Chemistry

Part 2 of 2
Chapters 12-21

Table of Contents

Preface

Welcome to *Chemistry*, an OpenStax resource. This textbook has been created with several goals in mind: accessibility, customization, and student engagement—all while encouraging students toward high levels of academic scholarship. Instructors and students alike will find that this textbook offers a strong foundation in chemistry in an accessible format.

About OpenStax

OpenStax is a non-profit organization committed to improving student access to quality learning materials. Our free textbooks go through a rigorous editorial publishing process. Our texts are developed and peer-reviewed by educators to ensure they are readable, accurate, and meet the scope and sequence requirements of today's college courses. Unlike traditional textbooks, OpenStax resources live online and are owned by the community of educators using them. Through our partnerships with companies and foundations committed to reducing costs for students, OpenStax is working to improve access to higher education for all. OpenStax is an initiative of Rice University and is made possible through the generous support of several philanthropic foundations. Since our launch in 2012 our texts have been used by millions of learners online and over 1,091 institutions worldwide.

About OpenStax's Resources

OpenStax resources provide quality academic instruction. Three key features set our materials apart from others: they can be customized by instructors for each class, they are a "living" resource that grows online through contributions from educators, and they are available free or for minimal cost.

Customization

OpenStax learning resources are designed to be customized for each course. Our textbooks provide a solid foundation on which instructors can build, and our resources are conceived and written with flexibility in mind. Instructors can select the sections most relevant to their curricula and create a textbook that speaks directly to the needs of their classes and student body. Teachers are encouraged to expand on existing examples by adding unique context via geographically localized applications and topical connections. *Chemistry* can be easily customized using our online platform (http://cnx.org/content/col11760/latest). Simply select the content most relevant to your current semester and create a textbook that speaks directly to the needs of your class. *Chemistry* is organized as a collection of sections that can be rearranged, modified, and enhanced through localized examples or to incorporate a specific theme of your course. This customization feature will ensure that your textbook truly reflects the goals of your course.

Curation

To broaden access and encourage community curation, *Chemistry* is "open source" licensed under a Creative Commons Attribution (CC-BY) license. The academic science community is invited to submit examples, emerging research, and other feedback to enhance and strengthen the material and keep it current and relevant for today's students.

Cost

Our textbooks are available for free online, and in low-cost print and e-book editions.

About *Chemistry*

Chemistry is designed for the two-semester general chemistry course. For many students, this course provides the foundation to a career in chemistry, while for others, this may be their only college-level science course. As such, this textbook provides an important opportunity for students to learn the core concepts of chemistry and understand how those concepts apply to their lives and the world around them. The text has been developed to meet the scope and sequence of most general chemistry courses. At the same time, the book includes a number of innovative features designed to enhance student learning. A strength of *Chemistry* is that instructors can customize the book, adapting it to the approach that works best in their classroom.

Coverage and Scope

Our *Chemistry* textbook adheres to the scope and sequence of most general chemistry courses nationwide. We strive to make chemistry, as a discipline, interesting and accessible to students. With this objective in mind, the content of this textbook has been developed and arranged to provide a logical progression from fundamental to more advanced concepts of chemical science. Topics are introduced within the context of familiar experiences whenever possible, treated with an appropriate rigor to satisfy the intellect of the learner, and reinforced in subsequent discussions of related content. The organization and pedagogical features were developed and vetted with feedback from chemistry educators dedicated to the project.

Chapter 1: Essential Ideas

Chapter 2: Atoms, Molecules, and Ions

Chapter 3: Composition of Substances and Solutions

Chapter 4: Stoichiometry of Chemical Reactions

Chapter 5: Thermochemistry

Chapter 6: Electronic Structures and Periodic Properties of Elements

Chapter 7: Chemical Bonding and Molecular Geometry

Chapter 8: Advanced Theories of Covalent Bonding

Chapter 9: Gases

Chapter 10: Liquids and Solids

Chapter 11: Solutions and Colloids

Chapter 12: Kinetics

Chapter 13: Fundamental Equilibrium Concepts

Chapter 14: Acid-Base Equilibria

Chapter 15: Equilibria of Other Reaction Classes

Chapter 16: Thermodynamics

Chapter 17: Electrochemistry

Chapter 18: Representative Metals, Metalloids, and Nonmetals

Chapter 19: Transition Metals and Coordination Chemistry

Chapter 20: Organic Chemistry

Chapter 21: Nuclear Chemistry

Pedagogical Foundation

Throughout *Chemistry*, you will find features that draw the students into scientific inquiry by taking selected topics a step further. Students and educators alike will appreciate discussions in these feature boxes.

Chemistry in Everyday Life ties chemistry concepts to everyday issues and real-world applications of science that students encounter in their lives. Topics include cell phones, solar thermal energy power plants, plastics recycling, and measuring blood pressure.

How Sciences Interconnect feature boxes discuss chemistry in context of its interconnectedness with other scientific disciplines. Topics include neurotransmitters, greenhouse gases and climate change, and proteins and enzymes.

Portrait of a Chemist features present a short bio and an introduction to the work of prominent figures from history and present day so that students can see the "face" of contributors in this field as well as science in action.

Comprehensive Art Program

Our art program is designed to enhance students' understanding of concepts through clear, effective illustrations, diagrams, and photographs.

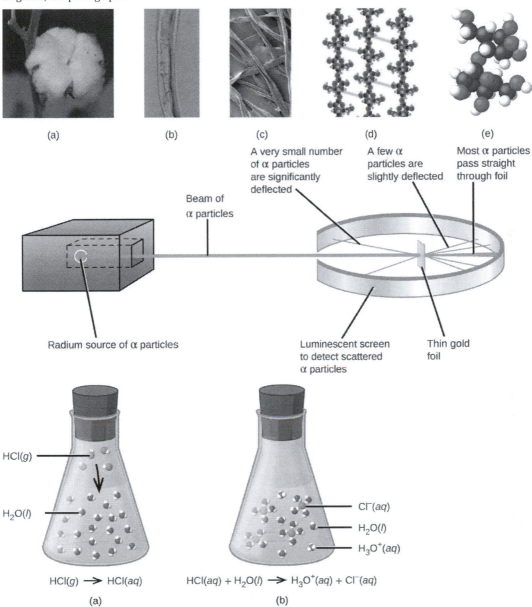

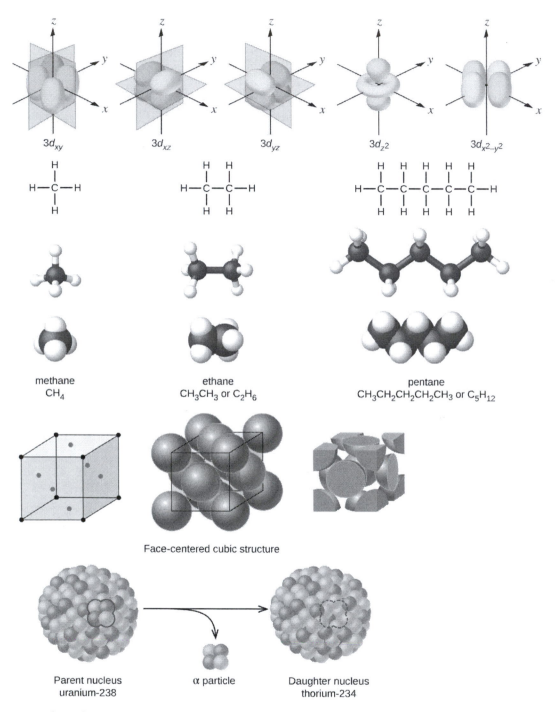

methane
CH_4

ethane
CH_3CH_3 or C_2H_6

pentane
$CH_3CH_2CH_2CH_2CH_3$ or C_5H_{12}

Face-centered cubic structure

Parent nucleus
uranium-238

α particle

Daughter nucleus
thorium-234

Interactives That Engage

Chemistry incorporates links to relevant interactive exercises and animations that help bring topics to life through our **Link to Learning** feature. Examples include:

PhET simulations

IUPAC data and interactives

TED talks

Assessments That Reinforce Key Concepts

In-chapter **Examples** walk students through problems by posing a question, stepping out a solution, and then asking students to practice the skill with a "Check Your Learning" component. The book also includes assessments at the end of each chapter so students can apply what they've learned through practice problems.

Atom-First Alternate Sequencing

Chemistry was conceived and written to fit a particular topical sequence, but it can be used flexibly to accommodate other course structures. Some instructors prefer to organize their course in a molecule-first or atom-first organization. For professors who use this approach, our OpenStax *Chemistry* textbook can be sequenced to fit this pedagogy. Please consider, however, that the chapters were not written to be completely independent, and that the proposed alternate sequence should be carefully considered for student preparation and textual consistency. We recommend these shifts in the table of contents structure if you plan to create a molecule/atom-first version of this text for your students:

Chapter 1: Essential Ideas

Chapter 2: Atoms, Molecules, and Ions

Chapter 6: Electronic Structure and Periodic Properties of Elements

Chapter 7: Chemical Bonding and Molecular Geometry

Chapter 8: Advanced Theories of Covalent Bonding

Chapter 3: Composition of Substances and Solutions

Chapter 4: Stoichiometry of Chemical Reactions

Chapter 5: Thermochemistry

Chapter 9: Gases

Chapter 10: Liquids and Solids

Chapter 11: Solutions and Colloids

Chapter 12: Kinetics

Chapter 13: Fundamental Equilibrium Concepts

Chapter 14: Acid-Base Equilibria

Chapter 15: Equilibria of Other Reaction Classes

Chapter 16: Thermodynamics

Chapter 17: Electrochemistry

Chapter 18: Representative Metals, Metalloids, and Nonmetals

Chapter 19: Transition Metals and Coordination Chemistry

Chapter 20: Organic Chemistry

Chapter 21: Nuclear Chemistry

Ancillaries

OpenStax projects offer an array of ancillaries for students and instructors. The following resources are available.

PowerPoint Slides

Instructor's Solution Manual

Our resources are continually expanding, so please visit http://openstaxcollege.org to view an up-to-date list of the Learning Resources for this title and to find information on accessing these resources.

About Our Team

Content Leads

Paul Flowers, PhD, University of North Carolina - Pembroke

Dr. Paul Flowers earned a BS in Chemistry from St. Andrews Presbyterian College in 1983 and a PhD in Analytical Chemistry from the University of Tennessee in 1988. After a one-year postdoctoral appointment at Los Alamos National Laboratory, he joined the University of North Carolina–Pembroke in the fall of 1989. Dr. Flowers teaches courses in general and analytical chemistry, and conducts experimental research involving the development of new devices and methods for microscale chemical analysis.

Klaus Theopold, PhD, University of Delaware

Dr. Klaus Theopold (born in Berlin, Germany) received his Vordiplom from the Universität Hamburg in 1977. He then decided to pursue his graduate studies in the United States, where he received his PhD in inorganic chemistry from UC Berkeley in 1982. After a year of postdoctoral research at MIT, he joined the faculty at Cornell University. In 1990, he moved to the University of Delaware, where he is a Professor in the Department of Chemistry and Biochemistry and serves as an Associate Director of the University's Center for Catalytic Science and Technology. Dr. Theopold regularly teaches graduate courses in inorganic and organometallic chemistry as well as General Chemistry.

Richard Langley, PhD, Stephen F. Austin State University

Dr. Richard Langley earned BS degrees in Chemistry and Mineralogy from Miami University of Ohio in the early 1970s and went on to receive his PhD in Chemistry from the University of Nebraska in 1977. After a postdoctoral fellowship at the Arizona State University Center for Solid State Studies, Dr. Langley taught in the University of Wisconsin system and participated in research at Argonne National Laboratory. Moving to Stephen F. Austin State University in 1982, Dr. Langley today serves as Professor of Chemistry. His areas of specialization are solid state chemistry, synthetic inorganic chemistry, fluorine chemistry, and chemical education.

Senior Contributing Author

William R. Robinson, PhD

Contributing Authors

Mark Blaser, Shasta College
Simon Bott, University of Houston
Donald Carpenetti, Craven Community College
Andrew Eklund, Alfred University
Emad El-Giar, University of Louisiana at Monroe
Don Frantz, Wilfrid Laurier University
Paul Hooker, Westminster College
Jennifer Look, Mercer University
George Kaminski, Worcester Polytechnic Institute
Carol Martinez, Central New Mexico Community College
Troy Milliken, Jackson State University
Vicki Moravec, Trine University
Jason Powell, Ferrum College
Thomas Sorensen, University of Wisconsin–Milwaukee
Allison Soult, University of Kentucky

Contributing Reviewers

Casey Akin, College Station Independent School District
Lara AL-Hariri, University of Massachusetts–Amherst
Sahar Atwa, University of Louisiana at Monroe

Todd Austell, University of North Carolina–Chapel Hill

Bobby Bailey, University of Maryland–University College

Robert Baker, Trinity College

Jeffrey Bartz, Kalamazoo College

Greg Baxley, Cuesta College

Ashley Beasley Green, National Institute of Standards and Technology

Patricia Bianconi, University of Massachusetts

Lisa Blank, Lyme Central School District

Daniel Branan, Colorado Community College System

Dorian Canelas, Duke University

Emmanuel Chang, York College

Carolyn Collins, College of Southern Nevada

Colleen Craig, University of Washington

Yasmine Daniels, Montgomery College–Germantown

Patricia Dockham, Grand Rapids Community College

Erick Fuoco, Richard J. Daley College

Andrea Geyer, University of Saint Francis

Daniel Goebbert, University of Alabama

John Goodwin, Coastal Carolina University

Stephanie Gould, Austin College

Patrick Holt, Bellarmine University

Kevin Kolack, Queensborough Community College

Amy Kovach, Roberts Wesleyan College

Judit Kovacs Beagle, University of Dayton

Krzysztof Kuczera, University of Kansas

Marcus Lay, University of Georgia

Pamela Lord, University of Saint Francis

Oleg Maksimov, Excelsior College

John Matson, Virginia Tech

Katrina Miranda, University of Arizona

Douglas Mulford, Emory University

Mark Ott, Jackson College

Adrienne Oxley, Columbia College

Richard Pennington, Georgia Gwinnett College

Rodney Powell, Coastal Carolina Community College

Jeanita Pritchett, Montgomery College–Rockville

Aheda Saber, University of Illinois at Chicago

Raymond Sadeghi, University of Texas at San Antonio

Nirmala Shankar, Rutgers University

Jonathan Smith, Temple University

Bryan Spiegelberg, Rider University

Ron Sternfels, Roane State Community College

Cynthia Strong, Cornell College

Kris Varazo, Francis Marion University

Victor Vilchiz, Virginia State University

Alex Waterson, Vanderbilt University

JuchaoYan, Eastern New Mexico University

Mustafa Yatin, Salem State University

Kazushige Yokoyama, State University of New York at Geneseo

Curtis Zaleski, Shippensburg University

Wei Zhang, University of Colorado–Boulder

Chapter 12

Kinetics

Figure 12.1   An agama lizard basks in the sun. As its body warms, the chemical reactions of its metabolism speed up.

Chapter Outline

Introduction

The lizard in the photograph is not simply enjoying the sunshine or working on its tan. The heat from the sun's rays is critical to the lizard's survival. A warm lizard can move faster than a cold one because the chemical reactions that allow its muscles to move occur more rapidly at higher temperatures. In the absence of warmth, the lizard is an easy meal for predators.

From baking a cake to determining the useful lifespan of a bridge, rates of chemical reactions play important roles in our understanding of processes that involve chemical changes. When planning to run a chemical reaction, we should ask at least two questions. The first is: "Will the reaction produce the desired products in useful quantities?" The second question is: "How rapidly will the reaction occur?" A reaction that takes 50 years to produce a product is about as useful as one that never gives a product at all. A third question is often asked when investigating reactions in greater detail: "What specific molecular-level processes take place as the reaction occurs?" Knowing the answer to this question is of practical importance when the yield or rate of a reaction needs to be controlled.

The study of chemical kinetics concerns the second and third questions—that is, the rate at which a reaction yields products and the molecular-scale means by which a reaction occurs. In this chapter, we will examine the factors that influence the rates of chemical reactions, the mechanisms by which reactions proceed, and the quantitative techniques used to determine and describe the rate at which reactions occur.

12.1 Chemical Reaction Rates

By the end of this section, you will be able to:

- Define chemical reaction rate
- Derive rate expressions from the balanced equation for a given chemical reaction
- Calculate reaction rates from experimental data

A *rate* is a measure of how some property varies with time. Speed is a familiar rate that expresses the distance traveled by an object in a given amount of time. Wage is a rate that represents the amount of money earned by a person working for a given amount of time. Likewise, the rate of a chemical reaction is a measure of how much reactant is consumed, or how much product is produced, by the reaction in a given amount of time.

The **rate of reaction** is the change in the amount of a reactant or product per unit time. Reaction rates are therefore determined by measuring the time dependence of some property that can be related to reactant or product amounts. Rates of reactions that consume or produce gaseous substances, for example, are conveniently determined by measuring changes in volume or pressure. For reactions involving one or more colored substances, rates may be monitored via measurements of light absorption. For reactions involving aqueous electrolytes, rates may be measured via changes in a solution's conductivity.

For reactants and products in solution, their relative amounts (concentrations) are conveniently used for purposes of expressing reaction rates. If we measure the concentration of hydrogen peroxide, H_2O_2, in an aqueous solution, we find that it changes slowly over time as the H_2O_2 decomposes, according to the equation:

$$2H_2O_2(aq) \longrightarrow 2H_2O(l) + O_2(g)$$

The rate at which the hydrogen peroxide decomposes can be expressed in terms of the rate of change of its concentration, as shown here:

$$
\begin{aligned}
\text{rate of decomposition of } H_2O_2 &= -\frac{\text{change in concentration of reactant}}{\text{time interval}} \\
&= -\frac{[H_2O_2]_{t_2} - [H_2O_2]_{t_1}}{t_2 - t_1} \\
&= -\frac{\Delta[H_2O_2]}{\Delta t}
\end{aligned}
$$

This mathematical representation of the change in species concentration over time is the **rate expression** for the reaction. The brackets indicate molar concentrations, and the symbol delta (Δ) indicates "change in." Thus, $[H_2O_2]_{t_1}$ represents the molar concentration of hydrogen peroxide at some time t_1; likewise, $[H_2O_2]_{t_2}$ represents the molar concentration of hydrogen peroxide at a later time t_2; and $\Delta[H_2O_2]$ represents the change in molar concentration of hydrogen peroxide during the time interval Δt (that is, $t_2 - t_1$). Since the reactant concentration decreases as the reaction proceeds, $\Delta[H_2O_2]$ is a negative quantity; we place a negative sign in front of the expression because reaction rates are, by convention, positive quantities. Figure 12.2 provides an example of data collected during the decomposition of H_2O_2.

Time (h)	$[H_2O_2]$ (mol L^{-1})	$\Delta[H_2O_2]$ (mol L^{-1})	Δt (h)	Rate of Decomposition, (mol/L/h)
0.00	1.000			
		−0.500	6.00	−0.0833
6.00	0.500			
		−0.250	6.00	−0.0417
12.00	0.250			
		−0.125	6.00	−0.0208
18.00	0.125			
		−0.062	6.00	−0.0103
24.00	0.0625			

Figure 12.2 The rate of decomposition of H_2O_2 in an aqueous solution decreases as the concentration of H_2O_2 decreases.

To obtain the tabulated results for this decomposition, the concentration of hydrogen peroxide was measured every 6 hours over the course of a day at a constant temperature of 40 °C. Reaction rates were computed for each time interval by dividing the change in concentration by the corresponding time increment, as shown here for the first 6-hour period:

$$\frac{-\Delta[H_2O_2]}{\Delta t} = \frac{-(0.500 \text{ mol/L} - 1.000 \text{ mol/L})}{(6.00 \text{ h} - 0.00 \text{ h})} = 0.0833 \text{ mol L}^{-1} \text{ h}^{-1}$$

Notice that the reaction rates vary with time, decreasing as the reaction proceeds. Results for the last 6-hour period yield a reaction rate of:

$$\frac{-\Delta[H_2O_2]}{\Delta t} = \frac{-(0.0625 \text{ mol/L} - 0.125 \text{ mol/L})}{(24.00 \text{ h} - 18.00 \text{ h})} = 0.0104 \text{ mol L}^{-1} \text{ h}^{-1}$$

This behavior indicates the reaction continually slows with time. Using the concentrations at the beginning and end of a time period over which the reaction rate is changing results in the calculation of an **average rate** for the reaction over this time interval. At any specific time, the rate at which a reaction is proceeding is known as its **instantaneous rate**. The instantaneous rate of a reaction at "time zero," when the reaction commences, is its **initial rate**. Consider the analogy of a car slowing down as it approaches a stop sign. The vehicle's initial rate—analogous to the beginning of a chemical reaction—would be the speedometer reading at the moment the driver begins pressing the brakes (t_0). A few moments later, the instantaneous rate at a specific moment—call it t_1—would be somewhat slower, as indicated by the speedometer reading at that point in time. As time passes, the instantaneous rate will continue to fall until it reaches zero, when the car (or reaction) stops. Unlike instantaneous speed, the car's average speed is not indicated by the speedometer; but it can be calculated as the ratio of the distance traveled to the time required to bring the vehicle to a complete stop (Δt). Like the decelerating car, the average rate of a chemical reaction will fall somewhere between its initial and final rates.

The instantaneous rate of a reaction may be determined one of two ways. If experimental conditions permit the measurement of concentration changes over very short time intervals, then average rates computed as described earlier provide reasonably good approximations of instantaneous rates. Alternatively, a graphical procedure may be used that, in effect, yields the results that would be obtained if short time interval measurements were possible. If we plot the concentration of hydrogen peroxide against time, the instantaneous rate of decomposition of H_2O_2 at any time t is given by the slope of a straight line that is tangent to the curve at that time (**Figure 12.3**). We can use calculus to evaluating the slopes of such tangent lines, but the procedure for doing so is beyond the scope of this chapter.

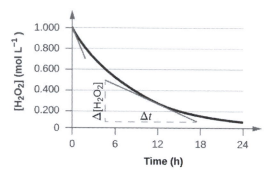

Figure 12.3 This graph shows a plot of concentration versus time for a 1.000 *M* solution of H_2O_2. The rate at any instant is equal to the opposite of the slope of a line tangential to this curve at that time. Tangents are shown at *t* = 0 h ("initial rate") and at *t* = 10 h ("instantaneous rate" at that particular time).

Chemistry in Everyday Life

Reaction Rates in Analysis: Test Strips for Urinalysis

Physicians often use disposable test strips to measure the amounts of various substances in a patient's urine (Figure 12.4). These test strips contain various chemical reagents, embedded in small pads at various locations along the strip, which undergo changes in color upon exposure to sufficient concentrations of specific substances. The usage instructions for test strips often stress that proper read time is critical for optimal results. This emphasis on read time suggests that kinetic aspects of the chemical reactions occurring on the test strip are important considerations.

The test for urinary glucose relies on a two-step process represented by the chemical equations shown here:

$$C_6H_{12}O_6 + O_2 \xrightarrow[\text{catalyst}]{} C_6H_{10}O_6 + H_2O_2$$

$$2H_2O_2 + 2I^- \xrightarrow[\text{catalyst}]{} I_2 + 2H_2O + O_2$$

The first equation depicts the oxidation of glucose in the urine to yield glucolactone and hydrogen peroxide. The hydrogen peroxide produced subsequently oxidizes colorless iodide ion to yield brown iodine, which may be visually detected. Some strips include an additional substance that reacts with iodine to produce a more distinct color change.

The two test reactions shown above are inherently very slow, but their rates are increased by special enzymes embedded in the test strip pad. This is an example of *catalysis*, a topic discussed later in this chapter. A typical glucose test strip for use with urine requires approximately 30 seconds for completion of the color-forming reactions. Reading the result too soon might lead one to conclude that the glucose concentration of the urine sample is lower than it actually is (a *false-negative* result). Waiting too long to assess the color change can lead to a *false positive* due to the slower (not catalyzed) oxidation of iodide ion by other substances found in urine.

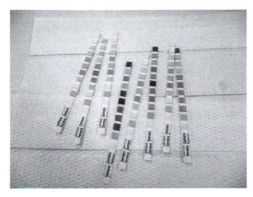

Figure 12.4 Test strips are commonly used to detect the presence of specific substances in a person's urine. Many test strips have several pads containing various reagents to permit the detection of multiple substances on a single strip. (credit: Iqbal Osman)

Relative Rates of Reaction

The rate of a reaction may be expressed in terms of the change in the amount of any reactant or product, and may be simply derived from the stoichiometry of the reaction. Consider the reaction represented by the following equation:

$$2NH_3(g) \longrightarrow N_2(g) + 3H_2(g)$$

The stoichiometric factors derived from this equation may be used to relate reaction rates in the same manner that they are used to related reactant and product amounts. The relation between the reaction rates expressed in terms of nitrogen production and ammonia consumption, for example, is:

$$-\frac{\Delta mol\ NH_3}{\Delta t} \times \frac{1\ mol\ N_2}{2\ mol\ NH_3} = \frac{\Delta mol\ N_2}{\Delta t}$$

We can express this more simply without showing the stoichiometric factor's units:

$$-\frac{1}{2}\frac{\Delta mol\ NH_3}{\Delta t} = \frac{\Delta mol\ N_2}{\Delta t}$$

Note that a negative sign has been added to account for the opposite signs of the two amount changes (the reactant amount is decreasing while the product amount is increasing). If the reactants and products are present in the same solution, the molar amounts may be replaced by concentrations:

$$-\frac{1}{2}\frac{\Delta[NH_3]}{\Delta t} = \frac{\Delta[N_2]}{\Delta t}$$

Similarly, the rate of formation of H_2 is three times the rate of formation of N_2 because three moles of H_2 form during the time required for the formation of one mole of N_2:

$$\frac{1}{3}\frac{\Delta[H_2]}{\Delta t} = \frac{\Delta[N_2]}{\Delta t}$$

Figure 12.5 illustrates the change in concentrations over time for the decomposition of ammonia into nitrogen and hydrogen at 1100 °C. We can see from the slopes of the tangents drawn at $t = 500$ seconds that the instantaneous rates of change in the concentrations of the reactants and products are related by their stoichiometric factors. The rate of hydrogen production, for example, is observed to be three times greater than that for nitrogen production:

$$\frac{2.91 \times 10^{-6}\ M/s}{9.71 \times 10^{-6}\ M/s} \approx 3$$

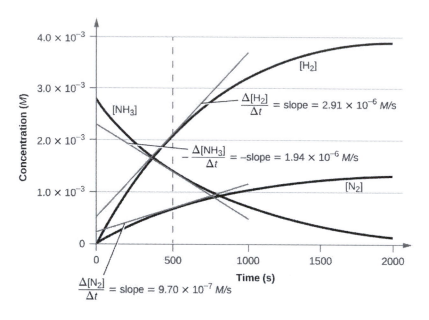

$$\frac{\Delta[N_2]}{\Delta t} = \text{slope} = 9.70 \times 10^{-7} \text{ M/s}$$

Figure 12.5 This graph shows the changes in concentrations of the reactants and products during the reaction $2NH_3 \longrightarrow 3N_2 + H_2$. The rates of change of the three concentrations are related by their stoichiometric factors, as shown by the different slopes of the tangents at $t = 500$ s.

Example 12.1

Expressions for Relative Reaction Rates

The first step in the production of nitric acid is the combustion of ammonia:

$$4NH_3(g) + 5O_2(g) \longrightarrow 4NO(g) + 6H_2O(g)$$

Write the equations that relate the rates of consumption of the reactants and the rates of formation of the products.

Solution

Considering the stoichiometry of this homogeneous reaction, the rates for the consumption of reactants and formation of products are:

$$-\frac{1}{4}\frac{\Delta[NH_3]}{\Delta t} = -\frac{1}{5}\frac{\Delta[O_2]}{\Delta t} = \frac{1}{4}\frac{\Delta[NO]}{\Delta t} = \frac{1}{6}\frac{\Delta[H_2O]}{\Delta t}$$

Check Your Learning

The rate of formation of Br_2 is 6.0×10^{-6} mol/L/s in a reaction described by the following net ionic equation:

$$5Br^- + BrO_3^- + 6H^+ \longrightarrow 3Br_2 + 3H_2O$$

Write the equations that relate the rates of consumption of the reactants and the rates of formation of the products.

Answer: $-\frac{1}{5}\frac{\Delta[Br^-]}{\Delta t} = -\frac{\Delta[BrO_3^-]}{\Delta t} = -\frac{1}{6}\frac{\Delta[H^+]}{\Delta t} = \frac{1}{3}\frac{\Delta[Br_2]}{\Delta t} = \frac{1}{3}\frac{\Delta[H_2O]}{\Delta t}$

Example 12.2

Reaction Rate Expressions for Decomposition of H_2O_2

The graph in Figure 12.3 shows the rate of the decomposition of H_2O_2 over time:

$$2H_2O_2 \longrightarrow 2H_2O + O_2$$

Based on these data, the instantaneous rate of decomposition of H_2O_2 at $t = 11.1$ h is determined to be 3.20×10^{-2} mol/L/h, that is:

$$-\frac{\Delta[H_2O_2]}{\Delta t} = 3.20 \times 10^{-2}\,\text{mol}\,L^{-1}h^{-1}$$

What is the instantaneous rate of production of H_2O and O_2?

Solution

Using the stoichiometry of the reaction, we may determine that:

$$-\frac{1}{2}\frac{\Delta[H_2O_2]}{\Delta t} = \frac{1}{2}\frac{\Delta[H_2O]}{\Delta t} = \frac{\Delta[O_2]}{\Delta t}$$

Therefore:

$$\frac{1}{2} \times 3.20 \times 10^{-2}\,\text{mol}\,L^{-1}h^{-1} = \frac{\Delta[O_2]}{\Delta t}$$

and

$$\frac{\Delta[O_2]}{\Delta t} = 1.60 \times 10^{-2}\,\text{mol}\,L^{-1}h^{-1}$$

Check Your Learning

If the rate of decomposition of ammonia, NH_3, at 1150 K is 2.10×10^{-6} mol/L/s, what is the rate of production of nitrogen and hydrogen?

Answer: 1.05×10^{-6} mol/L/s, N_2 and 3.15×10^{-6} mol/L/s, H_2.

12.2 Factors Affecting Reaction Rates

By the end of this section, you will be able to:

- Describe the effects of chemical nature, physical state, temperature, concentration, and catalysis on reaction rates

The rates at which reactants are consumed and products are formed during chemical reactions vary greatly. We can identify five factors that affect the rates of chemical reactions: the chemical nature of the reacting substances, the state of subdivision (one large lump versus many small particles) of the reactants, the temperature of the reactants, the concentration of the reactants, and the presence of a catalyst.

The Chemical Nature of the Reacting Substances

The rate of a reaction depends on the nature of the participating substances. Reactions that appear similar may have different rates under the same conditions, depending on the identity of the reactants. For example, when small pieces of the metals iron and sodium are exposed to air, the sodium reacts completely with air overnight, whereas the iron is barely affected. The active metals calcium and sodium both react with water to form hydrogen gas and a base. Yet calcium reacts at a moderate rate, whereas sodium reacts so rapidly that the reaction is almost explosive.

The State of Subdivision of the Reactants

Except for substances in the gaseous state or in solution, reactions occur at the boundary, or interface, between two phases. Hence, the rate of a reaction between two phases depends to a great extent on the surface contact between them. A finely divided solid has more surface area available for reaction than does one large piece of the same substance. Thus a liquid will react more rapidly with a finely divided solid than with a large piece of the same solid. For example, large pieces of iron react slowly with acids; finely divided iron reacts much more rapidly (Figure 12.6). Large pieces of wood smolder, smaller pieces burn rapidly, and saw dust burns explosively.

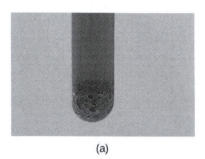 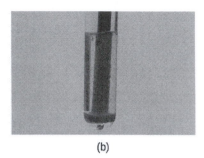

(a)   (b)

Figure 12.6   (a) Iron powder reacts rapidly with dilute hydrochloric acid and produces bubbles of hydrogen gas because the powder has a large total surface area: 2Fe(s) + 6HCl(aq) $\longrightarrow$ 2FeCl3(aq) + 3H2(g). (b) An iron nail reacts more slowly.

Link to Learning

Watch this video (http://openstaxcollege.org/l/16cesium) to see the reaction of cesium with water in slow motion and a discussion of how the state of reactants and particle size affect reaction rates.

Temperature of the Reactants

Chemical reactions typically occur faster at higher temperatures. Food can spoil quickly when left on the kitchen counter. However, the lower temperature inside of a refrigerator slows that process so that the same food remains fresh for days. We use a burner or a hot plate in the laboratory to increase the speed of reactions that proceed slowly at ordinary temperatures. In many cases, an increase in temperature of only 10 °C will approximately double the rate of a reaction in a homogeneous system.

Concentrations of the Reactants

The rates of many reactions depend on the concentrations of the reactants. Rates usually increase when the concentration of one or more of the reactants increases. For example, calcium carbonate ($CaCO_3$) deteriorates as a result of its reaction with the pollutant sulfur dioxide. The rate of this reaction depends on the amount of sulfur dioxide in the air (Figure 12.7). An acidic oxide, sulfur dioxide combines with water vapor in the air to produce sulfurous acid in the following reaction:

$$SO_2(g) + H_2O(g) \longrightarrow H_2SO_3(aq)$$

Calcium carbonate reacts with sulfurous acid as follows:

$$CaCO_3(s) + H_2SO_3(aq) \longrightarrow CaSO_3(aq) + CO_2(g) + H_2O(l)$$

In a polluted atmosphere where the concentration of sulfur dioxide is high, calcium carbonate deteriorates more rapidly than in less polluted air. Similarly, phosphorus burns much more rapidly in an atmosphere of pure oxygen than in air, which is only about 20% oxygen.

Figure 12.7 Statues made from carbonate compounds such as limestone and marble typically weather slowly over time due to the actions of water, and thermal expansion and contraction. However, pollutants like sulfur dioxide can accelerate weathering. As the concentration of air pollutants increases, deterioration of limestone occurs more rapidly. (credit: James P Fisher III)

Link to Learning

Phosphorous burns rapidly in air, but it will burn even more rapidly if the concentration of oxygen in is higher. Watch this video (http://openstaxcollege.org/l/16phosphor) to see an example.

The Presence of a Catalyst

Hydrogen peroxide solutions foam when poured onto an open wound because substances in the exposed tissues act as catalysts, increasing the rate of hydrogen peroxide's decomposition. However, in the absence of these catalysts (for example, in the bottle in the medicine cabinet) complete decomposition can take months. A **catalyst** is a substance that increases the rate of a chemical reaction by lowering the activation energy without itself being consumed by the reaction. Activation energy is the minimum amount of energy required for a chemical reaction to proceed in the forward direction. A catalyst increases the reaction rate by providing an alternative pathway or mechanism for the reaction to follow (Figure 12.8). Catalysis will be discussed in greater detail later in this chapter as it relates to mechanisms of reactions.

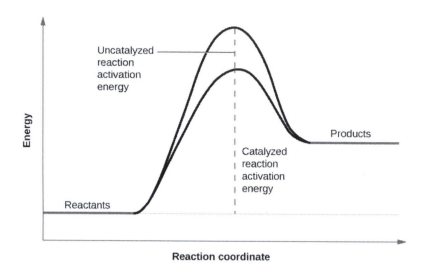

Reaction coordinate

Figure 12.8 The presence of a catalyst increases the rate of a reaction by lowering its activation energy.

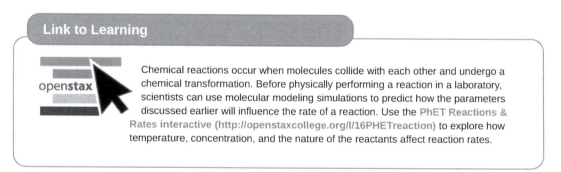

Link to Learning

Chemical reactions occur when molecules collide with each other and undergo a chemical transformation. Before physically performing a reaction in a laboratory, scientists can use molecular modeling simulations to predict how the parameters discussed earlier will influence the rate of a reaction. Use the PhET Reactions & Rates interactive (http://openstaxcollege.org/l/16PHETreaction) to explore how temperature, concentration, and the nature of the reactants affect reaction rates.

12.3 Rate Laws

By the end of this section, you will be able to:

- Explain the form and function of a rate law
- Use rate laws to calculate reaction rates
- Use rate and concentration data to identify reaction orders and derive rate laws

As described in the previous module, the rate of a reaction is affected by the concentrations of reactants. **Rate laws** or **rate equations** are mathematical expressions that describe the relationship between the rate of a chemical reaction and the concentration of its reactants. In general, a rate law (or differential rate law, as it is sometimes called) takes this form:

$$\text{rate} = k[A]^m[B]^n[C]^p \ldots$$

in which $[A]$, $[B]$, and $[C]$ represent the molar concentrations of reactants, and **k** is the **rate constant**, which is specific for a particular reaction at a particular temperature. The exponents m, n, and p are usually positive integers (although it is possible for them to be fractions or negative numbers). The rate constant k and the exponents m, n, and p must be determined experimentally by observing how the rate of a reaction changes as the concentrations of the reactants are changed. The rate constant k is independent of the concentration of A, B, or C, but it does vary with temperature and surface area.

The exponents in a rate law describe the effects of the reactant concentrations on the reaction rate and define the **reaction order**. Consider a reaction for which the rate law is:

$$\text{rate} = k[A]^m[B]^n$$

If the exponent m is 1, the reaction is first order with respect to A. If m is 2, the reaction is second order with respect to A. If n is 1, the reaction is first order in B. If n is 2, the reaction is second order in B. If m or n is zero, the reaction is zero order in A or B, respectively, and the rate of the reaction is not affected by the concentration of that reactant. The **overall reaction order** is the sum of the orders with respect to each reactant. If $m = 1$ and $n = 1$, the overall order of the reaction is second order ($m + n = 1 + 1 = 2$).

The rate law:

$$\text{rate} = k[H_2O_2]$$

describes a reaction that is first order in hydrogen peroxide and first order overall. The rate law:

$$\text{rate} = k[C_4H_6]^2$$

describes a reaction that is second order in C_4H_6 and second order overall. The rate law:

$$\text{rate} = k[H^+][OH^-]$$

describes a reaction that is first order in H^+, first order in OH^-, and second order overall.

Example 12.3

Writing Rate Laws from Reaction Orders

An experiment shows that the reaction of nitrogen dioxide with carbon monoxide:

$$NO_2(g) + CO(g) \longrightarrow NO(g) + CO_2(g)$$

is second order in NO_2 and zero order in CO at 100 °C. What is the rate law for the reaction?

Solution

The reaction will have the form:

$$\text{rate} = k[NO_2]^m[CO]^n$$

The reaction is second order in NO_2; thus $m = 2$. The reaction is zero order in CO; thus $n = 0$. The rate law is:

$$\text{rate} = k[NO_2]^2[CO]^0 = k[NO_2]^2$$

Remember that a number raised to the zero power is equal to 1, thus $[CO]^0 = 1$, which is why we can simply drop the concentration of CO from the rate equation: the rate of reaction is solely dependent on the concentration of NO_2. When we consider rate mechanisms later in this chapter, we will explain how a reactant's concentration can have no effect on a reaction despite being involved in the reaction.

Check Your Learning

The rate law for the reaction:

$$H_2(g) + 2NO(g) \longrightarrow N_2O(g) + H_2O(g)$$

has been determined to be rate $= k[NO]^2[H_2]$. What are the orders with respect to each reactant, and what is the overall order of the reaction?

Answer: order in NO = 2; order in H_2 = 1; overall order = 3

Check Your Learning

In a transesterification reaction, a triglyceride reacts with an alcohol to form an ester and glycerol. Many students learn about the reaction between methanol (CH_3OH) and ethyl acetate ($CH_3CH_2OCOCH_3$) as a sample reaction before studying the chemical reactions that produce biodiesel:

$$CH_3OH + CH_3CH_2OCOCH_3 \longrightarrow CH_3OCOCH_3 + CH_3CH_2OH$$

The rate law for the reaction between methanol and ethyl acetate is, under certain conditions, determined to be:

$$rate = k[CH_3OH]$$

What is the order of reaction with respect to methanol and ethyl acetate, and what is the overall order of reaction?

Answer: order in CH_3OH = 1; order in $CH_3CH_2OCOCH_3$ = 0; overall order = 1

It is sometimes helpful to use a more explicit algebraic method, often referred to as the **method of initial rates**, to determine the orders in rate laws. To use this method, we select two sets of rate data that differ in the concentration of only one reactant and set up a ratio of the two rates and the two rate laws. After canceling terms that are equal, we are left with an equation that contains only one unknown, the coefficient of the concentration that varies. We then solve this equation for the coefficient.

Example 12.4

Determining a Rate Law from Initial Rates

Ozone in the upper atmosphere is depleted when it reacts with nitrogen oxides. The rates of the reactions of nitrogen oxides with ozone are important factors in deciding how significant these reactions are in the formation of the ozone hole over Antarctica (Figure 12.9). One such reaction is the combination of nitric oxide, NO, with ozone, O_3:

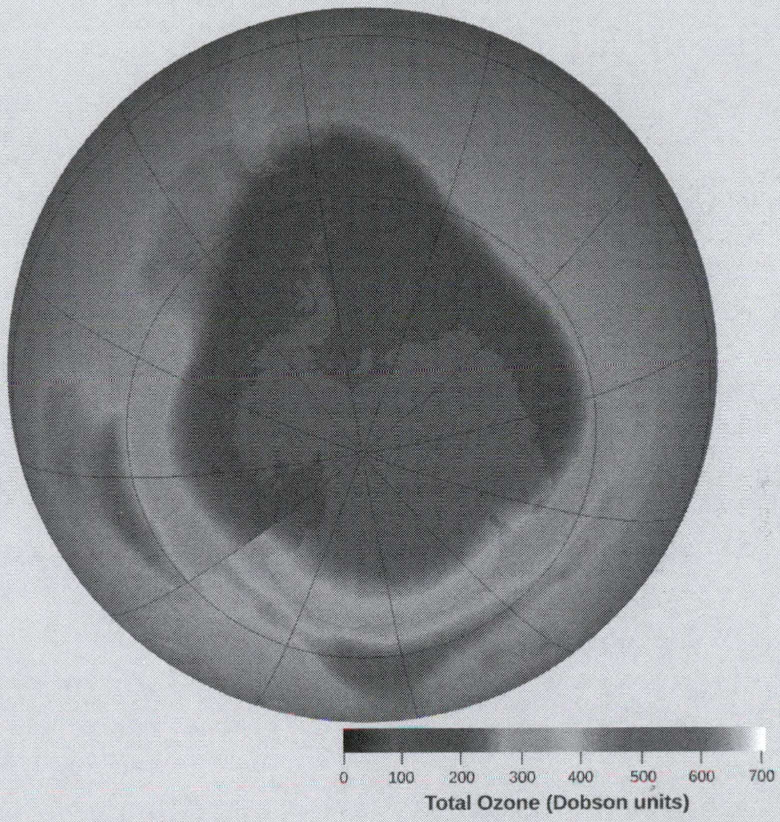

Figure 12.9 Over the past several years, the atmospheric ozone concentration over Antarctica has decreased during the winter. This map shows the decreased concentration as a purple area. (credit: modification of work by NASA)

$$NO(g) + O_3(g) \longrightarrow NO_2(g) + O_2(g)$$

This reaction has been studied in the laboratory, and the following rate data were determined at 25 °C.

Trial	[NO] (mol/L)	[O_3] (mol/L)	$\dfrac{\Delta[NO_2]}{\Delta t}$ (mol L^{-1} s^{-1})
1	1.00×10^{-6}	3.00×10^{-6}	6.60×10^{-5}
2	1.00×10^{-6}	6.00×10^{-6}	1.32×10^{-4}
3	1.00×10^{-6}	9.00×10^{-6}	1.98×10^{-4}
4	2.00×10^{-6}	9.00×10^{-6}	3.96×10^{-4}
5	3.00×10^{-6}	9.00×10^{-6}	5.94×10^{-4}

Determine the rate law and the rate constant for the reaction at 25 °C.

Solution

The rate law will have the form:

$$\text{rate} = k[\text{NO}]^m[\text{O}_3]^n$$

We can determine the values of m, n, and k from the experimental data using the following three-part process:

Step 1. *Determine the value of* m *from the data in which [NO] varies and [O_3] is constant.* In the last three experiments, [NO] varies while [O_3] remains constant. When [NO] doubles from trial 3 to 4, the rate doubles, and when [NO] triples from trial 3 to 5, the rate also triples. Thus, the rate is also directly proportional to [NO], and m in the rate law is equal to 1.

Step 2. *Determine the value of* n *from data in which [O_3] varies and [NO] is constant.* In the first three experiments, [NO] is constant and [O_3] varies. The reaction rate changes in direct proportion to the change in [O_3]. When [O_3] doubles from trial 1 to 2, the rate doubles; when [O_3] triples from trial 1 to 3, the rate increases also triples. Thus, the rate is directly proportional to [O_3], and n is equal to 1.The rate law is thus:

$$\text{rate} = k[\text{NO}]^1[\text{O}_3]^1 = k[\text{NO}][\text{O}_3]$$

Step 3. *Determine the value of* k *from one set of concentrations and the corresponding rate.*

$$
\begin{aligned}
k &= \frac{\text{rate}}{[\text{NO}][\text{O}_3]} \\[2mm]
&= \frac{6.60 \times 10^{-5}\ \cancel{\text{mol L}^{-1}}\text{s}^{-1}}{\left(1.00 \times 10^{-6}\ \cancel{\text{mol L}^{-1}}\right)\left(3.00 \times 10^{-6}\ \text{mol L}^{-1}\right)} \\[2mm]
&= 2.20 \times 10^7\ \text{L mol}^{-1}\text{s}^{-1}
\end{aligned}
$$

The large value of k tells us that this is a fast reaction that could play an important role in ozone depletion if [NO] is large enough.

Check Your Learning

Acetaldehyde decomposes when heated to yield methane and carbon monoxide according to the equation:

$$\text{CH}_3\text{CHO}(g) \longrightarrow \text{CH}_4(g) + \text{CO}(g)$$

Determine the rate law and the rate constant for the reaction from the following experimental data:

Trial	[CH$_3$CHO] (mol/L)	$-\dfrac{\Delta[\text{CH}_3\text{CHO}]}{\Delta t}$ (mol L^{-1} s^{-1})
1	1.75×10^{-3}	2.06×10^{-11}
2	3.50×10^{-3}	8.24×10^{-11}
3	7.00×10^{-3}	3.30×10^{-10}

Answer: rate $= k[\text{CH}_3\text{CHO}]^2$ with $k = 6.73 \times 10^{-6}$ L/mol/s

Example 12.5

Determining Rate Laws from Initial Rates

Using the initial rates method and the experimental data, determine the rate law and the value of the rate constant for this reaction:

$$2\text{NO}(g) + \text{Cl}_2(g) \longrightarrow 2\text{NOCl}(g)$$

Trial	[NO] (mol/L)	[Cl$_2$] (mol/L)	$-\dfrac{\Delta[NO]}{\Delta t}$ (mol L^{-1} s^{-1})
1	0.10	0.10	0.00300
2	0.10	0.15	0.00450
3	0.15	0.10	0.00675

Solution

The rate law for this reaction will have the form:

$$rate = k[NO]^m[Cl_2]^n$$

As in Example 12.4, we can approach this problem in a stepwise fashion, determining the values of m and n from the experimental data and then using these values to determine the value of k. In this example, however, we will use a different approach to determine the values of m and n:

Step 1. *Determine the value of* m *from the data in which [NO] varies and [Cl$_2$] is constant.* We can write the ratios with the subscripts x and y to indicate data from two different trials:

$$\frac{rate_x}{rate_y} = \frac{k[NO]_x^m[Cl_2]_x^n}{k[NO]_y^m[Cl_2]_y^n}$$

Using the third trial and the first trial, in which [Cl$_2$] does not vary, gives:

$$\frac{rate\ 3}{rate\ 1} = \frac{0.00675}{0.00300} = \frac{k(0.15)^m(0.10)^n}{k(0.10)^m(0.10)^n}$$

After canceling equivalent terms in the numerator and denominator, we are left with:

$$\frac{0.00675}{0.00300} = \frac{(0.15)^m}{(0.10)^m}$$

which simplifies to:

$$2.25 = (1.5)^m$$

We can use natural logs to determine the value of the exponent m:

$$\ln(2.25) = m\ln(1.5)$$
$$\frac{\ln(2.25)}{\ln(1.5)} = m$$
$$2 = m$$

We can confirm the result easily, since:

$$1.5^2 = 2.25$$

Step 2. *Determine the value of* n *from data in which [Cl$_2$] varies and [NO] is constant.*

$$\frac{rate\ 2}{rate\ 1} = \frac{0.00450}{0.00300} = \frac{k(0.10)^m(0.15)^n}{k(0.10)^m(0.10)^n}$$

Cancelation gives:

$$\frac{0.0045}{0.0030} = \frac{(0.15)^n}{(0.10)^n}$$

which simplifies to:

$$1.5 = (1.5)^n$$

Thus n must be 1, and the form of the rate law is:

$$Rate = k[NO]^m[Cl_2]^n = k[NO]^2[Cl_2]$$

Step 3. *Determine the numerical value of the rate constant* k *with appropriate units.* The units for the rate of a reaction are mol/L/s. The units for k are whatever is needed so that substituting into the rate law expression affords the appropriate units for the rate. In this example, the concentration units are mol^3/L^3. The units for k should be $mol^{-2}\, L^2/s$ so that the rate is in terms of mol/L/s.

To determine the value of k once the rate law expression has been solved, simply plug in values from the first experimental trial and solve for k:

$$0.00300\,\text{mol L}^{-1}\,\text{s}^{-1} = k\big(0.10\,\text{mol L}^{-1}\big)^2\big(0.10\,\text{mol L}^{-1}\big)^1$$

$$k = 3.0\,\text{mol}^{-2}\,\text{L}^2\,\text{s}^{-1}$$

Check Your Learning

Use the provided initial rate data to derive the rate law for the reaction whose equation is:

$$OCl^-(aq) + I^-(aq) \longrightarrow OI^-(aq) + Cl^-(aq)$$

Trial	[OCl⁻] (mol/L)	[I⁻] (mol/L)	Initial Rate (mol/L/s)
1	0.0040	0.0020	0.00184
2	0.0020	0.0040	0.00092
3	0.0020	0.0020	0.00046

Determine the rate law expression and the value of the rate constant k with appropriate units for this reaction.

Answer: $\dfrac{\text{rate 2}}{\text{rate 3}} = \dfrac{0.00092}{0.00046} = \dfrac{k(0.0020)^x(0.0040)^y}{k(0.0020)^x(0.0020)^y}$

$$2.00 = 2.00^y$$
$$y = 1$$

$$\frac{\text{rate 1}}{\text{rate 2}} = \frac{0.00184}{0.00092} = \frac{k(0.0040)^x(0.0020)^y}{k(0.0020)^x(0.0040)^y}$$

$$2.00 = \frac{2^x}{2^y}$$

$$2.00 = \frac{2^x}{2^1}$$

$$4.00 = 2^x$$

$$x = 2$$

Substituting the concentration data from trial 1 and solving for k yields:

$$\text{rate} = k[OCl^-]^2[I^-]^1$$

$$0.00184 = k(0.0040)^2(0.0020)^1$$

$$k = 5.75 \times 10^4\,\text{mol}^{-2}\,\text{L}^2\,\text{s}^{-1}$$

Reaction Order and Rate Constant Units

In some of our examples, the reaction orders in the rate law happen to be the same as the coefficients in the chemical equation for the reaction. This is merely a coincidence and very often not the case.

Rate laws may exhibit fractional orders for some reactants, and negative reaction orders are sometimes observed when an increase in the concentration of one reactant causes a decrease in reaction rate. A few examples illustrating these points are provided:

$$NO_2 + CO \longrightarrow NO + CO_2 \qquad \text{rate} = k[NO_2]^2$$
$$CH_3CHO \longrightarrow CH_4 + CO \qquad \text{rate} = k[CH_3CHO]^2$$
$$2N_2O_5 \longrightarrow 2NO_2 + O_2 \qquad \text{rate} = k[N_2O_5]$$
$$2NO_2 + F_2 \longrightarrow 2NO_2F \qquad \text{rate} = k[NO_2][F_2]$$
$$2NO_2Cl \longrightarrow 2NO_2 + Cl_2 \qquad \text{rate} = k[NO_2Cl]$$

It is important to note that *rate laws are determined by experiment only and are not reliably predicted by reaction stoichiometry.*

Reaction orders also play a role in determining the units for the rate constant k. In Example 12.4, a second-order reaction, we found the units for k to be $L \, mol^{-4} \, s^{-1}$, whereas in Example 12.5, a third order reaction, we found the units for k to be $mol^{-2} \, L^2/s$. More generally speaking, the units for the rate constant for a reaction of order $(m + n)$ are $mol^{1-(m+n)} \, L^{(m+n)-1} \, s^{-1}$. Table 12.1 summarizes the rate constant units for common reaction orders.

Rate Constants for Common Reaction Orders

Reaction Order	Units of k
$(m + n)$	$mol^{1-(m+n)} \, L^{(m+n)-1} \, s^{-1}$
zero	mol/L/s
first	s^{-1}
second	L/mol/s
third	$mol^{-2} \, L^2 \, s^{-1}$

Table 12.1

Note that the units in the table can also be expressed in terms of molarity (M) instead of mol/L. Also, units of time other than the second (such as minutes, hours, days) may be used, depending on the situation.

12.4 Integrated Rate Laws

By the end of this section, you will be able to:

- Explain the form and function of an integrated rate law
- Perform integrated rate law calculations for zero-, first-, and second-order reactions
- Define half-life and carry out related calculations
- Identify the order of a reaction from concentration/time data

The rate laws we have seen thus far relate the rate and the concentrations of reactants. We can also determine a second form of each rate law that relates the concentrations of reactants and time. These are called **integrated rate laws**. We can use an integrated rate law to determine the amount of reactant or product present after a period of time or to estimate the time required for a reaction to proceed to a certain extent. For example, an integrated rate law is used to determine the length of time a radioactive material must be stored for its radioactivity to decay to a safe level.

Using calculus, the differential rate law for a chemical reaction can be integrated with respect to time to give an equation that relates the amount of reactant or product present in a reaction mixture to the elapsed time of the reaction. This process can either be very straightforward or very complex, depending on the complexity of the differential rate law. For purposes of discussion, we will focus on the resulting integrated rate laws for first-, second-, and zero-order reactions.

First-Order Reactions

An equation relating the rate constant k to the initial concentration $[A]_0$ and the concentration $[A]_t$ present after any given time t can be derived for a first-order reaction and shown to be:

$$\ln\left(\frac{[A]_t}{[A]_0}\right) = -kt$$

or

$$\ln\left(\frac{[A]_0}{[A]_t}\right) = kt$$

or

$$[A] = [A]_0 e^{-kt}$$

Example 12.6

The Integrated Rate Law for a First-Order Reaction

The rate constant for the first-order decomposition of cyclobutane, C_4H_8 at 500 °C is $9.2 \times 10^{-3}\ s^{-1}$:

$$C_4H_8 \longrightarrow 2C_2H_4$$

How long will it take for 80.0% of a sample of C_4H_8 to decompose?

Solution

We use the integrated form of the rate law to answer questions regarding time:

$$\ln\left(\frac{[A]_0}{[A]}\right) = kt$$

There are four variables in the rate law, so if we know three of them, we can determine the fourth. In this case we know $[A]_0$, $[A]$, and k, and need to find t.

The initial concentration of C_4H_8, $[A]_0$, is not provided, but the provision that 80.0% of the sample has decomposed is enough information to solve this problem. Let x be the initial concentration, in which case the concentration after 80.0% decomposition is 20.0% of x or $0.200x$. Rearranging the rate law to isolate t and substituting the provided quantities yields:

$$t = \ln\frac{[x]}{[0.200x]} \times \frac{1}{k}$$

$$= \ln\frac{0.100\ \text{mol L}^{-1}}{0.020\ \text{mol L}^{-1}} \times \frac{1}{9.2 \times 10^{-3}\ s^{-1}}$$

$$= 1.609 \times \frac{1}{9.2 \times 10^{-3}\ s^{-1}}$$

$$= 1.7 \times 10^2\ s$$

Check Your Learning

Iodine-131 is a radioactive isotope that is used to diagnose and treat some forms of thyroid cancer. Iodine-131 decays to xenon-131 according to the equation:

$$\text{I-131} \longrightarrow \text{Xe-131} + \text{electron}$$

The decay is first-order with a rate constant of 0.138 d^{-1}. All radioactive decay is first order. How many days will it take for 90% of the iodine−131 in a 0.500 M solution of this substance to decay to Xe-131?

Answer: 16.7 days

We can use integrated rate laws with experimental data that consist of time and concentration information to determine the order and rate constant of a reaction. The integrated rate law can be rearranged to a standard linear equation format:

$$\ln[A] \;=\; (-k)(t) + \ln[A]_0$$
$$y \;=\; mx + b$$

A plot of $\ln[A]$ versus t for a first-order reaction is a straight line with a slope of $-k$ and an intercept of $\ln[A]_0$. If a set of rate data are plotted in this fashion but do *not* result in a straight line, the reaction is not first order in A.

Example 12.7

Determination of Reaction Order by Graphing

Show that the data in Figure 12.2 can be represented by a first-order rate law by graphing $\ln[H_2O_2]$ versus time. Determine the rate constant for the rate of decomposition of H_2O_2 from this data.

Solution

The data from Figure 12.2 with the addition of values of $\ln[H_2O_2]$ are given in Figure 12.10.

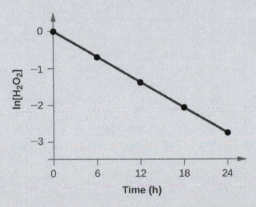

Figure 12.10   The linear relationship between the $\ln[H_2O_2]$ and time shows that the decomposition of hydrogen peroxide is a first-order reaction.

Trial	Time (h)	$[H_2O_2]$ (M)	$\ln[H_2O_2]$
1	0	1.000	0.0
2	6.00	0.500	−0.693
3	12.00	0.250	−1.386
4	18.00	0.125	−2.079
5	24.00	0.0625	−2.772

The plot of $\ln[H_2O_2]$ versus time is linear, thus we have verified that the reaction may be described by a first-order rate law.

The rate constant for a first-order reaction is equal to the negative of the slope of the plot of $\ln[H_2O_2]$ versus time where:

$$\text{slope} = \frac{\text{change in } y}{\text{change in } x} = \frac{\Delta y}{\Delta x} = \frac{\Delta \ln[H_2O_2]}{\Delta t}$$

In order to determine the slope of the line, we need two values of $\ln[H_2O_2]$ at different values of t (one near each end of the line is preferable). For example, the value of $\ln[H_2O_2]$ when t is 6.00 h is −0.693; the value when t = 12.00 h is −1.386:

$$\begin{aligned}
\text{slope} &= \frac{-1.386 - (-0.693)}{12.00 \text{ h} - 6.00 \text{ h}} \\
&= \frac{-0.693}{6.00 \text{ h}} \\
&= -1.155 \times 10^{-2} \text{ h}^{-1} \\
k &= -\text{slope} = -\left(-1.155 \times 10^{-1} \text{ h}^{-1}\right) = 1.155 \times 10^{-1} \text{ h}^{-1}
\end{aligned}$$

Check Your Learning

Graph the following data to determine whether the reaction $A \longrightarrow B + C$ is first order.

Trial	Time (s)	[A]
1	4.0	0.220
2	8.0	0.144
3	12.0	0.110
4	16.0	0.088
5	20.0	0.074

Answer: The plot of $\ln[A]$ vs. t is not a straight line. The equation is not first order:

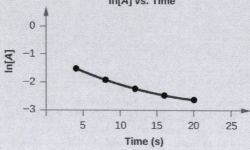

ln[A] vs. Time

Second-Order Reactions

The equations that relate the concentrations of reactants and the rate constant of second-order reactions are fairly complicated. We will limit ourselves to the simplest second-order reactions, namely, those with rates that are dependent upon just one reactant's concentration and described by the differential rate law:

$$\text{Rate} = k[A]^2$$

For these second-order reactions, the integrated rate law is:

$$\frac{1}{[A]} = kt + \frac{1}{[A]_0}$$

where the terms in the equation have their usual meanings as defined earlier.

Example 12.8

The Integrated Rate Law for a Second-Order Reaction

The reaction of butadiene gas (C_4H_6) with itself produces C_8H_{12} gas as follows:

$$2C_4H_6(g) \longrightarrow C_8H_{12}(g)$$

The reaction is second order with a rate constant equal to 5.76×10^{-2} L/mol/min under certain conditions. If the initial concentration of butadiene is 0.200 M, what is the concentration remaining after 10.0 min?

Solution

We use the integrated form of the rate law to answer questions regarding time. For a second-order reaction, we have:

$$\frac{1}{[A]} = kt + \frac{1}{[A]_0}$$

We know three variables in this equation: $[A]_0 = 0.200$ mol/L, $k = 5.76 \times 10^{-2}$ L/mol/min, and $t = 10.0$ min. Therefore, we can solve for $[A]$, the fourth variable:

$$\frac{1}{[A]} = \left(5.76 \times 10^{-2} \text{ L mol}^{-1} \text{ min}^{-1}\right)(10 \text{ min}) + \frac{1}{0.200 \text{ mol}^{-1}}$$

$$\frac{1}{[A]} = \left(5.76 \times 10^{-1} \text{ L mol}^{-1}\right) + 5.00 \text{ L mol}^{-1}$$

$$\frac{1}{[A]} = 5.58 \text{ L mol}^{-1}$$

$$[A] = 1.79 \times 10^{-1} \text{ mol L}^{-1}$$

Therefore 0.179 mol/L of butadiene remain at the end of 10.0 min, compared to the 0.200 mol/L that was originally present.

Check Your Learning

If the initial concentration of butadiene is 0.0200 M, what is the concentration remaining after 20.0 min?

Answer: 0.0196 mol/L

The integrated rate law for our second-order reactions has the form of the equation of a straight line:

$$\frac{1}{[A]} = kt + \frac{1}{[A]_0}$$
$$y = mx + b$$

A plot of $\frac{1}{[A]}$ versus t for a second-order reaction is a straight line with a slope of k and an intercept of $\frac{1}{[A]_0}$. If the plot is not a straight line, then the reaction is not second order.

Example 12.9 Determination of Reaction Order by Graphing

Test the data given to show whether the dimerization of C_4H_6 is a first- or a second-order reaction.

Solution

Trial	Time (s)	$[C_4H_6]$ (M)
1	0	1.00×10^{-2}
2	1600	5.04×10^{-3}
3	3200	3.37×10^{-3}
4	4800	2.53×10^{-3}
5	6200	2.08×10^{-3}

In order to distinguish a first-order reaction from a second-order reaction, we plot $\ln[C_4H_6]$ versus t and compare it with a plot of $\dfrac{1}{[C_4H_6]}$ versus t. The values needed for these plots follow.

Time (s)	$\dfrac{1}{[C_4H_6]}$ (M^{-1})	$\ln[C_4H_6]$
0	100	−4.605
1600	198	−5.289
3200	296	−5.692
4800	395	−5.978
6200	481	−6.175

The plots are shown in Figure 12.11. As you can see, the plot of $\ln[C_4H_6]$ versus t is not linear, therefore the reaction is not first order. The plot of $\dfrac{1}{[C_4H_6]}$ versus t is linear, indicating that the reaction is second order.

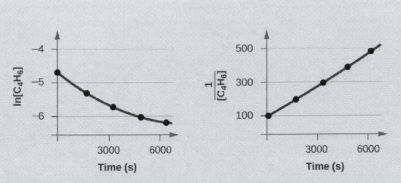

Figure 12.11 These two graphs show first- and second-order plots for the dimerization of C_4H_6. Since the first-order plot (left) is not linear, we know that the reaction is not first order. The linear trend in the second-order plot (right) indicates that the reaction follows second-order kinetics.

Check Your Learning

Does the following data fit a second-order rate law?

Trial	Time (s)	[A] (M)
1	5	0.952
2	10	0.625
3	15	0.465
4	20	0.370
5	25	0.308
6	35	0.230

Answer: Yes. The plot of $\frac{1}{[A]}$ vs. t is linear: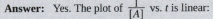

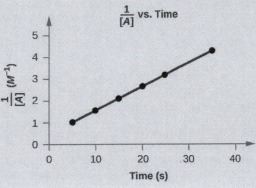

Zero-Order Reactions

For zero-order reactions, the differential rate law is:

$$\text{Rate} = k[A]^0 = k$$

A zero-order reaction thus exhibits a constant reaction rate, regardless of the concentration of its reactants.

The integrated rate law for a zero-order reaction also has the form of the equation of a straight line:

$$[A] = -kt + [A]_0$$
$$y = mx + b$$

A plot of $[A]$ versus t for a zero-order reaction is a straight line with a slope of $-k$ and an intercept of $[A]_0$. Figure 12.12 shows a plot of $[NH_3]$ versus t for the decomposition of ammonia on a hot tungsten wire and for the decomposition of ammonia on hot quartz (SiO_2). The decomposition of NH_3 on hot tungsten is zero order; the plot is a straight line. The decomposition of NH_3 on hot quartz is not zero order (it is first order). From the slope of the line for the zero-order decomposition, we can determine the rate constant:

$$\text{slope} = -k = 1.3110^{-6} \text{ mol/L/s}$$

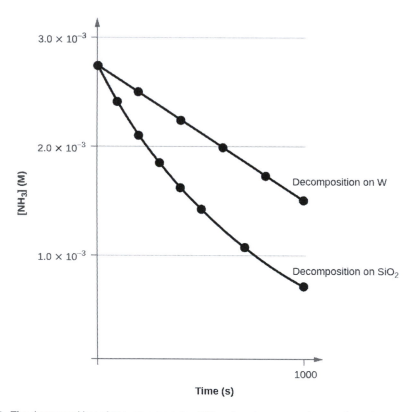

Figure 12.12   The decomposition of NH_3 on a tungsten (W) surface is a zero-order reaction, whereas on a quartz (SiO_2) surface, the reaction is first order.

The Half-Life of a Reaction

The **half-life of a reaction ($t_{1/2}$)** is the time required for one-half of a given amount of reactant to be consumed. In each succeeding half-life, half of the remaining concentration of the reactant is consumed. Using the decomposition of hydrogen peroxide (Figure 12.2) as an example, we find that during the first half-life (from 0.00 hours to 6.00 hours), the concentration of H_2O_2 decreases from 1.000 M to 0.500 M. During the second half-life (from 6.00 hours to 12.00 hours), it decreases from 0.500 *M* to 0.250 *M*; during the third half-life, it decreases from 0.250 *M* to 0.125 *M*. The concentration of H_2O_2 decreases by half during each successive period of 6.00 hours. The decomposition of hydrogen peroxide is a first-order reaction, and, as can be shown, the half-life of a first-order reaction is independent of the concentration of the reactant. However, half-lives of reactions with other orders depend on the concentrations of the reactants.

First-Order Reactions

We can derive an equation for determining the half-life of a first-order reaction from the alternate form of the integrated rate law as follows:

$$\ln \frac{[A]_0}{[A]} = kt$$

$$t = \ln \frac{[A]_0}{[A]} \times \frac{1}{k}$$

If we set the time *t* equal to the half-life, $t_{1/2}$, the corresponding concentration of *A* at this time is equal to one-half of its initial concentration. Hence, when $t = t_{1/2}$, $[A] = \frac{1}{2}[A]_0$.

Therefore:

$$t_{1/2} = \ln \frac{[A]_0}{\frac{1}{2}[A]_0} \times \frac{1}{k}$$

$$= \ln 2 \times \frac{1}{k} = 0.693 \times \frac{1}{k}$$

Thus:

$$t_{1/2} = \frac{0.693}{k}$$

We can see that the half-life of a first-order reaction is inversely proportional to the rate constant *k*. A fast reaction (shorter half-life) will have a larger *k*; a slow reaction (longer half-life) will have a smaller *k*.

Example 12.10

Calculation of a First-order Rate Constant using Half-Life

Calculate the rate constant for the first-order decomposition of hydrogen peroxide in water at 40 °C, using the data given in Figure 12.13.

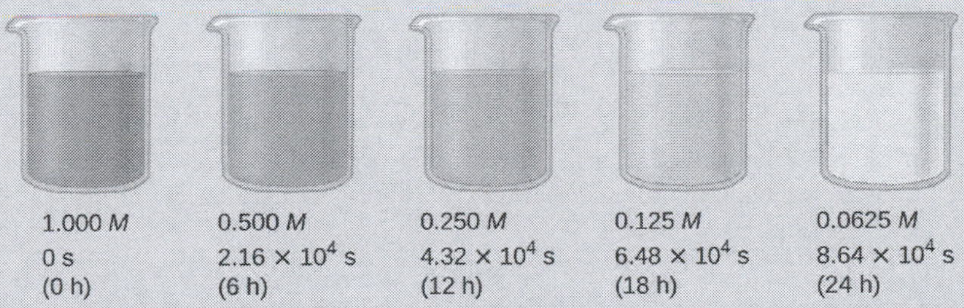

1.000 *M*	0.500 *M*	0.250 *M*	0.125 *M*	0.0625 *M*
0 s	2.16×10^4 s	4.32×10^4 s	6.48×10^4 s	8.64×10^4 s
(0 h)	(6 h)	(12 h)	(18 h)	(24 h)

Figure 12.13 The decomposition of H_2O_2 ($2H_2O_2 \longrightarrow 2H_2O + O_2$) at 40 °C is illustrated. The intensity of the color symbolizes the concentration of H_2O_2 at the indicated times; H_2O_2 is actually colorless.

Solution

The half-life for the decomposition of H_2O_2 is 2.16×10^4 s:

$$t_{1/2} = \frac{0.693}{k}$$

$$k = \frac{0.693}{t_{1/2}} = \frac{0.693}{2.16 \times 10^4 \text{ s}} = 3.21 \times 10^{-5} \text{ s}^{-1}$$

Check Your Learning

The first-order radioactive decay of iodine-131 exhibits a rate constant of 0.138 d^{-1}. What is the half-life for this decay?

Answer: 5.02 d.

Second-Order Reactions

We can derive the equation for calculating the half-life of a second order as follows:

$$\frac{1}{[A]} = kt + \frac{1}{[A]_0}$$

or

$$\frac{1}{[A]} - \frac{1}{[A]_0} = kt$$

If

$$t = t_{1/2}$$

then

$$[A] = \frac{1}{2}[A]_0$$

and we can write:

$$\begin{aligned} \frac{1}{\frac{1}{2}[A]_0} - \frac{1}{[A]_0} &= kt_{1/2} \\ 2[A]_0 - \frac{1}{[A]_0} &= kt_{1/2} \\ \frac{1}{[A]_0} &= kt_{1/2} \end{aligned}$$

Thus:

$$t_{1/2} = \frac{1}{k[A]_0}$$

For a second-order reaction, $t_{1/2}$ is inversely proportional to the concentration of the reactant, and the half-life increases as the reaction proceeds because the concentration of reactant decreases. Consequently, we find the use of the half-life concept to be more complex for second-order reactions than for first-order reactions. Unlike with first-order reactions, the rate constant of a second-order reaction cannot be calculated directly from the half-life unless the initial concentration is known.

Zero-Order Reactions

We can derive an equation for calculating the half-life of a zero order reaction as follows:

$$[A] = -kt + [A]_0$$

When half of the initial amount of reactant has been consumed $t = t_{1/2}$ and $[A] = \frac{[A]_0}{2}$. Thus:

$$\frac{[A]_0}{2} = -kt_{1/2} + [A]_0$$

$$kt_{1/2} = \frac{[A]_0}{2}$$

and

$$t_{1/2} = \frac{[A]_0}{2k}$$

The half-life of a zero-order reaction increases as the initial concentration increases.

Equations for both differential and integrated rate laws and the corresponding half-lives for zero-, first-, and second-order reactions are summarized in Table 12.2.

Summary of Rate Laws for Zero-, First-, and Second-Order Reactions

	Zero-Order	First-Order	Second-Order
rate law	rate = k	rate = $k[A]$	rate = $k[A]^2$
units of rate constant	$M\,s^{-1}$	s^{-1}	$M^{-1}\,s^{-1}$
integrated rate law	$[A] = -kt + [A]_0$	$\ln[A] = -kt + \ln[A]_0$	$\frac{1}{[A]} = kt + \left(\frac{1}{[A]_0}\right)$
plot needed for linear fit of rate data	$[A]$ vs. t	$\ln[A]$ vs. t	$\frac{1}{[A]}$ vs. t
relationship between slope of linear plot and rate constant	$k = -$slope	$k = -$slope	$k = +$slope
half-life	$t_{1/2} = \frac{[A]_0}{2k}$	$t_{1/2} = \frac{0.693}{k}$	$t_{1/2} = \frac{1}{[A]_0 k}$

Table 12.2

12.5 Collision Theory

By the end of this section, you will be able to:

- Use the postulates of collision theory to explain the effects of physical state, temperature, and concentration on reaction rates
- Define the concepts of activation energy and transition state
- Use the Arrhenius equation in calculations relating rate constants to temperature

We should not be surprised that atoms, molecules, or ions must collide before they can react with each other. Atoms must be close together to form chemical bonds. This simple premise is the basis for a very powerful theory that explains many observations regarding chemical kinetics, including factors affecting reaction rates.

Collision theory is based on the following postulates:

1. The rate of a reaction is proportional to the rate of reactant collisions:

$$\text{reaction rate} \propto \frac{\#\ \text{collisions}}{\text{time}}$$

2. The reacting species must collide in an orientation that allows contact between the atoms that will become bonded together in the product.

3. The collision must occur with adequate energy to permit mutual penetration of the reacting species' valence shells so that the electrons can rearrange and form new bonds (and new chemical species).

We can see the importance of the two physical factors noted in postulates 2 and 3, the orientation and energy of collisions, when we consider the reaction of carbon monoxide with oxygen:

$$2\,CO(g) + O_2(g) \longrightarrow 2\,CO_2(g)$$

Carbon monoxide is a pollutant produced by the combustion of hydrocarbon fuels. To reduce this pollutant, automobiles have catalytic converters that use a catalyst to carry out this reaction. It is also a side reaction of the combustion of gunpowder that results in muzzle flash for many firearms. If carbon monoxide and oxygen are present in sufficient quantity, the reaction is spontaneous at high temperature and pressure.

The first step in the gas-phase reaction between carbon monoxide and oxygen is a collision between the two molecules:

$$CO(g) + O_2(g) \longrightarrow CO_2(g) + O(g)$$

Although there are many different possible orientations the two molecules can have relative to each other, consider the two presented in Figure 12.14. In the first case, the oxygen side of the carbon monoxide molecule collides with the oxygen molecule. In the second case, the carbon side of the carbon monoxide molecule collides with the oxygen molecule. The second case is clearly more likely to result in the formation of carbon dioxide, which has a central carbon atom bonded to two oxygen atoms $(O = C = O)$. This is a rather simple example of how important the orientation of the collision is in terms of creating the desired product of the reaction.

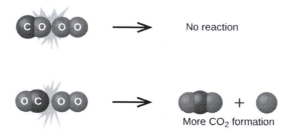

Figure 12.14 Illustrated are two collisions that might take place between carbon monoxide and oxygen molecules. The orientation of the colliding molecules partially determines whether a reaction between the two molecules will occur.

If the collision does take place with the correct orientation, there is still no guarantee that the reaction will proceed to form carbon dioxide. Every reaction requires a certain amount of activation energy for it to proceed in the forward direction, yielding an appropriate activated complex along the way. As Figure 12.15 demonstrates, even a collision with the correct orientation can fail to form the reaction product. In the study of reaction mechanisms, each of these three arrangements of atoms is called a proposed **activated complex** or **transition state**.

$$O=C\cdots O=O$$

$$O\cdots C\cdots O=O$$

$$O=C\cdots O\cdots O$$

Figure 12.15 Possible transition states (activated complexes) for carbon monoxide reacting with oxygen to form carbon dioxide. Solid lines represent covalent bonds, while dotted lines represent unstable orbital overlaps that may, or may not, become covalent bonds as product is formed. In the first two examples in this figure, the O=O double bond is not impacted; therefore, carbon dioxide cannot form. The third proposed transition state will result in the formation of carbon dioxide if the third "extra" oxygen atom separates from the rest of the molecule.

In most circumstances, it is impossible to isolate or identify a transition state or activated complex. In the reaction between carbon monoxide and oxygen to form carbon dioxide, activated complexes have only been observed spectroscopically in systems that utilize a heterogeneous catalyst. The gas-phase reaction occurs too rapidly to isolate any such chemical compound.

Collision theory explains why most reaction rates increase as concentrations increase. With an increase in the concentration of any reacting substance, the chances for collisions between molecules are increased because there are more molecules per unit of volume. More collisions mean a faster reaction rate, assuming the energy of the collisions is adequate.

Activation Energy and the Arrhenius Equation

The minimum energy necessary to form a product during a collision between reactants is called the **activation energy** (E_a). The kinetic energy of reactant molecules plays an important role in a reaction because the energy necessary to form a product is provided by a collision of a reactant molecule with another reactant molecule. (In single-reactant reactions, activation energy may be provided by a collision of the reactant molecule with the wall of the reaction vessel or with molecules of an inert contaminant.) If the activation energy is much larger than the average kinetic energy of the molecules, the reaction will occur slowly: Only a few fast-moving molecules will have enough energy to react. If the activation energy is much smaller than the average kinetic energy of the molecules, the fraction of molecules possessing the necessary kinetic energy will be large; most collisions between molecules will result in reaction, and the reaction will occur rapidly.

Figure 12.16 shows the energy relationships for the general reaction of a molecule of A with a molecule of B to form molecules of C and D:

$$A + B \longrightarrow C + D$$

The figure shows that the energy of the transition state is higher than that of the reactants A and B by an amount equal to E_a, the activation energy. Thus, the sum of the kinetic energies of A and B must be equal to or greater than E_a to reach the transition state. After the transition state has been reached, and as C and D begin to form, the system loses energy until its total energy is lower than that of the initial mixture. This lost energy is transferred to other molecules, giving them enough energy to reach the transition state. The forward reaction (that between molecules A and B) therefore tends to take place readily once the reaction has started. In Figure 12.16, ΔH represents the difference in enthalpy between the reactants (A and B) and the products (C and D). The sum of E_a and ΔH represents the activation energy for the reverse reaction:

$$C + D \longrightarrow A + B$$

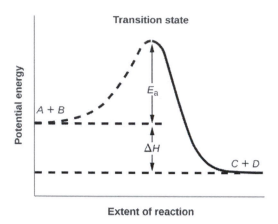

Figure 12.16 This graph shows the potential energy relationships for the reaction $A + B \longrightarrow C + D$. The dashed portion of the curve represents the energy of the system with a molecule of A and a molecule of B present, and the solid portion the energy of the system with a molecule of C and a molecule of D present. The activation energy for the forward reaction is represented by E_a. The activation energy for the reverse reaction is greater than that for the forward reaction by an amount equal to ΔH. The curve's peak is represented the transition state.

We can use the **Arrhenius equation** to relate the activation energy and the rate constant, k, of a given reaction:

$$k = Ae^{-E_a/RT}$$

In this equation, R is the ideal gas constant, which has a value 8.314 J/mol/K, T is temperature on the Kelvin scale, E_a is the activation energy in joules per mole, e is the constant 2.7183, and A is a constant called the **frequency factor**, which is related to the frequency of collisions and the orientation of the reacting molecules.

Both postulates of the collision theory of reaction rates are accommodated in the Arrhenius equation. The frequency factor A is related to the rate at which collisions having the correct *orientation* occur. The exponential term, $e^{-E_a/RT}$, is related to the fraction of collisions providing adequate *energy* to overcome the activation barrier of the reaction.

At one extreme, the system does not contain enough energy for collisions to overcome the activation barrier. In such cases, no reaction occurs. At the other extreme, the system has so much energy that every collision with the correct orientation can overcome the activation barrier, causing the reaction to proceed. In such cases, the reaction is nearly instantaneous.

The Arrhenius equation describes quantitatively much of what we have already discussed about reaction rates. For two reactions at the same temperature, the reaction with the higher activation energy has the lower rate constant and the slower rate. The larger value of E_a results in a smaller value for $e^{-E_a/RT}$, reflecting the smaller fraction of molecules with enough energy to react. Alternatively, the reaction with the smaller E_a has a larger fraction of molecules with enough energy to react. This will be reflected as a larger value of $e^{-E_a/RT}$, a larger rate constant, and a faster rate for the reaction. An increase in temperature has the same effect as a decrease in activation energy. A larger fraction of molecules has the necessary energy to react (Figure 12.17), as indicated by an increase in the value of $e^{-E_a/RT}$. The rate constant is also directly proportional to the frequency factor, A. Hence a change in conditions or reactants that increases the number of collisions with a favorable orientation for reaction results in an increase in A and, consequently, an increase in k.

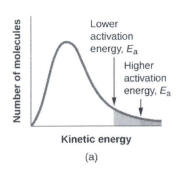
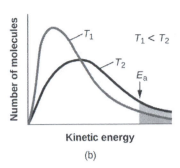

Number of molecules (y-axis, graph a)

Lower activation energy, E_a

Higher activation energy, E_a

Kinetic energy

(a)

Number of molecules (y-axis, graph b)

T_1

$T_1 < T_2$

T_2

E_a

Kinetic energy

(b)

Figure 12.17   (a) As the activation energy of a reaction decreases, the number of molecules with at least this much energy increases, as shown by the shaded areas. (b) At a higher temperature, T_2, more molecules have kinetic energies greater than E_a, as shown by the yellow shaded area.

A convenient approach to determining E_a for a reaction involves the measurement of k at different temperatures and using of an alternate version of the Arrhenius equation that takes the form of linear equation:

$$\ln k = \left(\frac{-E_a}{R}\right)\left(\frac{1}{T}\right) + \ln A$$
$$y = mx + b$$

Thus, a plot of $\ln k$ versus $\frac{1}{T}$ gives a straight line with the slope $\frac{-E_a}{R}$, from which E_a may be determined. The intercept gives the value of $\ln A$.

Example 12.11

Determination of E_a

The variation of the rate constant with temperature for the decomposition of HI(g) to $H_2(g)$ and $I_2(g)$ is given here. What is the activation energy for the reaction?

$$2HI(g) \longrightarrow H_2(g) + I_2(g)$$

T (K)	k (L/mol/s)
555	3.52×10^{-7}
575	1.22×10^{-6}
645	8.59×10^{-5}
700	1.16×10^{-3}
781	3.95×10^{-2}

Solution

Values of $\frac{1}{T}$ and $\ln k$ are:

$\frac{1}{T}$ (K^{-1})	ln k
1.80×10^{-3}	-14.860
1.74×10^{-3}	-13.617
1.55×10^{-3}	-9.362
1.43×10^{-3}	-6.759
1.28×10^{-3}	-3.231

Figure 12.18 is a graph of ln k versus $\frac{1}{T}$. To determine the slope of the line, we need two values of ln k, which are determined from the line at two values of $\frac{1}{T}$ (one near each end of the line is preferable). For example, the value of ln k determined from the line when $\frac{1}{T} = 1.25 \times 10^{-3}$ is -2.593; the value when $\frac{1}{T} = 1.78 \times 10^{-3}$ is -14.447.

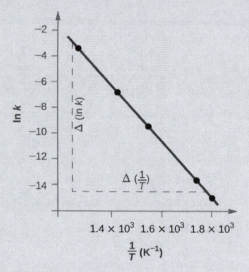

Figure 12.18 This graph shows the linear relationship between ln k and $\frac{1}{T}$ for the reaction $2HI \longrightarrow H_2 + I_2$ according to the Arrhenius equation.

The slope of this line is given by the following expression:

$$\text{Slope} = \frac{\Delta(\ln k)}{\Delta\left(\frac{1}{T}\right)}$$

$$= \frac{(-14.447) - (-2.593)}{\left(1.78 \times 10^{-3}\text{ K}^{-1}\right) - \left(1.25 \times 10^{-3}\text{ K}^{-1}\right)}$$

$$= \frac{-11.854}{0.53 \times 10^{-3}\text{ K}^{-1}} = 2.2 \times 10^4\text{ K}$$

$$= -\frac{E_a}{R}$$

Thus:

$$E_a = -\text{slope} \times R = -(-2.2 \times 10^4\text{ K} \times 8.314\text{ J mol}^{-1}\text{ K}^{-1})$$

$$E_a = 1.8 \times 10^5 \text{ J mol}$$

In many situations, it is possible to obtain a reasonable estimate of the activation energy without going through the entire process of constructing the Arrhenius plot. The Arrhenius equation:

$$\ln k = \left(\frac{-E_a}{R}\right)\left(\frac{1}{T}\right) + \ln A$$

can be rearranged as shown to give:

$$\frac{\Delta(\ln k)}{\Delta\left(\frac{1}{T}\right)} = -\frac{E_a}{R}$$

or

$$\ln\frac{k_1}{k_2} = \frac{E_a}{R}\left(\frac{1}{T_2} - \frac{1}{T_1}\right)$$

This equation can be rearranged to give a one-step calculation to obtain an estimate for the activation energy:

$$E_a = -R\left(\frac{\ln k_2 - \ln k_1}{\left(\frac{1}{T_2}\right) - \left(\frac{1}{T_1}\right)}\right)$$

Using the experimental data presented here, we can simply select two data entries. For this example, we select the first entry and the last entry:

T (K)	k (L/mol/s)	$\frac{1}{T}$ (K^{-1})	$\ln k$
555	3.52×10^{-7}	1.80×10^{-3}	-14.860
781	3.95×10^{-2}	1.28×10^{-3}	-3.231

After calculating $\frac{1}{T}$ and $\ln k$, we can substitute into the equation:

$$E_a = -8.314\text{ J mol}^{-1}\text{ K}^{-1}\left(\frac{-3.231 - (-14.860)}{1.28 \times 10^{-3}\text{ K}^{-1} - 1.80 \times 10^{-3}\text{ K}^{-1}}\right)$$

and the result is $E_a = 185,900$ J/mol.

This method is very effective, especially when a limited number of temperature-dependent rate constants are available for the reaction of interest.

Check Your Learning

The rate constant for the rate of decomposition of N_2O_5 to NO and O_2 in the gas phase is 1.66 L/mol/s at 650 K and 7.39 L/mol/s at 700 K:

$$2N_2O_5(g) \longrightarrow 4NO(g) + 3O_2(g)$$

Assuming the kinetics of this reaction are consistent with the Arrhenius equation, calculate the activation energy for this decomposition.

Answer: 113,000 J/mol

12.6 Reaction Mechanisms

By the end of this section, you will be able to:

- Distinguish net reactions from elementary reactions (steps)
- Identify the molecularity of elementary reactions
- Write a balanced chemical equation for a process given its reaction mechanism
- Derive the rate law consistent with a given reaction mechanism

A balanced equation for a chemical reaction indicates what is reacting and what is produced, but it reveals nothing about how the reaction actually takes place. The **reaction mechanism** (or reaction path) is the process, or pathway, by which a reaction occurs.

A chemical reaction usually occurs in steps, although it may not always be obvious to an observer. The decomposition of ozone, for example, appears to follow a mechanism with two steps:

$$O_3(g) \longrightarrow O_2(g) + O$$
$$O + O_3(g) \longrightarrow 2O_2(g)$$

We call each step in a reaction mechanism an **elementary reaction**. Elementary reactions occur exactly as they are written and cannot be broken down into simpler steps. Elementary reactions add up to the overall reaction, which, for the decomposition, is:

$$2O_3(g) \longrightarrow 3O_2(g)$$

Notice that the oxygen atom produced in the first step of this mechanism is consumed in the second step and therefore does not appear as a product in the overall reaction. Species that are produced in one step and consumed in a subsequent step are called **intermediates**.

While the overall reaction equation for the decomposition of ozone indicates that two molecules of ozone react to give three molecules of oxygen, the mechanism of the reaction does not involve the collision and reaction of two ozone molecules. Rather, it involves a molecule of ozone decomposing to an oxygen molecule and an intermediate oxygen atom; the oxygen atom then reacts with a second ozone molecule to give two oxygen molecules. These two elementary reactions occur exactly as they are shown in the reaction mechanism.

Unimolecular Elementary Reactions

The **molecularity** of an elementary reaction is the number of reactant species (atoms, molecules, or ions). For example, a **unimolecular reaction** involves the rearrangement of a *single* reactant species to produce one or more molecules of product:

$$A \longrightarrow \text{products}$$

The rate equation for a unimolecular reaction is:

$$\text{rate} = k[A]$$

A unimolecular reaction may be one of several elementary reactions in a complex mechanism. For example, the reaction:

$$O_3 \longrightarrow O_2 + O$$

illustrates a unimolecular elementary reaction that occurs as one part of a two-step reaction mechanism. However, some unimolecular reactions may have only a single reaction in the reaction mechanism. (In other words, an elementary reaction can also be an overall reaction in some cases.) For example, the gas-phase decomposition of cyclobutane, C_4H_8, to ethylene, C_2H_4, occurs via a unimolecular, single-step mechanism:

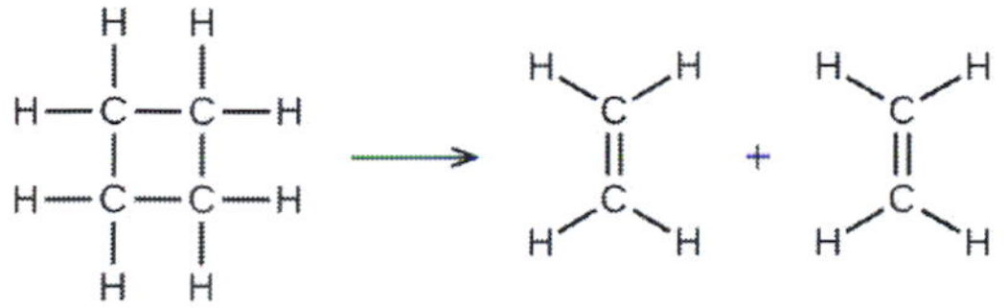

For these unimolecular reactions to occur, all that is required is the separation of parts of single reactant molecules into products.

Chemical bonds do not simply fall apart during chemical reactions. Energy is required to break chemical bonds. The activation energy for the decomposition of C_4H_8, for example, is 261 kJ per mole. This means that it requires 261 kilojoules to distort one mole of these molecules into activated complexes that decompose into products:

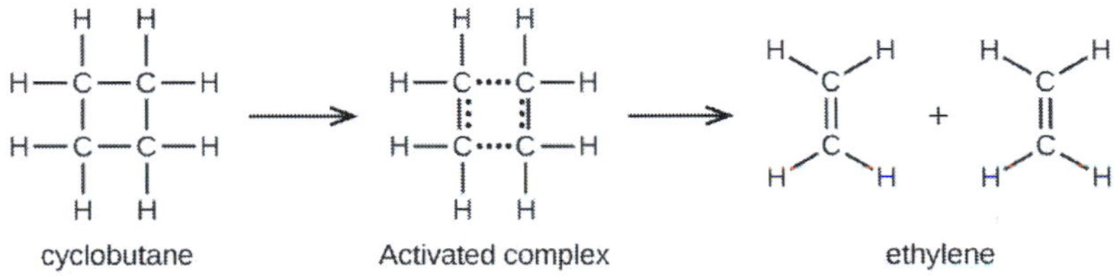

cyclobutane   Activated complex   ethylene

In a sample of C_4H_8, a few of the rapidly moving C_4H_8 molecules collide with other rapidly moving molecules and pick up additional energy. When the C_4H_8 molecules gain enough energy, they can transform into an activated complex, and the formation of ethylene molecules can occur. In effect, a particularly energetic collision knocks a C_4H_8 molecule into the geometry of the activated complex. However, only a small fraction of gas molecules travel at sufficiently high speeds with large enough kinetic energies to accomplish this. Hence, at any given moment, only a few molecules pick up enough energy from collisions to react.

The rate of decomposition of C_4H_8 is directly proportional to its concentration. Doubling the concentration of C_4H_8 in a sample gives twice as many molecules per liter. Although the fraction of molecules with enough energy to react remains the same, the total number of such molecules is twice as great. Consequently, there is twice as much C_4H_8 per liter, and the reaction rate is twice as fast:

$$\text{rate} = -\frac{\Delta[C_4H_8]}{\Delta t} = k[C_4H_8]$$

A similar relationship applies to any unimolecular elementary reaction; the reaction rate is directly proportional to the concentration of the reactant, and the reaction exhibits first-order behavior. The proportionality constant is the rate constant for the particular unimolecular reaction.

Bimolecular Elementary Reactions

The collision and combination of two molecules or atoms to form an activated complex in an elementary reaction is called a **bimolecular reaction**. There are two types of bimolecular elementary reactions:

$$A + B \longrightarrow \text{products}$$
$$\text{and}$$
$$2A \longrightarrow \text{products}$$

For the first type, in which the two reactant molecules are different, the rate law is first-order in A and first order in B:

$$\text{rate} = k[A][B]$$

For the second type, in which two identical molecules collide and react, the rate law is second order in A:

$$\text{rate} = k[A][A] = k[A]^2$$

Some chemical reactions have mechanisms that consist of a single bimolecular elementary reaction. One example is the reaction of nitrogen dioxide with carbon monoxide:

$$NO_2(g) + CO(g) \longrightarrow NO(g) + CO_2(g)$$

Another is the decomposition of two hydrogen iodide molecules to produce hydrogen, H_2, and iodine, I_2 Figure 12.19:

$$2HI(g) \longrightarrow H_2(g) + I_2(g)$$

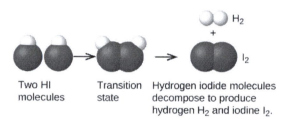

Two HI molecules   Transition state   Hydrogen iodide molecules decompose to produce hydrogen H_2 and iodine I_2.

Figure 12.19 The probable mechanism for the dissociation of two HI molecules to produce one molecule of H_2 and one molecule of I_2.

Bimolecular elementary reactions may also be involved as steps in a multistep reaction mechanism. The reaction of atomic oxygen with ozone is one example:

$$O(g) + O_3(g) \longrightarrow 2O_2(g)$$

Termolecular Elementary Reactions

An elementary **termolecular reaction** involves the simultaneous collision of three atoms, molecules, or ions. Termolecular elementary reactions are uncommon because the probability of three particles colliding simultaneously is less than one one-thousandth of the probability of two particles colliding. There are, however, a few established termolecular elementary reactions. The reaction of nitric oxide with oxygen appears to involve termolecular steps:

$$2NO + O_2 \longrightarrow 2NO_2$$
$$\text{rate} = k[NO]^2[O_2]$$

Likewise, the reaction of nitric oxide with chlorine appears to involve termolecular steps:

$$2NO + Cl_2 \longrightarrow 2NOCl$$
$$\text{rate} = k[NO]^2[Cl_2]$$

Relating Reaction Mechanisms to Rate Laws

It's often the case that one step in a multistep reaction mechanism is significantly slower than the others. Because a reaction cannot proceed faster than its slowest step, this step will limit the rate at which the overall reaction occurs. The slowest step is therefore called the **rate-limiting step** (or rate-determining step) of the reaction Figure 12.20.

Figure 12.20   A cattle chute is a nonchemical example of a rate-determining step. Cattle can only be moved from one holding pen to another as quickly as one animal can make its way through the chute. (credit: Loren Kerns)

As described earlier, rate laws may be derived directly from the chemical equations for elementary reactions. This is not the case, however, for ordinary chemical reactions. The balanced equations most often encountered represent the overall change for some chemical system, and very often this is the result of some multistep reaction mechanisms. In every case, we must determine the overall rate law from experimental data and deduce the mechanism from the rate law (and sometimes from other data). The reaction of NO_2 and CO provides an illustrative example:

$$NO_2(g) + CO(g) \longrightarrow CO_2(g) + NO(g)$$

For temperatures above 225 °C, the rate law has been found to be:

$$\text{rate} = k[NO_2][CO]$$

The reaction is first order with respect to NO_2 and first-order with respect to CO. This is consistent with a single-step bimolecular mechanism and it is *possible* that this is the mechanism for this reaction at high temperatures.

At temperatures below 225 °C, the reaction is described by a rate law that is second order with respect to NO_2:

$$\text{rate} = k[NO_2]^2$$

This is consistent with a mechanism that involves the following two elementary reactions, the first of which is slower and is therefore the rate-determining step:

$$NO_2(g) + NO_2(g) \longrightarrow NO_3(g) + NO(g) \text{ (slow)}$$
$$NO_3(g) + CO(g) \longrightarrow NO_2(g) + CO_2(g) \text{ (fast)}$$

The rate-determining step gives a rate law showing second-order dependence on the NO_2 concentration, and the sum of the two equations gives the net overall reaction.

In general, when the rate-determining (slower) step is the first step in a mechanism, the rate law for the overall reaction is the same as the rate law for this step. However, when the rate-determining step is preceded by a step involving an *equilibrium* reaction, the rate law for the overall reaction may be more difficult to derive.

An elementary reaction is at equilibrium when it proceeds in both the forward and reverse directions at equal rates. Consider the dimerization of NO to N_2O_2, with k_1 used to represent the rate constant of the forward reaction and k_{-1} used to represent the rate constant of the reverse reaction:

$$NO + NO \rightleftharpoons N_2O_2$$
$$\text{rate}_{forward} = \text{rate}_{reverse}$$
$$k_1[NO]^2 = k_{-1}[N_2O_2]$$

If N_2O_2 was an intermediate in a mechanism, this expression could be rearranged to represent the concentration of N_2O_2 in the overall rate law expression using algebraic manipulation:

$$\left(\frac{k_1[NO]^2}{k_{-1}}\right) = [N_2O_2]$$

However, once again, intermediates cannot be listed as part of the overall rate law expression, though they can be included in an individual elementary reaction of a mechanism. Example 12.12 will illustrate how to derive overall rate laws from mechanisms involving equilibrium steps preceding the rate-determining step.

Example 12.12

Deriving the Overall Rate Law Expression for a Multistep Reaction Mechanism

Nitryl chloride (NO_2Cl) decomposes to nitrogen dioxide (NO_2) and chlorine gas (Cl_2) according to the following mechanism:

1. $2NO_2Cl(g) \rightleftharpoons ClO_2(g) + N_2O(g) + ClO(g)$ (fast, k_1 represents the rate constant for the forward reaction and k_{-1} the rate constant for the reverse reaction)

2. $N_2O(g) + ClO_2(g) \rightleftharpoons NO_2(g) + NOCl(g)$ (fast, k_2 for the forward reaction, k_{-2} for the reverse reaction)

3. $NOCl + ClO \longrightarrow NO_2 + Cl_2$ (slow, k_3 the rate constant for the forward reaction)

Determine the overall reaction, write the rate law expression for each elementary reaction, identify any intermediates, and determine the overall rate law expression.

Solution

For the overall reaction, simply sum the three steps, cancel intermediates, and combine like formulas:

$$2NO_2Cl(g) \longrightarrow 2NO_2(g) + Cl_2(g)$$

Next, write the rate law expression for each elementary reaction. Remember that for elementary reactions that are part of a mechanism, the rate law expression can be derived directly from the stoichiometry:

$$k_1[NO_2Cl]_2 = k_{-1}[ClO_2][N_2O][ClO]$$
$$k_2[N_2O][ClO_2] = k_{-2}[NO_2][NOCl]$$
$$\text{Rate} = k_3[NOCl][ClO]$$

The third step, which is the slow step, is the rate-determining step. Therefore, the overall rate law expression could be written as Rate = k_3 [NOCl][ClO]. However, both NOCl and ClO are intermediates. Algebraic expressions must be used to represent [NOCl] and [ClO] such that no intermediates remain in the overall rate law expression.

Using elementary reaction 1, $[ClO] = \dfrac{k_1[NO_2Cl]^2}{k_{-1}[ClO_2][N_2O]}$.

Using elementary reaction 2, $[NOCl] = \dfrac{k_2[N_2O][ClO_2]}{k_{-2}[NO_2]}$.

Now substitute these algebraic expressions into the overall rate law expression and simplify:

$$\text{rate} = k_3 \left(\frac{k_2[N_2O][ClO_2]}{k_{-2}[NO_2]} \right) \left(\frac{k_1[NO_2Cl]^2}{k_{-1}[ClO_2][N_2O]} \right)$$

$$\text{rate} = \frac{k_3 k_2 k_1 [NO_2Cl]^2}{k_{-2} k_{-1} [NO_2]}$$

Notice that this rate law shows an *inverse* dependence on the concentration of one of the product species, consistent with the presence of an equilibrium step in the reaction mechanism.

Check Your Learning

Atomic chlorine in the atmosphere reacts with ozone in the following pair of elementary reactions:

$$Cl + O_3(g) \longrightarrow ClO(g) + O_2(g) \qquad \text{(rate constant } k_1\text{)}$$
$$ClO(g) + O \longrightarrow Cl(g) + O_2(g) \qquad \text{(rate constant } k_2\text{)}$$

Determine the overall reaction, write the rate law expression for each elementary reaction, identify any intermediates, and determine the overall rate law expression.

Answer: overall reaction: $O_3(g) + O \longrightarrow 2O_2(g)$

$\text{rate}_1 = k_1[O_3][Cl]$; $\text{rate}_2 = k_2[ClO][O]$

intermediate: $ClO(g)$

overall rate $= k_2 k_1 [O_3][Cl][O]$

12.7 Catalysis

By the end of this section, you will be able to:

- Explain the function of a catalyst in terms of reaction mechanisms and potential energy diagrams
- List examples of catalysis in natural and industrial processes

We have seen that the rate of many reactions can be accelerated by catalysts. A catalyst speeds up the rate of a reaction by lowering the activation energy; in addition, the catalyst is regenerated in the process. Several reactions that are thermodynamically favorable in the absence of a catalyst only occur at a reasonable rate when a catalyst is present. One such reaction is catalytic hydrogenation, the process by which hydrogen is added across an alkene C=C bond to afford the saturated alkane product. A comparison of the reaction coordinate diagrams (also known as energy diagrams) for catalyzed and uncatalyzed alkene hydrogenation is shown in Figure 12.21.

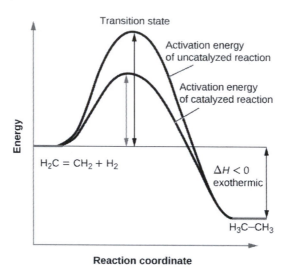

Figure 12.21 This graph compares the reaction coordinates for catalyzed and uncatalyzed alkene hydrogenation.

Catalysts function by providing an alternate reaction mechanism that has a lower activation energy than would be found in the absence of the catalyst. In some cases, the catalyzed mechanism may include additional steps, as depicted in the reaction diagrams shown in **Figure 12.22**. This lower activation energy results in an increase in rate as described by the Arrhenius equation. Note that a catalyst decreases the activation energy for both the forward and the reverse reactions and hence *accelerates both the forward and the reverse reactions*. Consequently, the presence of a catalyst will permit a system to reach equilibrium more quickly, but it has no effect on the position of the equilibrium as reflected in the value of its equilibrium constant (see the later chapter on chemical equilibrium).

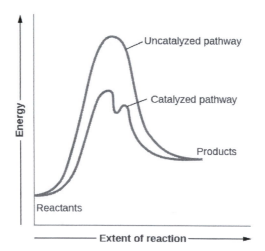

Figure 12.22 This potential energy diagram shows the effect of a catalyst on the activation energy. The catalyst provides a different reaction path with a lower activation energy. As shown, the catalyzed pathway involves a two-step mechanism (note the presence of two transition states) and an intermediate species (represented by the valley between the two transitions states).

Example 12.13

Using Reaction Diagrams to Compare Catalyzed Reactions

The two reaction diagrams here represent the same reaction: one without a catalyst and one with a catalyst. Identify which diagram suggests the presence of a catalyst, and determine the activation energy for the catalyzed reaction:

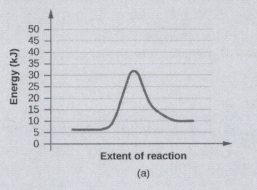
(a)

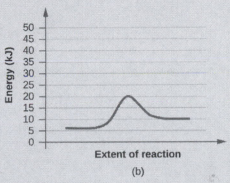
(b)

Solution

A catalyst does not affect the energy of reactant or product, so those aspects of the diagrams can be ignored; they are, as we would expect, identical in that respect. There is, however, a noticeable difference in the transition state, which is distinctly lower in diagram (b) than it is in (a). This indicates the use of a catalyst in diagram (b). The activation energy is the difference between the energy of the starting reagents and the transition state—a maximum on the reaction coordinate diagram. The reagents are at 6 kJ and the transition state is at 20 kJ, so the activation energy can be calculated as follows:

$$E_a = 20\,kJ - 6\,kJ = 14\,kJ$$

Check Your Learning

Determine which of the two diagrams here (both for the same reaction) involves a catalyst, and identify the activation energy for the catalyzed reaction:

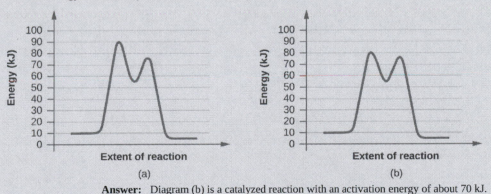

(a)   (b)

Answer: Diagram (b) is a catalyzed reaction with an activation energy of about 70 kJ.

Homogeneous Catalysts

A **homogeneous catalyst** is present in the same phase as the reactants. It interacts with a reactant to form an intermediate substance, which then decomposes or reacts with another reactant in one or more steps to regenerate the original catalyst and form product.

As an important illustration of homogeneous catalysis, consider the earth's ozone layer. Ozone in the upper atmosphere, which protects the earth from ultraviolet radiation, is formed when oxygen molecules absorb ultraviolet light and undergo the reaction:

$$3O_2(g) \xrightarrow{hv} 2O_3(g)$$

Ozone is a relatively unstable molecule that decomposes to yield diatomic oxygen by the reverse of this equation. This decomposition reaction is consistent with the following mechanism:

$$O_3 \longrightarrow O_2 + O$$
$$O + O_3 \longrightarrow 2O_2$$

The presence of nitric oxide, NO, influences the rate of decomposition of ozone. Nitric oxide acts as a catalyst in the following mechanism:

$$NO(g) + O_3(g) \longrightarrow NO_2(g) + O_2(g)$$
$$O_3(g) \longrightarrow O_2(g) + O(g)$$
$$NO_2(g) + O(g) \longrightarrow NO(g) + O_2(g)$$

The overall chemical change for the catalyzed mechanism is the same as:

$$2O_3(g) \longrightarrow 3O_2(g)$$

The nitric oxide reacts and is regenerated in these reactions. It is not permanently used up; thus, it acts as a catalyst. The rate of decomposition of ozone is greater in the presence of nitric oxide because of the catalytic activity of NO. Certain compounds that contain chlorine also catalyze the decomposition of ozone.

Portrait of a Chemist

Mario J. Molina

The 1995 Nobel Prize in Chemistry was shared by Paul J. Crutzen, Mario J. Molina (Figure 12.23), and F. Sherwood Rowland "for their work in atmospheric chemistry, particularly concerning the formation and decomposition of ozone."[1] Molina, a Mexican citizen, carried out the majority of his work at the Massachusetts Institute of Technology (MIT).

---

1. "The Nobel Prize in Chemistry 1995," Nobel Prize.org, accessed February 18, 2015, http://www.nobelprize.org/nobel_prizes/chemistry/laureates/1995/.

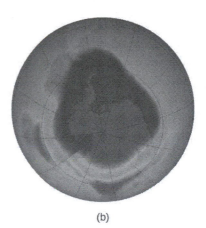

(a)   (b)

Figure 12.23   (a) Mexican chemist Mario Molina (1943 –) shared the Nobel Prize in Chemistry in 1995 for his research on (b) the Antarctic ozone hole. (credit a: courtesy of Mario Molina; credit b: modification of work by NASA)

In 1974, Molina and Rowland published a paper in the journal *Nature* (one of the major peer-reviewed publications in the field of science) detailing the threat of chlorofluorocarbon gases to the stability of the ozone layer in earth's upper atmosphere. The ozone layer protects earth from solar radiation by absorbing ultraviolet light. As chemical reactions deplete the amount of ozone in the upper atmosphere, a measurable "hole" forms above Antarctica, and an increase in the amount of solar ultraviolet radiation— strongly linked to the prevalence of skin cancers—reaches earth's surface. The work of Molina and Rowland was instrumental in the adoption of the Montreal Protocol, an international treaty signed in 1987 that successfully began phasing out production of chemicals linked to ozone destruction.

Molina and Rowland demonstrated that chlorine atoms from human-made chemicals can catalyze ozone destruction in a process similar to that by which NO accelerates the depletion of ozone. Chlorine atoms are generated when chlorocarbons or chlorofluorocarbons—once widely used as refrigerants and propellants—are photochemically decomposed by ultraviolet light or react with hydroxyl radicals. A sample mechanism is shown here using methyl chloride:

$$CH_3Cl + OH \longrightarrow Cl + \text{other products}$$

Chlorine radicals break down ozone and are regenerated by the following catalytic cycle:

$$Cl + O_3 \longrightarrow ClO + O_2$$
$$ClO + O \longrightarrow Cl + O_2$$
$$\text{overall Reaction: } O_3 + O \longrightarrow 2O_2$$

A single monatomic chlorine can break down thousands of ozone molecules. Luckily, the majority of atmospheric chlorine exists as the catalytically inactive forms Cl_2 and $ClONO_2$.

Since receiving his portion of the Nobel Prize, Molina has continued his work in atmospheric chemistry at MIT.

How Sciences Interconnect

Glucose-6-Phosphate Dehydrogenase Deficiency

Enzymes in the human body act as catalysts for important chemical reactions in cellular metabolism. As such, a deficiency of a particular enzyme can translate to a life-threatening disease. G6PD (glucose-6-phosphate dehydrogenase) deficiency, a genetic condition that results in a shortage of the enzyme glucose-6-phosphate dehydrogenase, is the most common enzyme deficiency in humans. This enzyme, shown in Figure 12.24, is the rate-limiting enzyme for the metabolic pathway that supplies NADPH to cells (Figure 12.25).

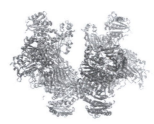

Figure 12.24   Glucose-6-phosphate dehydrogenase is a rate-limiting enzyme for the metabolic pathway that supplies NADPH to cells.

A disruption in this pathway can lead to reduced glutathione in red blood cells; once all glutathione is consumed, enzymes and other proteins such as hemoglobin are susceptible to damage. For example, hemoglobin can be metabolized to bilirubin, which leads to jaundice, a condition that can become severe. People who suffer from G6PD deficiency must avoid certain foods and medicines containing chemicals that can trigger damage their glutathione-deficient red blood cells.

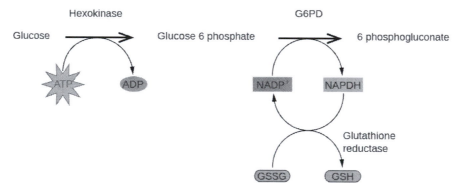

Figure 12.25   In the mechanism for the pentose phosphate pathway, G6PD catalyzes the reaction that regulates NAPDH, a co-enzyme that regulates glutathione, an antioxidant that protects red blood cells and other cells from oxidative damage.

Heterogeneous Catalysts

A **heterogeneous catalyst** is a catalyst that is present in a different phase (usually a solid) than the reactants. Such catalysts generally function by furnishing an active surface upon which a reaction can occur. Gas and liquid phase reactions catalyzed by heterogeneous catalysts occur on the surface of the catalyst rather than within the gas or liquid phase.

Heterogeneous catalysis has at least four steps:

1. Adsorption of the reactant onto the surface of the catalyst

2. Activation of the adsorbed reactant

3. Reaction of the adsorbed reactant

4. Diffusion of the product from the surface into the gas or liquid phase (desorption).

Any one of these steps may be slow and thus may serve as the rate determining step. In general, however, in the presence of the catalyst, the overall rate of the reaction is faster than it would be if the reactants were in the gas or liquid phase.

Figure 12.26 illustrates the steps that chemists believe to occur in the reaction of compounds containing a carbon–carbon double bond with hydrogen on a nickel catalyst. Nickel is the catalyst used in the hydrogenation of polyunsaturated fats and oils (which contain several carbon–carbon double bonds) to produce saturated fats and oils (which contain only carbon–carbon single bonds).

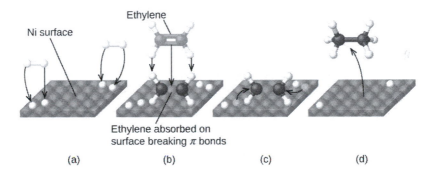

Figure 12.26   There are four steps in the catalysis of the reaction $C_2H_4 + H_2 \longrightarrow C_2H_6$ by nickel. (a) Hydrogen is adsorbed on the surface, breaking the H–H bonds and forming Ni–H bonds. (b) Ethylene is adsorbed on the surface, breaking the π-bond and forming Ni–C bonds. (c) Atoms diffuse across the surface and form new C–H bonds when they collide. (d) C_2H_6 molecules escape from the nickel surface, since they are not strongly attracted to nickel.

Other significant industrial processes that involve the use of heterogeneous catalysts include the preparation of sulfuric acid, the preparation of ammonia, the oxidation of ammonia to nitric acid, and the synthesis of methanol, CH_3OH. Heterogeneous catalysts are also used in the catalytic converters found on most gasoline-powered automobiles (Figure 12.27).

Chemistry in Everyday Life

Automobile Catalytic Converters

Scientists developed catalytic converters to reduce the amount of toxic emissions produced by burning gasoline in internal combustion engines. Catalytic converters take advantage of all five factors that affect the speed of chemical reactions to ensure that exhaust emissions are as safe as possible.

By utilizing a carefully selected blend of catalytically active metals, it is possible to effect complete combustion of all carbon-containing compounds to carbon dioxide while also reducing the output of nitrogen oxides. This is particularly impressive when we consider that one step involves adding more oxygen to the molecule and the other involves removing the oxygen (Figure 12.27).

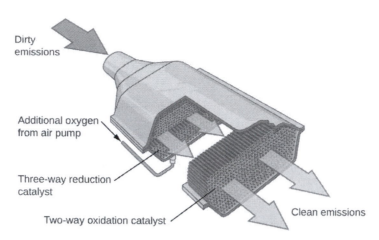

Dirty emissions

Additional oxygen from air pump

Three-way reduction catalyst

Two-way oxidation catalyst

Clean emissions

Figure 12.27 A catalytic converter allows for the combustion of all carbon-containing compounds to carbon dioxide, while at the same time reducing the output of nitrogen oxide and other pollutants in emissions from gasoline-burning engines.

Most modern, three-way catalytic converters possess a surface impregnated with a platinum-rhodium catalyst, which catalyzes the conversion nitric oxide into dinitrogen and oxygen as well as the conversion of carbon monoxide and hydrocarbons such as octane into carbon dioxide and water vapor:

$$2NO_2(g) \longrightarrow N_2(g) + 2O_2(g)$$
$$2CO(g) + O_2(g) \longrightarrow 2CO_2(g)$$
$$2C_8H_{18}(g) + 25O_2(g) \longrightarrow 16CO_2(g) + 18H_2O(g)$$

In order to be as efficient as possible, most catalytic converters are preheated by an electric heater. This ensures that the metals in the catalyst are fully active even before the automobile exhaust is hot enough to maintain appropriate reaction temperatures.

Link to Learning

The University of California at Davis' "ChemWiki" provides a thorough explanation (http://openstaxcollege.org/l/16catconvert) of how catalytic converters work.

How Sciences Interconnect

Enzyme Structure and Function

The study of enzymes is an important interconnection between biology and chemistry. Enzymes are usually proteins (polypeptides) that help to control the rate of chemical reactions between biologically important compounds, particularly those that are involved in cellular metabolism. Different classes of enzymes perform a variety of functions, as shown in Table 12.3.

Classes of Enzymes and Their Functions

Class	Function
oxidoreductases	redox reactions
transferases	transfer of functional groups
hydrolases	hydrolysis reactions
lyases	group elimination to form double bonds
isomerases	isomerization
ligases	bond formation with ATP hydrolysis

Table 12.3

Enzyme molecules possess an active site, a part of the molecule with a shape that allows it to bond to a specific substrate (a reactant molecule), forming an enzyme-substrate complex as a reaction intermediate. There are two models that attempt to explain how this active site works. The most simplistic model is referred to as the lock-and-key hypothesis, which suggests that the molecular shapes of the active site and substrate are complementary, fitting together like a key in a lock. The induced fit hypothesis, on the other hand, suggests that the enzyme molecule is flexible and changes shape to accommodate a bond with the substrate. This is not to suggest that an enzyme's active site is completely malleable, however. Both the lock-and-key model and the induced fit model account for the fact that enzymes can only bind with specific substrates, since in general a particular enzyme only catalyzes a particular reaction (Figure 12.28).

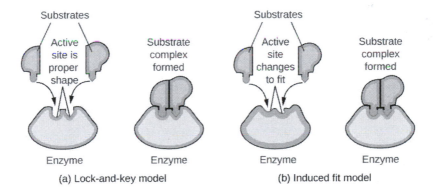

Figure 12.28   (a) According to the lock-and-key model, the shape of an enzyme's active site is a perfect fit for the substrate. (b) According to the induced fit model, the active site is somewhat flexible, and can change shape in order to bond with the substrate.

Link to Learning

The Royal Society of Chemistry (http://openstaxcollege.org/l/16enzymes) provides an excellent introduction to enzymes for students and teachers.

Key Terms

activated complex (also, transition state) unstable combination of reactant species representing the highest energy state of a reaction system

activation energy (E_a) energy necessary in order for a reaction to take place

Arrhenius equation mathematical relationship between the rate constant and the activation energy of a reaction

average rate rate of a chemical reaction computed as the ratio of a measured change in amount or concentration of substance to the time interval over which the change occurred

bimolecular reaction elementary reaction involving the collision and combination of two reactant species

catalyst substance that increases the rate of a reaction without itself being consumed by the reaction

collision theory model that emphasizes the energy and orientation of molecular collisions to explain and predict reaction kinetics

elementary reaction reaction that takes place precisely as depicted in its chemical equation

frequency factor (A) proportionality constant in the Arrhenius equation, related to the relative number of collisions having an orientation capable of leading to product formation

half-life of a reaction ($t_{1/2}$) time required for half of a given amount of reactant to be consumed

heterogeneous catalyst catalyst present in a different phase from the reactants, furnishing a surface at which a reaction can occur

homogeneous catalyst catalyst present in the same phase as the reactants

initial rate instantaneous rate of a chemical reaction at $t = 0$ s (immediately after the reaction has begun)

instantaneous rate rate of a chemical reaction at any instant in time, determined by the slope of the line tangential to a graph of concentration as a function of time

integrated rate law equation that relates the concentration of a reactant to elapsed time of reaction

intermediate molecule or ion produced in one step of a reaction mechanism and consumed in another

method of initial rates use of a more explicit algebraic method to determine the orders in a rate law

molecularity number of reactant species (atoms, molecules or ions) involved in an elementary reaction

overall reaction order sum of the reaction orders for each substance represented in the rate law

rate constant (k) proportionality constant in the relationship between reaction rate and concentrations of reactants

rate expression mathematical representation relating reaction rate to changes in amount, concentration, or pressure of reactant or product species per unit time

rate law (also, rate equation) mathematical equation showing the dependence of reaction rate on the rate constant and the concentration of one or more reactants

rate of reaction measure of the speed at which a chemical reaction takes place

rate-determining step (also, rate-limiting step) slowest elementary reaction in a reaction mechanism; determines the rate of the overall reaction

reaction mechanism stepwise sequence of elementary reactions by which a chemical change takes place

reaction order value of an exponent in a rate law, expressed as an ordinal number (for example, zero order for 0, first order for 1, second order for 2, and so on)

termolecular reaction elementary reaction involving the simultaneous collision and combination of three reactant species

unimolecular reaction elementary reaction involving the rearrangement of a single reactant species to produce one or more molecules of product

Key Equations

- relative reaction rates for $a\text{A} \longrightarrow b\text{B} = -\frac{1}{a}\frac{\Delta[A]}{\Delta t} = \frac{1}{b}\frac{\Delta[B]}{\Delta t}$

- integrated rate law for zero-order reactions: $[A] = -kt + [A]_0, \quad t_{1/2} = \frac{[A]_0}{2k}$

- integrated rate law for first-order reactions: $\ln[A] = -kt + \ln[A]_0, \quad t_{1/2} = \frac{0.693}{k}$

- integrated rate law for second-order reactions: $\frac{1}{[A]} = kt + \frac{1}{[A]_0}, \quad t_{1/2} = \frac{1}{[A]_0 k}$

- $k = Ae^{-E_a/RT}$

- $\ln k = \left(\frac{-E_a}{R}\right)\left(\frac{1}{T}\right) + \ln A$

- $\ln\frac{k_1}{k_2} = \frac{E_a}{R}\left(\frac{1}{T_2} - \frac{1}{T_1}\right)$

Summary

12.1 Chemical Reaction Rates
The rate of a reaction can be expressed either in terms of the decrease in the amount of a reactant or the increase in the amount of a product per unit time. Relations between different rate expressions for a given reaction are derived directly from the stoichiometric coefficients of the equation representing the reaction.

12.2 Factors Affecting Reaction Rates
The rate of a chemical reaction is affected by several parameters. Reactions involving two phases proceed more rapidly when there is greater surface area contact. If temperature or reactant concentration is increased, the rate of a given reaction generally increases as well. A catalyst can increase the rate of a reaction by providing an alternative pathway that causes the activation energy of the reaction to decrease.

12.3 Rate Laws
Rate laws provide a mathematical description of how changes in the amount of a substance affect the rate of a chemical reaction. Rate laws are determined experimentally and cannot be predicted by reaction stoichiometry. The order of reaction describes how much a change in the amount of each substance affects the overall rate, and the overall order of a reaction is the sum of the orders for each substance present in the reaction. Reaction orders are typically first order, second order, or zero order, but fractional and even negative orders are possible.

12.4 Integrated Rate Laws
Differential rate laws can be determined by the method of initial rates or other methods. We measure values for the initial rates of a reaction at different concentrations of the reactants. From these measurements, we determine the order of the reaction in each reactant. Integrated rate laws are determined by integration of the corresponding differential

rate laws. Rate constants for those rate laws are determined from measurements of concentration at various times during a reaction.

The half-life of a reaction is the time required to decrease the amount of a given reactant by one-half. The half-life of a zero-order reaction decreases as the initial concentration of the reactant in the reaction decreases. The half-life of a first-order reaction is independent of concentration, and the half-life of a second-order reaction decreases as the concentration increases.

12.5 Collision Theory

Chemical reactions require collisions between reactant species. These reactant collisions must be of proper orientation and sufficient energy in order to result in product formation. Collision theory provides a simple but effective explanation for the effect of many experimental parameters on reaction rates. The Arrhenius equation describes the relation between a reaction's rate constant and its activation energy, temperature, and dependence on collision orientation.

12.6 Reaction Mechanisms

The sequence of individual steps, or elementary reactions, by which reactants are converted into products during the course of a reaction is called the reaction mechanism. The overall rate of a reaction is determined by the rate of the slowest step, called the rate-determining step. Unimolecular elementary reactions have first-order rate laws, while bimolecular elementary reactions have second-order rate laws. By comparing the rate laws derived from a reaction mechanism to that determined experimentally, the mechanism may be deemed either incorrect or plausible.

12.7 Catalysis

Catalysts affect the rate of a chemical reaction by altering its mechanism to provide a lower activation energy. Catalysts can be homogenous (in the same phase as the reactants) or heterogeneous (a different phase than the reactants).

Exercises

12.1 Chemical Reaction Rates

1. What is the difference between average rate, initial rate, and instantaneous rate?

2. Ozone decomposes to oxygen according to the equation $2O_3(g) \longrightarrow 3O_2(g)$. Write the equation that relates the rate expressions for this reaction in terms of the disappearance of O_3 and the formation of oxygen.

3. In the nuclear industry, chlorine trifluoride is used to prepare uranium hexafluoride, a volatile compound of uranium used in the separation of uranium isotopes. Chlorine trifluoride is prepared by the reaction $Cl_2(g) + 3F_2(g) \longrightarrow 2ClF_3(g)$. Write the equation that relates the rate expressions for this reaction in terms of the disappearance of Cl_2 and F_2 and the formation of ClF_3.

4. A study of the rate of dimerization of C_4H_6 gave the data shown in the table:

$2C_4H_6 \longrightarrow C_8H_{12}$

Time (s)	0	1600	3200	4800	6200
[C₄H₆] (M)	1.00×10^{-2}	5.04×10^{-3}	3.37×10^{-3}	2.53×10^{-3}	2.08×10^{-3}

(a) Determine the average rate of dimerization between 0 s and 1600 s, and between 1600 s and 3200 s.

(b) Estimate the instantaneous rate of dimerization at 3200 s from a graph of time versus $[C_4H_6]$. What are the units of this rate?

(c) Determine the average rate of formation of C_8H_{12} at 1600 s and the instantaneous rate of formation at 3200 s from the rates found in parts (a) and (b).

5. A study of the rate of the reaction represented as $2A \longrightarrow B$ gave the following data:

Time (s)	0.0	5.0	10.0	15.0	20.0	25.0	35.0
[A] (M)	1.00	0.952	0.625	0.465	0.370	0.308	0.230

(a) Determine the average rate of disappearance of A between 0.0 s and 10.0 s, and between 10.0 s and 20.0 s.

(b) Estimate the instantaneous rate of disappearance of A at 15.0 s from a graph of time versus [A]. What are the units of this rate?

(c) Use the rates found in parts (a) and (b) to determine the average rate of formation of B between 0.00 s and 10.0 s, and the instantaneous rate of formation of B at 15.0 s.

6. Consider the following reaction in aqueous solution:

$5Br^-(aq) + BrO_3{}^-(aq) + 6H^+(aq) \longrightarrow 3Br_2(aq) + 3H_2O(l)$

If the rate of disappearance of $Br^-(aq)$ at a particular moment during the reaction is $3.5 \times 10^{-4} \, M \, s^{-1}$, what is the rate of appearance of $Br_2(aq)$ at that moment?

12.2 Factors Affecting Reaction Rates

7. Describe the effect of each of the following on the rate of the reaction of magnesium metal with a solution of hydrochloric acid: the molarity of the hydrochloric acid, the temperature of the solution, and the size of the pieces of magnesium.

8. Explain why an egg cooks more slowly in boiling water in Denver than in New York City. (Hint: Consider the effect of temperature on reaction rate and the effect of pressure on boiling point.)

9. Go to the PhET Reactions & Rates (http://openstaxcollege.org/l/16PHETreaction) interactive. Use the Single Collision tab to represent how the collision between monatomic oxygen (O) and carbon monoxide (CO) results in the breaking of one bond and the formation of another. Pull back on the red plunger to release the atom and observe the results. Then, click on "Reload Launcher" and change to "Angled shot" to see the difference.

(a) What happens when the angle of the collision is changed?

(b) Explain how this is relevant to rate of reaction.

10. In the PhET Reactions & Rates (http://openstaxcollege.org/l/16PHETreaction) interactive, use the "Many Collisions" tab to observe how multiple atoms and molecules interact under varying conditions. Select a molecule to pump into the chamber. Set the initial temperature and select the current amounts of each reactant. Select "Show bonds" under Options. How is the rate of the reaction affected by concentration and temperature?

11. In the PhET Reactions & Rates (http://openstaxcollege.org/l/16PHETreaction) interactive, on the Many Collisions tab, set up a simulation with 15 molecules of A and 10 molecules of BC. Select "Show Bonds" under Options.

(a) Leave the Initial Temperature at the default setting. Observe the reaction. Is the rate of reaction fast or slow?

(b) Click "Pause" and then "Reset All," and then enter 15 molecules of A and 10 molecules of BC once again. Select "Show Bonds" under Options. This time, increase the initial temperature until, on the graph, the total average energy line is completely above the potential energy curve. Describe what happens to the reaction.

12.3 Rate Laws

12. How do the rate of a reaction and its rate constant differ?

13. Doubling the concentration of a reactant increases the rate of a reaction four times. With this knowledge, answer the following questions:

(a) What is the order of the reaction with respect to that reactant?

(b) Tripling the concentration of a different reactant increases the rate of a reaction three times. What is the order of the reaction with respect to that reactant?

14. Tripling the concentration of a reactant increases the rate of a reaction nine times. With this knowledge, answer the following questions:

(a) What is the order of the reaction with respect to that reactant?

(b) Increasing the concentration of a reactant by a factor of four increases the rate of a reaction four times. What is the order of the reaction with respect to that reactant?

15. How much and in what direction will each of the following affect the rate of the reaction: $CO(g) + NO_2(g) \longrightarrow CO_2(g) + NO(g)$ if the rate law for the reaction is $rate = k[NO_2]^2$?

(a) Decreasing the pressure of NO_2 from 0.50 atm to 0.250 atm.

(b) Increasing the concentration of CO from 0.01 M to 0.03 M.

16. How will each of the following affect the rate of the reaction: $CO(g) + NO_2(g) \longrightarrow CO_2(g) + NO(g)$ if the rate law for the reaction is $rate = k[NO_2][CO]$?

(a) Increasing the pressure of NO_2 from 0.1 atm to 0.3 atm

(b) Increasing the concentration of CO from 0.02 M to 0.06 M.

17. Regular flights of supersonic aircraft in the stratosphere are of concern because such aircraft produce nitric oxide, NO, as a byproduct in the exhaust of their engines. Nitric oxide reacts with ozone, and it has been suggested that this could contribute to depletion of the ozone layer. The reaction $NO + O_3 \longrightarrow NO_2 + O_2$ is first order with respect to both NO and O_3 with a rate constant of 2.20×10^7 L/mol/s. What is the instantaneous rate of disappearance of NO when [NO] = 3.3×10^{-6} M and [O_3] = 5.9×10^{-7} M?

18. Radioactive phosphorus is used in the study of biochemical reaction mechanisms because phosphorus atoms are components of many biochemical molecules. The location of the phosphorus (and the location of the molecule it is bound in) can be detected from the electrons (beta particles) it produces:

$$^{32}_{15}P \longrightarrow ^{32}_{16}S + e^-$$

Rate = 4.85×10^{-2} day$^{-1}[^{32}P]$

What is the instantaneous rate of production of electrons in a sample with a phosphorus concentration of 0.0033 M?

19. The rate constant for the radioactive decay of ^{14}C is 1.21×10^{-4} year^{-1}. The products of the decay are nitrogen atoms and electrons (beta particles):

$$^{6}_{14}C \longrightarrow ^{6}_{14}N + e^-$$

rate = $k\left[^{6}_{14}C\right]$

What is the instantaneous rate of production of N atoms in a sample with a carbon-14 content of 6.5×10^{-9} M?

20. The decomposition of acetaldehyde is a second order reaction with a rate constant of 4.71×10^{-8} L/mol/s. What is the instantaneous rate of decomposition of acetaldehyde in a solution with a concentration of 5.55×10^{-4} M?

21. Alcohol is removed from the bloodstream by a series of metabolic reactions. The first reaction produces acetaldehyde; then other products are formed. The following data have been determined for the rate at which alcohol is removed from the blood of an average male, although individual rates can vary by 25–30%. Women metabolize alcohol a little more slowly than men:

[C$_2$H$_5$OH] (M)	4.4×10^{-2}	3.3×10^{-2}	2.2×10^{-2}
Rate (mol/L/h)	2.0×10^{-2}	2.0×10^{-2}	2.0×10^{-2}

Determine the rate equation, the rate constant, and the overall order for this reaction.

22. Under certain conditions the decomposition of ammonia on a metal surface gives the following data:

[NH$_3$] (M)	1.0 × 10^{-3}	2.0 × 10^{-3}	3.0 × 10^{-3}
Rate (mol/L/h^1)	1.5 × 10^{-6}	1.5 × 10^{-6}	1.5 × 10^{-6}

Determine the rate equation, the rate constant, and the overall order for this reaction.

23. Nitrosyl chloride, NOCl, decomposes to NO and Cl$_2$.

$$2NOCl(g) \longrightarrow 2NO(g) + Cl_2(g)$$

Determine the rate equation, the rate constant, and the overall order for this reaction from the following data:

[NOCl] (M)	0.10	0.20	0.30
Rate (mol/L/h)	8.0 × 10^{-10}	3.2 × 10^{-9}	7.2 × 10^{-9}

24. From the following data, determine the rate equation, the rate constant, and the order with respect to A for the reaction $A \longrightarrow 2C$.

[A] (M)	1.33 × 10^{-2}	2.66 × 10^{-2}	3.99 × 10^{-2}
Rate (mol/L/h)	3.80 × 10^{-7}	1.52 × 10^{-6}	3.42 × 10^{-6}

25. Nitrogen(II) oxide reacts with chlorine according to the equation:

$$2NO(g) + Cl_2(g) \longrightarrow 2NOCl(g)$$

The following initial rates of reaction have been observed for certain reactant concentrations:

[NO] (mol/L^1)	[Cl$_2$] (mol/L)	Rate (mol/L/h)
0.50	0.50	1.14
1.00	0.50	4.56
1.00	1.00	9.12

What is the rate equation that describes the rate's dependence on the concentrations of NO and Cl$_2$? What is the rate constant? What are the orders with respect to each reactant?

26. Hydrogen reacts with nitrogen monoxide to form dinitrogen monoxide (laughing gas) according to the equation: $H_2(g) + 2NO(g) \longrightarrow N_2O(g) + H_2O(g)$

Determine the rate equation, the rate constant, and the orders with respect to each reactant from the following data:

[NO] (M)	0.30	0.60	0.60
[H$_2$] (M)	0.35	0.35	0.70
Rate (mol/L/s)	2.835 × 10^{-3}	1.134 × 10^{-2}	2.268 × 10^{-2}

27. For the reaction $A \longrightarrow B + C$, the following data were obtained at 30 °C:

[A] (M)	0.230	0.356	0.557
Rate (mol/L/s)	4.17×10^{-4}	9.99×10^{-4}	2.44×10^{-3}

(a) What is the order of the reaction with respect to [A], and what is the rate equation?

(b) What is the rate constant?

28. For the reaction $Q \longrightarrow W + X$, the following data were obtained at 30 °C:

$[Q]_{initial}$ (M)	0.170	0.212	0.357
Rate (mol/L/s)	6.68×10^{-3}	1.04×10^{-2}	2.94×10^{-2}

(a) What is the order of the reaction with respect to [Q], and what is the rate equation?

(b) What is the rate constant?

29. The rate constant for the first-order decomposition at 45 °C of dinitrogen pentoxide, N_2O_5, dissolved in chloroform, $CHCl_3$, is 6.2×10^{-4} min^{-1}.
$$2N_2O_5 \longrightarrow 4NO_2 + O_2$$

What is the rate of the reaction when $[N_2O_5] = 0.40$ M?

30. The annual production of HNO_3 in 2013 was 60 million metric tons Most of that was prepared by the following sequence of reactions, each run in a separate reaction vessel.

(a) $4NH_3(g) + 5O_2(g) \longrightarrow 4NO(g) + 6H_2O(g)$

(b) $2NO(g) + O_2(g) \longrightarrow 2NO_2(g)$

(c) $3NO_2(g) + H_2O(l) \longrightarrow 2HNO_3(aq) + NO(g)$

The first reaction is run by burning ammonia in air over a platinum catalyst. This reaction is fast. The reaction in equation (c) is also fast. The second reaction limits the rate at which nitric acid can be prepared from ammonia. If equation (b) is second order in NO and first order in O_2, what is the rate of formation of NO_2 when the oxygen concentration is 0.50 M and the nitric oxide concentration is 0.75 M? The rate constant for the reaction is 5.8×10^{-6} $L^2/mol^2/s$.

31. The following data have been determined for the reaction:
$$I^- + OCl^- \longrightarrow IO^- + Cl^-$$

	1	2	3
$[I^-]_{initial}$ (M)	0.10	0.20	0.30
$[OCl^-]_{initial}$ (M)	0.050	0.050	0.010
Rate (mol/L/s)	3.05×10^{-4}	6.20×10^{-4}	1.83×10^{-4}

Determine the rate equation and the rate constant for this reaction.

12.4 Integrated Rate Laws

32. Describe how graphical methods can be used to determine the order of a reaction and its rate constant from a series of data that includes the concentration of A at varying times.

33. Use the data provided to graphically determine the order and rate constant of the following reaction:
$SO_2Cl_2 \longrightarrow SO_2 + Cl_2$

Time (s)	0	5.00×10^3	1.00×10^4	1.50×10^4
$[SO_2Cl_2]$ (M)	0.100	0.0896	0.0802	0.0719
Time (s)	2.50×10^4	3.00×10^4	4.00×10^4	
$[SO_2Cl_2]$ (M)	0.0577	0.0517	0.0415	

34. Use the data provided in a graphical method to determine the order and rate constant of the following reaction:
$2P \longrightarrow Q + W$

Time (s)	9.0	13.0	18.0	22.0	25.0
[P] (M)	1.077×10^{-3}	1.068×10^{-3}	1.055×10^{-3}	1.046×10^{-3}	1.039×10^{-3}

35. Pure ozone decomposes slowly to oxygen, $2O_3(g) \longrightarrow 3O_2(g)$. Use the data provided in a graphical method and determine the order and rate constant of the reaction.

Time (h)	0	2.0×10^3	7.6×10^3	1.00×10^4
$[O_3]$ (M)	1.00×10^{-5}	4.98×10^{-6}	2.07×10^{-6}	1.66×10^{-6}
Time (h)	1.23×10^4	1.43×10^4	1.70×10^4	
$[O_3]$ (M)	1.39×10^{-6}	1.22×10^{-6}	1.05×10^{-6}	

36. From the given data, use a graphical method to determine the order and rate constant of the following reaction:
$2X \longrightarrow Y + Z$

Time (s)	5.0	10.0	15.0	20.0	25.0	30.0	35.0	40.0
[X] (M)	0.0990	0.0497	0.0332	0.0249	0.0200	0.0166	0.0143	0.0125

37. What is the half-life for the first-order decay of phosphorus-32? $\left({}^{32}_{15}P \longrightarrow {}^{32}_{16}S + e^- \right)$ The rate constant for the decay is 4.85×10^{-2} day^{-1}.

38. What is the half-life for the first-order decay of carbon-14? $\left({}^{14}_{6}C \longrightarrow {}^{14}_{7}N + e^- \right)$ The rate constant for the decay is 1.21×10^{-4} year^{-1}.

39. What is the half-life for the decomposition of NOCl when the concentration of NOCl is 0.15 M? The rate constant for this second-order reaction is 8.0×10^{-8} L/mol/s.

40. What is the half-life for the decomposition of O_3 when the concentration of O_3 is 2.35×10^{-6} M? The rate constant for this second-order reaction is 50.4 L/mol/h.

41. The reaction of compound A to give compounds C and D was found to be second-order in A. The rate constant for the reaction was determined to be 2.42 L/mol/s. If the initial concentration is 0.500 mol/L, what is the value of $t_{1/2}$?

42. The half-life of a reaction of compound A to give compounds D and E is 8.50 min when the initial concentration of A is 0.150 mol/L. How long will it take for the concentration to drop to 0.0300 mol/L if the reaction is (a) first order with respect to A or (b) second order with respect to A?

43. Some bacteria are resistant to the antibiotic penicillin because they produce penicillinase, an enzyme with a molecular weight of 3×10^4 g/mol that converts penicillin into inactive molecules. Although the kinetics of enzyme-catalyzed reactions can be complex, at low concentrations this reaction can be described by a rate equation that is first order in the catalyst (penicillinase) and that also involves the concentration of penicillin. From the following data: 1.0 L of a solution containing 0.15 µg (0.15×10^{-6} g) of penicillinase, determine the order of the reaction with respect to penicillin and the value of the rate constant.

[Penicillin] (M)	Rate (mol/L/min)
2.0×10^{-6}	1.0×10^{-10}
3.0×10^{-6}	1.5×10^{-10}
4.0×10^{-6}	2.0×10^{-10}

44. Both technetium-99 and thallium-201 are used to image heart muscle in patients with suspected heart problems. The half-lives are 6 h and 73 h, respectively. What percent of the radioactivity would remain for each of the isotopes after 2 days (48 h)?

45. There are two molecules with the formula C_3H_6. Propene, $CH_3CH = CH_2$, is the monomer of the polymer polypropylene, which is used for indoor-outdoor carpets. Cyclopropane is used as an anesthetic:

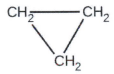

When heated to 499 °C, cyclopropane rearranges (isomerizes) and forms propene with a rate constant of 5.95×10^{-4} s^{-1}. What is the half-life of this reaction? What fraction of the cyclopropane remains after 0.75 h at 499.5 °C?

46. Fluorine-18 is a radioactive isotope that decays by positron emission to form oxygen-18 with a half-life of 109.7 min. (A positron is a particle with the mass of an electron and a single unit of positive charge; the equation is $_{9}^{18}F \longrightarrow {}_{8}^{18}O + e^-$.) Physicians use ^{18}F to study the brain by injecting a quantity of fluoro-substituted glucose into the blood of a patient. The glucose accumulates in the regions where the brain is active and needs nourishment.

(a) What is the rate constant for the decomposition of fluorine-18?

(b) If a sample of glucose containing radioactive fluorine-18 is injected into the blood, what percent of the radioactivity will remain after 5.59 h?

(c) How long does it take for 99.99% of the ^{18}F to decay?

47. Suppose that the half-life of steroids taken by an athlete is 42 days. Assuming that the steroids biodegrade by a first-order process, how long would it take for $\frac{1}{64}$ of the initial dose to remain in the athlete's body?

48. Recently, the skeleton of King Richard III was found under a parking lot in England. If tissue samples from the skeleton contain about 93.79% of the carbon-14 expected in living tissue, what year did King Richard III die? The half-life for carbon-14 is 5730 years.

49. Nitroglycerine is an extremely sensitive explosive. In a series of carefully controlled experiments, samples of the explosive were heated to 160 °C and their first-order decomposition studied. Determine the average rate constants for each experiment using the following data:

Initial [$C_3H_5N_3O_9$] (*M*)	4.88	3.52	2.29	1.81	5.33	4.05	2.95	1.72
t (s)	300	300	300	300	180	180	180	180
% Decomposed	52.0	52.9	53.2	53.9	34.6	35.9	36.0	35.4

50. For the past 10 years, the unsaturated hydrocarbon 1,3-butadiene $(CH_2 = CH - CH = CH_2)$ has ranked 38th among the top 50 industrial chemicals. It is used primarily for the manufacture of synthetic rubber. An isomer exists also as cyclobutene:

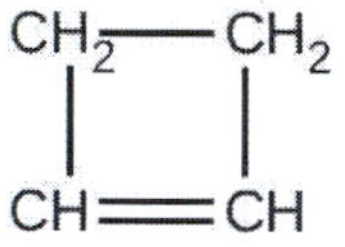

The isomerization of cyclobutene to butadiene is first-order and the rate constant has been measured as 2.0×10^{-4} s^{-1} at 150 °C in a 0.53-L flask. Determine the partial pressure of cyclobutene and its concentration after 30.0 minutes if an isomerization reaction is carried out at 150 °C with an initial pressure of 55 torr.

12.5 Collision Theory

51. Chemical reactions occur when reactants collide. What are two factors that may prevent a collision from producing a chemical reaction?

52. When every collision between reactants leads to a reaction, what determines the rate at which the reaction occurs?

53. What is the activation energy of a reaction, and how is this energy related to the activated complex of the reaction?

54. Account for the relationship between the rate of a reaction and its activation energy.

55. Describe how graphical methods can be used to determine the activation energy of a reaction from a series of data that includes the rate of reaction at varying temperatures.

56. How does an increase in temperature affect rate of reaction? Explain this effect in terms of the collision theory of the reaction rate.

57. The rate of a certain reaction doubles for every 10 °C rise in temperature.

(a) How much faster does the reaction proceed at 45 °C than at 25 °C?

(b) How much faster does the reaction proceed at 95 °C than at 25 °C?

58. In an experiment, a sample of $NaClO_3$ was 90% decomposed in 48 min. Approximately how long would this decomposition have taken if the sample had been heated 20 °C higher?

59. The rate constant at 325 °C for the decomposition reaction $C_4H_8 \longrightarrow 2C_2H_4$ is $6.1 \times 10^{-8} s^{-1}$, and the activation energy is 261 kJ per mole of C_4H_8. Determine the frequency factor for the reaction.

60. The rate constant for the decomposition of acetaldehyde, CH_3CHO, to methane, CH_4, and carbon monoxide, CO, in the gas phase is 1.1×10^{-2} L/mol/s at 703 K and 4.95 L/mol/s at 865 K. Determine the activation energy for this decomposition.

61. An elevated level of the enzyme alkaline phosphatase (ALP) in the serum is an indication of possible liver or bone disorder. The level of serum ALP is so low that it is very difficult to measure directly. However, ALP catalyzes a number of reactions, and its relative concentration can be determined by measuring the rate of one of these reactions under controlled conditions. One such reaction is the conversion of p-nitrophenyl phosphate (PNPP) to pnitrophenoxide ion (PNP) and phosphate ion. Control of temperature during the test is very important; the rate of the reaction increases 1.47 times if the temperature changes from 30 °C to 37 °C. What is the activation energy for the ALP–catalyzed conversion of PNPP to PNP and phosphate?

62. In terms of collision theory, to which of the following is the rate of a chemical reaction proportional?

(a) the change in free energy per second

(b) the change in temperature per second

(c) the number of collisions per second

(d) the number of product molecules

63. Hydrogen iodide, HI, decomposes in the gas phase to produce hydrogen, H_2, and iodine, I_2. The value of the rate constant, k, for the reaction was measured at several different temperatures and the data are shown here:

Temperature (K)	k (M^{-1} s^{-1})
555	6.23×10^{-7}
575	2.42×10^{-6}
645	1.44×10^{-4}
700	2.01×10^{-3}

What is the value of the activation energy (in kJ/mol) for this reaction?

64. The element Co exists in two oxidation states, Co(II) and Co(III), and the ions form many complexes. The rate at which one of the complexes of Co(III) was reduced by Fe(II) in water was measured. Determine the activation energy of the reaction from the following data:

T (K)	k (s^{-1})
293	0.054
298	0.100

65. The hydrolysis of the sugar sucrose to the sugars glucose and fructose,

$$C_{12}H_{22}O_{11} + H_2O \longrightarrow C_6H_{12}O_6 + C_6H_{12}O_6$$

follows a first-order rate equation for the disappearance of sucrose: Rate = $k[C_{12}H_{22}O_{11}]$ (The products of the reaction, glucose and fructose, have the same molecular formulas but differ in the arrangement of the atoms in their molecules.)

(a) In neutral solution, $k = 2.1 \times 10^{-11}$ s^{-1} at 27 °C and 8.5×10^{-11} s^{-1} at 37 °C. Determine the activation energy, the frequency factor, and the rate constant for this equation at 47 °C (assuming the kinetics remain consistent with the Arrhenius equation at this temperature).

(b) When a solution of sucrose with an initial concentration of 0.150 M reaches equilibrium, the concentration of sucrose is 1.65×10^{-7} M. How long will it take the solution to reach equilibrium at 27 °C in the absence of a catalyst? Because the concentration of sucrose at equilibrium is so low, assume that the reaction is irreversible.

(c) Why does assuming that the reaction is irreversible simplify the calculation in part (b)?

66. Use the PhET Reactions & Rates interactive simulation (http://openstaxcollege.org/l/ 16PHETreaction) to simulate a system. On the "Single collision" tab of the simulation applet, enable the "Energy view" by clicking the "+" icon. Select the first $A + BC \longrightarrow AB + C$ reaction (A is yellow, B is purple, and C is navy blue). Using the "straight shot" default option, try launching the A atom with varying amounts of energy. What changes when the Total Energy line at launch is below the transition state of the Potential Energy line? Why? What happens when it is above the transition state? Why?

67. Use the PhET Reactions & Rates interactive simulation (http://openstaxcollege.org/l/16PHETreaction) to simulate a system. On the "Single collision" tab of the simulation applet, enable the "Energy view" by clicking the "+" icon. Select the first $A + BC \longrightarrow AB + C$ reaction (A is yellow, B is purple, and C is navy blue). Using the "angled shot" option, try launching the A atom with varying angles, but with more Total energy than the transition state. What happens when the A atom hits the BC molecule from different directions? Why?

12.6 Reaction Mechanisms

68. Why are elementary reactions involving three or more reactants very uncommon?

69. In general, can we predict the effect of doubling the concentration of A on the rate of the overall reaction $A + B \longrightarrow C$? Can we predict the effect if the reaction is known to be an elementary reaction?

70. Define these terms:

(a) unimolecular reaction

(b) bimolecular reaction

(c) elementary reaction

(d) overall reaction

71. What is the rate equation for the elementary termolecular reaction $A + 2B \longrightarrow$ products? For $3A \longrightarrow$ products?

72. Given the following reactions and the corresponding rate laws, in which of the reactions might the elementary reaction and the overall reaction be the same?

(a)
$Cl_2 + CO \longrightarrow Cl_2CO$
rate $= k[Cl_2]^{3/2}[CO]$

(b)
$PCl_3 + Cl_2 \longrightarrow PCl_5$
rate $= k[PCl_3][Cl_2]$

(c)
$2NO + H_2 \longrightarrow N_2 + H_2O$
rate $= k[NO][H_2]$

(d)
$2NO + O_2 \longrightarrow 2NO_2$
rate $= k[NO]^2[O_2]$

(e)
$NO + O_3 \longrightarrow NO_2 + O_2$
rate $= k[NO][O_3]$

73. Write the rate equation for each of the following elementary reactions:

(a) $O_3 \xrightarrow{\text{sunlight}} O_2 + O$

(b) $O_3 + Cl \longrightarrow O_2 + ClO$

(c) $ClO + O \longrightarrow Cl + O_2$

(d) $O_3 + NO \longrightarrow NO_2 + O_2$

(e) $NO_2 + O \longrightarrow NO + O_2$

74. Nitrogen(II) oxide, NO, reacts with hydrogen, H_2, according to the following equation:
$2NO + 2H_2 \longrightarrow N_2 + 2H_2O$

What would the rate law be if the mechanism for this reaction were:

$2NO + H_2 \longrightarrow N_2 + H_2O_2$ (slow)

$H_2O_2 + H_2 \longrightarrow 2H_2O$ (fast)

75. Experiments were conducted to study the rate of the reaction represented by this equation.[2]

$2NO(g) + 2H_2(g) \longrightarrow N_2(g) + 2H_2O(g)$

Initial concentrations and rates of reaction are given here.

Experiment	Initial Concentration [NO] (mol/L)	Initial Concentration, [H₂] (mol/L)	Initial Rate of Formation of N₂ (mol/L min)
1	0.0060	0.0010	1.8×10^{-4}
2	0.0060	0.0020	3.6×10^{-4}
3	0.0010	0.0060	0.30×10^{-4}
4	0.0020	0.0060	1.2×10^{-4}

Consider the following questions:

(a) Determine the order for each of the reactants, NO and H_2, from the data given and show your reasoning.

(b) Write the overall rate law for the reaction.

(c) Calculate the value of the rate constant, k, for the reaction. Include units.

(d) For experiment 2, calculate the concentration of NO remaining when exactly one-half of the original amount of H_2 had been consumed.

(e) The following sequence of elementary steps is a proposed mechanism for the reaction.

Step 1: $NO + NO \rightleftharpoons N_2O_2$

Step 2: $N_2O_2 + H_2 \rightleftharpoons H_2O + N_2O$

Step 3: $N_2O + H_2 \rightleftharpoons N_2 + H_2O$

Based on the data presented, which of these is the rate determining step? Show that the mechanism is consistent with the observed rate law for the reaction and the overall stoichiometry of the reaction.

76. The reaction of CO with Cl_2 gives phosgene ($COCl_2$), a nerve gas that was used in World War I. Use the mechanism shown here to complete the following exercises:

$Cl_2(g) \rightleftharpoons 2Cl(g)$ (fast, k_1 represents the forward rate constant, k_{-1} the reverse rate constant)

$CO(g) + Cl(g) \longrightarrow COCl(g)$ (slow, k_2 the rate constant)

$COCl(g) + Cl(g) \longrightarrow COCl_2(g)$ (fast, k_3 the rate constant)

(a) Write the overall reaction.

(b) Identify all intermediates.

(c) Write the rate law for each elementary reaction.

(d) Write the overall rate law expression.

12.7 Catalysis

77. Account for the increase in reaction rate brought about by a catalyst.

---

2. This question is taken from the Chemistry Advanced Placement Examination and is used with the permission of the Educational Testing Service.

78. Compare the functions of homogeneous and heterogeneous catalysts.

79. Consider this scenario and answer the following questions: Chlorine atoms resulting from decomposition of chlorofluoromethanes, such as CCl_2F_2, catalyze the decomposition of ozone in the atmosphere. One simplified mechanism for the decomposition is:

$$O_3 \xrightarrow{\text{sunlight}} O_2 + O$$

$$O_3 + Cl \longrightarrow O_2 + ClO$$
$$ClO + O \longrightarrow Cl + O_2$$

(a) Explain why chlorine atoms are catalysts in the gas-phase transformation:
$$2O_3 \longrightarrow 3O_2$$

(b) Nitric oxide is also involved in the decomposition of ozone by the mechanism:

$$O_3 \xrightarrow{\text{sunlight}} O_2 + O$$

$$O_3 + NO \longrightarrow NO_2 + O_2$$
$$NO_2 + O \longrightarrow NO + O_2$$

Is NO a catalyst for the decomposition? Explain your answer.

80. For each of the following pairs of reaction diagrams, identify which of the pair is catalyzed:

(a)

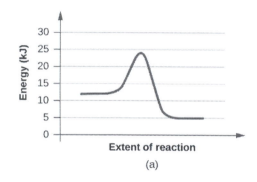

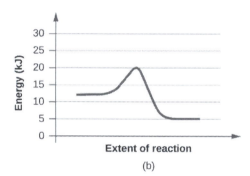

(b)

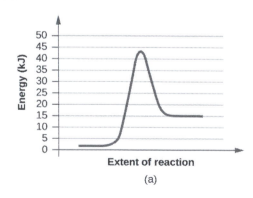

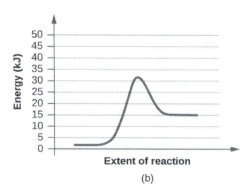

81. For each of the following pairs of reaction diagrams, identify which of the pairs is catalyzed:

(a)

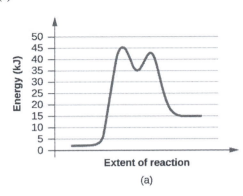

(a)   (b)

(b)

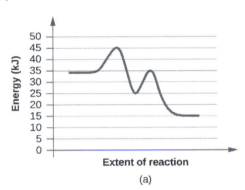

(a)   (b)

82. For each of the following reaction diagrams, estimate the activation energy (E_a) of the reaction:

(a)

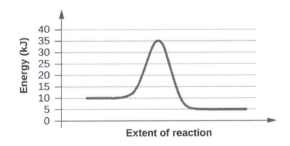

(b)

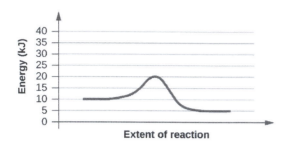

83. For each of the following reaction diagrams, estimate the activation energy (E_a) of the reaction:

(a)

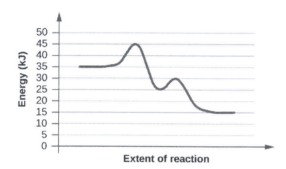

(b)

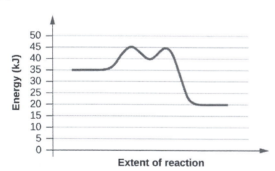

84. Based on the diagrams in Exercise 12.82, which of the reactions has the fastest rate? Which has the slowest rate?

85. Based on the diagrams in Exercise 12.83, which of the reactions has the fastest rate? Which has the slowest rate?

Chapter 13

Fundamental Equilibrium Concepts

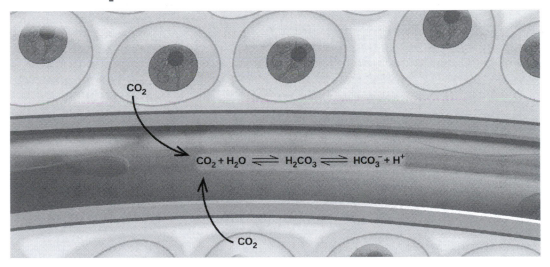

Figure 13.1 Movement of carbon dioxide through tissues and blood cells involves several equilibrium reactions.

Chapter Outline

Introduction

Imagine a beach populated with sunbathers and swimmers. As those basking in the sun get too hot and want to cool off, they head into the surf to swim. As the swimmers tire, they head to the beach to rest. If these two rates of transfer (sunbathers entering the water, swimmers leaving the water) are equal, the number of sunbathers and swimmers would be constant, or at equilibrium, although the identities of the people are constantly changing from sunbather to swimmer and back. An analogous situation occurs in chemical reactions. Reactions can occur in both directions simultaneously (reactants to products and products to reactants) and eventually reach a state of balance.

These balanced two-way reactions occur all around and even in us. For example, they occur in our blood, where the reaction between carbon dioxide and water forms carbonic acid $(HCO_3{}^-)$ (Figure 13.1). Human physiology is adapted to the amount of ionized products produced by this reaction $(HCO_3{}^-$ and $H^+)$. In this chapter, you will learn how to predict the position of the balance and the yield of a product of a reaction under specific conditions, how to change a reaction's conditions to increase or reduce yield, and how to evaluate an equilibrium system's reaction to disturbances.

13.1 Chemical Equilibria

By the end of this section, you will be able to:

- Describe the nature of equilibrium systems
- Explain the dynamic nature of a chemical equilibrium

A chemical reaction is usually written in a way that suggests it proceeds in one direction, the direction in which we read, but all chemical reactions are reversible, and both the forward and reverse reaction occur to one degree or another depending on conditions. In a chemical **equilibrium**, the forward and reverse reactions occur at equal rates, and the concentrations of products and reactants remain constant. If we run a reaction in a closed system so that the products cannot escape, we often find the reaction does not give a 100% yield of products. Instead, some reactants remain after the concentrations stop changing. At this point, when there is no further change in concentrations of reactants and products, we say the reaction is at equilibrium. A mixture of reactants and products is found at equilibrium.

For example, when we place a sample of dinitrogen tetroxide (N_2O_4, a colorless gas) in a glass tube, it forms nitrogen dioxide (NO_2, a brown gas) by the reaction

$$N_2O_4(g) \rightleftharpoons 2NO_2(g)$$

The color becomes darker as N_2O_4 is converted to NO_2. When the system reaches equilibrium, both N_2O_4 and NO_2 are present (Figure 13.2).

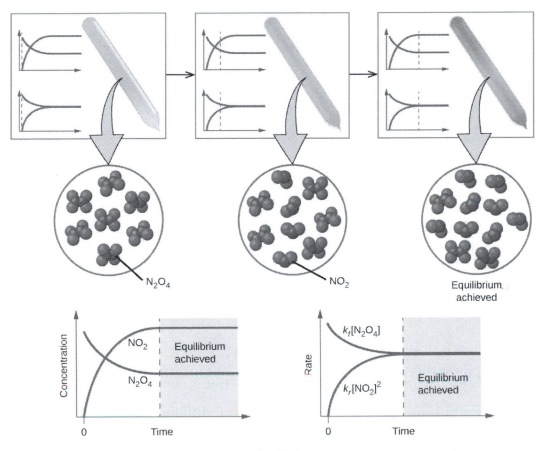

Figure 13.2 A mixture of NO_2 and N_2O_4 moves toward equilibrium. Colorless N_2O_4 reacts to form brown NO_2. As the reaction proceeds toward equilibrium, the color of the mixture darkens due to the increasing concentration of NO_2.

The formation of NO_2 from N_2O_4 is a **reversible reaction**, which is identified by the equilibrium arrow ($\rightleftharpoons$). All reactions are reversible, but many reactions, for all practical purposes, proceed in one direction until the reactants are exhausted and will reverse only under certain conditions. Such reactions are often depicted with a one-way arrow from reactants to products. Many other reactions, such as the formation of NO_2 from N_2O_4, are reversible under more easily obtainable conditions and, therefore, are named as such. In a reversible reaction, the reactants can combine to form products and the products can react to form the reactants. Thus, not only can N_2O_4 decompose to form NO_2, but the NO_2 produced can react to form N_2O_4. As soon as the forward reaction produces any NO_2, the reverse reaction begins and NO_2 starts to react to form N_2O_4. At equilibrium, the concentrations of N_2O_4 and NO_2 no longer change because the rate of formation of NO_2 is exactly equal to the rate of consumption of NO_2, and the rate of formation of N_2O_4 is exactly equal to the rate of consumption of N_2O_4. *Chemical equilibrium is a dynamic process*: As with the swimmers and the sunbathers, the numbers of each remain constant, yet there is a flux back and forth between them (Figure 13.3).

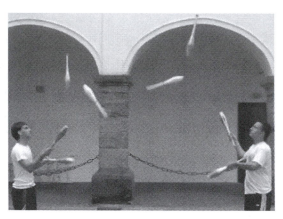

Figure 13.3 These jugglers provide an illustration of dynamic equilibrium. Each throws clubs to the other at the same rate at which he receives clubs from that person. Because clubs are thrown continuously in both directions, the number of clubs moving in each direction is constant, and the number of clubs each juggler has at a given time remains (roughly) constant.

In a chemical equilibrium, the forward and reverse reactions do not stop, rather they continue to occur at the same rate, leading to constant concentrations of the reactants and the products. Plots showing how the reaction rates and concentrations change with respect to time are shown in Figure 13.2.

We can detect a state of equilibrium because the concentrations of reactants and products do not appear to change. However, it is important that we verify that the absence of change is due to equilibrium and not to a reaction rate that is so slow that changes in concentration are difficult to detect.

We use a double arrow when writing an equation for a reversible reaction. Such a reaction may or may not be at equilibrium. For example, Figure 13.2 shows the reaction:

$$N_2O_4(g) \rightleftharpoons 2NO_2(g)$$

When we wish to speak about one particular component of a reversible reaction, we use a single arrow. For example, in the equilibrium shown in Figure 13.2, the rate of the forward reaction

$$2NO_2(g) \longrightarrow N_2O_4(g)$$

is equal to the rate of the backward reaction

$$N_2O_4(g) \longrightarrow 2NO_2(g)$$

Chemistry in Everyday Life

Equilibrium and Soft Drinks

The connection between chemistry and carbonated soft drinks goes back to 1767, when Joseph Priestley (1733–1804; mostly known today for his role in the discovery and identification of oxygen) discovered a method of infusing water with carbon dioxide to make carbonated water. In 1772, Priestly published a paper entitled "Impregnating Water with Fixed Air." The paper describes dripping oil of vitriol (today we call this sulfuric acid, but what a great way to describe sulfuric acid: "oil of vitriol" literally means "liquid nastiness") onto chalk (calcium carbonate). The resulting CO_2 falls into the container of water beneath the vessel in which the initial reaction takes place; agitation helps the gaseous CO_2 mix into the liquid water.

$$H_2SO_4(l) + CaCO_3(s) \longrightarrow CO_2(g) + H_2O(l) + CaSO_4(aq)$$

Carbon dioxide is slightly soluble in water. There is an equilibrium reaction that occurs as the carbon dioxide reacts with the water to form carbonic acid (H_2CO_3). Since carbonic acid is a weak acid, it can dissociate into protons (H^+) and hydrogen carbonate ions ($HCO_3{}^-$).

$$CO_2(aq) + H_2O(l) \rightleftharpoons H_2CO_3(aq) \rightleftharpoons HCO_3{}^-(aq) + H^+(aq)$$

Today, CO_2 can be pressurized into soft drinks, establishing the equilibrium shown above. Once you open the beverage container, however, a cascade of equilibrium shifts occurs. First, the CO_2 gas in the air space on top of the bottle escapes, causing the equilibrium between gas-phase CO_2 and dissolved or aqueous CO_2 to shift, lowering the concentration of CO_2 in the soft drink. Less CO_2 dissolved in the liquid leads to carbonic acid decomposing to dissolved CO_2 and H_2O. The lowered carbonic acid concentration causes a shift of the final equilibrium. As long as the soft drink is in an open container, the CO_2 bubbles up out of the beverage, releasing the gas into the air (Figure 13.4). With the lid off the bottle, the CO_2 reactions are no longer at equilibrium and will continue until no more of the reactants remain. This results in a soft drink with a much lowered CO_2 concentration, often referred to as "flat."

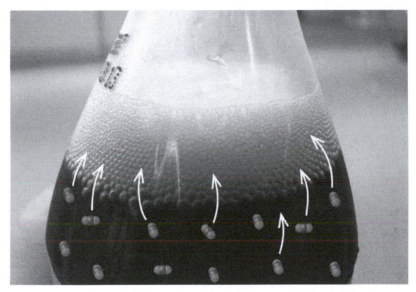

Figure 13.4   When a soft drink is opened, several equilibrium shifts occur. (credit: modification of work by "D Coetzee"/Flickr)

Let us consider the evaporation of bromine as a second example of a system at equilibrium.

$$Br_2(l) \rightleftharpoons Br_2(g)$$

An equilibrium can be established for a physical change—like this liquid to gas transition—as well as for a chemical reaction. Figure 13.5 shows a sample of liquid bromine at equilibrium with bromine vapor in a closed container. When we pour liquid bromine into an empty bottle in which there is no bromine vapor, some liquid evaporates, the amount of liquid decreases, and the amount of vapor increases. If we cap the bottle so no vapor escapes, the amount of liquid and vapor will eventually stop changing and an equilibrium between the liquid and the vapor will be established. If the bottle were not capped, the bromine vapor would escape and no equilibrium would be reached.

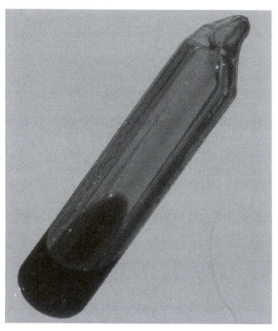

Figure 13.5 An equilibrium is pictured between liquid bromine, $Br_2(l)$, the dark liquid, and bromine vapor, $Br_2(g)$, the orange gas. Because the container is sealed, bromine vapor cannot escape and equilibrium is maintained. (credit: http://images-of-elements.com/bromine.php)

13.2 Equilibrium Constants

By the end of this section, you will be able to:

- Derive reaction quotients from chemical equations representing homogeneous and heterogeneous reactions
- Calculate values of reaction quotients and equilibrium constants, using concentrations and pressures
- Relate the magnitude of an equilibrium constant to properties of the chemical system

Now that we have a symbol $(\rightleftharpoons)$ to designate reversible reactions, we will need a way to express mathematically how the amounts of reactants and products affect the equilibrium of the system. A general equation for a reversible reaction may be written as follows:

$$m\text{A} + n\text{B} + \ \rightleftharpoons x\text{C} + y\text{D}$$

We can write the **reaction quotient (Q)** for this equation. When evaluated using concentrations, it is called Q_c. We use brackets to indicate molar concentrations of reactants and products.

$$Q_c = \frac{[\text{C}]^x[\text{D}]^y}{[\text{A}]^m[\text{B}]^n}$$

The reaction quotient is equal to the molar concentrations of the products of the chemical equation (multiplied together) over the reactants (also multiplied together), with each concentration raised to the power of the coefficient of that substance in the balanced chemical equation. For example, the reaction quotient for the reversible reaction $2NO_2(g) \rightleftharpoons N_2O_4(g)$ is given by this expression:

$$Q_c = \frac{[N_2O_4]}{[NO_2]^2}$$

Example 13.1

Writing Reaction Quotient Expressions

Write the expression for the reaction quotient for each of the following reactions:

(a) $3O_2(g) \rightleftharpoons 2O_3(g)$

(b) $N_2(g) + 3H_2(g) \rightleftharpoons 2NH_3(g)$

(c) $4NH_3(g) + 7O_2(g) \rightleftharpoons 4NO_2(g) + 6H_2O(g)$

Solution

(a) $Q_c = \dfrac{[O_3]^2}{[O_2]^3}$

(b) $Q_c = \dfrac{[NH_3]^2}{[N_2][H_2]^3}$

(c) $Q_c = \dfrac{[NO_2]^4[H_2O]^6}{[NH_3]^4[O_2]^7}$

Check Your Learning

Write the expression for the reaction quotient for each of the following reactions:

(a) $2SO_2(g) + O_2(g) \rightleftharpoons 2SO_3(g)$

(b) $C_4H_8(g) \rightleftharpoons 2C_2H_4(g)$

(c) $2C_4H_{10}(g) + 13O_2(g) \rightleftharpoons 8CO_2(g) + 10H_2O(g)$

Answer: (a) $Q_c = \dfrac{[SO_3]^2}{[SO_2]^2[O_2]}$; (b) $Q_c = \dfrac{[C_2H_4]^2}{[C_4H_8]}$; (c) $Q_c = \dfrac{[CO_2]^8[H_2O]^{10}}{[C_4H_{10}]^2[O_2]^{13}}$

The numeric value of Q_c for a given reaction varies; it depends on the concentrations of products and reactants present at the time when Q_c is determined. When pure reactants are mixed, Q_c is initially zero because there are no products present at that point. As the reaction proceeds, the value of Q_c increases as the concentrations of the products increase and the concentrations of the reactants simultaneously decrease (Figure 13.6). When the reaction reaches equilibrium, the value of the reaction quotient no longer changes because the concentrations no longer change.

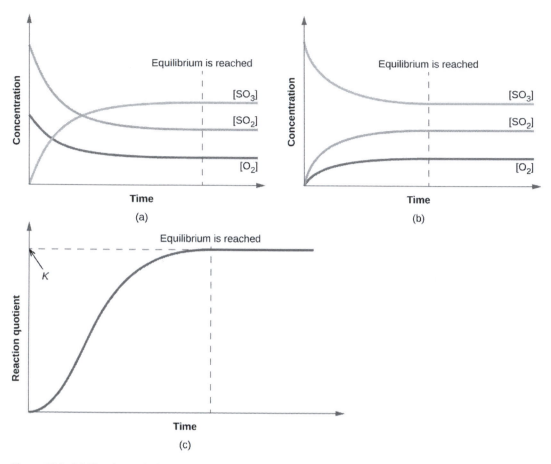

Figure 13.6 (a) The change in the concentrations of reactants and products is depicted as the $2SO_2(g) + O_2(g) \rightleftharpoons 2SO_3(g)$ reaction approaches equilibrium. (b) The change in concentrations of reactants and products is depicted as the reaction $2SO_3(g) \rightleftharpoons 2SO_2(g) + O_2(g)$ approaches equilibrium. (c) The graph shows the change in the value of the reaction quotient as the reaction approaches equilibrium.

When a mixture of reactants and products of a reaction reaches equilibrium at a given temperature, its reaction quotient always has the same value. This value is called the **equilibrium constant (K)** of the reaction at that temperature. As for the reaction quotient, when evaluated in terms of concentrations, it is noted as K_c.

That a reaction quotient always assumes the same value at equilibrium can be expressed as:

$$Q_c \text{ at equilibrium} = K_c = \frac{[C]^x[D]^y...}{[A]^m[B]^n...}$$

This equation is a mathematical statement of the **law of mass action**: When a reaction has attained equilibrium at a given temperature, the reaction quotient for the reaction always has the same value.

Example 13.2

Evaluating a Reaction Quotient

Gaseous nitrogen dioxide forms dinitrogen tetroxide according to this equation:

$$2NO_2(g) \rightleftharpoons N_2O_4(g)$$

When 0.10 mol NO_2 is added to a 1.0-L flask at 25 °C, the concentration changes so that at equilibrium, $[NO_2] = 0.016\ M$ and $[N_2O_4] = 0.042\ M$.

(a) What is the value of the reaction quotient before any reaction occurs?

(b) What is the value of the equilibrium constant for the reaction?

Solution

(a) Before any product is formed, $[NO_2] = \dfrac{0.10\ \text{mol}}{1.0\ \text{L}} = 0.10\ M$, and $[N_2O_4] = 0\ M$. Thus,

$$Q_c = \frac{[N_2O_4]}{[NO_2]^2} = \frac{0}{0.10^2} = 0$$

(b) At equilibrium, the value of the equilibrium constant is equal to the value of the reaction quotient. At equilibrium, $K_c = Q_c = \dfrac{[N_2O_4]}{[NO_2]^2} = \dfrac{0.042}{0.016^2} = 1.6 \times 10^2$. The equilibrium constant is 1.6×10^2.

Note that dimensional analysis would suggest the unit for this K_c value should be M^{-1}. However, it is common practice to omit units for K_c values computed as described here, since it is the magnitude of an equilibrium constant that relays useful information. As will be discussed later in this module, the rigorous approach to computing equilibrium constants uses dimensionless quantities derived from concentrations instead of actual concentrations, and so K_c values are truly unitless.

Check Your Learning

For the reaction $2SO_2(g) + O_2(g) \rightleftharpoons 2SO_3(g)$, the concentrations at equilibrium are $[SO_2] = 0.90\ M$, $[O_2] = 0.35\ M$, and $[SO_3] = 1.1\ M$. What is the value of the equilibrium constant, K_c?

Answer: $K_c = 4.3$

The magnitude of an equilibrium constant is a measure of the yield of a reaction when it reaches equilibrium. A large value for K_c indicates that equilibrium is attained only after the reactants have been largely converted into products. A small value of K_c—much less than 1—indicates that equilibrium is attained when only a small proportion of the reactants have been converted into products.

Once a value of K_c is known for a reaction, it can be used to predict directional shifts when compared to the value of Q_c. A system that is not at equilibrium will proceed in the direction that establishes equilibrium. The data in Figure 13.7 illustrate this. When heated to a consistent temperature, 800 °C, different starting mixtures of CO, H_2O, CO_2, and H_2 react to reach compositions adhering to the same equilibrium (the value of Q_c changes until it equals the value of K_c). This value is 0.640, the equilibrium constant for the reaction under these conditions.

$$CO(g) + H_2O(g) \rightleftharpoons CO_2(g) + H_2(g) \qquad K_c = 0.640 \qquad T = 800\ °C$$

It is important to recognize that an equilibrium can be established starting either from reactants or from products, or from a mixture of both. For example, equilibrium was established from Mixture 2 in Figure 13.7 when the products of the reaction were heated in a closed container. In fact, one technique used to determine whether a reaction is truly at equilibrium is to approach equilibrium starting with reactants in one experiment and starting with products in another. If the same value of the reaction quotient is observed when the concentrations stop changing in both experiments, then we may be certain that the system has reached equilibrium.

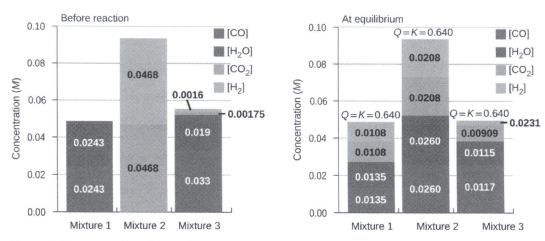

Figure 13.7 Concentrations of three mixtures are shown before and after reaching equilibrium at 800 °C for the so-called water gas shift reaction: $CO(g) + H_2O(g) \rightleftharpoons CO_2(g) + H_2(g)$.

Example 13.3

Predicting the Direction of Reaction

Given here are the starting concentrations of reactants and products for three experiments involving this reaction:

$$CO(g) + H_2O(g) \rightleftharpoons CO_2(g) + H_2(g)$$
$$K_c = 0.64$$

Determine in which direction the reaction proceeds as it goes to equilibrium in each of the three experiments shown.

Reactants/Products	Experiment 1	Experiment 2	Experiment 3
$[CO]_i$	0.0203 M	0.011 M	0.0094 M
$[H_2O]_i$	0.0203 M	0.0011 M	0.0025 M
$[CO_2]_i$	0.0040 M	0.037 M	0.0015 M
$[H_2]_i$	0.0040 M	0.046 M	0.0076 M

Solution

Experiment 1:

$$Q_c = \frac{[CO_2][H_2]}{[CO][H_2O]} = \frac{(0.0040)(0.0040)}{(0.0203)(0.0203)} = 0.039.$$

$Q_c < K_c$ (0.039 < 0.64)

The reaction will shift to the right.

Experiment 2:

$$Q_c = \frac{[CO_2][H_2]}{[CO][H_2O]} = \frac{(0.037)(0.046)}{(0.011)(0.0011)} = 1.4 \times 10^2$$

$Q_c > K_c$ (140 > 0.64)

The reaction will shift to the left.

Experiment 3:

$$Q_c = \frac{[CO_2][H_2]}{[CO][H_2O]} = \frac{(0.0015)(0.0076)}{(0.0094)(0.0025)} = 0.48$$

$Q_c < K_c$ (0.48 < 0.64)

The reaction will shift to the right.

Check Your Learning

Calculate the reaction quotient and determine the direction in which each of the following reactions will proceed to reach equilibrium.

(a) A 1.00-L flask containing 0.0500 mol of NO(g), 0.0155 mol of Cl2(g), and 0.500 mol of NOCl:

$$2NO(g) + Cl_2(g) \rightleftharpoons 2NOCl(g) \qquad K_c = 4.6 \times 10^4$$

(b) A 5.0-L flask containing 17 g of NH_3, 14 g of N_2, and 12 g of H_2:

$$N_2(g) + 3H_2(g) \rightleftharpoons 2NH_3(g) \qquad K_c = 0.060$$

(c) A 2.00-L flask containing 230 g of SO_3(g):

$$2SO_3(g) \rightleftharpoons 2SO_2(g) + O_2(g) \qquad K_c = 0.230$$

Answer: (a) $Q_c = 6.45 \times 10^3$, shifts right. (b) $Q_c = 0.12$, shifts left. (c) $Q_c = 0$, shifts right

In Example 13.2, it was mentioned that the common practice is to omit units when evaluating reaction quotients and equilibrium constants. It should be pointed out that using concentrations in these computations is a convenient but simplified approach that sometimes leads to results that seemingly conflict with the law of mass action. For example, equilibria involving aqueous ions often exhibit equilibrium constants that vary quite significantly (are *not* constant) at high solution concentrations. This may be avoided by computing K_c values using the *activities* of the reactants and products in the equilibrium system instead of their concentrations. The **activity** of a substance is a measure of its effective concentration under specified conditions. While a detailed discussion of this important quantity is beyond the scope of an introductory text, it is necessary to be aware of a few important aspects:

- Activities are dimensionless (unitless) quantities and are in essence "adjusted" concentrations.
- For relatively dilute solutions, a substance's activity and its molar concentration are roughly equal.
- Activities for pure condensed phases (solids and liquids) are equal to 1.

As a consequence of this last consideration, Q_c and K_c expressions do not contain terms for solids or liquids (being numerically equal to 1, these terms have no effect on the expression's value). Several examples of equilibria yielding such expressions will be encountered in this section.

Homogeneous Equilibria

A **homogeneous equilibrium** is one in which all of the reactants and products are present in a single solution (by definition, a homogeneous mixture). In this chapter, we will concentrate on the two most common types of homogeneous equilibria: those occurring in liquid-phase solutions and those involving exclusively gaseous species. Reactions between solutes in liquid solutions belong to one type of homogeneous equilibria. The chemical species involved can be molecules, ions, or a mixture of both. Several examples are provided here.

$$C_2H_2(aq) + 2Br_2(aq) \rightleftharpoons C_2H_2Br_4(aq) \qquad K_c = \frac{[C_2H_2Br_4]}{[C_2H_2][Br_2]^2}$$

$$I_2(aq) + I^-(aq) \rightleftharpoons I_3{}^-(aq) \qquad K_c = \frac{[I_3{}^-]}{[I_2][I^-]}$$

$$Hg_2{}^{2+}(aq) + NO_3{}^-(aq) + 3H_3O^+(aq) \rightleftharpoons 2Hg^{2+}(aq) + HNO_2(aq) + 4H_2O(l)$$

$$K_c = \frac{[Hg^{2+}]^2[HNO_2]}{[Hg_2{}^{2+}][NO_3{}^-][H_3O^+]^3}$$

$$HF(aq) + H_2O(l) \rightleftharpoons H_3O^+(aq) + F^-(aq) \qquad K_c = \frac{[H_3O^+][F^-]}{[HF]}$$

$$NH_3(aq) + H_2O(l) \rightleftharpoons NH_4{}^+(aq) + OH^-(aq) \qquad K_c = \frac{[NH_4{}^+][OH^-]}{[NH_3]}$$

In each of these examples, the equilibrium system is an aqueous solution, as denoted by the *aq* annotations on the solute formulas. Since $H_2O(l)$ is the solvent for these solutions, its concentration does not appear as a term in the K_c expression, as discussed earlier, even though it may also appear as a reactant or product in the chemical equation.

Reactions in which all reactants and products are gases represent a second class of homogeneous equilibria. We use molar concentrations in the following examples, but we will see shortly that partial pressures of the gases may be used as well.

$$C_2H_6(g) \rightleftharpoons C_2H_4(g) + H_2(g) \qquad K_c = \frac{[C_2H_4][H_2]}{[C_2H_6]}$$

$$3O_2(g) \rightleftharpoons 2O_3(g) \qquad K_c = \frac{[O_3]^2}{[O_2]^3}$$

$$N_2(g) + 3H_2(g) \rightleftharpoons 2NH_3(g) \qquad K_c = \frac{[NH_3]^2}{[N_2][H_2]^3}$$

$$C_3H_8(g) + 5O_2(g) \rightleftharpoons 3CO_2(g) + 4H_2O(g) \qquad K_c = \frac{[CO_2]^3[H_2O]^4}{[C_3H_8][O_2]^5}$$

Note that the concentration of $H_2O(g)$ has been included in the last example because water is not the solvent in this gas-phase reaction and its concentration (and activity) changes.

Whenever gases are involved in a reaction, the partial pressure of each gas can be used instead of its concentration in the equation for the reaction quotient because the partial pressure of a gas is directly proportional to its concentration at constant temperature. This relationship can be derived from the ideal gas equation, where M is the molar concentration of gas, $\frac{n}{V}$.

$$PV = nRT$$
$$P = \left(\frac{n}{V}\right)RT$$
$$= MRT$$

Thus, at constant temperature, the pressure of a gas is directly proportional to its concentration.

Using the partial pressures of the gases, we can write the reaction quotient for the system $C_2H_6(g) \rightleftharpoons C_2H_4(g) + H_2(g)$ by following the same guidelines for deriving concentration-based expressions:

$$Q_P = \frac{P_{C_2H_4}P_{H_2}}{P_{C_2H_6}}$$

In this equation we use Q_P to indicate a reaction quotient written with partial pressures: $P_{C_2H_6}$ is the partial pressure of C_2H_6; P_{H_2}, the partial pressure of H_2; and $P_{C_2H_6}$, the partial pressure of C_2H_4. At equilibrium:

$$K_P = Q_P = \frac{P_{C_2H_4}P_{H_2}}{P_{C_2H_6}}$$

The subscript P in the symbol K_P designates an equilibrium constant derived using partial pressures instead of concentrations. The equilibrium constant, K_P, is still a constant, but its numeric value may differ from the equilibrium constant found for the same reaction by using concentrations.

Conversion between a value for K_c, an equilibrium constant expressed in terms of concentrations, and a value for K_P, an equilibrium constant expressed in terms of pressures, is straightforward (a K or Q without a subscript could be either concentration or pressure).

The equation relating K_c and K_P is derived as follows. For the gas-phase reaction $mA + nB \rightleftharpoons xC + yD$:

$$
\begin{aligned}
K_P &= \frac{(P_C)^x (P_D)^y}{(P_A)^m (P_B)^n} \\
&= \frac{([C] \times RT)^x ([D] \times RT)^y}{([A] \times RT)^m ([B] \times RT)^n} \\
&= \frac{[C]^x [D]^y}{[A]^m [B]^n} \times \frac{(RT)^{x+y}}{(RT)^{m+n}} \\
&= K_c (RT)^{(x+y) - (m+n)} \\
&= K_c (RT)^{\Delta n}
\end{aligned}
$$

The relationship between K_c and K_P is

$$K_P = K_c (RT)^{\Delta n}$$

In this equation, Δn is the difference between the sum of the coefficients of the *gaseous* products and the sum of the coefficients of the *gaseous* reactants in the reaction (the change in moles of gas between the reactants and the products). For the gas-phase reaction $mA + nB \rightleftharpoons xC + yD$, we have

$$\Delta n = (x+y) - (m+n)$$

Example 13.4

Calculation of K_P

Write the equations for the conversion of K_c to K_P for each of the following reactions:

(a) $C_2H_6(g) \rightleftharpoons C_2H_4(g) + H_2(g)$

(b) $CO(g) + H_2O(g) \rightleftharpoons CO_2(g) + H_2(g)$

(c) $N_2(g) + 3H_2(g) \rightleftharpoons 2NH_3(g)$

(d) K_c is equal to 0.28 for the following reaction at 900 °C:

$$CS_2(g) + 4H_2(g) \rightleftharpoons CH_4(g) + 2H_2S(g)$$

What is K_P at this temperature?

Solution

(a) $\Delta n = (2) - (1) = 1$
$K_P = K_c (RT)^{\Delta n} = K_c (RT)^1 = K_c (RT)$

(b) $\Delta n = (2) - (2) = 0$
$K_P = K_c (RT)^{\Delta n} = K_c (RT)^0 = K_c$

(c) $\Delta n = (2) - (1 + 3) = -2$
$K_P = K_c (RT)^{\Delta n} = K_c (RT)^{-2} = \dfrac{K_c}{(RT)^2}$

(d) $K_P = K_c (RT)^{\Delta n} = (0.28)[(0.0821)(1173)]^{-2} = 3.0 \times 10^{-5}$

Check Your Learning

Write the equations for the conversion of K_c to K_P for each of the following reactions, which occur in the gas phase:

(a) $2SO_2(g) + O_2(g) \rightleftharpoons 2SO_3(g)$

(b) $N_2O_4(g) \rightleftharpoons 2NO_2(g)$

(c) $C_3H_8(g) + 5O_2(g) \rightleftharpoons 3CO_2(g) + 4H_2O(g)$

(d) At 227 °C, the following reaction has $K_c = 0.0952$:

$$CH_3OH(g) \rightleftharpoons CO(g) + 2H_2(g)$$

What would be the value of K_P at this temperature?

Answer: (a) $K_P = K_c (RT)^{-1}$; (b) $K_P = K_c (RT)$; (c) $K_P = K_c (RT)$; (d) 160 or 1.6×10^2

Heterogeneous Equilibria

A **heterogeneous equilibrium** is a system in which reactants and products are found in two or more phases. The phases may be any combination of solid, liquid, or gas phases, and solutions. When dealing with these equilibria, remember that solids and pure liquids do not appear in equilibrium constant expressions (the activities of pure solids, pure liquids, and solvents are 1).

Some heterogeneous equilibria involve chemical changes; for example:

$$PbCl_2(s) \rightleftharpoons Pb^{2+}(aq) + 2Cl^-(aq) \qquad K_c = [Pb^{2+}][Cl^-]^2$$
$$CaO(s) + CO_2(g) \rightleftharpoons CaCO_3(s) \qquad K_c = \frac{1}{[CO_2]}$$
$$C(s) + 2S(g) \rightleftharpoons CS_2(g) \qquad K_c = \frac{[CS_2]}{[S]^2}$$

Other heterogeneous equilibria involve phase changes, for example, the evaporation of liquid bromine, as shown in the following equation:

$$Br_2(l) \rightleftharpoons Br_2(g) \qquad K_c = [Br_2]$$

We can write equations for reaction quotients of heterogeneous equilibria that involve gases, using partial pressures instead of concentrations. Two examples are:

$$CaO(s) + CO_2(g) \rightleftharpoons CaCO_3(s) \qquad K_P = \frac{1}{P_{CO_2}}$$
$$C(s) + 2S(g) \rightleftharpoons CS_2(g) \qquad K_P = \frac{P_{CS_2}}{(P_S)^2}$$

13.3 Shifting Equilibria: Le Châtelier's Principle

By the end of this section, you will be able to:

- Describe the ways in which an equilibrium system can be stressed
- Predict the response of a stressed equilibrium using Le Châtelier's principle

As we saw in the previous section, reactions proceed in both directions (reactants go to products and products go to reactants). We can tell a reaction is at equilibrium if the reaction quotient (Q) is equal to the equilibrium constant (K). We next address what happens when a system at equilibrium is disturbed so that Q is no longer equal to K. If a system at equilibrium is subjected to a perturbance or **stress** (such as a change in concentration) the **position of equilibrium** changes. Since this stress affects the concentrations of the reactants and the products, the value of Q will no longer equal the value of K. To re-establish equilibrium, the system will either shift toward the products (if $Q < K$) or the reactants (if $Q > K$) until Q returns to the same value as K.

This process is described by **Le Châtelier's principle**: When a chemical system at equilibrium is disturbed, it returns to equilibrium by counteracting the disturbance. As described in the previous paragraph, the disturbance causes a change in Q; the reaction will shift to re-establish $Q = K$.

Predicting the Direction of a Reversible Reaction

Le Châtelier's principle can be used to predict changes in equilibrium concentrations when a system that is at equilibrium is subjected to a stress. However, if we have a mixture of reactants and products that have not yet reached equilibrium, the changes necessary to reach equilibrium may not be so obvious. In such a case, we can compare the values of Q and K for the system to predict the changes.

Effect of Change in Concentration on Equilibrium

A chemical system at equilibrium can be temporarily shifted out of equilibrium by adding or removing one or more of the reactants or products. The concentrations of both reactants and products then undergo additional changes to return the system to equilibrium.

The stress on the system in Figure 13.8 is the reduction of the equilibrium concentration of SCN^- (lowering the concentration of one of the reactants would cause Q to be larger than K). As a consequence, Le Châtelier's principle leads us to predict that the concentration of $Fe(SCN)^{2+}$ should decrease, increasing the concentration of SCN^- part way back to its original concentration, and increasing the concentration of Fe^{3+} above its initial equilibrium concentration.

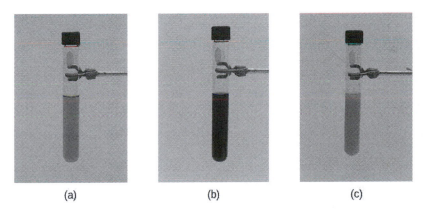

(a) (b) (c)

Figure 13.8 (a) The test tube contains 0.1 M Fe^{3+}. (b) Thiocyanate ion has been added to solution in (a), forming the red $Fe(SCN)^{2+}$ ion. $Fe^{3+}(aq) + SCN^-(aq) \rightleftharpoons Fe(SCN)^{2+}(aq)$. (c) Silver nitrate has been added to the solution in (b), precipitating some of the SCN^- as the white solid AgSCN. $Ag^+(aq) + SCN^-(aq) \rightleftharpoons AgSCN(s)$. The decrease in the SCN^- concentration shifts the first equilibrium in the solution to the left, decreasing the concentration (and lightening color) of the $Fe(SCN)^{2+}$. (credit: modification of work by Mark Ott)

The effect of a change in concentration on a system at equilibrium is illustrated further by the equilibrium of this chemical reaction:

$$H_2(g) + I_2(g) \rightleftharpoons 2HI(g) \qquad K_c = 50.0 \text{ at } 400\ °C$$

The numeric values for this example have been determined experimentally. A mixture of gases at 400 °C with $[H_2]$ = $[I_2]$ = 0.221 M and $[HI]$ = 1.563 M is at equilibrium; for this mixture, $Q_c = K_c = 50.0$. If H_2 is introduced into the system so quickly that its concentration doubles before it begins to react (new $[H_2] = 0.442\ M$), the reaction will shift so that a new equilibrium is reached, at which $[H_2] = 0.374\ M$, $[I_2] = 0.153\ M$, and $[HI] = 1.692\ M$. This gives:

$$Q_c = \frac{[HI]^2}{[H_2][I_2]} = \frac{(1.692)^2}{(0.374)(0.153)} = 50.0 = K_c$$

We have stressed this system by introducing additional H_2. The stress is relieved when the reaction shifts to the right, using up some (but not all) of the excess H_2, reducing the amount of uncombined I_2, and forming additional HI.

Effect of Change in Pressure on Equilibrium

Sometimes we can change the position of equilibrium by changing the pressure of a system. However, changes in pressure have a measurable effect only in systems in which gases are involved, and then only when the chemical reaction produces a change in the total number of gas molecules in the system. An easy way to recognize such a system is to look for different numbers of moles of gas on the reactant and product sides of the equilibrium. While evaluating pressure (as well as related factors like volume), it is important to remember that equilibrium constants are defined with regard to concentration (for K_c) or partial pressure (for K_P). Some changes to total pressure, like adding an inert gas that is not part of the equilibrium, will change the total pressure but not the partial pressures of the gases in the equilibrium constant expression. Thus, addition of a gas not involved in the equilibrium will not perturb the equilibrium.

Link to Learning

Check out this link (http://openstaxcollege.org/l/16equichange) to see a dramatic visual demonstration of how equilibrium changes with pressure changes.

As we increase the pressure of a gaseous system at equilibrium, either by decreasing the volume of the system or by adding more of one of the components of the equilibrium mixture, we introduce a stress by increasing the partial pressures of one or more of the components. In accordance with Le Châtelier's principle, a shift in the equilibrium that reduces the total number of molecules per unit of volume will be favored because this relieves the stress. The reverse reaction would be favored by a decrease in pressure.

Consider what happens when we increase the pressure on a system in which NO, O_2, and NO_2 are at equilibrium:

$$2NO(g) + O_2(g) \rightleftharpoons 2NO_2(g)$$

The formation of additional amounts of NO_2 decreases the total number of molecules in the system because each time two molecules of NO_2 form, a total of three molecules of NO and O_2 are consumed. This reduces the total pressure exerted by the system and reduces, but does not completely relieve, the stress of the increased pressure. On the other hand, a decrease in the pressure on the system favors decomposition of NO_2 into NO and O_2, which tends to restore the pressure.

Now consider this reaction:

$$N_2(g) + O_2(g) \rightleftharpoons 2NO(g)$$

Because there is no change in the total number of molecules in the system during reaction, a change in pressure does not favor either formation or decomposition of gaseous nitrogen monoxide.

Effect of Change in Temperature on Equilibrium

Changing concentration or pressure perturbs an equilibrium because the reaction quotient is shifted away from the equilibrium value. Changing the temperature of a system at equilibrium has a different effect: A change in temperature actually changes the value of the equilibrium constant. However, we can qualitatively predict the effect of the temperature change by treating it as a stress on the system and applying Le Châtelier's principle.

When hydrogen reacts with gaseous iodine, heat is evolved.

$$H_2(g) + I_2(g) \rightleftharpoons 2HI(g) \qquad\qquad \Delta H = -9.4\ kJ\ (exothermic)$$

Because this reaction is exothermic, we can write it with heat as a product.

$$H_2(g) + I_2(g) \rightleftharpoons 2HI(g) + heat$$

Increasing the temperature of the reaction increases the internal energy of the system. Thus, increasing the temperature has the effect of increasing the amount of one of the products of this reaction. The reaction shifts to the left to relieve the stress, and there is an increase in the concentration of H_2 and I_2 and a reduction in the concentration of HI. Lowering the temperature of this system reduces the amount of energy present, favors the production of heat, and favors the formation of hydrogen iodide.

When we change the temperature of a system at equilibrium, the equilibrium constant for the reaction changes. Lowering the temperature in the HI system increases the equilibrium constant: At the new equilibrium the concentration of HI has increased and the concentrations of H_2 and I_2 decreased. Raising the temperature decreases the value of the equilibrium constant, from 67.5 at 357 °C to 50.0 at 400 °C.

Temperature affects the equilibrium between NO_2 and N_2O_4 in this reaction

$$N_2O_4(g) \rightleftharpoons 2NO_2(g) \qquad\qquad \Delta H = 57.20\ kJ$$

The positive ΔH value tells us that the reaction is endothermic and could be written

$$heat + N_2O_4(g) \rightleftharpoons 2NO_2(g)$$

At higher temperatures, the gas mixture has a deep brown color, indicative of a significant amount of brown NO_2 molecules. If, however, we put a stress on the system by cooling the mixture (withdrawing energy), the equilibrium shifts to the left to supply some of the energy lost by cooling. The concentration of colorless N_2O_4 increases, and the concentration of brown NO_2 decreases, causing the brown color to fade.

Link to Learning

This interactive animation (http://openstaxcollege.org/l/16chatelier) allows you to apply Le Châtelier's principle to predict the effects of changes in concentration, pressure, and temperature on reactant and product concentrations.

Catalysts Do Not Affect Equilibrium

As we learned during our study of kinetics, a catalyst can speed up the rate of a reaction. Though this increase in reaction rate may cause a system to reach equilibrium more quickly (by speeding up the forward and reverse reactions), a catalyst has no effect on the value of an equilibrium constant nor on equilibrium concentrations.

The interplay of changes in concentration or pressure, temperature, and the lack of an influence of a catalyst on a chemical equilibrium is illustrated in the industrial synthesis of ammonia from nitrogen and hydrogen according to the equation

$$N_2(g) + 3H_2(g) \rightleftharpoons 2NH_3(g)$$

A large quantity of ammonia is manufactured by this reaction. Each year, ammonia is among the top 10 chemicals, by mass, manufactured in the world. About 2 billion pounds are manufactured in the United States each year.

Ammonia plays a vital role in our global economy. It is used in the production of fertilizers and is, itself, an important fertilizer for the growth of corn, cotton, and other crops. Large quantities of ammonia are converted to nitric acid, which plays an important role in the production of fertilizers, explosives, plastics, dyes, and fibers, and is also used in the steel industry.

Portrait of a Chemist

Fritz Haber

In the early 20th century, German chemist Fritz Haber (Figure 13.9) developed a practical process for converting diatomic nitrogen, which cannot be used by plants as a nutrient, to ammonia, a form of nitrogen that is easiest for plants to absorb.

$$N_2(g) + 3H_2(g) \rightleftharpoons 2NH_3(g)$$

The availability of nitrogen is a strong limiting factor to the growth of plants. Despite accounting for 78% of air, diatomic nitrogen (N_2) is nutritionally unavailable due the tremendous stability of the nitrogen-nitrogen triple bond. For plants to use atmospheric nitrogen, the nitrogen must be converted to a more bioavailable form (this conversion is called nitrogen fixation).

Haber was born in Breslau, Prussia (presently Wroclaw, Poland) in December 1868. He went on to study chemistry and, while at the University of Karlsruhe, he developed what would later be known as the Haber process: the catalytic formation of ammonia from hydrogen and atmospheric nitrogen under high temperatures and pressures. For this work, Haber was awarded the 1918 Nobel Prize in Chemistry for synthesis of ammonia from its elements. The Haber process was a boon to agriculture, as it allowed the production of fertilizers to no longer be dependent on mined feed stocks such as sodium nitrate. Currently, the annual production of synthetic nitrogen fertilizers exceeds 100 million tons and synthetic fertilizer production has increased the number of humans that arable land can support from 1.9 persons per hectare in 1908 to 4.3 in 2008.

Figure 13.9   The work of Nobel Prize recipient Fritz Haber revolutionized agricultural practices in the early 20th century. His work also affected wartime strategies, adding chemical weapons to the artillery.

In addition to his work in ammonia production, Haber is also remembered by history as one of the fathers of chemical warfare. During World War I, he played a major role in the development of poisonous gases used for trench warfare. Regarding his role in these developments, Haber said, "During peace time a scientist belongs to the World, but during war time he belongs to his country."[1] Haber defended the use of gas warfare against accusations that it was inhumane, saying that death was death, by whatever means it was inflicted. He stands as an example of the ethical dilemmas that face scientists in times of war and the double-edged nature of the sword of science.

Like Haber, the products made from ammonia can be multifaceted. In addition to their value for agriculture, nitrogen compounds can also be used to achieve destructive ends. Ammonium nitrate has also been used in explosives, including improvised explosive devices. Ammonium nitrate was one of the components of the bomb used in the attack on the Alfred P. Murrah Federal Building in downtown Oklahoma City on April 19, 1995.

It has long been known that nitrogen and hydrogen react to form ammonia. However, it became possible to manufacture ammonia in useful quantities by the reaction of nitrogen and hydrogen only in the early 20th century after the factors that influence its equilibrium were understood.

To be practical, an industrial process must give a large yield of product relatively quickly. One way to increase the yield of ammonia is to increase the pressure on the system in which N_2, H_2, and NH_3 are at equilibrium or are coming to equilibrium.

$$N_2(g) + 3H_2(g) \rightleftharpoons 2NH_3(g)$$

The formation of additional amounts of ammonia reduces the total pressure exerted by the system and somewhat reduces the stress of the increased pressure.

Although increasing the pressure of a mixture of N_2, H_2, and NH_3 will increase the yield of ammonia, at low temperatures, the rate of formation of ammonia is slow. At room temperature, for example, the reaction is so slow that if we prepared a mixture of N_2 and H_2, no detectable amount of ammonia would form during our lifetime. The formation of ammonia from hydrogen and nitrogen is an exothermic process:

$$N_2(g) + 3H_2(g) \longrightarrow 2NH_3(g) \qquad\qquad \Delta H = -92.2 \text{ kJ}$$

Thus, increasing the temperature to increase the rate lowers the yield. If we lower the temperature to shift the equilibrium to favor the formation of more ammonia, equilibrium is reached more slowly because of the large decrease of reaction rate with decreasing temperature.

Part of the rate of formation lost by operating at lower temperatures can be recovered by using a catalyst. The net effect of the catalyst on the reaction is to cause equilibrium to be reached more rapidly.

In the commercial production of ammonia, conditions of about 500 °C, 150–900 atm, and the presence of a catalyst are used to give the best compromise among rate, yield, and the cost of the equipment necessary to produce and contain high-pressure gases at high temperatures (Figure 13.10).

---

1. Herrlich, P. "The Responsibility of the Scientist: What Can History Teach Us About How Scientists Should Handle Research That Has the Potential to Create Harm?" *EMBO Reports* 14 (2013): 759–764.

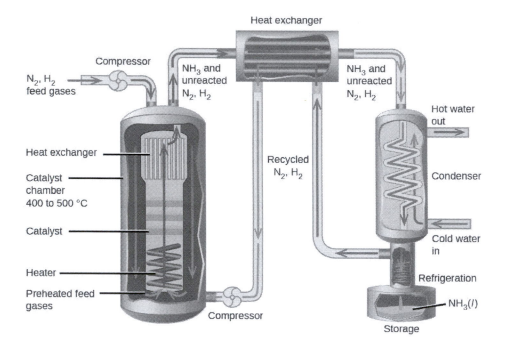

Figure 13.10 Commercial production of ammonia requires heavy equipment to handle the high temperatures and pressures required. This schematic outlines the design of an ammonia plant.

13.4 Equilibrium Calculations

By the end of this section, you will be able to:

- Write equations representing changes in concentration and pressure for chemical species in equilibrium systems
- Use algebra to perform various types of equilibrium calculations

We know that at equilibrium, the value of the reaction quotient of any reaction is equal to its equilibrium constant. Thus, we can use the mathematical expression for Q to determine a number of quantities associated with a reaction at equilibrium or approaching equilibrium. While we have learned to identify in which direction a reaction will shift to reach equilibrium, we want to extend that understanding to quantitative calculations. We do so by evaluating the ways that the concentrations of products and reactants change as a reaction approaches equilibrium, keeping in mind the stoichiometric ratios of the reaction. This algebraic approach to equilibrium calculations will be explored in this section.

Changes in concentrations or pressures of reactants and products occur as a reaction system approaches equilibrium. In this section we will see that we can relate these changes to each other using the coefficients in the balanced chemical equation describing the system. We use the decomposition of ammonia as an example.

On heating, ammonia reversibly decomposes into nitrogen and hydrogen according to this equation:

$$2NH_3(g) \rightleftharpoons N_2(g) + 3H_2(g)$$

If a sample of ammonia decomposes in a closed system and the concentration of N_2 increases by 0.11 M, the change in the N_2 concentration, $\Delta[N_2]$, the final concentration minus the initial concentration, is 0.11 M. The change is positive because the concentration of N_2 increases.

The change in the H_2 concentration, $\Delta[H_2]$, is also positive—the concentration of H_2 increases as ammonia decomposes. The chemical equation tells us that the change in the concentration of H_2 is three times the change in the concentration of N_2 because for each mole of N_2 produced, 3 moles of H_2 are produced.

$$\Delta[H_2] = 3 \times \Delta[N_2]$$
$$= 3 \times (0.11\ M) = 0.33\ M$$

The change in concentration of NH_3, $\Delta[NH_3]$, is twice that of $\Delta[N_2]$; the equation indicates that 2 moles of NH_3 must decompose for each mole of N_2 formed. However, the change in the NH_3 concentration is negative because the concentration of ammonia *decreases* as it decomposes.

$$\Delta[NH_3] = -2 \times \Delta[N_2] = -2 \times (0.11\ M) = -0.22\ M$$

We can relate these relationships directly to the coefficients in the equation

$$2NH_3(g) \quad\rightleftharpoons\quad N_2(g) \quad + \quad 3H_2(g)$$
$$\Delta[NH_3] = -2 \times \Delta[N_2] \qquad \Delta[N_2] = 0.11\ M \qquad \Delta[H_2] = 3 \times \Delta[N_2]$$

Note that all the changes on one side of the arrows are of the same sign and that all the changes on the other side of the arrows are of the opposite sign.

If we did not know the magnitude of the change in the concentration of N_2, we could represent it by the symbol x.

$$\Delta[N_2] = x$$

The changes in the other concentrations would then be represented as:

$$\Delta[H_2] = 3 \times \Delta[N_2] = 3x$$
$$\Delta[NH_3] = -2 \times \Delta[N_2] = -2x$$

The coefficients in the Δ terms are identical to those in the balanced equation for the reaction.

$$2NH_3(g) \quad\rightleftharpoons\quad N_2(g) \quad + \quad 3H_2(g)$$
$$-2x \qquad\qquad\quad x \qquad\qquad 3x$$

The simplest way for us to find the coefficients for the concentration changes in any reaction is to use the coefficients in the balanced chemical equation. The sign of the coefficient is positive when the concentration increases; it is negative when the concentration decreases.

Example 13.5

Determining Relative Changes in Concentration

Complete the changes in concentrations for each of the following reactions.

(a) $\quad C_2H_2(g) + \quad 2Br_2(g) \quad\rightleftharpoons\quad C_2H_2Br_4(g)$
$\qquad x \qquad\qquad\qquad \underline{\hspace{1.5cm}} \qquad\qquad \underline{\hspace{1.5cm}}$

(b) $\quad I_2(aq) + \quad I^-(aq) \quad\rightleftharpoons\quad I_3{}^-(aq)$
$\qquad \underline{\hspace{1.5cm}} \qquad \underline{\hspace{1.5cm}} \qquad x$

(c) $\quad C_3H_8(g) + \quad 5O_2(g) \quad\rightleftharpoons\quad 3CO_2(g) + \quad 4H_2O(g)$
$\qquad x \qquad\qquad\qquad \underline{\hspace{1.5cm}} \qquad\qquad \underline{\hspace{1.5cm}} \qquad \underline{\hspace{1.5cm}}$

Solution

(a) $\quad C_2H_2(g) + \quad 2Br_2(g) \quad\rightleftharpoons\quad C_2H_2Br_4(g)$
$\qquad x \qquad\qquad\quad 2x \qquad\qquad\qquad -x$

(b) $I_2(aq) + I^-(aq) \rightleftharpoons I_3^-(aq)$
    $-x$ $-x$ x

(c) $C_3H_8(g) + 5O_2(g) \rightleftharpoons 3CO_2(g) + 4H_2O(g)$
    x $5x$ $-3x$ $-4x$

Check Your Learning

Complete the changes in concentrations for each of the following reactions:

(a) $2SO_2(g) + O_2(g) \rightleftharpoons 2SO_3(g)$
    _____ x _____

(b) $C_4H_8(g) \rightleftharpoons 2C_2H_4(g)$
    _____ $-2x$

(c) $4NH_3(g) + 7H_2O(g) \rightleftharpoons 4NO_2(g) + 6H_2O(g)$
    _____   _____   _____   _____

Answer: (a) $2x, x, -2x$; (b) $x, -2x$; (c) $4x, 7x, -4x, -6x$ or $-4x, -7x, 4x, 6x$

Calculations Involving Equilibrium Concentrations

Because the value of the reaction quotient of any reaction at equilibrium is equal to its equilibrium constant, we can use the mathematical expression for Q_c (i.e., *the law of mass action*) to determine a number of quantities associated with a reaction at equilibrium. It may help if we keep in mind that $Q_c = K_c$ (at equilibrium) in all of these situations and that there are only three basic types of calculations:

1. **Calculation of an equilibrium constant**. If concentrations of reactants and products at equilibrium are known, the value of the equilibrium constant for the reaction can be calculated.

2. **Calculation of missing equilibrium concentrations**. If the value of the equilibrium constant and all of the equilibrium concentrations, except one, are known, the remaining concentration can be calculated.

3. **Calculation of equilibrium concentrations from initial concentrations**. If the value of the equilibrium constant and a set of concentrations of reactants and products that are not at equilibrium are known, the concentrations at equilibrium can be calculated.

A similar list could be generated using Q_P, K_P, and partial pressure. We will look at solving each of these cases in sequence.

Calculation of an Equilibrium Constant

Since the law of mass action is the only equation we have to describe the relationship between K_c and the concentrations of reactants and products, any problem that requires us to solve for K_c must provide enough information to determine the reactant and product concentrations at equilibrium. Armed with the concentrations, we can solve the equation for K_c, as it will be the only unknown.

Example 13.6 showed us how to determine the equilibrium constant of a reaction if we know the concentrations of reactants and products at equilibrium. The following example shows how to use the stoichiometry of the reaction and a combination of initial concentrations and equilibrium concentrations to determine an equilibrium constant. This technique, commonly called an ICE chart—for **I**nitial, **C**hange, and **E**quilibrium–will be helpful in solving many equilibrium problems. A chart is generated beginning with the equilibrium reaction in question. Underneath the reaction the initial concentrations of the reactants and products are listed—these conditions are usually provided in the problem and we consider no shift toward equilibrium to have happened. The next row of data is the change that occurs as the system shifts toward equilibrium—do not forget to consider the reaction stoichiometry as described in a previous section of this chapter. The last row contains the concentrations once equilibrium has been reached.

Example 13.6

Calculation of an Equilibrium Constant

Iodine molecules react reversibly with iodide ions to produce triiodide ions.

$$I_2(aq) + I^-(aq) \rightleftharpoons I_3^-(aq)$$

If a solution with the concentrations of I_2 and I^- both equal to 1.000×10^{-3} M before reaction gives an equilibrium concentration of I_2 of 6.61×10^{-4} M, what is the equilibrium constant for the reaction?

Solution

We will begin this problem by calculating the changes in concentration as the system goes to equilibrium. Then we determine the equilibrium concentrations and, finally, the equilibrium constant. First, we set up a table with the initial concentrations, the changes in concentrations, and the equilibrium concentrations using $-x$ as the change in concentration of I_2.

	I_2 +	I^- $\rightleftharpoons$	I_3^-
Initial concentration (M)	1.000×10^{-3}	1.000×10^{-3}	0
Change (M)	$-x$	$-x$	$+x$
Equilibrium concentration (M)	$[1.000 \times 10^{-3}]_i - x$	$[1.000 \times 10^{-3}]_i - x$	x

Since the equilibrium concentration of I_2 is given, we can solve for x. At equilibrium the concentration of I_2 is 6.61×10^{-4} M so that

$$1.000 \times 10^{-3} - x = 6.61 \times 10^{-4}$$
$$x = 1.000 \times 10^{-3} - 6.61 \times 10^{-4}$$
$$= 3.39 \times 10^{-4} \ M$$

Now we can fill in the table with the concentrations at equilibrium.

	I_2 +	I^- $\rightleftharpoons$	I_3^-
Initial concentration (M)	1.000×10^{-3}	1.000×10^{-3}	0
Change (M)	$-x = -3.39 \times 10^{-4}$	$-x$	$+x$
Equilibrium concentration (M)	6.61×10^{-4}	6.61×10^{-4}	3.39×10^{-4}

We now calculate the value of the equilibrium constant.

$$K_c = Q_c = \frac{[I_3^-]}{[I_2][I^-]}$$

$$= \frac{3.39 \times 10^{-4} \ M}{(6.61 \times 10^{-4} \ M)(6.61 \times 10^{-4} \ M)} = 776$$

Check Your Learning

Ethanol and acetic acid react and form water and ethyl acetate, the solvent responsible for the odor of some nail polish removers.

$$C_2H_5OH + CH_3CO_2H \rightleftharpoons CH_3CO_2C_2H_5 + H_2O$$

When 1 mol each of C_2H_5OH and CH_3CO_2H are allowed to react in 1 L of the solvent dioxane, equilibrium is established when $\frac{1}{3}$ mol of each of the reactants remains. Calculate the equilibrium constant for the reaction. (Note: Water is not a solvent in this reaction.)

Answer: $K_c = 4$

Calculation of a Missing Equilibrium Concentration

If we know the equilibrium constant for a reaction and know the concentrations at equilibrium of all reactants and products except one, we can calculate the missing concentration.

Example 13.7

Calculation of a Missing Equilibrium Concentration

Nitrogen oxides are air pollutants produced by the reaction of nitrogen and oxygen at high temperatures. At 2000 °C, the value of the equilibrium constant for the reaction, $N_2(g) + O_2(g) \rightleftharpoons 2NO(g)$, is 4.1×10^{-4}. Find the concentration of $NO(g)$ in an equilibrium mixture with air at 1 atm pressure at this temperature. In air, $[N_2] = 0.036$ mol/L and $[O_2]$ 0.0089 mol/L.

Solution

We are given all of the equilibrium concentrations except that of NO. Thus, we can solve for the missing equilibrium concentration by rearranging the equation for the equilibrium constant.

$$K_c = Q_c = \frac{[NO]^2}{[N_2][O_2]}$$
$$[NO]^2 = K_c[N_2][O_2]$$
$$[NO] = \sqrt{K_c[N_2][O_2]}$$
$$= \sqrt{(4.1 \times 10^{-4})(0.036)(0.0089)}$$
$$= \sqrt{1.31 \times 10^{-7}}$$
$$= 3.6 \times 10^{-4}$$

Thus [NO] is 3.6×10^{-4} mol/L at equilibrium under these conditions.

We can check our answer by substituting all equilibrium concentrations into the expression for the reaction quotient to see whether it is equal to the equilibrium constant.

$$Q_c = \frac{[NO]^2}{[N_2][O_2]}$$
$$= \frac{(3.6 \times 10^{-4})^2}{(0.036)(0.0089)}$$
$$Q_c = 4.0 \times 10^{-4} = K_c$$

The answer checks; our calculated value gives the equilibrium constant within the error associated with the significant figures in the problem.

Check Your Learning

The equilibrium constant for the reaction of nitrogen and hydrogen to produce ammonia at a certain temperature is 6.00×10^{-2}. Calculate the equilibrium concentration of ammonia if the equilibrium concentrations of nitrogen and hydrogen are 4.26 M and 2.09 M, respectively.

Answer: 1.53 mol/L

Calculation of Changes in Concentration

If we know the equilibrium constant for a reaction and a set of concentrations of reactants and products that are *not at equilibrium*, we can calculate the changes in concentrations as the system comes to equilibrium, as well as the new concentrations at equilibrium. The typical procedure can be summarized in four steps.

1. Determine the direction the reaction proceeds to come to equilibrium.

    a. Write a balanced chemical equation for the reaction.

    b. If the direction in which the reaction must proceed to reach equilibrium is not obvious, calculate Q_c from the initial concentrations and compare to K_c to determine the direction of change.

2. Determine the relative changes needed to reach equilibrium, then write the equilibrium concentrations in terms of these changes.

    a. Define the changes in the initial concentrations that are needed for the reaction to reach equilibrium. Generally, we represent the smallest change with the symbol x and express the other changes in terms of the smallest change.

    b. Define missing equilibrium concentrations in terms of the initial concentrations and the changes in concentration determined in (a).

3. Solve for the change and the equilibrium concentrations.

    a. Substitute the equilibrium concentrations into the expression for the equilibrium constant, solve for x, and check any assumptions used to find x.

    b. Calculate the equilibrium concentrations.

4. Check the arithmetic.

    a. Check the calculated equilibrium concentrations by substituting them into the equilibrium expression and determining whether they give the equilibrium constant.
    Sometimes a particular step may differ from problem to problem—it may be more complex in some problems and less complex in others. However, every calculation of equilibrium concentrations from a set of initial concentrations will involve these steps.
    In solving equilibrium problems that involve changes in concentration, sometimes it is convenient to set up an ICE table, as described in the previous section.

Example 13.8

Calculation of Concentration Changes as a Reaction Goes to Equilibrium

Under certain conditions, the equilibrium constant for the decomposition of $PCl_5(g)$ into $PCl_3(g)$ and $Cl_2(g)$ is 0.0211. What are the equilibrium concentrations of PCl_5, PCl_3, and Cl_2 if the initial concentration of PCl_5 was 1.00 *M*?

Solution

Use the stepwise process described earlier.

>   ***Step 1.*** *Determine the direction the reaction proceeds.*
>   The balanced equation for the decomposition of PCl_5 is
>
> $$PCl_5(g) \rightleftharpoons PCl_3(g) + Cl_2(g)$$
>
>   Because we have no products initially, $Q_c = 0$ and the reaction will proceed to the right.
>
>   ***Step 2.*** *Determine the relative changes needed to reach equilibrium, then write the equilibrium concentrations in terms of these changes.*

Let us represent the increase in concentration of PCl_3 by the symbol x. The other changes may be written in terms of x by considering the coefficients in the chemical equation.

$$PCl_5(g) \rightleftharpoons PCl_3(g) + Cl_2(g)$$
$${-x} \qquad {x} \qquad {x}$$

The changes in concentration and the expressions for the equilibrium concentrations are:

	PCl$_5$ $\rightleftharpoons$	PCl$_3$ +	Cl$_2$
Initial concentration (M)	1.00	0	0
Change (M)	$-x$	$+x$	$+x$
Equilibrium concentration (M)	$1.00 - x$	$0 + x = x$	$0 + x = x$

Step 3. *Solve for the change and the equilibrium concentrations.*

Substituting the equilibrium concentrations into the equilibrium constant equation gives

$$K_c = \frac{[PCl_3][Cl_2]}{[PCl_5]} = 0.0211$$
$$= \frac{(x)(x)}{(1.00 - x)}$$

This equation contains only one variable, x, the change in concentration. We can write the equation as a quadratic equation and solve for x using the quadratic formula.

$$0.0211 = \frac{(x)(x)}{(1.00 - x)}$$
$$0.0211(1.00 - x) = x^2$$
$$x^2 + 0.0211x - 0.0211 = 0$$

Appendix B shows us an equation of the form $ax^2 + bx + c = 0$ can be rearranged to solve for x:

$$x = \frac{-b \pm \sqrt{b^2 - 4ac}}{2a}$$

In this case, $a = 1$, $b = 0.0211$, and $c = -0.0211$. Substituting the appropriate values for a, b, and c yields:

$$x = \frac{-0.0211 \pm \sqrt{(0.0211)^2 - 4(1)(-0.0211)}}{2(1)}$$
$$= \frac{-0.0211 \pm \sqrt{(4.45 \times 10^{-4}) + (8.44 \times 10^{-2})}}{2}$$
$$= \frac{-0.0211 \pm 0.291}{2}$$

Hence

$$x = \frac{-0.0211 + 0.291}{2} = 0.135$$

or

$$x = \frac{-0.0211 - 0.291}{2} = -0.156$$

Quadratic equations often have two different solutions, one that is physically possible and one that is physically impossible (an extraneous root). In this case, the second solution (-0.156) is physically impossible because we know the change must be a positive number (otherwise we would end up with negative values for concentrations of the products). Thus, $x = 0.135\ M$.

The equilibrium concentrations are

$$[PCl_5] = 1.00 - 0.135 = 0.87\ M$$

$$[PCl_3] = x = 0.135\ M$$
$$[Cl_2] = x = 0.135\ M$$

Step 4.   *Check the arithmetic.*

Substitution into the expression for K_c (to check the calculation) gives

$$K_c = \frac{[PCl_3][Cl_2]}{[PCl_5]} = \frac{(0.135)(0.135)}{0.87} = 0.021$$

The equilibrium constant calculated from the equilibrium concentrations is equal to the value of K_c given in the problem (when rounded to the proper number of significant figures). Thus, the calculated equilibrium concentrations check.

Check Your Learning

Acetic acid, CH_3CO_2H, reacts with ethanol, C_2H_5OH, to form water and ethyl acetate, $CH_3CO_2C_2H_5$.

$$CH_3CO_2H + C_2H_5OH \rightleftharpoons CH_3CO_2C_2H_5 + H_2O$$

The equilibrium constant for this reaction with dioxane as a solvent is 4.0. What are the equilibrium concentrations when a mixture that is 0.15 M in CH_3CO_2H, 0.15 M in C_2H_5OH, 0.40 M in $CH_3CO_2C_2H_5$, and 0.40 M in H_2O are mixed in enough dioxane to make 1.0 L of solution?

Answer: $[CH_3CO_2H] = 0.36\ M$, $[C_2H_5OH] = 0.36\ M$, $[CH_3CO_2C_2H_5] = 0.17\ M$, $[H_2O] = 0.17\ M$

Check Your Learning

A 1.00-L flask is filled with 1.00 moles of H_2 and 2.00 moles of I_2. The value of the equilibrium constant for the reaction of hydrogen and iodine reacting to form hydrogen iodide is 50.5 under the given conditions. What are the equilibrium concentrations of H_2, I_2, and HI in moles/L?

$$H_2(g) + I_2(g) \rightleftharpoons 2HI(g)$$

Answer: $[H_2] = 0.06\ M$, $[I_2] = 1.06\ M$, $[HI] = 1.88\ M$

Sometimes it is possible to use chemical insight to find solutions to equilibrium problems without actually solving a quadratic (or more complicated) equation. First, however, it is useful to verify that equilibrium can be obtained starting from two extremes: all (or mostly) reactants and all (or mostly) products (similar to what was shown in Figure 13.7).

Consider the ionization of 0.150 M HA, a weak acid.

$$HA(aq) \rightleftharpoons H^+(aq) + A^-(aq) \qquad K_c = 6.80 \times 10^{-4}$$

The most obvious way to determine the equilibrium concentrations would be to start with only reactants. This could be called the "all reactant" starting point. Using x for the amount of acid ionized at equilibrium, this is the ICE table and solution.

	HA(aq) $\rightleftharpoons$ H⁺(aq) + A⁻(aq)		
Initial concentration (M)	0.150	0	0
Change (M)	−x	x	x
Equilibrium concentration (M)	0.150 − x	x	x

Setting up and solving the quadratic equation gives

$$K_c = \frac{[H^+][A^-]}{[HA]} = \frac{(x)(x)}{(0.150 - x)} = 6.80 \times 10^{-4}$$

$$x2 + 6.80 \times 10^{-4}\,x - 1.02 \times 10^{-4} = 0$$

$$x = \frac{-6.80 \times 10^{-4} \pm \sqrt{(6.80 \times 10^{-4})^2 - (4)(1)(-1.02 \times 10^{-4})}}{(2)(1)}$$

$$x = 0.00977\ M \text{ or} -0.0104\ M$$

Using the positive (physical) root, the equilibrium concentrations are

$$[HA] = 0.150 - x = 0.140\ M$$
$$[H^+] = [A^-] = x = 0.00977\ M$$

A less obvious way to solve the problem would be to assume all the HA ionizes first, then the system comes to equilibrium. This could be called the "all product" starting point. Assuming all of the HA ionizes gives

$$[HA] = 0.150 - 0.150 = 0\ M$$
$$[H^+] = 0 + 0.150 = 0.150\ M$$
$$[A^-] = 0 + 0.150 = 0.150\ M$$

Using these as initial concentrations and "y" to represent the concentration of HA at equilibrium, this is the ICE table for this starting point.

	HA(aq) $\rightleftharpoons$ H$^+$(aq) + A$^-$(aq)		
Initial concentration (M)	0	0.150	0.150
Change (M)	+y	−y	−y
Equilibrium concentration (M)	y	0.150 − y	0.150 − y

Setting up and solving the quadratic equation gives

$$K_c = \frac{[H^+][A^-]}{[HA]} = \frac{(0.150 - y)(0.150 - y)}{(y)} = 6.80 \times 10^{-4}$$

$$6.80 \times 10^{-4} y = 0.0225 - 0.300y + y^2$$

Retain a few extra significant figures to minimize rounding problems.

$$y^2 - 0.30068y + 0.022500 = 0$$

$$y = \frac{0.30068 \pm \sqrt{(0.30068)^2 - (4)(1)(0.022500)}}{(2)(1)}$$

$$y = \frac{0.30068 \pm 0.020210}{2}$$

Rounding each solution to three significant figures gives

$$y = 0.160\ M \qquad \text{or} \qquad y = 0.140\ M$$

Using the physically significant root (0.140 M) gives the equilibrium concentrations as

$$[HA] = y = 0.140\ M$$
$$[H^+] = 0.150 - y = 0.010\ M$$
$$[A^-] = 0.150 - y = 0.010\ M$$

Thus, the two approaches give the same results (to three *decimal places*), and show that *both* starting points lead to the same equilibrium conditions. The "all reactant" starting point resulted in a relatively small change (x) because the system was close to equilibrium, while the "all product" starting point had a relatively large change (y) that was nearly the size of the initial concentrations. It can be said that a system that starts "close" to equilibrium will require only a "small" change in conditions (x) to reach equilibrium.

Recall that a small K_c means that very little of the reactants form products and a large K_c means that most of the reactants form products. If the system can be arranged so it starts "close" to equilibrium, then if the change (x) is small compared to any initial concentrations, it can be neglected. Small is usually defined as resulting in an error of less than 5%. The following two examples demonstrate this.

Example 13.9

Approximate Solution Starting Close to Equilibrium

What are the concentrations at equilibrium of a 0.15 M solution of HCN?

$$HCN(aq) \rightleftharpoons H^+(aq) + CN^-(aq) \qquad K_c = 4.9 \times 10^{-10}$$

Solution

Using "x" to represent the concentration of each product at equilibrium gives this ICE table.

	HCN(aq) $\rightleftharpoons$ H$^+$(aq) + CN$^-$(aq)		
Initial concentration (M)	0.15	0	0
Change (M)	$-x$	x	x
Equilibrium concentration (M)	$0.15 - x$	x	x

The exact solution may be obtained using the quadratic formula with

$$K_c = \frac{(x)(x)}{0.15 - x}$$

solving

$$x^2 + 4.9 \times 10^{-10} - 7.35 \times 10^{-11} = 0$$
$$x = 8.56 \times 10^{-6} \ M \ (3 \text{ sig. figs.}) = 8.6 \times 10^{-6} \ M \ (2 \text{ sig. figs.})$$

Thus $[H^+] = [CN^-] = x = 8.6 \times 10^{-6} \ M$ and $[HCN] = 0.15 - x = 0.15 \ M$.

In this case, chemical intuition can provide a simpler solution. From the equilibrium constant and the initial conditions, x must be small compared to 0.15 M. More formally, if $x \ll 0.15$, then $0.15 - x \approx 0.15$. If this assumption is true, then it simplifies obtaining x

$$K_c = \frac{(x)(x)}{0.15 - x} \approx \frac{x^2}{0.15}$$
$$4.9 \times 10^{-10} = \frac{x^2}{0.15}$$
$$x^2 = (0.15)(4.9 \times 10^{-10}) = 7.4 \times 10^{-11}$$
$$x = \sqrt{7.4 \times 10^{-11}} = 8.6 \times 10^{-6} \ M$$

In this example, solving the exact (quadratic) equation and using approximations gave the same result to two significant figures. While most of the time the approximation is a bit different from the exact solution, as long as the error is less than 5%, the approximate solution is considered valid. In this problem, the 5% applies to IF $(0.15 - x) \approx 0.15 \ M$, so if

$$\frac{x}{0.15} \times 100\% = \frac{8.6 \times 10^{-6}}{0.15} \times 100\% = 0.006\%$$

is less than 5%, as it is in this case, the assumption is valid. The approximate solution is thus a valid solution.

Check Your Learning

What are the equilibrium concentrations in a 0.25 M NH$_3$ solution?

$$NH_3(aq) + H_2O(l) \rightleftharpoons NH_4^+(aq) + OH^-(aq) \qquad K_c = 1.8 \times 10^{-5}$$

Assume that x is much less than 0.25 M and calculate the error in your assumption.

Answer: $[OH^-] = [NH_4^+] = 0.0021 \ M$; $[NH_3] = 0.25 \ M$, error = 0.84%

The second example requires that the original information be processed a bit, but it still can be solved using a small x approximation.

Example 13.10

Approximate Solution After Shifting Starting Concentration

Copper(II) ions form a complex ion in the presence of ammonia

$$Cu^{2+}(aq) + 4NH_3(aq) \rightleftharpoons Cu(NH_3)_4{}^{2+}(aq) \qquad K_c = 5.0 \times 10^{13} = \frac{[Cu(NH_3)_4{}^{2+}]}{[Cu^{2+}(aq)][NH_3]^4}$$

If 0.010 mol Cu^{2+} is added to 1.00 L of a solution that is 1.00 M NH_3 what are the concentrations when the system comes to equilibrium?

Solution

The initial concentration of copper(II) is 0.010 M. The equilibrium constant is very large so it would be better to start with as much product as possible because "all products" is much closer to equilibrium than "all reactants." Note that Cu^{2+} is the limiting reactant; if all 0.010 M of it reacts to form product the concentrations would be

$$[Cu^{2+}] = 0.010 - 0.010 = 0\ M$$
$$[Cu(NH_3)_4{}^{2+}] = 0.010\ M$$
$$[NH_3] = 1.00 - 4 \times 0.010 = 0.96\ M$$

Using these "shifted" values as initial concentrations with x as the free copper(II) ion concentration at equilibrium gives this ICE table.

	$Cu^{2+}(aq)$ +	$4NH_3(aq) \rightleftharpoons$	$Cu(NH_3)_4{}^{2+}(aq)$
Initial concentration (M)	0	0.96	0.010
Change (M)	$+x$	$+4x$	$-x$
Equilibrium concentration (M)	x	$0.96 + 4x$	$0.010 - x$

Since we are starting close to equilibrium, x should be small so that

$$0.96 + 4x \approx 0.96\ M$$
$$0.010 - x \approx 0.010\ M$$
$$K_c = \frac{(0.010 - x)}{x(0.96 - 4x)^4} \approx \frac{(0.010)}{x(0.96)^4} = 5.0 \times 10^{13}$$
$$x = \frac{(0.010)}{K_c(0.96)^4} = 2.4 \times 10^{-16}\ M$$

Select the smallest concentration for the 5% rule.

$$\frac{2.4 \times 10^{-16}}{0.010} \times 100\% = 2 \times 10^{-12}\%$$

This is much less than 5%, so the assumptions are valid. The concentrations at equilibrium are

$$[Cu^{2+}] = x = 2.4 \times 10^{-16}\ M$$
$$[NH_3] = 0.96 - 4x = 0.96\ M$$
$$[Cu(NH_3)_4{}^{2+}] = 0.010 - x = 0.010\ M$$

By starting with the maximum amount of product, this system was near equilibrium and the change (x) was very small. With only a small change required to get to equilibrium, the equation for x was greatly simplified and gave a valid result well within the 5% error maximum.

Check Your Learning

What are the equilibrium concentrations when 0.25 mol Ni^{2+} is added to 1.00 L of 2.00 M NH_3 solution?

$$Ni^{2+}(aq) + 6NH_3(aq) \rightleftharpoons Ni(NH_3)_6{}^{2+}(aq) \qquad K_c = 5.5 \times 10^8$$

With such a large equilibrium constant, first form as much product as possible, then assume that only a small amount (x) of the product shifts left. Calculate the error in your assumption.

Answer: $[Ni(NH_3)_6{}^{2+}] = 0.25\ M$, $[NH_3] = 0.50\ M$, $[Ni^{2+}] = 2.9 \times 10^{-8}\ M$, error $= 1.2 \times 10^{-5}\%$

Key Terms

equilibrium in chemical reactions, the state in which the conversion of reactants into products and the conversion of products back into reactants occur simultaneously at the same rate; state of balance

equilibrium constant (*K*) value of the reaction quotient for a system at equilibrium

heterogeneous equilibria equilibria between reactants and products in different phases

homogeneous equilibria equilibria within a single phase

K_c equilibrium constant for reactions based on concentrations of reactants and products

K_P equilibrium constant for gas-phase reactions based on partial pressures of reactants and products

law of mass action when a reversible reaction has attained equilibrium at a given temperature, the reaction quotient remains constant

Le Châtelier's principle when a chemical system at equilibrium is disturbed, it returns to equilibrium by counteracting the disturbance

position of equilibrium concentrations or partial pressures of components of a reaction at equilibrium (commonly used to describe conditions before a disturbance)

reaction quotient (*Q*) ratio of the product of molar concentrations (or pressures) of the products to that of the reactants, each concentration (or pressure) being raised to the power equal to the coefficient in the equation

reversible reaction chemical reaction that can proceed in both the forward and reverse directions under given conditions

stress change to a reaction's conditions that may cause a shift in the equilibrium

Key Equations

- $Q = \dfrac{[C]^x[D]^y}{[A]^m[B]^n}$ where $mA + nB \rightleftharpoons xC + yD$

- $Q_P = \dfrac{(P_C)^x(P_D)^y}{(P_A)^m(P_B)^n}$ where $mA + nB \rightleftharpoons xC + yD$

- $P = MRT$

- $K_P = K_c\,(RT)^{\Delta n}$

Summary

13.1 Chemical Equilibria

A reaction is at equilibrium when the amounts of reactants or products no longer change. Chemical equilibrium is a dynamic process, meaning the rate of formation of products by the forward reaction is equal to the rate at which the products re-form reactants by the reverse reaction.

13.2 Equilibrium Constants

For any reaction that is at equilibrium, the reaction quotient Q is equal to the equilibrium constant K for the reaction. If a reactant or product is a pure solid, a pure liquid, or the solvent in a dilute solution, the concentration of this component does not appear in the expression for the equilibrium constant. At equilibrium, the values of the concentrations of the reactants and products are constant. Their particular values may vary depending on conditions, but the value of the reaction quotient will always equal K (K_c when using concentrations or K_P when using partial pressures).

A homogeneous equilibrium is an equilibrium in which all components are in the same phase. A heterogeneous equilibrium is an equilibrium in which components are in two or more phases. We can decide whether a reaction is at equilibrium by comparing the reaction quotient with the equilibrium constant for the reaction.

13.3 Shifting Equilibria: Le Châtelier's Principle

Systems at equilibrium can be disturbed by changes to temperature, concentration, and, in some cases, volume and pressure; volume and pressure changes will disturb equilibrium if the number of moles of gas is different on the reactant and product sides of the reaction. The system's response to these disturbances is described by Le Châtelier's principle: The system will respond in a way that counteracts the disturbance. Not all changes to the system result in a disturbance of the equilibrium. Adding a catalyst affects the rates of the reactions but does not alter the equilibrium, and changing pressure or volume will not significantly disturb systems with no gases or with equal numbers of moles of gas on the reactant and product side.

Effects of Disturbances of Equilibrium and K

Disturbance	Observed Change as Equilibrium is Restored	Direction of Shift	Effect on K
reactant added	added reactant is partially consumed	toward products	none
product added	added product is partially consumed	toward reactants	none
decrease in volume/ increase in gas pressure	pressure decreases	toward side with fewer moles of gas	none
increase in volume/ decrease in gas pressure	pressure increases	toward side with more moles of gas	none
temperature increase	heat is absorbed	toward products for endothermic, toward reactants for exothermic	changes
temperature decrease	heat is given off	toward reactants for endothermic, toward products for exothermic	changes

Table 13.1

13.4 Equilibrium Calculations

The ratios of the rate of change in concentrations of a reaction are equal to the ratios of the coefficients in the balanced chemical equation. The sign of the coefficient of X is positive when the concentration increases and negative when it decreases. We learned to approach three basic types of equilibrium problems. When given the concentrations of the reactants and products at equilibrium, we can solve for the equilibrium constant; when given the equilibrium constant and some of the concentrations involved, we can solve for the missing concentrations; and when given the equilibrium constant and the initial concentrations, we can solve for the concentrations at equilibrium.

Exercises

13.1 Chemical Equilibria

1. What does it mean to describe a reaction as "reversible"?

2. When writing an equation, how is a reversible reaction distinguished from a nonreversible reaction?

3. If a reaction is reversible, when can it be said to have reached equilibrium?

4. Is a system at equilibrium if the rate constants of the forward and reverse reactions are equal?

5. If the concentrations of products and reactants are equal, is the system at equilibrium?

13.2 Equilibrium Constants

6. Explain why there may be an infinite number of values for the reaction quotient of a reaction at a given temperature but there can be only one value for the equilibrium constant at that temperature.

7. Explain why an equilibrium between $Br_2(l)$ and $Br_2(g)$ would not be established if the container were not a closed vessel shown in Figure 13.5.

8. If you observe the following reaction at equilibrium, is it possible to tell whether the reaction started with pure NO_2 or with pure N_2O_4?
$$2NO_2(g) \rightleftharpoons N_2O_4(g)$$

9. Among the solubility rules previously discussed is the statement: All chlorides are soluble except Hg_2Cl_2, $AgCl$, $PbCl_2$, and $CuCl$.

(a) Write the expression for the equilibrium constant for the reaction represented by the equation
$AgCl(s) \rightleftharpoons Ag^+(aq) + Cl^-(aq)$. Is $K_c > 1$, < 1, or ≈ 1? Explain your answer.

(b) Write the expression for the equilibrium constant for the reaction represented by the equation
$Pb^{2+}(aq) + 2Cl^-(aq) \rightleftharpoons PbCl_2(s)$. Is $K_c > 1$, < 1, or ≈ 1? Explain your answer.

10. Among the solubility rules previously discussed is the statement: Carbonates, phosphates, borates, and arsenates—except those of the ammonium ion and the alkali metals—are insoluble.

(a) Write the expression for the equilibrium constant for the reaction represented by the equation
$CaCO_3(s) \rightleftharpoons Ca^{2+}(aq) + CO_3{}^-(aq)$. Is $K_c > 1$, < 1, or ≈ 1? Explain your answer.

(b) Write the expression for the equilibrium constant for the reaction represented by the equation
$3Ba^{2+}(aq) + 2PO_4{}^{3-}(aq) \rightleftharpoons Ba_3(PO_4)_2(s)$. Is $K_c > 1$, < 1, or ≈ 1? Explain your answer.

11. Benzene is one of the compounds used as octane enhancers in unleaded gasoline. It is manufactured by the catalytic conversion of acetylene to benzene: $3C_2H_2(g) \longrightarrow C_6H_6(g)$. Which value of K_c would make this reaction most useful commercially? $K_c \approx 0.01$, $K_c \approx 1$, or $K_c \approx 10$. Explain your answer.

12. Show that the complete chemical equation, the total ionic equation, and the net ionic equation for the reaction represented by the equation $KI(aq) + I_2(aq) \rightleftharpoons KI_3(aq)$ give the same expression for the reaction quotient. KI_3 is composed of the ions K^+ and $I_3{}^-$.

13. For a titration to be effective, the reaction must be rapid and the yield of the reaction must essentially be 100%. Is $K_c > 1$, < 1, or ≈ 1 for a titration reaction?

14. For a precipitation reaction to be useful in a gravimetric analysis, the product of the reaction must be insoluble. Is $K_c > 1$, < 1, or ≈ 1 for a useful precipitation reaction?

15. Write the mathematical expression for the reaction quotient, Q_c, for each of the following reactions:

(a) $CH_4(g) + Cl_2(g) \rightleftharpoons CH_3Cl(g) + HCl(g)$

(b) $N_2(g) + O_2(g) \rightleftharpoons 2NO(g)$

(c) $2SO_2(g) + O_2(g) \rightleftharpoons 2SO_3(g)$

(d) $BaSO_3(s) \rightleftharpoons BaO(s) + SO_2(g)$

(e) $P_4(g) + 5O_2(g) \rightleftharpoons P_4O_{10}(s)$

(f) $Br_2(g) \rightleftharpoons 2Br(g)$

(g) $CH_4(g) + 2O_2(g) \rightleftharpoons CO_2(g) + 2H_2O(l)$

(h) $CuSO_4 \cdot 5H_2O(s) \rightleftharpoons CuSO_4(s) + 5H_2O(g)$

16. Write the mathematical expression for the reaction quotient, Q_c, for each of the following reactions:

(a) $N_2(g) + 3H_2(g) \rightleftharpoons 2NH_3(g)$

(b) $4NH_3(g) + 5O_2(g) \rightleftharpoons 4NO(g) + 6H_2O(g)$

(c) $N_2O_4(g) \rightleftharpoons 2NO_2(g)$

(d) $CO_2(g) + H_2(g) \rightleftharpoons CO(g) + H_2O(g)$

(e) $NH_4Cl(s) \rightleftharpoons NH_3(g) + HCl(g)$

(f) $2Pb(NO_3)_2(s) \rightleftharpoons 2PbO(s) + 4NO_2(g) + O_2(g)$

(g) $2H_2(g) + O_2(g) \rightleftharpoons 2H_2O(l)$

(h) $S_8(g) \rightleftharpoons 8S(g)$

17. The initial concentrations or pressures of reactants and products are given for each of the following systems. Calculate the reaction quotient and determine the direction in which each system will proceed to reach equilibrium.

(a) $2NH_3(g) \rightleftharpoons N_2(g) + 3H_2(g)$ $K_c = 17$; $[NH_3] = 0.20\ M$, $[N_2] = 1.00\ M$, $[H_2] = 1.00\ M$

(b) $2NH_3(g) \rightleftharpoons N_2(g) + 3H_2(g)$ $K_P = 6.8 \times 10^4$; initial pressures: $NH_3 = 3.0$ atm, $N_2 = 2.0$ atm, $H_2 =$ 1.0 atm

(c) $2SO_3(g) \rightleftharpoons 2SO_2(g) + O_2(g)$ $K_c = 0.230$; $[SO_3] = 0.00\ M$, $[SO_2] = 1.00\ M$, $[O_2] = 1.00\ M$

(d) $2SO_3(g) \rightleftharpoons 2SO_2(g) + O_2(g)$ $K_P = 16.5$; initial pressures: $SO_3 = 1.00$ atm, $SO_2 = 1.00$ atm, $O_2 =$ 1.00 atm

(e) $2NO(g) + Cl_2(g) \rightleftharpoons 2NOCl(g)$ $K_c = 4.6 \times 10^4$; $[NO] = 1.00\ M$, $[Cl_2] = 1.00\ M$, $[NOCl] = 0\ M$

(f) $N_2(g) + O_2(g) \rightleftharpoons 2NO(g)$ $K_P = 0.050$; initial pressures: $NO = 10.0$ atm, $N_2 = O_2 = 5$ atm

18. The initial concentrations or pressures of reactants and products are given for each of the following systems. Calculate the reaction quotient and determine the direction in which each system will proceed to reach equilibrium.

(a) $2NH_3(g) \rightleftharpoons N_2(g) + 3H_2(g)$ $K_c = 17$; $[NH_3] = 0.50\ M$, $[N_2] = 0.15\ M$, $[H_2] = 0.12\ M$

(b) $2NH_3(g) \rightleftharpoons N_2(g) + 3H_2(g)$ $K_P = 6.8 \times 10^4$; initial pressures: $NH_3 = 2.00$ atm, $N_2 = 10.00$ atm, H_2 = 10.00 atm

(c) $2SO_3(g) \rightleftharpoons 2SO_2(g) + O_2(g)$ $K_c = 0.230$; $[SO_3] = 2.00\ M$, $[SO_2] = 2.00\ M$, $[O_2] = 2.00\ M$

(d) $2SO_3(g) \rightleftharpoons 2SO_2(g) + O_2(g)$ $K_P = 6.5$ atm; initial pressures: $SO_2 = 1.00$ atm, $O_2 = 1.130$ atm, SO_3 = 0 atm

(e) $2NO(g) + Cl_2(g) \rightleftharpoons 2NOCl(g)$ $K_P = 2.5 \times 10^3$; initial pressures: $NO = 1.00$ atm, $Cl_2 = 1.00$ atm, $NOCl = 0$ atm

(f) $N_2(g) + O_2(g) \rightleftharpoons 2NO(g)$ $K_c = 0.050$; $[N_2] = 0.100\ M$, $[O_2] = 0.200\ M$, $[NO] = 1.00\ M$

19. The following reaction has $K_P = 4.50 \times 10^{-5}$ at 720 K.
$N_2(g) + 3H_2(g) \rightleftharpoons 2NH_3(g)$

If a reaction vessel is filled with each gas to the partial pressures listed, in which direction will it shift to reach equilibrium? $P(NH_3) = 93$ atm, $P(N_2) = 48$ atm, and $P(H_2) = 52$

20. Determine if the following system is at equilibrium. If not, in which direction will the system need to shift to reach equilibrium?
$SO_2Cl_2(g) \rightleftharpoons SO_2(g) + Cl_2(g)$

$[SO_2Cl_2] = 0.12\ M$, $[Cl_2] = 0.16\ M$ and $[SO_2] = 0.050\ M$. K_c for the reaction is 0.078.

21. Which of the systems described in Exercise 13.15 give homogeneous equilibria? Which give heterogeneous equilibria?

22. Which of the systems described in Exercise 13.16 give homogeneous equilibria? Which give heterogeneous equilibria?

23. For which of the reactions in Exercise 13.15 does K_c (calculated using concentrations) equal K_P (calculated using pressures)?

24. For which of the reactions in Exercise 13.16 does K_c (calculated using concentrations) equal K_P (calculated using pressures)?

25. Convert the values of K_c to values of K_P or the values of K_P to values of K_c.

(a) $N_2(g) + 3H_2(g) \rightleftharpoons 2NH_3(g)$ $\qquad$ $K_c = 0.50$ at $400\ ^\circ C$

(b) $H_2 + I_2 \rightleftharpoons 2HI$ $\qquad$ $K_c = 50.2$ at $448\ ^\circ C$

(c) $Na_2SO_4 \cdot 10H_2O(s) \rightleftharpoons Na_2SO_4(s) + 10H_2O(g)$ $\qquad$ $K_P = 4.08 \times 10^{-25}$ at $25\ ^\circ C$

(d) $H_2O(l) \rightleftharpoons H_2O(g)$ $\qquad$ $K_P = 0.122$ at $50\ ^\circ C$

26. Convert the values of K_c to values of K_P or the values of K_P to values of K_c.

(a) $Cl_2(g) + Br_2(g) \rightleftharpoons 2BrCl(g)$ $\qquad$ $K_c = 4.7 \times 10^{-2}$ at $25\ ^\circ C$

(b) $2SO_2(g) + O_2(g) \rightleftharpoons 2SO_3(g)$ $\qquad$ $K_P = 48.2$ at $500\ ^\circ C$

(c) $CaCl_2 \cdot 6H_2O(s) \rightleftharpoons CaCl_2(s) + 6H_2O(g)$ $\qquad$ $K_P = 5.09 \times 10^{-44}$ at $25\ ^\circ C$

(d) $H_2O(l) \rightleftharpoons H_2O(g)$ $\qquad$ $K_P = 0.196$ at $60\ ^\circ C$

27. What is the value of the equilibrium constant expression for the change $H_2O(l) \rightleftharpoons H_2O(g)$ at $30\ ^\circ C$?

28. Write the expression of the reaction quotient for the ionization of HOCN in water.

29. Write the reaction quotient expression for the ionization of NH_3 in water.

30. What is the approximate value of the equilibrium constant K_P for the change $C_2H_5OC_2H_5(l) \rightleftharpoons C_2H_5OC_2H_5(g)$ at $25\ ^\circ C$. (Vapor pressure was described in the previous chapter on liquids and solids; refer back to this chapter to find the relevant information needed to solve this problem.)

13.3 Shifting Equilibria: Le Châtelier's Principle

31. The following equation represents a reversible decomposition:
$CaCO_3(s) \rightleftharpoons CaO(s) + CO_2(g)$

Under what conditions will decomposition in a closed container proceed to completion so that no $CaCO_3$ remains?

32. Explain how to recognize the conditions under which changes in pressure would affect systems at equilibrium.

33. What property of a reaction can we use to predict the effect of a change in temperature on the value of an equilibrium constant?

34. What would happen to the color of the solution in part (b) of Figure 13.8 if a small amount of NaOH were added and $Fe(OH)_3$ precipitated? Explain your answer.

35. The following reaction occurs when a burner on a gas stove is lit:
$CH_4(g) + 2O_2(g) \rightleftharpoons CO_2(g) + 2H_2O(g)$

Is an equilibrium among CH_4, O_2, CO_2, and H_2O established under these conditions? Explain your answer.

36. A necessary step in the manufacture of sulfuric acid is the formation of sulfur trioxide, SO_3, from sulfur dioxide, SO_2, and oxygen, O_2, shown here. At high temperatures, the rate of formation of SO_3 is higher, but the equilibrium amount (concentration or partial pressure) of SO_3 is lower than it would be at lower temperatures.
$2SO_2(g) + O_2(g) \longrightarrow 2SO_3(g)$

(a) Does the equilibrium constant for the reaction increase, decrease, or remain about the same as the temperature increases?

(b) Is the reaction endothermic or exothermic?

37. Suggest four ways in which the concentration of hydrazine, N_2H_4, could be increased in an equilibrium described by the following equation:

$$N_2(g) + 2H_2(g) \rightleftharpoons N_2H_4(g) \qquad \Delta H = 95 \text{ kJ}$$

38. Suggest four ways in which the concentration of PH_3 could be increased in an equilibrium described by the following equation:

$$P_4(g) + 6H_2(g) \rightleftharpoons 4PH_3(g) \qquad \Delta H = 110.5 \text{ kJ}$$

39. How will an increase in temperature affect each of the following equilibria? How will a decrease in the volume of the reaction vessel affect each?

(a) $2NH_3(g) \rightleftharpoons N_2(g) + 3H_2(g) \qquad \Delta H = 92 \text{ kJ}$

(b) $N_2(g) + O_2(g) \rightleftharpoons 2NO(g) \qquad \Delta H = 181 \text{ kJ}$

(c) $2O_3(g) \rightleftharpoons 3O_2(g) \qquad \Delta H = -285 \text{ kJ}$

(d) $CaO(s) + CO_2(g) \rightleftharpoons CaCO_3(s) \qquad \Delta H = -176 \text{ kJ}$

40. How will an increase in temperature affect each of the following equilibria? How will a decrease in the volume of the reaction vessel affect each?

(a) $2H_2O(g) \rightleftharpoons 2H_2(g) + O_2(g) \qquad \Delta H = 484 \text{ kJ}$

(b) $N_2(g) + 3H_2(g) \rightleftharpoons 2NH_3(g) \qquad \Delta H = -92.2 \text{ kJ}$

(c) $2Br(g) \rightleftharpoons Br_2(g) \qquad \Delta H = -224 \text{ kJ}$

(d) $H_2(g) + I_2(s) \rightleftharpoons 2HI(g) \qquad \Delta H = 53 \text{ kJ}$

41. Water gas is a 1:1 mixture of carbon monoxide and hydrogen gas and is called water gas because it is formed from steam and hot carbon in the following reaction: $H_2O(g) + C(s) \rightleftharpoons H_2(g) + CO(g)$. Methanol, a liquid fuel that could possibly replace gasoline, can be prepared from water gas and hydrogen at high temperature and pressure in the presence of a suitable catalyst.

(a) Write the expression for the equilibrium constant (K_c) for the reversible reaction

$$2H_2(g) + CO(g) \rightleftharpoons CH_3OH(g) \qquad \Delta H = -90.2 \text{ kJ}$$

(b) What will happen to the concentrations of H_2, CO, and CH_3OH at equilibrium if more H_2 is added?

(c) What will happen to the concentrations of H_2, CO, and CH_3OH at equilibrium if CO is removed?

(d) What will happen to the concentrations of H_2, CO, and CH_3OH at equilibrium if CH_3OH is added?

(e) What will happen to the concentrations of H_2, CO, and CH_3OH at equilibrium if the temperature of the system is increased?

(f) What will happen to the concentrations of H_2, CO, and CH_3OH at equilibrium if more catalyst is added?

42. Nitrogen and oxygen react at high temperatures.

(a) Write the expression for the equilibrium constant (K_c) for the reversible reaction

$$N_2(g) + O_2(g) \rightleftharpoons 2NO(g) \qquad \Delta H = 181 \text{ kJ}$$

(b) What will happen to the concentrations of N_2, O_2, and NO at equilibrium if more O_2 is added?

(c) What will happen to the concentrations of N_2, O_2, and NO at equilibrium if N_2 is removed?

(d) What will happen to the concentrations of N_2, O_2, and NO at equilibrium if NO is added?

(e) What will happen to the concentrations of N_2, O_2, and NO at equilibrium if the pressure on the system is increased by reducing the volume of the reaction vessel?

(f) What will happen to the concentrations of N_2, O_2, and NO at equilibrium if the temperature of the system is increased?

(g) What will happen to the concentrations of N_2, O_2, and NO at equilibrium if a catalyst is added?

43. Water gas, a mixture of H_2 and CO, is an important industrial fuel produced by the reaction of steam with red hot coke, essentially pure carbon.

(a) Write the expression for the equilibrium constant for the reversible reaction
$$C(s) + H_2O(g) \rightleftharpoons CO(g) + H_2(g) \qquad \Delta H = 131.30 \text{ kJ}$$

(b) What will happen to the concentration of each reactant and product at equilibrium if more C is added?

(c) What will happen to the concentration of each reactant and product at equilibrium if H_2O is removed?

(d) What will happen to the concentration of each reactant and product at equilibrium if CO is added?

(e) What will happen to the concentration of each reactant and product at equilibrium if the temperature of the system is increased?

44. Pure iron metal can be produced by the reduction of iron(III) oxide with hydrogen gas.

(a) Write the expression for the equilibrium constant (K_c) for the reversible reaction
$$Fe_2O_3(s) + 3H_2(g) \rightleftharpoons 2Fe(s) + 3H_2O(g) \qquad \Delta H = 98.7 \text{ kJ}$$

(b) What will happen to the concentration of each reactant and product at equilibrium if more Fe is added?

(c) What will happen to the concentration of each reactant and product at equilibrium if H_2O is removed?

(d) What will happen to the concentration of each reactant and product at equilibrium if H_2 is added?

(e) What will happen to the concentration of each reactant and product at equilibrium if the pressure on the system is increased by reducing the volume of the reaction vessel?

(f) What will happen to the concentration of each reactant and product at equilibrium if the temperature of the system is increased?

45. Ammonia is a weak base that reacts with water according to this equation:
$$NH_3(aq) + H_2O(l) \rightleftharpoons NH_4^+(aq) + OH^-(aq)$$

Will any of the following increase the percent of ammonia that is converted to the ammonium ion in water?

(a) Addition of NaOH

(b) Addition of HCl

(c) Addition of NH_4Cl

46. Acetic acid is a weak acid that reacts with water according to this equation:
$$CH_3CO_2H(aq) + H_2O(aq) \rightleftharpoons H_3O^+(aq) + CH_3CO_2^-(aq)$$

Will any of the following increase the percent of acetic acid that reacts and produces $CH_3CO_2^-$ ion?

(a) Addition of HCl

(b) Addition of NaOH

(c) Addition of $NaCH_3CO_2$

47. Suggest two ways in which the equilibrium concentration of Ag^+ can be reduced in a solution of Na^+, Cl^-, Ag^+, and NO_3^-, in contact with solid AgCl.
$$Na^+(aq) + Cl^-(aq) + Ag^+(aq) + NO_3^-(aq) \rightleftharpoons AgCl(s) + Na^+(aq) + NO_3^-(aq)$$
$$\Delta H = -65.9 \text{ kJ}$$

48. How can the pressure of water vapor be increased in the following equilibrium?
$$H_2O(l) \rightleftharpoons H_2O(g) \qquad \Delta H = 41 \text{ kJ}$$

49. Additional solid silver sulfate, a slightly soluble solid, is added to a solution of silver ion and sulfate ion at equilibrium with solid silver sulfate.

$$2Ag^+(aq) + SO_4{}^{2-}(aq) \rightleftharpoons Ag_2SO_4(s)$$

Which of the following will occur?

(a) Ag^+ or $SO_4{}^{2-}$ concentrations will not change.

(b) The added silver sulfate will dissolve.

(c) Additional silver sulfate will form and precipitate from solution as Ag^+ ions and $SO_4{}^{2-}$ ions combine.

(d) The Ag^+ ion concentration will increase and the $SO_4{}^{2-}$ ion concentration will decrease.

50. The amino acid alanine has two isomers, α-alanine and β-alanine. When equal masses of these two compounds are dissolved in equal amounts of a solvent, the solution of α-alanine freezes at the lowest temperature. Which form, α-alanine or β-alanine, has the larger equilibrium constant for ionization $(HX \rightleftharpoons H^+ + X^-)$?

13.4 Equilibrium Calculations

51. A reaction is represented by this equation: $A(aq) + 2B(aq) \rightleftharpoons 2C(aq)$ $\qquad K_c = 1 \times 10^3$

(a) Write the mathematical expression for the equilibrium constant.

(b) Using concentrations ≤1 M, make up two sets of concentrations that describe a mixture of A, B, and C at equilibrium.

52. A reaction is represented by this equation: $2W(aq) \rightleftharpoons X(aq) + 2Y(aq)$ $\qquad K_c = 5 \times 10^{-4}$

(a) Write the mathematical expression for the equilibrium constant.

(b) Using concentrations of ≤1 M, make up two sets of concentrations that describe a mixture of W, X, and Y at equilibrium.

53. What is the value of the equilibrium constant at 500 °C for the formation of NH_3 according to the following equation?

$$N_2(g) + 3H_2(g) \rightleftharpoons 2NH_3(g)$$

An equilibrium mixture of $NH_3(g)$, $H_2(g)$, and $N_2(g)$ at 500 °C was found to contain 1.35 M H_2, 1.15 M N_2, and 4.12 $\times$ 10^{-1} M NH_3.

54. Hydrogen is prepared commercially by the reaction of methane and water vapor at elevated temperatures.

$$CH_4(g) + H_2O(g) \rightleftharpoons 3H_2(g) + CO(g)$$

What is the equilibrium constant for the reaction if a mixture at equilibrium contains gases with the following concentrations: CH_4, 0.126 M; H_2O, 0.242 M; CO, 0.126 M; H_2 1.15 M, at a temperature of 760 °C?

55. A 0.72-mol sample of PCl_5 is put into a 1.00-L vessel and heated. At equilibrium, the vessel contains 0.40 mol of $PCl_3(g)$ and 0.40 mol of $Cl_2(g)$. Calculate the value of the equilibrium constant for the decomposition of PCl_5 to PCl_3 and Cl_2 at this temperature.

56. At 1 atm and 25 °C, NO_2 with an initial concentration of 1.00 M is 3.3×10^{-3}% decomposed into NO and O_2. Calculate the value of the equilibrium constant for the reaction.

$$2NO_2(g) \rightleftharpoons 2NO(g) + O_2(g)$$

57. Calculate the value of the equilibrium constant K_P for the reaction $2NO(g) + Cl_2(g) \rightleftharpoons 2NOCl(g)$ from these equilibrium pressures: NO, 0.050 atm; Cl_2, 0.30 atm; NOCl, 1.2 atm.

58. When heated, iodine vapor dissociates according to this equation:

$$I_2(g) \rightleftharpoons 2I(g)$$

At 1274 K, a sample exhibits a partial pressure of I_2 of 0.1122 atm and a partial pressure due to I atoms of 0.1378 atm. Determine the value of the equilibrium constant, K_P, for the decomposition at 1274 K.

59. A sample of ammonium chloride was heated in a closed container.

$$NH_4Cl(s) \rightleftharpoons NH_3(g) + HCl(g)$$

At equilibrium, the pressure of $NH_3(g)$ was found to be 1.75 atm. What is the value of the equilibrium constant K_P for the decomposition at this temperature?

60. At a temperature of 60 °C, the vapor pressure of water is 0.196 atm. What is the value of the equilibrium constant K_P for the transformation at 60 °C?

$$H_2O(l) \rightleftharpoons H_2O(g)$$

61. Complete the changes in concentrations (or pressure, if requested) for each of the following reactions.

(a)

$$2SO_3(g) \rightleftharpoons 2SO_2(g) + O_2(g)$$

___	___	$+x$
___	___	0.125 M

(b)

$$4NH_3(g) + 3O_2(g) \rightleftharpoons 2N_2(g) + 6H_2O(g)$$

___	$3x$	___	___
___	0.24 M	___	___

(c) Change in pressure:

$$2CH_4(g) \rightleftharpoons C_2H_2(g) + 3H_2(g)$$

___	x	___
___	25 torr	___

(d) Change in pressure:

$$CH_4(g) + H_2O(g) \rightleftharpoons CO(g) + 3H_2(g)$$

___	x	___	___
___	5 atm	___	___

(e)

$$NH_4Cl(s) \rightleftharpoons NH_3(g) + HCl(g)$$

	x	___
	1.03×10^{-4} M	___

(f) change in pressure:

$$Ni(s) + 4CO(g) \rightleftharpoons Ni(CO)_4(g)$$

	$4x$	___
	0.40 atm	___

62. Complete the changes in concentrations (or pressure, if requested) for each of the following reactions.

(a)

$$2H_2(g) + O_2(g) \rightleftharpoons 2H_2O(g)$$

___	___	$+2x$
___	___	1.50 M

(b)

$$CS_2(g) + 4H_2(g) \rightleftharpoons CH_4(g) + 2H_2S(g)$$

x	___	___	___
0.020 M	___	___	___

(c) Change in pressure:

$$H_2(g) + Cl_2(g) \rightleftharpoons 2HCl(g)$$
x ___ ___
1.50 atm ___ ___

(d) Change in pressure:

$$2NH_3(g) + 2O_2(g) \rightleftharpoons N_2O(g) + 3H_2O(g)$$
___ ___ ___ x
___ ___ ___ 60.6 torr

(e)

$$NH_4HS(s) \rightleftharpoons NH_3(g) + H_2S(g)$$
x ___
$9.8 \times 10^{-6} M$ ___

(f) Change in pressure:

$$Fe(s) + 5CO(g) \rightleftharpoons Fe(CO)_4(g)$$
___ x
___ 0.012 atm

63. Why are there no changes specified for Ni in Exercise 13.61, part (f)? What property of Ni does change?

64. Why are there no changes specified for NH₄HS in Exercise 13.62, part (e)? What property of NH₄HS does change?

65. Analysis of the gases in a sealed reaction vessel containing NH_3, N_2, and H_2 at equilibrium at 400 °C established the concentration of N_2 to be 1.2 M and the concentration of H_2 to be 0.24 M.

$$N_2(g) + 3H_2(g) \rightleftharpoons 2NH_3(g) \qquad K_c = 0.50 \text{ at } 400 \text{ °C}$$

Calculate the equilibrium molar concentration of NH_3.

66. Calculate the number of moles of HI that are at equilibrium with 1.25 mol of H_2 and 1.25 mol of I_2 in a 5.00−L flask at 448 °C.

$$H_2 + I_2 \rightleftharpoons 2HI \qquad K_c = 50.2 \text{ at } 448 \text{ °C}$$

67. What is the pressure of BrCl in an equilibrium mixture of Cl_2, Br_2, and BrCl if the pressure of Cl_2 in the mixture is 0.115 atm and the pressure of Br_2 in the mixture is 0.450 atm?

$$Cl_2(g) + Br_2(g) \rightleftharpoons 2BrCl(g) \qquad K_P = 4.7 \times 10^{-2}$$

68. What is the pressure of CO_2 in a mixture at equilibrium that contains 0.50 atm H_2, 2.0 atm of H_2O, and 1.0 atm of CO at 990 °C?

$$H_2(g) + CO_2(g) \rightleftharpoons H_2O(g) + CO(g) \qquad K_P = 1.6 \text{ at } 990 \text{ °C}$$

69. Cobalt metal can be prepared by reducing cobalt(II) oxide with carbon monoxide.

$$CoO(s) + CO(g) \rightleftharpoons Co(s) + CO_2(g) \qquad K_c = 4.90 \times 10^2 \text{ at } 550 \text{ °C}$$

What concentration of CO remains in an equilibrium mixture with $[CO_2] = 0.100 \ M$?

70. Carbon reacts with water vapor at elevated temperatures.

$$C(s) + H_2O(g) \rightleftharpoons CO(g) + H_2(g) \qquad K_c = 0.2 \text{ at } 1000 \text{ °C}$$

What is the concentration of CO in an equilibrium mixture with $[H_2O] = 0.500 \ M$ at 1000 °C?

71. Sodium sulfate 10−hydrate, $Na_2SO_4 \cdot 10H_2O$, dehydrates according to the equation

$$Na_2SO_4 \cdot 10H_2O(s) \rightleftharpoons Na_2SO_4(s) + 10H_2O(g) \qquad K_P = 4.08 \times 10^{-25} \text{ at } 25 \text{ °C}$$

What is the pressure of water vapor at equilibrium with a mixture of $Na_2SO_4 \cdot 10H_2O$ and $NaSO_4$?

72. Calcium chloride 6−hydrate, $CaCl_2 \cdot 6H_2O$, dehydrates according to the equation

$$CaCl_2 \cdot 6H_2O(s) \rightleftharpoons CaCl_2(s) + 6H_2O(g) \qquad\qquad K_P = 5.09 \times 10^{-44} \text{ at } 25 \text{ °C}$$

What is the pressure of water vapor at equilibrium with a mixture of $CaCl_2 \cdot 6H_2O$ and $CaCl_2$?

73. A student solved the following problem and found the equilibrium concentrations to be $[SO_2] = 0.590\ M$, $[O_2]$ = 0.0450 M, and $[SO_3] = 0.260\ M$. How could this student check the work without reworking the problem? The problem was: For the following reaction at 600 °C:

$$2SO_2(g) + O_2(g) \rightleftharpoons 2SO_3(g) \qquad\qquad K_c = 4.32$$

What are the equilibrium concentrations of all species in a mixture that was prepared with $[SO_3] = 0.500\ M$, $[SO_2]$ = 0 M, and $[O_2] = 0.350\ M$?

74. A student solved the following problem and found $[N_2O_4] = 0.16\ M$ at equilibrium. How could this student recognize that the answer was wrong without reworking the problem? The problem was: What is the equilibrium concentration of N_2O_4 in a mixture formed from a sample of NO_2 with a concentration of 0.10 M?

$$2NO_2(g) \rightleftharpoons N_2O_4(g) \qquad\qquad K_c = 160$$

75. Assume that the change in concentration of N_2O_4 is small enough to be neglected in the following problem.

(a) Calculate the equilibrium concentration of both species in 1.00 L of a solution prepared from 0.129 mol of N_2O_4 with chloroform as the solvent.

$$N_2O_4(g) \rightleftharpoons 2NO_2(g) \qquad\qquad K_c = 1.07 \times 10^{-5} \text{ in chloroform}$$

(b) Show that the change is small enough to be neglected.

76. Assume that the change in concentration of $COCl_2$ is small enough to be neglected in the following problem.

(a) Calculate the equilibrium concentration of all species in an equilibrium mixture that results from the decomposition of $COCl_2$ with an initial concentration of 0.3166 M.

$$COCl_2(g) \rightleftharpoons CO(g) + Cl_2(g) \qquad\qquad K_c = 2.2 \times 10^{-10}$$

(b) Show that the change is small enough to be neglected.

77. Assume that the change in pressure of H_2S is small enough to be neglected in the following problem.

(a) Calculate the equilibrium pressures of all species in an equilibrium mixture that results from the decomposition of H_2S with an initial pressure of 0.824 atm.

$$2H_2S(g) \rightleftharpoons 2H_2(g) + S_2(g) \qquad\qquad K_P = 2.2 \times 10^{-6}$$

(b) Show that the change is small enough to be neglected.

78. What are all concentrations after a mixture that contains $[H_2O] = 1.00\ M$ and $[Cl_2O] = 1.00\ M$ comes to equilibrium at 25 °C?

$$H_2O(g) + Cl_2O(g) \rightleftharpoons 2HOCl(g) \qquad\qquad K_c = 0.0900$$

79. What are the concentrations of PCl_5, PCl_3, and Cl_2 in an equilibrium mixture produced by the decomposition of a sample of pure PCl_5 with $[PCl_5] = 2.00\ M$?

$$PCl_5(g) \rightleftharpoons PCl_3(g) + Cl_2(g) \qquad\qquad K_c = 0.0211$$

80. Calculate the pressures of all species at equilibrium in a mixture of $NOCl$, NO, and Cl_2 produced when a sample of $NOCl$ with a pressure of 10.0 atm comes to equilibrium according to this reaction:

$$2NOCl(g) \rightleftharpoons 2NO(g) + Cl_2(g) \qquad\qquad K_P = 4.0 \times 10^{-4}$$

81. Calculate the equilibrium concentrations of NO, O_2, and NO_2 in a mixture at 250 °C that results from the reaction of 0.20 M NO and 0.10 M O_2. (Hint: K is large; assume the reaction goes to completion then comes back to equilibrium.)

$$2NO(g) + O_2(g) \rightleftharpoons 2NO_2(g) \qquad\qquad K_c = 2.3 \times 10^5 \text{ at } 250\,°C$$

82. Calculate the equilibrium concentrations that result when 0.25 M O_2 and 1.0 M HCl react and come to equilibrium.

$$4HCl(g) + O_2(g) \rightleftharpoons 2Cl_2(g) + 2H_2O(g) \qquad\qquad K_c = 3.1 \times 10^{13}$$

83. One of the important reactions in the formation of smog is represented by the equation

$$O_3(g) + NO(g) \rightleftharpoons NO_2(g) + O_2(g) \qquad\qquad K_P = 6.0 \times 10^{34}$$

What is the pressure of O_3 remaining after a mixture of O_3 with a pressure of 1.2×10^{-8} atm and NO with a pressure of 1.2×10^{-8} atm comes to equilibrium? (Hint: K_P is large; assume the reaction goes to completion then comes back to equilibrium.)

84. Calculate the pressures of NO, Cl_2, and NOCl in an equilibrium mixture produced by the reaction of a starting mixture with 4.0 atm NO and 2.0 atm Cl_2. (Hint: K_P is small; assume the reverse reaction goes to completion then comes back to equilibrium.)

$$2NO(g) + Cl_2(g) \rightleftharpoons 2NOCl(g) \qquad\qquad K_P = 2.5 \times 10^3$$

85. Calculate the number of grams of HI that are at equilibrium with 1.25 mol of H_2 and 63.5 g of iodine at 448 °C.

$$H_2 + I_2 \rightleftharpoons 2HI \qquad\qquad K_c = 50.2 \text{ at } 448\,°C$$

86. Butane exists as two isomers, n–butane and isobutane.

n-butane   isobutane

K_P = 2.5 at 25 °C

What is the pressure of isobutane in a container of the two isomers at equilibrium with a total pressure of 1.22 atm?

87. What is the minimum mass of $CaCO_3$ required to establish equilibrium at a certain temperature in a 6.50-L container if the equilibrium constant (K_c) is 0.050 for the decomposition reaction of $CaCO_3$ at that temperature?

$$CaCO_3(s) \rightleftharpoons CaO(s) + CO_2(g)$$

88. The equilibrium constant (K_c) for this reaction is 1.60 at 990 °C:

$$H_2(g) + CO_2(g) \rightleftharpoons H_2O(g) + CO(g)$$

Calculate the number of moles of each component in the final equilibrium mixture obtained from adding 1.00 mol of H_2, 2.00 mol of CO_2, 0.750 mol of H_2O, and 1.00 mol of CO to a 5.00-L container at 990 °C.

89. At 25 °C and at 1 atm, the partial pressures in an equilibrium mixture of N_2O_4 and NO_2 are $P_{N_2O_4} = 0.70$ atm and $P_{NO_2} = 0.30$ atm.

(a) Predict how the pressures of NO_2 and N_2O_4 will change if the total pressure increases to 9.0 atm. Will they increase, decrease, or remain the same?

(b) Calculate the partial pressures of NO_2 and N_2O_4 when they are at equilibrium at 9.0 atm and 25 °C.

90. In a 3.0-L vessel, the following equilibrium partial pressures are measured: N_2, 190 torr; H_2, 317 torr; NH_3, 1.00×10^3 torr.

$$N_2(g) + 3H_2(g) \rightleftharpoons 2NH_3(g)$$

(a) How will the partial pressures of H_2, N_2, and NH_3 change if H_2 is removed from the system? Will they increase, decrease, or remain the same?

(b) Hydrogen is removed from the vessel until the partial pressure of nitrogen, at equilibrium, is 250 torr. Calculate the partial pressures of the other substances under the new conditions.

91. The equilibrium constant (K_c) for this reaction is 5.0 at a given temperature.

$$CO(g) + H_2O(g) \rightleftharpoons CO_2(g) + H_2(g)$$

(a) On analysis, an equilibrium mixture of the substances present at the given temperature was found to contain 0.20 mol of CO, 0.30 mol of water vapor, and 0.90 mol of H_2 in a liter. How many moles of CO_2 were there in the equilibrium mixture?

(b) Maintaining the same temperature, additional H_2 was added to the system, and some water vapor was removed by drying. A new equilibrium mixture was thereby established containing 0.40 mol of CO, 0.30 mol of water vapor, and 1.2 mol of H_2 in a liter. How many moles of CO_2 were in the new equilibrium mixture? Compare this with the quantity in part (a), and discuss whether the second value is reasonable. Explain how it is possible for the water vapor concentration to be the same in the two equilibrium solutions even though some vapor was removed before the second equilibrium was established.

92. Antimony pentachloride decomposes according to this equation:

$$SbCl_5(g) \rightleftharpoons SbCl_3(g) + Cl_2(g)$$

An equilibrium mixture in a 5.00-L flask at 448 °C contains 3.85 g of $SbCl_5$, 9.14 g of $SbCl_3$, and 2.84 g of Cl_2. How many grams of each will be found if the mixture is transferred into a 2.00-L flask at the same temperature?

93. Consider the reaction between H_2 and O_2 at 1000 K

$$2H_2(g) + O_2(g) \rightleftharpoons 2H_2O(g) \qquad K_P = \frac{(P_{H_2O})^2}{(P_{O_2})(P_{H_2})^3} = 1.33 \times 10^{20}$$

If 0.500 atm of H_2 and 0.500 atm of O_2 are allowed to come to equilibrium at this temperature, what are the partial pressures of the components?

94. An equilibrium is established according to the following equation

$$Hg_2{}^{2+}(aq) + NO_3{}^-(aq) + 3H^+(aq) \rightleftharpoons 2Hg^{2+}(aq) + HNO_2(aq) + H_2O(l) \qquad K_c = 4.6$$

What will happen in a solution that is 0.20 M each in $Hg_2{}^{2+}$, $NO_3{}^-$, H^+, Hg^{2+}, and HNO_2?

(a) $Hg_2{}^{2+}$ will be oxidized and $NO_3{}^-$ reduced.

(b) $Hg_2{}^{2+}$ will be reduced and $NO_3{}^-$ oxidized.

(c) Hg^{2+} will be oxidized and HNO_2 reduced.

(d) Hg^{2+} will be reduced and HNO_2 oxidized.

(e) There will be no change because all reactants and products have an activity of 1.

95. Consider the equilibrium

$$4NO_2(g) + 6H_2O(g) \rightleftharpoons 4NH_3(g) + 7O_2(g)$$

(a) What is the expression for the equilibrium constant (K_c) of the reaction?

(b) How must the concentration of NH_3 change to reach equilibrium if the reaction quotient is less than the equilibrium constant?

(c) If the reaction were at equilibrium, how would a decrease in pressure (from an increase in the volume of the reaction vessel) affect the pressure of NO_2?

(d) If the change in the pressure of NO_2 is 28 torr as a mixture of the four gases reaches equilibrium, how much will the pressure of O_2 change?

96. The binding of oxygen by hemoglobin (Hb), giving oxyhemoglobin (HbO_2), is partially regulated by the concentration of H_3O^+ and dissolved CO_2 in the blood. Although the equilibrium is complicated, it can be summarized as

$$HbO_2(aq) + H_3O^+(aq) + CO_2(g) \rightleftharpoons CO_2-Hb-H^+ + O_2(g) + H_2O(l)$$

(a) Write the equilibrium constant expression for this reaction.

(b) Explain why the production of lactic acid and CO_2 in a muscle during exertion stimulates release of O_2 from the oxyhemoglobin in the blood passing through the muscle.

97. The hydrolysis of the sugar sucrose to the sugars glucose and fructose follows a first-order rate equation for the disappearance of sucrose.

$$C_{12}H_{22}O_{11}(aq) + H_2O(l) \longrightarrow C_6H_{12}O_6(aq) + C_6H_{12}O_6(aq)$$

Rate = $k[C_{12}H_{22}O_{11}]$

In neutral solution, $k = 2.1 \times 10^{-11}$/s at 27 °C. (As indicated by the rate constant, this is a very slow reaction. In the human body, the rate of this reaction is sped up by a type of catalyst called an enzyme.) (Note: That is not a mistake in the equation—the products of the reaction, glucose and fructose, have the same molecular formulas, $C_6H_{12}O_6$, but differ in the arrangement of the atoms in their molecules). The equilibrium constant for the reaction is 1.36×10^5 at 27 °C. What are the concentrations of glucose, fructose, and sucrose after a 0.150 M aqueous solution of sucrose has reached equilibrium? Remember that the activity of a solvent (the effective concentration) is 1.

98. The density of trifluoroacetic acid vapor was determined at 118.1 °C and 468.5 torr, and found to be 2.784 g/L. Calculate K_c for the association of the acid.

$$2CF_3CO_2H(g) \rightleftharpoons CF_3C \overset{\displaystyle O-H\cdots O}{\underset{\displaystyle O\cdots H-O}{\Big\langle}} CF_3C(g)$$

99. Liquid N_2O_3 is dark blue at low temperatures, but the color fades and becomes greenish at higher temperatures as the compound decomposes to NO and NO_2. At 25 °C, a value of $K_P = 1.91$ has been established for this decomposition. If 0.236 moles of N_2O_3 are placed in a 1.52-L vessel at 25 °C, calculate the equilibrium partial pressures of $N_2O_3(g)$, $NO_2(g)$, and $NO(g)$.

100. A 1.00-L vessel at 400 °C contains the following equilibrium concentrations: N_2, 1.00 M; H_2, 0.50 M; and NH_3, 0.25 M. How many moles of hydrogen must be removed from the vessel to increase the concentration of nitrogen to 1.1 M?

101. A 0.010 M solution of the weak acid HA has an osmotic pressure (see chapter on solutions and colloids) of 0.293 atm at 25 °C. A 0.010 M solution of the weak acid HB has an osmotic pressure of 0.345 atm under the same conditions.

(a) Which acid has the larger equilibrium constant for ionization

HA $[HA(aq) \rightleftharpoons A^-(aq) + H^+(aq)]$ or HB $[HB(aq) \rightleftharpoons H^+(aq) + B^-(aq)]$?

(b) What are the equilibrium constants for the ionization of these acids?

(Hint: Remember that each solution contains three dissolved species: the weak acid (HA or HB), the conjugate base (A^- or B^-), and the hydrogen ion (H^+). Remember that osmotic pressure (like all colligative properties) is related to the total number of solute particles. Specifically for osmotic pressure, those concentrations are described by molarities.)

Chapter 14

Acid-Base Equilibria

Figure 14.1 Sinkholes such as this are the result of reactions between acidic groundwaters and basic rock formations, like limestone. (credit: modification of work by Emil Kehnel)

Chapter Outline

14.1 Brønsted-Lowry Acids and Bases

14.2 pH and pOH

14.3 Relative Strengths of Acids and Bases

14.4 Hydrolysis of Salt Solutions

14.5 Polyprotic Acids

14.6 Buffers

14.7 Acid-Base Titrations

Introduction

In our bodies, in our homes, and in our industrial society, acids and bases play key roles. Proteins, enzymes, blood, genetic material, and other components of living matter contain both acids and bases. We seem to like the sour taste of acids; we add them to soft drinks, salad dressings, and spices. Many foods, including citrus fruits and some vegetables, contain acids. Cleaners in our homes contain acids or bases. Acids and bases play important roles in the chemical industry. Currently, approximately 36 million metric tons of sulfuric acid are produced annually in the United States alone. Huge quantities of ammonia (8 million tons), urea (10 million tons), and phosphoric acid (10 million tons) are also produced annually.

This chapter will illustrate the chemistry of acid-base reactions and equilibria, and provide you with tools for quantifying the concentrations of acids and bases in solutions.

14.1 Brønsted-Lowry Acids and Bases

By the end of this section, you will be able to:

- Identify acids, bases, and conjugate acid-base pairs according to the Brønsted-Lowry definition
- Write equations for acid and base ionization reactions
- Use the ion-product constant for water to calculate hydronium and hydroxide ion concentrations
- Describe the acid-base behavior of amphiprotic substances

Acids and bases have been known for a long time. When Robert Boyle characterized them in 1680, he noted that acids dissolve many substances, change the color of certain natural dyes (for example, they change litmus from blue to red), and lose these characteristic properties after coming into contact with alkalis (bases). In the eighteenth century, it was recognized that acids have a sour taste, react with limestone to liberate a gaseous substance (now known to be CO_2), and interact with alkalis to form neutral substances. In 1815, Humphry Davy contributed greatly to the development of the modern acid-base concept by demonstrating that hydrogen is the essential constituent of acids. Around that same time, Joseph Louis Gay-Lussac concluded that acids are substances that can neutralize bases and that these two classes of substances can be defined only in terms of each other. The significance of hydrogen was reemphasized in 1884 when Carl Axel Arrhenius defined an acid as a compound that dissolves in water to yield hydrogen cations (now recognized to be hydronium ions) and a base as a compound that dissolves in water to yield hydroxide anions.

In an earlier chapter on chemical reactions, we defined acids and bases as Arrhenius did: We identified an acid as a compound that dissolves in water to yield hydronium ions (H_3O^+) and a base as a compound that dissolves in water to yield hydroxide ions (OH^-). This definition is not wrong; it is simply limited.

Later, we extended the definition of an acid or a base using the more general definition proposed in 1923 by the Danish chemist Johannes Brønsted and the English chemist Thomas Lowry. Their definition centers on the proton, H^+. A proton is what remains when a normal hydrogen atom, $_1^1 H$, loses an electron. A compound that donates a proton to another compound is called a **Brønsted-Lowry acid**, and a compound that accepts a proton is called a **Brønsted-Lowry base**. An acid-base reaction is the transfer of a proton from a proton donor (acid) to a proton acceptor (base). In a subsequent chapter of this text we will introduce the most general model of acid-base behavior introduced by the American chemist G. N. Lewis.

Acids may be compounds such as HCl or H_2SO_4, organic acids like acetic acid (CH_3COOH) or ascorbic acid (vitamin C), or H_2O. Anions (such as HSO_4^-, $H_2PO_4^-$, HS^-, and HCO_3^-) and cations (such as H_3O^+, NH_4^+, and $[Al(H_2O)_6]^{3+}$) may also act as acids. Bases fall into the same three categories. Bases may be neutral molecules (such as H_2O, NH_3, and CH_3NH_2), anions (such as OH^-, HS^-, HCO_3^-, CO_3^{2-}, F^-, and PO_4^{3-}), or cations (such as $[Al(H_2O)_5 OH]^{2+}$). The most familiar bases are ionic compounds such as NaOH and $Ca(OH)_2$, which contain the hydroxide ion, OH^-. The hydroxide ion in these compounds accepts a proton from acids to form water:

$$H^+ + OH^- \longrightarrow H_2O$$

We call the product that remains after an acid donates a proton the **conjugate base** of the acid. This species is a base because it can accept a proton (to re-form the acid):

$$\text{acid} \rightleftharpoons \text{proton} + \text{conjugate base}$$
$$HF \rightleftharpoons H^+ + F^-$$
$$H_2SO_4 \rightleftharpoons H^+ + HSO_4^-$$
$$H_2O \rightleftharpoons H^+ + OH^-$$
$$HSO_4^- \rightleftharpoons H^+ + SO_4^{2-}$$
$$NH_4^+ \rightleftharpoons H^+ + NH_3$$

We call the product that results when a base accepts a proton the base's **conjugate acid**. This species is an acid because it can give up a proton (and thus re-form the base):

$$\text{base} + \text{proton} \;\rightleftharpoons\; \text{conjugate acid}$$
$$OH^- + H^+ \;\rightleftharpoons\; H_2O$$
$$H_2O + H^+ \;\rightleftharpoons\; H_3O^+$$
$$NH_3 + H^+ \;\rightleftharpoons\; NH_4{}^+$$
$$S^{2-} + H^+ \;\rightleftharpoons\; HS^-$$
$$CO_3{}^{2-} + H^+ \;\rightleftharpoons\; HCO_3{}^-$$
$$F^- + H^+ \;\rightleftharpoons\; HF$$

In these two sets of equations, the behaviors of acids as proton donors and bases as proton acceptors are represented in isolation. In reality, all acid-base reactions involve the *transfer* of protons between acids and bases. For example, consider the acid-base reaction that takes place when ammonia is dissolved in water. A water molecule (functioning as an acid) transfers a proton to an ammonia molecule (functioning as a base), yielding the conjugate base of water, OH^-, and the conjugate acid of ammonia, $NH_4{}^+$:

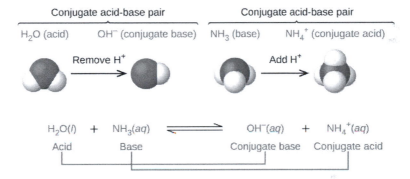

The reaction between a Brønsted-Lowry acid and water is called **acid ionization**. For example, when hydrogen fluoride dissolves in water and ionizes, protons are transferred from hydrogen fluoride molecules to water molecules, yielding hydronium ions and fluoride ions:

$$HF + H_2O \;\rightleftharpoons\; H_3O^+ + F^-$$
$$\text{Acid} \quad\;\; \text{Base} \qquad\qquad \text{Acid} \quad\;\; \text{Base}$$

$$K = \frac{[H_3O^+][F^-]}{[HF]}$$

When we add a base to water, a **base ionization** reaction occurs in which protons are transferred from water molecules to base molecules. For example, adding ammonia to water yields hydroxide ions and ammonium ions:

$$H_2O + C_5NH_5 \;\rightleftharpoons\; C_5NH_6{}^+ + OH^-$$
$$\text{Acid} \quad\;\; \text{Base} \qquad\qquad\quad \text{Acid} \qquad\quad \text{Base}$$

$$K = \frac{[C_5NH_6{}^+][OH^-]}{[C_5NH_5]}$$

Notice that both these ionization reactions are represented as equilibrium processes. The relative extent to which these acid and base ionization reactions proceed is an important topic treated in a later section of this chapter. In the preceding paragraphs we saw that water can function as either an acid or a base, depending on the nature of the solute dissolved in it. In fact, in pure water or in any aqueous solution, water acts both as an acid and a base. A very small fraction of water molecules donate protons to other water molecules to form hydronium ions and hydroxide ions:

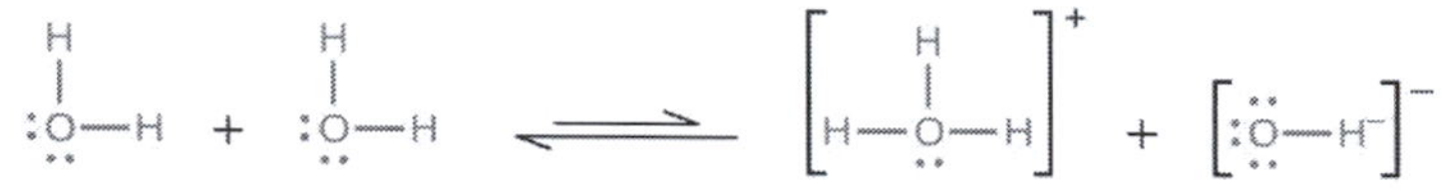

$$\underset{\text{Acid}}{H_2O} + \underset{\text{Base}}{H_2O} \rightleftharpoons \underset{\text{Acid}}{H_3O^+} + \underset{\text{Base}}{OH^-}$$

This type of reaction, in which a substance ionizes when one molecule of the substance reacts with another molecule of the same substance, is referred to as **autoionization**.

Pure water undergoes autoionization to a very slight extent. Only about two out of every 10^9 molecules in a sample of pure water are ionized at 25 °C. The equilibrium constant for the ionization of water is called the **ion-product constant for water (K_w)**:

$$H_2O(l) + H_2O(l) \rightleftharpoons H_3O^+(aq) + OH^-(aq) \qquad K_w = [H_3O^+][OH^-]$$

The slight ionization of pure water is reflected in the small value of the equilibrium constant; at 25 °C, K_w has a value of 1.0×10^{-14}. The process is endothermic, and so the extent of ionization and the resulting concentrations of hydronium ion and hydroxide ion increase with temperature. For example, at 100 °C, the value for K_w is about 5.1×10^{-13}, roughly 50 times larger than the value at 25 °C.

Example 14.1

Ion Concentrations in Pure Water

What are the hydronium ion concentration and the hydroxide ion concentration in pure water at 25 °C?

Solution

The autoionization of water yields the same number of hydronium and hydroxide ions. Therefore, in pure water, $[H_3O^+] = [OH^-]$. At 25 °C:

$$K_w = [H_3O^+][OH^-] = [H_3O^+]^{2+} = [OH^-]^{2+} = 1.0 \times 10^{-14}$$

So:

$$[H_3O^+] = [OH^-] = \sqrt{1.0 \times 10^{-14}} = 1.0 \times 10^{-7}\ M$$

The hydronium ion concentration and the hydroxide ion concentration are the same, and we find that both equal $1.0 \times 10^{-7}\ M$.

Check Your Learning

The ion product of water at 80 °C is 2.4×10^{-13}. What are the concentrations of hydronium and hydroxide ions in pure water at 80 °C?

Answer: $[H_3O^+] = [OH^-] = 4.9 \times 10^{-7}\ M$

It is important to realize that the autoionization equilibrium for water is established in all aqueous solutions. Adding an acid or base to water will not change the position of the equilibrium. Example 14.2 demonstrates the quantitative aspects of this relation between hydronium and hydroxide ion concentrations.

Example 14.2

The Inverse Proportionality of [H₃O⁺] and [OH⁻]

A solution of carbon dioxide in water has a hydronium ion concentration of 2.0×10^{-6} M. What is the concentration of hydroxide ion at 25 °C?

Solution

We know the value of the ion-product constant for water at 25 °C:

$$2H_2O(l) \rightleftharpoons H_3O^+(aq) + OH^-(aq) \qquad K_w = [H_3O^+][OH^-] = 1.0 \times 10^{-14}$$

Thus, we can calculate the missing equilibrium concentration.

Rearrangement of the K_w expression yields that [OH⁻] is directly proportional to the inverse of [H₃O⁺]:

$$[OH^-] = \frac{K_w}{[H_3O^+]} = \frac{1.0 \times 10^{-14}}{2.0 \times 10^{-6}} = 5.0 \times 10^{-9}$$

The hydroxide ion concentration in water is reduced to 5.0×10^{-9} M as the hydrogen ion concentration increases to 2.0×10^{-6} M. This is expected from Le Châtelier's principle; the autoionization reaction shifts to the left to reduce the stress of the increased hydronium ion concentration and the [OH⁻] is reduced relative to that in pure water.

A check of these concentrations confirms that our arithmetic is correct:

$$K_w = [H_3O^+][OH^-] = (2.0 \times 10^{-6})(5.0 \times 10^{-9}) = 1.0 \times 10^{-14}$$

Check Your Learning

What is the hydronium ion concentration in an aqueous solution with a hydroxide ion concentration of 0.001 M at 25 °C?

Answer: $[H_3O^+] = 1 \times 10^{-11}$ M

Amphiprotic Species

Like water, many molecules and ions may either gain or lose a proton under the appropriate conditions. Such species are said to be **amphiprotic**. Another term used to describe such species is **amphoteric**, which is a more general term for a species that may act either as an acid or a base by any definition (not just the Brønsted-Lowry one). Consider for example the bicarbonate ion, which may either donate or accept a proton as shown here:

$$HCO_3^-(aq) + H_2O(l) \rightleftharpoons CO_3^{2-}(aq) + H_3O^+(aq)$$
$$HCO_3^-(aq) + H_2O(l) \rightleftharpoons H_2CO_3(aq) + OH^-(aq)$$

Example 14.3

Representing the Acid-Base Behavior of an Amphoteric Substance

Write separate equations representing the reaction of HSO_3^-

(a) as an acid with OH⁻

(b) as a base with HI

Solution

(a) $HSO_3^-(aq) + OH^-(aq) \rightleftharpoons SO_3^{2-}(aq) + H_2O(l)$

(b) $HSO_3^-(aq) + HI(aq) \rightleftharpoons H_2SO_3(aq) + I^-(aq)$

Check Your Learning

Write separate equations representing the reaction of $H_2PO_4^-$

(a) as a base with HBr

(b) as an acid with OH^-

Answer: (a) $H_2PO_4^-(aq) + HBr(aq) \rightleftharpoons H_3PO_4(aq) + Br^-(aq)$; (b) $H_2PO_4^-(aq) + OH^-(aq) \rightleftharpoons HPO_4^{2-}(aq) + H_2O(l)$

14.2 pH and pOH

By the end of this section, you will be able to:

- Explain the characterization of aqueous solutions as acidic, basic, or neutral
- Express hydronium and hydroxide ion concentrations on the pH and pOH scales
- Perform calculations relating pH and pOH

As discussed earlier, hydronium and hydroxide ions are present both in pure water and in all aqueous solutions, and their concentrations are inversely proportional as determined by the ion product of water (K_w). The concentrations of these ions in a solution are often critical determinants of the solution's properties and the chemical behaviors of its other solutes, and specific vocabulary has been developed to describe these concentrations in relative terms. A solution is **neutral** if it contains equal concentrations of hydronium and hydroxide ions; **acidic** if it contains a greater concentration of hydronium ions than hydroxide ions; and **basic** if it contains a lesser concentration of hydronium ions than hydroxide ions.

A common means of expressing quantities, the values of which may span many orders of magnitude, is to use a logarithmic scale. One such scale that is very popular for chemical concentrations and equilibrium constants is based on the p-function, defined as shown where "X" is the quantity of interest and "log" is the base-10 logarithm:

$$pX = -\log X$$

The **pH** of a solution is therefore defined as shown here, where $[H_3O^+]$ is the molar concentration of hydronium ion in the solution:

$$pH = -\log[H_3O^+]$$

Rearranging this equation to isolate the hydronium ion molarity yields the equivalent expression:

$$[H_3O^+] = 10^{-pH}$$

Likewise, the hydroxide ion molarity may be expressed as a p-function, or **pOH**:

$$pOH = -\log[OH^-]$$

or

$$[OH^-] = 10^{-pOH}$$

Finally, the relation between these two ion concentration expressed as p-functions is easily derived from the K_w expression:

$$K_w = [H_3O^+][OH^-]$$
$$-\log K_w = -\log([H_3O^+][OH^-]) = -\log[H_3O^+] + -\log[OH^-]$$
$$pK_w = pH + pOH$$

At 25 °C, the value of K_w is 1.0×10^{-14}, and so:

$$14.00 = pH + pOH$$

As was shown in Example 14.1, the hydronium ion molarity in pure water (or any neutral solution) is 1.0×10^{-7} M at 25 °C. The pH and pOH of a neutral solution at this temperature are therefore:

$$pH = -\log[H_3O^+] = -\log(1.0 \times 10^{-7}) = 7.00$$
$$pOH = -\log[OH^-] = -\log(1.0 \times 10^{-7}) = 7.00$$

And so, *at this temperature*, acidic solutions are those with hydronium ion molarities greater than 1.0×10^{-7} M and hydroxide ion molarities less than 1.0×10^{-7} M (corresponding to pH values less than 7.00 and pOH values greater than 7.00). Basic solutions are those with hydronium ion molarities less than 1.0×10^{-7} M and hydroxide ion molarities greater than 1.0×10^{-7} M (corresponding to pH values greater than 7.00 and pOH values less than 7.00).

Since the autoionization constant K_w is temperature dependent, these correlations between pH values and the acidic/neutral/basic adjectives will be different at temperatures other than 25 °C. For example, the "Check Your Learning" exercise accompanying Example 14.1 showed the hydronium molarity of pure water at 80 °C is 4.9×10^{-7} M, which corresponds to pH and pOH values of:

$$pH = -\log[H_3O^+] = -\log(4.9 \times 10^{-7}) = 6.31$$
$$pOH = -\log[OH^-] = -\log(4.9 \times 10^{-7}) = 6.31$$

At this temperature, then, neutral solutions exhibit pH = pOH = 6.31, acidic solutions exhibit pH less than 6.31 and pOH greater than 6.31, whereas basic solutions exhibit pH greater than 6.31 and pOH less than 6.31. This distinction can be important when studying certain processes that occur at nonstandard temperatures, such as enzyme reactions in warm-blooded organisms. Unless otherwise noted, references to pH values are presumed to be those at standard temperature (25 °C) (Table 14.1).

Summary of Relations for Acidic, Basic and Neutral Solutions

Classification	Relative Ion Concentrations	pH at 25 °C
acidic	$[H_3O^+] > [OH^-]$	pH < 7
neutral	$[H_3O^+] = [OH^-]$	pH = 7
basic	$[H_3O^+] < [OH^-]$	pH > 7

Table 14.1

Figure 14.2 shows the relationships between $[H_3O^+]$, $[OH^-]$, pH, and pOH, and gives values for these properties at standard temperatures for some common substances.

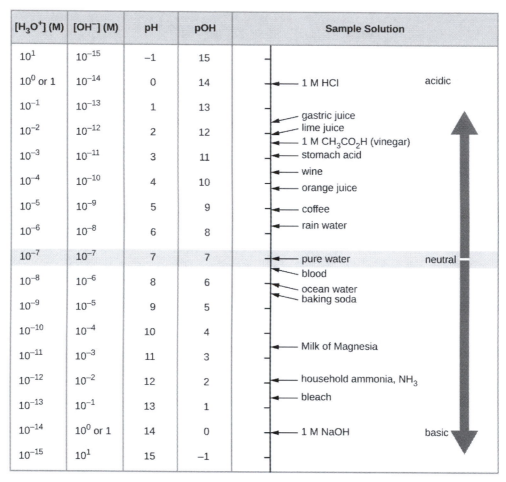

[H$_3$O$^+$] (M)	[OH$^-$] (M)	pH	pOH	Sample Solution
10^1	10^{-15}	−1	15	
10^0 or 1	10^{-14}	0	14	1 M HCl acidic
10^{-1}	10^{-13}	1	13	gastric juice
10^{-2}	10^{-12}	2	12	lime juice 1 M CH$_3$CO$_2$H (vinegar)
10^{-3}	10^{-11}	3	11	stomach acid
10^{-4}	10^{-10}	4	10	wine orange juice
10^{-5}	10^{-9}	5	9	coffee
10^{-6}	10^{-8}	6	8	rain water
10^{-7}	10^{-7}	7	7	pure water neutral
10^{-8}	10^{-6}	8	6	blood
10^{-9}	10^{-5}	9	5	ocean water baking soda
10^{-10}	10^{-4}	10	4	
10^{-11}	10^{-3}	11	3	Milk of Magnesia
10^{-12}	10^{-2}	12	2	household ammonia, NH$_3$
10^{-13}	10^{-1}	13	1	bleach
10^{-14}	10^0 or 1	14	0	1 M NaOH basic
10^{-15}	10^1	15	−1	

Figure 14.2 The pH and pOH scales represent concentrations of [H$_3$O$^+$] and OH$^-$, respectively. The pH and pOH values of some common substances at standard temperature (25 °C) are shown in this chart.

Example 14.4

Calculation of pH from [H$_3$O$^+$]

What is the pH of stomach acid, a solution of HCl with a hydronium ion concentration of 1.2×10^{-3} *M*?

Solution

$$pH = -\log[H_3O^+]$$
$$= -\log(1.2 \times 10^{-3})$$
$$= -(-2.92) = 2.92$$

(The use of logarithms is explained in Appendix B. Recall that, as we have done here, when taking the log of a value, keep as many decimal places in the result as there are significant figures in the value.)

Check Your Learning

Water exposed to air contains carbonic acid, H_2CO_3, due to the reaction between carbon dioxide and water:

$$CO_2(aq) + H_2O(l) \rightleftharpoons H_2CO_3(aq)$$

Air-saturated water has a hydronium ion concentration caused by the dissolved CO_2 of $2.0 \times 10^{-6}\ M$, about 20-times larger than that of pure water. Calculate the pH of the solution at 25 °C.

Answer: 5.70

Example 14.5

Calculation of Hydronium Ion Concentration from pH

Calculate the hydronium ion concentration of blood, the pH of which is 7.3 (slightly alkaline).

Solution

$$pH = -\log[H_3O^+] = 7.3$$

$$\log[H_3O^+] = -7.3$$

$$[H_3O^+] = 10^{-7.3}\ \text{or}\ [H_3O^+] = \text{antilog of} -7.3$$

$$[H_3O^+] = 5 \times 10^{-8}\ M$$

(On a calculator take the antilog, or the "inverse" log, of −7.3, or calculate $10^{-7.3}$.)

Check Your Learning

Calculate the hydronium ion concentration of a solution with a pH of −1.07.

Answer: $12\ M$

How Sciences Interconnect

Environmental Science

Normal rainwater has a pH between 5 and 6 due to the presence of dissolved CO_2 which forms carbonic acid:

$$H_2O(l) + CO_2(g) \longrightarrow H_2CO_3(aq)$$
$$H_2CO_3(aq) \rightleftharpoons H^+(aq) + HCO_3^-(aq)$$

Acid rain is rainwater that has a pH of less than 5, due to a variety of nonmetal oxides, including CO_2, SO_2, SO_3, NO, and NO_2 being dissolved in the water and reacting with it to form not only carbonic acid, but sulfuric acid and nitric acid. The formation and subsequent ionization of sulfuric acid are shown here:

$$H_2O(l) + SO_3(g) \longrightarrow H_2SO_4(aq)$$
$$H_2SO_4(aq) \longrightarrow H^+(aq) + HSO_4^-(aq)$$

Carbon dioxide is naturally present in the atmosphere because we and most other organisms produce it as a waste product of metabolism. Carbon dioxide is also formed when fires release carbon stored in vegetation or when we burn wood or fossil fuels. Sulfur trioxide in the atmosphere is naturally produced by volcanic activity, but it also stems from burning fossil fuels, which have traces of sulfur, and from the process of "roasting" ores of metal sulfides in metal-refining processes. Oxides of nitrogen are formed in internal combustion engines where the high temperatures make it possible for the nitrogen and oxygen in air to chemically combine.

Acid rain is a particular problem in industrial areas where the products of combustion and smelting are released into the air without being stripped of sulfur and nitrogen oxides. In North America and Europe until the 1980s, it was responsible for the destruction of forests and freshwater lakes, when the acidity of the rain actually killed trees, damaged soil, and made lakes uninhabitable for all but the most acid-tolerant species. Acid rain also corrodes statuary and building facades that are made of marble and limestone (Figure 14.3). Regulations limiting the amount of sulfur and nitrogen oxides that can be released into the atmosphere by industry and automobiles have reduced the severity of acid damage to both natural and manmade environments in North America and Europe. It is now a growing problem in industrial areas of China and India.

For further information on acid rain, visit this website (http://openstaxcollege.org/l/16EPA) hosted by the US Environmental Protection Agency.

(a)

(b)

Figure 14.3   (a) Acid rain makes trees more susceptible to drought and insect infestation, and depletes nutrients in the soil. (b) It also is corrodes statues that are carved from marble or limestone. (credit a: modification of work by Chris M Morris; credit b: modification of work by "Eden, Janine and Jim"/Flickr)

Example 14.6

Calculation of pOH

What are the pOH and the pH of a 0.0125-M solution of potassium hydroxide, KOH?

Solution

Potassium hydroxide is a highly soluble ionic compound and completely dissociates when dissolved in dilute solution, yielding $[OH^-] = 0.0125\ M$:

$$pOH = -\log[OH^-] = -\log 0.0125$$
$$= -(-1.903) = 1.903$$

The pH can be found from the pOH:

$$pH + pOH = 14.00$$
$$pH = 14.00 - pOH = 14.00 - 1.903 = 12.10$$

Check Your Learning

The hydronium ion concentration of vinegar is approximately $4 \times 10^{-3}\ M$. What are the corresponding values of pOH and pH?

Answer: pOH = 11.6, pH = 2.4

The acidity of a solution is typically assessed experimentally by measurement of its pH. The pOH of a solution is not usually measured, as it is easily calculated from an experimentally determined pH value. The pH of a solution can be directly measured using a pH meter (Figure 14.4).

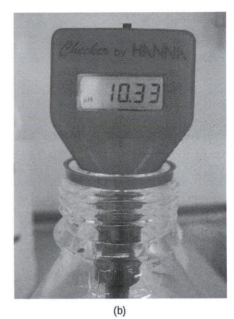

(a)   (b)

Figure 14.4   (a) A research-grade pH meter used in a laboratory can have a resolution of 0.001 pH units, an accuracy of ± 0.002 pH units, and may cost in excess of $1000. (b) A portable pH meter has lower resolution (0.01 pH units), lower accuracy (± 0.2 pH units), and a far lower price tag. (credit b: modification of work by Jacopo Werther)

The pH of a solution may also be visually estimated using colored indicators (Figure 14.5).

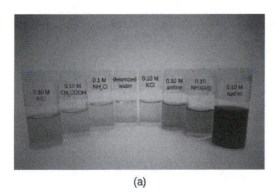
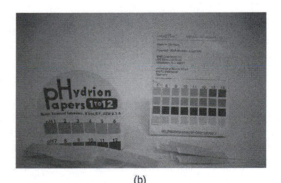

(a)   (b)

Figure 14.5   (a) A universal indicator assumes a different color in solutions of different pH values. Thus, it can be added to a solution to determine the pH of the solution. The eight vials each contain a universal indicator and 0.1-M solutions of progressively weaker acids: HCl (pH = l), CH_3CO_2H (pH = 3), and NH_4Cl (pH = 5), deionized water, a neutral substance (pH = 7); and 0.1-M solutions of the progressively stronger bases: KCl (pH = 7), aniline, $C_6H_5NH_2$ (pH = 9), NH_3 (pH = 11), and NaOH (pH = 13). (b) pH paper contains a mixture of indicators that give different colors in solutions of differing pH values. (credit: modification of work by Sahar Atwa)

14.3 Relative Strengths of Acids and Bases

By the end of this section, you will be able to:

- Assess the relative strengths of acids and bases according to their ionization constants
- Rationalize trends in acid–base strength in relation to molecular structure
- Carry out equilibrium calculations for weak acid–base systems

We can rank the strengths of acids by the extent to which they ionize in aqueous solution. The reaction of an acid with water is given by the general expression:

$$HA(aq) + H_2O(l) \rightleftharpoons H_3O^+(aq) + A^-(aq)$$

Water is the base that reacts with the acid HA, A^- is the conjugate base of the acid HA, and the hydronium ion is the conjugate acid of water. A strong acid yields 100% (or very nearly so) of H_3O^+ and A^- when the acid ionizes in water; Figure 14.6 lists several strong acids. A weak acid gives small amounts of H_3O^+ and A^-.

6 Strong Acids		6 Strong Bases	
$HClO_4$	perchloric acid	LiOH	lithium hydroxide
HCl	hydrochloric acid	NaOH	sodium hydroxide
HBr	hydrobromic acid	KOH	potassium hydroxide
HI	hydroiodic acid	$Ca(OH)_2$	calcium hydroxide
HNO_3	nitric acid	$Sr(OH)_2$	strontium hydroxide
H_2SO_4	sulfuric acid	$Ba(OH)_2$	barium hydroxide

Figure 14.6   Some of the common strong acids and bases are listed here.

The relative strengths of acids may be determined by measuring their equlibrium constants in aqueous solutions. In solutions of the same concentration, stronger acids ionize to a greater extent, and so yield higher concentrations of hydronium ions than do weaker acids. The equilibrium constant for an acid is called the **acid-ionization constant, K_a**. For the reaction of an acid HA:

$$HA(aq) + H_2O(l) \rightleftharpoons H_3O^+(aq) + A^-(aq),$$

we write the equation for the ionization constant as:

$$K_a = \frac{[H_3O^+][A^-]}{[HA]}$$

where the concentrations are those at equilibrium. Although water is a reactant in the reaction, it is the solvent as well, so we do not include $[H_2O]$ in the equation. The larger the K_a of an acid, the larger the concentration of H_3O^+ and A^- relative to the concentration of the nonionized acid, HA. Thus a stronger acid has a larger ionization constant than does a weaker acid. The ionization constants increase as the strengths of the acids increase. (A table of ionization constants of weak acids appears in Appendix H, with a partial listing in Table 14.2.)

The following data on acid-ionization constants indicate the order of acid strength $CH_3CO_2H < HNO_2 < HSO_4{}^-$:

$$CH_3CO_2H(aq) + H_2O(l) \rightleftharpoons H_3O^+(aq) + CH_3CO_2{}^-(aq) \qquad K_a = 1.8 \times 10^{-5}$$

$$HNO_2(aq) + H_2O(l) \rightleftharpoons H_3O^+(aq) + NO_2^-(aq) \qquad\qquad K_a = 4.6 \times 10^{-4}$$

$$HSO_4^-(aq) + H_2O(aq) \rightleftharpoons H_3O^+(aq) + SO_4^{2-}(aq) \qquad\qquad K_a = 1.2 \times 10^{-2}$$

Another measure of the strength of an acid is its percent ionization. The **percent ionization** of a weak acid is the ratio of the concentration of the ionized acid to the initial acid concentration, times 100:

$$\% \text{ ionization} = \frac{[H_3O^+]_{eq}}{[HA]_0} \times 100$$

Because the ratio includes the initial concentration, the percent ionization for a solution of a given weak acid varies depending on the original concentration of the acid, and actually decreases with increasing acid concentration.

Example 14.7

Calculation of Percent Ionization from pH

Calculate the percent ionization of a 0.125-M solution of nitrous acid (a weak acid), with a pH of 2.09.

Solution

The percent ionization for an acid is:

$$\frac{[H_3O^+]_{eq}}{[HNO_2]_0} \times 100$$

The chemical equation for the dissociation of the nitrous acid is: $HNO_2(aq) + H_2O(l) \rightleftharpoons NO_2^-(aq) + H_3O^+(aq)$. Since $10^{-pH} = [H_3O^+]$, we find that $10^{-2.09} = 8.1 \times 10^{-3}\ M$, so that percent ionization is:

$$\frac{8.1 \times 10^{-3}}{0.125} \times 100 = 6.5\%$$

Remember, the logarithm 2.09 indicates a hydronium ion concentration with only two significant figures.

Check Your Learning

Calculate the percent ionization of a 0.10-M solution of acetic acid with a pH of 2.89.

Answer: 1.3% ionized

We can rank the strengths of bases by their tendency to form hydroxide ions in aqueous solution. The reaction of a Brønsted-Lowry base with water is given by:

$$B(aq) + H_2O(l) \rightleftharpoons HB^+(aq) + OH^-(aq)$$

Water is the acid that reacts with the base, HB^+ is the conjugate acid of the base B, and the hydroxide ion is the conjugate base of water. A strong base yields 100% (or very nearly so) of OH^- and HB^+ when it reacts with water; Figure 14.6 lists several strong bases. A weak base yields a small proportion of hydroxide ions. Soluble ionic hydroxides such as NaOH are considered strong bases because they dissociate completely when dissolved in water.

Link to Learning

View the simulation (http://openstaxcollege.org/l/16AcidBase) of strong and weak acids and bases at the molecular level.

As we did with acids, we can measure the relative strengths of bases by measuring their **base-ionization constant** **(K_b)** in aqueous solutions. In solutions of the same concentration, stronger bases ionize to a greater extent, and so yield higher hydroxide ion concentrations than do weaker bases. A stronger base has a larger ionization constant than does a weaker base. For the reaction of a base, B:

$$B(aq) + H_2O(l) \rightleftharpoons HB^+(aq) + OH^-(aq),$$

we write the equation for the ionization constant as:

$$K_b = \frac{[HB^+][OH^-]}{[B]}$$

where the concentrations are those at equilibrium. Again, we do not include [H_2O] in the equation because water is the solvent. The chemical reactions and ionization constants of the three bases shown are:

$$NO_2^-(aq) + H_2O(l) \rightleftharpoons HNO_2(aq) + OH^-(aq) \qquad K_b = 2.22 \times 10^{-11}$$
$$CH_3CO_2^-(aq) + H_2O(l) \rightleftharpoons CH_3CO_2H(aq) + OH^-(aq) \qquad K_b = 5.6 \times 10^{-10}$$
$$NH_3(aq) + H_2O(l) \rightleftharpoons NH_4^+(aq) + OH^-(aq) \qquad K_b = 1.8 \times 10^{-5}$$

A table of ionization constants of weak bases appears in Appendix I (with a partial list in Table 14.3). As with acids, percent ionization can be measured for basic solutions, but will vary depending on the base ionization constant and the initial concentration of the solution.

Consider the ionization reactions for a conjugate acid-base pair, HA − A⁻:

$$HA(aq) + H_2O(l) \rightleftharpoons H_3O^+(aq) + A^-(aq) \qquad K_a = \frac{[H_3O^+][A^-]}{[HA]}$$

$$A^-(aq) + H_2O(l) \rightleftharpoons OH^-(aq) + HA(aq) \qquad K_b = \frac{[HA][OH]}{[A^-]}$$

Adding these two chemical equations yields the equation for the autoionization for water:

$$\cancel{HA(aq)} + H_2O(l) + \cancel{A^-(aq)} + H_2O(l) \rightleftharpoons H_3O^+(aq) + \cancel{A^-(aq)} + OH^-(aq) + \cancel{HA(aq)}$$
$$2H_2O(l) \rightleftharpoons H_3O^+(aq) + OH^-(aq)$$

As shown in the previous chapter on equilibrium, the K expression for a chemical equation derived from adding two or more other equations is the mathematical product of the other equations' K expressions. Multiplying the mass-action expressions together and cancelling common terms, we see that:

$$K_a \times K_b = \frac{[H_3O^+][A^-]}{[HA]} \times \frac{[HA][OH^-]}{[A^-]} = [H_3O^+][OH^-] = K_w$$

For example, the acid ionization constant of acetic acid (CH_3COOH) is 1.8×10^{-5}, and the base ionization constant of its conjugate base, acetate ion (CH_3COO^-), is 5.6×10^{-10}. The product of these two constants is indeed equal to K_w:

$$K_a \times K_b = (1.8 \times 10^{-5}) \times (5.6 \times 10^{-10}) = 1.0 \times 10^{-14} = K_w$$

The extent to which an acid, HA, donates protons to water molecules depends on the strength of the conjugate base, A⁻, of the acid. If A⁻ is a strong base, any protons that are donated to water molecules are recaptured by A⁻. Thus there is relatively little A⁻ and H_3O^+ in solution, and the acid, HA, is weak. If A⁻ is a weak base, water binds the protons more strongly, and the solution contains primarily A⁻ and H_3O^+—the acid is strong. Strong acids form very weak conjugate bases, and weak acids form stronger conjugate bases (Figure 14.7).

Relative acid strength

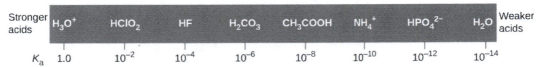

Relative conjugate base strength

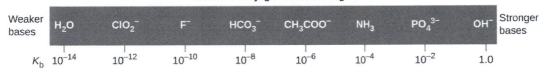

Figure 14.7 This diagram shows the relative strengths of conjugate acid-base pairs, as indicated by their ionization constants in aqueous solution.

Figure 14.8 lists a series of acids and bases in order of the decreasing strengths of the acids and the corresponding increasing strengths of the bases. The acid and base in a given row are conjugate to each other.

Acid					Base		
perchloric acid	$HClO_4$				ClO_4^-	perchlorate ion	
sulfuric acid	H_2SO_4	Undergo complete acid ionization in water	Do not undergo base ionization in water		HSO_4^-	hydrogen sulfate ion	
hydrogen iodide	HI				I^-	iodide ion	
hydrogen bromide	HBr				Br^-	bromide ion	
hydrogen chloride	HCl				Cl^-	chloride ion	
nitric acid	HNO_3				NO_3^-	nitrate ion	
hydronium ion	H_3O^+				H_2O	water	
hydrogen sulfate ion	HSO_4^-				SO_4^{2-}	sulfate ion	
phosphoric acid	H_3PO_4				$H_2PO_4^-$	dihydrogen phosphate ion	
hydrogen fluoride	HF				F^-	fluoride ion	
nitrous acid	HNO_2				NO_2^-	nitrite ion	
acetic acid	CH_3CO_2H				$CH_3CO_2^-$	acetate ion	
carbonic acid	H_2CO_3				HCO_3^-	hydrogen carbonate ion	
hydrogen sulfide	H_2S				HS^-	hydrogen sulfide ion	
ammonium ion	NH_4^+				HN_3	ammonia	
hydrogen cyanide	HCN				CN^-	cyanide ion	
hydrogen carbonate ion	HCO_3^-				CO_3^{2-}	carbonate ion	
water	H_2O				OH^-	hydroxide ion	
hydrogen sulfide ion	HS^-	Do not undergo acid ionization in water	Undergo complete base ionization in water		S^{2-}	sulfide ion	
ethanol	C_2H_5OH				$C_2H_5O^-$	ethoxide ion	
ammonia	NH_3				NH_2^-	amide ion	
hydrogen	H_2				H^-	hydride ion	
methane	CH_4				CH_3^-	methide ion	

Increasing acid strength (left arrow, bottom to top). Increasing base strength (right arrow, top to bottom).

Figure 14.8   The chart shows the relative strengths of conjugate acid-base pairs.

The first six acids in Figure 14.8 are the most common strong acids. These acids are completely dissociated in aqueous solution. The conjugate bases of these acids are weaker bases than water. When one of these acids dissolves in water, their protons are completely transferred to water, the stronger base.

Those acids that lie between the hydronium ion and water in Figure 14.8 form conjugate bases that can compete with water for possession of a proton. Both hydronium ions and nonionized acid molecules are present in equilibrium in a solution of one of these acids. Compounds that are weaker acids than water (those found below water in the column of acids) in Figure 14.8 exhibit no observable acidic behavior when dissolved in water. Their conjugate bases are stronger than the hydroxide ion, and if any conjugate base were formed, it would react with water to re-form the acid.

The extent to which a base forms hydroxide ion in aqueous solution depends on the strength of the base relative to that of the hydroxide ion, as shown in the last column in Figure 14.8. A strong base, such as one of those lying below hydroxide ion, accepts protons from water to yield 100% of the conjugate acid and hydroxide ion. Those bases lying between water and hydroxide ion accept protons from water, but a mixture of the hydroxide ion and the base results. Bases that are weaker than water (those that lie above water in the column of bases) show no observable basic behavior in aqueous solution.

Example 14.8

The Product $K_a \times K_b = K_w$

Use the K_b for the nitrite ion, NO_2^-, to calculate the K_a for its conjugate acid.

Solution

K_b for NO_2^- is given in this section as 2.22×10^{-11}. The conjugate acid of NO_2^- is HNO_2; K_a for HNO_2 can be calculated using the relationship:

$$K_a \times K_b = 1.0 \times 10^{-14} = K_w$$

Solving for K_a, we get:

$$K_a = \frac{K_w}{K_b} = \frac{1.0 \times 10^{-14}}{2.22 \times 10^{-11}} = 4.5 \times 10^{-4}$$

This answer can be verified by finding the K_a for HNO_2 in Appendix H.

Check Your Learning

We can determine the relative acid strengths of NH_4^+ and HCN by comparing their ionization constants. The ionization constant of HCN is given in Appendix H as 4×10^{-10}. The ionization constant of NH_4^+ is not listed, but the ionization constant of its conjugate base, NH_3, is listed as 1.8×10^{-5}. Determine the ionization constant of NH_4^+, and decide which is the stronger acid, HCN or NH_4^+.

Answer: NH_4^+ is the slightly stronger acid (K_a for $NH_4^+ = 5.6 \times 10^{-10}$).

The Ionization of Weak Acids and Weak Bases

Many acids and bases are weak; that is, they do not ionize fully in aqueous solution. A solution of a weak acid in water is a mixture of the nonionized acid, hydronium ion, and the conjugate base of the acid, with the nonionized acid present in the greatest concentration. Thus, a weak acid increases the hydronium ion concentration in an aqueous solution (but not as much as the same amount of a strong acid).

Acetic acid, CH_3CO_2H, is a weak acid. When we add acetic acid to water, it ionizes to a small extent according to the equation:

$$CH_3CO_2H(aq) + H_2O(l) \rightleftharpoons H_3O^+(aq) + CH_3CO_2^-(aq),$$

giving an equilibrium mixture with most of the acid present in the nonionized (molecular) form. This equilibrium, like other equilibria, is dynamic; acetic acid molecules donate hydrogen ions to water molecules and form hydronium ions and acetate ions at the same rate that hydronium ions donate hydrogen ions to acetate ions to reform acetic acid molecules and water molecules. We can tell by measuring the pH of an aqueous solution of known concentration that only a fraction of the weak acid is ionized at any moment (Figure 14.9). The remaining weak acid is present in the nonionized form.

For acetic acid, at equilibrium:

$$K_a = \frac{[H_3O^+][CH_3CO_2^-]}{[CH_3CO_2H]} = 1.8 \times 10^{-5}$$

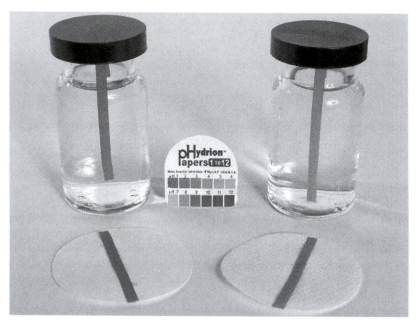

Figure 14.9 pH paper indicates that a 0.l-*M* solution of HCl (beaker on left) has a pH of 1. The acid is fully ionized and $[H_3O^+]$ = 0.1 *M*. A 0.1-*M* solution of CH_3CO_2H (beaker on right) is has a pH of 3 ($[H_3O^+]$ = 0.001 *M*) because the weak acid CH_3CO_2H is only partially ionized. In this solution, $[H_3O^+]$ < $[CH_3CO_2H]$. (credit: modification of work by Sahar Atwa)

Ionization Constants of Some Weak Acids

Ionization Reaction	K_a at 25 °C
$HSO_4^- + H_2O \rightleftharpoons H_3O^+ + SO_4^{2-}$	1.2×10^{-2}
$HF + H_2O \rightleftharpoons H_3O^+ + F^-$	7.2×10^{-4}
$HNO_2 + H_2O \rightleftharpoons H_3O^+ + NO_2^-$	4.5×10^{-4}
$HNCO + H_2O \rightleftharpoons H_3O^+ + NCO^-$	3.46×10^{-4}
$HCO_2H + H_2O \rightleftharpoons H_3O^+ + HCO_2^-$	1.8×10^{-4}
$CH_3CO_2H + H_2O \rightleftharpoons H_3O^+ + CH_3CO_2^-$	1.8×10^{-5}
$HCIO + H_2O \rightleftharpoons H_3O^+ + CIO^-$	3.5×10^{-8}
$HBrO + H_2O \rightleftharpoons H_3O^+ + BrO^-$	2×10^{-9}
$HCN + H_2O \rightleftharpoons H_3O^+ + CN^-$	4×10^{-10}

Table 14.2

Table 14.2 gives the ionization constants for several weak acids; additional ionization constants can be found in Appendix H.

At equilibrium, a solution of a weak base in water is a mixture of the nonionized base, the conjugate acid of the weak base, and hydroxide ion with the nonionized base present in the greatest concentration. Thus, a weak base increases the hydroxide ion concentration in an aqueous solution (but not as much as the same amount of a strong base).

For example, a solution of the weak base trimethylamine, $(CH_3)_3N$, in water reacts according to the equation:

$$(CH_3)_3N(aq) + H_2O(l) \rightleftharpoons (CH_3)_3NH^+(aq) + OH^-(aq),$$

giving an equilibrium mixture with most of the base present as the nonionized amine. This equilibrium is analogous to that described for weak acids.

We can confirm by measuring the pH of an aqueous solution of a weak base of known concentration that only a fraction of the base reacts with water (Figure 14.10). The remaining weak base is present as the unreacted form. The equilibrium constant for the ionization of a weak base, K_b, is called the ionization constant of the weak base, and is equal to the reaction quotient when the reaction is at equilibrium. For trimethylamine, at equilibrium:

$$K_b = \frac{[(CH_3)_3NH^+][OH^-]}{[(CH_3)_3N]}$$

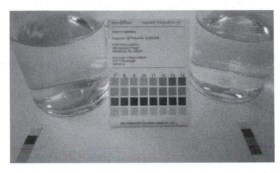

Figure 14.10 pH paper indicates that a 0.1-*M* solution of NH_3 (left) is weakly basic. The solution has a pOH of 3 ($[OH^-]$ = 0.001 *M*) because the weak base NH_3 only partially reacts with water. A 0.1-*M* solution of NaOH (right) has a pOH of 1 because NaOH is a strong base. (credit: modification of work by Sahar Atwa)

The ionization constants of several weak bases are given in Table 14.3 and in Appendix I.

Ionization Constants of Some Weak Bases

Ionization Reaction	K_b at 25 °C
$(CH_3)_2NH + H_2O \rightleftharpoons (CH_3)_2NH_2^+ + OH^-$	7.4×10^{-4}
$CH_3NH_2 + H_2O \rightleftharpoons CH_3NH_3^+ + OH^-$	4.4×10^{-4}
$(CH_3)_3N + H_2O \rightleftharpoons (CH_3)_3NH^+ + OH^-$	7.4×10^{-5}
$NH_3 + H_2O \rightleftharpoons NH_4^+ + OH^-$	1.8×10^{-5}
$C_6H_5NH_2 + H_2O \rightleftharpoons C_6N_5NH_3^+ + OH^-$	4.6×10^{-10}

Table 14.3

Example 14.9

Determination of K_a from Equilibrium Concentrations

Acetic acid is the principal ingredient in vinegar (Figure 14.11); that's why it tastes sour. At equilibrium, a solution contains $[CH_3CO_2H] = 0.0787\ M$ and $[H_3O^+] = [CH_3CO_2{}^-] = 0.00118\ M$. What is the value of K_a for acetic acid?

Figure 14.11 Vinegar is a solution of acetic acid, a weak acid. (credit: modification of work by "HomeSpot HQ"/Flickr)

Solution

We are asked to calculate an equilibrium constant from equilibrium concentrations. At equilibrium, the value of the equilibrium constant is equal to the reaction quotient for the reaction:

$$CH_3CO_2H(aq) + H_2O(l) \rightleftharpoons H_3O^+(aq) + CH_3CO_2{}^-(aq)$$

$$K_a = \frac{[H_3O^+][CH_3CO_2{}^-]}{[CH_3CO_2H]} = \frac{(0.00118)(0.00118)}{0.0787} = 1.77 \times 10^{-5}$$

Check Your Learning

What is the equilibrium constant for the ionization of the $HSO_4{}^-$ ion, the weak acid used in some household cleansers:

$$HSO_4{}^-(aq) + H_2O(l) \rightleftharpoons H_3O^+(aq) + SO_4{}^{2-}(aq)$$

In one mixture of $NaHSO_4$ and Na_2SO_4 at equilibrium, $[H_3O^+] = 0.027\ M$; $[HSO_4{}^-] = 0.29\ M$; and $[SO_4{}^{2-}] = 0.13\ M$.

Answer: K_a for $HSO_4{}^- = 1.2 \times 10^{-2}$

Example 14.10

Determination of K_b from Equilibrium Concentrations

Caffeine, $C_8H_{10}N_4O_2$ is a weak base. What is the value of K_b for caffeine if a solution at equilibrium has $[C_8H_{10}N_4O_2] = 0.050\ M$, $[C_8H_{10}N_4O_2H^+] = 5.0 \times 10^{-3}\ M$, and $[OH^-] = 2.5 \times 10^{-3}\ M$?

Solution

At equilibrium, the value of the equilibrium constant is equal to the reaction quotient for the reaction:

$$C_8H_{10}N_4O_2(aq) + H_2O(l) \rightleftharpoons C_8H_{10}N_4O_2H^+(aq) + OH^-(aq)$$

$$K_b = \frac{[C_8H_{10}N_4O_2H^+][OH^-]}{[C_8H_{10}N_4O_2]} = \frac{(5.0 \times 10^{-3})(2.5 \times 10^{-3})}{0.050} = 2.5 \times 10^{-4}$$

Check Your Learning

What is the equilibrium constant for the ionization of the $HPO_4{}^{2-}$ ion, a weak base:

$$HPO_4{}^{2-}(aq) + H_2O(l) \rightleftharpoons H_2PO_4{}^-(aq) + OH^-(aq)$$

In a solution containing a mixture of NaH_2PO_4 and Na_2HPO_4 at equilibrium, $[OH^-] = 1.3 \times 10^{-6}\ M$; $[H_2PO_4{}^-] = 0.042\ M$; and $[HPO_4{}^{2-}] = 0.341\ M$.

Answer: K_b for $HPO_4{}^{2-} = 1.6 \times 10^{-7}$

Example 14.11

Determination of K_a or K_b from pH

The pH of a 0.0516-M solution of nitrous acid, HNO_2, is 2.34. What is its K_a?

$$HNO_2(aq) + H_2O(l) \rightleftharpoons H_3O^+(aq) + NO_2{}^-(aq)$$

Solution

We determine an equilibrium constant starting with the initial concentrations of HNO_2, H_3O^+, and $NO_2{}^-$ as well as one of the final concentrations, the concentration of hydronium ion at equilibrium. (Remember that pH is simply another way to express the concentration of hydronium ion.)

We can solve this problem with the following steps in which x is a change in concentration of a species in the reaction:

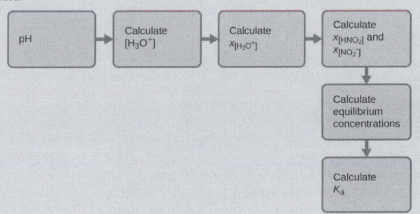

We can summarize the various concentrations and changes as shown here (the concentration of water does not appear in the expression for the equilibrium constant, so we do not need to consider its concentration):

	HNO_2	+ H_2O ⇌	H_3O^+	+ NO_2^-
Initial concentration (M)	0.0516		~0	0
Change (M)	−x	+	x	x
Equilibrium concentration (M)	$[HNO_2]_i + (-x) =$ 0.0516 + (−x)		$[H_3O]^+ + x[NO_2]^-$ + x ~0 + x	0.0046

To get the various values in the ICE (Initial, Change, Equilibrium) table, we first calculate $[H_3O^+]$, the equilibrium concentration of H_3O^+, from the pH:

$$[H_3O^+] = 10^{-2.34} = 0.0046\ M$$

The change in concentration of H_3O^+, $x_{[H_3O^+]}$, is the difference between the equilibrium concentration of H_3O^+, which we determined from the pH, and the initial concentration, $[H_3O^+]_i$. The initial concentration of H_3O^+ is its concentration in pure water, which is so much less than the final concentration that we approximate it as zero (~0).

The change in concentration of NO_2^- is equal to the change in concentration of $[H_3O^+]$. For each 1 mol of H_3O^+ that forms, 1 mol of NO_2^- forms. The equilibrium concentration of HNO_2 is equal to its initial concentration plus the change in its concentration.

Now we can fill in the ICE table with the concentrations at equilibrium, as shown here:

	HNO_2	+ H_2O ⇌	H_3O^+	+ NO_2^-
Initial concentration (M)	0.0516		~0	0
Change (M)	−x	+	x = 0.0046	x = 0.0046
Equilibrium concentration (M)	0.0470		0.0046	0.0046

Finally, we calculate the value of the equilibrium constant using the data in the table:

$$K_a = \frac{[H_3O^+][NO_2^-]}{[HNO_2]} = \frac{(0.0046)(0.0046)}{(0.0470)} = 4.5 \times 10^{-4}$$

Check Your Learning.

The pH of a solution of household ammonia, a 0.950-M solution of NH_3, is 11.612. What is K_b for NH_3.

Answer: $K_b = 1.8 \times 10^{-5}$

Example 14.12

Equilibrium Concentrations in a Solution of a Weak Acid

Formic acid, HCO_2H, is the irritant that causes the body's reaction to ant stings (Figure 14.12).

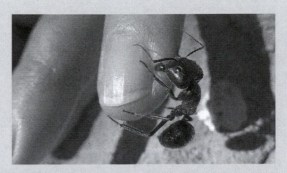

Figure 14.12   The pain of an ant's sting is caused by formic acid. (credit: John Tann)

What is the concentration of hydronium ion and the pH in a 0.534-M solution of formic acid?

$$HCO_2H(aq) + H_2O(l) \rightleftharpoons H_3O^+(aq) + HCO_2{}^-(aq) \qquad\qquad K_a = 1.8 \times 10^{-4}$$

Solution

> **Step 1.** *Determine x and equilibrium concentrations.* The equilibrium expression is:
> $$HCO_2H(aq) + H_2O(l) \rightleftharpoons H_3O^+(aq) + HCO_2{}^-(aq)$$

The concentration of water does not appear in the expression for the equilibrium constant, so we do not need to consider its change in concentration when setting up the ICE table.

The table shows initial concentrations (concentrations before the acid ionizes), changes in concentration, and equilibrium concentrations follows (the data given in the problem appear in color):

	HCO$_2$H +	H$_2$O $\rightleftharpoons$	H$_3$O$^+$
Initial concentration (M)	0.534	~0	0
Change (M)	$-x$	x	x
Equilibrium concentration (M)	$0.534 + (-x)$	$0 + x = x$	$0 + x = x$

> **Step 2.** *Solve for x and the equilibrium concentrations.* At equilibrium:
> $$K_a = 1.8 \times 10^{-4} = \frac{[H_3O^+][HCO_2{}^-]}{[HCO_2H]}$$
>
> $$= \frac{(x)(x)}{0.534 - x} = 1.8 \times 10^{-4}$$

Now solve for x. Because the initial concentration of acid is reasonably large and K_a is very small, we assume that $x \ll 0.534$, which *permits* us to simplify the denominator term as $(0.534 - x) = 0.534$. This gives:

$$K_a = 1.8 \times 10^{-4} = \frac{x^{2+}}{0.534}$$

Solve for x as follows:

$$x^{2+} = 0.534 \times \left(1.8 \times 10^{-4}\right) = 9.6 \times 10^{-5}$$
$$x = \sqrt{9.6 \times 10^{-5}}$$
$$= 9.8 \times 10^{-3}$$

To check the assumption that x is small compared to 0.534, we calculate:

$$\frac{x}{0.534} = \frac{9.8 \times 10^{-3}}{0.534} = 1.8 \times 10^{-2} \ (1.8\% \ \text{of } 0.534)$$

x is less than 5% of the initial concentration; the assumption is valid.

We find the equilibrium concentration of hydronium ion in this formic acid solution from its initial concentration and the change in that concentration as indicated in the last line of the table:

$$[H_3O^+] = {\sim}0 + x = 0 + 9.8 \times 10^{-3} \ M.$$
$$= 9.8 \times 10^{-3} \ M$$

The pH of the solution can be found by taking the negative log of the $[H_3O^+]$, so:

$$-\log\!\left(9.8 \times 10^{-3}\right) = 2.01$$

Check Your Learning

Only a small fraction of a weak acid ionizes in aqueous solution. What is the percent ionization of acetic acid in a 0.100-M solution of acetic acid, CH_3CO_2H?

$$CH_3CO_2H(aq) + H_2O(l) \rightleftharpoons H_3O^+(aq) + CH_3CO_2{}^-(aq) \qquad K_a = 1.8 \times 10^{-5}$$

(Hint: Determine $[CH_3CO_2{}^-]$ at equilibrium.) Recall that the percent ionization is the fraction of acetic acid that is ionized $\times$ 100, or $\dfrac{[CH_3CO_2{}^-]}{[CH_3CO_2H]_{\text{initial}}} \times 100$.

Answer: percent ionization = 1.3%

The following example shows that the concentration of products produced by the ionization of a weak base can be determined by the same series of steps used with a weak acid.

Example 14.13

Equilibrium Concentrations in a Solution of a Weak Base

Find the concentration of hydroxide ion in a 0.25-M solution of trimethylamine, a weak base:

$$(CH_3)_3N(aq) + H_2O(l) \rightleftharpoons (CH_3)_3NH^+(aq) + OH^-(aq) \qquad K_b = 7.4 \times 10^{-5}$$

Solution

This problem requires that we calculate an equilibrium concentration by determining concentration changes as the ionization of a base goes to equilibrium. The solution is approached in the same way as that for the ionization of formic acid in Example 14.12. The reactants and products will be different and the numbers will be different, but the logic will be the same:

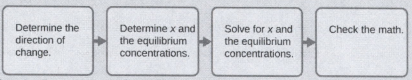

Step 1. *Determine x and equilibrium concentrations.* The table shows the changes and concentrations:

	$(CH_3)_3N$ + H_2O ⇌ $(CH_3)_3NH^+$ + OH^-			
Initial concentration (M)	0.25		0	~0
Change (M)	−x		x	x
Equilibrium concentration (M)	0.25 + (−x)		0 + x	~0 + x

Step 2. *Solve for x and the equilibrium concentrations.* At equilibrium:

$$K_b = \frac{[(CH_3)_3NH^+][OH^-]}{[(CH_3)_3N]} = \frac{(x)(x)}{0.25 - x} = 7.4 \times 10^{-5}$$

If we assume that x is small relative to 0.25, then we can replace (0.25 − x) in the preceding equation with 0.25. Solving the simplified equation gives:

$$x = 4.3 \times 10^{-3}$$

This change is less than 5% of the initial concentration (0.25), so the assumption is justified. Recall that, for this computation, x is equal to the equilibrium concentration of *hydroxide ion* in the solution (see earlier tabulation):

$$[OH^-] = \sim0 + x = x = 4.3 \times 10^{-3}\ M$$

$$= 4.3 \times 10^{-3}\ M$$

Then calculate pOH as follows:

$$pOH = -\log(4.3 \times 10^{-3}) = 2.37$$

Using the relation introduced in the previous section of this chapter:

$$pH + pOH = pK_w = 14.00$$

permits the computation of pH:

$$pH = 14.00 - pOH = 14.00 - 2.37 = 11.63$$

Step 3. *Check the work.* A check of our arithmetic shows that $K_b = 7.4 \times 10^{-5}$.

Check Your Learning

(a) Show that the calculation in Step 2 of this example gives an x of 4.3 × 10^{-3} and the calculation in Step 3 shows $K_b = 7.4 \times 10^{-5}$.

(b) Find the concentration of hydroxide ion in a 0.0325-*M* solution of ammonia, a weak base with a K_b of 1.76 × 10^{-5}. Calculate the percent ionization of ammonia, the fraction ionized × 100, or $\frac{[NH_4^+]}{[NH_3]} \times 100$

Answer: 7.56 × 10^{-4} *M*, 2.33%

Some weak acids and weak bases ionize to such an extent that the simplifying assumption that x is small relative to the initial concentration of the acid or base is inappropriate. As we solve for the equilibrium concentrations in such cases, we will see that we cannot neglect the change in the initial concentration of the acid or base, and we must solve the equilibrium equations by using the quadratic equation.

Example 14.14

Equilibrium Concentrations in a Solution of a Weak Acid

Sodium bisulfate, $NaHSO_4$, is used in some household cleansers because it contains the HSO_4^- ion, a weak acid. What is the pH of a 0.50-M solution of HSO_4^-?

$$HSO_4^-(aq) + H_2O(l) \rightleftharpoons H_3O^+(aq) + SO_4^{2-}(aq) \qquad K_a = 1.2 \times 10^{-2}$$

Solution

We need to determine the equilibrium concentration of the hydronium ion that results from the ionization of HSO_4^- so that we can use $[H_3O^+]$ to determine the pH. As in the previous examples, we can approach the solution by the following steps:

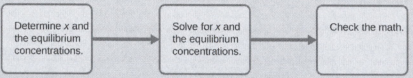

Step 1. *Determine x and equilibrium concentrations.* This table shows the changes and concentrations:

	HSO_4^-	+ H_2O	$\rightleftharpoons$	H_3O^+	+ SO_4^{2-}
Initial concentration (M)	0.50			~0	0
Change (M)	$-x$			x	x
Equilibrium concentration (M)	$0.50 + (-x) =$ $0.50 - x$			$0 + x = x$	$0 + x = x$

Step 2. *Solve for x and the concentrations.* As we begin solving for x, we will find this is more complicated than in previous examples. As we discuss these complications we should not lose track of the fact that it is still the purpose of this step to determine the value of x.
At equilibrium:

$$K_a = 1.2 \times 10^{-2} = \frac{[H_3O^+][SO_4^{2-}]}{[HSO_4^-]} = \frac{(x)(x)}{0.50 - x}$$

If we assume that x is small and approximate $(0.50 - x)$ as 0.50, we find:

$$x = 7.7 \times 10^{-2}$$

When we check the assumption, we calculate:

$$\frac{x}{[HSO_4^-]_i}$$

$$\frac{x}{0.50} = \frac{7.7 \times 10^{-2}}{0.50} = 0.15\ (15\%)$$

The value of x is not less than 5% of 0.50, so the assumption is not valid. We need the quadratic formula to find x.
The equation:

$$K_a = 1.2 \times 10^{-2} = \frac{(x)(x)}{0.50 - x}$$

gives

$$6.0 \times 10^{-3} - 1.2 \times 10^{-2} x = x^{2+}$$

or

$$x^{2+} + 1.2 \times 10^{-2} x - 6.0 \times 10^{-3} = 0$$

This equation can be solved using the quadratic formula. For an equation of the form

$$ax^{2+} + bx + c = 0,$$

x is given by the equation:

$$x = \frac{-b \pm \sqrt{b^{2+} - 4ac}}{2a}$$

In this problem, $a = 1$, $b = 1.2 \times 10^{-3}$, and $c = -6.0 \times 10^{-3}$.
Solving for x gives a negative root (which cannot be correct since concentration cannot be negative) and a positive root:

$$x = 7.2 \times 10^{-2}$$

Now determine the hydronium ion concentration and the pH:

$$[H_3O^+] = \sim 0 + x = 0 + 7.2 \times 10^{-2} \ M$$

$$= 7.2 \times 10^{-2} \ M$$

The pH of this solution is:

$$pH = -\log[H_3O^+] = -\log 7.2 \times 10^{-2} = 1.14$$

Check Your Learning

(a) Show that the quadratic formula gives $x = 7.2 \times 10^{-2}$.

(b) Calculate the pH in a 0.010-M solution of caffeine, a weak base:

$$C_8H_{10}N_4O_2(aq) + H_2O(l) \rightleftharpoons C_8H_{10}N_4O_2H^+(aq) + OH^-(aq) \qquad K_b = 2.5 \times 10^{-4}$$

(Hint: It will be necessary to convert $[OH^-]$ to $[H_3O^+]$ or pOH to pH toward the end of the calculation.)

Answer: pH 11.16

The Relative Strengths of Strong Acids and Bases

Strong acids, such as HCl, HBr, and HI, all exhibit the same strength in water. The water molecule is such a strong base compared to the conjugate bases Cl^-, Br^-, and I^- that ionization of these strong acids is essentially complete in aqueous solutions. In solvents less basic than water, we find HCl, HBr, and HI differ markedly in their tendency to give up a proton to the solvent. For example, when dissolved in ethanol (a weaker base than water), the extent of ionization increases in the order HCl < HBr < HI, and so HI is demonstrated to be the strongest of these acids. The inability to discern differences in strength among strong acids dissolved in water is known as the **leveling effect of water**.

Water also exerts a leveling effect on the strengths of strong bases. For example, the oxide ion, O^{2-}, and the amide ion, NH_2^-, are such strong bases that they react completely with water:

$$O^{2-}(aq) + H_2O(l) \longrightarrow OH^-(aq) + OH^-(aq)$$
$$NH_2^-(aq) + H_2O(l) \longrightarrow NH_3(aq) + OH^-(aq)$$

Thus, O^{2-} and NH_2^- appear to have the same base strength in water; they both give a 100% yield of hydroxide ion.

Effect of Molecular Structure on Acid-Base Strength

In the absence of any leveling effect, the acid strength of binary compounds of hydrogen with nonmetals (A) increases as the H-A bond strength decreases down a group in the periodic table. For group 7A, the order of increasing acidity is HF < HCl < HBr < HI. Likewise, for group 6A, the order of increasing acid strength is $H_2O < H_2S < H_2Se < H_2Te$.

Across a row in the periodic table, the acid strength of binary hydrogen compounds increases with increasing electronegativity of the nonmetal atom because the polarity of the H-A bond increases. Thus, the order of increasing acidity (for removal of one proton) across the second row is $CH_4 < NH_3 < H_2O < HF$; across the third row, it is $SiH_4 < PH_3 < H_2S < HCl$ (see Figure 14.13).

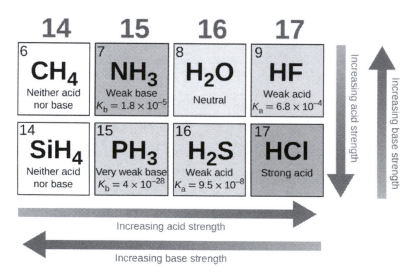

Figure 14.13 As you move from left to right and down the periodic table, the acid strength increases. As you move from right to left and up, the base strength increases.

Compounds containing oxygen and one or more hydroxyl (OH) groups can be acidic, basic, or amphoteric, depending on the position in the periodic table of the central atom E, the atom bonded to the hydroxyl group. Such compounds have the general formula $O_nE(OH)_m$, and include sulfuric acid, $O_2S(OH)_2$, sulfurous acid, $OS(OH)_2$, nitric acid, O_2NOH, perchloric acid, O_3ClOH, aluminum hydroxide, $Al(OH)_3$, calcium hydroxide, $Ca(OH)_2$, and potassium hydroxide, KOH:

If the central atom, E, has a low electronegativity, its attraction for electrons is low. Little tendency exists for the central atom to form a strong covalent bond with the oxygen atom, and bond *a* between the element and oxygen is more readily broken than bond *b* between oxygen and hydrogen. Hence bond *a* is ionic, hydroxide ions are released to the solution, and the material behaves as a base—this is the case with $Ca(OH)_2$ and KOH. Lower electronegativity is characteristic of the more metallic elements; hence, the metallic elements form ionic hydroxides that are by definition basic compounds.

If, on the other hand, the atom E has a relatively high electronegativity, it strongly attracts the electrons it shares with the oxygen atom, making bond *a* relatively strongly covalent. The oxygen-hydrogen bond, bond *b*, is thereby weakened because electrons are displaced toward E. Bond *b* is polar and readily releases hydrogen ions to the solution, so the material behaves as an acid. High electronegativities are characteristic of the more nonmetallic elements. Thus, nonmetallic elements form covalent compounds containing acidic –OH groups that are called **oxyacids**.

Increasing the oxidation number of the central atom E also increases the acidity of an oxyacid because this increases the attraction of E for the electrons it shares with oxygen and thereby weakens the O-H bond. Sulfuric acid, H_2SO_4, or $O_2S(OH)_2$ (with a sulfur oxidation number of +6), is more acidic than sulfurous acid, H_2SO_3, or $OS(OH)_2$ (with a sulfur oxidation number of +4). Likewise nitric acid, HNO_3, or O_2NOH (N oxidation number = +5), is more acidic than nitrous acid, HNO_2, or $ONOH$ (N oxidation number = +3). In each of these pairs, the oxidation number of the central atom is larger for the stronger acid (Figure 14.14).

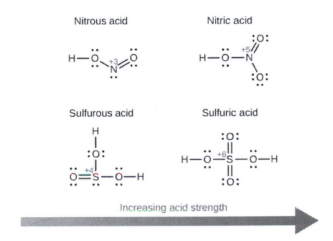

Figure 14.14 As the oxidation number of the central atom E increases, the acidity also increases.

Hydroxy compounds of elements with intermediate electronegativities and relatively high oxidation numbers (for example, elements near the diagonal line separating the metals from the nonmetals in the periodic table) are usually amphoteric. This means that the hydroxy compounds act as acids when they react with strong bases and as bases when they react with strong acids. The amphoterism of aluminum hydroxide, which commonly exists as the hydrate $Al(H_2O)_3(OH)_3$, is reflected in its solubility in both strong acids and strong bases. In strong bases, the relatively insoluble hydrated aluminum hydroxide, $Al(H_2O)_3(OH)_3$, is converted into the soluble ion, $[Al(H_2O)_2(OH)_4]^-$, by reaction with hydroxide ion:

$$Al(H_2O)_3(OH)_3(aq) + OH^-(aq) \rightleftharpoons H_2O(l) + [Al(H_2O)_2(OH)_4]^-(aq)$$

In this reaction, a proton is transferred from one of the aluminum-bound H_2O molecules to a hydroxide ion in solution. The $Al(H_2O)_3(OH)_3$ compound thus acts as an acid under these conditions. On the other hand, when dissolved in strong acids, it is converted to the soluble ion $[Al(H_2O)_6]^{3+}$ by reaction with hydronium ion:

$$3H_3O^+(aq) + Al(H_2O)_3(OH)_3(aq) \rightleftharpoons Al(H_2O)_6{}^{3+}(aq) + 3H_2O(l)$$

In this case, protons are transferred from hydronium ions in solution to $Al(H_2O)_3(OH)_3$, and the compound functions as a base.

14.4 Hydrolysis of Salt Solutions

By the end of this section, you will be able to:

- Predict whether a salt solution will be acidic, basic, or neutral
- Calculate the concentrations of the various species in a salt solution
- Describe the process that causes solutions of certain metal ions to be acidic

As we have seen in the section on chemical reactions, when an acid and base are mixed, they undergo a neutralization reaction. The word "neutralization" seems to imply that a stoichiometrically equivalent solution of an acid and a base would be neutral. This is sometimes true, but the salts that are formed in these reactions may have acidic or basic properties of their own, as we shall now see.

Acid-Base Neutralization

A solution is neutral when it contains equal concentrations of hydronium and hydroxide ions. When we mix solutions of an acid and a base, an acid-base neutralization reaction occurs. However, even if we mix stoichiometrically equivalent quantities, we may find that the resulting solution is not neutral. It could contain either an excess of hydronium ions or an excess of hydroxide ions because the nature of the salt formed determines whether the solution is acidic, neutral, or basic. The following four situations illustrate how solutions with various pH values can arise following a neutralization reaction using stoichiometrically equivalent quantities:

1. A strong acid and a strong base, such as HCl(aq) and NaOH(aq) will react to form a neutral solution since the conjugate partners produced are of negligible strength (see Figure 14.8):

$$HCl(aq) + NaOH(aq) \rightleftharpoons NaCl(aq) + H_2O(l)$$

2. A strong acid and a weak base yield a weakly acidic solution, not because of the strong acid involved, but because of the conjugate acid of the weak base.

3. A weak acid and a strong base yield a weakly basic solution. A solution of a weak acid reacts with a solution of a strong base to form the conjugate base of the weak acid and the conjugate acid of the strong base. The conjugate acid of the strong base is a weaker acid than water and has no effect on the acidity of the resulting solution. However, the conjugate base of the weak acid is a weak base and ionizes slightly in water. This increases the amount of hydroxide ion in the solution produced in the reaction and renders it slightly basic.

4. A weak acid plus a weak base can yield either an acidic, basic, or neutral solution. This is the most complex of the four types of reactions. When the conjugate acid and the conjugate base are of unequal strengths, the solution can be either acidic or basic, depending on the relative strengths of the two conjugates. Occasionally the weak acid and the weak base will have the *same* strength, so their respective conjugate base and acid will have the same strength, and the solution will be neutral. To predict whether a particular combination will be acidic, basic or neutral, tabulated K values of the conjugates must be compared.

> ### Chemistry in Everyday Life
>
> ### Stomach Antacids
>
> Our stomachs contain a solution of roughly 0.03 *M* HCl, which helps us digest the food we eat. The burning sensation associated with heartburn is a result of the acid of the stomach leaking through the muscular valve

at the top of the stomach into the lower reaches of the esophagus. The lining of the esophagus is not protected from the corrosive effects of stomach acid the way the lining of the stomach is, and the results can be very painful. When we have heartburn, it feels better if we reduce the excess acid in the esophagus by taking an antacid. As you may have guessed, antacids are bases. One of the most common antacids is calcium carbonate, $CaCO_3$. The reaction,

$$CaCO_3(s) + 2HCl(aq) \rightleftharpoons CaCl_2(aq) + H_2O(l) + CO_2(g)$$

not only neutralizes stomach acid, it also produces $CO_2(g)$, which may result in a satisfying belch.

Milk of Magnesia is a suspension of the sparingly soluble base magnesium hydroxide, $Mg(OH)_2$. It works according to the reaction:

$$Mg(OH)_2(s) \rightleftharpoons Mg^{2+}(aq) + 2OH^-(aq)$$

The hydroxide ions generated in this equilibrium then go on to react with the hydronium ions from the stomach acid, so that :

$$H_3O^+ + OH^- \rightleftharpoons 2H_2O(l)$$

This reaction does not produce carbon dioxide, but magnesium-containing antacids can have a laxative effect.

Several antacids have aluminum hydroxide, $Al(OH)_3$, as an active ingredient. The aluminum hydroxide tends to cause constipation, and some antacids use aluminum hydroxide in concert with magnesium hydroxide to balance the side effects of the two substances.

Chemistry in Everyday Life

Culinary Aspects of Chemistry

Cooking is essentially synthetic chemistry that happens to be safe to eat. There are a number of examples of acid-base chemistry in the culinary world. One example is the use of baking soda, or sodium bicarbonate in baking. $NaHCO_3$ is a base. When it reacts with an acid such as lemon juice, buttermilk, or sour cream in a batter, bubbles of carbon dioxide gas are formed from decomposition of the resulting carbonic acid, and the batter "rises." Baking powder is a combination of sodium bicarbonate, and one or more acid salts that react when the two chemicals come in contact with water in the batter.

Many people like to put lemon juice or vinegar, both of which are acids, on cooked fish (Figure 14.15). It turns out that fish have volatile amines (bases) in their systems, which are neutralized by the acids to yield involatile ammonium salts. This reduces the odor of the fish, and also adds a "sour" taste that we seem to enjoy.

$$CH_3COOH + NH_2CH_2CH_2CH_2CH_2NH_2 \longrightarrow CH_3COO^- + NH_3^+CH_2CH_2CH_2CH_2NH_2$$

Acetic acid +   Putrescine $\longrightarrow$ Acetate ion +   Putrescinium ion

Figure 14.15 A neutralization reaction takes place between citric acid in lemons or acetic acid in vinegar, and the bases in the flesh of fish.

Pickling is a method used to preserve vegetables using a naturally produced acidic environment. The vegetable, such as a cucumber, is placed in a sealed jar submerged in a brine solution. The brine solution favors the growth of beneficial bacteria and suppresses the growth of harmful bacteria. The beneficial bacteria feed on starches in the cucumber and produce lactic acid as a waste product in a process called fermentation. The lactic acid eventually increases the acidity of the brine to a level that kills any harmful bacteria, which require a basic environment. Without the harmful bacteria consuming the cucumbers they are able to last much longer than if they were unprotected. A byproduct of the pickling process changes the flavor of the vegetables with the acid making them taste sour.

Salts of Weak Bases and Strong Acids

When we neutralize a weak base with a strong acid, the product is a salt containing the conjugate acid of the weak base. This conjugate acid is a weak acid. For example, ammonium chloride, NH_4Cl, is a salt formed by the reaction of the weak base ammonia with the strong acid HCl:

$$NH_3(aq) + HCl(aq) \longrightarrow NH_4Cl(aq)$$

A solution of this salt contains ammonium ions and chloride ions. The chloride ion has no effect on the acidity of the solution since HCl is a strong acid. Chloride is a very weak base and will not accept a proton to a measurable extent. However, the ammonium ion, the conjugate acid of ammonia, reacts with water and increases the hydronium ion concentration:

$$NH_4^+(aq) + H_2O(l) \rightleftharpoons H_3O^+(aq) + NH_3(aq)$$

The equilibrium equation for this reaction is simply the ionization constant. K_a, for the acid NH_4^+ :

$$\frac{[H_3O^+][NH_3]}{[NH_4^+]} = K_a$$

We will not find a value of K_a for the ammonium ion in Appendix H. However, it is not difficult to determine K_a for NH_4^+ from the value of the ionization constant of water, K_w, and K_b, the ionization constant of its conjugate base, NH_3, using the following relationship:

$$K_w = K_a \times K_b$$

This relation holds for any base and its conjugate acid or for any acid and its conjugate base.

Example 14.15

The pH of a Solution of a Salt of a Weak Base and a Strong Acid

Aniline is an amine that is used to manufacture dyes. It is isolated as aniline hydrochloride, $[C_6H_5NH_3^+]Cl$, a salt prepared by the reaction of the weak base aniline and hydrochloric acid. What is the pH of a 0.233 M solution of aniline hydrochloride?

$$C_6H_5NH_3^+(aq) + H_2O(l) \rightleftharpoons H_3O^+(aq) + C_6H_5NH_2(aq)$$

Solution

The new step in this example is to determine K_a for the $C_6H_5NH_3^+$ ion. The $C_6H_5NH_3^+$ ion is the conjugate acid of a weak base. The value of K_a for this acid is not listed in Appendix H, but we can determine it from the value of K_b for aniline, $C_6H_5NH_2$, which is given as 4.6×10^{-10} (Table 14.3 and Appendix I):

$$K_a \text{ (for } C_6H_5NH_3^+) \times K_b \text{ (for } C_6H_5NH_2) = K_w = 1.0 \times 10^{-14}$$

$$K_a \text{ (for } C_6H_5NH_3^+) = \frac{K_w}{K_b \text{ (for } C_6H_5NH_2)} = \frac{1.0 \times 10^{-14}}{4.6 \times 10^{-10}} = 2.2 \times 10^{-5}$$

Now we have the ionization constant and the initial concentration of the weak acid, the information necessary to determine the equilibrium concentration of H_3O^+, and the pH:

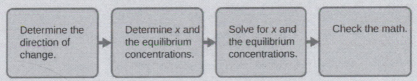

With these steps we find $[H_3O^+] = 2.3 \times 10^{-3} M$ and pH = 2.64

Check Your Learning

(a) Do the calculations and show that the hydronium ion concentration for a 0.233-M solution of $C_6H_5NH_3^+$ is 2.3×10^{-3} and the pH is 2.64.

(b) What is the hydronium ion concentration in a 0.100-M solution of ammonium nitrate, NH_4NO_3, a salt composed of the ions NH_4^+ and NO_3^-. Use the data in Table 14.3 to determine K_b for the ammonium ion. Which is the stronger acid $C_6H_5NH_3^+$ or NH_4^+?

Answer: (a) K_a (for NH_4^+) = 5.6×10^{-10}, $[H_3O^+] = 7.5 \times 10^{-6} M$; (b) $C_6H_5NH_3^+$ is the stronger acid.

Salts of Weak Acids and Strong Bases

When we neutralize a weak acid with a strong base, we get a salt that contains the conjugate base of the weak acid. This conjugate base is usually a weak base. For example, sodium acetate, $NaCH_3CO_2$, is a salt formed by the reaction of the weak acid acetic acid with the strong base sodium hydroxide:

$$CH_3CO_2H(aq) + NaOH(aq) \longrightarrow NaCH_3CO_2(aq) + H_2O(aq)$$

A solution of this salt contains sodium ions and acetate ions. The sodium ion, as the conjugate acid of a strong base, has no effect on the acidity of the solution. However, the acetate ion, the conjugate base of acetic acid, reacts with water and increases the concentration of hydroxide ion:

$$CH_3CO_2^-(aq) + H_2O(l) \rightleftharpoons CH_3CO_2H(aq) + OH^-(aq)$$

The equilibrium equation for this reaction is the ionization constant, K_b, for the base $CH_3CO_2^-$. The value of K_b can be calculated from the value of the ionization constant of water, K_w, and K_a, the ionization constant of the conjugate acid of the anion using the equation:

$$K_w = K_a \times K_b$$

For the acetate ion and its conjugate acid we have:

$$K_b \text{ (for } CH_3CO_2^-) = \frac{K_w}{K_a \text{ (for } CH_3CO_2H)} = \frac{1.0 \times 10^{-14}}{1.8 \times 10^{-5}} = 5.6 \times 10^{-10}$$

Some handbooks do not report values of K_b. They only report ionization constants for acids. If we want to determine a K_b value using one of these handbooks, we must look up the value of K_a for the conjugate acid and convert it to a K_b value.

Example 14.16

Equilibrium in a Solution of a Salt of a Weak Acid and a Strong Base

Determine the acetic acid concentration in a solution with $[CH_3CO_2^-] = 0.050\,M$ and $[OH^-] = 2.5 \times 10^{-6}\,M$ at equilibrium. The reaction is:

$$CH_3CO_2^-(aq) + H_2O(l) \rightleftharpoons CH_3CO_2H(aq) + OH^-(aq)$$

Solution

We are given two of three equilibrium concentrations and asked to find the missing concentration. If we can find the equilibrium constant for the reaction, the process is straightforward.

The acetate ion behaves as a base in this reaction; hydroxide ions are a product. We determine K_b as follows:

$$K_b \text{ (for } CH_3CO_2^-) = \frac{K_w}{K_a \text{ (for } CH_3CO_2H)} = \frac{1.0 \times 10^{-14}}{1.8 \times 10^{-5}} = 5.6 \times 10^{-10}$$

Now find the missing concentration:

$$K_b = \frac{[CH_3CO_2H][OH^-]}{[CH_3CO_2^-]} = 5.6 \times 10^{-10}$$

$$= \frac{[CH_3CO_2H](2.5 \times 10^{-6})}{(0.050)} = 5.6 \times 10^{-10}$$

Solving this equation we get $[CH_3CO_2H] = 1.1 \times 10^{-5}\,M$.

Check Your Learning

What is the pH of a 0.083-M solution of CN^-? Use 4.0×10^{-10} as K_a for HCN. Hint: We will probably need to convert pOH to pH or find $[H_3O^+]$ using $[OH^-]$ in the final stages of this problem.

Answer: 11.16

Equilibrium in a Solution of a Salt of a Weak Acid and a Weak Base

In a solution of a salt formed by the reaction of a weak acid and a weak base, to predict the pH, we must know both the K_a of the weak acid and the K_b of the weak base. If $K_a > K_b$, the solution is acidic, and if $K_b > K_a$, the solution is basic.

Example 14.17

Determining the Acidic or Basic Nature of Salts

Determine whether aqueous solutions of the following salts are acidic, basic, or neutral:

(a) KBr

(b) $NaHCO_3$

(c) NH_4Cl

(d) Na_2HPO_4

(e) NH_4F

Solution

Consider each of the ions separately in terms of its effect on the pH of the solution, as shown here:

(a) The K^+ cation and the Br^- anion are both spectators, since they are the cation of a strong base (KOH) and the anion of a strong acid (HBr), respectively. The solution is neutral.

(b) The Na^+ cation is a spectator, and will not affect the pH of the solution; while the HCO_3^- anion is amphiprotic, it could either behave as an acid or a base. The K_a of HCO_3^- is 4.7×10^{-11}, so the K_b of its conjugate base is $\dfrac{1.0 \times 10^{-14}}{4.7 \times 10^{-11}} = 2.1 \times 10^{-4}$.

Since $K_b \gg K_a$, the solution is basic.

(c) The NH_4^+ ion is acidic and the Cl^- ion is a spectator. The solution will be acidic.

(d) The Na^+ ion is a spectator, while the HPO_4^- ion is amphiprotic, with a K_a of 3.6×10^{-13}

so that the K_b of its conjugate base is $\dfrac{1.0 \times 10^{-14}}{3.6 \times 10^{-13}} = 2.8 \times 10^{-2}$. Because $K_b \gg K_a$, the solution is basic.

(e) The NH_4^+ ion is listed as being acidic, and the F^- ion is listed as a base, so we must directly compare the K_a and the K_b of the two ions. K_a of NH_4^+ is 5.6×10^{-10}, which seems very small, yet the K_b of F^- is 1.4×10^{-11}, so the solution is acidic, since $K_a > K_b$.

Check Your Learning

Determine whether aqueous solutions of the following salts are acidic, basic, or neutral:

(a) K_2CO_3

(b) $CaCl_2$

(c) KH_2PO_4

(d) $(NH_4)_2CO_3$

(e) $AlBr_3$

Answer: (a) basic; (b) neutral; (c) basic; (d) basic; (e) acidic

The Ionization of Hydrated Metal Ions

If we measure the pH of the solutions of a variety of metal ions we will find that these ions act as weak acids when in solution. The aluminum ion is an example. When aluminum nitrate dissolves in water, the aluminum ion reacts with water to give a hydrated aluminum ion, $Al(H_2O)_6^{3+}$, dissolved in bulk water. What this means is that the aluminum ion has the strongest interactions with the six closest water molecules (the so-called first solvation shell), even though it does interact with the other water molecules surrounding this $Al(H_2O)_6^{3+}$ cluster as well:

$$Al(NO_3)_3(s) + 6H_2O(l) \longrightarrow Al(H_2O)_6^{3+}(aq) + 3NO_3^-(aq)$$

We frequently see the formula of this ion simply as "$Al^{3+}(aq)$", without explicitly noting the six water molecules that are the closest ones to the aluminum ion and just describing the ion as being solvated in water (hydrated). This is similar to the simplification of the formula of the hydronium ion, H_3O^+ to H^+. However, in this case, the hydrated aluminum ion is a weak acid (Figure 14.16) and donates a proton to a water molecule. Thus, the hydration becomes important and we may use formulas that show the extent of hydration:

$$Al(H_2O)_6^{3+}(aq) + H_2O(l) \rightleftharpoons H_3O^+(aq) + Al(H_2O)_5(OH)^{2+}(aq) \qquad K_a = 1.4 \times 10^{-5}$$

As with other polyprotic acids, the hydrated aluminum ion ionizes in stages, as shown by:

$$Al(H_2O)_6^{3+}(aq) + H_2O(l) \rightleftharpoons H_3O^+(aq) + Al(H_2O)_5(OH)^{2+}(aq)$$
$$Al(H_2O)_5(OH)^{2+}(aq) + H_2O(l) \rightleftharpoons H_3O^+(aq) + Al(H_2O)_4(OH)_2^+(aq)$$
$$Al(H_2O)_4(OH)_2^+(aq) + H_2O(l) \rightleftharpoons H_3O^+(aq) + Al(H_2O)_3(OH)_3(aq)$$

Note that some of these aluminum species are exhibiting amphiprotic behavior, since they are acting as acids when they appear on the right side of the equilibrium expressions and as bases when they appear on the left side.

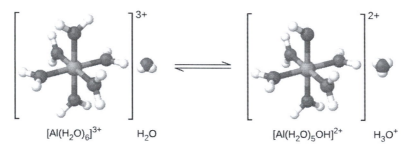

$[Al(H_2O)_6]^{3+}$ H_2O $[Al(H_2O)_5OH]^{2+}$ H_3O^+

Figure 14.16 When an aluminum ion reacts with water, the hydrated aluminum ion becomes a weak acid.

However, the ionization of a cation carrying more than one charge is usually not extensive beyond the first stage. Additional examples of the first stage in the ionization of hydrated metal ions are:

$$Fe(H_2O)_6^{3+}(aq) + H_2O(l) \rightleftharpoons H_3O^+(aq) + Fe(H_2O)_5(OH)^{2+}(aq) \qquad K_a = 2.74$$
$$Cu(H_2O)_6^{2+}(aq) + H_2O(l) \rightleftharpoons H_3O^+(aq) + Cu(H_2O)_5(OH)^+(aq) \qquad K_a = {\sim}6.3$$
$$Zn(H_2O)_4^{2+}(aq) + H_2O(l) \rightleftharpoons H_3O^+(aq) + Zn(H_2O)_3(OH)^+(aq) \qquad K_a = 9.6$$

Example 14.18

Hydrolysis of $[Al(H_2O)_6]^{3+}$

Calculate the pH of a 0.10-M solution of aluminum chloride, which dissolves completely to give the hydrated aluminum ion $[Al(H_2O)_6]^{3+}$ in solution.

Solution

In spite of the unusual appearance of the acid, this is a typical acid ionization problem.

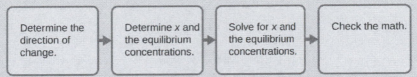

Step 1. *Determine the direction of change.* The equation for the reaction and K_a are:

$$Al(H_2O)_6^{3+}(aq) + H_2O(l) \rightleftharpoons H_3O^+(aq) + Al(H_2O)_5(OH)^{2+}(aq) \qquad K_a = 1.4 \times 10^{-5}$$

The reaction shifts to the right to reach equilibrium.

Step 2. *Determine x and equilibrium concentrations.* Use the table:

	$Al(H_2O)_6^{3+} + H_2O \rightleftharpoons H_3O^+ + Al(H_2O)_5(OH)^{2+}$		
Initial concentration (M)	0.10	~0	0
Change (M)	$-x$	x	x
Equilibrium constant (M)	$0.10 - x$	x	x

Step 3. *Solve for x and the equilibrium concentrations.* Substituting the expressions for the equilibrium concentrations into the equation for the ionization constant yields:

$$K_a = \frac{[H_3O^+][Al(H_2O)_5(OH)^{2+}]}{[Al(H_2O)_6^{3+}]}$$

$$= \frac{(x)(x)}{0.10 - x} = 1.4 \times 10^{-5}$$

Solving this equation gives:

$$x = 1.2 \times 10^{-3} \ M$$

From this we find:

$$[H_3O^+] = 0 + x = 1.2 \times 10^{-3} \ M$$

$$pH = -\log[H_3O^+] = 2.92 \ (\text{an acidic solution})$$

Step 4. *Check the work.* The arithmetic checks; when $1.2 \times 10^{-3} \ M$ is substituted for x, the result $= K_a$.

Check Your Learning

What is $[Al(H_2O)_5(OH)^{2+}]$ in a 0.15-M solution of $Al(NO_3)_3$ that contains enough of the strong acid HNO_3 to bring $[H_3O^+]$ to 0.10 M?

Answer: $2.1 \times 10^{-5} M$

The constants for the different stages of ionization are not known for many metal ions, so we cannot calculate the extent of their ionization. However, practically all hydrated metal ions other than those of the alkali metals ionize to give acidic solutions. Ionization increases as the charge of the metal ion increases or as the size of the metal ion decreases.

14.5 Polyprotic Acids

By the end of this section, you will be able to:

- Extend previously introduced equilibrium concepts to acids and bases that may donate or accept more than one proton

We can classify acids by the number of protons per molecule that they can give up in a reaction. Acids such as HCl, HNO_3, and HCN that contain one ionizable hydrogen atom in each molecule are called **monoprotic acids**. Their reactions with water are:

$$HCl(aq) + H_2O(l) \longrightarrow H_3O^+(aq) + Cl^-(aq)$$
$$HNO_3(aq) + H_2O(l) \longrightarrow H_3O^+(aq) + NO_3{}^-(aq)$$
$$HCN(aq) + H_2O(l) \longrightarrow H_3O^+(aq) + CN^-(aq)$$

Even though it contains four hydrogen atoms, acetic acid, CH_3CO_2H, is also monoprotic because only the hydrogen atom from the carboxyl group (COOH) reacts with bases:

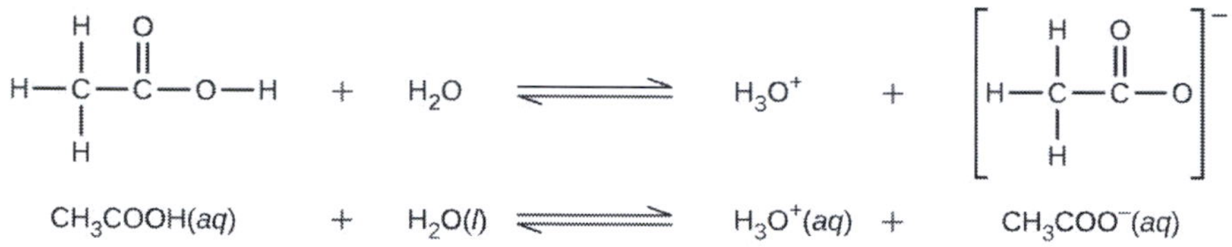

$$CH_3COOH(aq) + H_2O(l) \rightleftharpoons H_3O^+(aq) + CH_3COO^-(aq)$$

Similarly, monoprotic bases are bases that will accept a single proton.

Diprotic acids contain two ionizable hydrogen atoms per molecule; ionization of such acids occurs in two steps. The first ionization always takes place to a greater extent than the second ionization. For example, sulfuric acid, a strong acid, ionizes as follows:

First ionization: $H_2SO_4(aq) + H_2O(l) \rightleftharpoons H_3O^+(aq) + HSO_4{}^-(aq)$ $K_{a1} =$ more than 10^2; complete dissociation

Second ionization: $HSO_4{}^-(aq) + H_2O(l) \rightleftharpoons H_3O^+(aq) + SO_4{}^{2-}(aq)$ $K_{a2} = 1.2 \times 10^{-2}$

This **stepwise ionization** process occurs for all polyprotic acids. When we make a solution of a weak diprotic acid, we get a solution that contains a mixture of acids. Carbonic acid, H_2CO_3, is an example of a weak diprotic acid. The first ionization of carbonic acid yields hydronium ions and bicarbonate ions in small amounts.

First ionization:
$$H_2CO_3(aq) + H_2O(l) \rightleftharpoons H_3O^+(aq) + HCO_3{}^-(aq) \qquad K_{H_2CO_3} = \frac{[H_3O^+][HCO_3{}^-]}{[H_2CO_3]} = 4.3 \times 10^{-7}$$

The bicarbonate ion can also act as an acid. It ionizes and forms hydronium ions and carbonate ions in even smaller quantities.

Second ionization:
$$HCO_3{}^-(aq) + H_2O(l) \rightleftharpoons H_3O^+(aq) + CO_3{}^{2-}(aq) \qquad K_{HCO_3{}^-} = \frac{[H_3O^+][CO_3{}^{2-}]}{[HCO_3{}^-]} = 4.7 \times 10^{-11}$$

$K_{H_2CO_3}$ is larger than $K_{HCO_3{}^-}$ by a factor of 10^4, so H_2CO_3 is the dominant producer of hydronium ion in the solution. This means that little of the $HCO_3{}^-$ formed by the ionization of H_2CO_3 ionizes to give hydronium ions (and carbonate ions), and the concentrations of H_3O^+ and $HCO_3{}^-$ are practically equal in a pure aqueous solution of H_2CO_3.

If the first ionization constant of a weak diprotic acid is larger than the second by a factor of at least 20, it is appropriate to treat the first ionization separately and calculate concentrations resulting from it before calculating concentrations of species resulting from subsequent ionization. This can simplify our work considerably because we can determine the concentration of H_3O^+ and the conjugate base from the first ionization, then determine the concentration of the conjugate base of the second ionization in a solution with concentrations determined by the first ionization.

Example 14.19

Ionization of a Diprotic Acid

When we buy soda water (carbonated water), we are buying a solution of carbon dioxide in water. The solution is acidic because CO_2 reacts with water to form carbonic acid, H_2CO_3. What are $[H_3O^+]$, $[HCO_3{}^-]$, and $[CO_3{}^{2-}]$ in a saturated solution of CO_2 with an initial $[H_2CO_3] = 0.033\ M$?

$$H_2CO_3(aq) + H_2O(l) \rightleftharpoons H_3O^+(aq) + HCO_3{}^-(aq) \qquad K_{a1} = 4.3 \times 10^{-7}$$
$$HCO_3{}^-(aq) + H_2O(l) \rightleftharpoons H_3O^+(aq) + CO_3{}^{2-}(aq) \qquad K_{a2} = 4.7 \times 10^{-11}$$

Solution

As indicated by the ionization constants, H_2CO_3 is a much stronger acid than $HCO_3{}^-$, so H_2CO_3 is the dominant producer of hydronium ion in solution. Thus there are two parts in the solution of this problem: (1) Using the customary four steps, we determine the concentration of H_3O^+ and $HCO_3{}^-$ produced by ionization of H_2CO_3. (2) Then we determine the concentration of $CO_3{}^{2-}$ in a solution with the concentration of H_3O^+ and $HCO_3{}^-$ determined in (1). To summarize:

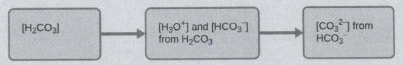

Step 1. Determine the concentrations of H_3O^+ and $HCO_3{}^-$.

$$H_2CO_3(aq) + H_2O(l) \rightleftharpoons H_3O^+(aq) + HCO_3{}^-(aq) \qquad K_{a1} = 4.3 \times 10^{-7}$$

As for the ionization of any other weak acid:

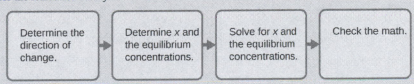

An abbreviated table of changes and concentrations shows:

	H_2CO_3 + H_2O $\rightleftharpoons$ H_3O^+ + $HCO_3{}^-$		
Initial concentration (M)	0.033	~0	0
Change (M)	−x	x	x
Equilibrium constant (M)	0.033 − x	x	x

Substituting the equilibrium concentrations into the equilibrium gives us:

$$K_{H_2CO_3} = \frac{[H_3O^+][HCO_3{}^-]}{[H_2CO_3]} = \frac{(x)(x)}{0.033 - x} = 4.3 \times 10^{-7}$$

Solving the preceding equation making our standard assumptions gives:

$$x = 1.2 \times 10^{-4}$$

Thus:

$$[H_2CO_3] = 0.033 \ M$$

$$[H_3O^+] = [HCO_3^-] = 1.2 \times 10^{-4} \ M$$

Step 2. *Determine the concentration of* CO_3^{2-} *in a solution at equilibrium with* $[H_3O^+]$ *and* $[HCO_3^-]$ *both equal to* 1.2×10^{-4} M.

$$HCO_3^-(aq) + H_2O(l) \rightleftharpoons H_3O^+(aq) + CO_3^{2-}(aq)$$

$$K_{HCO_3^-} = \frac{[H_3O^+][CO_3^{2-}]}{[HCO_3^-]} = \frac{(1.2 \times 10^{-4})[CO_3^{2-}]}{1.2 \times 10^{-4}}$$

$$[CO_3^{2-}] = \frac{(4.7 \times 10^{-11})(1.2 \times 10^{-4})}{1.2 \times 10^{-4}} = 4.7 \times 10^{-11} \ M$$

To summarize: In part 1 of this example, we found that the H_2CO_3 in a 0.033-M solution ionizes slightly and at equilibrium $[H_2CO_3] = 0.033 \ M$; $[H_3O^+] = 1.2 \times 10^{-4}$; and $[HCO_3^-] = 1.2 \times 10^{-4} \ M$. In part 2, we determined that $[CO_3^{2-}] = 4.7 \times 10^{-11} \ M$.

Check Your Learning

The concentration of H_2S in a saturated aqueous solution at room temperature is approximately 0.1 M. Calculate $[H_3O^+]$, $[HS^-]$, and $[S^{2-}]$ in the solution:

$$H_2S(aq) + H_2O(l) \rightleftharpoons H_3O^+(aq) + HS^-(aq) \qquad K_{a1} = 1.0 \times 10^{-7}$$
$$HS^-(aq) + H_2O(l) \rightleftharpoons H_3O^+(aq) + S^{2-}(aq) \qquad K_{a2} = 1.0 \times 10^{-19}$$

Answer: $[H_2S] = 0.1 \ M$; $[H_3O^+] = [HS^-] = 0.0001 \ M$; $[S^{2-}] = 1 \times 10^{-19} \ M$

We note that the concentration of the sulfide ion is the same as K_{a2}. This is due to the fact that each subsequent dissociation occurs to a lesser degree (as acid gets weaker).

A **triprotic acid** is an acid that has three dissociable protons that undergo stepwise ionization: Phosphoric acid is a typical example:

First ionization: $H_3PO_4(aq) + H_2O(l) \rightleftharpoons H_3O^+(aq) + H_2PO_4^-(aq)$ $\qquad K_{a1} = 7.5 \times 10^{-3}$

Second ionization: $H_2PO_4^-(aq) + H_2O(l) \rightleftharpoons H_3O^+(aq) + HPO_4^{2-}(aq)$ $\qquad K_{a2} = 6.3 \times 10^{-8}$

Third ionization: $HPO_4^{2-}(aq) + H_2O(l) \rightleftharpoons H_3O^+(aq) + PO_4^{3-}(aq)$ $\qquad K_{a3} = 3.6 \times 10^{-13}$

As with the diprotic acids, the differences in the ionization constants of these reactions tell us that in each successive step the degree of ionization is significantly weaker. This is a general characteristic of polyprotic acids and successive ionization constants often differ by a factor of about 10^5 to 10^6.

This set of three dissociation reactions may appear to make calculations of equilibrium concentrations in a solution of H_3PO_4 complicated. However, because the successive ionization constants differ by a factor of 10^5 to 10^6, the calculations can be broken down into a series of parts similar to those for diprotic acids.

Polyprotic bases can accept more than one hydrogen ion in solution. The carbonate ion is an example of a **diprotic base**, since it can accept up to two protons. Solutions of alkali metal carbonates are quite alkaline, due to the reactions:

$$H_2O(l) + CO_3{}^{2-}(aq) \rightleftharpoons HCO_3{}^-(aq) + OH^-(aq) \quad \text{and} \quad H_2O(l) + HCO_3{}^-(aq) \rightleftharpoons H_2CO_3(aq) + OH^-(aq)$$

14.6 Buffers

By the end of this section, you will be able to:

- Describe the composition and function of acid–base buffers
- Calculate the pH of a buffer before and after the addition of added acid or base

A mixture of a weak acid and its conjugate base (or a mixture of a weak base and its conjugate acid) is called a buffer solution, or a **buffer**. Buffer solutions resist a change in pH when small amounts of a strong acid or a strong base are added (Figure 14.17). A solution of acetic acid and sodium acetate ($CH_3COOH + CH_3COONa$) is an example of a buffer that consists of a weak acid and its salt. An example of a buffer that consists of a weak base and its salt is a solution of ammonia and ammonium chloride ($NH_3(aq) + NH_4Cl(aq)$).

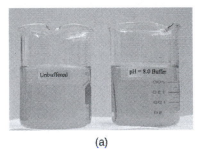
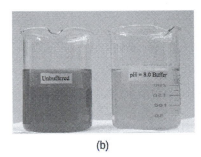

(a)   (b)

Figure 14.17   (a) The buffered solution on the left and the unbuffered solution on the right have the same pH (pH 8); they are basic, showing the yellow color of the indicator methyl orange at this pH. (b) After the addition of 1 mL of a 0.01-*M* HCl solution, the buffered solution has not detectably changed its pH but the unbuffered solution has become acidic, as indicated by the change in color of the methyl orange, which turns red at a pH of about 4. (credit: modification of work by Mark Ott)

How Buffers Work

A mixture of acetic acid and sodium acetate is acidic because the K_a of acetic acid is greater than the K_b of its conjugate base acetate. It is a buffer because it contains both the weak acid and its salt. Hence, it acts to keep the hydronium ion concentration (and the pH) almost constant by the addition of either a small amount of a strong acid or a strong base. If we add a base such as sodium hydroxide, the hydroxide ions react with the few hydronium ions present. Then more of the acetic acid reacts with water, restoring the hydronium ion concentration almost to its original value:

$$CH_3CO_2H(aq) + H_2O(l) \longrightarrow H_3O^+(aq) + CH_3CO_2{}^-(aq)$$

The pH changes very little. If we add an acid such as hydrochloric acid, most of the hydronium ions from the hydrochloric acid combine with acetate ions, forming acetic acid molecules:

$$H_3O^+(aq) + CH_3CO_2{}^-(aq) \longrightarrow CH_3CO_2H(aq) + H_2O(l)$$

Thus, there is very little increase in the concentration of the hydronium ion, and the pH remains practically unchanged (Figure 14.18).

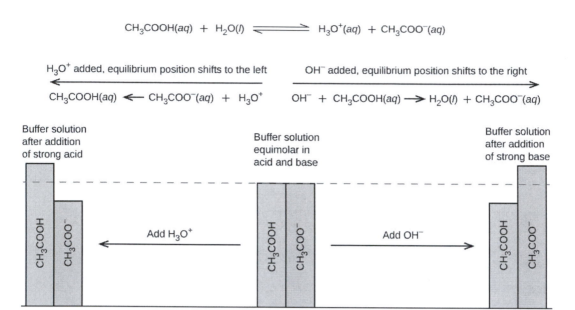

Figure 14.18 This diagram shows the buffer action of these reactions.

A mixture of ammonia and ammonium chloride is basic because the K_b for ammonia is greater than the K_a for the ammonium ion. It is a buffer because it also contains the salt of the weak base. If we add a base (hydroxide ions), ammonium ions in the buffer react with the hydroxide ions to form ammonia and water and reduce the hydroxide ion concentration almost to its original value:

$$NH_4{}^+(aq) + OH^-(aq) \longrightarrow NH_3(aq) + H_2O(l)$$

If we add an acid (hydronium ions), ammonia molecules in the buffer mixture react with the hydronium ions to form ammonium ions and reduce the hydronium ion concentration almost to its original value:

$$H_3O^+(aq) + NH_3(aq) \longrightarrow NH_4{}^+(aq) + H_2O(l)$$

The three parts of the following example illustrate the change in pH that accompanies the addition of base to a buffered solution of a weak acid and to an unbuffered solution of a strong acid.

Example 14.20

pH Changes in Buffered and Unbuffered Solutions

Acetate buffers are used in biochemical studies of enzymes and other chemical components of cells to prevent pH changes that might change the biochemical activity of these compounds.

(a) Calculate the pH of an acetate buffer that is a mixture with 0.10 M acetic acid and 0.10 M sodium acetate.

Solution

To determine the pH of the buffer solution we use a typical equilibrium calculation (as illustrated in earlier Examples):

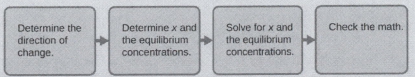

Step 1. *Determine the direction of change.* The equilibrium in a mixture of H_3O^+, $CH_3CO_2^-$, and CH_3CO_2H is:

$$CH_3CO_2H(aq) + H_2O(l) \rightleftharpoons H_3O^+(aq) + CH_3CO_2^-(aq)$$

The equilibrium constant for CH_3CO_2H is not given, so we look it up in Appendix H: $K_a = 1.8 \times 10^{-5}$. With $[CH_3CO_2H] = [CH_3CO_2^-] = 0.10\ M$ and $[H_3O^+] = \sim 0\ M$, the reaction shifts to the right to form H_3O^+.

Step 2. *Determine* x *and equilibrium concentrations.* A table of changes and concentrations follows:

	$[CH_3CO_2H]\ +\ [H_2O] \rightleftharpoons H_3O^+\ +\ [CH_3CO_2^-]$		
Initial concentration (*M*)	0.10	~0	0.10
Change (*M*)	−x	x	x
Equilibrium constant (*M*)	0.10 − x	x	0.10 + x

Step 3. *Solve for x and the equilibrium concentrations.* We find:

$$x = 1.8 \times 10^{-5}\ M$$

and

$$[H_3O^+] = 0 + x = 1.8 \times 10^{-5}\ M$$

Thus:

$$pH = -\log[H_3O^+] = -\log(1.8 \times 10^{-5})$$

$$= 4.74$$

Step 4. *Check the work.* If we calculate all calculated equilibrium concentrations, we find that the equilibrium value of the reaction coefficient, $Q = K_a$.

(b) Calculate the pH after 1.0 mL of 0.10 *M* NaOH is added to 100 mL of this buffer, giving a solution with a volume of 101 mL.

First, we calculate the concentrations of an intermediate mixture resulting from the complete reaction between the acid in the buffer and the added base. Then we determine the concentrations of the mixture at the new equilibrium:

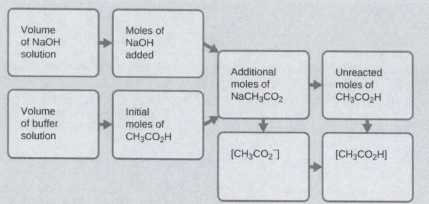

Step 1. *Determine the moles of NaOH.* One milliliter (0.0010 L) of 0.10 M NaOH contains:

$$0.0010 \, \text{L} \times \left(\frac{0.10 \text{ mol NaOH}}{1 \, \text{L}} \right) = 1.0 \times 10^{-4} \text{ mol NaOH}$$

Step 2. *Determine the moles of CH_2CO_2H.* Before reaction, 0.100 L of the buffer solution contains:

$$0.100 \, \text{L} \times \left(\frac{0.100 \text{ mol CH}_3\text{CO}_2\text{H}}{1 \, \text{L}} \right) = 1.00 \times 10^{-2} \text{ mol CH}_3\text{CO}_2\text{H}$$

Step 3. *Solve for the amount of $NaCH_3CO_2$ produced.* The 1.0×10^{-4} mol of NaOH neutralizes 1.0×10^{-4} mol of CH_3CO_2H, leaving:

$$(1.0 \times 10^{-2}) - (0.01 \times 10^{-2}) = 0.99 \times 10^{-2} \text{ mol CH}_3\text{CO}_2\text{H}$$

and producing 1.0×10^{-4} mol of $NaCH_3CO_2$. This makes a total of:

$$(1.0 \times 10^{-2}) + (0.01 \times 10^{-2}) = 1.01 \times 10^{-2} \text{ mol NaCH}_3\text{CO}_2$$

Step 4. *Find the molarity of the products.* After reaction, CH_3CO_2H and $NaCH_3CO_2$ are contained in 101 mL of the intermediate solution, so:

$$[CH_3CO_2H] = \frac{9.9 \times 10^{-3} \text{ mol}}{0.101 \text{ L}} = 0.098 \, M$$

$$[NaCH_3CO_2] = \frac{1.01 \times 10^{-2} \text{ mol}}{0.101 \text{ L}} = 0.100 \, M$$

Now we calculate the pH after the intermediate solution, which is 0.098 M in CH_3CO_2H and 0.100 M in $NaCH_3CO_2$, comes to equilibrium. The calculation is very similar to that in part (a) of this example:

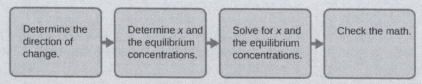

This series of calculations gives a pH = 4.75. Thus the addition of the base barely changes the pH of the solution (Figure 14.17).

(c) For comparison, calculate the pH after 1.0 mL of 0.10 M NaOH is added to 100 mL of a solution of an unbuffered solution with a pH of 4.74 (a 1.8×10^{-5}-M solution of HCl). The volume of the final solution is 101 mL.

Solution

This 1.8×10^{-5}-M solution of HCl has the same hydronium ion concentration as the 0.10-M solution of acetic acid-sodium acetate buffer described in part (a) of this example. The solution contains:

$$0.100 \, L \times \left(\frac{1.8 \times 10^{-5} \, mol \, HCl}{1 \, L} \right) = 1.8 \times 10^{-6} \, mol \, HCl$$

As shown in part (b), 1 mL of 0.10 M NaOH contains 1.0×10^{-4} mol of NaOH. When the NaOH and HCl solutions are mixed, the HCl is the limiting reagent in the reaction. All of the HCl reacts, and the amount of NaOH that remains is:

$$(1.0 \times 10^{-4}) - (1.8 \times 10^{-6}) = 9.8 \times 10^{-5} \, M$$

The concentration of NaOH is:

$$\frac{9.8 \times 10^{-5} \, M \, NaOH}{0.101 \, L} = 9.7 \times 10^{-4} \, M$$

The pOH of this solution is:

$$pOH = -\log[OH^-] = -\log\!\left(9.7 \times 10^{-4}\right) = 3.01$$

The pH is:

$$pH = 14.00 - pOH = 10.99$$

The pH changes from 4.74 to 10.99 in this unbuffered solution. This compares to the change of 4.74 to 4.75 that occurred when the same amount of NaOH was added to the buffered solution described in part (b).

Check Your Learning

Show that adding 1.0 mL of 0.10 M HCl changes the pH of 100 mL of a 1.8×10^{-5} M HCl solution from 4.74 to 3.00.

Answer: Initial pH of 1.8×10^{-5} M HCl; pH $= -\log[H_3O^+] = -\log[1.8 \times 10^{-5}] = 4.74$
Moles of H_3O^+ in 100 mL 1.8×10^{-5} M HCl; 1.8×10^{-5} moles/L $\times$ 0.100 L $= 1.8 \times 10^{-6}$
Moles of H_3O^+ added by addition of 1.0 mL of 0.10 M HCl: 0.10 moles/L $\times$ 0.0010 L $= 1.0 \times 10^{-4}$ moles;
final pH after addition of 1.0 mL of 0.10 M HCl:

$$pH = -\log[H_3O^+] = -\log\!\left(\frac{total \, moles \, H_3O^+}{total \, volume} \right) = -\log\!\left(\frac{1.0 \times 10^{-4} \, mol + 1.8 \times 10^{-6} \, mol}{101 \, mL\!\left(\frac{1 \, L}{1000 \, mL} \right)} \right) = 3.00$$

If we add an acid or a base to a buffer that is a mixture of a weak base and its salt, the calculations of the changes in pH are analogous to those for a buffer mixture of a weak acid and its salt.

Buffer Capacity

Buffer solutions do not have an unlimited capacity to keep the pH relatively constant (Figure 14.19). If we add so much base to a buffer that the weak acid is exhausted, no more buffering action toward the base is possible. On the other hand, if we add an excess of acid, the weak base would be exhausted, and no more buffering action toward any additional acid would be possible. In fact, we do not even need to exhaust all of the acid or base in a buffer to overwhelm it; its buffering action will diminish rapidly as a given component nears depletion.

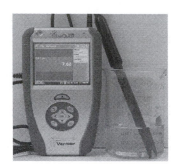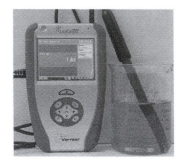

Figure 14.19   The indicator color (methyl orange) shows that a small amount of acid added to a buffered solution of pH 8 (beaker on the left) has little affect on the buffered system (middle beaker). However, a large amount of acid exhausts the buffering capacity of the solution and the pH changes dramatically (beaker on the right). (credit: modification of work by Mark Ott)

The **buffer capacity** is the amount of acid or base that can be added to a given volume of a buffer solution before the pH changes significantly, usually by one unit. Buffer capacity depends on the amounts of the weak acid and its conjugate base that are in a buffer mixture. For example, 1 L of a solution that is 1.0 M in acetic acid and 1.0 M in sodium acetate has a greater buffer capacity than 1 L of a solution that is 0.10 M in acetic acid and 0.10 M in sodium acetate even though both solutions have the same pH. The first solution has more buffer capacity because it contains more acetic acid and acetate ion.

Selection of Suitable Buffer Mixtures

There are two useful rules of thumb for selecting buffer mixtures:

1. A good buffer mixture should have about equal concentrations of both of its components. A buffer solution has generally lost its usefulness when one component of the buffer pair is less than about 10% of the other. Figure 14.20 shows an acetic acid-acetate ion buffer as base is added. The initial pH is 4.74. A change of 1 pH unit occurs when the acetic acid concentration is reduced to 11% of the acetate ion concentration.

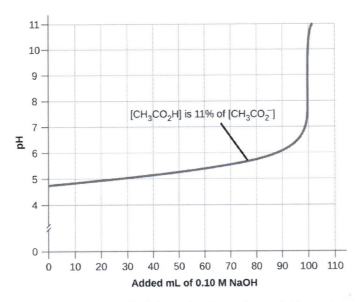

Figure 14.20 The graph, an illustration of buffering action, shows change of pH as an increasing amount of a 0.10-*M* NaOH solution is added to 100 mL of a buffer solution in which, initially, $[CH_3CO_2H] = 0.10\ M$ and $[CH_3CO_2{}^-] = 0.10\ M$.

2. Weak acids and their salts are better as buffers for pHs less than 7; weak bases and their salts are better as buffers for pHs greater than 7.

Blood is an important example of a buffered solution, with the principal acid and ion responsible for the buffering action being carbonic acid, H_2CO_3, and the bicarbonate ion, $HCO_3{}^-$. When an excess of hydrogen ion enters the blood stream, it is removed primarily by the reaction:

$$H_3O^+(aq) + HCO_3{}^-(aq) \longrightarrow H_2CO_3(aq) + H_2O(l)$$

When an excess of the hydroxide ion is present, it is removed by the reaction:

$$OH^-(aq) + H_2CO_3(aq) \longrightarrow HCO_3{}^-(aq) + H_2O(l)$$

The pH of human blood thus remains very near 7.35, that is, slightly basic. Variations are usually less than 0.1 of a pH unit. A change of 0.4 of a pH unit is likely to be fatal.

The Henderson-Hasselbalch Equation

The ionization-constant expression for a solution of a weak acid can be written as:

$$K_a = \frac{[H_3O^+][A^-]}{[HA]}$$

Rearranging to solve for $[H_3O^+]$, we get:

$$[H_3O^+] = K_a \times \frac{[HA]}{[A^-]}$$

Taking the negative logarithm of both sides of this equation, we arrive at:

$$-\log[H_3O^+] = -\log K_a - \log\frac{[HA]}{[A^-]},$$

which can be written as

$$pH = pK_a + \log \frac{[A^-]}{[HA]}$$

where pK_a is the negative of the common logarithm of the ionization constant of the weak acid ($pK_a = -\log K_a$). This equation relates the pH, the ionization constant of a weak acid, and the concentrations of the weak acid and its salt in a buffered solution. Scientists often use this expression, called the **Henderson-Hasselbalch equation**, to calculate the pH of buffer solutions. It is important to note that the "x is small" assumption must be valid to use this equation.

Portrait of a Chemist

Lawrence Joseph Henderson and Karl Albert Hasselbalch

Lawrence Joseph Henderson (1878–1942) was an American physician, biochemist and physiologist, to name only a few of his many pursuits. He obtained a medical degree from Harvard and then spent 2 years studying in Strasbourg, then a part of Germany, before returning to take a lecturer position at Harvard. He eventually became a professor at Harvard and worked there his entire life. He discovered that the acid-base balance in human blood is regulated by a buffer system formed by the dissolved carbon dioxide in blood. He wrote an equation in 1908 to describe the carbonic acid-carbonate buffer system in blood. Henderson was broadly knowledgeable; in addition to his important research on the physiology of blood, he also wrote on the adaptations of organisms and their fit with their environments, on sociology and on university education. He also founded the Fatigue Laboratory, at the Harvard Business School, which examined human physiology with specific focus on work in industry, exercise, and nutrition.

In 1916, Karl Albert Hasselbalch (1874–1962), a Danish physician and chemist, shared authorship in a paper with Christian Bohr in 1904 that described the Bohr effect, which showed that the ability of hemoglobin in the blood to bind with oxygen was inversely related to the acidity of the blood and the concentration of carbon dioxide. The pH scale was introduced in 1909 by another Dane, Sørensen, and in 1912, Hasselbalch published measurements of the pH of blood. In 1916, Hasselbalch expressed Henderson's equation in logarithmic terms, consistent with the logarithmic scale of pH, and thus the Henderson-Hasselbalch equation was born.

How Sciences Interconnect

Medicine: The Buffer System in Blood

The normal pH of human blood is about 7.4. The carbonate buffer system in the blood uses the following equilibrium reaction:

$$CO_2(g) + 2H_2O(l) \rightleftharpoons H_2CO_3(aq) \rightleftharpoons HCO_3^-(aq) + H_3O^+(aq)$$

The concentration of carbonic acid, H_2CO_3 is approximately 0.0012 M, and the concentration of the hydrogen carbonate ion, HCO_3^-, is around 0.024 M. Using the Henderson-Hasselbalch equation and the pK_a of carbonic acid at body temperature, we can calculate the pH of blood:

$$pH = pK_a + \log \frac{[\text{base}]}{[\text{acid}]} = 6.1 + \log \frac{0.024}{0.0012} = 7.4$$

The fact that the H_2CO_3 concentration is significantly lower than that of the HCO_3^- ion may seem unusual, but this imbalance is due to the fact that most of the by-products of our metabolism that enter our bloodstream are acidic. Therefore, there must be a larger proportion of base than acid, so that the capacity of the buffer will not be exceeded.

Lactic acid is produced in our muscles when we exercise. As the lactic acid enters the bloodstream, it is neutralized by the HCO_3^- ion, producing H_2CO_3. An enzyme then accelerates the breakdown of the excess carbonic acid to carbon dioxide and water, which can be eliminated by breathing. In fact, in addition to the regulating effects of the carbonate buffering system on the pH of blood, the body uses breathing to regulate blood pH. If the pH of the blood decreases too far, an increase in breathing removes CO_2 from the blood through the lungs driving the equilibrium reaction such that $[H_3O^+]$ is lowered. If the blood is too alkaline, a lower breath rate increases CO_2 concentration in the blood, driving the equilibrium reaction the other way, increasing $[H^+]$ and restoring an appropriate pH.

Link to Learning

View information (http://openstaxcollege.org/l/16BufferSystem) on the buffer system encountered in natural waters.

14.7 Acid-Base Titrations

By the end of this section, you will be able to:

- Interpret titration curves for strong and weak acid-base systems
- Compute sample pH at important stages of a titration
- Explain the function of acid-base indicators

As seen in the chapter on the stoichiometry of chemical reactions, titrations can be used to quantitatively analyze solutions for their acid or base concentrations. In this section, we will explore the changes in the concentrations of the acidic and basic species present in a solution during the process of a titration.

Titration Curve

Previously, when we studied acid-base reactions in solution, we focused only on the point at which the acid and base were stoichiometrically equivalent. No consideration was given to the pH of the solution before, during, or after the neutralization.

Example 14.21

Calculating pH for Titration Solutions: Strong Acid/Strong Base

A titration is carried out for 25.00 mL of 0.100 M HCl (strong acid) with 0.100 M of a strong base NaOH the titration curve is shown in Figure 14.21. Calculate the pH at these volumes of added base solution:

(a) 0.00 mL

(b) 12.50 mL

(c) 25.00 mL

(d) 37.50 mL

Solution

Since HCl is a strong acid, we can assume that all of it dissociates. The initial concentration of H_3O^+ is $[H_3O^+]_0 = 0.100\ M$. When the base solution is added, it also dissociates completely, providing OH^- ions. The H_3O^+ and OH^- ions neutralize each other, so only those of the two that were in excess remain, and their concentration determines the pH. Thus, the solution is initially acidic (pH < 7), but eventually all the hydronium ions present from the original acid are neutralized, and the solution becomes neutral. As more base is added, the solution turns basic.

The total initial amount of the hydronium ions is:

$$n(H^+)_0 = [H_3O^+]_0 \times 0.02500\ L = 0.002500\ mol$$

Once X mL of the 0.100-M base solution is added, the number of moles of the OH^- ions introduced is:

$$n(OH^-)_0 = 0.100\ M \times X\ mL \times \left(\frac{1\ L}{1000\ mL}\right)$$

The total volume becomes: $V = (25.00\ mL + X\ mL)\left(\frac{1\ L}{1000\ mL}\right)$

The number of moles of H_3O^+ becomes:

$$n(H^+) = n(H^+)_0 - n(OH^-)_0 = 0.002500\ mol - 0.100\ M \times X\ mL \times \left(\frac{1\ L}{1000\ mL}\right)$$

The concentration of H_3O^+ is:

$$[H_3O^+] = \frac{n(H^+)}{V} = \frac{0.002500\ mol - 0.100\ M \times X\ mL \times \left(\frac{1\ L}{1000\ mL}\right)}{(25.00\ mL + X\ mL)\left(\frac{1\ L}{1000\ mL}\right)}$$

$$= \frac{0.002500\ mol \times \left(\frac{1000\ mL}{1\ L}\right) - 0.100\ M \times X\ mL}{25.00\ mL + X\ mL}$$

$$pH = -\log([H_3O^+])$$

The preceding calculations work if $n(H^+)_0 - n(OH^-)_0 > 0$ and so $n(H^+) > 0$. When $n(H^+)_0 = n(OH^-)_0$, the H_3O^+ ions from the acid and the OH^- ions from the base mutually neutralize. At this point, the only hydronium ions left are those from the autoionization of water, and there are no OH^- particles to neutralize them. Therefore, in this case:

$$[H_3O^+] = [OH^-],\ [H_3O^+] = K_w = 1.0 \times 10^{-14};\ [H_3O^+] = 1.0 \times 10^{-7}$$
$$pH = -\log(1.0 \times 10^{-7}) = 7.00$$

Finally, when $n(OH^-)_0 > n(H^+)_0$, there are not enough H_3O^+ ions to neutralize all the OH^- ions, and instead of $n(H^+) = n(H^+)_0 - n(OH^-)_0$, we calculate: $n(OH^-) = n(OH^-)_0 - n(H^+)_0$

In this case:

$$[OH^-] = \frac{n(OH^-)}{V} = \frac{0.100\ M \times X\ mL \times \left(\frac{1\ L}{1000\ mL}\right) - 0.002500\ mol}{(25.00\ mL + X\ mL)\left(\frac{1\ L}{1000\ mL}\right)}$$

$$= \frac{0.100\ M \times X\ mL - 0.002500\ mol \times \left(\frac{1000\ mL}{1\ L}\right)}{25.00\ mL + X\ mL}$$

$$pH = 14 - pOH = 14 + \log([OH^-])$$

Let us now consider the four specific cases presented in this problem:

(a) X = 0 mL

$$[H_3O^+] = \frac{n(H^+)}{V} = \frac{0.002500 \text{ mol} \times \left(\frac{1000 \text{ mL}}{1 \text{ L}}\right)}{25.00 \text{ mL}} = 0.1 \ M$$

$pH = -\log(0.100) = 1.000$

(b) X = 12.50 mL

$$[H_3O^+] = \frac{n(H^+)}{V} = \frac{0.002500 \text{ mol} \times \left(\frac{1000 \text{ mL}}{1 \text{ L}}\right) - 0.100 \ M \times 12.50 \text{ mL}}{25.00 \text{ mL} + 12.50 \text{ mL}} = 0.0333 \ M$$

$pH = -\log(0.0333) = 1.477$

(c) X = 25.00 mL

Since the volumes and concentrations of the acid and base solutions are the same: $n(H^+)_0 = n(OH^-)_0$, and pH = 7.000, as described earlier.

(d) X = 37.50 mL

In this case:

$$n(OH^-)_0 > n(H^+)_0$$

$$[OH^-] = \frac{n(OH^-)}{V} = \frac{0.100 \ M \times 35.70 \text{ mL} - 0.002500 \text{ mol} \times \left(\frac{1000 \text{ mL}}{1 \text{ L}}\right)}{25.00 \text{ mL} + 37.50 \text{ mL}} = 0.0200 \ M$$

$pH = 14 - pOH = 14 + \log([OH^-]) = 14 + \log(0.0200) = 12.30$

Check Your Learning

Calculate the pH for the strong acid/strong base titration between 50.0 mL of 0.100 M $HNO_3(aq)$ and 0.200 M NaOH (titrant) at the listed volumes of added base: 0.00 mL, 15.0 mL, 25.0 mL, and 40.0 mL.

Answer: 0.00: 1.000; 15.0: 1.5111; 25.0: 7; 40.0: 12.523

In the example, we calculated pH at four points during a titration. Table 14.4 shows a detailed sequence of changes in the pH of a strong acid and a weak acid in a titration with NaOH.

pH Values in the Titrations of a Strong Acid with a Strong Base and of a Weak Acid with a Strong Base

Volume of 0.100 M NaOH Added (mL)	Moles of NaOH Added	pH Values 0.100 M HCl[1]	pH Values 0.100 M CH_3CO_2H[2]
0.0	0.0	1.00	2.87
5.0	0.00050	1.18	4.14
10.0	0.00100	1.37	4.57
15.0	0.00150	1.60	4.92
20.0	0.00200	1.95	5.35
22.0	0.00220	2.20	5.61
24.0	0.00240	2.69	6.13
24.5	0.00245	3.00	6.44
24.9	0.00249	3.70	7.14

Table 14.4

---

1. Titration of 25.00 mL of 0.100 M HCl (0.00250 mol of HCl) with 0.100 M NaOH.
2. Titration of 25.00 mL of 0.100 M CH_3CO_2H (0.00250 mol of CH_3CO_2H) with 0.100 M NaOH.

pH Values in the Titrations of a Strong Acid with a Strong Base and of a Weak Acid with a Strong Base

Volume of 0.100 *M* NaOH Added (mL)	Moles of NaOH Added	pH Values 0.100 *M* HCl[3]	pH Values 0.100 *M* CH₃CO₂H[4]
25.0	0.00250	7.00	8.72
25.1	0.00251	10.30	10.30
25.5	0.00255	11.00	11.00
26.0	0.00260	11.29	11.29
28.0	0.00280	11.75	11.75
30.0	0.00300	11.96	11.96
35.0	0.00350	12.22	12.22
40.0	0.00400	12.36	12.36
45.0	0.00450	12.46	12.46
50.0	0.00500	12.52	12.52

Table 14.4

The simplest acid-base reactions are those of a strong acid with a strong base. Table 14.4 shows data for the titration of a 25.0-mL sample of 0.100 *M* hydrochloric acid with 0.100 *M* sodium hydroxide. The values of the pH measured after successive additions of small amounts of NaOH are listed in the first column of this table, and are graphed in Figure 14.21, in a form that is called a **titration curve**. The pH increases slowly at first, increases rapidly in the middle portion of the curve, and then increases slowly again. The point of inflection (located at the midpoint of the vertical part of the curve) is the equivalence point for the titration. It indicates when equivalent quantities of acid and base are present. For the titration of a strong acid with a strong base, the equivalence point occurs at a pH of 7.00 and the points on the titration curve can be calculated using solution stoichiometry (Table 14.4 and Figure 14.21).

---

3. Titration of 25.00 mL of 0.100 *M* HCl (0.00250 mol of HCl) with 0.100 *M* NaOH.
4. Titration of 25.00 mL of 0.100 *M* CH₃CO₂H (0.00250 mol of CH₃CO₂H) with 0.100 *M* NaOH.

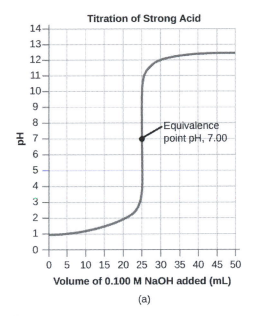
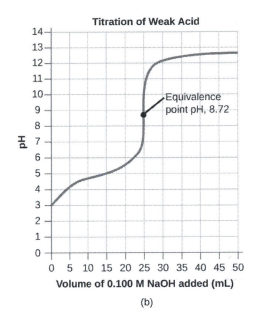

Figure 14.21 (a) The titration curve for the titration of 25.00 mL of 0.100 M CH₃CO₂H (weak acid) with 0.100 M NaOH (strong base) has an equivalence point of 7.00 pH. (b) The titration curve for the titration of 25.00 mL of 0.100 M HCl (strong acid) with 0.100 M NaOH (strong base) has an equivalence point of 8.72 pH.

The titration of a weak acid with a strong base (or of a weak base with a strong acid) is somewhat more complicated than that just discussed, but it follows the same general principles. Let us consider the titration of 25.0 mL of 0.100 M acetic acid (a weak acid) with 0.100 M sodium hydroxide and compare the titration curve with that of the strong acid. Table 14.4 gives the pH values during the titration, Figure 14.21 shows the titration curve.

Although the initial volume and molarity of the acids are the same, there are important differences between the two titration curves. The titration curve for the weak acid begins at a higher value (less acidic) and maintains higher pH values up to the equivalence point. This is because acetic acid is a weak acid, which is only partially ionized. The pH at the equivalence point is also higher (8.72 rather than 7.00) due to the hydrolysis of acetate, a weak base that raises the pH:

$$CH_3CO_2^-(aq) + H_2O(l) \rightleftharpoons CH_3CO_2H(l) + OH^-(aq)$$

After the equivalence point, the two curves are identical because the pH is dependent on the excess of hydroxide ion in both cases.

Example 14.22

Titration of a Weak Acid with a Strong Base

The titration curve shown in Figure 14.23 is for the titration of 25.00 mL of 0.100 M CH₃CO₂H with 0.100 M NaOH. The reaction can be represented as:

$$CH_3CO_2H + OH^- \longrightarrow CH_3CO_2^- + H_2O$$

(a) What is the initial pH before any amount of the NaOH solution has been added? $K_a = 1.8 \times 10^{-5}$ for CH₃CO₂H.

(b) Find the pH after 25.00 mL of the NaOH solution have been added.

(c) Find the pH after 12.50 mL of the NaOH solution has been added.

(d) Find the pH after 37.50 mL of the NaOH solution has been added.

Solution

(a) Assuming that the dissociated amount is small compared to 0.100 M, we find that:

$$K_a = \frac{[H_3O^+][CH_3CO_2{}^-]}{[CH_3CO_2H]} \approx \frac{[H_3O^+]^2}{[CH_3CO_2H]_0}, \qquad \text{and}$$

$$[H_3O^+] = \sqrt{K_a \times [CH_3CO_2H]} = \sqrt{1.8 \times 10^{-5} \times 0.100} = 1.3 \times 10^{-3}$$

$$pH = -\log(1.3 \times 10^{-3}) = 2.87$$

(b) After 25.00 mL of NaOH are added, the number of moles of NaOH and CH_3CO_2H are equal because the amounts of the solutions and their concentrations are the same. All of the CH_3CO_2H has been converted to $CH_3CO_2{}^-$. The concentration of the $CH_3CO_2{}^-$ ion is:

$$\frac{0.00250 \text{ mol}}{0.0500 \text{ L}} = 0.0500 \text{ M} CH_3CO_2{}^-$$

The equilibrium that must be focused on now is the basicity equilibrium for $CH_3CO_2{}^-$:

$$CH_3CO_2{}^-(aq) + H_2O(l) \rightleftharpoons CH_3CO_2H(aq) + OH^-(aq)$$

so we must determine K_b for the base by using the ion product constant for water:

$$K_b = \frac{[CH_3CO_2H][OH^-]}{[CH_3CO_2{}^-]}$$

$$K_a = \frac{[CH_3CO_2{}^-][H^+]}{[CH_3CO_2H]}, \quad \text{so} \quad \frac{[CH_3CO_2H]}{[CH_3CO_2{}^-]} = \frac{[H^+]}{K_a}.$$

Since $K_w = [H^+][OH^-]$:

$$K_b = \frac{[H^+][OH^-]}{K_a} = \frac{K_w}{K_a} = \frac{1.0 \times 10^{-14}}{1.8 \times 10^{-5}} = 5.6 \times 10^{-10}$$

Let us denote the concentration of each of the products of this reaction, CH_3CO_2H and OH^-, as x. Using the assumption that x is small compared to 0.0500 M, $K_b = \frac{x^2}{0.0500 \, M}$, and then:

$$x = [OH^-] = 5.3 \times 10^{-6}$$

$$pOH = -\log(5.3 \times 10^{-6}) = 5.28$$

$$pH = 14.00 - 5.28 = 8.72$$

Note that the pH at the equivalence point of this titration is significantly greater than 7.

(c) In (a), 25.00 mL of the NaOH solution was added, and so practically all the CH_3CO_2H was converted into $CH_3CO_2{}^-$. In this case, only 12.50 mL of the base solution has been introduced, and so only half of all the CH_3CO_2H is converted into $CH_3CO_2{}^-$. The total initial number of moles of CH_3CO_2H is 0.02500L $\times$ 0.100 M = 0.00250 mol, and so after adding the NaOH, the numbers of moles of CH_3CO_2H and $CH_3CO_2{}^-$ are both approximately equal to $\frac{0.00250 \text{ mol}}{2} = 0.00125$ mol, and their concentrations are the same.

Since the amount of the added base is smaller than the original amount of the acid, the equivalence point has not been reached, the solution remains a buffer, and we can use the Henderson-Hasselbalch equation:

$$pH = pK_a + \log\frac{[Base]}{[Acid]} = -\log(K_a) + \log\frac{[CH_3CO_2{}^-]}{[CH_3CO_2H]} = -\log(1.8 \times 10^{-5}) + \log(1)$$

(as the concentrations of $CH_3CO_2^-$ and CH_3CO_2H are the same)

Thus:

$$pH = -\log(1.8 \times 10^{-5}) = 4.74$$

(the pH = the pK_a at the halfway point in a titration of a weak acid)

(d) After 37.50 mL of NaOH is added, the amount of NaOH is 0.03750 L $\times$ 0.100 M = 0.003750 mol NaOH. Since this is past the equivalence point, the excess hydroxide ions will make the solution basic, and we can again use stoichiometric calculations to determine the pH:

$$[OH^-] = \frac{(0.003750 \text{ mol} - 0.00250 \text{ mol})}{0.06250 \text{ L}} = 2.00 \times 10^{-2} M$$

So:

$$pOH = -\log(2.00 \times 10^{-2}) = 1.70, \text{ and } pH = 14.00 - 1.70 = 12.30$$

Note that this result is the same as for the strong acid-strong base titration example provided, since the amount of the strong base added moves the solution past the equivalence point.

Check Your Learning

Calculate the pH for the weak acid/strong base titration between 50.0 mL of 0.100 M HCOOH(aq) (formic acid) and 0.200 M NaOH (titrant) at the listed volumes of added base: 0.00 mL, 15.0 mL, 25.0 mL, and 30.0 mL.

Answer: 0.00 mL: 2.37; 15.0 mL: 3.92; 25.00 mL: 8.29; 30.0 mL: 12.097

Acid-Base Indicators

Certain organic substances change color in dilute solution when the hydronium ion concentration reaches a particular value. For example, phenolphthalein is a colorless substance in any aqueous solution with a hydronium ion concentration greater than 5.0 $\times$ 10^{-9} M (pH < 8.3). In more basic solutions where the hydronium ion concentration is less than 5.0 $\times$ 10^{-9} M (pH > 8.3), it is red or pink. Substances such as phenolphthalein, which can be used to determine the pH of a solution, are called **acid-base indicators**. Acid-base indicators are either weak organic acids or weak organic bases.

The equilibrium in a solution of the acid-base indicator methyl orange, a weak acid, can be represented by an equation in which we use HIn as a simple representation for the complex methyl orange molecule:

$$HIn(aq) + H_2O(l) \ \rightleftharpoons \ H_3O^+(aq) + In^-(aq)$$
$$\text{red} \qquad\qquad\qquad\qquad\qquad \text{yellow}$$

$$K_a = \frac{[H_3O^+][In^-]}{[HIn]} = 4.0 \times 10^{-4}$$

The anion of methyl orange, In^-, is yellow, and the nonionized form, HIn, is red. When we add acid to a solution of methyl orange, the increased hydronium ion concentration shifts the equilibrium toward the nonionized red form, in accordance with Le Châtelier's principle. If we add base, we shift the equilibrium towards the yellow form. This behavior is completely analogous to the action of buffers.

An indicator's color is the visible result of the ratio of the concentrations of the two species In^- and HIn. If most of the indicator (typically about 60−90% or more) is present as In^-, then we see the color of the In^- ion, which would be yellow for methyl orange. If most is present as HIn, then we see the color of the HIn molecule: red for methyl orange. For methyl orange, we can rearrange the equation for K_a and write:

$$\frac{[In^-]}{[HIn]} = \frac{[\text{substance with yellow color}]}{[\text{substance with red color}]} = \frac{K_a}{[H_3O^+]}$$

This shows us how the ratio of $\frac{[In^-]}{[HIn]}$ varies with the concentration of hydronium ion.

The above expression describing the indicator equilibrium can be rearranged:

$$\frac{[H_3O^+]}{K_a} = \frac{[HIn]}{[In^-]}$$

$$\log\left(\frac{[H_3O^+]}{K_a}\right) = \log\left(\frac{[HIn]}{[In^-]}\right)$$

$$\log([H_3O^+]) - \log(K_a) = -\log\left(\frac{[In^-]}{[HIn]}\right)$$

$$-pH + pK_a = -\log\left(\frac{[In^-]}{[HIn]}\right)$$

$$pH = pKa + \log\left(\frac{[In^-]}{[HIn]}\right) \text{ or } pH = pK_a + \log\left(\frac{[base]}{[acid]}\right)$$

The last formula is the same as the Henderson-Hasselbalch equation, which can be used to describe the equilibrium of indicators.

When $[H_3O^+]$ has the same numerical value as K_a, the ratio of $[In^-]$ to $[HIn]$ is equal to 1, meaning that 50% of the indicator is present in the red form (HIn) and 50% is in the yellow ionic form (In^-), and the solution appears orange in color. When the hydronium ion concentration increases to 8×10^{-4} M (a pH of 3.1), the solution turns red. No change in color is visible for any further increase in the hydronium ion concentration (decrease in pH). At a hydronium ion concentration of 4×10^{-5} M (a pH of 4.4), most of the indicator is in the yellow ionic form, and a further decrease in the hydronium ion concentration (increase in pH) does not produce a visible color change. The pH range between 3.1 (red) and 4.4 (yellow) is the **color-change interval** of methyl orange; the pronounced color change takes place between these pH values.

There are many different acid-base indicators that cover a wide range of pH values and can be used to determine the approximate pH of an unknown solution by a process of elimination. Universal indicators and pH paper contain a mixture of indicators and exhibit different colors at different pHs. Figure 14.22 presents several indicators, their colors, and their color-change intervals.

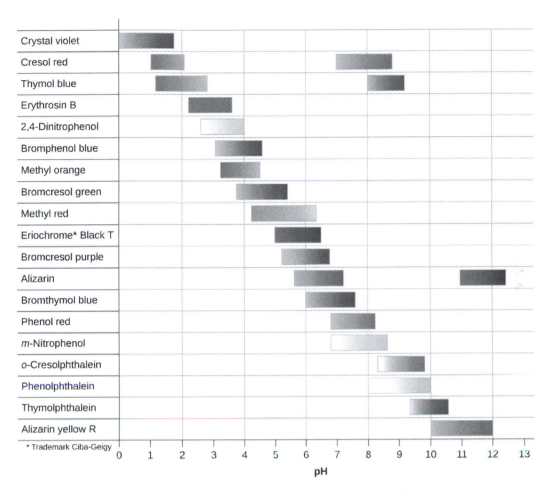

Figure 14.22   This chart illustrates the ranges of color change for several acid-base indicators.

Titration curves help us pick an indicator that will provide a sharp color change at the equivalence point. The best selection would be an indicator that has a color change interval that brackets the pH at the equivalence point of the titration.

The color change intervals of three indicators are shown in Figure 14.23. The equivalence points of both the titration of the strong acid and of the weak acid are located in the color-change interval of phenolphthalein. We can use it for titrations of either strong acid with strong base or weak acid with strong base.

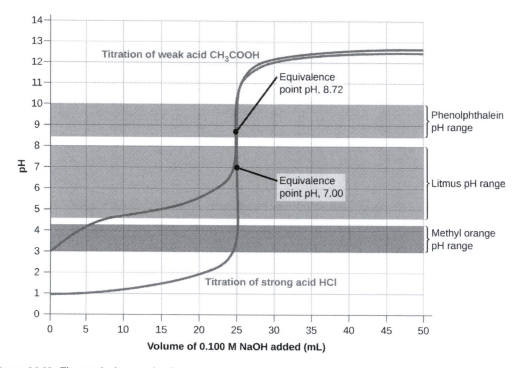

Figure 14.23 The graph shows a titration curve for the titration of 25.00 mL of 0.100 M CH_3CO_2H (weak acid) with 0.100 M NaOH (strong base) and the titration curve for the titration of HCl (strong acid) with NaOH (strong base). The pH ranges for the color change of phenolphthalein, litmus, and methyl orange are indicated by the shaded areas.

Litmus is a suitable indicator for the HCl titration because its color change brackets the equivalence point. However, we should not use litmus for the CH_3CO_2H titration because the pH is within the color-change interval of litmus when only about 12 mL of NaOH has been added, and it does not leave the range until 25 mL has been added. The color change would be very gradual, taking place during the addition of 13 mL of NaOH, making litmus useless as an indicator of the equivalence point.

We could use methyl orange for the HCl titration, but it would not give very accurate results: (1) It completes its color change slightly before the equivalence point is reached (but very close to it, so this is not too serious); (2) it changes color, as Figure 14.23 shows, during the addition of nearly 0.5 mL of NaOH, which is not so sharp a color change as that of litmus or phenolphthalein; and (3) it goes from yellow to orange to red, making detection of a precise endpoint much more challenging than the colorless to pink change of phenolphthalein. Figure 14.23 shows us that methyl orange would be completely useless as an indicator for the CH_3CO_2H titration. Its color change begins after about 1 mL of NaOH has been added and ends when about 8 mL has been added. The color change is completed long before the equivalence point (which occurs when 25.0 mL of NaOH has been added) is reached and hence provides no indication of the equivalence point.

We base our choice of indicator on a calculated pH, the pH at the equivalence point. At the equivalence point, equimolar amounts of acid and base have been mixed, and the calculation becomes that of the pH of a solution of the salt resulting from the titration.

Key Terms

acid ionization reaction involving the transfer of a proton from an acid to water, yielding hydronium ions and the conjugate base of the acid

acid ionization constant (K_a) equilibrium constant for the ionization of a weak acid

acid-base indicator organic acid or base whose color changes depending on the pH of the solution it is in

acidic describes a solution in which $[H_3O^+] > [OH^-]$

amphiprotic species that may either gain or lose a proton in a reaction

amphoteric species that can act as either an acid or a base

autoionization reaction between identical species yielding ionic products; for water, this reaction involves transfer of protons to yield hydronium and hydroxide ions

base ionization reaction involving the transfer of a proton from water to a base, yielding hydroxide ions and the conjugate acid of the base

base ionization constant (K_b) equilibrium constant for the ionization of a weak base

basic describes a solution in which $[H_3O^+] < [OH^-]$

Brønsted-Lowry acid proton donor

Brønsted-Lowry base proton acceptor

buffer mixture of a weak acid or a weak base and the salt of its conjugate; the pH of a buffer resists change when small amounts of acid or base are added

buffer capacity amount of an acid or base that can be added to a volume of a buffer solution before its pH changes significantly (usually by one pH unit)

color-change interval range in pH over which the color change of an indicator takes place

conjugate acid substance formed when a base gains a proton

conjugate base substance formed when an acid loses a proton

diprotic acid acid containing two ionizable hydrogen atoms per molecule. A diprotic acid ionizes in two steps

diprotic base base capable of accepting two protons. The protons are accepted in two steps

Henderson-Hasselbalch equation equation used to calculate the pH of buffer solutions

ion-product constant for water (K_w) equilibrium constant for the autoionization of water

leveling effect of water any acid stronger than H_3O^+, or any base stronger than OH^- will react with water to form H_3O^+, or OH^-, respectively; water acts as a base to make all strong acids appear equally strong, and it acts as an acid to make all strong bases appear equally strong

monoprotic acid acid containing one ionizable hydrogen atom per molecule

neutral describes a solution in which $[H_3O^+] = [OH^-]$

oxyacid compound containing a nonmetal and one or more hydroxyl groups

percent ionization ratio of the concentration of the ionized acid to the initial acid concentration, times 100

pH logarithmic measure of the concentration of hydronium ions in a solution

pOH logarithmic measure of the concentration of hydroxide ions in a solution

stepwise ionization process in which an acid is ionized by losing protons sequentially

titration curve plot of the pH of a solution of acid or base versus the volume of base or acid added during a titration

triprotic acid acid that contains three ionizable hydrogen atoms per molecule; ionization of triprotic acids occurs in three steps

Key Equations

- $K_w = [H_3O^+][OH^-] = 1.0 \times 10^{-14}$ (at 25 °C)
- $pH = -\log[H_3O^+]$
- $pOH = -\log[OH^-]$
- $[H_3O^+] = 10^{-pH}$
- $[OH^-] = 10^{-pOH}$
- $pH + pOH = pK_w = 14.00$ at 25 °C
- $K_a = \dfrac{[H_3O^+][A^-]}{[HA]}$
- $K_b = \dfrac{[HB^+][OH^-]}{[B]}$
- $K_a \times K_b = 1.0 \times 10^{-14} = K_w$
- Percent ionization $= \dfrac{[H_3O^+]_{eq}}{[HA]_0} \times 100$
- $pK_a = -\log K_a$
- $pK_b = -\log K_b$
- $pH = pK_a + \log \dfrac{[A^-]}{[HA]}$

Summary

14.1 Brønsted-Lowry Acids and Bases

A compound that can donate a proton (a hydrogen ion) to another compound is called a Brønsted-Lowry acid. The compound that accepts the proton is called a Brønsted-Lowry base. The species remaining after a Brønsted-Lowry acid has lost a proton is the conjugate base of the acid. The species formed when a Brønsted-Lowry base gains a proton is the conjugate acid of the base. Thus, an acid-base reaction occurs when a proton is transferred from an acid to a base, with formation of the conjugate base of the reactant acid and formation of the conjugate acid of the reactant base. Amphiprotic species can act as both proton donors and proton acceptors. Water is the most important amphiprotic species. It can form both the hydronium ion, H_3O^+, and the hydroxide ion, OH^- when it undergoes autoionization:

$$2H_2O(l) \rightleftharpoons H_3O^+(aq) + OH^-(aq)$$

The ion product of water, K_w is the equilibrium constant for the autoionization reaction:

$$K_w = [H_2O^+][OH^-] = 1.0 \times 10^{-14} \text{ at } 25 \text{ °C}$$

14.2 pH and pOH

The concentration of hydronium ion in a solution of an acid in water is greater than 1.0×10^{-7} M at 25 °C. The concentration of hydroxide ion in a solution of a base in water is greater than 1.0×10^{-7} M at 25 °C. The concentration of H_3O^+ in a solution can be expressed as the pH of the solution; pH = $-\log H_3O^+$. The concentration of OH^- can be expressed as the pOH of the solution: pOH = $-\log[OH^-]$. In pure water, pH = 7.00 and pOH = 7.00

14.3 Relative Strengths of Acids and Bases

The strengths of Brønsted-Lowry acids and bases in aqueous solutions can be determined by their acid or base ionization constants. Stronger acids form weaker conjugate bases, and weaker acids form stronger conjugate bases. Thus strong acids are completely ionized in aqueous solution because their conjugate bases are weaker bases than water. Weak acids are only partially ionized because their conjugate bases are strong enough to compete successfully with water for possession of protons. Strong bases react with water to quantitatively form hydroxide ions. Weak bases give only small amounts of hydroxide ion. The strengths of the binary acids increase from left to right across a period of the periodic table ($CH_4 < NH_3 < H_2O < HF$), and they increase down a group (HF < HCl < HBr < HI). The strengths of oxyacids that contain the same central element increase as the oxidation number of the element increases ($H_2SO_3 < H_2SO_4$). The strengths of oxyacids also increase as the electronegativity of the central element increases [$H_2SeO_4 < H_2SO_4$].

14.4 Hydrolysis of Salt Solutions

The characteristic properties of aqueous solutions of Brønsted-Lowry acids are due to the presence of hydronium ions; those of aqueous solutions of Brønsted-Lowry bases are due to the presence of hydroxide ions. The neutralization that occurs when aqueous solutions of acids and bases are combined results from the reaction of the hydronium and hydroxide ions to form water. Some salts formed in neutralization reactions may make the product solutions slightly acidic or slightly basic.

Solutions that contain salts or hydrated metal ions have a pH that is determined by the extent of the hydrolysis of the ions in the solution. The pH of the solutions may be calculated using familiar equilibrium techniques, or it may be qualitatively determined to be acidic, basic, or neutral depending on the relative K_a and K_b of the ions involved.

14.5 Polyprotic Acids

An acid that contains more than one ionizable proton is a polyprotic acid. The protons of these acids ionize in steps. The differences in the acid ionization constants for the successive ionizations of the protons in a polyprotic acid usually vary by roughly five orders of magnitude. As long as the difference between the successive values of K_a of the acid is greater than about a factor of 20, it is appropriate to break down the calculations of the concentrations of the ions in solution into a series of steps.

14.6 Buffers

A solution containing a mixture of an acid and its conjugate base, or of a base and its conjugate acid, is called a buffer solution. Unlike in the case of an acid, base, or salt solution, the hydronium ion concentration of a buffer solution does not change greatly when a small amount of acid or base is added to the buffer solution. The base (or acid) in the buffer reacts with the added acid (or base).

14.7 Acid-Base Titrations

A titration curve is a graph that relates the change in pH of an acidic or basic solution to the volume of added titrant. The characteristics of the titration curve are dependent on the specific solutions being titrated. The pH of the solution at the equivalence point may be greater than, equal to, or less than 7.00. The choice of an indicator for a given titration depends on the expected pH at the equivalence point of the titration, and the range of the color change of the indicator.

Exercises

14.1 Brønsted-Lowry Acids and Bases

1. Write equations that show NH_3 as both a conjugate acid and a conjugate base.

2. Write equations that show $H_2PO_4{}^-$ acting both as an acid and as a base.

3. Show by suitable net ionic equations that each of the following species can act as a Brønsted-Lowry acid:

(a) H_3O^+

(b) HCl

(c) NH_3

(d) CH_3CO_2H

(e) $NH_4{}^+$

(f) $HSO_4{}^-$

4. Show by suitable net ionic equations that each of the following species can act as a Brønsted-Lowry acid:

(a) HNO_3

(b) $PH_4{}^+$

(c) H_2S

(d) CH_3CH_2COOH

(e) $H_2PO_4{}^-$

(f) HS^-

5. Show by suitable net ionic equations that each of the following species can act as a Brønsted-Lowry base:

(a) H_2O

(b) OH^-

(c) NH_3

(d) CN^-

(e) S^{2-}

(f) $H_2PO_4{}^-$

6. Show by suitable net ionic equations that each of the following species can act as a Brønsted-Lowry base:

(a) HS^-

(b) $PO_4{}^{3-}$

(c) $NH_2{}^-$

(d) C_2H_5OH

(e) O^{2-}

(f) $H_2PO_4{}^-$

7. What is the conjugate acid of each of the following? What is the conjugate base of each?

(a) OH^-

(b) H_2O

(c) $HCO_3{}^-$

(d) NH_3

(e) HSO_4^-

(f) H_2O_2

(g) HS^-

(h) $H_5N_2^+$

8. What is the conjugate acid of each of the following? What is the conjugate base of each?

(a) H_2S

(b) $H_2PO_4^-$

(c) PH_3

(d) HS^-

(e) HSO_3^-

(f) $H_3O_2^+$

(g) H_4N_2

(h) CH_3OH

9. Identify and label the Brønsted-Lowry acid, its conjugate base, the Brønsted-Lowry base, and its conjugate acid in each of the following equations:

(a) $HNO_3 + H_2O \longrightarrow H_3O^+ + NO_3^-$

(b) $CN^- + H_2O \longrightarrow HCN + OH^-$

(c) $H_2SO_4 + Cl^- \longrightarrow HCl + HSO_4^-$

(d) $HSO_4^- + OH^- \longrightarrow SO_4^{2-} + H_2O$

(e) $O^{2-} + H_2O \longrightarrow 2OH^-$

(f) $[Cu(H_2O)_3(OH)]^+ + [Al(H_2O)_6]^{3+} \longrightarrow [Cu(H_2O)_4]^{2+} + [Al(H_2O)_5(OH)]^{2+}$

(g) $H_2S + NH_2^- \longrightarrow HS^- + NH_3$

10. Identify and label the Brønsted-Lowry acid, its conjugate base, the Brønsted-Lowry base, and its conjugate acid in each of the following equations:

(a) $NO_2^- + H_2O \longrightarrow HNO_2 + OH^-$

(b) $HBr + H_2O \longrightarrow H_3O^+ + Br^-$

(c) $HS^- + H_2O \longrightarrow H_2S + OH^-$

(d) $H_2PO_4^- + OH^- \longrightarrow HPO_4^{2-} + H_2O$

(e) $H_2PO_4^- + HCl \longrightarrow H_3PO_4 + Cl^-$

(f) $[Fe(H_2O)_5(OH)]^{2+} + [Al(H_2O)_6]^{3+} \longrightarrow [Fe(H_2O)_6]^{3+} + [Al(H_2O)_5(OH)]^{2+}$

(g) $CH_3OH + H^- \longrightarrow CH_3O^- + H_2$

11. What are amphiprotic species? Illustrate with suitable equations.

12. State which of the following species are amphiprotic and write chemical equations illustrating the amphiprotic character of these species:

(a) H_2O

(b) $H_2PO_4^-$

(c) S^{2-}

(d) $CO_3{}^{2-}$

(e) $HSO_4{}^-$

13. State which of the following species are amphiprotic and write chemical equations illustrating the amphiprotic character of these species.

(a) NH_3

(b) $HPO_4{}^-$

(c) Br^-

(d) $NH_4{}^+$

(e) $ASO_4{}^{3-}$

14. Is the self ionization of water endothermic or exothermic? The ionization constant for water (K_w) is 2.9 × 10^{-14} at 40 °C and 9.6 × 10^{-14} at 60 °C.

14.2 pH and pOH

15. Explain why a sample of pure water at 40 °C is neutral even though $[H_3O^+]$ = 1.7 × 10^{-7} M. K_w is 2.9 × 10^{-14} at 40 °C.

16. The ionization constant for water (K_w) is 2.9 × 10^{-14} at 40 °C. Calculate $[H_3O^+]$, $[OH^-]$, pH, and pOH for pure water at 40 °C.

17. The ionization constant for water (K_w) is 9.614 × 10^{-14} at 60 °C. Calculate $[H_3O^+]$, $[OH^-]$, pH, and pOH for pure water at 60 °C.

18. Calculate the pH and the pOH of each of the following solutions at 25 °C for which the substances ionize completely:

(a) 0.200 *M* HCl

(b) 0.0143 *M* NaOH

(c) 3.0 *M* HNO_3

(d) 0.0031 *M* $Ca(OH)_2$

19. Calculate the pH and the pOH of each of the following solutions at 25 °C for which the substances ionize completely:

(a) 0.000259 *M* $HClO_4$

(b) 0.21 *M* NaOH

(c) 0.000071 *M* $Ba(OH)_2$

(d) 2.5 *M* KOH

20. What are the pH and pOH of a solution of 2.0 M HCl, which ionizes completely?

21. What are the hydronium and hydroxide ion concentrations in a solution whose pH is 6.52?

22. Calculate the hydrogen ion concentration and the hydroxide ion concentration in wine from its pH. See Figure 14.2 for useful information.

23. Calculate the hydronium ion concentration and the hydroxide ion concentration in lime juice from its pH. See Figure 14.2 for useful information.

24. The hydronium ion concentration in a sample of rainwater is found to be 1.7 × 10^{-6} M at 25 °C. What is the concentration of hydroxide ions in the rainwater?

25. The hydroxide ion concentration in household ammonia is 3.2 × 10^{-3} M at 25 °C. What is the concentration of hydronium ions in the solution?

14.3 Relative Strengths of Acids and Bases

26. Explain why the neutralization reaction of a strong acid and a weak base gives a weakly acidic solution.

27. Explain why the neutralization reaction of a weak acid and a strong base gives a weakly basic solution.

28. Use this list of important industrial compounds (and Figure 14.8) to answer the following questions regarding: CaO, $Ca(OH)_2$, CH_3CO_2H, CO_2, HCl, H_2CO_3, HF, HNO_2, HNO_3, H_3PO_4, H_2SO_4, NH_3, $NaOH$, Na_2CO_3.

(a) Identify the strong Brønsted-Lowry acids and strong Brønsted-Lowry bases.

(b) List those compounds in (a) that can behave as Brønsted-Lowry acids with strengths lying between those of H_3O^+ and H_2O.

(c) List those compounds in (a) that can behave as Brønsted-Lowry bases with strengths lying between those of H_2O and OH^-.

29. The odor of vinegar is due to the presence of acetic acid, CH_3CO_2H, a weak acid. List, in order of descending concentration, all of the ionic and molecular species present in a 1-M aqueous solution of this acid.

30. Household ammonia is a solution of the weak base NH_3 in water. List, in order of descending concentration, all of the ionic and molecular species present in a 1-M aqueous solution of this base.

31. Explain why the ionization constant, K_a, for H_2SO_4 is larger than the ionization constant for H_2SO_3.

32. Explain why the ionization constant, K_a, for HI is larger than the ionization constant for HF.

33. Gastric juice, the digestive fluid produced in the stomach, contains hydrochloric acid, HCl. Milk of Magnesia, a suspension of solid $Mg(OH)_2$ in an aqueous medium, is sometimes used to neutralize excess stomach acid. Write a complete balanced equation for the neutralization reaction, and identify the conjugate acid-base pairs.

34. Nitric acid reacts with insoluble copper(II) oxide to form soluble copper(II) nitrate, $Cu(NO_3)_2$, a compound that has been used to prevent the growth of algae in swimming pools. Write the balanced chemical equation for the reaction of an aqueous solution of HNO_3 with CuO.

35. What is the ionization constant at 25 °C for the weak acid $CH_3NH_3{}^+$, the conjugate acid of the weak base CH_3NH_2, $K_b = 4.4 \times 10^{-4}$.

36. What is the ionization constant at 25 °C for the weak acid $(CH_3)_2NH_2{}^+$, the conjugate acid of the weak base $(CH_3)_2NH$, $K_b = 7.4 \times 10^{-4}$?

37. Which base, CH_3NH_2 or $(CH_3)_2NH$, is the strongest base? Which conjugate acid, $(CH_3)_2NH_2{}^+$ or $(CH_3)_2NH$, is the strongest acid?

38. Which is the stronger acid, $NH_4{}^+$ or HBrO?

39. Which is the stronger base, $(CH_3)_3N$ or $H_2BO_3{}^-$?

40. Predict which acid in each of the following pairs is the stronger and explain your reasoning for each.

(a) H_2O or HF

(b) $B(OH)_3$ or $Al(OH)_3$

(c) $HSO_3{}^-$ or $HSO_4{}^-$

(d) NH_3 or H_2S

(e) H_2O or H_2Te

41. Predict which compound in each of the following pairs of compounds is more acidic and explain your reasoning for each.

(a) $HSO_4{}^-$ or $HSeO_4{}^-$

(b) NH_3 or H_2O

(c) PH_3 or HI

(d) NH_3 or PH_3

(e) H_2S or HBr

42. Rank the compounds in each of the following groups in order of increasing acidity or basicity, as indicated, and explain the order you assign.

(a) acidity: HCl, HBr, HI

(b) basicity: H_2O, OH^-, H^-, Cl^-

(c) basicity: $Mg(OH)_2$, $Si(OH)_4$, $ClO_3(OH)$ (Hint: Formula could also be written as $HClO_4$).

(d) acidity: HF, H_2O, NH_3, CH_4

43. Rank the compounds in each of the following groups in order of increasing acidity or basicity, as indicated, and explain the order you assign.

(a) acidity: $NaHSO_3$, $NaHSeO_3$, $NaHSO_4$

(b) basicity: BrO_2^-, ClO_2^-, IO_2^-

(c) acidity: HOCl, HOBr, HOI

(d) acidity: HOCl, HOClO, $HOClO_2$, $HOClO_3$

(e) basicity: NH_2^-, HS^-, HTe^-, PH_2^-

(f) basicity: BrO^-, BrO_2^-, BrO_3^-, BrO_4^-

44. Both HF and HCN ionize in water to a limited extent. Which of the conjugate bases, F^- or CN^-, is the stronger base? See Table 14.3.

45. The active ingredient formed by aspirin in the body is salicylic acid, $C_6H_4OH(CO_2H)$. The carboxyl group ($-CO_2H$) acts as a weak acid. The phenol group (an OH group bonded to an aromatic ring) also acts as an acid but a much weaker acid. List, in order of descending concentration, all of the ionic and molecular species present in a 0.001-M aqueous solution of $C_6H_4OH(CO_2H)$.

46. What do we represent when we write:
$$CH_3CO_2H(aq) + H_2O(l) \rightleftharpoons H_3O^+(aq) + CH_3CO_2^-(aq)?$$

47. Explain why equilibrium calculations are not necessary to determine ionic concentrations in solutions of certain strong electrolytes such as NaOH and HCl. Under what conditions are equilibrium calculations necessary as part of the determination of the concentrations of all ions of some other strong electrolytes in solution?

48. Are the concentrations of hydronium ion and hydroxide ion in a solution of an acid or a base in water directly proportional or inversely proportional? Explain your answer.

49. What two common assumptions can simplify calculation of equilibrium concentrations in a solution of a weak acid?

50. What two common assumptions can simplify calculation of equilibrium concentrations in a solution of a weak base?

51. Which of the following will increase the percent of NH_3 that is converted to the ammonium ion in water (Hint: Use LeChâtelier's principle.)?

(a) addition of NaOH

(b) addition of HCl

(c) addition of NH_4Cl

52. Which of the following will increase the percent of HF that is converted to the fluoride ion in water?

(a) addition of NaOH

(b) addition of HCl

(c) addition of NaF

53. What is the effect on the concentrations of NO_2^-, HNO_2, and OH^- when the following are added to a solution of KNO_2 in water:

(a) HCl

(b) HNO_2

(c) NaOH

(d) NaCl

(e) KNO

The equation for the equilibrium is:
$$NO_2^-(aq) + H_2O(l) \rightleftharpoons HNO_2(aq) + OH^-(aq)$$

54. What is the effect on the concentration of hydrofluoric acid, hydronium ion, and fluoride ion when the following are added to separate solutions of hydrofluoric acid?

(a) HCl

(b) KF

(c) NaCl

(d) KOH

(e) HF

The equation for the equilibrium is:
$$HF(aq) + H_2O(l) \rightleftharpoons H_3O^+(aq) + F^-(aq)$$

55. Why is the hydronium ion concentration in a solution that is 0.10 M in HCl and 0.10 M in HCOOH determined by the concentration of HCl?

56. From the equilibrium concentrations given, calculate K_a for each of the weak acids and K_b for each of the weak bases.

(a) CH_3CO_2H: $[H_3O^+]$ = 1.34 × 10^{-3} M;
$[CH_3CO_2^-]$ = 1.34 × 10^{-3} M;

$[CH_3CO_2H]$ = 9.866 × 10^{-2} M;

(b) ClO^-: $[OH^-]$ = 4.0 × 10^{-4} M;

$[HClO]$ = 2.38 × 10^{-5} M;

$[ClO^-]$ = 0.273 M;

(c) HCO_2H: $[HCO_2H]$ = 0.524 M;
$[H_3O^+]$ = 9.8 × 10^{-3} M;
$[HCO_2^-]$ = 9.8 × 10^{-3} M;

(d) $C_6H_5NH_3^+$: $[C_6H_5NH_3^+]$ = 0.233 M;

$[C_6H_5NH_2]$ = 2.3 × 10^{-3} M;
$[H_3O^+]$ = 2.3 × 10^{-3} M

57. From the equilibrium concentrations given, calculate K_a for each of the weak acids and K_b for each of the weak bases.

(a) NH_3: $[OH^-]$ = 3.1 × 10^{-3} M;
$[NH_4^+]$ = 3.1 × 10^{-3} M;

$[NH_3]$ = 0.533 M;

(b) HNO_2: $[H_3O^+]$ = 0.011 M;
$[NO_2^-]$ = 0.0438 M;

$[HNO_2]$ = 1.07 M;

(c) $(CH_3)_3N$: $[(CH_3)_3N] = 0.25\ M$;
$[(CH_3)_3NH^+] = 4.3 \times 10^{-3}\ M$;

$[OH^-] = 4.3 \times 10^{-3}\ M$;

(d) NH_4^+: $[NH_4^+] = 0.100\ M$;

$[NH_3] = 7.5 \times 10^{-6}\ M$;
$[H_3O^+] = 7.5 \times 10^{-6}\ M$

58. Determine K_b for the nitrite ion, NO_2^-. In a 0.10-M solution this base is 0.0015% ionized.

59. Determine K_a for hydrogen sulfate ion, HSO_4^-. In a 0.10-M solution the acid is 29% ionized.

60. Calculate the ionization constant for each of the following acids or bases from the ionization constant of its conjugate base or conjugate acid:

(a) F^-

(b) NH_4^+

(c) AsO_4^{3-}

(d) $(CH_3)_2NH_2^+$

(e) NO_2^-

(f) $HC_2O_4^-$ (as a base)

61. Calculate the ionization constant for each of the following acids or bases from the ionization constant of its conjugate base or conjugate acid:

(a) HTe^- (as a base)

(b) $(CH_3)_3NH^+$

(c) $HAsO_4^{3-}$ (as a base)

(d) HO_2^- (as a base)

(e) $C_6H_5NH_3^+$

(f) HSO_3^- (as a base)

62. For which of the following solutions must we consider the ionization of water when calculating the pH or pOH?

(a) $3 \times 10^{-8}\ M\ HNO_3$

(b) 0.10 g HCl in 1.0 L of solution

(c) 0.00080 g NaOH in 0.50 L of solution

(d) $1 \times 10^{-7}\ M\ Ca(OH)_2$

(e) 0.0245 $M\ KNO_3$

63. Even though both NH_3 and $C_6H_5NH_2$ are weak bases, NH_3 is a much stronger acid than $C_6H_5NH_2$. Which of the following is correct at equilibrium for a solution that is initially 0.10 M in NH_3 and 0.10 M in $C_6H_5NH_2$?

(a) $[OH^-] = [NH_4^+]$

(b) $[NH_4^+] = [C_6H_5NH_3^+]$

(c) $[OH^-] = [C_6H_5NH_3^+]$

(d) $[NH_3] = [C_6H_5NH_2]$

(e) both a and b are correct

64. Calculate the equilibrium concentration of the nonionized acids and all ions in a solution that is 0.25 M in HCO_2H and 0.10 M in $HClO$.

65. Calculate the equilibrium concentration of the nonionized acids and all ions in a solution that is 0.134 M in HNO_2 and 0.120 M in $HBrO$.

66. Calculate the equilibrium concentration of the nonionized bases and all ions in a solution that is 0.25 M in CH_3NH_2 and 0.10 M in C_5H_5N ($K_b = 1.7 \times 10^{-9}$).

67. Calculate the equilibrium concentration of the nonionized bases and all ions in a solution that is 0.115 M in NH_3 and 0.100 M in $C_6H_5NH_2$.

68. Using the K_a values in Appendix H, place $Al(H_2O)_6^{3+}$ in the correct location in Figure 14.8.

69. Calculate the concentration of all solute species in each of the following solutions of acids or bases. Assume that the ionization of water can be neglected, and show that the change in the initial concentrations can be neglected. Ionization constants can be found in Appendix H and Appendix I.

(a) 0.0092 M HClO, a weak acid

(b) 0.0784 M $C_6H_5NH_2$, a weak base

(c) 0.0810 M HCN, a weak acid

(d) 0.11 M $(CH_3)_3N$, a weak base

(e) 0.120 M $Fe(H_2O)_6^{2+}$ a weak acid, $K_a = 1.6 \times 10^{-7}$

70. Propionic acid, $C_2H_5CO_2H$ ($K_a = 1.34 \times 10^{-5}$), is used in the manufacture of calcium propionate, a food preservative. What is the hydronium ion concentration in a 0.698-M solution of $C_2H_5CO_2H$?

71. White vinegar is a 5.0% by mass solution of acetic acid in water. If the density of white vinegar is 1.007 g/cm^3, what is the pH?

72. The ionization constant of lactic acid, $CH_3CH(OH)CO_2H$, an acid found in the blood after strenuous exercise, is 1.36×10^{-4}. If 20.0 g of lactic acid is used to make a solution with a volume of 1.00 L, what is the concentration of hydronium ion in the solution?

73. Nicotine, $C_{10}H_{14}N_2$, is a base that will accept two protons ($K_1 = 7 \times 10^{-7}$, $K_2 = 1.4 \times 10^{-11}$). What is the concentration of each species present in a 0.050-M solution of nicotine?

74. The pH of a 0.20-M solution of HF is 1.92. Determine K_a for HF from these data.

75. The pH of a 0.15-M solution of HSO_4^- is 1.43. Determine K_a for HSO_4^- from these data.

76. The pH of a 0.10-M solution of caffeine is 11.16. Determine K_b for caffeine from these data:
$$C_8H_{10}N_4O_2(aq) + H_2O(l) \rightleftharpoons C_8H_{10}N_4O_2H^+(aq) + OH^-(aq)$$

77. The pH of a solution of household ammonia, a 0.950 M solution of NH_3, is 11.612. Determine K_b for NH_3 from these data.

14.4 Hydrolysis of Salt Solutions

78. Determine whether aqueous solutions of the following salts are acidic, basic, or neutral:

(a) $Al(NO_3)_3$

(b) RbI

(c) $KHCO_2$

(d) CH_3NH_3Br

79. Determine whether aqueous solutions of the following salts are acidic, basic, or neutral:

(a) $FeCl_3$

(b) K_2CO_3

(c) NH_4Br

(d) $KClO_4$

80. Novocaine, $C_{13}H_{21}O_2N_2Cl$, is the salt of the base procaine and hydrochloric acid. The ionization constant for procaine is 7×10^{-6}. Is a solution of novocaine acidic or basic? What are $[H_3O^+]$, $[OH^-]$, and pH of a 2.0% solution by mass of novocaine, assuming that the density of the solution is 1.0 g/mL.

14.5 Polyprotic Acids

81. Which of the following concentrations would be practically equal in a calculation of the equilibrium concentrations in a 0.134-M solution of H_2CO_3, a diprotic acid: $[H_3O^+]$, $[OH^-]$, $[H_2CO_3]$, $[HCO_3^-]$, $[CO_3^{2-}]$? No calculations are needed to answer this question.

82. Calculate the concentration of each species present in a 0.050-M solution of H_2S.

83. Calculate the concentration of each species present in a 0.010-M solution of phthalic acid, $C_6H_4(CO_2H)_2$.

$$C_6H_4(CO_2H)_2(aq) + H_2O(l) \rightleftharpoons H_3O^+(aq) + C_6H_4(CO_2H)(CO_2)^-(aq) \qquad K_a = 1.1 \times 10^{-3}$$

$$C_6H_4(CO_2H)(CO_2)(aq) + H_2O(l) \rightleftharpoons H_3O^+(aq) + C_6H_4(CO_2)_2^{2-}(aq) \qquad K_a = 3.9 \times 10^{-6}$$

84. Salicylic acid, $HOC_6H_4CO_2H$, and its derivatives have been used as pain relievers for a long time. Salicylic acid occurs in small amounts in the leaves, bark, and roots of some vegetation (most notably historically in the bark of the willow tree). Extracts of these plants have been used as medications for centuries. The acid was first isolated in the laboratory in 1838.

(a) Both functional groups of salicylic acid ionize in water, with $K_a = 1.0 \times 10^{-3}$ for the—CO_2H group and 4.2×10^{-13} for the $-OH$ group. What is the pH of a saturated solution of the acid (solubility = 1.8 g/L).

(b) Aspirin was discovered as a result of efforts to produce a derivative of salicylic acid that would not be irritating to the stomach lining. Aspirin is acetylsalicylic acid, $CH_3CO_2C_6H_4CO_2H$. The $-CO_2H$ functional group is still present, but its acidity is reduced, $K_a = 3.0 \times 10^{-4}$. What is the pH of a solution of aspirin with the same concentration as a saturated solution of salicylic acid (See Part a).

(c) Under some conditions, aspirin reacts with water and forms a solution of salicylic acid and acetic acid:
$$CH_3CO_2C_6H_4CO_2H(aq) + H_2O(l) \longrightarrow HOC_6H_4CO_2H(aq) + CH_3CO_2H(aq)$$

i. Which of the acids, salicylic acid or acetic acid, produces more hydronium ions in such a solution?

ii. What are the concentrations of molecules and ions in a solution produced by the hydrolysis of 0.50 g of aspirin dissolved in enough water to give 75 mL of solution?

85. The ion HTe^- is an amphiprotic species; it can act as either an acid or a base.

(a) What is K_a for the acid reaction of HTe^- with H_2O?

(b) What is K_b for the reaction in which HTe^- functions as a base in water?

(c) Demonstrate whether or not the second ionization of H_2Te can be neglected in the calculation of $[HTe^-]$ in a 0.10 M solution of H_2Te.

14.6 Buffers

86. Explain why a buffer can be prepared from a mixture of NH_4Cl and NaOH but not from NH_3 and NaOH.

87. Explain why the pH does not change significantly when a small amount of an acid or a base is added to a solution that contains equal amounts of the acid H_3PO_4 and a salt of its conjugate base NaH_2PO_4.

88. Explain why the pH does not change significantly when a small amount of an acid or a base is added to a solution that contains equal amounts of the base NH_3 and a salt of its conjugate acid NH_4Cl.

89. What is $[H_3O^+]$ in a solution of 0.25 M CH_3CO_2H and 0.030 M $NaCH_3CO_2$?
$$CH_3CO_2H(aq) + H_2O(l) \rightleftharpoons H_3O^+(aq) + CH_3CO_2^-(aq) \qquad K_a = 1.8 \times 10^{-5}$$

90. What is $[H_3O^+]$ in a solution of 0.075 M HNO_2 and 0.030 M $NaNO_2$?
$$HNO_2(aq) + H_2O(l) \rightleftharpoons H_3O^+(aq) + NO_2^-(aq) \qquad K_a = 4.5 \times 10^{-5}$$

91. What is $[OH^-]$ in a solution of 0.125 M CH_3NH_2 and 0.130 M CH_3NH_3Cl?
$$CH_3NH_2(aq) + H_2O(l) \rightleftharpoons CH_3NH_3^+(aq) + OH^-(aq) \qquad K_b = 4.4 \times 10^{-4}$$

92. What is $[OH^-]$ in a solution of 1.25 M NH_3 and 0.78 M NH_4NO_3?

$$NH_3(aq) + H_2O(l) \rightleftharpoons NH_4^+(aq) + OH^-(aq) \qquad\qquad K_b = 1.8 \times 10^{-5}$$

93. What concentration of NH_4NO_3 is required to make $[OH^-] = 1.0 \times 10^{-5}$ in a 0.200-M solution of NH_3?

94. What concentration of NaF is required to make $[H_3O^+] = 2.3 \times 10^{-4}$ in a 0.300-M solution of HF?

95. What is the effect on the concentration of acetic acid, hydronium ion, and acetate ion when the following are added to an acidic buffer solution of equal concentrations of acetic acid and sodium acetate:

(a) HCl

(b) KCH_3CO_2

(c) NaCl

(d) KOH

(e) CH_3CO_2H

96. What is the effect on the concentration of ammonia, hydroxide ion, and ammonium ion when the following are added to a basic buffer solution of equal concentrations of ammonia and ammonium nitrate:

(a) KI

(b) NH_3

(c) HI

(d) NaOH

(e) NH_4Cl

97. What will be the pH of a buffer solution prepared from 0.20 mol NH_3, 0.40 mol NH_4NO_3, and just enough water to give 1.00 L of solution?

98. Calculate the pH of a buffer solution prepared from 0.155 mol of phosphoric acid, 0.250 mole of KH_2PO_4, and enough water to make 0.500 L of solution.

99. How much solid $NaCH_3CO_2\cdot3H_2O$ must be added to 0.300 L of a 0.50-M acetic acid solution to give a buffer with a pH of 5.00? (Hint: Assume a negligible change in volume as the solid is added.)

100. What mass of NH_4Cl must be added to 0.750 L of a 0.100-M solution of NH_3 to give a buffer solution with a pH of 9.26? (Hint: Assume a negligible change in volume as the solid is added.)

101. A buffer solution is prepared from equal volumes of 0.200 M acetic acid and 0.600 M sodium acetate. Use 1.80×10^{-5} as K_a for acetic acid.

(a) What is the pH of the solution?

(b) Is the solution acidic or basic?

(c) What is the pH of a solution that results when 3.00 mL of 0.034 M HCl is added to 0.200 L of the original buffer?

102. A 5.36–g sample of NH_4Cl was added to 25.0 mL of 1.00 M NaOH and the resulting solution diluted to 0.100 L.

(a) What is the pH of this buffer solution?

(b) Is the solution acidic or basic?

(c) What is the pH of a solution that results when 3.00 mL of 0.034 M HCl is added to the solution?

103. Which acid in Table 14.2 is most appropriate for preparation of a buffer solution with a pH of 3.1? Explain your choice.

104. Which acid in Table 14.2 is most appropriate for preparation of a buffer solution with a pH of 3.7? Explain your choice.

105. Which base in Table 14.3 is most appropriate for preparation of a buffer solution with a pH of 10.65? Explain your choice.

106. Which base in Table 14.3 is most appropriate for preparation of a buffer solution with a pH of 9.20? Explain your choice.

107. Saccharin, $C_7H_4NSO_3H$, is a weak acid ($K_a = 2.1 \times 10^{-2}$). If 0.250 L of diet cola with a buffered pH of 5.48 was prepared from 2.00×10^{-3} g of sodium saccharide, $Na(C_7H_4NSO_3)$, what are the final concentrations of saccharine and sodium saccharide in the solution?

108. What is the pH of 1.000 L of a solution of 100.0 g of glutamic acid ($C_5H_9NO_4$, a diprotic acid; $K_1 = 8.5 \times 10^{-5}$, $K_2 = 3.39 \times 10^{-10}$) to which has been added 20.0 g of NaOH during the preparation of monosodium glutamate, the flavoring agent? What is the pH when exactly 1 mol of NaOH per mole of acid has been added?

14.7 Acid-Base Titrations

109. Explain how to choose the appropriate acid-base indicator for the titration of a weak base with a strong acid.

110. Explain why an acid-base indicator changes color over a range of pH values rather than at a specific pH.

111. Why can we ignore the contribution of water to the concentrations of H_3O^+ in the solutions of following acids:

0.0092 M HClO, a weak acid

0.0810 M HCN, a weak acid

0.120 M $Fe(H_2O)_6^{2+}$ a weak acid, $K_a = 1.6 \times 10^{-7}$

but not the contribution of water to the concentration of OH^-?

112. We can ignore the contribution of water to the concentration of OH^- in a solution of the following bases:

0.0784 M $C_6H_5NH_2$, a weak base

0.11 M $(CH_3)_3N$, a weak base

but not the contribution of water to the concentration of H_3O^+?

113. Draw a curve for a series of solutions of HF. Plot $[H_3O^+]_{total}$ on the vertical axis and the total concentration of HF (the sum of the concentrations of both the ionized and nonionized HF molecules) on the horizontal axis. Let the total concentration of HF vary from 1×10^{-10} M to 1×10^{-2} M.

114. Draw a curve similar to that shown in Figure 14.23 for a series of solutions of NH_3. Plot $[OH^-]$ on the vertical axis and the total concentration of NH_3 (both ionized and nonionized NH_3 molecules) on the horizontal axis. Let the total concentration of NH_3 vary from 1×10^{-10} M to 1×10^{-2} M.

115. Calculate the pH at the following points in a titration of 40 mL (0.040 L) of 0.100 M barbituric acid ($K_a = 9.8 \times 10^{-5}$) with 0.100 M KOH.

(a) no KOH added

(b) 20 mL of KOH solution added

(c) 39 mL of KOH solution added

(d) 40 mL of KOH solution added

(e) 41 mL of KOH solution added

116. The indicator dinitrophenol is an acid with a K_a of 1.1×10^{-4}. In a 1.0×10^{-4}-M solution, it is colorless in acid and yellow in base. Calculate the pH range over which it goes from 10% ionized (colorless) to 90% ionized (yellow).

Chapter 15

Equilibria of Other Reaction Classes

Figure 15.1 The mineral fluorite (CaF_2) is deposited through a precipitation process. Note that pure fluorite is colorless, and that the color in this sample is due to the presence of other metals in the crystal.

Chapter Outline

15.1 Precipitation and Dissolution

15.2 Lewis Acids and Bases

15.3 Multiple Equilibria

Introduction

In Figure 15.1, we see a close-up image of the mineral fluorite, which is commonly used as a semiprecious stone in many types of jewelry because of its striking appearance. These solid deposits of fluorite are formed through a process called hydrothermal precipitation. In this process, the fluorite remains dissolved in solution, usually in hot water heated by volcanic activity deep below the earth, until conditions arise that allow the mineral to come out of solution and form a deposit. These deposit-forming conditions can include a change in temperature of the solution, availability of new locations to form a deposit such as a rock crevice, contact between the solution and a reactive substance such as certain types of rock, or a combination of any of these factors.

We previously learned about aqueous solutions and their importance, as well as about solubility rules. While this gives us a picture of solubility, that picture is not complete if we look at the rules alone. Solubility equilibrium, which we will explore in this chapter, is a more complex topic that allows us to determine the extent to which a slightly soluble ionic solid will dissolve, and the conditions under which precipitation (such as the fluorite deposit in Figure 15.1) will occur.

15.1 Precipitation and Dissolution

By the end of this section, you will be able to:

- Write chemical equations and equilibrium expressions representing solubility equilibria
- Carry out equilibrium computations involving solubility, equilibrium expressions, and solute concentrations

The preservation of medical laboratory blood samples, mining of sea water for magnesium, formulation of over-the-counter medicines such as Milk of Magnesia and antacids, and treating the presence of hard water in your home's water supply are just a few of the many tasks that involve controlling the equilibrium between a slightly soluble ionic solid and an aqueous solution of its ions.

In some cases, we want to prevent dissolution from occurring. Tooth decay, for example, occurs when the calcium hydroxylapatite, which has the formula $Ca_5(PO_4)_3(OH)$, in our teeth dissolves. The dissolution process is aided when bacteria in our mouths feast on the sugars in our diets to produce lactic acid, which reacts with the hydroxide ions in the calcium hydroxylapatite. Preventing the dissolution prevents the decay. On the other hand, sometimes we want a substance to dissolve. We want the calcium carbonate in a chewable antacid to dissolve because the $CO_3{}^{2-}$ ions produced in this process help soothe an upset stomach.

In this section, we will find out how we can control the dissolution of a slightly soluble ionic solid by the application of Le Châtelier's principle. We will also learn how to use the equilibrium constant of the reaction to determine the concentration of ions present in a solution.

The Solubility Product Constant

Silver chloride is what's known as a sparingly soluble ionic solid (Figure 15.2). Recall from the solubility rules in an earlier chapter that halides of Ag^+ are not normally soluble. However, when we add an excess of solid AgCl to water, it dissolves to a small extent and produces a mixture consisting of a very dilute solution of Ag^+ and Cl^- ions in equilibrium with undissolved silver chloride:

$$AgCl(s) \underset{\text{precipitation}}{\overset{\text{dissolution}}{\rightleftharpoons}} Ag^+(aq) + Cl^-(aq)$$

This equilibrium, like other equilibria, is dynamic; some of the solid AgCl continues to dissolve, but at the same time, Ag^+ and Cl^- ions in the solution combine to produce an equal amount of the solid. At equilibrium, the opposing processes have equal rates.

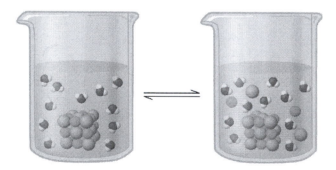

Figure 15.2 Silver chloride is a sparingly soluble ionic solid. When it is added to water, it dissolves slightly and produces a mixture consisting of a very dilute solution of Ag^+ and Cl^- ions in equilibrium with undissolved silver chloride.

The equilibrium constant for the equilibrium between a slightly soluble ionic solid and a solution of its ions is called the **solubility product (K_{sp})** of the solid. Recall from the chapter on solutions and colloids that we use an ion's concentration as an approximation of its activity in a dilute solution. For silver chloride, at equilibrium:

$$AgCl(s) \rightleftharpoons Ag^+(aq) + Cl^-(aq) \qquad K_{sp} = [Ag^+(aq)][Cl^-(aq)]$$

When looking at dissolution reactions such as this, the solid is listed as a reactant, whereas the ions are listed as products. The solubility product constant, as with every equilibrium constant expression, is written as the product of the concentrations of each of the ions, raised to the power of their stoichiometric coefficients. Here, the solubility product constant is equal to Ag^+ and Cl^- when a solution of silver chloride is in equilibrium with undissolved AgCl. There is no denominator representing the reactants in this equilibrium expression since the reactant is a pure solid; therefore [AgCl] does not appear in the expression for K_{sp}.

Some common solubility products are listed in Table 15.1 according to their K_{sp} values, whereas a more extensive compilation of products appears in Appendix J. Each of these equilibrium constants is much smaller than 1 because the compounds listed are only slightly soluble. A small K_{sp} represents a system in which the equilibrium lies to the left, so that relatively few hydrated ions would be present in a saturated solution.

Common Solubility Products by Decreasing Equilibrium Constants

Substance	K_{sp} at 25 °C
CuCl	1.2×10^{-6}
CuBr	6.27×10^{-9}
AgI	1.5×10^{-16}
PbS	7×10^{-29}
Al(OH)$_3$	2×10^{-32}
Fe(OH)$_3$	4×10^{-38}

Table 15.1

Example 15.1

Writing Equations and Solubility Products

Write the ionic equation for the dissolution and the solubility product expression for each of the following slightly soluble ionic compounds:

(a) AgI, silver iodide, a solid with antiseptic properties

(b) CaCO$_3$, calcium carbonate, the active ingredient in many over-the-counter chewable antacids

(c) Mg(OH)$_2$, magnesium hydroxide, the active ingredient in Milk of Magnesia

(d) Mg(NH$_4$)PO$_4$, magnesium ammonium phosphate, an essentially insoluble substance used in tests for magnesium

(e) Ca$_5$(PO$_4$)$_3$OH, the mineral apatite, a source of phosphate for fertilizers

(Hint: When determining how to break (d) and (e) up into ions, refer to the list of polyatomic ions in the section on chemical nomenclature.)

Solution

(a) $AgI(s) \rightleftharpoons Ag^+(aq) + I^-(aq)$ $\qquad K_{sp} = [Ag^+][I^-]$

(b) $CaCO_3(s) \rightleftharpoons Ca^{2+}(aq) + CO_3{}^{2-}(aq)$ $\qquad K_{sp} = [Ca^{2+}][CO_3{}^{2-}]$

(c) $Mg(OH)_2(s) \rightleftharpoons Mg^{2+}(aq) + 2OH^-(aq)$ $\qquad K_{sp} = [Mg^{2+}][OH^-]^2$

(d) $Mg(NH_4)PO_4(s) \rightleftharpoons Mg^{2+}(aq) + NH_4{}^+(aq) + PO_4{}^{3-}(aq)$ $\qquad K_{sp} = [Mg^{2+}][NH_4{}^+][PO_4{}^{3-}]$

(e) $Ca_5(PO_4)3OH(s) \rightleftharpoons 5Ca^{2+}(aq) + 3PO_4{}^{3-}(aq) + OH^-(aq)$ $\qquad K_{sp} = [Ca^{2+}]^5[PO_4{}^{3-}]^3[OH^-]$

Check Your Learning

Write the ionic equation for the dissolution and the solubility product for each of the following slightly soluble compounds:

(a) $BaSO_4$

(b) Ag_2SO_4

(c) $Al(OH)_3$

(d) $Pb(OH)Cl$

Answer: (a) $BaSO_4(s) \rightleftharpoons Ba^{2+}(aq) + SO_4{}^{2-}(aq)$ $\qquad K_{sp} = [Ba^{2+}][SO_4{}^{2-}]$; (b) $Ag_2SO_4(s) \rightleftharpoons 2Ag^+(aq) + SO_4{}^{2-}(aq)$ $\qquad K_{sp} = [Ag^+]^2[SO_4{}^{2-}]$; (c) $Al(OH)_3(s) \rightleftharpoons Al^{2+}(aq) + 3OH^-(aq)$ $\qquad K_{sp} = [Al^{3+}][OH^-]^3$; (d) $Pb(OH)Cl(s) \rightleftharpoons Pb^{2+}(aq) + OH^-(aq) + Cl^-(aq)$ $\qquad K_{sp} = [Pb^{2+}][OH^-][Cl^-]$

Now we will extend the discussion of K_{sp} and show how the solubility product constant is determined from the solubility of its ions, as well as how K_{sp} can be used to determine the molar solubility of a substance.

K_{sp} and Solubility

Recall that the definition of *solubility* is the maximum possible concentration of a solute in a solution at a given temperature and pressure. We can determine the solubility product of a slightly soluble solid from that measure of its solubility at a given temperature and pressure, provided that the only significant reaction that occurs when the solid dissolves is its dissociation into solvated ions, that is, the only equilibrium involved is:

$$M_pX_q(s) \rightleftharpoons pM^{m+}(aq) + qX^{n-}(aq)$$

In this case, we calculate the solubility product by taking the solid's solubility expressed in units of moles per liter (mol/L), known as its **molar solubility**.

Example 15.2

Calculation of K_{sp} from Equilibrium Concentrations

We began the chapter with an informal discussion of how the mineral fluorite (Figure 15.1) is formed. Fluorite, CaF_2, is a slightly soluble solid that dissolves according to the equation:

$$CaF_2(s) \rightleftharpoons Ca^{2+}(aq) + 2F^-(aq)$$

The concentration of Ca^{2+} in a saturated solution of CaF_2 is 2.15×10^{-4} M; therefore, that of F^- is 4.30×10^{-4} M, that is, twice the concentration of Ca^{2+}. What is the solubility product of fluorite?

Solution

First, write out the K_{sp} expression, then substitute in concentrations and solve for K_{sp}:

$$CaF_2(s) \rightleftharpoons Ca^{2+}(aq) + 2F^-(aq)$$

A saturated solution is a solution at equilibrium with the solid. Thus:

$$K_{sp} = [Ca^{2+}][F^-]^2 = (2.1 \times 10^{-4})(4.2 \times 10^{-4})^2 = 3.7 \times 10^{-11}$$

As with other equilibrium constants, we do not include units with K_{sp}.

Check Your Learning

In a saturated solution that is in contact with solid $Mg(OH)_2$, the concentration of Mg^{2+} is 1.31×10^{-4} M. What is the solubility product for $Mg(OH)_2$?

$$Mg(OH)_2(s) \rightleftharpoons Mg^{2+}(aq) + 2OH^-(aq)$$

Answer: 8.99×10^{-12}

Example 15.3

Determination of Molar Solubility from K_{sp}

The K_{sp} of copper(I) bromide, CuBr, is 6.3×10^{-9}. Calculate the molar solubility of copper bromide.

Solution

The solubility product constant of copper(I) bromide is 6.3×10^{-9}.

The reaction is:

$$CuBr(s) \rightleftharpoons Cu^+(aq) + Br^-(aq)$$

First, write out the solubility product equilibrium constant expression:

$$K_{sp} = [Cu^+][Br^-]$$

Create an ICE table (as introduced in the chapter on fundamental equilibrium concepts), leaving the CuBr column empty as it is a solid and does not contribute to the K_{sp}:

	CuBr $\rightleftharpoons$	Cu$^+$ +	Br$^-$
Initial concentration (M)		0	0
Change (M)		x	x
Equilibrium concentration (M)		$0 + x = x$	$0 + x = x$

At equilibrium:

$$K_{sp} = [Cu^+][Br^-]$$
$$6.3 \times 10^{-9} = (x)(x) = x^2$$
$$x = \sqrt{(6.3 \times 10^{-9})} = 7.9 \times 10^{-5}$$

Therefore, the molar solubility of CuBr is 7.9×10^{-5} M.

Check Your Learning

The K_{sp} of AgI is 1.5×10^{-16}. Calculate the molar solubility of silver iodide.

Answer: 1.2×10^{-8} M

Example 15.4

Determination of Molar Solubility from K_{sp}, Part II

The K_{sp} of calcium hydroxide, $Ca(OH)_2$, is 1.3×10^{-6}. Calculate the molar solubility of calcium hydroxide.

Solution

The solubility product constant of calcium hydroxide is 1.3×10^{-6}.

The reaction is:

$$Ca(OH)_2(s) \rightleftharpoons Ca^{2+}(aq) + 2OH^-(aq)$$

First, write out the solubility product equilibrium constant expression:

$$K_{sp} = [Ca^{2+}][OH^-]^2$$

Create an ICE table, leaving the $Ca(OH)_2$ column empty as it is a solid and does not contribute to the K_{sp}:

	$Ca(OH)_2 \rightleftharpoons$	Ca^{2+} +	$2OH^-$
Initial concentration (M)		0	0
Change (M)		x	$2x$
Equilibrium concentration (M)		$0 + x = x$	$0 + 2x = 2x$

At equilibrium:

$$K_{sp} = [Ca^{2+}][OH^-]^2$$
$$1.3 \times 10^{-6} = (x)(2x)^2 = (x)(4x^2) = 4x^3$$
$$x = \sqrt[3]{\frac{1.3 \times 10^{-6}}{4}} = 6.9 \times 10^{-3}$$

Therefore, the molar solubility of $Ca(OH)_2$ is 1.3×10^{-2} M.

Check Your Learning

The K_{sp} of PbI_2 is 1.4×10^{-8}. Calculate the molar solubility of lead(II) iodide.

Answer: 1.5×10^{-3} M

Note that solubility is not always given as a molar value. When the solubility of a compound is given in some unit other than moles per liter, we must convert the solubility into moles per liter (i.e., molarity) in order to use it in the solubility product constant expression. Example 15.5 shows how to perform those unit conversions before determining the solubility product equilibrium.

Example 15.5

Determination of K_{sp} from Gram Solubility

Many of the pigments used by artists in oil-based paints (Figure 15.3) are sparingly soluble in water. For example, the solubility of the artist's pigment chrome yellow, $PbCrO_4$, is 4.6×10^{-6} g/L. Determine the solubility product equilibrium constant for $PbCrO_4$.

Figure 15.3 Oil paints contain pigments that are very slightly soluble in water. In addition to chrome yellow ($PbCrO_4$), examples include Prussian blue ($Fe_7(CN)_{18}$), the reddish-orange color vermilion (HgS), and green color veridian (Cr_2O_3). (credit: Sonny Abesamis)

Solution

We are given the solubility of $PbCrO_4$ in grams per liter. If we convert this solubility into moles per liter, we can find the equilibrium concentrations of Pb^{2+} and CrO_4^{2-}, then K_{sp}:

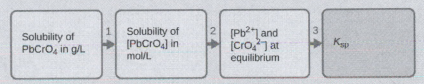

Step 1. *Use the molar mass of $PbCrO_4$ $\left(\dfrac{323.2 \text{ g}}{1 \text{ mol}}\right)$ to convert the solubility of $PbCrO_4$ in grams per liter into moles per liter:*

$$[PbCrO_4] = \frac{4.6 \times 10^{-6} \text{ g PbCrO}_4}{1 \text{ L}} \times \frac{1 \text{ mol PbCrO}_4}{323.2 \text{ g PbCrO}_4}$$

$$= \frac{1.4 \times 10^{-8} \text{ mol PbCrO}_4}{1 \text{ L}}$$

$$= 1.4 \times 10^{-8} M$$

Step 2. *The chemical equation for the dissolution indicates that 1 mol of $PbCrO_4$ gives 1 mol of $Pb^{2+}(aq)$ and 1 mol of $CrO_4^{2-}(aq)$:*

$$PbCrO_4(s) \rightleftharpoons Pb^{2+}(aq) + CrO_4^{2-}(aq)$$

Thus, both $[Pb^{2+}]$ and $[CrO_4^{2-}]$ are equal to the molar solubility of $PbCrO_4$:

$$[Pb^{2+}] = [CrO_4^{2-}] = 1.4 \times 10^{-8} M$$

Step 3. *Solve.* $K_{sp} = [Pb^{2+}][CrO_4^{2-}] = (1.4 \times 10^{-8})(1.4 \times 10^{-8}) = 2.0 \times 10^{-16}$

Check Your Learning

The solubility of TlCl [thallium(I) chloride], an intermediate formed when thallium is being isolated from ores, is 3.46 grams per liter at 20 °C. What is its solubility product?

Answer: 1.7×10^{-4}

Example 15.6

Calculating the Solubility of Hg$_2$Cl$_2$

Calomel, Hg$_2$Cl$_2$, is a compound composed of the diatomic ion of mercury(I), Hg$_2^{2+}$, and chloride ions, Cl$^-$. Although most mercury compounds are now known to be poisonous, eighteenth-century physicians used calomel as a medication. Their patients rarely suffered any mercury poisoning from the treatments because calomel is quite insoluble:

$$Hg_2Cl_2(s) \rightleftharpoons Hg_2^{2+}(aq) + 2Cl^-(aq) \qquad K_{sp} = 1.1 \times 10^{-18}$$

Calculate the molar solubility of Hg$_2$Cl$_2$.

Solution

The molar solubility of Hg$_2$Cl$_2$ is equal to the concentration of Hg$_2^{2+}$ ions because for each 1 mol of Hg$_2$Cl$_2$ that dissolves, 1 mol of Hg$_2^{2+}$ forms:

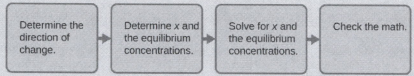

Step 1. *Determine the direction of change.* Before any Hg$_2$Cl$_2$ dissolves, Q is zero, and the reaction will shift to the right to reach equilibrium.

Step 2. *Determine x and equilibrium concentrations.* Concentrations and changes are given in the following ICE table:

	Hg$_2$Cl$_2$ $\rightleftharpoons$ Hg$_2^{2+}$	+	2Cl$^-$
Initial concentration (M)	0		0
Change (M)	x		2x
Equilibrium concentration (M)	0 + x = x		0 + 2x = 2x

Note that the change in the concentration of Cl$^-$ (2x) is twice as large as the change in the concentration of Hg$_2^{2+}$ (x) because 2 mol of Cl$^-$ forms for each 1 mol of Hg$_2^{2+}$ that forms. Hg$_2$Cl$_2$ is a pure solid, so it does not appear in the calculation.

Step 3. *Solve for x and the equilibrium concentrations.* We substitute the equilibrium concentrations into the expression for K_{sp} and calculate the value of x:

$$K_{sp} = [Hg_2^{2+}][Cl^-]^2$$
$$1.1 \times 10^{-18} = (x)(2x)^2$$
$$4x^3 = 1.1 \times 10^{-18}$$
$$x = \sqrt[3]{\left(\frac{1.1 \times 10^{-18}}{4}\right)} = 6.5 \times 10^{-7} \ M$$
$$[Hg_2^{2+}] = 6.5 \times 10^{-7} \ M = 6.5 \times 10^{-7} \ M$$
$$[Cl^-] = 2x = 2(6.5 \times 10^{-7}) = 1.3 \times 10^{-6} \ M$$

The molar solubility of Hg$_2$Cl$_2$ is equal to [Hg$_2^{2+}$], or 6.5 $\times$ 10^{-7} M.

Step 4. *Check the work.* At equilibrium, Q = K_{sp}:
$$Q = [Hg_2^{2+}][Cl^-]^2 = (6.5 \times 10^{-7})(1.3 \times 10^{-6})^2 = 1.1 \times 10^{-18}$$

The calculations check.

Check Your Learning

Determine the molar solubility of MgF_2 from its solubility product: $K_{sp} = 6.4 \times 10^{-9}$.

Answer: $1.2 \times 10^{-3}\ M$

Tabulated K_{sp} values can also be compared to reaction quotients calculated from experimental data to tell whether a solid will precipitate in a reaction under specific conditions: Q equals K_{sp} at equilibrium; if Q is less than K_{sp}, the solid will dissolve until Q equals K_{sp}; if Q is greater than K_{sp}, precipitation will occur at a given temperature until Q equals K_{sp}.

How Sciences Interconnect

Using Barium Sulfate for Medical Imaging

Various types of medical imaging techniques are used to aid diagnoses of illnesses in a noninvasive manner. One such technique utilizes the ingestion of a barium compound before taking an X-ray image. A suspension of barium sulfate, a chalky powder, is ingested by the patient. Since the K_{sp} of barium sulfate is 1.1×10^{-10}, very little of it dissolves as it coats the lining of the patient's intestinal tract. Barium-coated areas of the digestive tract then appear on an X-ray as white, allowing for greater visual detail than a traditional X-ray (Figure 15.4).

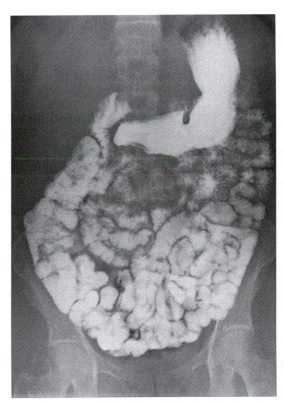

Figure 15.4 The suspension of barium sulfate coats the intestinal tract, which allows for greater visual detail than a traditional X-ray. (credit modification of work by "glitzy queen00"/Wikimedia Commons)

Further diagnostic testing can be done using barium sulfate and fluoroscopy. In fluoroscopy, a continuous X-ray is passed through the body so the doctor can monitor, on a TV or computer screen, the barium sulfate's movement as it passes through the digestive tract. Medical imaging using barium sulfate can be used to diagnose acid reflux disease, Crohn's disease, and ulcers in addition to other conditions.

Visit this website (http://openstaxcollege.org/l/16barium) for more information on how barium is used in medical diagnoses and which conditions it is used to diagnose.

Predicting Precipitation

The equation that describes the equilibrium between solid calcium carbonate and its solvated ions is:

$$CaCO_3(s) \rightleftharpoons Ca^{2+}(aq) + CO_3{}^{2-}(aq)$$

We can establish this equilibrium either by adding solid calcium carbonate to water or by mixing a solution that contains calcium ions with a solution that contains carbonate ions. If we add calcium carbonate to water, the solid will dissolve until the concentrations are such that the value of the reaction quotient $(Q = [Ca^{2+}][CO_3{}^{2-}])$ is equal to the solubility product $(K_{sp} = 8.7 \times 10^{-9})$. If we mix a solution of calcium nitrate, which contains Ca^{2+} ions, with a solution of sodium carbonate, which contains $CO_3{}^{2-}$ ions, the slightly soluble ionic solid $CaCO_3$ will precipitate, provided that the concentrations of Ca^{2+} and $CO_3{}^{2-}$ ions are such that Q is greater than K_{sp} for the mixture. The reaction shifts to the left and the concentrations of the ions are reduced by formation of the solid until the value of Q equals K_{sp}. A saturated solution in equilibrium with the undissolved solid will result. If the concentrations are such that Q is less than K_{sp}, then the solution is not saturated and no precipitate will form.

We can compare numerical values of Q with K_{sp} to predict whether precipitation will occur, as Example 15.7 shows. (Note: Since all forms of equilibrium constants are temperature dependent, we will assume a room temperature environment going forward in this chapter unless a different temperature value is explicitly specified.)

Example 15.7

Precipitation of Mg(OH)$_2$

The first step in the preparation of magnesium metal is the precipitation of $Mg(OH)_2$ from sea water by the addition of lime, $Ca(OH)_2$, a readily available inexpensive source of OH^- ion:

$$Mg(OH)_2(s) \rightleftharpoons Mg^{2+}(aq) + 2OH^-(aq) \qquad K_{sp} = 8.9 \times 10^{-12}$$

The concentration of $Mg^{2+}(aq)$ in sea water is 0.0537 M. Will $Mg(OH)_2$ precipitate when enough $Ca(OH)_2$ is added to give a $[OH^-]$ of 0.0010 M?

Solution

This problem asks whether the reaction:

$$Mg(OH)_2(s) \rightleftharpoons Mg^{2+}(aq) + 2OH^-(aq)$$

shifts to the left and forms solid $Mg(OH)_2$ when $[Mg^{2+}] = 0.0537$ M and $[OH^-] = 0.0010$ M. The reaction shifts to the left if Q is greater than K_{sp}. Calculation of the reaction quotient under these conditions is shown here:

$$Q = [Mg^{2+}][OH^-]^2 = (0.0537)(0.0010)^2 = 5.4 \times 10^{-8}$$

Because Q is greater than K_{sp} ($Q = 5.4 \times 10^{-8}$ is larger than $K_{sp} = 8.9 \times 10^{-12}$), we can expect the reaction to shift to the left and form solid magnesium hydroxide. $Mg(OH)_2(s)$ forms until the concentrations of magnesium ion and hydroxide ion are reduced sufficiently so that the value of Q is equal to K_{sp}.

Check Your Learning

Use the solubility product in Appendix J to determine whether $CaHPO_4$ will precipitate from a solution with $[Ca^{2+}] = 0.0001$ M and $[HPO_4{}^{2-}] = 0.001$ M.

Answer: No precipitation of $CaHPO_4$; $Q = 1 \times 10^{-7}$, which is less than K_{sp}

Example 15.8

Precipitation of AgCl upon Mixing Solutions

Does silver chloride precipitate when equal volumes of a 2.0×10^{-4}-M solution of $AgNO_3$ and a 2.0×10^{-4}-M solution of $NaCl$ are mixed?

(Note: The solution also contains Na^+ and $NO_3{}^-$ ions, but when referring to solubility rules, one can see that sodium nitrate is very soluble and cannot form a precipitate.)

Solution

The equation for the equilibrium between solid silver chloride, silver ion, and chloride ion is:

$$AgCl(s) \rightleftharpoons Ag^+(aq) + Cl^-(aq)$$

The solubility product is 1.6×10^{-10} (see Appendix J).

$AgCl$ will precipitate if the reaction quotient calculated from the concentrations in the mixture of $AgNO_3$ and $NaCl$ is greater than K_{sp}. The volume doubles when we mix equal volumes of $AgNO_3$ and $NaCl$ solutions, so each concentration is reduced to half its initial value. Consequently, immediately upon mixing, $[Ag^+]$ and $[Cl^-]$ are both equal to:

$$\frac{1}{2}(2.0 \times 10^{-4})\,M = 1.0 \times 10^{-4}\,M$$

The reaction quotient, Q, is *momentarily* greater than K_{sp} for $AgCl$, so a supersaturated solution is formed:

$$Q = [Ag^+][Cl^-] = (1.0 \times 10^{-4})(1.0 \times 10^{-4}) = 1.0 \times 10^{-8} > K_{sp}$$

Since supersaturated solutions are unstable, $AgCl$ will precipitate from the mixture until the solution returns to equilibrium, with Q equal to K_{sp}.

Check Your Learning

Will $KClO_4$ precipitate when 20 mL of a 0.050-M solution of K^+ is added to 80 mL of a 0.50-M solution of $ClO_4{}^-$? (Remember to calculate the new concentration of each ion after mixing the solutions before plugging into the reaction quotient expression.)

Answer: No, $Q = 4.0 \times 10^{-3}$, which is less than $K_{sp} = 1.05 \times 10^{-2}$

In the previous two examples, we have seen that $Mg(OH)_2$ or $AgCl$ precipitate when Q is greater than K_{sp}. In general, when a solution of a soluble salt of the M^{m+} ion is mixed with a solution of a soluble salt of the X^{n-} ion, the solid, M_pX_q precipitates if the value of Q for the mixture of M^{m+} and X^{n-} is greater than K_{sp} for M_pX_q. Thus, if we know the concentration of one of the ions of a slightly soluble ionic solid and the value for the solubility product of the solid, then we can calculate the concentration that the other ion must exceed for precipitation to begin. To simplify the calculation, we will assume that precipitation begins when the reaction quotient becomes equal to the solubility product constant.

Example 15.9

Precipitation of Calcium Oxalate

Blood will not clot if calcium ions are removed from its plasma. Some blood collection tubes contain salts of the oxalate ion, $C_2O_4{}^{2-}$, for this purpose (Figure 15.5). At sufficiently high concentrations, the calcium and oxalate ions form solid, $CaC_2O_4 \cdot H_2O$ (which also contains water bound in the solid). The concentration of Ca^{2+} in a sample of blood serum is 2.2×10^{-3} M. What concentration of $C_2O_4{}^{2-}$ ion must be established before $CaC_2O_4 \cdot H_2O$ begins to precipitate?

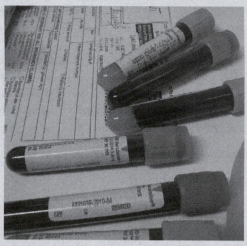

Figure 15.5 Anticoagulants can be added to blood that will combine with the Ca^{2+} ions in blood serum and prevent the blood from clotting. (credit: modification of work by Neeta Lind)

Solution

The equilibrium expression is:

$$CaC_2O_4(s) \rightleftharpoons Ca^{2+}(aq) + C_2O_4{}^{2-}(aq)$$

For this reaction:

$$K_{sp} = [Ca^{2+}][C_2O_4{}^{2-}] = 1.96 \times 10^{-8}$$

(see Appendix J)

CaC_2O_4 does not appear in this expression because it is a solid. Water does not appear because it is the solvent.

Solid CaC_2O_4 does not begin to form until Q equals K_{sp}. Because we know K_{sp} and $[Ca^{2+}]$, we can solve for the concentration of $C_2O_4{}^{2-}$ that is necessary to produce the first trace of solid:

$$Q = K_{sp} = [Ca^{2+}][C_2O_4{}^{2-}] = 1.96 \times 10^{-8}$$

$$(2.2 \times 10^{-3})[C_2O_4{}^{2-}] = 1.96 \times 10^{-8}$$

$$[C_2O_4{}^{2-}] = \frac{1.96 \times 10^{-8}}{2.2 \times 10^{-3}} = 8.9 \times 10^{-6}$$

A concentration of $[C_2O_4{}^{2-}] = 8.9 \times 10^{-6}\ M$ is necessary to initiate the precipitation of CaC_2O_4 under these conditions.

Check Your Learning

If a solution contains 0.0020 mol of $CrO_4{}^{2-}$ per liter, what concentration of Ag^+ ion must be reached by adding solid $AgNO_3$ before Ag_2CrO_4 begins to precipitate? Neglect any increase in volume upon adding the solid silver nitrate.

Answer: $4.5 \times 10^{-9}\ M$

It is sometimes useful to know the concentration of an ion that remains in solution after precipitation. We can use the solubility product for this calculation too: If we know the value of K_{sp} and the concentration of one ion in solution, we can calculate the concentration of the second ion remaining in solution. The calculation is of the same type as that in Example 15.9—calculation of the concentration of a species in an equilibrium mixture from the concentrations of the other species and the equilibrium constant. However, the concentrations are different; we are calculating concentrations after precipitation is complete, rather than at the start of precipitation.

Example 15.10

Concentrations Following Precipitation

Clothing washed in water that has a manganese $[Mn^{2+}(aq)]$ concentration exceeding 0.1 mg/L (1.8×10^{-6} M) may be stained by the manganese upon oxidation, but the amount of Mn^{2+} in the water can be reduced by adding a base. If a person doing laundry wishes to add a buffer to keep the pH high enough to precipitate the manganese as the hydroxide, $Mn(OH)_2$, what pH is required to keep $[Mn^{2+}]$ equal to $1.8 \times 10^{-6}\ M$?

Solution

The dissolution of $Mn(OH)_2$ is described by the equation:

$$Mn(OH)_2(s) \rightleftharpoons Mn^{2+}(aq) + 2OH^-(aq) \qquad K_{sp} = 2 \times 10^{-3}$$

We need to calculate the concentration of OH^- when the concentration of Mn^{2+} is $1.8 \times 10^{-6}\ M$. From that, we calculate the pH. At equilibrium:

$$K_{sp} = [Mn^{2+}][OH^-]^2$$

or

$$(1.8 \times 10^{-6})[OH^-]^2 = 2 \times 10^{-3}$$

so

$$[OH^-] = 3.3 \times 10^{-4}\ M$$

Now we calculate the pH from the pOH:

$$pOH = -\log[OH^-] = -\log(3.3 \times 10 - 4) = 3.48$$
$$pH = 14.00 - pOH = 14.00 - 3.80 = 10.52$$

If the person doing laundry adds a base, such as the sodium silicate (Na_4SiO_4) in some detergents, to the wash water until the pH is raised to 10.52, the manganese ion will be reduced to a concentration of $1.8 \times 10^{-6}\ M$; at that concentration or less, the ion will not stain clothing.

Check Your Learning

The first step in the preparation of magnesium metal is the precipitation of $Mg(OH)_2$ from sea water by the addition of $Ca(OH)_2$. The concentration of $Mg^{2+}(aq)$ in sea water is 5.37×10^{-2} M. Calculate the pH at which $[Mg^{2+}]$ is diminished to 1.0×10^{-5} M by the addition of $Ca(OH)_2$.

Answer: 10.97

Due to their light sensitivity, mixtures of silver halides are used in fiber optics for medical lasers, in photochromic eyeglass lenses (glass lenses that automatically darken when exposed to sunlight), and—before the advent of digital photography—in photographic film. Even though AgCl ($K_{sp} = 1.6 \times 10^{-10}$), AgBr ($K_{sp} = 5.0 \times 10^{-13}$), and AgI ($K_{sp} = 1.5 \times 10^{-16}$) are each quite insoluble, we cannot prepare a homogeneous solid mixture of them by adding Ag^+ to a solution of Cl^-, Br^-, and I^-; essentially all of the AgI will precipitate before any of the other solid halides form because of its smaller value for K_{sp}. However, we can prepare a homogeneous mixture of the solids by slowly adding a solution of Cl^-, Br^-, and I^- to a solution of Ag^+.

When two anions form slightly soluble compounds with the same cation, or when two cations form slightly soluble compounds with the same anion, the less soluble compound (usually, the compound with the smaller K_{sp}) generally precipitates first when we add a precipitating agent to a solution containing both anions (or both cations). When the K_{sp} values of the two compounds differ by two orders of magnitude or more (e.g., 10^{-2} vs. 10^{-4}), almost all of the less soluble compound precipitates before any of the more soluble one does. This is an example of **selective precipitation**, where a reagent is added to a solution of dissolved ions causing one of the ions to precipitate out before the rest.

Chemistry in Everyday Life

The Role of Precipitation in Wastewater Treatment

Solubility equilibria are useful tools in the treatment of wastewater carried out in facilities that may treat the municipal water in your city or town (Figure 15.6). Specifically, selective precipitation is used to remove contaminants from wastewater before it is released back into natural bodies of water. For example, phosphate ions $(PO_4{}^{2-})$ are often present in the water discharged from manufacturing facilities. An abundance of phosphate causes excess algae to grow, which impacts the amount of oxygen available for marine life as well as making water unsuitable for human consumption.

Figure 15.6 Wastewater treatment facilities, such as this one, remove contaminants from wastewater before the water is released back into the natural environment. (credit: "eutrophication&hypoxia"/Wikimedia Commons)

One common way to remove phosphates from water is by the addition of calcium hydroxide, known as lime, $Ca(OH)_2$. The lime is converted into calcium carbonate, a strong base, in the water. As the water is made more basic, the calcium ions react with phosphate ions to produce hydroxylapatite, $Ca_5(PO4)_3(OH)$, which then precipitates out of the solution:

$$5Ca^{2+} + 3PO_4{}^{3-} + OH^- \rightleftharpoons Ca_{10}(PO_4)_6 \cdot (OH)_2(s)$$

The precipitate is then removed by filtration and the water is brought back to a neutral pH by the addition of CO_2 in a recarbonation process. Other chemicals can also be used for the removal of phosphates by precipitation, including iron(III) chloride and aluminum sulfate.

View this site (http://openstaxcollege.org/l/16Wastewater) for more information on how phosphorus is removed from wastewater.

Selective precipitation can also be used in qualitative analysis. In this method, reagents are added to an unknown chemical mixture in order to induce precipitation. Certain reagents cause specific ions to precipitate out; therefore, the addition of the reagent can be used to determine whether the ion is present in the solution.

Link to Learning

View this simulation (http://openstaxcollege.org/l/16solublesalts) to study the process of salts dissolving and forming saturated solutions and precipitates for specific compounds, or compounds for which you select the charges on the ions and the K_{sp}

Example 15.11

Precipitation of Silver Halides

A solution contains 0.0010 mol of KI and 0.10 mol of KCl per liter. $AgNO_3$ is gradually added to this solution. Which forms first, solid AgI or solid AgCl?

Solution

The two equilibria involved are:

$$AgCl(s) \rightleftharpoons Ag^+(aq) + Cl^-(aq) \qquad K_{sp} = 1.6 \times 10^{-10}$$
$$AgI(s) \rightleftharpoons Ag^+(aq) + I^-(aq) \qquad K_{sp} = 1.5 \times 10^{-16}$$

If the solution contained about *equal* concentrations of Cl^- and I^-, then the silver salt with the smallest K_{sp} (AgI) would precipitate first. The concentrations are not equal, however, so we should find the $[Ag^+]$ at which AgCl begins to precipitate and the $[Ag^+]$ at which AgI begins to precipitate. The salt that forms at the lower $[Ag^+]$ precipitates first.

For AgI: AgI precipitates when Q equals K_{sp} for AgI (1.5×10^{-16}). When $[I^-] = 0.0010\ M$:

$$Q = [Ag^+][I^-] = [Ag^+](0.0010) = 1.5 \times 10^{-16}$$
$$[Ag^+] = \frac{1.8 \times 10^{-10}}{0.10} = 1.6 \times 10^{-9}$$

AgI begins to precipitate when $[Ag^+]$ is 1.5×10^{-13} M.

For AgCl: AgCl precipitates when Q equals K_{sp} for AgCl (1.6×10^{-10}). When $[Cl^-] = 0.10$ M:

$$Q_{sp} = [Ag^+][Cl^-] = [Ag^+](0.10) = 1.6 \times 10^{-10}$$

$$[Ag^+] = \frac{1.8 \times 10^{-10}}{0.10} = 1.6 \times 10^{-9} \ M$$

AgCl begins to precipitate when $[Ag^+]$ is 1.6×10^{-9} M.

AgI begins to precipitate at a lower $[Ag^+]$ than AgCl, so AgI begins to precipitate first.

Check Your Learning

If silver nitrate solution is added to a solution which is 0.050 M in both Cl^- and Br^- ions, at what $[Ag^+]$ would precipitation begin, and what would be the formula of the precipitate?

Answer: $[Ag^+] = 1.0 \times 10^{-11}$ M; AgBr precipitates first

Common Ion Effect

As we saw when we discussed buffer solutions, the hydronium ion concentration of an aqueous solution of acetic acid decreases when the strong electrolyte sodium acetate, $NaCH_3CO_2$, is added. We can explain this effect using Le Châtelier's principle. The addition of acetate ions causes the equilibrium to shift to the left, decreasing the concentration of H_3O^+ to compensate for the increased acetate ion concentration. This increases the concentration of CH_3CO_2H:

$$CH_3CO_2H + H_2O \rightleftharpoons H_3O^+ + CH_3CO_2^-$$

Because sodium acetate and acetic acid have the acetate ion in common, the influence on the equilibrium is called the **common ion effect**.

The common ion effect can also have a direct effect on solubility equilibria. Suppose we are looking at the reaction where silver iodide is dissolved:

$$AgI(s) \rightleftharpoons Ag^+(aq) + I^-(aq)$$

If we were to add potassium iodide (KI) to this solution, we would be adding a substance that shares a common ion with silver iodide. Le Châtelier's principle tells us that when a change is made to a system at equilibrium, the reaction will shift to counteract that change. In this example, there would be an excess of iodide ions, so the reaction would shift toward the left, causing more silver iodide to precipitate out of solution.

Link to Learning

View this simulation (http://openstaxcollege.org/l/16commonion) to see how the common ion effect work with different concentrations of salts.

Example 15.12

Common Ion Effect

Calculate the molar solubility of cadmium sulfide (CdS) in a 0.010-M solution of cadmium bromide (CdBr$_2$). The K_{sp} of CdS is 1.0×10^{-28}.

Solution

The first thing you should notice is that the cadmium sulfide is dissolved in a solution that contains cadmium ions. We need to use an ICE table to set up this problem and include the CdBr$_2$ concentration as a contributor of cadmium ions:

$$CdS(s) \rightleftharpoons Cd^{2+}(aq) + S^{2-}(aq)$$

	CdS $\rightleftharpoons$	Cd^{2+} +	S^{2-}
Initial concentration (*M*)		0.010	0
Change (*M*)		*x*	*x*
Equilibrium concentration (*M*)		0.010 + *x*	0 + *x* = *x*

$$K_{sp} = [Cd^{2+}][S^{2-}] = 1.0 \times 10^{-28}$$
$$(0.010 + x)(x) = 1.0 \times 10^{-28}$$
$$x^2 + 0.010x - 1.0 \times 10^{-28} = 0$$

We can solve this equation using the quadratic formula, but we can also make an assumption to make this calculation much simpler. Since the K_{sp} value is so small compared with the cadmium concentration, we can assume that the change between the initial concentration and the equilibrium concentration is negligible, so that $0.010 + x \sim 0.010$. Going back to our K_{sp} expression, we would now get:

$$K_{sp} = [Cd^{2+}][S^{2-}] = 1.0 \times 10^{-28}$$
$$(0.010)(x) = 1.0 \times 10^{-28}$$
$$x = 1.0 \times 10^{-26}$$

Therefore, the molar solubility of CdS in this solution is 1.0×10^{-26} M.

Check Your Learning

Calculate the molar solubility of aluminum hydroxide, Al(OH)$_3$, in a 0.015-M solution of aluminum nitrate, Al(NO$_3$)$_3$. The K_{sp} of Al(OH)$_3$ is 2×10^{-32}.

Answer: 1×10^{-10} M

15.2 Lewis Acids and Bases

By the end of this section, you will be able to:

- Explain the Lewis model of acid-base chemistry
- Write equations for the formation of adducts and complex ions
- Perform equilibrium calculations involving formation constants

In 1923, G. N. Lewis proposed a generalized definition of acid-base behavior in which acids and bases are identified by their ability to accept or to donate a pair of electrons and form a coordinate covalent bond.

A **coordinate covalent bond** (or dative bond) occurs when one of the atoms in the bond provides both bonding electrons. For example, a coordinate covalent bond occurs when a water molecule combines with a hydrogen ion to form a hydronium ion. A coordinate covalent bond also results when an ammonia molecule combines with a hydrogen ion to form an ammonium ion. Both of these equations are shown here.

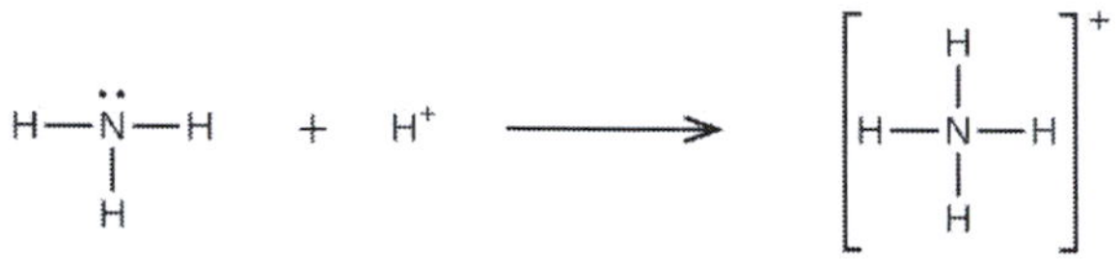

A **Lewis acid** is any species (molecule or ion) that can accept a pair of electrons, and a **Lewis base** is any species (molecule or ion) that can donate a pair of electrons.

A Lewis acid-base reaction occurs when a base donates a pair of electrons to an acid. A **Lewis acid-base adduct**, a compound that contains a coordinate covalent bond between the Lewis acid and the Lewis base, is formed. The following equations illustrate the general application of the Lewis concept.

The boron atom in boron trifluoride, BF_3, has only six electrons in its valence shell. Being short of the preferred octet, BF_3 is a very good Lewis acid and reacts with many Lewis bases; a fluoride ion is the Lewis base in this reaction, donating one of its lone pairs:

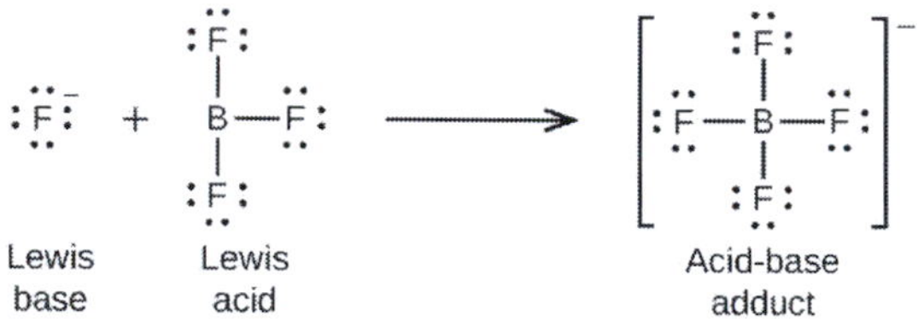

In the following reaction, each of two ammonia molecules, Lewis bases, donates a pair of electrons to a silver ion, the Lewis acid:

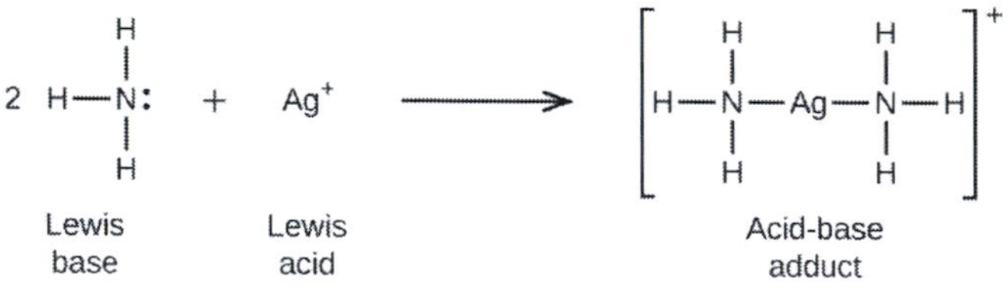

Nonmetal oxides act as Lewis acids and react with oxide ions, Lewis bases, to form oxyanions:

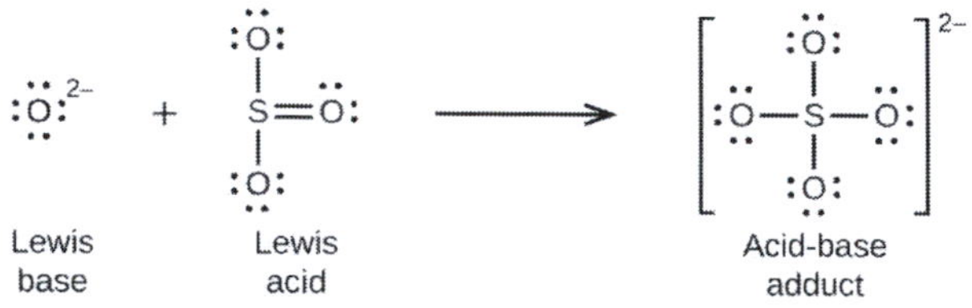

Many Lewis acid-base reactions are displacement reactions in which one Lewis base displaces another Lewis base from an acid-base adduct, or in which one Lewis acid displaces another Lewis acid:

The last displacement reaction shows how the reaction of a Brønsted-Lowry acid with a base fits into the Lewis concept. A Brønsted-Lowry acid such as HCl is an acid-base adduct according to the Lewis concept, and proton transfer occurs because a more stable acid-base adduct is formed. Thus, although the definitions of acids and bases in the two theories are quite different, the theories overlap considerably.

Many slightly soluble ionic solids dissolve when the concentration of the metal ion in solution is decreased through the formation of complex (polyatomic) ions in a Lewis acid-base reaction. For example, silver chloride dissolves in a solution of ammonia because the silver ion reacts with ammonia to form the **complex ion** $Ag(NH_3)_2^+$. The Lewis structure of the $Ag(NH_3)_2^+$ ion is:

The equations for the dissolution of $AgCl$ in a solution of NH_3 are:

$$AgCl(s) \longrightarrow Ag^+(aq) + Cl^-(aq)$$
$$Ag^+(aq) + 2NH_3(aq) \longrightarrow Ag(NH_3)_2^+(aq)$$
$$\text{Net: } AgCl(s) + 2NH_3(aq) \longrightarrow Ag(NH_3)_2^+(aq) + Cl^-(aq)$$

Aluminum hydroxide dissolves in a solution of sodium hydroxide or another strong base because of the formation of the complex ion $Al(OH)_4^-$. The Lewis structure of the $Al(OH)_4^-$ ion is:

The equations for the dissolution are:

$$Al(OH)_3(s) \longrightarrow Al^{3+}(aq) + 3OH^-(aq)$$

$$Al^{3+}(aq) + 4OH^-(aq) \longrightarrow Al(OH)_4{}^-(aq)$$
$$\text{Net: } Al(OH)_3(s) + OH^-(aq) \longrightarrow Al(OH)_4{}^-(aq)$$

Mercury(II) sulfide dissolves in a solution of sodium sulfide because HgS reacts with the S^{2-} ion:

$$HgS(s) \longrightarrow Hg^{2+}(aq) + S^{2-}(aq)$$
$$Hg^{2+}(aq) + 2S^{2-}(aq) \longrightarrow HgS_2{}^{2-}(aq)$$
$$\text{Net: } HgS(s) + S^{2-}(aq) \longrightarrow HgS_2{}^{2-}(aq)$$

A complex ion consists of a central atom, typically a transition metal cation, surrounded by ions, or molecules called **ligands**. These ligands can be neutral molecules like H_2O or NH_3, or ions such as CN^- or OH^-. Often, the ligands act as Lewis bases, donating a pair of electrons to the central atom. The ligands aggregate themselves around the central atom, creating a new ion with a charge equal to the sum of the charges and, most often, a transitional metal ion. This more complex arrangement is why the resulting ion is called a *complex ion*. The complex ion formed in these reactions cannot be predicted; it must be determined experimentally. The types of bonds formed in complex ions are called coordinate covalent bonds, as electrons from the ligands are being shared with the central atom. Because of this, complex ions are sometimes referred to as coordination complexes. This will be studied further in upcoming chapters.

The equilibrium constant for the reaction of the components of a complex ion to form the complex ion in solution is called a **formation constant (K_f)** (sometimes called a stability constant). For example, the complex ion $Cu(CN)_2{}^-$ is shown here:

$$:N \equiv C - Cu - C \equiv N:$$

It forms by the reaction:

$$Cu^+(aq) + 2CN^-(aq) \rightleftharpoons Cu(CN)_2{}^-(aq)$$

At equilibrium:

$$K_f = Q = \frac{[Cu(CN)_2{}^-]}{[Cu^+][CN^-]^2}$$

The inverse of the formation constant is the **dissociation constant (K_d)**, the equilibrium constant for the *decomposition* of a complex ion into its components in solution. We will work with dissociation constants further in the exercises for this section. Appendix K and Table 15.2 are tables of formation constants. In general, the larger the formation constant, the more stable the complex; however, as in the case of K_{sp} values, the stoichiometry of the compound must be considered.

Common Complex Ions by Decreasing Formulation Constants

Substance	K_f at 25 °C
$[Cd(CN)_4]^{2-}$	1.3×10^7
$Ag(NH_3)_2{}^+$	1.7×10^7
$[AlF_6]^{3-}$	7×10^{19}

Table 15.2

As an example of dissolution by complex ion formation, let us consider what happens when we add aqueous ammonia to a mixture of silver chloride and water. Silver chloride dissolves slightly in water, giving a small concentration of Ag^+ ($[Ag^+] = 1.3 \times 10^{-5}$ M):

$$AgCl(s) \rightleftharpoons Ag^+(aq) + Cl^-(aq)$$

However, if NH_3 is present in the water, the complex ion, $Ag(NH_3)_2{}^+$, can form according to the equation:

$$Ag^+(aq) + 2NH_3(aq) \rightleftharpoons Ag(NH_3)_2{}^+(aq)$$

with

$$K_f = \frac{[Ag(NH_3)_2{}^+]}{[Ag^+][NH_3]^2} = 1.6 \times 10^7$$

The large size of this formation constant indicates that most of the free silver ions produced by the dissolution of AgCl combine with NH_3 to form $Ag(NH_3)_2{}^+$. As a consequence, the concentration of silver ions, $[Ag^+]$, is reduced, and the reaction quotient for the dissolution of silver chloride, $[Ag^+][Cl^-]$, falls below the solubility product of AgCl:

$$Q = [Ag^+][Cl^-] < K_{sp}$$

More silver chloride then dissolves. If the concentration of ammonia is great enough, all of the silver chloride dissolves.

Example 15.13

Dissociation of a Complex Ion

Calculate the concentration of the silver ion in a solution that initially is 0.10 M with respect to $Ag(NH_3)_2{}^+$.

Solution

We use the familiar path to solve this problem:

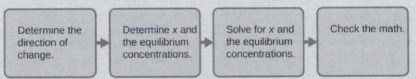

Step 1. *Determine the direction of change.* The complex ion $Ag(NH_3)_2{}^+$ is in equilibrium with its components, as represented by the equation:

$$Ag^+(aq) + 2NH_3(aq) \rightleftharpoons Ag(NH_3)_2{}^+(aq)$$

We write the equilibrium as a formation reaction because Appendix K lists formation constants for complex ions. Before equilibrium, the reaction quotient is larger than the equilibrium constant [$K_f = 1.6 \times 10^7$, and $Q = \frac{0.10}{0 \times 0}$, it is infinitely large], so the reaction shifts to the left to reach equilibrium.

Step 2. *Determine x and equilibrium concentrations.* We let the change in concentration of Ag^+ be x. Dissociation of 1 mol of $Ag(NH_3)_2{}^+$ gives 1 mol of Ag^+ and 2 mol of NH_3, so the change in $[NH_3]$ is 2x and that of $Ag(NH_3)_2{}^+$ is –x. In summary:

	Ag^+	+	$2NH_3$	$\rightleftharpoons$	$Ag(NH_3)_2{}^+$
Initial concentration (M)	0		0		0.10
Change (M)	x		2x		–x
Equilibrium concentration (M)	0 + x		0 + 2x		0.10 – x

Step 3. *Solve for x and the equilibrium concentrations.* At equilibrium:

$$K_f = \frac{[Ag(NH_3)_2{}^+]}{[Ag^+][NH_3]^2}$$

$$1.6 \times 10^7 = \frac{0.10 - x}{(x)(2x)^2}$$

Both Q and K_f are much larger than 1, so let us assume that the changes in concentrations needed to reach equilibrium are small. Thus $0.10 - x$ is approximated as 0.10:

$$1.6 \times 10^7 = \frac{0.10 - x}{(x)(2x)^2}$$

$$x^3 = \frac{0.10}{4(1.6 \times 10^7)} = 1.6 \times 10^{-9}$$

$$x = \sqrt[3]{1.6 \times 10^{-19}} = 1.2 \times 10^{-3}$$

Because only 1.2% of the $Ag(NH_3)_2{}^+$ dissociates into Ag^+ and NH_3, the assumption that x is small is justified.

Now we determine the equilibrium concentrations:

$$[Ag^+] = 0 + x = 1.2 \times 10^{-3} \ M$$
$$[NH_3] = 0 + 2x = 2.4 \times 10^{-3} \ M$$
$$[Ag(NH_3)_2{}^+] = 0.10 - x = 0.10 - 0.0012 = 0.099$$

The concentration of free silver ion in the solution is 0.0012 M.

Step 4.   *Check the work.* The value of Q calculated using the equilibrium concentrations is equal to K_f within the error associated with the significant figures in the calculation.

Check Your Learning

Calculate the silver ion concentration, $[Ag^+]$, of a solution prepared by dissolving 1.00 g of $AgNO_3$ and 10.0 g of KCN in sufficient water to make 1.00 L of solution. (Hint: Because $Q < K_f$, assume the reaction goes to completion then calculate the $[Ag^+]$ produced by dissociation of the complex.)

Answer: $3 \times 10^{-21} \ M$

15.3 Multiple Equilibria

By the end of this section, you will be able to:

- Describe examples of systems involving two (or more) simultaneous chemical equilibria
- Calculate reactant and product concentrations for multiple equilibrium systems
- Compare dissolution and weak electrolyte formation

There are times when one equilibrium reaction does not adequately describe the system being studied. Sometimes we have more than one type of equilibrium occurring at once (for example, an acid-base reaction and a precipitation reaction).

The ocean is a unique example of a system with **multiple equilibria**, or multiple states of solubility equilibria working simultaneously. Carbon dioxide in the air dissolves in sea water, forming carbonic acid (H_2CO_3). The carbonic acid then ionizes to form hydrogen ions and bicarbonate ions ($HCO_3{}^-$), which can further ionize into more hydrogen ions and carbonate ions ($CO_3{}^{2-}$):

$$CO_2(g) \rightleftharpoons CO_2(aq)$$
$$CO_2(aq) + H_2O \rightleftharpoons H_2CO_3(aq)$$
$$H_2CO_3(aq) \rightleftharpoons H^+(aq) + HCO_3{}^-(aq)$$

$$HCO_3{}^-(aq) \rightleftharpoons H^+(aq) + CO_3{}^{2-}(aq)$$

The excess H^+ ions make seawater more acidic. Increased ocean acidification can then have negative impacts on reef-building coral, as they cannot absorb the calcium carbonate they need to grow and maintain their skeletons (Figure 15.7). This in turn disrupts the local biosystem that depends upon the health of the reefs for its survival. If enough local reefs are similarly affected, the disruptions to sea life can be felt globally. The world's oceans are presently in the midst of a period of intense acidification, believed to have begun in the mid-nineteenth century, and which is now accelerating at a rate faster than any change to oceanic pH in the last 20 million years.

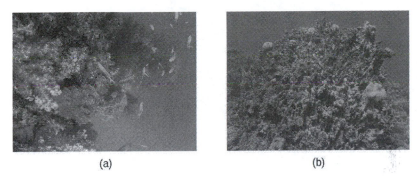

(a)   (b)

Figure 15.7   Healthy coral reefs (a) support a dense and diverse array of sea life across the ocean food chain. But when coral are unable to adequately build and maintain their calcium carbonite skeletons because of excess ocean acidification, the unhealthy reef (b) is only capable of hosting a small fraction of the species as before, and the local food chain starts to collapse. (credit a: modification of work by NOAA Photo Library; credit b: modification of work by "prilfish"/Flickr)

Link to Learning

Learn more about ocean acidification (http://openstaxcollege.org/l/16acidicocean) and how it affects other marine creatures.

This site (http://openstaxcollege.org/l/16coralreef) has detailed information about how ocean acidification specifically affects coral reefs.

Slightly soluble solids derived from weak acids generally dissolve in strong acids, unless their solubility products are extremely small. For example, we can dissolve $CuCO_3$, FeS, and $Ca_3(PO_4)_2$ in HCl because their basic anions react to form weak acids (H_2CO_3, H_2S, and $H_2PO_4{}^-$). The resulting decrease in the concentration of the anion results in a shift of the equilibrium concentrations to the right in accordance with Le Châtelier's principle.

Of particular relevance to us is the dissolution of hydroxylapatite, $Ca_5(PO_4)_3OH$, in acid. Apatites are a class of calcium phosphate minerals (Figure 15.8); a biological form of hydroxylapatite is found as the principal mineral in the enamel of our teeth. A mixture of hydroxylapatite and water (or saliva) contains an equilibrium mixture of solid $Ca_5(PO_4)_3OH$ and dissolved Ca^{2+}, $PO_4{}^{3-}$, and OH^- ions:

$$Ca_5(PO_4)_3 OH(s) \longrightarrow 5Ca^{2+}(aq) + 3PO_4{}^{3-}(aq) + OH^-(aq)$$

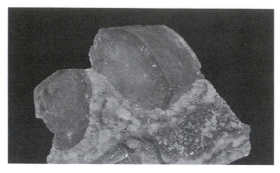

Figure 15.8 Crystal of the mineral hydroxylapatite, $Ca_5(PO_4)_3OH$, is shown here. Pure apatite is white, but like many other minerals, this sample is colored because of the presence of impurities.

When exposed to acid, phosphate ions react with hydronium ions to form hydrogen phosphate ions and ultimately, phosphoric acid:

$$PO_4^{3-}(aq) + H_3O^+ \longrightarrow H_2PO_4^{2-} + H_2O$$
$$PO_4^{2-}(aq) + H_3O^+ \longrightarrow H_2PO_4^- + H_2O$$
$$H_2PO_4^- + H_3O^+ \longrightarrow H_3PO_4 + H_2O$$

Hydroxide ion reacts to form water:

$$OH^-(aq) + H_3O^+ \longrightarrow 2H_2O$$

These reactions decrease the phosphate and hydroxide ion concentrations, and additional hydroxylapatite dissolves in an acidic solution in accord with Le Châtelier's principle. Our teeth develop cavities when acid waste produced by bacteria growing on them causes the hydroxylapatite of the enamel to dissolve. Fluoride toothpastes contain sodium fluoride, NaF, or stannous fluoride [more properly named tin(II) fluoride], SnF_2. They function by replacing the OH^- ion in hydroxylapatite with F^- ion, producing fluorapatite, $Ca_5(PO_4)_3F$:

$$NaF + Ca_5(PO_4)_3OH \rightleftharpoons Ca_5(PO_4)_3F + Na^+ + OH^-$$

The resulting $Ca_5(PO_4)_3F$ is slightly less soluble than $Ca_5(PO_4)_3OH$, and F^- is a weaker base than OH^-. Both of these factors make the fluorapatite more resistant to attack by acids than hydroxylapatite. See the Chemistry in Everyday Life feature on the role of fluoride in preventing tooth decay for more information.

Chemistry in Everyday Life

Role of Fluoride in Preventing Tooth Decay

As we saw previously, fluoride ions help protect our teeth by reacting with hydroxylapatite to form fluorapatite, $Ca_5(PO_4)_3F$. Since it lacks a hydroxide ion, fluorapatite is more resistant to attacks by acids in our mouths and is thus less soluble, protecting our teeth. Scientists discovered that naturally fluorinated water could be beneficial to your teeth, and so it became common practice to add fluoride to drinking water. Toothpastes and mouthwashes also contain amounts of fluoride (Figure 15.9).

Figure 15.9   Fluoride, found in many toothpastes, helps prevent tooth decay (credit: Kerry Ceszyk).

Unfortunately, excess fluoride can negate its advantages. Natural sources of drinking water in various parts of the world have varying concentrations of fluoride, and places where that concentration is high are prone to certain health risks when there is no other source of drinking water. The most serious side effect of excess fluoride is the bone disease, skeletal fluorosis. When excess fluoride is in the body, it can cause the joints to stiffen and the bones to thicken. It can severely impact mobility and can negatively affect the thyroid gland. Skeletal fluorosis is a condition that over 2.7 million people suffer from across the world. So while fluoride can protect our teeth from decay, the US Environmental Protection Agency sets a maximum level of 4 ppm (4 mg/ L) of fluoride in drinking water in the US. Fluoride levels in water are not regulated in all countries, so fluorosis is a problem in areas with high levels of fluoride in the groundwater.

When acid rain attacks limestone or marble, which are calcium carbonates, a reaction occurs that is similar to the acid attack on hydroxylapatite. The hydronium ion from the acid rain combines with the carbonate ion from calcium carbonates and forms the hydrogen carbonate ion, a weak acid:

$$H_3O^+(aq) + CO_3{}^{2-}(aq) \longrightarrow HCO_3{}^-(aq) + H_2O(l)$$

Calcium hydrogen carbonate, $Ca(HCO_3)_2$, is soluble, so limestone and marble objects slowly dissolve in acid rain.

If we add calcium carbonate to a concentrated acid, hydronium ion reacts with the carbonate ion according to the equation:

$$2H_3O^+(aq) + CO_3{}^{2-}(aq) \longrightarrow H_2CO_3(aq) + 2H_2O(l)$$

(Acid rain is usually not sufficiently acidic to cause this reaction; however, laboratory acids are.) The solution may become saturated with the weak electrolyte carbonic acid, which is unstable, and carbon dioxide gas can be evolved:

$$H_2CO_3(aq) \longrightarrow CO_2(g) + H_2O(l)$$

These reactions decrease the carbonate ion concentration, and additional calcium carbonate dissolves. If enough acid is present, the concentration of carbonate ion is reduced to such a low level that the reaction quotient for the dissolution of calcium carbonate remains less than the solubility product of calcium carbonate, even after all of the calcium carbonate has dissolved.

Example 15.14

Prevention of Precipitation of Mg(OH)₂

Calculate the concentration of ammonium ion that is required to prevent the precipitation of $Mg(OH)_2$ in a solution with $[Mg^{2+}] = 0.10\ M$ and $[NH_3] = 0.10\ M$.

Solution

Two equilibria are involved in this system:

Reaction (1): $Mg(OH)_2(s) \rightleftharpoons Mg^{2+}(aq) + 2OH^-(aq)$; $K_{sp} = 1.5 \times 10^{-11}$

Reaction (2): $NH_3(aq) + H_2O(l) \rightleftharpoons NH_4^+(aq) + OH^-(aq)$ $K_{sp} = 1.8 \times 10^{-5}$

To prevent the formation of solid $Mg(OH)_2$, we must adjust the concentration of OH^- so that the reaction quotient for Equation (1), $Q = [Mg^{2+}][OH^-]^2$, is less than K_{sp} for $Mg(OH)_2$. (To simplify the calculation, we determine the concentration of OH^- when $Q = K_{sp}$.) $[OH^-]$ can be reduced by the addition of NH_4^+, which shifts Reaction (2) to the left and reduces $[OH^-]$.

Step 1. *We determine the $[OH^-]$ at which $Q = K_{sp}$ when $[Mg^{2+}] = 0.10\ M$:*

$$Q = [Mg^{2+}][OH^-]^2 = (0.10)[OH^-]^2 = 1.5 \times 10^{-11}$$

$$[OH^-] = 1.2 \times 10^{-5}\ M$$

Solid $Mg(OH)_2$ will not form in this solution when $[OH^-]$ is less than $1.2 \times 10^{-5}\ M$.

Step 2. *We calculate the $[NH_4^+]$ needed to decrease $[OH^-]$ to $1.2 \times 10^{-5}\ M$ when $[NH_3] = 0.10$.*

$$K_b = \frac{[NH_4^+][OH^-]}{[NH_3]} = \frac{[NH_4^+](1.2 \times 10^{-5})}{0.10} = 1.8 \times 10^{-5}$$

$$[NH_4^+] = 0.15\ M$$

When $[NH_4^+]$ equals $0.15\ M$, $[OH^-]$ will be $1.2 \times 10^{-5}\ M$. Any $[NH_4^+]$ greater than $0.15\ M$ will reduce $[OH^-]$ below $1.2 \times 10^{-5}\ M$ and prevent the formation of $Mg(OH)_2$.

Check Your Learning

Consider the two equilibria:

$$ZnS(s) \rightleftharpoons Zn^{2+}(aq) + S^{2-}(aq) \qquad K_{sp} = 1 \times 10^{-27}$$

$$2H_2O(l) + H_2S(aq) \rightleftharpoons 2H_3O^+(aq) + S^{2-}(aq) \qquad K = 1.0 \times 10^{-26}$$

and calculate the concentration of hydronium ion required to prevent the precipitation of ZnS in a solution that is $0.050\ M$ in Zn^{2+} and saturated with H_2S (0.10 M H_2S).

Answer: $[H_3O^+] > 0.2\ M$ ($[S^{2-}]$ is less than $2 \times 10^{-26}\ M$ and precipitation of ZnS does not occur.)

Therefore, precise calculations of the solubility of solids from the solubility product are limited to cases in which the only significant reaction occurring when the solid dissolves is the formation of its ions.

Example 15.15

Multiple Equilibria

Unexposed silver halides are removed from photographic film when they react with sodium thiosulfate ($Na_2S_2O_3$, called hypo) to form the complex ion $Ag(S_2O_3)_2^{3-}$ ($K_f = 4.7 \times 10^{13}$). The reaction with silver bromide is:

What mass of $Na_2S_2O_3$ is required to prepare 1.00 L of a solution that will dissolve 1.00 g of AgBr by the formation of $Ag(S_2O_3)_2^{3-}$?

Solution

Two equilibria are involved when AgBr dissolves in a solution containing the $S_2O_3^{2-}$ ion:

Reaction (1): $AgBr(s) \rightleftharpoons Ag^+(aq) + Br^-(aq)$ $\qquad$ $K_{sp} = 3.3 \times 10^{-13}$

Reaction (2): $Ag^+(aq) + S_2O_3^{2-}(aq) \rightleftharpoons Ag(S_2O_3)_2^{3-}(aq)$ $\qquad$ $K_f = 4.7 \times 10^{13}$

In order for 1.00 g of AgBr to dissolve, the $[Ag^+]$ in the solution that results must be low enough for Q for Reaction (1) to be smaller than K_{sp} for this reaction. We reduce $[Ag^+]$ by adding $S_2O_3^{2-}$ and thus cause Reaction (2) to shift to the right. We need the following steps to determine what mass of $Na_2S_2O_3$ is needed to provide the necessary $S_2O_3^{2-}$.

> **Step 1.** *We calculate the $[Br^-]$ produced by the complete dissolution of 1.00 g of AgBr (5.33 $\times$ 10^{-3} mol AgBr) in 1.00 L of solution:*
>
> $$[Br^-] = 5.33 \times 10^{-3} \ M$$
>
> **Step 2.** *We use $[Br^-]$ and K_{sp} to determine the maximum possible concentration of Ag^+ that can be present without causing reprecipitation of AgBr:*
>
> $$[Ag^+] = 6.2 \times 10^{-11} \ M$$
>
> **Step 3.** *We determine the $[S_2O_3^{2-}]$ required to make $[Ag^+] = 6.2 \times 10^{-11}$ M after the remaining Ag^+ ion has reacted with $S_2O_3^{2-}$ according to the equation:*
>
> $$Ag^+ + 2S_2O_3^{2-} \rightleftharpoons Ag(S_2O_3)_2^{3-} \qquad K_f = 4.7 \times 10^{13}$$
>
> Because 5.33×10^{-3} mol of AgBr dissolves:
>
> $$(5.33 \times 10^{-3}) - (6.2 \times 10^{-11}) = 5.33 \times 10^{-3} \ mol \ Ag(S_2O_3)_2^{3-}$$
>
> Thus, at equilibrium: $[Ag(S_2O_3)_2^{3-}] = 5.33 \times 10^{-3}$ M, $[Ag^+] = 6.2 \times 10^{-11}$ M, and $Q = K_f = 4.7 \times 10^{13}$:
>
> $$K_f = \frac{[Ag(S_2O_3)_2^{3-}]}{[Ag^+][S_2O_3^{2-}]^2} = 4.7 \times 10^{13}$$
>
> $$[S_2O_3^{2-}] = 1.4 \times 10^{-3} \ M$$
>
> When $[S_2O_3^{2-}]$ is 1.4×10^{-3} M, $[Ag^+]$ is 6.2×10^{-11} M and all AgBr remains dissolved.
>
> **Step 4.** *We determine the total number of moles of $S_2O_3^{2-}$ that must be added to the solution.* This equals the amount that reacts with Ag^+ to form $Ag(S_2O_3)_2^{3-}$ plus the amount of free $S_2O_3^{2-}$ in solution at equilibrium. To form 5.33×10^{-3} mol of $Ag(S_2O_3)_2^{3-}$ requires $2 \times$ (5.33×10^{-3}) mol of $S_2O_3^{2-}$. In addition, 1.4×10^{-3} mol of unreacted $S_2O_3^{2-}$ is present (Step 3). Thus, the total amount of $S_2O_3^{2-}$ that must be added is:
>
> $$2 \times (5.33 \times 10^{-3} \ mol \ S_2O_3^{2-}) + 1.4 \times 10^{-3} \ mol \ S_2O_3^{2-} = 1.21 \times 10^{-2} \ mol \ S_2O_3^{2-}$$
>
> **Step 5.** *We determine the mass of $Na_2S_2O_3$ required to give 1.21×10^{-2} mol $S_2O_3^{2-}$ using the molar mass of $Na_2S_2O_3$:*
>
> $$1.21 \times 10^{-2} \ mol \ S_2O_3^{2-} \times \frac{158.1 \ g \ Na_2S_2O_3}{1 \ mol \ Na_2S_2O_3} = 1.9 \ g \ Na_2S_2O_3$$

Thus, 1.00 L of a solution prepared from 1.9 g $Na_2S_2O_3$ dissolves 1.0 g of AgBr.

Check Your Learning

AgCl(s), silver chloride, is well known to have a very low solubility: $Ag(s) \rightleftharpoons Ag^+(aq) + Cl^-(aq)$, K_{sp} = 1.77 × 10^{-10}. Adding ammonia significantly increases the solubility of AgCl because a complex ion is formed: $Ag^+(aq) + 2NH_3(aq) \rightleftharpoons Ag(NH_3)_2^+(aq)$, K_f = 1.7 × 10^7. What mass of NH$_3$ is required to prepare 1.00 L of solution that will dissolve 2.00 g of AgCl by formation of $Ag(NH_3)_2^+$?

Answer: 1.00 L of a solution prepared with 4.84 g NH$_3$ dissolves 2.0 g of AgCl.

Dissolution versus Weak Electrolyte Formation

We can determine how to shift the concentration of ions in the equilibrium between a slightly soluble solid and a solution of its ions by applying Le Châtelier's principle. For example, one way to control the concentration of manganese(II) ion, Mn^{2+}, in a solution is to adjust the pH of the solution and, consequently, to manipulate the equilibrium between the slightly soluble solid manganese(II) hydroxide, manganese(II) ion, and hydroxide ion:

$$Mn(OH)_2(s) \rightleftharpoons Mn^{2+}(aq) + 2OH^-(aq) \qquad K_{sp} = [Mn^{2+}][OH^-]^2$$

This could be important to a laundry because clothing washed in water that has a manganese concentration exceeding 0.1 mg per liter may be stained by the manganese. We can reduce the concentration of manganese by increasing the concentration of hydroxide ion. We could add, for example, a small amount of NaOH or some other base such as the silicates found in many laundry detergents. As the concentration of OH$^-$ ion increases, the equilibrium responds by shifting to the left and reducing the concentration of Mn^{2+} ion while increasing the amount of solid $Mn(OH)_2$ in the equilibrium mixture, as predicted by Le Châtelier's principle.

Example 15.16

Solubility Equilibrium of a Slightly Soluble Solid

What is the effect on the amount of solid $Mg(OH)_2$ that dissolves and the concentrations of Mg^{2+} and OH$^-$ when each of the following are added to a mixture of solid $Mg(OH)_2$ in water at equilibrium?

(a) MgCl$_2$

(b) KOH

(c) an acid

(d) NaNO$_3$

(e) Mg(OH)$_2$

Solution

The equilibrium among solid $Mg(OH)_2$ and a solution of Mg^{2+} and OH$^-$ is:

$$Mg(OH)_2(s) \rightleftharpoons Mg^{2+}(aq) + 2OH^-(aq)$$

(a) The reaction shifts to the left to relieve the stress produced by the additional Mg^{2+} ion, in accordance with Le Châtelier's principle. In quantitative terms, the added Mg^{2+} causes the reaction quotient to be larger than the solubility product ($Q > K_{sp}$), and $Mg(OH)_2$ forms until the reaction quotient again equals K_{sp}. At the new equilibrium, [OH$^-$] is less and [Mg^{2+}] is greater than in the solution of $Mg(OH)_2$ in pure water. More solid $Mg(OH)_2$ is present.

(b) The reaction shifts to the left to relieve the stress of the additional OH$^-$ ion. $Mg(OH)_2$ forms until the reaction quotient again equals K_{sp}. At the new equilibrium, [OH$^-$] is greater and [Mg^{2+}] is less than in the solution of $Mg(OH)_2$ in pure water. More solid $Mg(OH)_2$ is present.

(c) The concentration of OH^- is reduced as the OH^- reacts with the acid. The reaction shifts to the right to relieve the stress of less OH^- ion. In quantitative terms, the decrease in the OH^- concentration causes the reaction quotient to be smaller than the solubility product ($Q < K_{sp}$), and additional $Mg(OH)_2$ dissolves until the reaction quotient again equals K_{sp}. At the new equilibrium, $[OH^-]$ is less and $[Mg^{2+}]$ is greater than in the solution of $Mg(OH)_2$ in pure water. More $Mg(OH)_2$ is dissolved.

(d) $NaNO_3$ contains none of the species involved in the equilibrium, so we should expect that it has no appreciable effect on the concentrations of Mg^{2+} and OH^-. (As we have seen previously, dissolved salts change the activities of the ions of an electrolyte. However, the salt effect is generally small, and we shall neglect the slight errors that may result from it.)

(e) The addition of solid $Mg(OH)_2$ has no effect on the solubility of $Mg(OH)_2$ or on the concentration of Mg^{2+} and OH^-. The concentration of $Mg(OH)_2$ does not appear in the equation for the reaction quotient:

$$Q = [Mg^{2+}][OH^-]^2$$

Thus, changing the amount of solid magnesium hydroxide in the mixture has no effect on the value of Q, and no shift is required to restore Q to the value of the equilibrium constant.

Check Your Learning

What is the effect on the amount of solid $NiCO_3$ that dissolves and the concentrations of Ni^{2+} and $CO_3{}^{2-}$ when each of the following are added to a mixture of the slightly soluble solid $NiCO_3$ and water at equilibrium?

(a) $Ni(NO_3)_2$

(b) $KClO_4$

(c) $NiCO_3$

(d) K_2CO_3

(e) HNO_3 (reacts with carbonate giving $HCO_3{}^-$ or H_2O and CO_2)

Answer: (a) mass of $NiCO_3(s)$ increases, $[Ni^{2+}]$ increases, $[CO_3{}^{2-}]$ decreases; (b) no appreciable effect; (c) no effect except to increase the amount of solid $NiCO_3$; (d) mass of $NiCO_3(s)$ increases, $[Ni^{2+}]$ decreases, $[CO_3{}^{2-}]$ increases; (e) mass of $NiCO_3(s)$ decreases, $[Ni^{2+}]$ increases, $[CO_3{}^{2-}]$ decreases

Key Terms

common ion effect effect on equilibrium when a substance with an ion in common with the dissolved species is added to the solution; causes a decrease in the solubility of an ionic species, or a decrease in the ionization of a weak acid or base

complex ion ion consisting of a transition metal central atom and surrounding molecules or ions called ligands

coordinate covalent bond (also, dative bond) bond formed when one atom provides both electrons in a shared pair

dissociation constant (K_d) equilibrium constant for the decomposition of a complex ion into its components in solution

formation constant (K_f) (also, stability constant) equilibrium constant for the formation of a complex ion from its components in solution

Lewis acid any species that can accept a pair of electrons and form a coordinate covalent bond

Lewis acid-base adduct compound or ion that contains a coordinate covalent bond between a Lewis acid and a Lewis base

Lewis base any species that can donate a pair of electrons and form a coordinate covalent bond

ligand molecule or ion that surrounds a transition metal and forms a complex ion; ligands act as Lewis bases

molar solubility solubility of a compound expressed in units of moles per liter (mol/L)

multiple equilibrium system characterized by more than one state of balance between a slightly soluble ionic solid and an aqueous solution of ions working simultaneously

selective precipitation process in which ions are separated using differences in their solubility with a given precipitating reagent

solubility product (K_{sp}) equilibrium constant for the dissolution of a slightly soluble electrolyte

Key Equations

- $M_pX_q(s) \rightleftharpoons pM^{m+}(aq) + qX^{n-}(aq)$ $K_{sp} = [M^{m+}]^p[X^{n-}]^q$

Summary

15.1 Precipitation and Dissolution

The equilibrium constant for an equilibrium involving the precipitation or dissolution of a slightly soluble ionic solid is called the solubility product, K_{sp}, of the solid. When we have a heterogeneous equilibrium involving the slightly soluble solid M_pX_q and its ions M^{m+} and X^{n-}:

$$M_pX_q(s) \rightleftharpoons pM^{m+}(aq) + qX^{n-}(aq)$$

We write the solubility product expression as:

$$K_{sp} = [M^{m+}]^p[X^{n-}]^q$$

The solubility product of a slightly soluble electrolyte can be calculated from its solubility; conversely, its solubility can be calculated from its K_{sp}, provided the only significant reaction that occurs when the solid dissolves is the formation of its ions.

A slightly soluble electrolyte begins to precipitate when the magnitude of the reaction quotient for the dissolution reaction exceeds the magnitude of the solubility product. Precipitation continues until the reaction quotient equals the solubility product.

A reagent can be added to a solution of ions to allow one ion to selectively precipitate out of solution. The common ion effect can also play a role in precipitation reactions. In the presence of an ion in common with one of the ions in the solution, Le Châtelier's principle applies and more precipitate comes out of solution so that the molar solubility is reduced.

15.2 Lewis Acids and Bases

G.N. Lewis proposed a definition for acids and bases that relies on an atom's or molecule's ability to accept or donate electron pairs. A Lewis acid is a species that can accept an electron pair, whereas a Lewis base has an electron pair available for donation to a Lewis acid. Complex ions are examples of Lewis acid-base adducts. In a complex ion, we have a central atom, often consisting of a transition metal cation, which acts as a Lewis acid, and several neutral molecules or ions surrounding them called ligands that act as Lewis bases. Complex ions form by sharing electron pairs to form coordinate covalent bonds. The equilibrium reaction that occurs when forming a complex ion has an equilibrium constant associated with it called a formation constant, K_f. This is often referred to as a stability constant, as it represents the stability of the complex ion. Formation of complex ions in solution can have a profound effect on the solubility of a transition metal compound.

15.3 Multiple Equilibria

Several systems we encounter consist of multiple equilibria, systems where two or more equilibria processes are occurring simultaneously. Some common examples include acid rain, fluoridation, and dissolution of carbon dioxide in sea water. When looking at these systems, we need to consider each equilibrium separately and then combine the individual equilibrium constants into one solubility product or reaction quotient expression using the tools from the first equilibrium chapter. Le Châtelier's principle also must be considered, as each reaction in a multiple equilibria system will shift toward reactants or products based on what is added to the initial reaction and how it affects each subsequent equilibrium reaction.

Exercises

15.1 Precipitation and Dissolution

1. Complete the changes in concentrations for each of the following reactions:

(a) $AgI(s) \longrightarrow Ag^+(aq) + I^-(aq)$
$ x \quad \underline{}$

(b) $CaCO_3(s) \longrightarrow Ca^{2+}(aq) + CO_3{}^{2-}(aq)$
$ \underline{} \quad x$

(c) $Mg(OH)_2(s) \longrightarrow Mg^{2+}(aq) + 2OH^-(aq)$
$ x \quad \underline{}$

(d) $Mg_3(PO_4)_2(s) \longrightarrow 3Mg^{2+}(aq) + 2PO_4{}^{3-}(aq)$
$ x\underline{}$

(e) $Ca_5(PO_4)_3OH(s) \longrightarrow 5Ca^{2+}(aq) + 3PO_4{}^{3-}(aq) + OH^-(aq)$
$ \underline{} \quad \underline{} \quad x$

2. Complete the changes in concentrations for each of the following reactions:

(a) $BaSO_4(s) \longrightarrow Ba^{2+}(aq) + SO_4{}^{2-}(aq)$
$ x \quad \underline{}$

(b) $Ag_2SO_4(s) \longrightarrow 2Ag^+(aq) + SO_4{}^{2-}(aq)$

$\underline{\hspace{1.5cm}}$ x

(c) $Al(OH)_3(s) \longrightarrow Al^{3+}(aq) + 3OH^-(aq)$

x $\underline{\hspace{1.5cm}}$

(d) $Pb(OH)Cl(s) \longrightarrow Pb^{2+}(aq) + OH^-(aq) + Cl^-(aq)$

$\underline{\hspace{1.5cm}}$ x $\underline{\hspace{1.5cm}}$

(e) $Ca_3(AsO_4)_2(s) \longrightarrow 3Ca^{2+}(aq) + 2AsO_4{}^{3-}(aq)$

$3x$ $\underline{\hspace{1.5cm}}$

3. How do the concentrations of Ag^+ and $CrO_4{}^{2-}$ in a saturated solution above 1.0 g of solid Ag_2CrO_4 change when 100 g of solid Ag_2CrO_4 is added to the system? Explain.

4. How do the concentrations of Pb^{2+} and S^{2-} change when K_2S is added to a saturated solution of PbS?

5. What additional information do we need to answer the following question: How is the equilibrium of solid silver bromide with a saturated solution of its ions affected when the temperature is raised?

6. Which of the following slightly soluble compounds has a solubility greater than that calculated from its solubility product because of hydrolysis of the anion present: $CoSO_3$, CuI, $PbCO_3$, $PbCl_2$, Tl_2S, $KClO_4$?

7. Which of the following slightly soluble compounds has a solubility greater than that calculated from its solubility product because of hydrolysis of the anion present: $AgCl$, $BaSO_4$, CaF_2, Hg_2I_2, $MnCO_3$, ZnS, PbS?

8. Write the ionic equation for dissolution and the solubility product (K_{sp}) expression for each of the following slightly soluble ionic compounds:

(a) $PbCl_2$

(b) Ag_2S

(c) $Sr_3(PO_4)_2$

(d) $SrSO_4$

9. Write the ionic equation for the dissolution and the K_{sp} expression for each of the following slightly soluble ionic compounds:

(a) LaF_3

(b) $CaCO_3$

(c) Ag_2SO_4

(d) $Pb(OH)_2$

10. The *Handbook of Chemistry and Physics (http://openstaxcollege.org/l/16Handbook)* gives solubilities of the following compounds in grams per 100 mL of water. Because these compounds are only slightly soluble, assume that the volume does not change on dissolution and calculate the solubility product for each.

(a) $BaSiF_6$, 0.026 g/100 mL (contains $SiF_6{}^{2-}$ ions)

(b) $Ce(IO_3)_4$, 1.5×10^{-2} g/100 mL

(c) $Gd_2(SO_4)_3$, 3.98 g/100 mL

(d) $(NH_4)_2PtBr_6$, 0.59 g/100 mL (contains $PtBr_6{}^{2-}$ ions)

11. The *Handbook of Chemistry and Physics (http://openstaxcollege.org/l/16Handbook)* gives solubilities of the following compounds in grams per 100 mL of water. Because these compounds are only slightly soluble, assume that the volume does not change on dissolution and calculate the solubility product for each.

(a) $BaSeO_4$, 0.0118 g/100 mL

(b) $Ba(BrO_3)_2 \cdot H_2O$, 0.30 g/100 mL

(c) $NH_4MgAsO_4 \cdot 6H_2O$, 0.038 g/100 mL

(d) $La_2(MoO_4)_3$, 0.00179 g/100 mL

12. Use solubility products and predict which of the following salts is the most soluble, in terms of moles per liter, in pure water: CaF_2, Hg_2Cl_2, PbI_2, or $Sn(OH)_2$.

13. Assuming that no equilibria other than dissolution are involved, calculate the molar solubility of each of the following from its solubility product:

(a) $KHC_4H_4O_6$

(b) PbI_2

(c) $Ag_4[Fe(CN)_6]$, a salt containing the $Fe(CN)_4{}^-$ ion

(d) Hg_2I_2

14. Assuming that no equilibria other than dissolution are involved, calculate the molar solubility of each of the following from its solubility product:

(a) Ag_2SO_4

(b) $PbBr_2$

(c) AgI

(d) $CaC_2O_4 \cdot H_2O$

15. Assuming that no equilibria other than dissolution are involved, calculate the concentration of all solute species in each of the following solutions of salts in contact with a solution containing a common ion. Show that changes in the initial concentrations of the common ions can be neglected.

(a) $AgCl(s)$ in 0.025 M $NaCl$

(b) $CaF_2(s)$ in 0.00133 M KF

(c) $Ag_2SO_4(s)$ in 0.500 L of a solution containing 19.50 g of K_2SO_4

(d) $Zn(OH)_2(s)$ in a solution buffered at a pH of 11.45

16. Assuming that no equilibria other than dissolution are involved, calculate the concentration of all solute species in each of the following solutions of salts in contact with a solution containing a common ion. Show that changes in the initial concentrations of the common ions can be neglected.

(a) $TlCl(s)$ in 1.250 M HCl

(b) $PbI_2(s)$ in 0.0355 M CaI_2

(c) $Ag_2CrO_4(s)$ in 0.225 L of a solution containing 0.856 g of K_2CrO_4

(d) $Cd(OH)_2(s)$ in a solution buffered at a pH of 10.995

17. Assuming that no equilibria other than dissolution are involved, calculate the concentration of all solute species in each of the following solutions of salts in contact with a solution containing a common ion. Show that it is not appropriate to neglect the changes in the initial concentrations of the common ions.

(a) $TlCl(s)$ in 0.025 M $TlNO_3$

(b) $BaF_2(s)$ in 0.0313 M KF

(c) MgC_2O_4 in 2.250 L of a solution containing 8.156 g of $Mg(NO_3)_2$

(d) $Ca(OH)_2(s)$ in an unbuffered solution initially with a pH of 12.700

18. Explain why the changes in concentrations of the common ions in Exercise 15.17 can be neglected.

19. Explain why the changes in concentrations of the common ions in Exercise 15.18 cannot be neglected.

20. Calculate the solubility of aluminum hydroxide, $Al(OH)_3$, in a solution buffered at pH 11.00.

21. Refer to Appendix J for solubility products for calcium salts. Determine which of the calcium salts listed is most soluble in moles per liter and which is most soluble in grams per liter.

22. Most barium compounds are very poisonous; however, barium sulfate is often administered internally as an aid in the X-ray examination of the lower intestinal tract (Figure 15.4). This use of $BaSO_4$ is possible because of its low solubility. Calculate the molar solubility of $BaSO_4$ and the mass of barium present in 1.00 L of water saturated with $BaSO_4$.

23. Public Health Service standards for drinking water set a maximum of 250 mg/L (2.60×10^{-3} M) of $SO_4{}^{2-}$ because of its cathartic action (it is a laxative). Does natural water that is saturated with $CaSO_4$ ("gyp" water) as a result or passing through soil containing gypsum, $CaSO_4 \cdot 2H_2O$, meet these standards? What is $SO_4{}^{2-}$ in such water?

24. Perform the following calculations:

(a) Calculate $[Ag^+]$ in a saturated aqueous solution of AgBr.

(b) What will $[Ag^+]$ be when enough KBr has been added to make $[Br^-] = 0.050$ M?

(c) What will $[Br^-]$ be when enough $AgNO_3$ has been added to make $[Ag^+] = 0.020$ M?

25. The solubility product of $CaSO_4 \cdot 2H_2O$ is 2.4×10^{-5}. What mass of this salt will dissolve in 1.0 L of 0.010 M $SO4^{2-}$?

26. Assuming that no equilibria other than dissolution are involved, calculate the concentrations of ions in a saturated solution of each of the following (see Appendix J for solubility products).

(a) TlCl

(b) BaF_2

(c) Ag_2CrO_4

(d) $CaC_2O_4 \cdot H_2O$

(e) the mineral anglesite, $PbSO_4$

27. Assuming that no equilibria other than dissolution are involved, calculate the concentrations of ions in a saturated solution of each of the following (see Appendix J for solubility products):

(a) AgI

(b) Ag_2SO_4

(c) $Mn(OH)_2$

(d) $Sr(OH)_2 \cdot 8H_2O$

(e) the mineral brucite, $Mg(OH)_2$

28. The following concentrations are found in mixtures of ions in equilibrium with slightly soluble solids. From the concentrations given, calculate K_{sp} for each of the slightly soluble solids indicated:

(a) AgBr: $[Ag^+] = 5.7 \times 10^{-7}$ M, $[Br^-] = 5.7 \times 10^{-7}$ M

(b) $CaCO_3$: $[Ca^{2+}] = 5.3 \times 10^{-3}$ M, $[CO_3{}^{2-}] = 9.0 \times 10^{-7}$ M

(c) PbF_2: $[Pb^{2+}] = 2.1 \times 10^{-3}$ M, $[F^-] = 4.2 \times 10^{-3}$ M

(d) Ag_2CrO_4: $[Ag^+] = 5.3 \times 10^{-5}$ M, 3.2×10^{-3} M

(e) InF_3: $[In^{3+}] = 2.3 \times 10^{-3}$ M, $[F^-] = 7.0 \times 10^{-3}$ M

29. The following concentrations are found in mixtures of ions in equilibrium with slightly soluble solids. From the concentrations given, calculate K_{sp} for each of the slightly soluble solids indicated:

(a) TlCl: $[Tl^+] = 1.21 \times 10^{-2}$ M, $[Cl^-] = 1.2 \times 10^{-2}$ M

(b) $Ce(IO_3)_4$: $[Ce^{4+}] = 1.8 \times 10^{-4}\ M$, $[IO_3{}^-] = 2.6 \times 10^{-13}\ M$

(c) $Gd_2(SO_4)_3$: $[Gd^{3+}] = 0.132\ M$, $[SO_4{}^{2-}] = 0.198\ M$

(d) Ag_2SO_4: $[Ag^+] = 2.40 \times 10^{-2}\ M$, $[SO_4{}^{2-}] = 2.05 \times 10^{-2}\ M$

(e) $BaSO_4$: $[Ba^{2+}] = 0.500\ M$, $[SO_4{}^{2-}] = 2.16 \times 10^{-10}\ M$

30. Which of the following compounds precipitates from a solution that has the concentrations indicated? (See Appendix J for K_{sp} values.)

(a) $KClO_4$: $[K^+] = 0.01\ M$, $[ClO_4{}^-] = 0.01\ M$

(b) K_2PtCl_6: $[K^+] = 0.01\ M$, $[PtCl_6{}^{2-}] = 0.01\ M$

(c) PbI_2: $[Pb^{2+}] = 0.003\ M$, $[I^-] = 1.3 \times 10^{-3}\ M$

(d) Ag_2S: $[Ag^+] = 1 \times 10^{-10}\ M$, $[S^{2-}] = 1 \times 10^{-13}\ M$

31. Which of the following compounds precipitates from a solution that has the concentrations indicated? (See Appendix J for K_{sp} values.)

(a) $CaCO_3$: $[Ca^{2+}] = 0.003\ M$, $[CO_3{}^{2-}] = 0.003\ M$

(b) $Co(OH)_2$: $[Co^{2+}] = 0.01\ M$, $[OH^-] = 1 \times 10^{-7}\ M$

(c) $CaHPO_4$: $[Ca^{2+}] = 0.01\ M$, $[HPO_4{}^{2-}] = 2 \times 10^{-6}\ M$

(d) $Pb_3(PO_4)_2$: $[Pb^{2+}] = 0.01\ M$, $[PO_4{}^{3-}] = 1 \times 10^{-13}\ M$

32. Calculate the concentration of Tl^+ when $TlCl$ just begins to precipitate from a solution that is $0.0250\ M$ in Cl^-.

33. Calculate the concentration of sulfate ion when $BaSO_4$ just begins to precipitate from a solution that is $0.0758\ M$ in Ba^{2+}.

34. Calculate the concentration of Sr^{2+} when SrF_2 starts to precipitate from a solution that is $0.0025\ M$ in F^-.

35. Calculate the concentration of $PO_4{}^{3-}$ when Ag_3PO_4 starts to precipitate from a solution that is $0.0125\ M$ in Ag^+.

36. Calculate the concentration of F^- required to begin precipitation of CaF_2 in a solution that is $0.010\ M$ in Ca^{2+}.

37. Calculate the concentration of Ag^+ required to begin precipitation of Ag_2CO_3 in a solution that is $2.50 \times 10^{-6}\ M$ in $CO_3{}^{2-}$.

38. What $[Ag^+]$ is required to reduce $[CO_3{}^{2-}]$ to $8.2 \times 10^{-4}\ M$ by precipitation of Ag_2CO_3?

39. What $[F^-]$ is required to reduce $[Ca^{2+}]$ to $1.0 \times 10^{-4}\ M$ by precipitation of CaF_2?

40. A volume of $0.800\ L$ of a 2×10^{-4}-M $Ba(NO_3)_2$ solution is added to $0.200\ L$ of $5 \times 10^{-4}\ M$ Li_2SO_4. Does $BaSO_4$ precipitate? Explain your answer.

41. Perform these calculations for nickel(II) carbonate. (a) With what volume of water must a precipitate containing $NiCO_3$ be washed to dissolve $0.100\ g$ of this compound? Assume that the wash water becomes saturated with $NiCO_3$ ($K_{sp} = 1.36 \times 10^{-7}$).

(b) If the $NiCO_3$ were a contaminant in a sample of $CoCO_3$ ($K_{sp} = 1.0 \times 10^{-12}$), what mass of $CoCO_3$ would have been lost? Keep in mind that both $NiCO_3$ and $CoCO_3$ dissolve in the same solution.

42. Iron concentrations greater than $5.4 \times 10^{-6}\ M$ in water used for laundry purposes can cause staining. What $[OH^-]$ is required to reduce $[Fe^{2+}]$ to this level by precipitation of $Fe(OH)_2$?

43. A solution is $0.010\ M$ in both Cu^{2+} and Cd^{2+}. What percentage of Cd^{2+} remains in the solution when 99.9% of the Cu^{2+} has been precipitated as CuS by adding sulfide?

44. A solution is $0.15\ M$ in both Pb^{2+} and Ag^+. If Cl^- is added to this solution, what is $[Ag^+]$ when $PbCl_2$ begins to precipitate?

45. What reagent might be used to separate the ions in each of the following mixtures, which are 0.1 M with respect to each ion? In some cases it may be necessary to control the pH. (Hint: Consider the K_{sp} values given in Appendix J.)

(a) Hg_2^{2+} and Cu^{2+}

(b) SO_4^{2-} and Cl^-

(c) Hg^{2+} and Co^{2+}

(d) Zn^{2+} and Sr^{2+}

(e) Ba^{2+} and Mg^{2+}

(f) CO_3^{2-} and OH^-

46. A solution contains 1.0×10^{-5} mol of KBr and 0.10 mol of KCl per liter. $AgNO_3$ is gradually added to this solution. Which forms first, solid AgBr or solid AgCl?

47. A solution contains 1.0×10^{-2} mol of KI and 0.10 mol of KCl per liter. $AgNO_3$ is gradually added to this solution. Which forms first, solid AgI or solid AgCl?

48. The calcium ions in human blood serum are necessary for coagulation (Figure 15.5). Potassium oxalate, $K_2C_2O_4$, is used as an anticoagulant when a blood sample is drawn for laboratory tests because it removes the calcium as a precipitate of $CaC_2O_4 \cdot H_2O$. It is necessary to remove all but 1.0% of the Ca^{2+} in serum in order to prevent coagulation. If normal blood serum with a buffered pH of 7.40 contains 9.5 mg of Ca^{2+} per 100 mL of serum, what mass of $K_2C_2O_4$ is required to prevent the coagulation of a 10 mL blood sample that is 55% serum by volume? (All volumes are accurate to two significant figures. Note that the volume of serum in a 10-mL blood sample is 5.5 mL. Assume that the K_{sp} value for CaC_2O_4 in serum is the same as in water.)

49. About 50% of urinary calculi (kidney stones) consist of calcium phosphate, $Ca_3(PO_4)_2$. The normal mid range calcium content excreted in the urine is 0.10 g of Ca^{2+} per day. The normal mid range amount of urine passed may be taken as 1.4 L per day. What is the maximum concentration of phosphate ion that urine can contain before a calculus begins to form?

50. The pH of normal urine is 6.30, and the total phosphate concentration $([PO_4^{3-}] + [HPO_4^{2-}] + [H_2PO_4^-] + [H_3PO_4])$ is 0.020 M. What is the minimum concentration of Ca^{2+} necessary to induce kidney stone formation? (See Exercise 15.49 for additional information.)

51. Magnesium metal (a component of alloys used in aircraft and a reducing agent used in the production of uranium, titanium, and other active metals) is isolated from sea water by the following sequence of reactions:

$$Mg^{2+}(aq) + Ca(OH)_2(aq) \longrightarrow Mg(OH)_2(s) + Ca^{2+}(aq)$$

$$Mg(OH)_2(s) + 2HCl(aq) \longrightarrow MgCl_2(s) + 2H_2O(l)$$

$$MgCl_2(l) \xrightarrow{\text{electrolysis}} Mg(s) + Cl_2(g)$$

Sea water has a density of 1.026 g/cm^3 and contains 1272 parts per million of magnesium as $Mg^{2+}(aq)$ by mass. What mass, in kilograms, of $Ca(OH)_2$ is required to precipitate 99.9% of the magnesium in 1.00×10^3 L of sea water?

52. Hydrogen sulfide is bubbled into a solution that is 0.10 M in both Pb^{2+} and Fe^{2+} and 0.30 M in HCl. After the solution has come to equilibrium it is saturated with H_2S ($[H_2S] = 0.10$ M). What concentrations of Pb^{2+} and Fe^{2+} remain in the solution? For a saturated solution of H_2S we can use the equilibrium:

$$H_2S(aq) + 2H_2O(l) \rightleftharpoons 2H_3O^+(aq) + S^{2-}(aq) \qquad K = 1.0 \times 10^{-26}$$

(Hint: The $[H_3O^+]$ changes as metal sulfides precipitate.)

53. Perform the following calculations involving concentrations of iodate ions:

(a) The iodate ion concentration of a saturated solution of $La(IO_3)_3$ was found to be 3.1×10^{-3} mol/L. Find the K_{sp}.

(b) Find the concentration of iodate ions in a saturated solution of $Cu(IO_3)_2$ ($K_{sp} = 7.4 \times 10^{-8}$).

54. Calculate the molar solubility of AgBr in 0.035 M NaBr ($K_{sp} = 5 \times 10^{-13}$).

55. How many grams of $Pb(OH)_2$ will dissolve in 500 mL of a 0.050-M $PbCl_2$ solution ($K_{sp} = 1.2 \times 10^{-15}$)?

56. Use the simulation (http://openstaxcollege.org/l/16solublesalts) from the earlier Link to Learning to complete the following exercise:. Using 0.01 g CaF_2, give the K_{sp} values found in a 0.2-M solution of each of the salts. Discuss why the values change as you change soluble salts.

57. How many grams of Milk of Magnesia, $Mg(OH)_2$ (s) (58.3 g/mol), would be soluble in 200 mL of water. $K_{sp} = 7.1 \times 10^{-12}$. Include the ionic reaction and the expression for K_{sp} in your answer. ($K_w = 1 \times 10^{-14} = [H_3O^+][OH^-]$)

58. Two hypothetical salts, LM_2 and LQ, have the same molar solubility in H_2O. If K_{sp} for LM_2 is 3.20×10^{-5}, what is the K_{sp} value for LQ?

59. Which of the following carbonates will form first? Which of the following will form last? Explain.

(a) $MgCO_3$ $\quad\quad$ $K_{sp} = 3.5 \times 10^{-8}$

(b) $CaCO_3$ $\quad\quad$ $K_{sp} = 4.2 \times 10^{-7}$

(c) $SrCO_3$ $\quad\quad$ $K_{sp} = 3.9 \times 10^{-9}$

(d) $BaCO_3$ $\quad\quad$ $K_{sp} = 4.4 \times 10^{-5}$

(e) $MnCO_3$ $\quad\quad$ $K_{sp} = 5.1 \times 10^{-9}$

60. How many grams of $Zn(CN)_2$(s) (117.44 g/mol) would be soluble in 100 mL of H_2O? Include the balanced reaction and the expression for K_{sp} in your answer. The K_{sp} value for $Zn(CN)_2$(s) is 3.0×10^{-16}.

15.2 Lewis Acids and Bases

61. Under what circumstances, if any, does a sample of solid AgCl completely dissolve in pure water?

62. Explain why the addition of NH_3 or HNO_3 to a saturated solution of Ag_2CO_3 in contact with solid Ag_2CO_3 increases the solubility of the solid.

63. Calculate the cadmium ion concentration, $[Cd^{2+}]$, in a solution prepared by mixing 0.100 L of 0.0100 M $Cd(NO_3)_2$ with 1.150 L of 0.100 NH_3(aq).

64. Explain why addition of NH_3 or HNO_3 to a saturated solution of $Cu(OH)_2$ in contact with solid $Cu(OH)_2$ increases the solubility of the solid.

65. Sometimes equilibria for complex ions are described in terms of dissociation constants, K_d. For the complex ion $AlF_6{}^{3-}$ the dissociation reaction is:

$$AlF_6{}^{3-} \rightleftharpoons Al^{3+} + 6F^- \text{ and } K_d = \frac{[Al^{3+}][F^-]^6}{[AlF_6{}^{3-}]} = 2 \times 10^{-24}$$

Calculate the value of the formation constant, K_f, for $AlF_6{}^{3-}$.

66. Using the value of the formation constant for the complex ion $Co(NH_3)_6{}^{2+}$, calculate the dissociation constant.

67. Using the dissociation constant, $K_d = 7.8 \times 10^{-18}$, calculate the equilibrium concentrations of Cd^{2+} and CN^- in a 0.250-M solution of $Cd(CN)_4{}^{2-}$.

68. Using the dissociation constant, $K_d = 3.4 \times 10^{-15}$, calculate the equilibrium concentrations of Zn^{2+} and OH^- in a 0.0465-M solution of $Zn(OH)_4{}^{2-}$.

69. Using the dissociation constant, $K_d = 2.2 \times 10^{-34}$, calculate the equilibrium concentrations of Co^{3+} and NH_3 in a 0.500-M solution of $Co(NH_3)_6{}^{3+}$.

70. Using the dissociation constant, $K_d = 1 \times 10^{-44}$, calculate the equilibrium concentrations of Fe^{3+} and CN^- in a 0.333 M solution of $Fe(CN)_6{}^{3-}$.

71. Calculate the mass of potassium cyanide ion that must be added to 100 mL of solution to dissolve 2.0×10^{-2} mol of silver cyanide, AgCN.

72. Calculate the minimum concentration of ammonia needed in 1.0 L of solution to dissolve 3.0×10^{-3} mol of silver bromide.

73. A roll of 35-mm black and white photographic film contains about 0.27 g of unexposed AgBr before developing. What mass of $Na_2S_2O_3 \cdot 5H_2O$ (sodium thiosulfate pentahydrate or hypo) in 1.0 L of developer is required to dissolve the AgBr as $Ag(S_2O_3)_2{}^{3-}$ ($K_f = 4.7 \times 10^{13}$)?

74. We have seen an introductory definition of an acid: An acid is a compound that reacts with water and increases the amount of hydronium ion present. In the chapter on acids and bases, we saw two more definitions of acids: a compound that donates a proton (a hydrogen ion, H^+) to another compound is called a Brønsted-Lowry acid, and a Lewis acid is any species that can accept a pair of electrons. Explain why the introductory definition is a macroscopic definition, while the Brønsted-Lowry definition and the Lewis definition are microscopic definitions.

75. Write the Lewis structures of the reactants and product of each of the following equations, and identify the Lewis acid and the Lewis base in each:

(a) $CO_2 + OH^- \longrightarrow HCO_3{}^-$

(b) $B(OH)_3 + OH^- \longrightarrow B(OH)_4{}^-$

(c) $I^- + I_2 \longrightarrow I_3{}^-$

(d) $AlCl_3 + Cl^- \longrightarrow AlCl_4{}^-$ (use Al-Cl single bonds)

(e) $O^{2-} + SO_3 \longrightarrow SO_4{}^{2-}$

76. Write the Lewis structures of the reactants and product of each of the following equations, and identify the Lewis acid and the Lewis base in each:

(a) $CS_2 + SH^- \longrightarrow HCS_3{}^-$

(b) $BF_3 + F^- \longrightarrow BF_4{}^-$

(c) $I^- + SnI_2 \longrightarrow SnI_3{}^-$

(d) $Al(OH)_3 + OH^- \longrightarrow Al(OH)_4{}^-$

(e) $F^- + SO_3 \longrightarrow SFO_3{}^-$

77. Using Lewis structures, write balanced equations for the following reactions:

(a) $HCl(g) + PH_3(g) \longrightarrow$

(b) $H_3O^+ + CH_3{}^- \longrightarrow$

(c) $CaO + SO_3 \longrightarrow$

(d) $NH_4{}^+ + C_2H_5O^- \longrightarrow$

78. Calculate $[HgCl_4{}^{2-}]$ in a solution prepared by adding 0.0200 mol of NaCl to 0.250 L of a 0.100-M $HgCl_2$ solution.

79. In a titration of cyanide ion, 28.72 mL of 0.0100 M $AgNO_3$ is added before precipitation begins. [The reaction of Ag^+ with CN^- goes to completion, producing the $Ag(CN)_2{}^-$ complex.] Precipitation of solid AgCN takes place when excess Ag^+ is added to the solution, above the amount needed to complete the formation of $Ag(CN)_2{}^-$. How many grams of NaCN were in the original sample?

80. What are the concentrations of Ag^+, CN^-, and $Ag(CN)_2{}^-$ in a saturated solution of AgCN?

81. In dilute aqueous solution HF acts as a weak acid. However, pure liquid HF (boiling point = 19.5 °C) is a strong acid. In liquid HF, HNO_3 acts like a base and accepts protons. The acidity of liquid HF can be increased by adding one of several inorganic fluorides that are Lewis acids and accept F^- ion (for example, BF_3 or SbF_5). Write balanced chemical equations for the reaction of pure HNO_3 with pure HF and of pure HF with BF_3.

82. The simplest amino acid is glycine, $H_2NCH_2CO_2H$. The common feature of amino acids is that they contain the functional groups: an amine group, $-NH_2$, and a carboxylic acid group, $-CO_2H$. An amino acid can function as either an acid or a base. For glycine, the acid strength of the carboxyl group is about the same as that of acetic acid, CH_3CO_2H, and the base strength of the amino group is slightly greater than that of ammonia, NH_3.

(a) Write the Lewis structures of the ions that form when glycine is dissolved in 1 *M* HCl and in 1 *M* KOH.

(b) Write the Lewis structure of glycine when this amino acid is dissolved in water. (Hint: Consider the relative base strengths of the $-NH_2$ and $-CO_2^-$ groups.)

83. Boric acid, H_3BO_3, is not a Brønsted-Lowry acid but a Lewis acid.

(a) Write an equation for its reaction with water.

(b) Predict the shape of the anion thus formed.

(c) What is the hybridization on the boron consistent with the shape you have predicted?

15.3 Multiple Equilibria

84. A saturated solution of a slightly soluble electrolyte in contact with some of the solid electrolyte is said to be a system in equilibrium. Explain. Why is such a system called a heterogeneous equilibrium?

85. Calculate the equilibrium concentration of Ni^{2+} in a 1.0-*M* solution $[Ni(NH_3)_6](NO_3)_2$.

86. Calculate the equilibrium concentration of Zn^{2+} in a 0.30-*M* solution of $Zn(CN)_4^{2-}$.

87. Calculate the equilibrium concentration of Cu^{2+} in a solution initially with 0.050 *M* Cu^{2+} and 1.00 *M* NH_3.

88. Calculate the equilibrium concentration of Zn^{2+} in a solution initially with 0.150 *M* Zn^{2+} and 2.50 *M* CN^-.

89. Calculate the Fe^{3+} equilibrium concentration when 0.0888 mole of $K_3[Fe(CN)_6]$ is added to a solution with 0.0.00010 *M* CN^-.

90. Calculate the Co^{2+} equilibrium concentration when 0.100 mole of $[Co(NH_3)_6](NO_3)_2$ is added to a solution with 0.025 *M* NH_3. Assume the volume is 1.00 L.

91. The equilibrium constant for the reaction $Hg^{2+}(aq) + 2Cl^-(aq) \rightleftharpoons HgCl_2(aq)$ is 1.6 × 10^{13}. Is $HgCl_2$ a strong electrolyte or a weak electrolyte? What are the concentrations of Hg^{2+} and Cl^- in a 0.015-*M* solution of $HgCl_2$?

92. Calculate the molar solubility of $Sn(OH)_2$ in a buffer solution containing equal concentrations of NH_3 and NH_4^+.

93. Calculate the molar solubility of $Al(OH)_3$ in a buffer solution with 0.100 *M* NH_3 and 0.400 *M* NH_4^+.

94. What is the molar solubility of CaF_2 in a 0.100-*M* solution of HF? K_a for HF = 7.2 × 10^{-4}.

95. What is the molar solubility of $BaSO_4$ in a 0.250-*M* solution of $NaHSO_4$? K_a for $HSO_4^- = 1.2 × 10^{-2}$.

96. What is the molar solubility of $Tl(OH)_3$ in a 0.10-*M* solution of NH_3?

97. What is the molar solubility of $Pb(OH)_2$ in a 0.138-*M* solution of CH_3NH_2?

98. A solution of 0.075 *M* $CoBr_2$ is saturated with H_2S ($[H_2S] = 0.10$ *M*). What is the minimum pH at which CoS begins to precipitate?

$$CoS(s) \rightleftharpoons Co^{2+}(aq) + S^{2-}(aq) \qquad K_{sp} = 4.5 × 10^{-27}$$

$$H_2S(aq) + 2H_2O(l) \rightleftharpoons 2H_3O^+(aq) + S^{2-}(aq) \qquad K = 1.0 × 10^{-26}$$

99. A 0.125-*M* solution of $Mn(NO_3)_2$ is saturated with H_2S ($[H_2S] = 0.10$ *M*). At what pH does MnS begin to precipitate?

$$MnS(s) \rightleftharpoons Mn^{2+}(aq) + S^{2-}(aq) \qquad K_{sp} = 4.3 × 10^{-22}$$

$$H_2S(aq) + 2H_2O(l) \rightleftharpoons 2H_3O^+(aq) + S^{2-}(aq) \qquad K = 1.0 × 10^{-26}$$

100. Calculate the molar solubility of BaF_2 in a buffer solution containing 0.20 M HF and 0.20 M NaF.

101. Calculate the molar solubility of $CdCO_3$ in a buffer solution containing 0.115 M Na_2CO_3 and 0.120 M $NaHCO_3$

102. To a 0.10-M solution of $Pb(NO_3)_2$ is added enough HF(g) to make [HF] = 0.10 M.

(a) Does PbF_2 precipitate from this solution? Show the calculations that support your conclusion.

(b) What is the minimum pH at which PbF_2 precipitates?

103. Calculate the concentration of Cd^{2+} resulting from the dissolution of $CdCO_3$ in a solution that is 0.010 M in H_2CO_3.

104. Both AgCl and AgI dissolve in NH_3.

(a) What mass of AgI dissolves in 1.0 L of 1.0 M NH_3?

(b) What mass of AgCl dissolves in 1.0 L of 1.0 M NH_3?

105. Calculate the volume of 1.50 M CH_3CO_2H required to dissolve a precipitate composed of 350 mg each of $CaCO_3$, $SrCO_3$, and $BaCO_3$.

106. Even though $Ca(OH)_2$ is an inexpensive base, its limited solubility restricts its use. What is the pH of a saturated solution of $Ca(OH)_2$?

107. What mass of NaCN must be added to 1 L of 0.010 M $Mg(NO_3)_2$ in order to produce the first trace of $Mg(OH)_2$?

108. Magnesium hydroxide and magnesium citrate function as mild laxatives when they reach the small intestine. Why do magnesium hydroxide and magnesium citrate, two very different substances, have the same effect in your small intestine. (Hint: The contents of the small intestine are basic.)

109. The following question is taken from a Chemistry Advanced Placement Examination and is used with the permission of the Educational Testing Service.

Solve the following problem:

$$MgF_2(s) \rightleftharpoons Mg^{2+}(aq) + 2F^-(aq)$$

In a saturated solution of MgF_2 at 18 °C, the concentration of Mg^{2+} is 1.21×10^{-3} M. The equilibrium is represented by the preceding equation.

(a) Write the expression for the solubility-product constant, K_{sp}, and calculate its value at 18 °C.

(b) Calculate the equilibrium concentration of Mg^{2+} in 1.000 L of saturated MgF_2 solution at 18 °C to which 0.100 mol of solid KF has been added. The KF dissolves completely. Assume the volume change is negligible.

(c) Predict whether a precipitate of MgF_2 will form when 100.0 mL of a 3.00×10^{-3}-M solution of $Mg(NO_3)_2$ is mixed with 200.0 mL of a 2.00×10^{-3}-M solution of NaF at 18 °C. Show the calculations to support your prediction.

(d) At 27 °C the concentration of Mg^{2+} in a saturated solution of MgF_2 is 1.17×10^{-3} M. Is the dissolving of MgF_2 in water an endothermic or an exothermic process? Give an explanation to support your conclusion.

110. Which of the following compounds, when dissolved in a 0.01-M solution of $HClO_4$, has a solubility greater than in pure water: CuCl, $CaCO_3$, MnS, $PbBr_2$, CaF_2? Explain your answer.

111. Which of the following compounds, when dissolved in a 0.01-M solution of $HClO_4$, has a solubility greater than in pure water: AgBr, BaF_2, $Ca_3(PO_4)_3$, ZnS, PbI_2? Explain your answer.

112. What is the effect on the amount of solid $Mg(OH)_2$ that dissolves and the concentrations of Mg^{2+} and OH^- when each of the following are added to a mixture of solid $Mg(OH)_2$ and water at equilibrium?

(a) $MgCl_2$

(b) KOH

(c) $HClO_4$

(d) $NaNO_3$

(e) $Mg(OH)_2$

113. What is the effect on the amount of $CaHPO_4$ that dissolves and the concentrations of Ca^{2+} and HPO_4^- when each of the following are added to a mixture of solid $CaHPO_4$ and water at equilibrium?

(a) $CaCl_2$

(b) HCl

(c) $KClO_4$

(d) $NaOH$

(e) $CaHPO_4$

114. Identify all chemical species present in an aqueous solution of $Ca_3(PO_4)_2$ and list these species in decreasing order of their concentrations. (Hint: Remember that the PO_4^{3-} ion is a weak base.)

115. A volume of 50 mL of 1.8 M NH_3 is mixed with an equal volume of a solution containing 0.95 g of $MgCl_2$. What mass of NH_4Cl must be added to the resulting solution to prevent the precipitation of $Mg(OH)_2$?

Chapter 16

Thermodynamics

Figure 16.1 Geysers are a dramatic display of thermodynamic principles in nature. As water inside the earth heats up, it rises to the surface through small channels. Pressure builds up until the water turns to steam, and steam is expelled forcefully through a hole at the surface. (credit: modification of work by Yellowstone National Park)

Chapter Outline

16.1 Spontaneity

16.2 Entropy

16.3 The Second and Third Laws of Thermodynamics

16.4 Free Energy

Introduction

Among the many capabilities of chemistry is its ability to predict if a process will occur under specified conditions. Thermodynamics, the study of relationships between the energy and work associated with chemical and physical processes, provides this predictive ability. Previous chapters in this text have described various applications of thermochemistry, an important aspect of thermodynamics concerned with the heat flow accompanying chemical reactions and phase transitions. This chapter will introduce additional thermodynamic concepts, including those that enable the prediction of any chemical or physical changes under a given set of conditions.

16.1 Spontaneity

By the end of this section, you will be able to:

• Distinguish between spontaneous and nonspontaneous processes

• Describe the dispersal of matter and energy that accompanies certain spontaneous processes

In this section, consider the differences between two types of changes in a system: Those that occur spontaneously and those that occur by force. In doing so, we'll gain an understanding as to why some systems are naturally inclined to change in one direction under certain conditions and how relatively quickly or slowly that natural change proceeds.

We'll also gain insight into how the spontaneity of a process affects the distribution of energy and matter within the system.

Spontaneous and Nonspontaneous Processes

Processes have a natural tendency to occur in one direction under a given set of conditions. Water will naturally flow downhill, but uphill flow requires outside intervention such as the use of a pump. Iron exposed to the earth's atmosphere will corrode, but rust is not converted to iron without intentional chemical treatment. A **spontaneous process** is one that occurs naturally under certain conditions. A **nonspontaneous process**, on the other hand, will not take place unless it is "driven" by the continual input of energy from an external source. A process that is spontaneous in one direction under a particular set of conditions is nonspontaneous in the reverse direction. At room temperature and typical atmospheric pressure, for example, ice will spontaneously melt, but water will not spontaneously freeze.

The spontaneity of a process is *not* correlated to the speed of the process. A spontaneous change may be so rapid that it is essentially instantaneous or so slow that it cannot be observed over any practical period of time. To illustrate this concept, consider the decay of radioactive isotopes, a topic more thoroughly treated in the chapter on nuclear chemistry. Radioactive decay is by definition a spontaneous process in which the nuclei of unstable isotopes emit radiation as they are converted to more stable nuclei. All the decay processes occur spontaneously, but the rates at which different isotopes decay vary widely. Technetium-99m is a popular radioisotope for medical imaging studies that undergoes relatively rapid decay and exhibits a half-life of about six hours. Uranium-238 is the most abundant isotope of uranium, and its decay occurs much more slowly, exhibiting a half-life of more than four billion years (Figure 16.2).

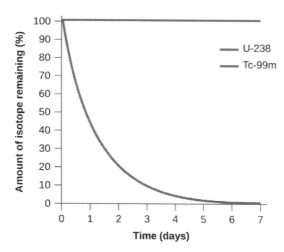

Figure 16.2 Both U-238 and Tc-99m undergo spontaneous radioactive decay, but at drastically different rates. Over the course of one week, essentially all of a Tc-99m sample and none of a U-238 sample will have decayed.

As another example, consider the conversion of diamond into graphite (Figure 16.3).

$$C(s, \ diamond) \longrightarrow C(s, \ graphite)$$

The phase diagram for carbon indicates that graphite is the stable form of this element under ambient atmospheric pressure, while diamond is the stable allotrope at very high pressures, such as those present during its geologic formation. Thermodynamic calculations of the sort described in the last section of this chapter indicate that the conversion of diamond to graphite at ambient pressure occurs spontaneously, yet diamonds are observed to exist, and persist, under these conditions. Though the process is spontaneous under typical ambient conditions, its rate is extremely slow, and so for all practical purposes diamonds are indeed "forever." Situations such as these emphasize the important distinction between the thermodynamic and the kinetic aspects of a process. In this particular case, diamonds are said to be *thermodynamically unstable* but *kinetically stable* under ambient conditions.

C (diamond)   C (graphite)

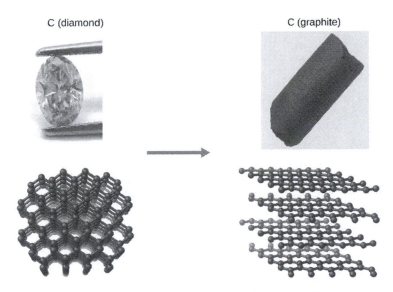

Figure 16.3 The conversion of carbon from the diamond allotrope to the graphite allotrope is spontaneous at ambient pressure, but its rate is immeasurably slow at low to moderate temperatures. This process is known as *graphitization*, and its rate can be increased to easily measurable values at temperatures in the 1000–2000 K range. (credit "diamond" photo: modification of work by "Fancy Diamonds"/Flickr; credit "graphite" photo: modificaton of work by images-of-elements.com/carbon.php)

Dispersal of Matter and Energy

As we extend our discussion of thermodynamic concepts toward the objective of predicting spontaneity, consider now an isolated system consisting of two flasks connected with a closed valve. Initially there is an ideal gas on the left and a vacuum on the right (Figure 16.4). When the valve is opened, the gas spontaneously expands to fill both flasks. Recalling the definition of pressure-volume work from the chapter on thermochemistry, note that no work has been done because the pressure in a vacuum is zero.

$$w = -P\Delta V = 0 \qquad (P = 0 \text{ in a vaccum})$$

Note as well that since the system is isolated, no heat has been exchanged with the surroundings ($q = 0$). The first law of thermodynamics confirms that there has been no change in the system's internal energy as a result of this process.

$$\Delta U = q + w = 0 + 0 = 0$$

The spontaneity of this process is therefore not a consequence of any change in energy that accompanies the process. Instead, the driving force appears to be related to the *greater, more uniform dispersal of matter* that results when the gas is allowed to expand. Initially, the system was comprised of one flask containing matter and another flask containing nothing. After the spontaneous process took place, the matter was distributed both more widely (occupying twice its original volume) and more uniformly (present in equal amounts in each flask).

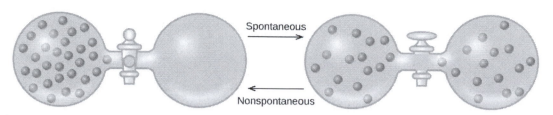

Figure 16.4 An isolated system consists of an ideal gas in one flask that is connected by a closed valve to a second flask containing a vacuum. Once the valve is opened, the gas spontaneously becomes evenly distributed between the flasks.

Now consider two objects at different temperatures: object X at temperature T_X and object Y at temperature T_Y, with $T_X > T_Y$ (Figure 16.5). When these objects come into contact, heat spontaneously flows from the hotter object (X) to the colder one (Y). This corresponds to a loss of thermal energy by X and a gain of thermal energy by Y.

$$q_X < 0 \qquad \text{and} \qquad q_Y = -q_X > 0$$

From the perspective of this two-object system, there was no net gain or loss of thermal energy, rather the available thermal energy was redistributed among the two objects. This spontaneous process resulted in a *more uniform dispersal of energy*.

Figure 16.5 When two objects at different temperatures come in contact, heat spontaneously flows from the hotter to the colder object.

As illustrated by the two processes described, an important factor in determining the spontaneity of a process is the extent to which it changes the dispersal or distribution of matter and/or energy. In each case, a spontaneous process took place that resulted in a more uniform distribution of matter or energy.

Example 16.1

Redistribution of Matter during a Spontaneous Process

Describe how matter is redistributed when the following spontaneous processes take place:

(a) A solid sublimes.

(b) A gas condenses.

(c) A drop of food coloring added to a glass of water forms a solution with uniform color.

Solution

(a)   (b)   (c)

Figure 16.6   (credit a: modification of work by Jenny Downing; credit b: modification of work by "Fuzzy Gerdes"/Flickr; credit c: modification of work by Sahar Atwa)

(a) Sublimation is the conversion of a solid (relatively high density) to a gas (much lesser density). This process yields a much greater dispersal of matter, since the molecules will occupy a much greater volume after the solid-to-gas transition.

(b) Condensation is the conversion of a gas (relatively low density) to a liquid (much greater density). This process yields a much lesser dispersal of matter, since the molecules will occupy a much lesser volume after the solid-to-gas transition.

(c) The process in question is dilution. The food dye molecules initially occupy a much smaller volume (the drop of dye solution) than they occupy once the process is complete (in the full glass of water). The process therefore entails a greater dispersal of matter. The process may also yield a more uniform dispersal of matter, since the initial state of the system involves two regions of different dye concentrations (high in the drop, zero in the water), and the final state of the system contains a single dye concentration throughout.

Check Your Learning

Describe how matter and/or energy is redistributed when you empty a canister of compressed air into a room.

Answer: This is also a dilution process, analogous to example (c). It entails both a greater and more uniform dispersal of matter as the compressed air in the canister is permitted to expand into the lower-pressure air of the room.

16.2 Entropy

By the end of this section, you will be able to:

- Define entropy
- Explain the relationship between entropy and the number of microstates
- Predict the sign of the entropy change for chemical and physical processes

In 1824, at the age of 28, Nicolas Léonard Sadi Carnot (Figure 16.7) published the results of an extensive study regarding the efficiency of steam heat engines. In a later review of Carnot's findings, Rudolf Clausius introduced a new thermodynamic property that relates the spontaneous heat flow accompanying a process to the temperature at which the process takes place. This new property was expressed as the ratio of the *reversible* heat (q_{rev}) and the kelvin temperature (T). The term **reversible process** refers to a process that takes place at such a slow rate that it is always at equilibrium and its direction can be changed (it can be "reversed") by an infinitesimally small change is some condition. Note that the idea of a reversible process is a formalism required to support the development of various thermodynamic concepts; no real processes are truly reversible, rather they are classified as *irreversible*.

(a)   (b)

Figure 16.7 (a) Nicholas Léonard Sadi Carnot's research into steam-powered machinery and (b) Rudolf Clausius's later study of those findings led to groundbreaking discoveries about spontaneous heat flow processes.

Similar to other thermodynamic properties, this new quantity is a state function, and so its change depends only upon the initial and final states of a system. In 1865, Clausius named this property **entropy (S)** and defined its change for any process as the following:

$$\Delta S = \frac{q_{rev}}{T}$$

The entropy change for a real, irreversible process is then equal to that for the theoretical reversible process that involves the same initial and final states.

Entropy and Microstates

Following the work of Carnot and Clausius, Ludwig Boltzmann developed a molecular-scale statistical model that related the entropy of a system to the number of *microstates* possible for the system. A **microstate (W)** is a specific configuration of the locations and energies of the atoms or molecules that comprise a system like the following:

$$S = k \ln W$$

Here k is the Boltzmann constant and has a value of 1.38×10^{-23} J/K.

As for other state functions, the change in entropy for a process is the difference between its final (S_f) and initial (S_i) values:

$$\Delta S = S_f - S_i = k \ln W_f - k \ln W_i = k \ln \frac{W_f}{W_i}$$

For processes involving an increase in the number of microstates, $W_f > W_i$, the entropy of the system increases, $\Delta S > 0$. Conversely, processes that reduce the number of microstates, $W_f < W_i$, yield a decrease in system entropy, $\Delta S < 0$. This molecular-scale interpretation of entropy provides a link to the probability that a process will occur as illustrated in the next paragraphs.

Consider the general case of a system comprised of N particles distributed among n boxes. The number of microstates possible for such a system is n^N. For example, distributing four particles among two boxes will result in 2^4 = 16 different microstates as illustrated in Figure 16.8. Microstates with equivalent particle arrangements (not considering individual particle identities) are grouped together and are called *distributions*. The probability that a system will exist with its components in a given distribution is proportional to the number of microstates within the distribution. Since entropy increases logarithmically with the number of microstates, *the most probable distribution is therefore the one of greatest entropy.*

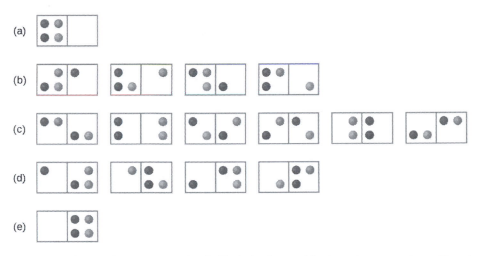

Figure 16.8 The sixteen microstates associated with placing four particles in two boxes are shown. The microstates are collected into five distributions—(a), (b), (c), (d), and (e)—based on the numbers of particles in each box.

For this system, the most probable configuration is one of the six microstates associated with distribution (c) where the particles are evenly distributed between the boxes, that is, a configuration of two particles in each box. The probability of finding the system in this configuration is or $\frac{6}{16}$ or $\frac{3}{8}$. The least probable configuration of the system is one in which all four particles are in one box, corresponding to distributions (a) and (d), each with a probability of $\frac{1}{16}$. The probability of finding all particles in only one box (either the left box or right box) is then $\left(\frac{1}{16} + \frac{1}{16}\right) = \frac{2}{16}$ or $\frac{1}{8}$.

As you add more particles to the system, the number of possible microstates increases exponentially (2^N). A macroscopic (laboratory-sized) system would typically consist of moles of particles ($N \sim 10^{23}$), and the corresponding number of microstates would be staggeringly huge. Regardless of the number of particles in the system, however, the distributions in which roughly equal numbers of particles are found in each box are always the most probable configurations.

The previous description of an ideal gas expanding into a vacuum (Figure 16.4) is a macroscopic example of this particle-in-a-box model. For this system, the most probable distribution is confirmed to be the one in which the matter is most uniformly dispersed or distributed between the two flasks. The spontaneous process whereby the gas contained initially in one flask expands to fill both flasks equally therefore yields an increase in entropy for the system.

A similar approach may be used to describe the spontaneous flow of heat. Consider a system consisting of two objects, each containing two particles, and two units of energy (represented as "*") in Figure 16.9. The hot object is comprised of particles **A** and **B** and initially contains both energy units. The cold object is comprised of particles **C** and **D**, which initially has no energy units. Distribution (a) shows the three microstates possible for the initial state of the system, with both units of energy contained within the hot object. If one of the two energy units is transferred, the result is distribution (b) consisting of four microstates. If both energy units are transferred, the result is distribution (c) consisting of three microstates. And so, we may describe this system by a total of ten microstates. The probability that the heat does not flow when the two objects are brought into contact, that is, that the system remains in distribution (a), is $\frac{3}{10}$. More likely is the flow of heat to yield one of the other two distribution, the combined probability being $\frac{7}{10}$. The most likely result is the flow of heat to yield the uniform dispersal of energy represented by distribution (b), the probability of this configuration being $\frac{4}{10}$. As for the previous example of matter dispersal, extrapolating this treatment to macroscopic collections of particles dramatically increases the probability of the uniform distribution relative to the other distributions. This supports the common observation that placing hot and cold objects in contact results in spontaneous heat flow that ultimately equalizes the objects' temperatures. And, again, this spontaneous process is also characterized by an increase in system entropy.

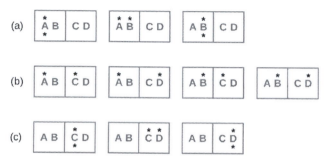

Figure 16.9 This shows a microstate model describing the flow of heat from a hot object to a cold object. (a) Before the heat flow occurs, the object comprised of particles **A** and **B** contains both units of energy and as represented by a distribution of three microstates. (b) If the heat flow results in an even dispersal of energy (one energy unit transferred), a distribution of four microstates results. (c) If both energy units are transferred, the resulting distribution has three microstates.

Example 16.2

Determination of ΔS

Consider the system shown here. What is the change in entropy for a process that converts the system from distribution (a) to (c)?

Solution

We are interested in the following change:

The initial number of microstates is one, the final six:

$$\Delta S = k \ln \frac{W_c}{W_a} = 1.38 \times 10^{-23} \text{ J/K} \times \ln \frac{6}{1} = 2.47 \times 10^{-23} \text{ J/K}$$

The sign of this result is consistent with expectation; since there are more microstates possible for the final state than for the initial state, the change in entropy should be positive.

Predicting the Sign of ΔS

The relationships between entropy, microstates, and matter/energy dispersal described previously allow us to make generalizations regarding the relative entropies of substances and to predict the sign of entropy changes for chemical and physical processes. Consider the phase changes illustrated in Figure 16.10. In the solid phase, the atoms or molecules are restricted to nearly fixed positions with respect to each other and are capable of only modest oscillations about these positions. With essentially fixed locations for the system's component particles, the number of microstates is relatively small. In the liquid phase, the atoms or molecules are free to move over and around each other, though they remain in relatively close proximity to one another. This increased freedom of motion results in a greater variation in possible particle locations, so the number of microstates is correspondingly greater than for the solid. As a result, $S_{liquid} > S_{solid}$ and the process of converting a substance from solid to liquid (melting) is characterized by an increase in entropy, $\Delta S > 0$. By the same logic, the reciprocal process (freezing) exhibits a decrease in entropy, $\Delta S < 0$.

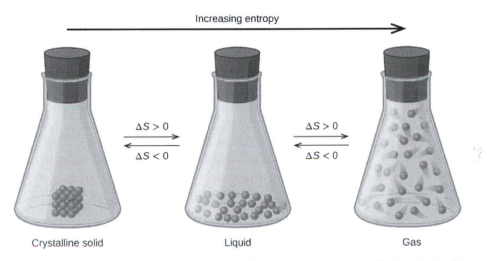

Increasing entropy

$\Delta S > 0$

$\Delta S < 0$

$\Delta S > 0$

$\Delta S < 0$

Crystalline solid

Liquid

Gas

Figure 16.10 The entropy of a substance increases ($\Delta S > 0$) as it transforms from a relatively ordered solid, to a less-ordered liquid, and then to a still less-ordered gas. The entropy decreases ($\Delta S < 0$) as the substance transforms from a gas to a liquid and then to a solid.

Now consider the vapor or gas phase. The atoms or molecules occupy a *much* greater volume than in the liquid phase; therefore each atom or molecule can be found in many more locations than in the liquid (or solid) phase. Consequently, for any substance, $S_{gas} > S_{liquid} > S_{solid}$, and the processes of vaporization and sublimation likewise involve increases in entropy, $\Delta S > 0$. Likewise, the reciprocal phase transitions, condensation and deposition, involve decreases in entropy, $\Delta S < 0$.

According to kinetic-molecular theory, the temperature of a substance is proportional to the average kinetic energy of its particles. Raising the temperature of a substance will result in more extensive vibrations of the particles in solids and more rapid translations of the particles in liquids and gases. At higher temperatures, the distribution of kinetic energies among the atoms or molecules of the substance is also broader (more dispersed) than at lower temperatures. Thus, the entropy for any substance increases with temperature (Figure 16.11).

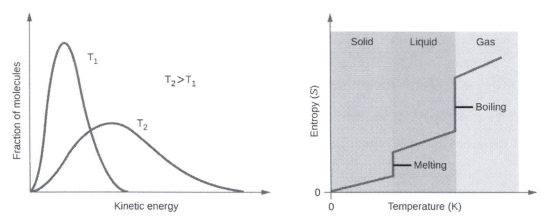

Figure 16.11 Entropy increases as the temperature of a substance is raised, which corresponds to the greater spread of kinetic energies. When a substance melts or vaporizes, it experiences a significant increase in entropy.

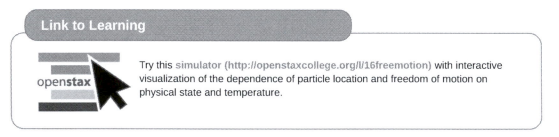

Link to Learning

Try this simulator (http://openstaxcollege.org/l/16freemotion) with interactive visualization of the dependence of particle location and freedom of motion on physical state and temperature.

The entropy of a substance is influenced by structure of the particles (atoms or molecules) that comprise the substance. With regard to atomic substances, heavier atoms possess greater entropy at a given temperature than lighter atoms, which is a consequence of the relation between a particle's mass and the spacing of quantized translational energy levels (which is a topic beyond the scope of our treatment). For molecules, greater numbers of atoms (regardless of their masses) increase the ways in which the molecules can vibrate and thus the number of possible microstates and the system entropy.

Finally, variations in the types of particles affects the entropy of a system. Compared to a pure substance, in which all particles are identical, the entropy of a mixture of two or more different particle types is greater. This is because of the additional orientations and interactions that are possible in a system comprised of nonidentical components. For example, when a solid dissolves in a liquid, the particles of the solid experience both a greater freedom of motion and additional interactions with the solvent particles. This corresponds to a more uniform dispersal of matter and energy and a greater number of microstates. The process of dissolution therefore involves an increase in entropy, $\Delta S > 0$.

Considering the various factors that affect entropy allows us to make informed predictions of the sign of ΔS for various chemical and physical processes as illustrated in Example 16.3.

Example 16.3

Predicting the Sign of ΔS

Predict the sign of the entropy change for the following processes. Indicate the reason for each of your predictions.

(a) One mole liquid water at room temperature $\longrightarrow$ one mole liquid water at 50 °C

(b) $Ag^+(aq) + Cl^-(aq) \longrightarrow AgCl(s)$

(c) $C_6H_6(l) + \frac{15}{2}O_2(g) \longrightarrow 6CO_2(g) + 3H_2O(l)$

(d) $NH_3(s) \longrightarrow NH_3(l)$

Solution

(a) positive, temperature increases

(b) negative, reduction in the number of ions (particles) in solution, decreased dispersal of matter

(c) negative, net decrease in the amount of gaseous species

(d) positive, phase transition from solid to liquid, net increase in dispersal of matter

Check Your Learning

Predict the sign of the enthalpy change for the following processes. Give a reason for your prediction.

(a) $NaNO_3(s) \longrightarrow Na^+(aq) + NO_3^-(aq)$

(b) the freezing of liquid water

(c) $CO_2(s) \longrightarrow CO_2(g)$

(d) $CaCO(s) \longrightarrow CaO(s) + CO_2(g)$

Answer: (a) Positive; The solid dissolves to give an increase of mobile ions in solution. (b) Negative; The liquid becomes a more ordered solid. (c) Positive; The relatively ordered solid becomes a gas. (d) Positive; There is a net production of one mole of gas.

16.3 The Second and Third Laws of Thermodynamics

By the end of this section, you will be able to:

- State and explain the second and third laws of thermodynamics
- Calculate entropy changes for phase transitions and chemical reactions under standard conditions

The Second Law of Thermodynamics

In the quest to identify a property that may reliably predict the spontaneity of a process, we have identified a very promising candidate: entropy. Processes that involve an increase in entropy *of the system* ($\Delta S > 0$) are very often spontaneous; however, examples to the contrary are plentiful. By expanding consideration of entropy changes to include *the surroundings*, we may reach a significant conclusion regarding the relation between this property and spontaneity. In thermodynamic models, the system and surroundings comprise everything, that is, the universe, and so the following is true:

$$\Delta S_{univ} = \Delta S_{sys} + \Delta S_{surr}$$

To illustrate this relation, consider again the process of heat flow between two objects, one identified as the system and the other as the surroundings. There are three possibilities for such a process:

1. The objects are at different temperatures, and heat flows from the hotter to the cooler object. *This is always observed to occur spontaneously.* Designating the hotter object as the system and invoking the definition of entropy yields the following:

$$\Delta S_{sys} = \frac{-q_{rev}}{T_{sys}} \quad \text{and} \quad \Delta S_{surr} = \frac{q_{rev}}{T_{surr}}$$

The arithmetic signs of q_{rev} denote the loss of heat by the system and the gain of heat by the surroundings. Since $T_{sys} > T_{surr}$ in this scenario, the magnitude of the entropy change for the surroundings will be greater than that for the system, and so the sum of ΔS_{sys} and ΔS_{surr} will yield a positive value for ΔS_{univ}. *This process involves an increase in the entropy of the universe.*

2. The objects are at different temperatures, and heat flows from the cooler to the hotter object. *This is never observed to occur spontaneously.* Again designating the hotter object as the system and invoking the definition of entropy yields the following:

$$\Delta S_{sys} = \frac{q_{rev}}{T_{sys}} \quad \text{and} \quad \Delta S_{surr} = \frac{-q_{rev}}{T_{surr}}$$

The arithmetic signs of q_{rev} denote the gain of heat by the system and the loss of heat by the surroundings. The magnitude of the entropy change for the surroundings will again be greater than that for the system, but in this case, the signs of the heat changes will yield a negative value for ΔS_{univ}. *This process involves a decrease in the entropy of the universe.*

3. The temperature difference between the objects is infinitesimally small, $T_{sys} \approx T_{surr}$, and so the heat flow is thermodynamically reversible. See the previous section's discussion). In this case, the system and surroundings experience entropy changes that are equal in magnitude and therefore sum to yield a value of zero for ΔS_{univ}. *This process involves no change in the entropy of the universe.*

These results lead to a profound statement regarding the relation between entropy and spontaneity known as the **second law of thermodynamics**: *all spontaneous changes cause an increase in the entropy of the universe.* A summary of these three relations is provided in Table 16.1.

The Second Law of Thermodynamics

$\Delta S_{univ} > 0$	spontaneous
$\Delta S_{univ} < 0$	nonspontaneous (spontaneous in opposite direction)
$\Delta S_{univ} = 0$	reversible (system is at equilibrium)

Table 16.1

For many realistic applications, the surroundings are vast in comparison to the system. In such cases, the heat gained or lost by the surroundings as a result of some process represents a very small, nearly infinitesimal, fraction of its total thermal energy. For example, combustion of a fuel in air involves transfer of heat from a system (the fuel and oxygen molecules undergoing reaction) to surroundings that are infinitely more massive (the earth's atmosphere). As a result, q_{surr} is a good approximation of q_{rev}, and the second law may be stated as the following:

$$\Delta S_{univ} = \Delta S_{sys} + \Delta S_{surr} = \Delta S_{sys} + \frac{q_{surr}}{T}$$

We may use this equation to predict the spontaneity of a process as illustrated in Example 16.4.

Example 16.4

Will Ice Spontaneously Melt?

The entropy change for the process

$$H_2O(s) \longrightarrow H_2O(l)$$

is 22.1 J/K and requires that the surroundings transfer 6.00 kJ of heat to the system. Is the process spontaneous at −10.00 °C? Is it spontaneous at +10.00 °C?

Solution

We can assess the spontaneity of the process by calculating the entropy change of the universe. If ΔS_{univ} is positive, then the process is spontaneous. At both temperatures, $\Delta S_{sys} = 22.1$ J/K and $q_{surr} = -6.00$ kJ.

At −10.00 °C (263.15 K), the following is true:

$$\Delta S_{univ} = \Delta S_{sys} + \Delta S_{surr} = \Delta S_{sys} + \frac{q_{surr}}{T}$$

$$= 22.1 \text{ J/K} + \frac{-6.00 \times 10^3 \text{ J}}{263.15 \text{ K}} = -0.7 \text{ J/K}$$

$S_{univ} < 0$, so melting is nonspontaneous (*not* spontaneous) at −10.0 °C.

At 10.00 °C (283.15 K), the following is true:

$$\Delta S_{univ} = \Delta S_{sys} + \frac{q_{surr}}{T}$$

$$= 22.1 \text{ J/K} + \frac{-6.00 \times 10^3 \text{ J}}{283.15 \text{ K}} = +0.9 \text{ J/K}$$

$S_{univ} > 0$, so melting *is* spontaneous at 10.00 °C.

Check Your Learning

Using this information, determine if liquid water will spontaneously freeze at the same temperatures. What can you say about the values of S_{univ}?

Answer: Entropy is a state function, and freezing is the opposite of melting. At −10.00 °C spontaneous, +0.7 J/K; at +10.00 °C nonspontaneous, −0.9 J/K.

The Third Law of Thermodynamics

The previous section described the various contributions of matter and energy dispersal that contribute to the entropy of a system. With these contributions in mind, consider the entropy of a pure, perfectly crystalline solid possessing no kinetic energy (that is, at a temperature of absolute zero, 0 K). This system may be described by a single microstate, as its purity, perfect crystallinity and complete lack of motion means there is but one possible location for each identical atom or molecule comprising the crystal ($W = 1$). According to the Boltzmann equation, the entropy of this system is zero.

$$S = k \ln W = k \ln(1) = 0$$

This limiting condition for a system's entropy represents the **third law of thermodynamics**: *the entropy of a pure, perfect crystalline substance at 0 K is zero.*

We can make careful calorimetric measurements to determine the temperature dependence of a substance's entropy and to derive absolute entropy values under specific conditions. **Standard entropies** are given the label S°_{298} for values determined for one mole of substance at a pressure of 1 bar and a temperature of 298 K. The **standard entropy change (ΔS°)** for any process may be computed from the standard entropies of its reactant and product species like the following:

$$\Delta S^\circ = \sum \nu S^\circ_{298}(\text{products}) - \sum \nu S^\circ_{298}(\text{reactants})$$

Here, ν represents stoichiometric coefficients in the balanced equation representing the process. For example, ΔS° for the following reaction at room temperature

$$mA + nB \longrightarrow xC + yD,$$

is computed as the following:

$$= \left[xS^{\circ}_{298}(C) + yS^{\circ}_{298}(D) \right] - \left[mS^{\circ}_{298}(A) + nS^{\circ}_{298}(B) \right]$$

Table 16.2 lists some standard entropies at 298.15 K. You can find additional standard entropies in Appendix G.

Standard Entropies (at 298.15 K, 1 atm)

Substance	S°_{298} (J mol^{-1} K^{-1})
carbon	
C(s, graphite)	5.740
C(s, diamond)	2.38
CO(g)	197.7
CO$_2$(g)	213.8
CH$_4$(g)	186.3
C$_2$H$_4$(g)	219.5
C$_2$H$_6$(g)	229.5
CH$_3$OH(l)	126.8
C$_2$H$_5$OH(l)	160.7
hydrogen	
H$_2$(g)	130.57
H(g)	114.6
H$_2$O(g)	188.71
H$_2$O(l)	69.91
HCl(g)	186.8
H$_2$S(g)	205.7
oxygen	
O$_2$(g)	205.03

Table 16.2

Example 16.5

Determination of ΔS°

Calculate the standard entropy change for the following process:

$$H_2O(g) \longrightarrow H_2O(l)$$

Solution

The value of the standard entropy change at room temperature, ΔS°_{298}, is the difference between the standard entropy of the product, $H_2O(l)$, and the standard entropy of the reactant, $H_2O(g)$.

$$\Delta S_{298}^{\circ} = S_{298}^{\circ}(H_2O(l)) - S_{298}^{\circ}(H_2O(g))$$
$$= (70.0 \text{ J mol}^{-1}\text{ K}^{-1}) - (188.8 \text{ Jmol}^{-1}\text{ K}^{-1}) = -118.8 \text{ J mol}^{-1}\text{ K}^{-1}$$

The value for ΔS_{298}° is negative, as expected for this phase transition (condensation), which the previous section discussed.

Check Your Learning

Calculate the standard entropy change for the following process:

$$H_2(g) + C_2H_4(g) \longrightarrow C_2H_6(g)$$

Answer: $-120.6 \text{ J mol}^{-1}\text{ K}^{-1}$

Example 16.6

Determination of ΔS°

Calculate the standard entropy change for the combustion of methanol, CH_3OH:

$$2CH_3OH(l) + 3O_2(g) \longrightarrow 2CO_2(g) + 4H_2O(l)$$

Solution

The value of the standard entropy change is equal to the difference between the standard entropies of the products and the entropies of the reactants scaled by their stoichiometric coefficients.

$$\Delta S^{\circ} = \Delta S_{298}^{\circ} = \sum \nu S_{298}^{\circ}(\text{products}) - \sum \nu S_{298}^{\circ}(\text{reactants})$$
$$[2S_{298}^{\circ}(CO_2(g)) + 4S_{298}^{\circ}(H_2O(l))] - [2S_{298}^{\circ}(CH_3OH(l)) + 3S_{298}^{\circ}(O_2(g))]$$
$$= \{[2(213.8) + 4 \times 70.0] - [2(126.8) + 3(205.03)]\} = -161.1 \text{ J/mol·K}$$

Check Your Learning

Calculate the standard entropy change for the following reaction:

$$Ca(OH)_2(s) \longrightarrow CaO(s) + H_2O(l)$$

Answer: 24.7 J/mol·K

16.4 Free Energy

By the end of this section, you will be able to:

- Define Gibbs free energy, and describe its relation to spontaneity
- Calculate free energy change for a process using free energies of formation for its reactants and products
- Calculate free energy change for a process using enthalpies of formation and the entropies for its reactants and products
- Explain how temperature affects the spontaneity of some processes
- Relate standard free energy changes to equilibrium constants

One of the challenges of using the second law of thermodynamics to determine if a process is spontaneous is that we must determine the entropy change for the system *and* the entropy change for the surroundings. An alternative approach involving a new thermodynamic property defined in terms of system properties only was introduced in the late nineteenth century by American mathematician Josiah Willard Gibbs. This new property is called the **Gibbs free**

energy change (*G*) (or simply the *free energy*), and it is defined in terms of a system's enthalpy and entropy as the following:

$$G = H - TS$$

Free energy is a state function, and at constant temperature and pressure, the **standard free energy change ($\Delta G°$)** may be expressed as the following:

$$\Delta G = \Delta H - T\Delta S$$

(For simplicity's sake, the subscript "sys" will be omitted henceforth.)

We can understand the relationship between this system property and the spontaneity of a process by recalling the previously derived second law expression:

$$\Delta S_{univ} = \Delta S + \frac{q_{surr}}{T}$$

The first law requires that $q_{surr} = -q_{sys}$, and at constant pressure $q_{sys} = \Delta H$, and so this expression may be rewritten as the following:

$$\Delta S_{univ} = \Delta S - \frac{\Delta H}{T}$$

ΔH is the enthalpy change *of the system*. Multiplying both sides of this equation by $-T$, and rearranging yields the following:

$$-T\Delta S_{univ} = \Delta H - T\Delta S$$

Comparing this equation to the previous one for free energy change shows the following relation:

$$\Delta G = -T\Delta S_{univ}$$

The free energy change is therefore a reliable indicator of the spontaneity of a process, being directly related to the previously identified spontaneity indicator, ΔS_{univ}. Table 16.3 summarizes the relation between the spontaneity of a process and the arithmetic signs of these indicators.

Relation between Process Spontaneity and Signs of Thermodynamic Properties

$\Delta S_{univ} > 0$	$\Delta G < 0$	spontaneous
$\Delta S_{univ} < 0$	$\Delta G > 0$	nonspontaneous
$\Delta S_{univ} = 0$	$\Delta G = 0$	reversible (at equilibrium)

Table 16.3

Calculating Free Energy Change

Free energy is a state function, so its value depends only on the conditions of the initial and final states of the system that have undergone some change. A convenient and common approach to the calculation of free energy changes for physical and chemical reactions is by use of widely available compilations of standard state thermodynamic data. One method involves the use of standard enthalpies and entropies to compute standard free energy changes according to the following relation as demonstrated in Example 16.7.

$$\Delta G° = \Delta H° - T\Delta S°$$

Example 16.7

Evaluation of $\Delta G°$ Change from $\Delta H°$ and $\Delta S°$

Use standard enthalpy and entropy data from Appendix G to calculate the standard free energy change for the vaporization of water at room temperature (298 K). What does the computed value for $\Delta G°$ say about the spontaneity of this process?

Solution

The process of interest is the following:

$$H_2O(l) \longrightarrow H_2O(g)$$

The standard change in free energy may be calculated using the following equation:

$$\Delta G°_{298} = \Delta H° - T\Delta S°$$

From Appendix G, here is the data:

Substance	$\Delta H°_f$ (kJ/mol)	$S°_{298}$ (J/K·mol)
$H_2O(l)$	−286.83	70.0
$H_2O(g)$	−241.82	188.8

Combining at 298 K:

$$\Delta H° = \Delta H°_{298} = \Delta H°_f\ (H_2O(g)) - \Delta H°_f\ (H_2O(l))$$
$$= [-241.82\ \text{kJ} - (-285.83)]\ \text{kJ/mol} = 44.01\ \text{kJ/mol}$$

$$\Delta S° = \Delta S°_{298} = S°_{298}(H_2O(g)) - S°_{298}(H_2O(l))$$
$$= 188.8\ \text{J/mol·K} - 70.0\ \text{J/K} = 118.8\ \text{J/mol·K}$$

$$\Delta G° = \Delta H° - T\Delta S°$$

Converting everything into kJ and combining at 298 K:

$$\Delta G°_{298} = \Delta H° - T\Delta S°$$
$$= 44.01\ \text{kJ/mol} - (298\ \text{K} \times 118.8\ \text{J/mol·K}) \times \frac{1\ \text{kJ}}{1000\ \text{J}}$$

$$44.01\ \text{kJ/mol} - 35.4\ \text{kJ/mol} = 8.6\ \text{kJ/mol}$$

At 298 K (25 °C) $\Delta G°_{298} > 0$, and so boiling is nonspontaneous (*not* spontaneous).

Check Your Learning

Use standard enthalpy and entropy data from Appendix G to calculate the standard free energy change for the reaction shown here (298 K). What does the computed value for $\Delta G°$ say about the spontaneity of this process?

$$C_2H_6(g) \longrightarrow H_2(g) + C_2H_4(g)$$

Answer: $\Delta G°_{298} = 102.0$ kJ/mol; the reaction is nonspontaneous (*not* spontaneous) at 25 °C.

Free energy changes may also use the **standard free energy of formation** $(\Delta G°_f)$, for each of the reactants and products involved in the reaction. The standard free energy of formation is the free energy change that accompanies the formation of one mole of a substance from its elements in their standard states. Similar to the standard enthalpies of formation, $\Delta G°_f$ is by definition zero for elemental substances under standard state conditions. The approach to computing the free energy change for a reaction using this approach is the same as that demonstrated previously for enthalpy and entropy changes. For the reaction

$$mA + nB \longrightarrow xC + yD,$$

the standard free energy change at room temperature may be calculated as

$$\Delta G_{298}^\circ = \Delta G^\circ = \sum \nu \, \Delta G_{298}^\circ (\text{products}) - \sum \nu \, \Delta G_{298}^\circ (\text{reactants})$$

$$= \left[x\Delta G_f^\circ \ (\text{C}) + y\Delta G_f^\circ \ (\text{D}) \right] - \left[m\Delta G_f^\circ \ (\text{A}) + n\Delta G_f^\circ \ (\text{B}) \right].$$

Example 16.8

Calculation of ΔG_{298}°

Consider the decomposition of yellow mercury(II) oxide.

$$HgO(s, \ \text{yellow}) \longrightarrow Hg(l) + \frac{1}{2}O_2(g)$$

Calculate the standard free energy change at room temperature, ΔG_{298}°, using (a) standard free energies of formation and (b) standard enthalpies of formation and standard entropies. Do the results indicate the reaction to be spontaneous or nonspontaneous under standard conditions?

Solution

The required data are available in Appendix G and are shown here.

Compound	ΔG_f° (kJ/mol)	ΔH_f° (kJ/mol)	S_{298}° (J/K·mol)
HgO (s, yellow)	−58.43	−90.46	71.13
Hg(l)	0	0	75.9
$O_2(g)$	0	0	205.2

(a) Using free energies of formation:

$$\Delta G_{298}^\circ = \sum \nu GS_{298}^\circ (\text{products}) - \sum \nu \Delta G_{298}^\circ (\text{reactants})$$

$$= \left[1\Delta G_{298}^\circ \, Hg(l) + \frac{1}{2}\Delta G_{298}^\circ \, O_2(g) \right] - 1\Delta G_{298}^\circ \, HgO(s, \ \text{yellow})$$

$$= \left[1 \, mol(0 \ kJ/mol) + \frac{1}{2} \, mol(0 \ kJ/mol) \right] - 1 \, mol(-58.43 \ kJ/mol) = 58.43 \ kJ/mol$$

(b) Using enthalpies and entropies of formation:

$$\Delta H_{298}^\circ = \sum \nu \Delta H_{298}^\circ (\text{products}) - \sum \nu \Delta H_{298}^\circ (\text{reactants})$$

$$= \left[1\Delta H_{298}^\circ \, Hg(l) + \frac{1}{2}\Delta H_{298}^\circ \, O_2(g) \right] - 1\Delta H_{298}^\circ \, HgO(s, \ \text{yellow})$$

$$= \left[1 \, mol(0 \ kJ/mol) + \frac{1}{2} \, mol(0 \ kJ/mol) \right] - 1 \, mol(-90.46 \ kJ/mol) = 90.46 \ kJ/mol$$

$$\Delta S_{298}^\circ = \sum \nu \Delta S_{298}^\circ (\text{products}) - \sum \nu \Delta S_{298}^\circ (\text{reactants})$$

$$= \left[1\Delta S_{298}^\circ \, Hg(l) + \frac{1}{2}\Delta S_{298}^\circ \, O_2(g) \right] - 1\Delta S_{298}^\circ \, HgO(s, \ \text{yellow})$$

$$= \left[1 \, mol \, (75.9 \ J/mol \ K) + \frac{1}{2} \, mol(205.2 \ J/mol \ K) \right] - 1 \, mol(71.13 \ J/mol \ K) = 107.4 \ J/mol \ K$$

$$\Delta G^\circ = \Delta H^\circ - T\Delta S^\circ = 90.46 \ kJ - 298.15 \ K \times 107.4 \ J/K\cdot mol \times \frac{1 \ kJ}{1000 \ J}$$

$$\Delta G^\circ = (90.46 - 32.01) \ kJ/mol = 58.45 \ kJ/mol$$

Both ways to calculate the standard free energy change at 25 °C give the same numerical value (to three significant figures), and both predict that the process is nonspontaneous (*not* spontaneous) at room temperature.

Check Your Learning

Calculate $\Delta G°$ using (a) free energies of formation and (b) enthalpies of formation and entropies (Appendix G). Do the results indicate the reaction to be spontaneous or nonspontaneous at 25 °C?

$$C_2H_4(g) \longrightarrow H_2(g) + C_2H_2(g)$$

Answer: −141.5 kJ/mol, nonspontaneous

Temperature Dependence of Spontaneity

As was previously demonstrated in this chapter's section on entropy, the spontaneity of a process may depend upon the temperature of the system. Phase transitions, for example, will proceed spontaneously in one direction or the other depending upon the temperature of the substance in question. Likewise, some chemical reactions can also exhibit temperature dependent spontaneities. To illustrate this concept, the equation relating free energy change to the enthalpy and entropy changes for the process is considered:

$$\Delta G = \Delta H - T\Delta S$$

The spontaneity of a process, as reflected in the arithmetic sign of its free energy change, is then determined by the signs of the enthalpy and entropy changes and, in some cases, the absolute temperature. Since T is the absolute (kelvin) temperature, it can only have positive values. Four possibilities therefore exist with regard to the signs of the enthalpy and entropy changes:

1. **Both ΔH and ΔS are positive.** This condition describes an endothermic process that involves an increase in system entropy. In this case, ΔG will be negative if the magnitude of the $T\Delta S$ term is greater than ΔH. If the $T\Delta S$ term is less than ΔH, the free energy change will be positive. Such a process is *spontaneous at high temperatures and nonspontaneous at low temperatures*.

2. **Both ΔH and ΔS are negative.** This condition describes an exothermic process that involves a decrease in system entropy. In this case, ΔG will be negative if the magnitude of the $T\Delta S$ term is less than ΔH. If the $T\Delta S$ term's magnitude is greater than ΔH, the free energy change will be positive. Such a process is *spontaneous at low temperatures and nonspontaneous at high temperatures*.

3. **ΔH is positive and ΔS is negative.** This condition describes an endothermic process that involves a decrease in system entropy. In this case, ΔG will be positive regardless of the temperature. Such a process is *nonspontaneous at all temperatures*.

4. **ΔH is negative and ΔS is positive.** This condition describes an exothermic process that involves an increase in system entropy. In this case, ΔG will be negative regardless of the temperature. Such a process is *spontaneous at all temperatures*.

These four scenarios are summarized in Figure 16.12.

Summary of the Four Scenarios for Enthalpy and Entropy Changes

	$\Delta H > 0$ (endothermic)	$\Delta H < 0$ (exothermic)
$\Delta S > 0$ (increase in entropy)	$\Delta G < 0$ at high temperature $\Delta G > 0$ at low temperature Process is spontaneous at high temperature	$\Delta G < 0$ at any temperature Process is spontaneous at any temperature
$\Delta S < 0$ (decrease in entropy)	$\Delta G > 0$ at any temperature Process is nonspontaneous at any temperature	$\Delta G < 0$ at low temperature $\Delta G > 0$ at high temperature Process is spontaneous at low temperature

Figure 16.12   There are four possibilities regarding the signs of enthalpy and entropy changes.

Example 16.9

Predicting the Temperature Dependence of Spontaneity

The incomplete combustion of carbon is described by the following equation:

$$2C(s) + O_2(g) \longrightarrow 2CO(g)$$

How does the spontaneity of this process depend upon temperature?

Solution

Combustion processes are exothermic ($\Delta H < 0$). This particular reaction involves an increase in entropy due to the accompanying increase in the amount of gaseous species (net gain of one mole of gas, $\Delta S > 0$). The reaction is therefore spontaneous ($\Delta G < 0$) at all temperatures.

Check Your Learning

Popular chemical hand warmers generate heat by the air-oxidation of iron:

$$4Fe(s) + 3O_2(g) \longrightarrow 2Fe_2O_3(s)$$

How does the spontaneity of this process depend upon temperature?

Answer: ΔH and ΔS are negative; the reaction is spontaneous at low temperatures.

When considering the conclusions drawn regarding the temperature dependence of spontaneity, it is important to keep in mind what the terms "high" and "low" mean. Since these terms are adjectives, the temperatures in question are deemed high or low relative to some reference temperature. A process that is nonspontaneous at one temperature but spontaneous at another will necessarily undergo a change in "spontaneity" (as reflected by its ΔG) as temperature varies. This is clearly illustrated by a graphical presentation of the free energy change equation, in which ΔG is plotted on the y axis versus T on the x axis:

$$\Delta G = \Delta H - T\Delta S$$
$$y = b + mx$$

Such a plot is shown in Figure 16.13. A process whose enthalpy and entropy changes are of the same arithmetic sign will exhibit a temperature-dependent spontaneity as depicted by the two yellow lines in the plot. Each line crosses from one spontaneity domain (positive or negative ΔG) to the other at a temperature that is characteristic of the process in question. This temperature is represented by the x-intercept of the line, that is, the value of T for which ΔG is zero:

$$\Delta G = 0 = \Delta H - T\Delta S$$
$$T = \frac{\Delta H}{\Delta S}$$

And so, saying a process is spontaneous at "high" or "low" temperatures means the temperature is above or below, respectively, that temperature at which ΔG for the process is zero. As noted earlier, this condition describes a system at equilibrium.

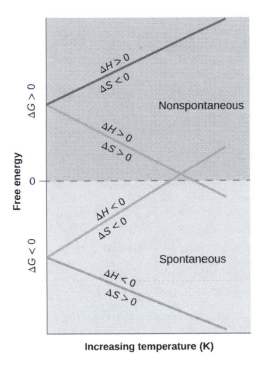

Figure 16.13 These plots show the variation in ΔG with temperature for the four possible combinations of arithmetic sign for ΔH and ΔS.

Example 16.10

Equilibrium Temperature for a Phase Transition

As defined in the chapter on liquids and solids, the boiling point of a liquid is the temperature at which its solid and liquid phases are in equilibrium (that is, when vaporization and condensation occur at equal rates). Use the information in Appendix G to estimate the boiling point of water.

Solution

The process of interest is the following phase change:

$$H_2O(l) \longrightarrow H_2O(g)$$

When this process is at equilibrium, $\Delta G = 0$, so the following is true:

$$0 = \Delta H° - T\Delta S° \qquad \text{or} \qquad T = \frac{\Delta H°}{\Delta S°}$$

Using the standard thermodynamic data from Appendix G,

$$\Delta H° = \Delta H_f° \ (H_2O(g)) - \Delta H_f° \ (H_2O(l))$$
$$= -241.82 \ \text{kJ/mol} - (-285.83 \ \text{kJ/mol}) = 44.01 \ \text{kJ/mol}$$
$$\Delta S° = \Delta S_{298}° (H_2O(g)) - \Delta S_{298}° (H_2O(l))$$
$$= 188.8 \ \text{J/K·mol} - 70.0 \ \text{J/K·mol} = 118.8 \ \text{J/K·mol}$$
$$T = \frac{\Delta H°}{\Delta S°} = \frac{44.01 \times 10^3 \ \text{J/mol}}{118.8 \ \text{J/K·mol}} = 370.5 \ \text{K} = 97.3 \ °\text{C}$$

The accepted value for water's normal boiling point is 373.2 K (100.0 °C), and so this calculation is in reasonable agreement. Note that the values for enthalpy and entropy changes data used were derived from standard data at 298 K (Appendix G). If desired, you could obtain more accurate results by using enthalpy and entropy changes determined at (or at least closer to) the actual boiling point.

Check Your Learning

Use the information in Appendix G to estimate the boiling point of CS_2.

Answer: 313 K (accepted value 319 K)

Free Energy and Equilibrium

The free energy change for a process may be viewed as a measure of its driving force. A negative value for ΔG represents a finite driving force for the process in the forward direction, while a positive value represents a driving force for the process in the reverse direction. When ΔG is zero, the forward and reverse driving forces are equal, and so the process occurs in both directions at the same rate (the system is at equilibrium).

In the chapter on equilibrium the *reaction quotient*, Q, was introduced as a convenient measure of the status of an equilibrium system. Recall that Q is the numerical value of the mass action expression for the system, and that you may use its value to identify the direction in which a reaction will proceed in order to achieve equilibrium. When Q is lesser than the equilibrium constant, K, the reaction will proceed in the forward direction until equilibrium is reached and Q = K. Conversely, if Q < K, the process will proceed in the reverse direction until equilibrium is achieved.

The free energy change for a process taking place with reactants and products present under nonstandard conditions, ΔG, is related to the standard free energy change, $\Delta G°$, according to this equation:

$$\Delta G = \Delta G° + RT \ln Q$$

R is the gas constant (8.314 J/K mol), T is the kelvin or absolute temperature, and Q is the reaction quotient. We may use this equation to predict the spontaneity for a process under any given set of conditions as illustrated in Example 16.11.

Example 16.11

Calculating ΔG under Nonstandard Conditions

What is the free energy change for the process shown here under the specified conditions?

$T = 25 \ °C$, $P_{N_2} = 0.870 \ \text{atm}$, $P_{H_2} = 0.250 \ \text{atm}$, and $P_{NH_3} = 12.9 \ \text{atm}$

$$2NH_3(g) \longrightarrow 3H_2(g) + N_2(g) \qquad \Delta G° = 33.0 \ \text{kJ/mol}$$

Solution

The equation relating free energy change to standard free energy change and reaction quotient may be used directly:

$$\Delta G = \Delta G° + RT \ln Q = 33.0 \ \frac{\text{kJ}}{\text{mol}} + \left(8.314 \ \frac{\text{J}}{\text{mol K}} \times 298 \ \text{K} \times \ln \frac{(0.250^3) \times 0.870}{12.9^2} \right) = 9680 \ \frac{\text{J}}{\text{mol}} \ \text{or} \ 9.68 \ \text{kJ/mol}$$

Since the computed value for ΔG is positive, the reaction is nonspontaneous under these conditions.

Check Your Learning

Calculate the free energy change for this same reaction at 875 °C in a 5.00 L mixture containing 0.100 mol of each gas. Is the reaction spontaneous under these conditions?

Answer: $\Delta G = -136$ kJ; yes

For a system at equilibrium, $Q = K$ and $\Delta G = 0$, and the previous equation may be written as

$$0 = \Delta G° + RT \ln K \qquad \text{(at equilibrium)}$$

$$\Delta G° = -RT \ln K \qquad \text{or} \qquad K = e^{-\frac{\Delta G°}{RT}}$$

This form of the equation provides a useful link between these two essential thermodynamic properties, and it can be used to derive equilibrium constants from standard free energy changes and vice versa. The relations between standard free energy changes and equilibrium constants are summarized in Table 16.4.

Relations between Standard Free Energy Changes and Equilibrium Constants

K	ΔG°	Comments
> 1	< 0	Products are more abundant at equilibrium.
< 1	> 0	Reactants are more abundant at equilibrium.
= 1	= 0	Reactants and products are equally abundant at equilibrium.

Table 16.4

Example 16.12

Calculating an Equilibrium Constant using Standard Free Energy Change

Given that the standard free energies of formation of $Ag^+(aq)$, $Cl^-(aq)$, and $AgCl(s)$ are 77.1 kJ/mol, −131.2 kJ/mol, and −109.8 kJ/mol, respectively, calculate the solubility product, K_{sp}, for AgCl.

Solution

The reaction of interest is the following:

$$AgCl(s) \rightleftharpoons Ag^+(aq) + Cl^-(aq) \qquad K_{sp} = [Ag^+][Cl^-]$$

The standard free energy change for this reaction is first computed using standard free energies of formation for its reactants and products:

$$\Delta G° = \Delta G°_{298} = \left[\Delta G°_f \ (Ag^+(aq)) + \Delta G°_f \ (Cl^-(aq))\right] - \left[\Delta G°_f \ (AgCl(s))\right]$$
$$= [77.1 \text{ kJ/mol} - 131.2 \text{ kJ/mol}] - [-109.8 \text{ kJ/mol}] = 55.7 \text{ kJ/mol}$$

The equilibrium constant for the reaction may then be derived from its standard free energy change:

$$K_{sp} = e^{-\frac{\Delta G°}{RT}} = \exp\left(-\frac{\Delta G°}{RT}\right) = \exp\left(-\frac{55.7 \times 10^3 \text{ J/mol}}{8.314 \text{ J/mol·K} \times 298.15 \text{ K}}\right) = \exp(-22.470) = e^{-22.470} = 1.74 \times 10^{-10}$$

This result is in reasonable agreement with the value provided in Appendix J.

Check Your Learning

Use the thermodynamic data provided in Appendix G to calculate the equilibrium constant for the dissociation of dinitrogen tetroxide at 25 °C.

$$2NO_2(g) \rightleftharpoons N_2O_4(g)$$

Answer: $K = 6.9$

To further illustrate the relation between these two essential thermodynamic concepts, consider the observation that reactions spontaneously proceed in a direction that ultimately establishes equilibrium. As may be shown by plotting the free energy change versus the extent of the reaction (for example, as reflected in the value of Q), equilibrium is established when the system's free energy is minimized (Figure 16.14). If a system is present with reactants and products present in nonequilibrium amounts ($Q \neq K$), the reaction will proceed spontaneously in the direction necessary to establish equilibrium.

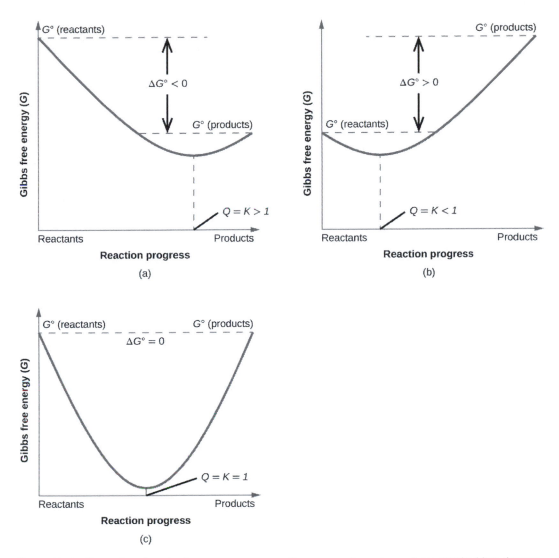

Figure 16.14 These plots show the free energy versus reaction progress for systems whose standard free changes are (a) negative, (b) positive, and (c) zero. Nonequilibrium systems will proceed spontaneously in whatever direction is necessary to minimize free energy and establish equilibrium.

Key Terms

entropy (S) state function that is a measure of the matter and/or energy dispersal within a system, determined by the number of system microstates often described as a measure of the disorder of the system

Gibbs free energy change (G) thermodynamic property defined in terms of system enthalpy and entropy; all spontaneous processes involve a decrease in G

microstate (W) possible configuration or arrangement of matter and energy within a system

nonspontaneous process process that requires continual input of energy from an external source

reversible process process that takes place so slowly as to be capable of reversing direction in response to an infinitesimally small change in conditions; hypothetical construct that can only be approximated by real processes removed

second law of thermodynamics entropy of the universe increases for a spontaneous process

spontaneous change process that takes place without a continuous input of energy from an external source

standard entropy (S°) entropy for a substance at 1 bar pressure; tabulated values are usually determined at 298.15 K and denoted S°_{298}

standard entropy change (ΔS°) change in entropy for a reaction calculated using the standard entropies, usually at room temperature and denoted ΔS°_{298}

standard free energy change (ΔG°) change in free energy for a process occurring under standard conditions (1 bar pressure for gases, 1 M concentration for solutions)

standard free energy of formation (ΔG°_f) change in free energy accompanying the formation of one mole of substance from its elements in their standard states

third law of thermodynamics entropy of a perfect crystal at absolute zero (0 K) is zero

Key Equations

- $\Delta S = \dfrac{q_{rev}}{T}$

- $S = k \ln W$

- $\Delta S = k \ln \dfrac{W_f}{W_i}$

- $\Delta S^\circ = \Delta S^\circ_{298} = \sum \nu S^\circ_{298} \text{ (products)} - \sum \nu S^\circ_{298} \text{ (reactants)}$

- $\Delta S = \dfrac{q_{rev}}{T}$

- $\Delta S_{univ} = \Delta S_{sys} + \Delta S_{surr}$

- $\Delta S_{univ} = \Delta S_{sys} + \Delta S_{surr} = \Delta S_{sys} + \dfrac{q_{surr}}{T}$

- $\Delta G = \Delta H - T\Delta S$

- $\Delta G = \Delta G^\circ + RT \ln Q$

- $\Delta G^\circ = -RT \ln K$

Summary

16.1 Spontaneity

Chemical and physical processes have a natural tendency to occur in one direction under certain conditions. A spontaneous process occurs without the need for a continual input of energy from some external source, while a nonspontaneous process requires such. Systems undergoing a spontaneous process may or may not experience a gain or loss of energy, but they will experience a change in the way matter and/or energy is distributed within the system.

16.2 Entropy

Entropy (S) is a state function that can be related to the number of microstates for a system (the number of ways the system can be arranged) and to the ratio of reversible heat to kelvin temperature. It may be interpreted as a measure of the dispersal or distribution of matter and/or energy in a system, and it is often described as representing the "disorder" of the system.

For a given substance, $S_{solid} < S_{liquid} < S_{gas}$ in a given physical state at a given temperature, entropy is typically greater for heavier atoms or more complex molecules. Entropy increases when a system is heated and when solutions form. Using these guidelines, the sign of entropy changes for some chemical reactions may be reliably predicted.

16.3 The Second and Third Laws of Thermodynamics

The second law of thermodynamics states that a spontaneous process increases the entropy of the universe, $S_{univ} > 0$. If $\Delta S_{univ} < 0$, the process is nonspontaneous, and if $\Delta S_{univ} = 0$, the system is at equilibrium. The third law of thermodynamics establishes the zero for entropy as that of a perfect, pure crystalline solid at 0 K. With only one possible microstate, the entropy is zero. We may compute the standard entropy change for a process by using standard entropy values for the reactants and products involved in the process.

16.4 Free Energy

Gibbs free energy (G) is a state function defined with regard to system quantities only and may be used to predict the spontaneity of a process. A negative value for ΔG indicates a spontaneous process; a positive ΔG indicates a nonspontaneous process; and a ΔG of zero indicates that the system is at equilibrium. A number of approaches to the computation of free energy changes are possible.

Exercises

16.1 Spontaneity

1. What is a spontaneous reaction?

2. What is a nonspontaneous reaction?

3. Indicate whether the following processes are spontaneous or nonspontaneous.

(a) Liquid water freezing at a temperature below its freezing point

(b) Liquid water freezing at a temperature above its freezing point

(c) The combustion of gasoline

(d) A ball thrown into the air

(e) A raindrop falling to the ground

(f) Iron rusting in a moist atmosphere

4. A helium-filled balloon spontaneously deflates overnight as He atoms diffuse through the wall of the balloon. Describe the redistribution of matter and/or energy that accompanies this process.

5. Many plastic materials are organic polymers that contain carbon and hydrogen. The oxidation of these plastics in air to form carbon dioxide and water is a spontaneous process; however, plastic materials tend to persist in the environment. Explain.

16.2 Entropy

6. In Figure 16.8 all possible distributions and microstates are shown for four different particles shared between two boxes. Determine the entropy change, ΔS, if the particles are initially evenly distributed between the two boxes, but upon redistribution all end up in Box (b).

7. In Figure 16.8 all of the possible distributions and microstates are shown for four different particles shared between two boxes. Determine the entropy change, ΔS, for the system when it is converted from distribution (b) to distribution (d).

8. How does the process described in the previous item relate to the system shown in Figure 16.4?

9. Consider a system similar to the one in Figure 16.8, except that it contains six particles instead of four. What is the probability of having all the particles in only one of the two boxes in the case? Compare this with the similar probability for the system of four particles that we have derived to be equal to $\frac{1}{8}$. What does this comparison tell us about even larger systems?

10. Consider the system shown in Figure 16.9. What is the change in entropy for the process where the energy is initially associated only with particle A, but in the final state the energy is distributed between two different particles?

11. Consider the system shown in Figure 16.9. What is the change in entropy for the process where the energy is initially associated with particles A and B, and the energy is distributed between two particles in different boxes (one in A-B, the other in C-D)?

12. Arrange the following sets of systems in order of increasing entropy. Assume one mole of each substance and the same temperature for each member of a set.

(a) $H_2(g)$, $HBrO_4(g)$, $HBr(g)$

(b) $H_2O(l)$, $H_2O(g)$, $H_2O(s)$

(c) $He(g)$, $Cl_2(g)$, $P_4(g)$

13. At room temperature, the entropy of the halogens increases from I_2 to Br_2 to Cl_2. Explain.

14. Consider two processes: sublimation of $I_2(s)$ and melting of $I_2(s)$ (Note: the latter process can occur at the same temperature but somewhat higher pressure).

$I_2(s) \longrightarrow I_2(g)$

$I_2(s) \longrightarrow I_2(l)$

Is ΔS positive or negative in these processes? In which of the processes will the magnitude of the entropy change be greater?

15. Indicate which substance in the given pairs has the higher entropy value. Explain your choices.

(a) $C_2H_5OH(l)$ or $C_3H_7OH(l)$

(b) $C_2H_5OH(l)$ or $C_2H_5OH(g)$

(c) $2H(g)$ or $H(g)$

16. Predict the sign of the entropy change for the following processes.

(a) An ice cube is warmed to near its melting point.

(b) Exhaled breath forms fog on a cold morning.

(c) Snow melts.

17. Predict the sign of the entropy change for the following processes. Give a reason for your prediction.

(a) $Pb^{2+}(aq) + S^{2-}(aq) \longrightarrow PbS(s)$

(b) $2Fe(s) + 3O_2(g) \longrightarrow Fe_2O_3(s)$

(c) $2C_6H_{14}(l) + 19O_2(g) \longrightarrow 14H_2O(g) + 12CO_2(g)$

18. Write the balanced chemical equation for the combustion of methane, $CH_4(g)$, to give carbon dioxide and water vapor. Explain why it is difficult to predict whether ΔS is positive or negative for this chemical reaction.

19. Write the balanced chemical equation for the combustion of benzene, $C_6H_6(l)$, to give carbon dioxide and water vapor. Would you expect ΔS to be positive or negative in this process?

16.3 The Second and Third Laws of Thermodynamics

20. What is the difference between ΔS, $\Delta S°$, and $\Delta S°_{298}$ for a chemical change?

21. Calculate $\Delta S°_{298}$ for the following changes.

(a) $SnCl_4(l) \longrightarrow SnCl_4(g)$

(b) $CS_2(g) \longrightarrow CS_2(l)$

(c) $Cu(s) \longrightarrow Cu(g)$

(d) $H_2O(l) \longrightarrow H_2O(g)$

(e) $2H_2(g) + O_2(g) \longrightarrow 2H_2O(l)$

(f) $2HCl(g) + Pb(s) \longrightarrow PbCl_2(s) + H_2(g)$

(g) $Zn(s) + CuSO_4(s) \longrightarrow Cu(s) + ZnSO_4(s)$

22. Determine the entropy change for the combustion of liquid ethanol, C_2H_5OH, under standard state conditions to give gaseous carbon dioxide and liquid water.

23. Determine the entropy change for the combustion of gaseous propane, C_3H_8, under standard state conditions to give gaseous carbon dioxide and water.

24. "Thermite" reactions have been used for welding metal parts such as railway rails and in metal refining. One such thermite reaction is $Fe_2O_3(s) + 2Al(s) \longrightarrow Al_2O_3(s) + 2Fe(s)$. Is the reaction spontaneous at room temperature under standard conditions? During the reaction, the surroundings absorb 851.8 kJ/mol of heat.

25. Using the relevant $S°_{298}$ values listed in Appendix G, calculate $S°_{298}$ for the following changes:

(a) $N_2(g) + 3H_2(g) \longrightarrow 2NH_3(g)$

(b) $N_2(g) + \frac{5}{2}O_2(g) \longrightarrow N_2O_5(g)$

26. From the following information, determine $\Delta S°_{298}$ for the following:

$N(g) + O(g) \longrightarrow NO(g) \qquad \Delta S°_{298} = ?$

$N_2(g) + O_2(g) \longrightarrow 2NO(g) \qquad \Delta S°_{298} = 24.8 \text{ J/K}$

$N_2(g) \longrightarrow 2N(g) \qquad \Delta S°_{298} = 115.0 \text{ J/K}$

$O_2(g) \longrightarrow 2O(g) \qquad \Delta S°_{298} = 117.0 \text{ J/K}$

27. By calculating ΔS_{univ} at each temperature, determine if the melting of 1 mole of $NaCl(s)$ is spontaneous at 500 °C and at 700 °C.

$$S°_{NaCl(s)} = 72.11\frac{J}{mol \cdot K} \qquad S°_{NaCl(l)} = 95.06\frac{J}{mol \cdot K} \qquad \Delta H°_{fusion} = 27.95 \text{ kJ/mol}$$

What assumptions are made about the thermodynamic information (entropy and enthalpy values) used to solve this problem?

28. Use the standard entropy data in Appendix G to determine the change in entropy for each of the reactions listed in Exercise 16.33. All are run under standard state conditions and 25 °C.

29. Use the standard entropy data in Appendix G to determine the change in entropy for each of the reactions listed in Exercise 16.34. All are run under standard state conditions and 25 °C.

16.4 Free Energy

30. What is the difference between ΔG, $\Delta G°$, and $\Delta G°_{298}$ for a chemical change?

31. A reactions has $\Delta H^{\circ}_{298} = 100$ kJ/mol and $\Delta S^{\circ}_{298} = 250$ J/mol·K. Is the reaction spontaneous at room temperature? If not, under what temperature conditions will it become spontaneous?

32. Explain what happens as a reaction starts with $\Delta G < 0$ (negative) and reaches the point where $\Delta G = 0$.

33. Use the standard free energy of formation data in Appendix G to determine the free energy change for each of the following reactions, which are run under standard state conditions and 25 °C. Identify each as either spontaneous or nonspontaneous at these conditions.

(a) $MnO_2(s) \longrightarrow Mn(s) + O_2(g)$

(b) $H_2(g) + Br_2(l) \longrightarrow 2HBr(g)$

(c) $Cu(s) + S(g) \longrightarrow CuS(s)$

(d) $2LiOH(s) + CO_2(g) \longrightarrow Li_2CO_3(s) + H_2O(g)$

(e) $CH_4(g) + O_2(g) \longrightarrow C(s,\ graphite) + 2H_2O(g)$

(f) $CS_2(g) + 3Cl_2(g) \longrightarrow CCl_4(g) + S_2Cl_2(g)$

34. Use the standard free energy data in Appendix G to determine the free energy change for each of the following reactions, which are run under standard state conditions and 25 °C. Identify each as either spontaneous or nonspontaneous at these conditions.

(a) $C(s, graphite) + O_2(g) \longrightarrow CO_2(g)$

(b) $O_2(g) + N_2(g) \longrightarrow 2NO(g)$

(c) $2Cu(s) + S(g) \longrightarrow Cu_2S(s)$

(d) $CaO(s) + H_2O(l) \longrightarrow Ca(OH)_2(s)$

(e) $Fe_2O_3(s) + 3CO(g) \longrightarrow 2Fe(s) + 3CO_2(g)$

(f) $CaSO_4 \cdot 2H_2O(s) \longrightarrow CaSO_4(s) + 2H_2O(g)$

35. Given:
$$P_4(s) + 5O_2(g) \longrightarrow P_4O_{10}(s) \qquad \Delta G^{\circ}_{298} = -2697.0 \text{ kJ/mol}$$
$$2H_2(g) + O_2(g) \longrightarrow 2H_2O(g) \qquad \Delta G^{\circ}_{298} = -457.18 \text{ kJ/mol}$$
$$6H_2O(g) + P_4O_{10}(g) \longrightarrow 4H_3PO_4(l) \qquad \Delta G^{\circ}_{298} = -428.66 \text{ kJ/mol}$$

(a) Determine the standard free energy of formation, ΔG°_f , for phosphoric acid.

(b) How does your calculated result compare to the value in Appendix G? Explain.

36. Is the formation of ozone ($O_3(g)$) from oxygen ($O_2(g)$) spontaneous at room temperature under standard state conditions?

37. Consider the decomposition of red mercury(II) oxide under standard state conditions.
$$2HgO(s,\ red) \longrightarrow 2Hg(l) + O_2(g)$$

(a) Is the decomposition spontaneous under standard state conditions?

(b) Above what temperature does the reaction become spontaneous?

38. Among other things, an ideal fuel for the control thrusters of a space vehicle should decompose in a spontaneous exothermic reaction when exposed to the appropriate catalyst. Evaluate the following substances under standard state conditions as suitable candidates for fuels.

(a) Ammonia: $2NH_3(g) \longrightarrow N_2(g) + 3H_2(g)$

(b) Diborane: $B_2H_6(g) \longrightarrow 2B(g) + 3H_2(g)$

(c) Hydrazine: $N_2H_4(g) \longrightarrow N_2(g) + 2H_2(g)$

(d) Hydrogen peroxide: $H_2O_2(l) \longrightarrow H_2O(g) + \frac{1}{2}O_2(g)$

39. Calculate $\Delta G°$ for each of the following reactions from the equilibrium constant at the temperature given.

(a) $N_2(g) + O_2(g) \longrightarrow 2NO(g)$ T = 2000 °C $K_p = 4.1 \times 10^{-4}$

(b) $H_2(g) + I_2(g) \longrightarrow 2HI(g)$ T = 400 °C $K_p = 50.0$

(c) $CO_2(g) + H_2(g) \longrightarrow CO(g) + H_2O(g)$ T = 980 °C $K_p = 1.67$

(d) $CaCO_3(s) \longrightarrow CaO(s) + CO_2(g)$ T = 900 °C $K_p = 1.04$

(e) $HF(aq) + H_2O(l) \longrightarrow H_3O^+(aq) + F^-(aq)$ T = 25 °C $K_p = 7.2 \times 10^{-4}$

(f) $AgBr(s) \longrightarrow Ag^+(aq) + Br^-(aq)$ T = 25 °C $K_p = 3.3 \times 10^{-13}$

40. Calculate $\Delta G°$ for each of the following reactions from the equilibrium constant at the temperature given.

(a) $Cl_2(g) + Br_2(g) \longrightarrow 2BrCl(g)$ T = 25 °C $K_p = 4.7 \times 10^{-2}$

(b) $2SO_2(g) + O_2(g) \rightleftharpoons 2SO_3(g)$ T = 500 °C $K_p = 48.2$

(c) $H_2O(l) \rightleftharpoons H_2O(g)$ T = 60 °C $K_p = 0.196$ atm

(d) $CoO(s) + CO(g) \rightleftharpoons Co(s) + CO_2(g)$ T = 550 °C $K_p = 4.90 \times 10^2$

(e) $CH_3NH_2(aq) + H_2O(l) \longrightarrow CH_3NH_3{}^+(aq) + OH^-(aq)$ T = 25 °C $K_p = 4.4 \times 10^{-4}$

(f) $PbI_2(s) \longrightarrow Pb^{2+}(aq) + 2I^-(aq)$ T = 25 °C $K_p = 8.7 \times 10^{-9}$

41. Calculate the equilibrium constant at 25 °C for each of the following reactions from the value of $\Delta G°$ given.

(a) $O_2(g) + 2F_2(g) \longrightarrow 2OF_2(g)$ $\Delta G° = -9.2$ kJ

(b) $I_2(s) + Br_2(l) \longrightarrow 2IBr(g)$ $\Delta G° = 7.3$ kJ

(c) $2LiOH(s) + CO_2(g) \longrightarrow Li_2CO_3(s) + H_2O(g)$ $\Delta G° = -79$ kJ

(d) $N_2O_3(g) \longrightarrow NO(g) + NO_2(g)$ $\Delta G° = -1.6$ kJ

(e) $SnCl_4(l) \longrightarrow SnCl_4(l)$ $\Delta G° = 8.0$ kJ

42. Calculate the equilibrium constant at 25 °C for each of the following reactions from the value of $\Delta G°$ given.

(a) $I_2(s) + Cl_2(g) \longrightarrow 2ICl(g)$ $\Delta G° = -10.88$ kJ

(b) $H_2(g) + I_2(s) \longrightarrow 2HI(g)$ $\Delta G° = 3.4$ kJ

(c) $CS_2(g) + 3Cl_2(g) \longrightarrow CCl_4(g) + S_2Cl_2(g)$ $\Delta G° = -39$ kJ

(d) $2SO_2(g) + O_2(g) \longrightarrow 2SO_3(g)$ $\Delta G° = -141.82$ kJ

(e) $CS_2(g) \longrightarrow CS_2(l)$ $\Delta G° = -1.88$ kJ

43. Calculate the equilibrium constant at the temperature given.

(a) $O_2(g) + 2F_2(g) \longrightarrow 2F_2O(g)$ (T = 100 °C)

(b) $I_2(s) + Br_2(l) \longrightarrow 2IBr(g)$ (T = 0.0 °C)

(c) $2LiOH(s) + CO_2(g) \longrightarrow Li_2CO_3(s) + H_2O(g)$ (T = 575 °C)

(d) $N_2O_3(g) \longrightarrow NO(g) + NO_2(g)$ (T = -10.0 °C)

(e) $SnCl_4(l) \longrightarrow SnCl_4(g)$ (T = 200 °C)

44. Calculate the equilibrium constant at the temperature given.

(a) $I_2(s) + Cl_2(g) \longrightarrow 2ICl(g)$ (T = 100 °C)

(b) $H_2(g) + I_2(s) \longrightarrow 2HI(g)$ (T = 0.0 °C)

(c) $CS_2(g) + 3Cl_2(g) \longrightarrow CCl_4(g) + S_2Cl_2(g)$ (T = 125 °C)

(d) $2SO_2(g) + O_2(g) \longrightarrow 2SO_3(g)$ (T = 675 °C)

(e) $CS_2(g) \longrightarrow CS_2(l)$ (T = 90 °C)

45. Consider the following reaction at 298 K:
$$N_2O_4(g) \rightleftharpoons 2NO_2(g) \qquad K_P = 0.142$$

What is the standard free energy change at this temperature? Describe what happens to the initial system, where the reactants and products are in standard states, as it approaches equilibrium.

46. Determine the normal boiling point (in kelvin) of dichloroethane, CH_2Cl_2. Find the actual boiling point using the Internet or some other source, and calculate the percent error in the temperature. Explain the differences, if any, between the two values.

47. Under what conditions is $N_2O_3(g) \longrightarrow NO(g) + NO_2(g)$ spontaneous?

48. At room temperature, the equilibrium constant (K_w) for the self-ionization of water is 1.00×10^{-14}. Using this information, calculate the standard free energy change for the aqueous reaction of hydrogen ion with hydroxide ion to produce water. (Hint: The reaction is the reverse of the self-ionization reaction.)

49. Hydrogen sulfide is a pollutant found in natural gas. Following its removal, it is converted to sulfur by the reaction $2H_2S(g) + SO_2(g) \rightleftharpoons \frac{3}{8}S_8(s, \text{ rhombic}) + 2H_2O(l)$. What is the equilibrium constant for this reaction? Is the reaction endothermic or exothermic?

50. Consider the decomposition of $CaCO_3(s)$ into $CaO(s)$ and $CO_2(g)$. What is the equilibrium partial pressure of CO_2 at room temperature?

51. In the laboratory, hydrogen chloride ($HCl(g)$) and ammonia ($NH_3(g)$) often escape from bottles of their solutions and react to form the ammonium chloride ($NH_4Cl(s)$), the white glaze often seen on glassware. Assuming that the number of moles of each gas that escapes into the room is the same, what is the maximum partial pressure of HCl and NH_3 in the laboratory at room temperature? (Hint: The partial pressures will be equal and are at their maximum value when at equilibrium.)

52. Benzene can be prepared from acetylene. $3C_2H_2(g) \rightleftharpoons C_6H_6(g)$. Determine the equilibrium constant at 25 °C and at 850 °C. Is the reaction spontaneous at either of these temperatures? Why is all acetylene not found as benzene?

53. Carbon dioxide decomposes into CO and O_2 at elevated temperatures. What is the equilibrium partial pressure of oxygen in a sample at 1000 °C for which the initial pressure of CO_2 was 1.15 atm?

54. Carbon tetrachloride, an important industrial solvent, is prepared by the chlorination of methane at 850 K.
$$CH_4(g) + 4Cl_2(g) \longrightarrow CCl_4(g) + 4HCl(g)$$

What is the equilibrium constant for the reaction at 850 K? Would the reaction vessel need to be heated or cooled to keep the temperature of the reaction constant?

55. Acetic acid, CH_3CO_2H, can form a dimer, $(CH_3CO_2H)_2$, in the gas phase.
$$2CH_3CO_2H(g) \longrightarrow (CH_3CO_2H)_2(g)$$

The dimer is held together by two hydrogen bonds with a total strength of 66.5 kJ per mole of dimer.

At 25 °C, the equilibrium constant for the dimerization is 1.3×10^3 (pressure in atm). What is $\Delta S°$ for the reaction?

56. Nitric acid, HNO_3, can be prepared by the following sequence of reactions:

$$4NH_3(g) + 5O_2(g) \longrightarrow 4NO(g) + 6H_2O(g)$$

$$2NO(g) + O_2(g) \longrightarrow 2NO_2(g)$$

$$3NO_2(g) + H_2O(l) \longrightarrow 2HNO_3(l) + NO(g)$$

How much heat is evolved when 1 mol of $NH_3(g)$ is converted to $HNO_3(l)$? Assume standard states at 25 °C.

57. Determine ΔG for the following reactions.

(a) Antimony pentachloride decomposes at 448 °C. The reaction is:

$$SbCl_5(g) \longrightarrow SbCl_3(g) + Cl_2(g)$$

An equilibrium mixture in a 5.00 L flask at 448 °C contains 3.85 g of $SbCl_5$, 9.14 g of $SbCl_3$, and 2.84 g of Cl_2.

(b) Chlorine molecules dissociate according to this reaction:

$$Cl_2(g) \longrightarrow 2Cl(g)$$

1.00% of Cl_2 molecules dissociate at 975 K and a pressure of 1.00 atm.

58. Given that the ΔG_f° for $Pb^{2+}(aq)$ and $Cl^-(aq)$ is −24.3 kJ/mole and −131.2 kJ/mole respectively, determine the solubility product, K_{sp}, for $PbCl_2(s)$.

59. Determine the standard free energy change, ΔG_f°, for the formation of $S^{2-}(aq)$ given that the ΔG_f° for $Ag^+(aq)$ and $Ag_2S(s)$ are 77.1 k/mole and −39.5 kJ/mole respectively, and the solubility product for $Ag_2S(s)$ is 8 × 10^{-51}.

60. Determine the standard enthalpy change, entropy change, and free energy change for the conversion of diamond to graphite. Discuss the spontaneity of the conversion with respect to the enthalpy and entropy changes. Explain why diamond spontaneously changing into graphite is not observed.

61. The evaporation of one mole of water at 298 K has a standard free energy change of 8.58 kJ.

$$H_2O(l) \rightleftharpoons H_2O(g) \qquad \Delta G_{298}^\circ = 8.58 \text{ kJ}$$

(a) Is the evaporation of water under standard thermodynamic conditions spontaneous?

(b) Determine the equilibrium constant, K_P, for this physical process.

(c) By calculating ΔG, determine if the evaporation of water at 298 K is spontaneous when the partial pressure of water, P_{H_2O}, is 0.011 atm.

(d) If the evaporation of water were always nonspontaneous at room temperature, wet laundry would never dry when placed outside. In order for laundry to dry, what must be the value of P_{H_2O} in the air?

62. In glycolysis, the reaction of glucose (Glu) to form glucose-6-phosphate (G6P) requires ATP to be present as described by the following equation:

$$\text{Glu} + \text{ATP} \longrightarrow \text{G6P} + \text{ADP} \qquad \Delta G_{298}^\circ = -17 \text{ kJ}$$

In this process, ATP becomes ADP summarized by the following equation:

$$\text{ATP} \longrightarrow \text{ADP} \qquad \Delta G_{298}^\circ = -30 \text{ kJ}$$

Determine the standard free energy change for the following reaction, and explain why ATP is necessary to drive this process:

$$\text{Glu} \longrightarrow \text{G6P} \qquad \Delta G_{298}^\circ = ?$$

63. One of the important reactions in the biochemical pathway glycolysis is the reaction of glucose-6-phosphate (G6P) to form fructose-6-phosphate (F6P):

$$\text{G6P} \rightleftharpoons \text{F6P} \qquad \Delta G_{298}^\circ = 1.7 \text{ kJ}$$

(a) Is the reaction spontaneous or nonspontaneous under standard thermodynamic conditions?

(b) Standard thermodynamic conditions imply the concentrations of G6P and F6P to be 1 M, however, in a typical cell, they are not even close to these values. Calculate ΔG when the concentrations of G6P and F6P are 120 μM and

28 μ*M* respectively, and discuss the spontaneity of the forward reaction under these conditions. Assume the temperature is 37 °C.

64. Without doing a numerical calculation, determine which of the following will reduce the free energy change for the reaction, that is, make it less positive or more negative, when the temperature is increased. Explain.

(a) $N_2(g) + 3H_2(g) \longrightarrow 2NH_3(g)$

(b) $HCl(g) + NH_3(g) \longrightarrow NH_4Cl(s)$

(c) $(NH_4)_2Cr_2O_7(s) \longrightarrow Cr_2O_3(s) + 4H_2O(g) + N_2(g)$

(d) $2Fe(s) + 3O_2(g) \longrightarrow Fe_2O_3(s)$

65. When ammonium chloride is added to water and stirred, it dissolves spontaneously and the resulting solution feels cold. Without doing any calculations, deduce the signs of Δ*G*, Δ*H*, and Δ*S* for this process, and justify your choices.

66. An important source of copper is from the copper ore, chalcocite, a form of copper(I) sulfide. When heated, the Cu_2S decomposes to form copper and sulfur described by the following equation:
$Cu_2S(s) \longrightarrow Cu(s) + S(s)$

(a) Determine ΔG°_{298} for the decomposition of $Cu_2S(s)$.

(b) The reaction of sulfur with oxygen yields sulfur dioxide as the only product. Write an equation that describes this reaction, and determine ΔG°_{298} for the process.

(c) The production of copper from chalcocite is performed by roasting the Cu_2S in air to produce the Cu. By combining the equations from Parts (a) and (b), write the equation that describes the roasting of the chalcocite, and explain why coupling these reactions together makes for a more efficient process for the production of the copper.

67. What happens to ΔG°_{298} (becomes more negative or more positive) for the following chemical reactions when the partial pressure of oxygen is increased?

(a) $S(s) + O_2(g) \longrightarrow SO_2(g)$

(b) $2SO_2(g) + O_2(g) \longrightarrow SO_3(g)$

(c) $HgO(s) \longrightarrow Hg(l) + O_2(g)$

Chapter 17

Electrochemistry

Figure 17.1 Electric vehicles contain batteries that can be recharged, thereby using electric energy to bring about a chemical change and vice versa. (credit: modification of work by Robert Couse-Baker)

Chapter Outline

Introduction

Electrochemistry deals with chemical reactions that produce electricity and the changes associated with the passage of electrical current through matter. The reactions involve electron transfer, and so they are oxidation-reduction (or redox) reactions. Many metals may be purified or electroplated using electrochemical methods. Devices such as automobiles, smartphones, electronic tablets, watches, pacemakers, and many others use batteries for power. Batteries use chemical reactions that produce electricity spontaneously and that can be converted into useful work. All electrochemical systems involve the transfer of electrons in a reacting system. In many systems, the reactions occur in a region known as the cell, where the transfer of electrons occurs at electrodes.

17.1 Balancing Oxidation-Reduction Reactions

By the end of this section, you will be able to:

- Define electrochemistry and a number of important associated terms
- Split oxidation-reduction reactions into their oxidation half-reactions and reduction half-reactions
- Produce balanced oxidation-reduction equations for reactions in acidic or basic solution
- Identify oxidizing agents and reducing agents

Electricity refers to a number of phenomena associated with the presence and flow of electric charge. Electricity includes such diverse things as lightning, static electricity, the current generated by a battery as it discharges, and many other influences on our daily lives. The flow or movement of charge is an electric current (Figure 17.2). Electrons or ions may carry the charge. The elementary unit of charge is the charge of a proton, which is equal in magnitude to the charge of an electron. The SI unit of charge is the coulomb (C) and the charge of a proton is 1.602×10^{-19} C. The presence of an electric charge generates an electric field. Electric **current** is the rate of flow of charge. The SI unit for electrical current is the SI base unit called the ampere (A), which is a flow rate of 1 coulomb of charge per second (1 A = 1 C/s). An electric current flows in a path, called an electric **circuit**. In most chemical systems, it is necessary to maintain a closed path for current to flow. The flow of charge is generated by an electrical potential difference, or potential, between two points in the circuit. **Electrical potential** is the ability of the electric field to do work on the charge. The SI unit of electrical potential is the volt (V). When 1 coulomb of charge moves through a potential difference of 1 volt, it gains or loses 1 joule (J) of energy. Table 17.1 summarizes some of this information about electricity.

Common Electrical Terms

Quantity	Definition	Measure or Unit
Electric charge	Charge on a proton	1.602×10^{-19} C
Electric current	The movement of charge	ampere = A = 1 C/s
Electric potential	The force trying to move the charge	volt = V = J/C
Electric field	The force acting upon other charges in the vicinity	

Table 17.1

Figure 17.2 Electricity-related phenomena include lightning, accumulation of static electricity, and current produced by a battery. (credit left: modification of work by Thomas Bresson; credit middle: modification of work by Chris Darling; credit right: modification of work by Windell Oskay)

Electrochemistry studies oxidation-reduction reactions, which were first discussed in an earlier chapter, where we learned that oxidation was the loss of electrons and reduction was the gain of electrons. The reactions discussed tended to be rather simple, and conservation of mass (atom counting by type) and deriving a correctly balanced chemical equation were relatively simple. In this section, we will concentrate on the half-reaction method for balancing oxidation-reduction reactions. The use of half-reactions is important partly for balancing more complicated reactions and partly because many aspects of electrochemistry are easier to discuss in terms of half-reactions. There are alternate methods of balancing these reactions; however, there are no good alternatives to half-reactions for discussing what is occurring in many systems. The **half-reaction method** splits oxidation-reduction reactions into their oxidation "half" and reduction "half" to make finding the overall equation easier.

Electrochemical reactions frequently occur in solutions, which could be acidic, basic, or neutral. When balancing oxidation-reduction reactions, the nature of the solution may be important. It helps to see this in an actual problem. Consider the following unbalanced oxidation-reduction reaction in acidic solution:

$$MnO_4^-(aq) + Fe^{2+}(aq) \longrightarrow Mn^{2+}(aq) + Fe^{3+}(aq)$$

We can start by collecting the species we have so far into an unbalanced oxidation half-reaction and an unbalanced **reduction half-reaction**. Each of these half-reactions contain the same element in two different oxidation states. The Fe^{2+} has lost an electron to become Fe^{3+}; therefore, the iron underwent oxidation. The reduction is not as obvious; however, the manganese gained five electrons to change from Mn^{7+} to Mn^{2+}.

$$\text{oxidation (unbalanced):} \quad Fe^{2+}(aq) \longrightarrow Fe^{3+}(aq)$$
$$\text{reduction (unbalanced):} \ MnO_4^-(aq) \longrightarrow Mn^{2+}(aq)$$

In acidic solution, there are hydrogen ions present, which are often useful in balancing half-reactions. It may be necessary to use the hydrogen ions directly or as a reactant that may react with oxygen to generate water. Hydrogen ions are very important in acidic solutions where the reactants or products contain hydrogen and/or oxygen. In this example, the oxidation half-reaction involves neither hydrogen nor oxygen, so hydrogen ions are not necessary to the balancing. However, the reduction half-reaction does involve oxygen. It is necessary to use hydrogen ions to convert this oxygen to water.

$$\text{charge not balanced:} \ MnO_4^-(aq) + 8H^+(aq) \longrightarrow Mn^{2+}(aq) + 4H_2O(l)$$

The situation is different in basic solution because the hydrogen ion concentration is lower and the hydroxide ion concentration is higher. After finishing this example, we will examine how basic solutions differ from acidic solutions. A neutral solution may be treated as acidic or basic, though treating it as acidic is usually easier.

The iron atoms in the oxidation half-reaction are balanced (mass balance); however, the charge is unbalanced, since the charges on the ions are not equal. It is necessary to use electrons to balance the charge. The way to balance the charge is by *adding* electrons to one side of the equation. Adding a single electron on the right side gives a balanced oxidation half-reaction:

$$\text{oxidation (balanced):} \ Fe^{2+}(aq) \longrightarrow Fe^{3+}(aq) + e^-$$

You should check the half-reaction for the number of each atom type and the total charge on each side of the equation. The charges include the actual charges of the ions times the number of ions and the charge on an electron times the number of electrons.

$$\text{Fe: Does } (1 \times 1) = (1 \times 1)? \text{ Yes.}$$
$$\text{Charge: Does } [1 \times (+2)] = [1 \times (+3) + 1 \times (-1)]? \text{ Yes.}$$

If the atoms and charges balance, the half-reaction is balanced. In oxidation half-reactions, electrons appear as products (on the right). As discussed in the earlier chapter, since iron underwent oxidation, iron is the reducing agent.

Now return to the reduction half-reaction equation:

$$\text{reduction (unbalanced):} \ MnO_4^-(aq) + 8H^+(aq) \longrightarrow Mn^{2+}(aq) + 4H_2O(l)$$

The atoms are balanced (mass balance), so it is now necessary to check for charge balance. The total charge on the left of the reaction arrow is $[(-1) \times (1) + (8) \times (+1)]$, or +7, while the total charge on the right side is $[(1) \times (+2) +$

(4) × (0)], or +2. The difference between +7 and +2 is five; therefore, it is necessary to add five electrons to the left side to achieve charge balance.

$$\text{Reduction (balanced): } MnO_4{}^-(aq) + 8H^+(aq) + 5e^- \longrightarrow Mn^{2+}(aq) + 4H_2O(l)$$

You should check this half-reaction for each atom type and for the charge, as well:

Mn: Does $(1 \times 1) = (1 \times 1)$? Yes.
H: Does $(8 \times 1) = (4 \times 2)$? Yes.
O: Does $(1 \times 4) = (4 \times 1)$? Yes.
Charge: Does $[1 \times (-1) + 8 \times (+1) + 5 \times (-1)] = [1 \times (+2)]$? Yes.

Now that this half-reaction is balanced, it is easy to see it involves reduction because electrons were gained when $MnO_4{}^-$ was reduced to Mn^{2+}. In all reduction half-reactions, electrons appear as reactants (on the left side). As discussed in the earlier chapter, the species that was reduced, $MnO_4{}^-$ in this case, is also called the oxidizing agent. We now have two balanced half-reactions.

$$\text{oxidation: } \qquad\qquad\qquad Fe^{2+}(aq) \longrightarrow Fe^{3+}(aq) + e^-$$
$$\text{reduction: } MnO_4{}^-(aq) + 8H^+(aq) + 5e^- \longrightarrow Mn^{2+}(aq) + 4H_2O(l)$$

It is now necessary to combine the two halves to produce a whole reaction. The key to combining the half-reactions is the electrons. The electrons lost during oxidation must go somewhere. These electrons go to cause reduction. The number of electrons transferred from the oxidation half-reaction to the reduction half-reaction must be equal. There can be no missing or excess electrons. In this example, the oxidation half-reaction generates one electron, while the reduction half-reaction requires five. The lowest common multiple of one and five is five; therefore, it is necessary to multiply every term in the oxidation half-reaction by five and every term in the reduction half-reaction by one. (In this case, the multiplication of the reduction half-reaction generates no change; however, this will not always be the case.) The multiplication of the two half-reactions by the appropriate factor followed by addition of the two halves gives

$$\text{oxidation: } \qquad\qquad\qquad 5 \times (Fe^{2+}(aq) \longrightarrow Fe^{3+}(aq) + e^-)$$
$$\text{reduction: } \quad MnO_4{}^-(aq) + 8H^+(aq) + 5e^- \longrightarrow Mn^{2+}(aq) + 4H_2O(l)$$
$$\overline{\text{overall: } 5Fe^{2+}(aq) + MnO_4{}^-(aq) + 8H^+(aq) \longrightarrow 5Fe^{3+}(aq) + Mn^{2+}(aq) + 4H_2O(l)}$$

The electrons do not appear in the final answer because the oxidation electrons are the same electrons as the reduction electrons and they "cancel." Carefully check each side of the overall equation to verify everything was combined correctly:

Fe: Does $(5 \times 1) = (5 \times 1)$? Yes.
Mn: Does $(1 \times 1) = (1 \times 1)$? Yes.
H: Does $(8 \times 1) = (4 \times 2)$? Yes.
O: Does $(1 \times 4) = (4 \times 1)$? Yes.
Charge: Does $[5 \times (+2) + 1 \times (-1) + 8 \times (+1)] = [5 \times (+3) + 1 \times (+2)]$? Yes.

Everything checks, so this is the overall equation in acidic solution. If something does not check, the most common error occurs during the multiplication of the individual half-reactions.

Now suppose we wanted the solution to be basic. Recall that basic solutions have excess hydroxide ions. Some of these hydroxide ions will react with hydrogen ions to produce water. The simplest way to generate the balanced overall equation in basic solution is to start with the balanced equation in acidic solution, then "convert" it to the equation for basic solution. However, it is necessary to exercise caution when doing this, as many reactants behave differently under basic conditions and many metal ions will precipitate as the metal hydroxide. We just produced the following reaction, which we want to change to a basic reaction:

$$5Fe^{2+}(aq) + MnO_4{}^-(aq) + 8H^+(aq) \longrightarrow 5Fe^{3+}(aq) + Mn^{2+}(aq) + 4H_2O(l)$$

However, under basic conditions, MnO_4^- normally reduces to MnO_2 and iron will be present as either $Fe(OH)_2$ or $Fe(OH)_3$. For these reasons, under basic conditions, this reaction will be

$$3Fe(OH)_2(s) + MnO_4^-(aq) + 2H_2O(l) \longrightarrow 3Fe(OH)_3(s) + MnO_2(s) + OH^-(aq)$$

(Under very basic conditions MnO_4^- will reduce to MnO_4^{2-}, instead of MnO_2.)

It is still possible to balance any oxidation-reduction reaction as an acidic reaction and then, when necessary, convert the equation to a basic reaction. This will work if the acidic and basic reactants and products are the same or if the basic reactants and products are used before the conversion from acidic or basic. There are very few examples in which the acidic and basic reactions will involve the same reactants and products. However, balancing a basic reaction as acidic and then converting to basic will work. To convert to a basic reaction, it is necessary to add the same number of hydroxide ions to each side of the equation so that all the hydrogen ions (H^+) are removed and mass balance is maintained. Hydrogen ion combines with hydroxide ion (OH^-) to produce water.

Let us now try a basic equation. We will start with the following basic reaction:

$$Cl^-(aq) + MnO_4^-(aq) \longrightarrow ClO_3^-(aq) + MnO_2(s)$$

Balancing this as acid gives

$$Cl^-(aq) + 2MnO_4^-(aq) + 2H^+(aq) \longrightarrow ClO_3^-(aq) + 2MnO_2(s) + H_2O(l)$$

In this case, it is necessary to add two hydroxide ions to each side of the equation to convert the two hydrogen ions on the left into water:

$$Cl^-(aq) + 2MnO_4^-(aq) + (2H^+ + 2OH^-)(aq) \longrightarrow ClO_3^-(aq) + 2MnO_2(s) + H_2O(l) + 2OH^-(aq)$$
$$Cl^-(aq) + 2MnO_4^-(aq) + (2H_2O)(l) \longrightarrow ClO_3^-(aq) + 2MnO_2(s) + H_2O(l) + 2OH^-(aq)$$

Note that both sides of the equation show water. Simplifying should be done when necessary, and gives the desired equation. In this case, it is necessary to remove one H_2O from each side of the reaction arrows.

$$Cl^-(aq) + 2MnO_4^-(aq) + H_2O(l) \longrightarrow ClO_3^-(aq) + 2MnO_2(s) + 2OH^-(aq)$$

Again, check each side of the overall equation to make sure there are no errors:

Cl: Does $(1 \times 1) = (1 \times 1)$? Yes.
Mn: Does $(2 \times 1) = (2 \times 1)$? Yes.
H: Does $(1 \times 2) = (2 \times 1)$? Yes.
O: Does $(2 \times 4 + 1 \times 1) = (3 \times 1 + 2 \times 2 + 2 \times 1)$? Yes.
Charge: Does $[1 \times (-1) + 2 \times (-1)] = [1 \times (-1) + 2 \times (-1)]$? Yes.

Everything checks, so this is the overall equation in basic solution.

Example 17.1

Balancing Acidic Oxidation-Reduction Reactions

Balance the following reaction equation in acidic solution:

$$MnO_4^-(aq) + Cr^{3+}(aq) \longrightarrow Mn^{2+}(aq) + Cr_2O_7^{2-}(aq)$$

Solution

This is an oxidation-reduction reaction, so start by collecting the species given into an unbalanced oxidation half-reaction and an unbalanced reduction half-reaction.

oxidation (unbalanced): $Cr^{3+}(aq) \longrightarrow Cr_2O_7^{2-}(aq)$
reduction (unbalanced): $MnO_4^-(aq) \longrightarrow Mn^{2+}(aq)$

Starting with the oxidation half-reaction, we can balance the chromium

$$\text{oxidation (unbalanced): } 2Cr^{3+}(aq) \longrightarrow Cr_2O_7{}^{2-}(aq)$$

In acidic solution, we can use or generate hydrogen ions (H^+). Adding seven water molecules to the left side provides the necessary oxygen; the "left over" hydrogen appears as 14 H^+ on the right:

$$\text{oxidation (unbalanced): } 2Cr^{3+}(aq) + 7H_2O(l) \longrightarrow Cr_2O_7{}^{2-}(aq) + 14H^+(aq)$$

The left side of the equation has a total charge of [2 × (+3) = +6], and the right side a total charge of [−2 + 14 × (+1) = +12]. The difference is six; adding six electrons to the right side produces a mass- and charge-balanced oxidation half-reaction (in acidic solution):

$$\text{oxidation (balanced): } 2Cr^{3+}(aq) + 7H_2O(l) \longrightarrow Cr_2O_7{}^{2-}(aq) + 14H^+(aq) + 6e^-$$

Checking the half-reaction:

> Cr: Does (2 × 1) = (1 × 2)? Yes.
> H: Does (7 × 2) = (14 × 1)? Yes.
> O: Does (7 × 1) = (1 × 7)? Yes.
> Charge: Does [2 × (+3)] = [1 × (−2) + 14 × (+1) + 6 × (−1)]? Yes.

Now work on the reduction. It is necessary to convert the four oxygen atoms in the permanganate into four water molecules. To do this, add eight H^+ to convert the oxygen into four water molecules:

$$\text{reduction (unbalanced): } MnO_4{}^-(aq) + 8H^+(aq) \longrightarrow Mn^{2+}(aq) + 4H_2O(l)$$

Then add five electrons to the left side to balance the charge:

$$\text{reduction (balanced): } MnO_4{}^-(aq) + 8H^+(aq) + 5e^- \longrightarrow Mn^{2+}(aq) + 4H_2O(l)$$

Make sure to check the half-reaction:

> Mn: Does (1 × 1) = (1 × 1)? Yes.
> H: Does (8 × 1) = (4 × 2)? Yes.
> O: Does (1 × 4) = (4 × 1)? Yes.
> Charge: Does [1 × (−1) + 8 × (+1) + 5 × (−1)] = [1 × (+2)]? Yes.

Collecting what we have so far:

$$\text{oxidation: } \quad 2Cr^{3+}(aq) + 7H_2O(l) \longrightarrow Cr_2O_7{}^{2-}(aq) + 14H^+(aq) + 6e^-$$

$$\text{reduction: } MnO_4{}^-(aq) + 8H^+(aq) + 5e^- \longrightarrow Mn^{2+}(aq) + 4H_2O(l)$$

The least common multiple for the electrons is 30, so multiply the oxidation half-reaction by five, the reduction half-reaction by six, combine, and simplify:

$$10Cr^{3+}(aq) + 35H_2O(l) + 6MnO_4{}^-(aq) + 48H^+(aq) \longrightarrow 5Cr_2O_7{}^{2-}(aq) + 70H^+(aq) + 6Mn^{2+}(aq) + 24H_2O(l)$$

$$10Cr^{3+}(aq) + 11H_2O(l) + 6MnO_4{}^-(aq) \longrightarrow 5\,Cr_2O_7{}^{2-}(aq) + 22H^+(aq) + 6Mn^{2+}(aq)$$

Checking each side of the equation:

> Mn: Does (6 × 1) = (6 × 1)? Yes.
> Cr: Does (10 × 1) = (5 × 2)? Yes.
> H: Does (11 × 2) = (22 × 1)? Yes.
> O: Does (11 × 1 + 6 × 4) = (5 × 7)? Yes.
> Charge: Does [10 × (+3) + 6 × (−1)] = [5 × (−2) + 22 × (+1) + 6 × (+2)]? Yes.

This is the balanced equation in acidic solution.

Check your learning

Balance the following equation in acidic solution:

$$Hg_2^{2+} + Ag \longrightarrow Hg + Ag^+$$

$$\textbf{Answer:} \quad Hg_2^{2+}(aq) + 2Ag(s) \longrightarrow 2Hg(l) + 2Ag^+(aq)$$

Example 17.2

Balancing Basic Oxidation-Reduction Reactions

Balance the following reaction equation in basic solution:

$$MnO_4^{-}(aq) + Cr(OH)_3(s) \longrightarrow MnO_2(s) + CrO_4^{2-}(aq)$$

Solution

This is an oxidation-reduction reaction, so start by collecting the species given into an unbalanced oxidation half-reaction and an unbalanced reduction half-reaction

$$\text{oxidation (unbalanced): } Cr(OH)_3(s) \longrightarrow CrO_4^{2-}(aq)$$

$$\text{reduction (unbalanced): } MnO_4^{-}(aq) \longrightarrow MnO_2(s)$$

Starting with the oxidation half-reaction, we can balance the chromium

$$\text{oxidation (unbalanced): } Cr(OH)_3(s) \longrightarrow CrO_4^{2-}(aq)$$

In acidic solution, we can use or generate hydrogen ions (H^+). Adding one water molecule to the left side provides the necessary oxygen; the "left over" hydrogen appears as five H^+ on the right side:

$$\text{oxidation (unbalanced): } Cr(OH)_3(s) + H_2O(l) \longrightarrow CrO_4^{2-}(aq) + 5H^+(aq)$$

The left side of the equation has a total charge of [0], and the right side a total charge of $[-2 + 5 \times (+1) = +3]$. The difference is three, adding three electrons to the right side produces a mass- and charge-balanced oxidation half-reaction (in acidic solution):

$$\text{oxidation (balanced): } Cr(OH)_3(s) + H_2O(l) \longrightarrow CrO_4^{2-}(aq) + 5H^+(aq) + 3e^-$$

Checking the half-reaction:

Cr: Does $(1 \times 1) = (1 \times 1)$? Yes.
H: Does $(1 \times 3 + 1 \times 2) = (5 \times 1)$? Yes.
O: Does $(1 \times 3 + 1 \times 1) = (4 \times 1)$? Yes.
Charge: Does $[0 = [1 \times (-2) + 5 \times (+1) + 3 \times (-1)]$? Yes.

Now work on the reduction. It is necessary to convert the four O atoms in the MnO_4^- minus the two O atoms in MnO_2 into two water molecules. To do this, add four H^+ to convert the oxygen into two water molecules:

$$\text{reduction (unbalanced): } MnO_4^{-}(aq) + 4H^+(aq) \longrightarrow MnO_2(s) + 2H_2O(l)$$

Then add three electrons to the left side to balance the charge:

$$\text{reduction (balanced): } MnO_4^{-}(aq) + 4H^+(aq) + 3e^- \longrightarrow MnO_2(s) + 2H_2O(l)$$

Make sure to check the half-reaction:

Mn: Does $(1 \times 1) = (1 \times 1)$? Yes.
H: Does $(4 \times 1) = (2 \times 2)$? Yes.
O: Does $(1 \times 4) = (1 \times 2 + 2 \times 1)$? Yes.
Charge: Does $[1 \times (-1) + 4 \times (+1) + 3 \times (-1)] = [0]$? Yes.

Collecting what we have so far:

$$\text{oxidation:} \qquad Cr(OH)_3(s) + H_2O(l) \longrightarrow CrO_4{}^{2-}(aq) + 5H^+(aq) + 3e^-$$

$$\text{reduction:} \; MnO_4{}^-(aq) + 4H^+(aq) + 3e^- \longrightarrow MnO_2(s) + 2H_2O(l)$$

In this case, both half reactions involve the same number of electrons; therefore, simply add the two half-reactions together.

$$MnO_4{}^-(aq) + 4H^+(aq) + Cr(OH)_3(s) + H_2O(l) \longrightarrow CrO_4{}^{2-}(aq) + MnO_2(s) + 2H_2O(l) + 5H^+(aq)$$

$$MnO_4{}^-(aq) + Cr(OH)_3(s) \longrightarrow CrO_4{}^{2-}(aq) + MnO_2(s) + H_2O(l) + H^+(aq)$$

Checking each side of the equation:

Mn: Does $(1 \times 1) = (1 \times 1)$? Yes.

Cr: Does $(1 \times 1) = (1 \times 1)$? Yes.

H: Does $(1 \times 3) = (2 \times 1 + 1 \times 1)$? Yes.

O: Does $(1 \times 4 + 1 \times 3) = (1 \times 4 + 1 \times 2 + 1 \times 1)$? Yes.

Charge: Does $[1 \times (-1)] = [1 \times (-2) + 1 \times (+1)]$? Yes.

This is the balanced equation in acidic solution. For a basic solution, add one hydroxide ion to each side and simplify:

$$OH^-(aq) + MnO_4{}^-(aq) + Cr(OH)_3(s) \longrightarrow CrO_4{}^{2-}(aq) + MnO_2(s) + H_2O(l) + (H^+ + OH^-)(aq)$$

$$OH^-(aq) + MnO_4{}^-(aq) + Cr(OH)_3(s) \longrightarrow CrO_4{}^{2-}(aq) + MnO_2(s) + 2H_2O(l)$$

Checking each side of the equation:

Mn: Does$(1 \times 1) = (1 \times 1)$? Yes.

Cr: Does $(1 \times 1) = (1 \times 1)$? Yes.

H: Does $(1 \times 1 + 1 \times 3) = (2 \times 2)$? Yes.

O: Does $(1 \times 1 + 1 \times 4 + 1 \times 3) = (1 \times 4 + 1 \times 2 + 2 \times 1)$? Yes.

Charge: Does $[1 \times (-1) + 1 \times (-1)] = [1 \times (-2)]$? Yes.

This is the balanced equation in basic solution.

Check Your Learning

Balance the following in the type of solution indicated.

(a) $H_2 + Cu^{2+} \longrightarrow Cu$ (acidic solution)

(b) $H_2 + Cu(OH)_2 \longrightarrow Cu$ (basic solution)

(c) $Fe + Ag^+ \longrightarrow Fe^{2+} + Ag$

(d) Identify the oxidizing agents in reactions (a), (b), and (c).

(e) Identify the reducing agents in reactions (a), (b), and (c).

Answer: (a) $H_2(g) + Cu^{2+}(aq) \longrightarrow 2H^+(aq) + Cu(s)$; (b) $H_2(g) + Cu(OH)_2(s) \longrightarrow 2H_2O(l) + Cu(s)$; (c) $Fe(s) + 2Ag^+(aq) \longrightarrow Fe^{2+}(aq) + 2Ag(s)$; (d) oxidizing agent = species reduced: Cu^{2+}, $Cu(OH)_2$, Ag^+

(e) reducing agent = species oxidized: H_2, H_2, Fe.

17.2 Galvanic Cells

By the end of this section, you will be able to:

- Use cell notation to describe galvanic cells
- Describe the basic components of galvanic cells

Galvanic cells, also known as **voltaic cells**, are electrochemical cells in which spontaneous oxidation-reduction reactions produce electrical energy. In writing the equations, it is often convenient to separate the oxidation-reduction reactions into half-reactions to facilitate balancing the overall equation and to emphasize the actual chemical transformations.

Consider what happens when a clean piece of copper metal is placed in a solution of silver nitrate (Figure 17.3). As soon as the copper metal is added, silver metal begins to form and copper ions pass into the solution. The blue color of the solution on the far right indicates the presence of copper ions. The reaction may be split into its two half-reactions. Half-reactions separate the oxidation from the reduction, so each can be considered individually.

$$
\begin{array}{lll}
\text{oxidation:} & Cu(s) \longrightarrow Cu^{2+}(aq) + 2e^- & \\
\text{reduction:} & 2 \times (Ag^+(aq) + e^- \longrightarrow Ag(s)) & \quad\text{or}\quad & 2Ag^+(aq) + 2e^- \longrightarrow 2Ag(s) \\
\hline
\text{overall:} & 2Ag^+(aq) + Cu(s) \longrightarrow 2Ag(s) + Cu^{2+}(aq) &
\end{array}
$$

The equation for the reduction half-reaction had to be doubled so the number electrons "gained" in the reduction half-reaction equaled the number of electrons "lost" in the oxidation half-reaction.

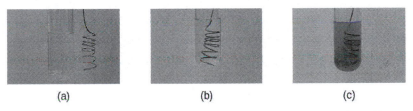

(a) (b) (c)

Figure 17.3 When a clean piece of copper metal is placed into a clear solution of silver nitrate (a), an oxidation-reduction reaction occurs that results in the exchange of Cu^{2+} for Ag^+ ions in solution. As the reaction proceeds (b), the solution turns blue (c) because of the copper ions present, and silver metal is deposited on the copper strip as the silver ions are removed from solution. (credit: modification of work by Mark Ott)

Galvanic or voltaic cells involve spontaneous electrochemical reactions in which the half-reactions are separated (Figure 17.4) so that current can flow through an external wire. The beaker on the left side of the figure is called a half-cell, and contains a 1 M solution of copper(II) nitrate [$Cu(NO_3)_2$] with a piece of copper metal partially submerged in the solution. The copper metal is an electrode. The copper is undergoing oxidation; therefore, the copper electrode is the **anode**. The anode is connected to a voltmeter with a wire and the other terminal of the voltmeter is connected to a silver electrode by a wire. The silver is undergoing reduction; therefore, the silver electrode is the **cathode**. The half-cell on the right side of the figure consists of the silver electrode in a 1 M solution of silver nitrate ($AgNO_3$). At this point, no current flows—that is, no significant movement of electrons through the wire occurs because the circuit is open. The circuit is closed using a salt bridge, which transmits the current with moving ions. The salt bridge consists of a concentrated, nonreactive, electrolyte solution such as the sodium nitrate ($NaNO_3$) solution used in this example. As electrons flow from left to right through the electrode and wire, nitrate ions (anions) pass through the porous plug on the left into the copper(II) nitrate solution. This keeps the beaker on the left electrically neutral by neutralizing the charge on the copper(II) ions that are produced in the solution as the copper metal is oxidized. At the same time, the nitrate ions are moving to the left, sodium ions (cations) move to the right, through the porous plug, and into the silver nitrate solution on the right. These added cations "replace" the silver ions that are removed from the solution as they were reduced to silver metal, keeping the beaker on the right electrically neutral. Without the salt bridge, the compartments would not remain electrically neutral and no significant current would flow. However, if the two compartments are in direct contact, a salt bridge is not necessary. The instant the circuit is completed, the voltmeter reads +0.46 V, this is called the **cell potential**. The cell potential is created when the two dissimilar metals are connected, and is a measure of the energy per unit charge available from the oxidation-reduction reaction. The volt is the derived SI unit for electrical potential

$$\text{volt} = V = \frac{J}{C}$$

In this equation, A is the current in amperes and C the charge in coulombs. Note that volts must be multiplied by the charge in coulombs (C) to obtain the energy in joules (J).

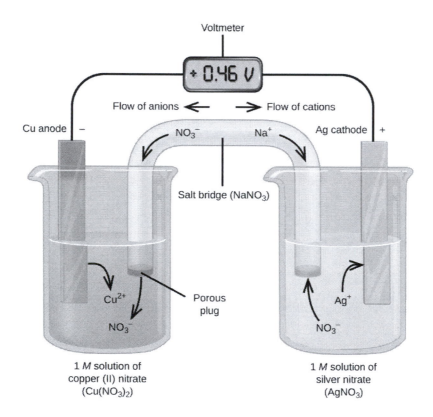

Figure 17.4 In this standard galvanic cell, the half-cells are separated; electrons can flow through an external wire and become available to do electrical work.

When the electrochemical cell is constructed in this fashion, a positive cell potential indicates a spontaneous reaction *and* that the electrons are flowing from the left to the right. There is a lot going on in Figure 17.4, so it is useful to summarize things for this system:

- Electrons flow from the anode to the cathode: left to right in the standard galvanic cell in the figure.

- The electrode in the left half-cell is the anode because oxidation occurs here. The name refers to the flow of anions in the salt bridge toward it.

- The electrode in the right half-cell is the cathode because reduction occurs here. The name refers to the flow of cations in the salt bridge toward it.

- Oxidation occurs at the anode (the left half-cell in the figure).

- Reduction occurs at the cathode (the right half-cell in the figure).

- The cell potential, +0.46 V, in this case, results from the inherent differences in the nature of the materials used to make the two half-cells.

- The salt bridge must be present to close (complete) the circuit and both an oxidation and reduction must occur for current to flow.

There are many possible galvanic cells, so a shorthand notation is usually used to describe them. The **cell notation** (sometimes called a cell diagram) provides information about the various species involved in the reaction. This notation also works for other types of cells. A vertical line, $|$, denotes a phase boundary and a double line, $\|$, the salt bridge. Information about the anode is written to the left, followed by the anode solution, then the salt bridge (when present), then the cathode solution, and, finally, information about the cathode to the right. The cell notation for the galvanic cell in Figure 17.4 is then

$$Cu(s) \mid Cu^{2+}(aq, 1\ M) \parallel Ag^{+}(aq, 1\ M) \mid Ag(s)$$

Note that spectator ions are not included and that the simplest form of each half-reaction was used. When known, the initial concentrations of the various ions are usually included.

One of the simplest cells is the Daniell cell. It is possible to construct this battery by placing a copper electrode at the bottom of a jar and covering the metal with a copper sulfate solution. A zinc sulfate solution is floated on top of the copper sulfate solution; then a zinc electrode is placed in the zinc sulfate solution. Connecting the copper electrode to the zinc electrode allows an electric current to flow. This is an example of a cell without a salt bridge, and ions may flow across the interface between the two solutions.

Some oxidation-reduction reactions involve species that are poor conductors of electricity, and so an electrode is used that does not participate in the reactions. Frequently, the electrode is platinum, gold, or graphite, all of which are inert to many chemical reactions. One such system is shown in Figure 17.5. Magnesium undergoes oxidation at the anode on the left in the figure and hydrogen ions undergo reduction at the cathode on the right. The reaction may be summarized as

$$\begin{aligned} \text{oxidation:} & \quad Mg(s) \longrightarrow Mg^{2+}(aq) + 2e^{-} \\ \text{reduction:} & \quad 2H^{+}(aq) + 2e^{-} \longrightarrow H_2(g) \\ \hline \text{overall:} & \quad Mg(s) + 2H^{+}(aq) \longrightarrow Mg^{2+}(aq) + H_2(g) \end{aligned}$$

The cell used an inert platinum wire for the cathode, so the cell notation is

$$Mg(s) \mid Mg^{2+}(aq) \parallel H^{+}(aq) \mid H_2(g) \mid Pt(s)$$

The magnesium electrode is an **active electrode** because it participates in the oxidation-reduction reaction. **Inert electrodes**, like the platinum electrode in Figure 17.5, do not participate in the oxidation-reduction reaction and are present so that current can flow through the cell. Platinum or gold generally make good inert electrodes because they are chemically unreactive.

Example 17.3

Using Cell Notation

Consider a galvanic cell consisting of

$$2Cr(s) + 3Cu^{2+}(aq) \longrightarrow 2Cr^{3+}(aq) + 3Cu(s)$$

Write the oxidation and reduction half-reactions and write the reaction using cell notation. Which reaction occurs at the anode? The cathode?

Solution

By inspection, Cr is oxidized when three electrons are lost to form Cr^{3+}, and Cu^{2+} is reduced as it gains two electrons to form Cu. Balancing the charge gives

$$\text{oxidation:} \quad 2Cr(s) \longrightarrow 2Cr^{3+}(aq) + 6e^-$$
$$\text{reduction: } 3Cu^{2+}(aq) + 6e^- \longrightarrow 3Cu(s)$$
$$\overline{\text{overall: } 2Cr(s) + 3Cu^{2+}(aq) \longrightarrow 2Cr^{3+}(aq) + 3Cu(s)}$$

Cell notation uses the simplest form of each of the equations, and starts with the reaction at the anode. No concentrations were specified so: $Cr(s) \mid Cr^{3+}(aq) \parallel Cu^{2+}(aq) \mid Cu(s)$. Oxidation occurs at the anode and reduction at the cathode.

Using Cell Notation

Consider a galvanic cell consisting of

$$5Fe^{2+}(aq) + MnO_4{}^-(aq) + 8H^+(aq) \longrightarrow 5Fe^{3+}(aq) + Mn^{2+}(aq) + 4H_2O(l)$$

Write the oxidation and reduction half-reactions and write the reaction using cell notation. Which reaction occurs at the anode? The cathode?

Solution

By inspection, Fe^{2+} undergoes oxidation when one electron is lost to form Fe^{3+}, and MnO_4^- is reduced as it gains five electrons to form Mn^{2+}. Balancing the charge gives

$$\text{oxidation:} \quad 5(Fe^{2+}(aq) \longrightarrow Fe^{3+}(aq) + e^-)$$
$$\text{reduction:} \quad MnO_4{}^-(aq) + 8H^+(aq) + 5e^- \longrightarrow Mn^{2+}(aq) + 4H_2O(l)$$
$$\overline{\text{overall: } 5Fe^{2+}(aq) + MnO_4{}^-(aq) + 8H^+(aq) \longrightarrow 5Fe^{3+}(aq) + Mn^{2+}(aq) + 4H_2O(l)}$$

Cell notation uses the simplest form of each of the equations, and starts with the reaction at the anode. It is necessary to use an inert electrode, such as platinum, because there is no metal present to conduct the electrons from the anode to the cathode. No concentrations were specified so: $Pt(s) \mid Fe^{2+}(aq), Fe^{3+}(aq) \parallel MnO_4{}^-(aq), H^+(aq), Mn^{2+}(aq) \mid Pt(s)$. Oxidation occurs at the anode and reduction at the cathode.

Check Your Learning

Use cell notation to describe the galvanic cell where copper(II) ions are reduced to copper metal and zinc metal is oxidized to zinc ions.

Answer: From the information given in the problem:

$$\text{anode (oxidation):} \quad Zn(s) \longrightarrow Zn^{2+}(aq) + 2e^-$$
$$\text{cathode (reduction): } Cu^{2+}(aq) + 2e^- \longrightarrow Cu(s)$$
$$\overline{\text{overall:} \quad Zn(s) + Cu^{2+}(aq) \longrightarrow Zn^{2+}(aq) + Cu(s)}$$

Using cell notation:
$$Zn(s) \mid Zn^{2+}(aq) \parallel Cu^{2+}(aq) \mid Cu(s).$$

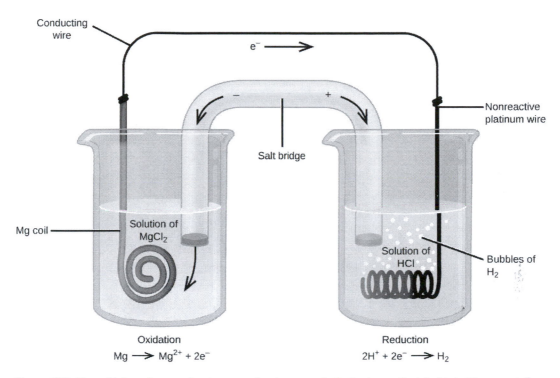

Figure 17.5 The oxidation of magnesium to magnesium ion occurs in the beaker on the left side in this apparatus; the reduction of hydrogen ions to hydrogen occurs in the beaker on the right. A nonreactive, or inert, platinum wire allows electrons from the left beaker to move into the right beaker. The overall reaction is:

$Mg + 2H^+ \longrightarrow Mg^{2+} + H_2,$ which is represented in cell notation as:

$Mg(s) \mid Mg^{2+}(aq) \parallel H^+(aq) \mid H_2(g) \mid Pt(s).$

17.3 Standard Reduction Potentials

By the end of this section, you will be able to:

- Determine standard cell potentials for oxidation-reduction reactions
- Use standard reduction potentials to determine the better oxidizing or reducing agent from among several possible choices

The cell potential in Figure 17.4 (+0.46 V) results from the difference in the electrical potentials for each electrode. While it is impossible to determine the electrical potential of a single electrode, we can assign an electrode the value of zero and then use it as a reference. The electrode chosen as the zero is shown in Figure 17.6 and is called the **standard hydrogen electrode (SHE)**. The SHE consists of 1 atm of hydrogen gas bubbled through a 1 M HCl solution, usually at room temperature. Platinum, which is chemically inert, is used as the electrode. The reduction half-reaction chosen as the reference is

$$2H^+(aq, 1\ M) + 2e^- \rightleftharpoons H_2(g,\ 1\ atm) \qquad E° = 0\ V$$

$E°$ is the standard reduction potential. The superscript "°" on the E denotes standard conditions (1 bar or 1 atm for gases, 1 M for solutes). The voltage is defined as zero for all temperatures.

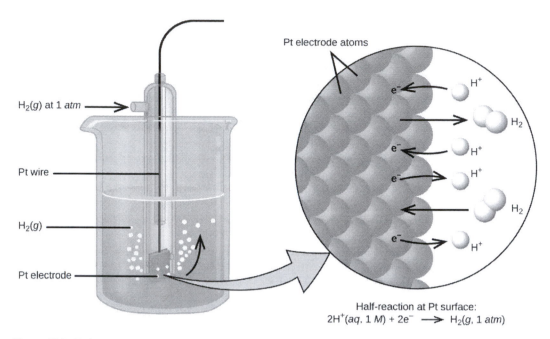

Pt electrode atoms

$H_2(g)$ at 1 *atm*

Pt wire

$H_2(g)$

Pt electrode

Half-reaction at Pt surface:
$2H^+(aq, 1\ M) + 2e^- \longrightarrow H_2(g, 1\ atm)$

Figure 17.6 Hydrogen gas at 1 atm is bubbled through 1 *M* HCl solution. Platinum, which is inert to the action of the 1 *M* HCl, is used as the electrode. Electrons on the surface of the electrode combine with H^+ in solution to produce hydrogen gas.

A galvanic cell consisting of a SHE and Cu^{2+}/Cu half-cell can be used to determine the standard reduction potential for Cu^{2+} (**Figure 17.7**). In cell notation, the reaction is

$$Pt(s)\ |\ H_2(g,\ 1\ atm)\ |\ H^+(aq,\ 1\ M)\ \|\ Cu^{2+}(aq,\ 1\ M)\ |\ Cu(s)$$

Electrons flow from the anode to the cathode. The reactions, which are reversible, are

Anode (oxidation): $\quad\quad\quad\quad H_2(g) \longrightarrow 2H^+(aq) + 2e^-$

Cathode (reduction): $Cu^{2+}(aq) + 2e^- \longrightarrow Cu(s)$

Overall: $\quad\quad\quad\quad\quad Cu^{2+}(aq) + H_2(g) \longrightarrow 2H^+(aq) + Cu(s)$

The standard reduction potential can be determined by subtracting the standard reduction potential for the reaction occurring at the anode from the standard reduction potential for the reaction occurring at the cathode. The minus sign is necessary because oxidation is the reverse of reduction.

$$E^\circ_{cell} = E^\circ_{cathode} - E^\circ_{anode}$$
$$+0.34\ V = E^\circ_{Cu^{2+}/Cu} - E^\circ_{H^+/H_2} = E^\circ_{Cu^{2+}/Cu} - 0 = E^\circ_{Cu^{2+}/Cu}$$

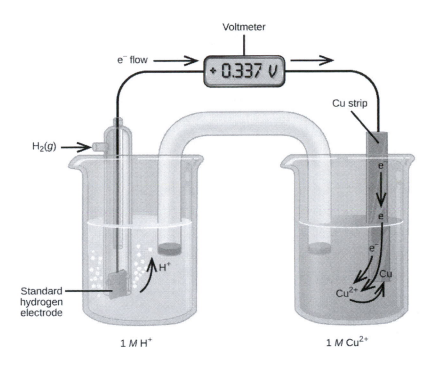

Figure 17.7 A galvanic cell can be used to determine the standard reduction potential of Cu^{2+}.

Using the SHE as a reference, other standard reduction potentials can be determined. Consider the cell shown in Figure 17.8, where

$$Pt(s) \mid H_2(g,\ 1\ atm) \mid H^+(aq, 1\ M) \parallel Ag^+(aq, 1\ M) \mid Ag(s)$$

Electrons flow from left to right, and the reactions are

anode (oxidation): $H_2(g) \longrightarrow 2H^+(aq) + 2e^-$

cathode (reduction): $2Ag^+(aq) + 2e^- \longrightarrow 2Ag(s)$

overall: $2Ag^+(aq) + H_2(g) \longrightarrow 2H^+(aq) + 2Ag(s)$

The standard reduction potential can be determined by subtracting the standard reduction potential for the reaction occurring at the anode from the standard reduction potential for the reaction occurring at the cathode. The minus sign is needed because oxidation is the reverse of reduction.

$$E^\circ_{cell} = E^\circ_{cathode} - E^\circ_{anode}$$
$$+0.80\ V = E^\circ_{Ag^+/Ag} - E^\circ_{H^+/H_2} = E^\circ_{Ag^+/Ag} - 0 = E^\circ_{Ag^+/Ag}$$

It is important to note that the potential is *not* doubled for the cathode reaction.

The SHE is rather dangerous and rarely used in the laboratory. Its main significance is that it established the zero for standard reduction potentials. Once determined, standard reduction potentials can be used to determine the **standard cell potential**, E°_{cell}, for any cell. For example, for the cell shown in Figure 17.4,

$$Cu(s) \mid Cu^{2+}(aq,\ 1\ M) \parallel Ag^+(aq,\ 1\ M) \mid Ag(s)$$

anode (oxidation): $Cu(s) \longrightarrow Cu^{2+}(aq) + 2e^-$

cathode (reduction): $2Ag^+(aq) + 2e^- \longrightarrow 2Ag(s)$

overall: $Cu(s) + 2Ag^+(aq) \longrightarrow Cu^{2+}(aq) + 2Ag(s)$

$$E^\circ_{cell} = E^\circ_{cathode} - E^\circ_{anode} = E^\circ_{Ag^+/Ag} - E^\circ_{Cu^{2+}/Cu} = 0.80 \text{ V} - 0.34 \text{ V} = 0.46 \text{ V}$$

Again, note that when calculating E°_{cell}, standard reduction potentials always remain the same even when a half-reaction is multiplied by a factor. Standard reduction potentials for selected reduction reactions are shown in Table 17.2. A more complete list is provided in Appendix L.

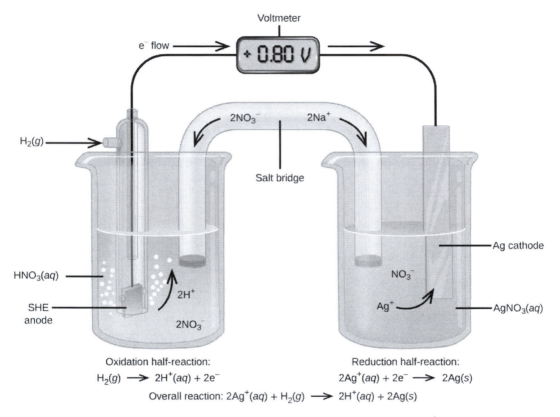

Oxidation half-reaction:

$H_2(g) \longrightarrow 2H^+(aq) + 2e^-$

Reduction half-reaction:

$2Ag^+(aq) + 2e^- \longrightarrow 2Ag(s)$

Overall reaction: $2Ag^+(aq) + H_2(g) \longrightarrow 2H^+(aq) + 2Ag(s)$

Figure 17.8 A galvanic cell can be used to determine the standard reduction potential of Ag^+. The SHE on the left is the anode and assigned a standard reduction potential of zero.

Selected Standard Reduction Potentials at 25 °C

Half-Reaction	E° (V)
$F_2(g) + 2e^- \longrightarrow 2F^-(aq)$	+2.866
$PbO_2(s) + SO_4{}^{2-}(aq) + 4H^+(aq) + 2e^- \longrightarrow PbSO_4(s) + 2H_2O(l)$	+1.69
$MnO_4{}^-(aq) + 8H^+(aq) + 5e^- \longrightarrow Mn^{2+}(aq) + 4H_2O(l)$	+1.507

Table 17.2

Selected Standard Reduction Potentials at 25 °C

Half-Reaction	$E°$ (V)
$Au^{3+}(aq) + 3e^- \longrightarrow Au(s)$	+1.498
$Cl_2(g) + 2e^- \longrightarrow 2Cl^-(aq)$	+1.35827
$O_2(g) + 4H^+(aq) + 4e^- \longrightarrow 2H_2O(l)$	+1.229
$Pt^{2+}(aq) + 2e^- \longrightarrow Pt(s)$	+1.20
$Br_2(aq) + 2e^- \longrightarrow 2Br^-(aq)$	+1.0873
$Ag^+(aq) + e^- \longrightarrow Ag(s)$	+0.7996
$Hg_2^{2+}(aq) + 2e^- \longrightarrow 2Hg(l)$	+0.7973
$Fe^{3+}(aq) + e^- \longrightarrow Fe^{2+}(aq)$	+0.771
$MnO_4^-(aq) + 2H_2O(l) + 3e^- \longrightarrow MnO_2(s) + 4OH^-(aq)$	+0.558
$I_2(s) + 2e^- \longrightarrow 2I^-(aq)$	+0.5355
$NiO_2(s) + 2H_2O(l) + 2e^- \longrightarrow Ni(OH)_2(s) + 2OH^-(aq)$	+0.49
$Cu^{2+}(aq) + 2e^- \longrightarrow Cu(s)$	+0.337
$Hg_2Cl_2(s) + 2e^- \longrightarrow 2Hg(l) + 2Cl^-(aq)$	+0.26808
$AgCl(s) + 2e^- \longrightarrow Ag(s) + Cl^-(aq)$	+0.22233
$Sn^{4+}(aq) + 2e^- \longrightarrow Sn^{2+}(aq)$	+0.151
$2H^+(aq) + 2e^- \longrightarrow H_2(g)$	0.00
$Pb^{2+}(aq) + 2e^- \longrightarrow Pb(s)$	−0.126
$Sn^{2+}(aq) + 2e^- \longrightarrow Sn(s)$	−0.1262
$Ni^{2+}(aq) + 2e^- \longrightarrow Ni(s)$	−0.257
$Co^{2+}(aq) + 2e^- \longrightarrow Co(s)$	−0.28
$PbSO_4(s) + 2e^- \longrightarrow Pb(s) + SO_4^{2-}(aq)$	−0.3505
$Cd^{2+}(aq) + 2e^- \longrightarrow Cd(s)$	−0.4030
$Fe^{2+}(aq) + 2e^- \longrightarrow Fe(s)$	−0.447
$Cr^{3+}(aq) + 3e^- \longrightarrow Cr(s)$	−0.744
$Mn^{2+}(aq) + 2e^- \longrightarrow Mn(s)$	−1.185
$Zn(OH)_2(s) + 2e^- \longrightarrow Zn(s) + 2OH^-(aq)$	−1.245

Table 17.2

Selected Standard Reduction Potentials at 25 °C

Half-Reaction	$E°$ (V)
$Zn^{2+}(aq) + 2e^- \longrightarrow Zn(s)$	−0.7618
$Al^{3+}(aq) + 3e^- \longrightarrow Al(s)$	−1.662
$Mg^2(aq) + 2e^- \longrightarrow Mg(s)$	−2.372
$Na^+(aq) + e^- \longrightarrow Na(s)$	−2.71
$Ca^{2+}(aq) + 2e^- \longrightarrow Ca(s)$	−2.868
$Ba^{2+}(aq) + 2e^- \longrightarrow Ba(s)$	−2.912
$K^+(aq) + e^- \longrightarrow K(s)$	−2.931
$Li^+(aq) + e^- \longrightarrow Li(s)$	−3.04

Table 17.2

Tables like this make it possible to determine the standard cell potential for many oxidation-reduction reactions.

Example 17.4

Cell Potentials from Standard Reduction Potentials

What is the standard cell potential for a galvanic cell that consists of Au^{3+}/Au and Ni^{2+}/Ni half-cells? Identify the oxidizing and reducing agents.

Solution

Using Table 17.2, the reactions involved in the galvanic cell, both written as reductions, are

$$Au^{3+}(aq) + 3e^- \longrightarrow Au(s) \qquad E°_{Au^{3+}/Au} = +1.498 \text{ V}$$
$$Ni^{2+}(aq) + 2e^- \longrightarrow Ni(s) \qquad E°_{Ni^{2+}/Ni} = -0.257 \text{ V}$$

Galvanic cells have positive cell potentials, and all the reduction reactions are reversible. The reaction at the anode will be the half-reaction with the smaller or more negative standard reduction potential. Reversing the reaction at the anode (to show the oxidation) but *not* its standard reduction potential gives:

Anode (oxidation): $\qquad Ni(s) \longrightarrow Ni^{2+}(aq) + 2e^- \qquad E°_{anode} = E°_{Ni^{2+}/Ni} = -0.257 \text{ V}$

Cathode (reduction): $Au^{3+}(aq) + 3e^- \longrightarrow Au(s) \qquad E°_{cathode} = E°_{Au^{3+}/Au} = +1.498 \text{ V}$

The least common factor is six, so the overall reaction is

$$3Ni(s) + 2Au^{3+}(aq) \longrightarrow 3Ni^{2+}(aq) + 2Au(s)$$

The reduction potentials are *not* scaled by the stoichiometric coefficients when calculating the cell potential, and the unmodified standard reduction potentials must be used.

$$E°_{cell} = E°_{cathode} - E°_{anode} = 1.498 \text{ V} - (-0.257 \text{ V}) = 1.755 \text{ V}$$

From the half-reactions, Ni is oxidized, so it is the reducing agent, and Au^{3+} is reduced, so it is the oxidizing agent.

Check Your Learning

A galvanic cell consists of a Mg electrode in 1 M $Mg(NO_3)_2$ solution and a Ag electrode in 1 M $AgNO_3$ solution. Calculate the standard cell potential at 25 °C.

Answer:

$$Mg(s) + 2Ag^+(aq) \longrightarrow Mg^{2+}(aq) + 2Ag(s) \qquad E°_{cell} = 0.7996 \text{ V} - (-2.372 \text{ V}) = 3.172 \text{ V}$$

17.4 The Nernst Equation

By the end of this section, you will be able to:

- Relate cell potentials to free energy changes
- Use the Nernst equation to determine cell potentials at nonstandard conditions
- Perform calculations that involve converting between cell potentials, free energy changes, and equilibrium constants

We will now extend electrochemistry by determining the relationship between $E°_{cell}$ and the thermodynamics quantities such as $\Delta G°$ (Gibbs free energy) and K (the equilibrium constant). In galvanic cells, chemical energy is converted into electrical energy, which can do work. The electrical work is the product of the charge transferred multiplied by the potential difference (voltage):

$$\text{electrical work} = \text{volts} \times (\text{charge in coulombs}) = \text{J}$$

The charge on 1 mole of electrons is given by **Faraday's constant** (F)

$$F = \frac{6.022 \times 10^{23}\,e^-}{\text{mol}} \times \frac{1.602 \times 10^{-19}\,C}{e^-} = 9.648 \times 10^4\,\frac{C}{\text{mol}} = 9.684 \times 10^4\,\frac{J}{\text{V·mol}}$$

$$\text{total charge} = (\text{number of moles of } e^-) \times F = nF$$

In this equation, n is the number of moles of electrons for the *balanced* oxidation-reduction reaction. The measured cell potential is the **maximum** potential the cell can produce and is related to the **electrical work** (w_{ele}) by

$$E_{cell} = \frac{-w_{ele}}{nF} \qquad \text{or} \qquad w_{ele} = -nFE_{cell}$$

The negative sign for the work indicates that the electrical work is done by the system (the galvanic cell) on the surroundings. In an earlier chapter, the free energy was defined as the energy that was available to do work. In particular, the change in free energy was defined in terms of the maximum work (w_{max}), which, for electrochemical systems, is w_{ele}.

$$\Delta G = w_{max} = w_{ele}$$
$$\Delta G = -nFE_{cell}$$

We can verify the signs are correct when we realize that n and F are positive constants and that galvanic cells, which have positive cell potentials, involve spontaneous reactions. Thus, spontaneous reactions, which have $\Delta G < 0$, must have $E_{cell} > 0$. If all the reactants and products are in their standard states, this becomes

$$\Delta G° = -nFE°_{cell}$$

This provides a way to relate standard cell potentials to equilibrium constants, since

$$\Delta G° = -RT \ln K$$

$$-nFE°_{cell} = -RT \ln K \qquad \text{or} \qquad E°_{cell} = \frac{RT}{nF} \ln K$$

Most of the time, the electrochemical reactions are run at standard temperature (298.15 K). Collecting terms at this temperature yields

$$E^\circ_{cell} = \frac{RT}{nF} \ln K = \frac{\left(8.314\frac{J}{K \cdot mol}\right)(298.15K)}{n \times 96,485 \text{ C/V·mol}} \ln K = \frac{0.0257 \text{ V}}{n} \ln K$$

where n is the number of moles of electrons. For historical reasons, the logarithm in equations involving cell potentials is often expressed using base 10 logarithms (log), which changes the constant by a factor of 2.303:

$$E^\circ_{cell} = \frac{0.0592 \text{ V}}{n} \log K$$

Thus, if ΔG°, K, or E°_{cell} is known or can be calculated, the other two quantities can be readily determined. The relationships are shown graphically in Figure 17.9.

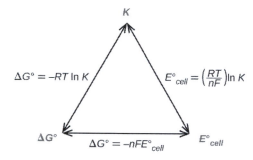

Figure 17.9 The relationships between ΔG°, K, and E°_{cell}. Given any one of the three quantities, the other two can be calculated, so any of the quantities could be used to determine whether a process was spontaneous.

Given any one of the quantities, the other two can be calculated.

Example 17.5

Equilibrium Constants, Standard Cell Potentials, and Standard Free Energy Changes

What is the standard free energy change and equilibrium constant for the following reaction at 25 °C?

$$2Ag^+(aq) + Fe(s) \rightleftharpoons 2Ag(s) + Fe^{2+}(aq)$$

Solution

The reaction involves an oxidation-reduction reaction, so the standard cell potential can be calculated using the data in Appendix L.

anode (oxidation): $\qquad$ $Fe(s) \longrightarrow Fe^{2+}(aq) + 2e^-$ $\qquad$ $E^\circ_{Fe^{2+}/Fe} = -0.447 \text{ V}$

cathode (reduction): $2 \times (Ag^+(aq) + e^- \longrightarrow Ag(s))$ $\qquad$ $E^\circ_{Ag^+/Ag} = 0.7996 \text{ V}$

$E^\circ_{cell} = E^\circ_{cathode} - E^\circ_{anode} = E^\circ_{Ag^+/Ag} - E^\circ_{Fe^{2+}/Fe} = +1.247 \text{ V}$

Remember that the cell potential for the cathode is not multiplied by two when determining the standard cell potential. With $n = 2$, the equilibrium constant is then

$$E^\circ_{cell} = \frac{0.0592 \text{ V}}{n} \log K$$
$$K = 10^{n \times E^\circ_{cell}/0.0592 \text{ V}}$$

$$K = 10^{2 \times 1.247 \text{ V}/0.0592 \text{ V}}$$
$$K = 10^{42.128}$$
$$K = 1.3 \times 10^{42}$$

The two equilibrium constants differ slightly due to rounding in the constants 0.0257 V and 0.0592 V. The standard free energy is then

$$\Delta G° = -nFE°_{cell}$$

$$\Delta G° = -2 \times 96,485 \, \frac{\text{J}}{\text{V·mol}} \times 1.247 \text{ V} = -240.6 \, \frac{\text{kJ}}{\text{mol}}$$

Check your answer: A positive standard cell potential means a spontaneous reaction, so the standard free energy change should be negative, and an equilibrium constant should be >1.

Check Your Learning

What is the standard free energy change and the equilibrium constant for the following reaction at room temperature? Is the reaction spontaneous?

$$Sn(s) + 2Cu^{2+}(aq) \rightleftharpoons Sn^{2+}(aq) + 2Cu^{+}(aq)$$

Answer: Spontaneous; $n = 2$; $E°_{cell} = +0.291$ V; $\Delta G° = -56.2 \, \frac{\text{kJ}}{\text{mol}}$; $K = 6.8 \times 10^9$.

Now that the connection has been made between the free energy and cell potentials, nonstandard concentrations follow. Recall that

$$\Delta G = \Delta G° + RT \ln Q$$

where Q is the reaction quotient (see the chapter on equilibrium fundamentals). Converting to cell potentials:

$$-nFE_{cell} = -nFE°_{cell} + RT \ln Q \qquad \text{or} \qquad E_{cell} = E°_{cell} - \frac{RT}{nF} \ln Q$$

This is the **Nernst equation**. At standard temperature (298.15 K), it is possible to write the above equations as

$$E_{cell} = E°_{cell} - \frac{0.0257 \text{ V}}{n} \ln Q \qquad \text{or} \qquad E_{cell} = E°_{cell} - \frac{0.0592 \text{ V}}{n} \log Q$$

If the temperature is not 273.15 K, it is necessary to recalculate the value of the constant. With the Nernst equation, it is possible to calculate the cell potential at nonstandard conditions. This adjustment is necessary because potentials determined under different conditions will have different values.

Example 17.6

Cell Potentials at Nonstandard Conditions

Consider the following reaction at room temperature:

$$Co(s) + Fe^{2+}(aq, \ 1.94 \, M) \longrightarrow Co^{2+}(aq, 0.15 \, M) + Fe(s)$$

Is the process spontaneous?

Solution

There are two ways to solve the problem. If the thermodynamic information in Appendix G were available, you could calculate the free energy change. If the free energy change is negative, the process is spontaneous. The other approach, which we will use, requires information like that given in Appendix L. Using those data, the cell potential can be determined. If the cell potential is positive, the process is spontaneous. Collecting information from Appendix L and the problem,

Anode (oxidation): $\qquad$ $Co(s) \longrightarrow Co^{2+}(aq) + 2e^-$ $\qquad$ $E^\circ_{Co^{2+}/Co} = -0.28$ V

Cathode (reduction): $Fe^{2+}(aq) + 2e^- \longrightarrow Fe(s)$ $\qquad$ $E^\circ_{Fe^{2+}/Fe} = -0.447$ V

$E^\circ_{cell} = E^\circ_{cathode} - E^\circ_{anode} = -0.447$ V $- (-0.28$ V$) = -0.17$ V

The process is not spontaneous under standard conditions. Using the Nernst equation and the concentrations stated in the problem and $n = 2$,

$$Q = \frac{[Co^{2+}]}{[Fe^{2+}]} = \frac{0.15\,M}{1.94\,M} = 0.077$$

$$E_{cell} = E^\circ_{cell} - \frac{0.0592\ V}{n} \log Q$$

$$E_{cell} = -0.17\ V - \frac{0.0592\ V}{2} \log 0.077$$

$$E_{cell} = -0.17\ V + 0.033\ V = -0.014\ V$$

The process is (still) nonspontaneous.

Check Your Learning

What is the cell potential for the following reaction at room temperature?

$$Al(s) \mid Al^{3+}(aq,\ 0.15\ M) \parallel Cu^{2+}(aq,\ 0.025\ M) \mid Cu(s)$$

What are the values of n and Q for the overall reaction? Is the reaction spontaneous under these conditions?

Answer: $n = 6$; $Q = 1440$; $E_{cell} = +1.97$ V, spontaneous.

Finally, we will take a brief look at a special type of cell called a **concentration cell**. In a concentration cell, the electrodes are the same material and the half-cells differ only in concentration. Since one or both compartments is not standard, the cell potentials will be unequal; therefore, there will be a potential difference, which can be determined with the aid of the Nernst equation.

Example 17.7

Concentration Cells

What is the cell potential of the concentration cell described by

$$Zn(s) \mid Zn^{2+}(aq,\ 0.10\ M) \parallel Zn^{2+}(aq,\ 0.50\ M) \mid Zn(s)$$

Solution

From the information given:

Anode: $\qquad$ $Zn(s) \longrightarrow Zn^{2+}(aq,\ 0.10\ M) + 2e^-$ $\qquad$ $E^\circ_{anode} = -0.7618$ V

Cathode: $Zn^{2+}(aq,\ 0.50\ M) + 2e^- \longrightarrow Zn(s)$ $\qquad$ $E^\circ_{cathode} = -0.7618$ V

Overall: $\qquad$ $Zn^{2+}(aq,\ 0.50\ M) \longrightarrow Zn^{2+}(aq,\ 0.10\ M)$ $\qquad$ $E^\circ_{cell} = 0.000$ V

The standard cell potential is zero because the anode and cathode involve the same reaction; only the concentration of Zn^{2+} changes. Substituting into the Nernst equation,

$$E_{cell} = 0.000\ V - \frac{0.0592\ V}{2} \log \frac{0.10}{0.50} = +0.021\ V$$

and the process is spontaneous at these conditions.

Check your answer: In a concentration cell, the standard cell potential will always be zero. To get a positive cell potential (spontaneous process) the reaction quotient Q must be <1. $Q < 1$ in this case, so the process is spontaneous.

Check Your Learning

What value of Q for the previous concentration cell would result in a voltage of 0.10 V? If the concentration of zinc ion at the cathode was 0.50 M, what was the concentration at the anode?

Answer: $Q = 0.00042$; $[Zn^{2+}]_{cat} = 2.1 \times 10^{-4} M$.

17.5 Batteries and Fuel Cells

By the end of this section, you will be able to:

- Classify batteries as primary or secondary
- List some of the characteristics and limitations of batteries
- Provide a general description of a fuel cell

A **battery** is an electrochemical cell or series of cells that produces an electric current. In principle, any galvanic cell could be used as a battery. An ideal battery would never run down, produce an unchanging voltage, and be capable of withstanding environmental extremes of heat and humidity. Real batteries strike a balance between ideal characteristics and practical limitations. For example, the mass of a car battery is about 18 kg or about 1% of the mass of an average car or light-duty truck. This type of battery would supply nearly unlimited energy if used in a smartphone, but would be rejected for this application because of its mass. Thus, no single battery is "best" and batteries are selected for a particular application, keeping things like the mass of the battery, its cost, reliability, and current capacity in mind. There are two basic types of batteries: primary and secondary. A few batteries of each type are described next.

Link to Learning

Visit this site (http://openstaxcollege.org/l/16batteries) to learn more about batteries.

Primary Batteries

Primary batteries are single-use batteries because they cannot be recharged. A common primary battery is the **dry cell** (Figure 17.10). The dry cell is a zinc-carbon battery. The zinc can serves as both a container and the negative electrode. The positive electrode is a rod made of carbon that is surrounded by a paste of manganese(IV) oxide, zinc chloride, ammonium chloride, carbon powder, and a small amount of water. The reaction at the anode can be represented as the ordinary oxidation of zinc:

$$Zn(s) \longrightarrow Zn^{2+}(aq) + 2e^- \qquad E^\circ_{Zn^{2+}/Zn} = -0.7618 \text{ V}$$

The reaction at the cathode is more complicated, in part because more than one reaction occurs. The series of reactions that occurs at the cathode is approximately

$$2MnO_2(s) + 2NH_4Cl(aq) + 2e^- \longrightarrow Mn_2O_3(s) + 2NH_3(aq) + H_2O(l) + 2Cl^-$$

The overall reaction for the zinc–carbon battery can be represented as $2MnO_2(s) + 2NH_4Cl(aq) + Zn(s) \longrightarrow Zn^{2+}(aq) + Mn_2O_3(s) + 2NH_3(aq) + H_2O(l) + 2Cl^-$ with an overall cell potential which is initially about 1.5 V, but decreases as the battery is used. It is important to remember that the voltage delivered by a battery is the same regardless of the size of a battery. For this reason, D, C, A, AA, and AAA batteries all have the same voltage rating. However, larger batteries can deliver more moles of electrons. As the zinc container oxidizes, its contents eventually leak out, so this type of battery should not be left in any electrical device for extended periods.

Figure 17.10   The diagram shows a cross section of a flashlight battery, a zinc-carbon dry cell.

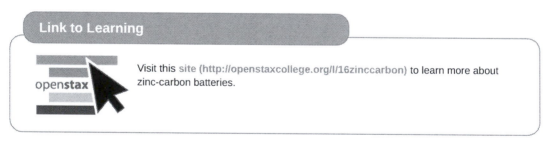

Link to Learning

Visit this site (http://openstaxcollege.org/l/16zinccarbon) to learn more about zinc-carbon batteries.

Alkaline batteries (Figure 17.11) were developed in the 1950s partly to address some of the performance issues with zinc–carbon dry cells. They are manufactured to be exact replacements for zinc-carbon dry cells. As their name suggests, these types of batteries use alkaline electrolytes, often potassium hydroxide. The reactions are

anode: $Zn(s) + 2OH^-(aq) \longrightarrow ZnO(s) + H_2O(l) + 2e^-$ $\qquad E°_{anode} = -1.28$ V

cathode: $2MnO_2(s) + H_2O(l) + 2e^- \longrightarrow Mn_2O_3(s) + 2OH^-(aq)$ $\qquad E°_{cathode} = +0.15$ V

overall: $Zn(s) + 2MnO_2(s) \longrightarrow ZnO(s) + Mn_2O_3(s)$ $\qquad E°_{cell} = +1.43$ V

An alkaline battery can deliver about three to five times the energy of a zinc-carbon dry cell of similar size. Alkaline batteries are prone to leaking potassium hydroxide, so these should also be removed from devices for long-term storage. While some alkaline batteries are rechargeable, most are not. Attempts to recharge an alkaline battery that is not rechargeable often leads to rupture of the battery and leakage of the potassium hydroxide electrolyte.

Metal top cover (+)

MnO$_2$ powder mixed
with coal dust (cathode)

Porous separator

zinc powder with
potassium hydroxide
(anode)

Anode current collector

steel can, cathode
current collector

Plastic cover

Metal bottom cover (−)

Figure 17.11   Alkaline batteries were designed as direct replacements for zinc-carbon (dry cell) batteries.

Link to Learning

Visit this site (http://openstaxcollege.org/l/16alkaline) to learn more about
alkaline batteries.

Secondary Batteries

Secondary batteries are rechargeable. These are the types of batteries found in devices such as smartphones, electronic tablets, and automobiles.

Nickel-cadmium, or NiCd, batteries (Figure 17.12) consist of a nickel-plated cathode, cadmium-plated anode, and a potassium hydroxide electrode. The positive and negative plates, which are prevented from shorting by the separator, are rolled together and put into the case. This is a "jelly-roll" design and allows the NiCd cell to deliver much more current than a similar-sized alkaline battery. The reactions are

$$\text{anode:} \qquad\qquad \text{Cd}(s) + 2\text{OH}^-(aq) \longrightarrow \text{Cd(OH)}_2(s) + 2e^-$$
$$\text{cathode: } \text{NiO}_2(s) + 2\text{H}_2\text{O}(l) + 2e^- \longrightarrow \text{Ni(OH)}_2(s) + 2\text{OH}^-(aq)$$
$$\overline{\text{overall: } \text{Cd}(s) + \text{NiO}_2(s) + 2\text{H}_2\text{O}(l) \longrightarrow \text{Cd(OH)}_2(s) + \text{Ni(OH)}_2(s)}$$

The voltage is about 1.2 V to 1.25 V as the battery discharges. When properly treated, a NiCd battery can be recharged about 1000 times. Cadmium is a toxic heavy metal so NiCd batteries should never be opened or put into the regular trash.

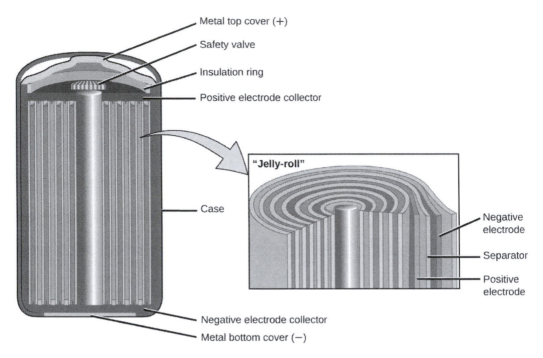

Figure 17.12   NiCd batteries use a "jelly-roll" design that significantly increases the amount of current the battery can deliver as compared to a similar-sized alkaline battery.

Link to Learning

Visit this site (http://openstaxcollege.org/l/16NiCdrecharge) for more information about nickel cadmium rechargeable batteries.

Lithium ion batteries (Figure 17.13) are among the most popular rechargeable batteries and are used in many portable electronic devices. The reactions are

$$\text{anode:} \qquad\qquad LiCoO_2 \;\rightleftharpoons\; Li_{x-1}CoO_2 + x\,Li^+ + x\,e^-$$

$$\text{cathode:}\; x\,Li^+ + x\,e^- + x\,C_6 \;\rightleftharpoons\; x\,LiC_6$$

$$\text{overall:} \qquad LiCoO_2 + x\,C_6 \;\rightleftharpoons\; Li_{x-1}CoO_2 + x\,LiC_6$$

With the coefficients representing moles, x is no more than about 0.5 moles. The battery voltage is about 3.7 V. Lithium batteries are popular because they can provide a large amount current, are lighter than comparable batteries of other types, produce a nearly constant voltage as they discharge, and only slowly lose their charge when stored.

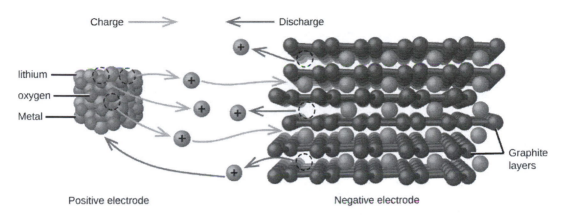

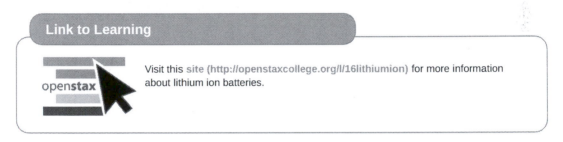

Figure 17.13   In a lithium ion battery, charge flows between the electrodes as the lithium ions move between the anode and cathode.

Link to Learning

Visit this site (http://openstaxcollege.org/l/16lithiumion) for more information about lithium ion batteries.

The **lead acid battery** (Figure 17.14) is the type of secondary battery used in your automobile. It is inexpensive and capable of producing the high current required by automobile starter motors. The reactions for a lead acid battery are

anode: $\quad Pb(s) + HSO_4{}^-(aq) \longrightarrow PbSO_4(s) + H^+(aq) + 2e^-$

cathode: $PbO_2(s) + HSO_4{}^-(aq) + 3H^+(aq) + 2e^- \longrightarrow PbSO_4(s) + 2H_2O(l)$

overall: $\quad Pb(s) + PbO_2(s) + 2H_2SO_4(aq) \longrightarrow 2PbSO_4(s) + 2H_2O(l)$

Each cell produces 2 V, so six cells are connected in series to produce a 12-V car battery. Lead acid batteries are heavy and contain a caustic liquid electrolyte, but are often still the battery of choice because of their high current density. Since these batteries contain a significant amount of lead, they must always be disposed of properly.

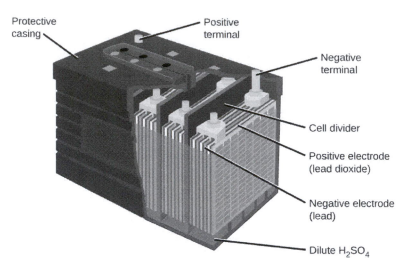

Figure 17.14 The lead acid battery in your automobile consists of six cells connected in series to give 12 V. Their low cost and high current output makes these excellent candidates for providing power for automobile starter motors.

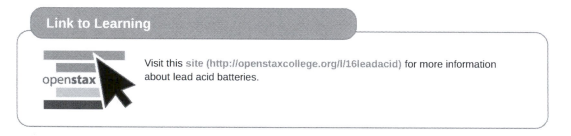

Link to Learning

Visit this site (http://openstaxcollege.org/l/16leadacid) for more information about lead acid batteries.

Fuel Cells

A **fuel cell** is a device that converts chemical energy into electrical energy. Fuel cells are similar to batteries but require a continuous source of fuel, often hydrogen. They will continue to produce electricity as long as fuel is available. Hydrogen fuel cells have been used to supply power for satellites, space capsules, automobiles, boats, and submarines (Figure 17.15).

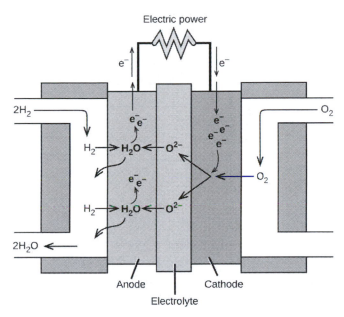

Figure 17.15 In this hydrogen fuel-cell schematic, oxygen from the air reacts with hydrogen, producing water and electricity.

In a hydrogen fuel cell, the reactions are

$$\text{anode: } 2H_2 + 2O^{2-} \longrightarrow 2H_2O + 4e^-$$
$$\text{cathode: } O_2 + 4e^- \longrightarrow 2O^{2-}$$
$$\overline{\text{overall: } 2H_2 + O_2 \longrightarrow 2H_2O}$$

The voltage is about 0.9 V. The efficiency of fuel cells is typically about 40% to 60%, which is higher than the typical internal combustion engine (25% to 35%) and, in the case of the hydrogen fuel cell, produces only water as exhaust. Currently, fuel cells are rather expensive and contain features that cause them to fail after a relatively short time.

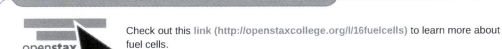

Link to Learning

Check out this link (http://openstaxcollege.org/l/16fuelcells) to learn more about fuel cells.

17.6 Corrosion

By the end of this section, you will be able to:

- Define corrosion
- List some of the methods used to prevent or slow corrosion

Corrosion is usually defined as the degradation of metals due to an electrochemical process. The formation of rust on iron, tarnish on silver, and the blue-green patina that develops on copper are all examples of corrosion. The total cost of corrosion in the United States is significant, with estimates in excess of half a trillion dollars a year.

Chemistry in Everyday Life

Statue of Liberty: Changing Colors

The Statue of Liberty is a landmark every American recognizes. The Statue of Liberty is easily identified by its height, stance, and unique blue-green color (Figure 17.16). When this statue was first delivered from France, its appearance was not green. It was brown, the color of its copper "skin." So how did the Statue of Liberty change colors? The change in appearance was a direct result of corrosion. The copper that is the primary component of the statue slowly underwent oxidation from the air. The oxidation-reduction reactions of copper metal in the environment occur in several steps. Copper metal is oxidized to copper(I) oxide (Cu_2O), which is red, and then to copper(II) oxide, which is black

$$2Cu(s) + \frac{1}{2}O_2(g) \longrightarrow Cu_2O(s) \qquad \text{(red)}$$

$$Cu_2O(s) + \frac{1}{2}O_2(g) \longrightarrow 2CuO(s) \qquad \text{(black)}$$

Coal, which was often high in sulfur, was burned extensively in the early part of the last century. As a result, sulfur trioxide, carbon dioxide, and water all reacted with the CuO

$$2CuO(s) + CO_2(g) + H_2O(l) \longrightarrow Cu_2CO_3(OH)_2(s) \qquad \text{(green)}$$

$$3CuO(s) + 2CO_2(g) + H_2O(l) \longrightarrow Cu_2(CO_3)_2(OH)_2(s) \qquad \text{(blue)}$$

$$4CuO(s) + SO_3(g) + 3H_2O(l) \longrightarrow Cu_4SO_4(OH)_6(s) \qquad \text{(green)}$$

These three compounds are responsible for the characteristic blue-green patina seen today. Fortunately, formation of the patina created a protective layer on the surface, preventing further corrosion of the copper skin. The formation of the protective layer is a form of passivation, which is discussed further in a later chapter.

(a)   (b)

Figure 17.16   (a) The Statue of Liberty is covered with a copper skin, and was originally brown, as shown in this painting. (b) Exposure to the elements has resulted in the formation of the blue-green patina seen today.

Perhaps the most familiar example of corrosion is the formation of rust on iron. Iron will rust when it is exposed to oxygen and water. The main steps in the rusting of iron appear to involve the following (Figure 17.17). Once exposed to the atmosphere, iron rapidly oxidizes.

$$\text{anode: Fe}(s) \longrightarrow \text{Fe}^{2+}(aq) + 2e^- \qquad E^\circ_{\text{Fe}^{2+}/\text{Fe}} = -0.44 \text{ V}$$

The electrons reduce oxygen in the air in acidic solutions.

$$\text{cathode: O}_2(g) + 2\text{H}^+(aq) + 4e^- \longrightarrow 2\text{H}_2\text{O}(l) \qquad E^\circ_{\text{O}_2/\text{O}^2} = +1.23 \text{ V}$$

$$\text{overall: 2Fe}(s) + \text{O}_2(g) + 4\text{H}^+(aq) \longrightarrow 2\text{Fe}^{2+}(aq) + 2\text{H}_2\text{O}(l) \qquad E^\circ_{\text{cell}} = +1.67 \text{ V}$$

What we call rust is hydrated iron(III) oxide, which forms when iron(II) ions react further with oxygen.

$$4\text{Fe}^{2+}(aq) + \text{O}_2(g) + (4 + 2x)\,\text{H}_2\text{O}(l) \longrightarrow 2\text{Fe}_2\text{O}_3 \cdot x\text{H}_2\text{O}(s) + 8\text{H}^+(aq)$$

The number of water molecules is variable, so it is represented by x. Unlike the patina on copper, the formation of rust does not create a protective layer and so corrosion of the iron continues as the rust flakes off and exposes fresh iron to the atmosphere.

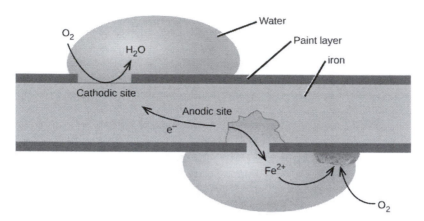

Figure 17.17 Once the paint is scratched on a painted iron surface, corrosion occurs and rust begins to form. The speed of the spontaneous reaction is increased in the presence of electrolytes, such as the sodium chloride used on roads to melt ice and snow or in salt water.

One way to keep iron from corroding is to keep it painted. The layer of paint prevents the water and oxygen necessary for rust formation from coming into contact with the iron. As long as the paint remains intact, the iron is protected from corrosion.

Other strategies include alloying the iron with other metals. For example, stainless steel is mostly iron with a bit of chromium. The chromium tends to collect near the surface, where it forms an oxide layer that protects the iron.

Zinc-plated or **galvanized iron** uses a different strategy. Zinc is more easily oxidized than iron because zinc has a lower reduction potential. Since zinc has a lower reduction potential, it is a more active metal. Thus, even if the zinc coating is scratched, the zinc will still oxidize before the iron. This suggests that this approach should work with other active metals.

Another important way to protect metal is to make it the cathode in a galvanic cell. This is **cathodic protection** and can be used for metals other than just iron. For example, the rusting of underground iron storage tanks and pipes can be prevented or greatly reduced by connecting them to a more active metal such as zinc or magnesium (Figure 17.18). This is also used to protect the metal parts in water heaters. The more active metals (lower reduction potential) are called **sacrificial anodes** because as they get used up as they corrode (oxidize) at the anode. The metal being protected serves as the cathode, and so does not oxidize (corrode). When the anodes are properly monitored and periodically replaced, the useful lifetime of the iron storage tank can be greatly extended.

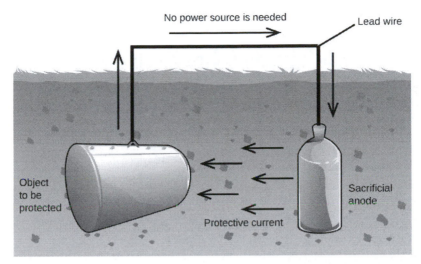

Figure 17.18 One way to protect an underground iron storage tank is through cathodic protection. Using an active metal like zinc or magnesium for the anode effectively makes the storage tank the cathode, preventing it from corroding (oxidizing).

17.7 Electrolysis

By the end of this section, you will be able to:

- Describe electrolytic cells and their relationship to galvanic cells
- Perform various calculations related to electrolysis

In galvanic cells, chemical energy is converted into electrical energy. The opposite is true for electrolytic cells. In **electrolytic cells**, electrical energy causes nonspontaneous reactions to occur in a process known as **electrolysis**. The charging electric car pictured in Figure 17.1 at the beginning of this chapter shows one such process. Electrical energy is converted into the chemical energy in the battery as it is charged. Once charged, the battery can be used to power the automobile.

The same principles are involved in electrolytic cells as in galvanic cells. We will look at three electrolytic cells and the quantitative aspects of electrolysis.

The Electrolysis of Molten Sodium Chloride

In molten sodium chloride, the ions are free to migrate to the electrodes of an electrolytic cell. A simplified diagram of the cell commercially used to produce sodium metal and chlorine gas is shown in Figure 17.19. Sodium is a strong reducing agent and chlorine is used to purify water, and is used in antiseptics and in paper production. The reactions are

$$\text{anode:} \quad 2Cl^-(l) \longrightarrow Cl_2(g) + 2e^- \qquad E^\circ_{Cl_2/Cl^-} = +1.3 \text{ V}$$

$$\text{cathode:} \quad Na^+(l) + e^- \longrightarrow Na(l) \qquad E^\circ_{Na^+/Na} = -2.7 \text{ V}$$

$$\text{overall: } 2Na^+(l) + 2Cl^-(l) \longrightarrow 2Na(l) + Cl_2(g) \qquad E^\circ_{cell} = -4.0 \text{ V}$$

The power supply (battery) must supply a minimum of 4 V, but, in practice, the applied voltages are typically higher because of inefficiencies in the process itself.

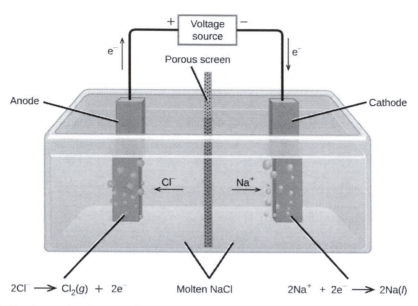

Figure 17.19 Passing an electric current through molten sodium chloride decomposes the material into sodium metal and chlorine gas. Care must be taken to keep the products separated to prevent the spontaneous formation of sodium chloride.

The Electrolysis of Water

It is possible to split water into hydrogen and oxygen gas by electrolysis. Acids are typically added to increase the concentration of hydrogen ion in solution (Figure 17.20). The reactions are

$$\text{anode:} \qquad 2H_2O(l) \longrightarrow O_2(g) + 4H^+(aq) + 4e^- \qquad\qquad E^\circ_{anode} = +1.229 \text{ V}$$

$$\text{cathode: } 2H^+(aq) + 2e^- \longrightarrow H_2(g) \qquad E^\circ_{cathode} = 0 \text{ V}$$

$$\text{overall:} \qquad 2H_2O(l) \longrightarrow 2H_2(g) + O_2(g) \qquad E^\circ_{cell} = -1.229 \text{ V}$$

Note that the sulfuric acid is not consumed and that the volume of hydrogen gas produced is twice the volume of oxygen gas produced. The minimum applied voltage is 1.229 V.

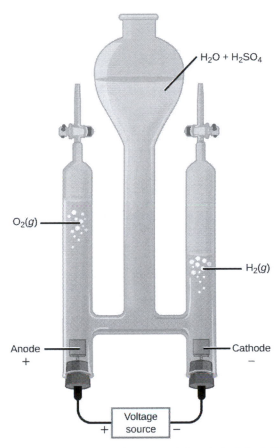

Figure 17.20 Water decomposes into oxygen and hydrogen gas during electrolysis. Sulfuric acid was added to increase the concentration of hydrogen ions and the total number of ions in solution, but does not take part in the reaction. The volume of hydrogen gas collected is twice the volume of oxygen gas collected, due to the stoichiometry of the reaction.

The Electrolysis of Aqueous Sodium Chloride

The electrolysis of aqueous sodium chloride is the more common example of electrolysis because more than one species can be oxidized and reduced. Considering the anode first, the possible reactions are

$$\text{(i) } 2Cl^-(aq) \longrightarrow Cl_2(g) + 2\,e^- \qquad E^\circ_{anode} = +1.35827 \text{ V}$$
$$\text{(ii) } 2H_2O(l) \longrightarrow O_2(g) + 4H^+(aq) + 4e^- \qquad E^\circ_{anode} = +1.229 \text{ V}$$

These values suggest that water should be oxidized at the anode because a smaller potential would be needed—using reaction (ii) for the oxidation would give a less-negative cell potential. When the experiment is run, it turns out chlorine, not oxygen, is produced at the anode. The unexpected process is so common in electrochemistry that it has been given the name overpotential. The **overpotential** is the difference between the theoretical cell voltage and the actual voltage that is necessary to cause electrolysis. It turns out that the overpotential for oxygen is rather high and effectively makes the reduction potential more positive. As a result, under normal conditions, chlorine gas is what actually forms at the anode.

Now consider the cathode. Three reductions could occur:

$$\text{(iii) } 2H^+(aq) + 2e^- \longrightarrow H_2(g) \qquad E^{\circ}_{cathode} = 0 \text{ V}$$

$$\text{(iv) } 2H_2O(l) + 2e^- \longrightarrow H_2(g) + 2OH^-(aq) \qquad E^{\circ}_{cathode} = -0.8277 \text{ V}$$

$$\text{(v) } Na^+(aq) + e^- \longrightarrow Na(s) \qquad E^{\circ}_{cathode} = -2.71 \text{ V}$$

Reaction (v) is ruled out because it has such a negative reduction potential. Under standard state conditions, reaction (iii) would be preferred to reaction (iv). However, the pH of a sodium chloride solution is 7, so the concentration of hydrogen ions is only 1×10^{-7} M. At such low concentrations, reaction (iii) is unlikely and reaction (iv) occurs. The overall reaction is then

$$\text{overall: } 2H_2O(l) + 2Cl^-(aq) \longrightarrow H_2(g) + Cl_2(g) + 2OH^-(aq) \qquad E^{\circ}_{cell} = -2.186 \text{ V}$$

As the reaction proceeds, hydroxide ions replace chloride ions in solution. Thus, sodium hydroxide can be obtained by evaporating the water after the electrolysis is complete. Sodium hydroxide is valuable in its own right and is used for things like oven cleaner, drain opener, and in the production of paper, fabrics, and soap.

Chemistry in Everyday Life

Electroplating

An important use for electrolytic cells is in **electroplating**. Electroplating results in a thin coating of one metal on top of a conducting surface. Reasons for electroplating include making the object more corrosion resistant, strengthening the surface, producing a more attractive finish, or for purifying metal. The metals commonly used in electroplating include cadmium, chromium, copper, gold, nickel, silver, and tin. Common consumer products include silver-plated or gold-plated tableware, chrome-plated automobile parts, and jewelry. We can get an idea of how this works by investigating how silver-plated tableware is produced (Figure 17.21).

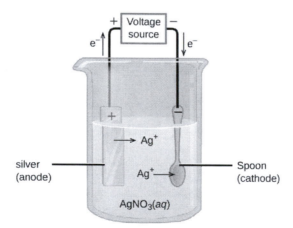

Figure 17.21 The spoon, which is made of an inexpensive metal, is connected to the negative terminal of the voltage source and acts as the cathode. The anode is a silver electrode. Both electrodes are immersed in a silver nitrate solution. When a steady current is passed through the solution, the net result is that silver metal is removed from the anode and deposited on the cathode.

In the figure, the anode consists of a silver electrode, shown on the left. The cathode is located on the right and is the spoon, which is made from inexpensive metal. Both electrodes are immersed in a solution of silver nitrate. As the potential is increased, current flows. Silver metal is lost at the anode as it goes into solution.

$$\text{anode: } Ag(s) \longrightarrow Ag^+(aq) + e^-$$

> The mass of the cathode increases as silver ions from the solution are deposited onto the spoon
>
> $$\text{cathode: } Ag^+(aq) + e^- \longrightarrow Ag(s)$$
>
> The net result is the transfer of silver metal from the anode to the cathode. The quality of the object is usually determined by the thickness of the deposited silver and the rate of deposition.

Quantitative Aspects of Electrolysis

The amount of current that is allowed to flow in an electrolytic cell is related to the number of moles of electrons. The number of moles of electrons can be related to the reactants and products using stoichiometry. Recall that the SI unit for current (I) is the ampere (A), which is the equivalent of 1 coulomb per second ($1 \text{ A} = 1 \frac{C}{s}$). The total charge ($Q$, in coulombs) is given by

$$Q = I \times t = n \times F$$

Where t is the time in seconds, n the number of moles of electrons, and F is the Faraday constant.

Moles of electrons can be used in stoichiometry problems. The time required to deposit a specified amount of metal might also be requested, as in the second of the following examples.

Example 17.8

Converting Current to Moles of Electrons

In one process used for electroplating silver, a current of 10.23 A was passed through an electrolytic cell for exactly 1 hour. How many moles of electrons passed through the cell? What mass of silver was deposited at the cathode from the silver nitrate solution?

Solution

Faraday's constant can be used to convert the charge (Q) into moles of electrons (n). The charge is the current (I) multiplied by the time

$$n = \frac{Q}{F} = \frac{\frac{10.23 \text{ C}}{s} \times 1 \text{ hr} \times \frac{60 \text{ min}}{hr} \times \frac{60 \text{ s}}{min}}{96,485 \text{ C/mol e}^-} = \frac{36,830 \text{ C}}{96,485 \text{ C/mol e}^-} = 0.3817 \text{ mol e}^-$$

From the problem, the solution contains $AgNO_3$, so the reaction at the cathode involves 1 mole of electrons for each mole of silver

$$\text{cathode: } Ag^+(aq) + e^- \longrightarrow Ag(s)$$

The atomic mass of silver is 107.9 g/mol, so

$$\text{mass Ag} = 0.3817 \text{ mol e}^- \times \frac{1 \text{ mol Ag}}{1 \text{ mol e}^-} \times \frac{107.9 \text{ g Ag}}{1 \text{ mol Ag}} = 41.19 \text{ g Ag}$$

Check your answer: From the stoichiometry, 1 mole of electrons would produce 1 mole of silver. Less than one-half a mole of electrons was involved and less than one-half a mole of silver was produced.

Check Your Learning

Aluminum metal can be made from aluminum ions by electrolysis. What is the half-reaction at the cathode? What mass of aluminum metal would be recovered if a current of 2.50×10^3 A passed through the solution for 15.0 minutes? Assume the yield is 100%.

Answer: $Al^{3+}(aq) + 3 e^- \longrightarrow Al(s)$; 7.77 mol Al = 210.0 g Al.

Example 17.9

Time Required for Deposition

In one application, a 0.010-mm layer of chromium must be deposited on a part with a total surface area of 3.3 m^2 from a solution of containing chromium(III) ions. How long would it take to deposit the layer of chromium if the current was 33.46 A? The density of chromium (metal) is 7.19 g/cm^3.

Solution

This problem brings in a number of topics covered earlier. An outline of what needs to be done is:

- If the total charge can be determined, the time required is just the charge divided by the current
- The total charge can be obtained from the amount of Cr needed and the stoichiometry
- The amount of Cr can be obtained using the density and the volume Cr required
- The volume Cr required is the thickness times the area

Solving in steps, and taking care with the units, the volume of Cr required is

$$\text{volume} = \left(0.010 \text{ mm} \times \frac{1 \text{ cm}}{10 \text{ mm}}\right) \times \left(3.3 \text{ m}^2 \times \left(\frac{10,000 \text{ cm}^2}{1 \text{ m}^2}\right)\right) = 33 \text{ cm}^3$$

Cubic centimeters were used because they match the volume unit used for the density. The amount of Cr is then

$$\text{mass} = \text{volume} \times \text{density} = 33 \text{ cm}^3 \times \frac{7.19 \text{ g}}{\text{cm}^3} = 237 \text{ g Cr}$$

$$\text{mol Cr} = 237 \text{ g Cr} \times \frac{1 \text{ mol Cr}}{52.00 \text{ g Cr}} = 4.56 \text{ mol Cr}$$

Since the solution contains chromium(III) ions, 3 moles of electrons are required per mole of Cr. The total charge is then

$$Q = 4.56 \text{ mol Cr} \times \frac{3 \text{ mol e}^-}{1 \text{ mol Cr}} \times \frac{96485 \text{ C}}{\text{mol e}^-} = 1.32 \times 10^6 \text{ C}$$

The time required is then

$$t = \frac{Q}{I} = \frac{1.32 \times 10^6 \text{ C}}{33.46 \text{ C/s}} = 3.95 \times 10^4 \text{ s} = 11.0 \text{ hr}$$

Check your answer: In a long problem like this, a single check is probably not enough. Each of the steps gives a reasonable number, so things are probably correct. Pay careful attention to unit conversions and the stoichiometry.

Check Your Learning

What mass of zinc is required to galvanize the top of a 3.00 m × 5.50 m sheet of iron to a thickness of 0.100 mm of zinc? If the zinc comes from a solution of $Zn(NO_3)_2$ and the current is 25.5 A, how long will it take to galvanize the top of the iron? The density of zinc is 7.140 g/cm^3.

Answer: 231 g Zn required 446 minutes.

Key Terms

active electrode electrode that participates in the oxidation-reduction reaction of an electrochemical cell; the mass of an active electrode changes during the oxidation-reduction reaction

alkaline battery primary battery that uses an alkaline (often potassium hydroxide) electrolyte; designed to be an exact replacement for the dry cell, but with more energy storage and less electrolyte leakage than typical dry cell

anode electrode in an electrochemical cell at which oxidation occurs; information about the anode is recorded on the left side of the salt bridge in cell notation

battery galvanic cell or series of cells that produces a current; in theory, any galvanic cell

cathode electrode in an electrochemical cell at which reduction occurs; information about the cathode is recorded on the right side of the salt bridge in cell notation

cathodic protection method of protecting metal by using a sacrificial anode and effectively making the metal that needs protecting the cathode, thus preventing its oxidation

cell notation shorthand way to represent the reactions in an electrochemical cell

cell potential difference in electrical potential that arises when dissimilar metals are connected; the driving force for the flow of charge (current) in oxidation-reduction reactions

circuit path taken by a current as it flows because of an electrical potential difference

concentration cell galvanic cell in which the two half-cells are the same except for the concentration of the solutes; spontaneous when the overall reaction is the dilution of the solute

corrosion degradation of metal through an electrochemical process

current flow of electrical charge; the SI unit of charge is the coulomb (C) and current is measured in amperes $\left(1 \text{ A} = 1\frac{\text{C}}{\text{s}}\right)$

dry cell primary battery, also called a zinc-carbon battery; can be used in any orientation because it uses a paste as the electrolyte; tends to leak electrolyte when stored

electrical potential energy per charge; in electrochemical systems, it depends on the way the charges are distributed within the system; the SI unit of electrical potential is the volt $\left(1 \text{ V} = 1\frac{\text{J}}{\text{C}}\right)$

electrical work (w_{ele}) negative of total charge times the cell potential; equal to w_{max} for the system, and so equals the free energy change (ΔG)

electrolysis process using electrical energy to cause a nonspontaneous process to occur

electrolytic cell electrochemical cell in which electrolysis is used; electrochemical cell with negative cell potentials

electroplating depositing a thin layer of one metal on top of a conducting surface

Faraday's constant (F) charge on 1 mol of electrons; $F = 96,485$ C/mol e$^-$

fuel cell devices that produce an electrical current as long as fuel and oxidizer are continuously added; more efficient than internal combustion engines

galvanic cell electrochemical cell that involves a spontaneous oxidation-reduction reaction; electrochemical cells with positive cell potentials; also called a voltaic cell

galvanized iron method for protecting iron by covering it with zinc, which will oxidize before the iron; zinc-plated iron

half-reaction method method that produces a balanced overall oxidation-reduction reaction by splitting the reaction into an oxidation "half" and reduction "half," balancing the two half-reactions, and then combining the oxidation half-reaction and reduction half-reaction in such a way that the number of electrons generated by the oxidation is exactly canceled by the number of electrons required by the reduction

inert electrode electrode that allows current to flow, but that does not otherwise participate in the oxidation-reduction reaction in an electrochemical cell; the mass of an inert electrode does not change during the oxidation-reduction reaction; inert electrodes are often made of platinum or gold because these metals are chemically unreactive.

lead acid battery secondary battery that consists of multiple cells; the lead acid battery found in automobiles has six cells and a voltage of 12 V

lithium ion battery very popular secondary battery; uses lithium ions to conduct current and is light, rechargeable, and produces a nearly constant potential as it discharges

Nernst equation equation that relates the logarithm of the reaction quotient (Q) to nonstandard cell potentials; can be used to relate equilibrium constants to standard cell potentials

nickel-cadmium battery (NiCd battery) secondary battery that uses cadmium, which is a toxic heavy metal; heavier than lithium ion batteries, but with similar performance characteristics

overpotential difference between the theoretical potential and actual potential in an electrolytic cell; the "extra" voltage required to make some nonspontaneous electrochemical reaction to occur

oxidation half-reaction the "half" of an oxidation-reduction reaction involving oxidation; the half-reaction in which electrons appear as products; balanced when each atom type, as well as the charge, is balanced

primary battery single-use nonrechargeable battery

reduction half-reaction the "half" of an oxidation-reduction reaction involving reduction; the half-reaction in which electrons appear as reactants; balanced when each atom type, as well as the charge, is balanced

sacrificial anode more active, inexpensive metal used as the anode in cathodic protection; frequently made from magnesium or zinc

secondary battery battery that can be recharged

standard cell potential ($E°_{cell}$) the cell potential when all reactants and products are in their standard states (1 bar or 1 atm or gases; 1 M for solutes), usually at 298.15 K; can be calculated by subtracting the standard reduction potential for the half-reaction at the anode from the standard reduction potential for the half-reaction occurring at the cathode

standard hydrogen electrode (SHE) the electrode consists of hydrogen gas bubbling through hydrochloric acid over an inert platinum electrode whose reduction at standard conditions is assigned a value of 0 V; the reference point for standard reduction potentials

standard reduction potential ($E°$) the value of the reduction under standard conditions (1 bar or 1 atm for gases; 1 M for solutes) usually at 298.15 K; tabulated values used to calculate standard cell potentials

voltaic cell another name for a galvanic cell

Key Equations

- $E°_{cell} = E°_{cathode} - E°_{anode}$

- $E°_{cell} = \dfrac{RT}{nF} \ln K$

- $E°_{cell} = \dfrac{0.0257 \text{ V}}{n} \ln K = \dfrac{0.0592 \text{ V}}{n} \log K$ (at 298.15 K)

- $E_{cell} = E°_{cell} - \dfrac{RT}{nF} \ln Q$ (Nernst equation)

- $E_{cell} = E°_{cell} - \dfrac{0.0257 \text{ V}}{n} \ln Q = E°_{cell} - \dfrac{0.0592 \text{ V}}{n} \log Q$ (at 298.15 K)

- $\Delta G = -nFE_{cell}$

- $\Delta G° = -nFE°_{cell}$

- $w_{ele} = w_{max} = -nFE_{cell}$

- $Q = I \times t = n \times F$

Summary

17.1 Balancing Oxidation-Reduction Reactions

An electric current consists of moving charge. The charge may be in the form of electrons or ions. Current flows through an unbroken or closed circular path called a circuit. The current flows through a conducting medium as a result of a difference in electrical potential between two points in a circuit. Electrical potential has the units of energy per charge. In SI units, charge is measured in coulombs (C), current in amperes $\left(A = \dfrac{C}{s} \right)$, and electrical potential in volts $\left(V = \dfrac{J}{C} \right)$.

Oxidation is the loss of electrons, and the species that is oxidized is also called the reducing agent. Reduction is the gain of electrons, and the species that is reduced is also called the oxidizing agent. Oxidation-reduction reactions can be balanced using the half-reaction method. In this method, the oxidation-reduction reaction is split into an oxidation half-reaction and a reduction half-reaction. The oxidation half-reaction and reduction half-reaction are then balanced separately. Each of the half-reactions must have the same number of each type of atom on both sides of the equation *and* show the same total charge on each side of the equation. Charge is balanced in oxidation half-reactions by adding electrons as products; in reduction half-reactions, charge is balanced by adding electrons as reactants. The total number of electrons gained by reduction must exactly equal the number of electrons lost by oxidation when combining the two half-reactions to give the overall balanced equation. Balancing oxidation-reduction reaction equations in aqueous solutions frequently requires that oxygen or hydrogen be added or removed from a reactant. In acidic solution, hydrogen is added by adding hydrogen ion (H^+) and removed by producing hydrogen ion; oxygen is removed by adding hydrogen ion and producing water, and added by adding water and producing hydrogen ion. A balanced equation in basic solution can be obtained by first balancing the equation in acidic solution, and then adding hydroxide ion to each side of the balanced equation in such numbers that all the hydrogen ions are converted to water.

17.2 Galvanic Cells

Electrochemical cells typically consist of two half-cells. The half-cells separate the oxidation half-reaction from the reduction half-reaction and make it possible for current to flow through an external wire. One half-cell, normally depicted on the left side in a figure, contains the anode. Oxidation occurs at the anode. The anode is connected to the cathode in the other half-cell, often shown on the right side in a figure. Reduction occurs at the cathode. Adding a salt bridge completes the circuit allowing current to flow. Anions in the salt bridge flow toward the anode and cations in the salt bridge flow toward the cathode. The movement of these ions completes the circuit and keeps each half-cell electrically neutral. Electrochemical cells can be described using cell notation. In this notation, information about the reaction at the anode appears on the left and information about the reaction at the cathode on the right. The salt bridge is represented by a double line, ‖. The solid, liquid, or aqueous phases within a half-cell are separated by a single line, $|$. The phase and concentration of the various species is included after the species name. Electrodes that participate in the oxidation-reduction reaction are called active electrodes. Electrodes that do not participate in the oxidation-reduction reaction but are there to allow current to flow are inert electrodes. Inert electrodes are often made from platinum or gold, which are unchanged by many chemical reactions.

17.3 Standard Reduction Potentials

Assigning the potential of the standard hydrogen electrode (SHE) as zero volts allows the determination of standard reduction potentials, $E°$, for half-reactions in electrochemical cells. As the name implies, standard reduction potentials use standard states (1 bar or 1 atm for gases; 1 M for solutes, often at 298.15 K) and are written as reductions (where electrons appear on the left side of the equation). The reduction reactions are reversible, so standard cell potentials can be calculated by subtracting the standard reduction potential for the reaction at the anode from the standard reduction for the reaction at the cathode. When calculating the standard cell potential, the standard reduction potentials are *not* scaled by the stoichiometric coefficients in the balanced overall equation.

17.4 The Nernst Equation

Electrical work (w_{ele}) is the negative of the product of the total charge (Q) and the cell potential (E_{cell}). The total charge can be calculated as the number of moles of electrons (n) times the Faraday constant ($F = 96,485$ C/mol e$^-$). Electrical work is the maximum work that the system can produce and so is equal to the change in free energy. Thus, anything that can be done with or to a free energy change can also be done to or with a cell potential. The Nernst equation relates the cell potential at nonstandard conditions to the logarithm of the reaction quotient. Concentration cells exploit this relationship and produce a positive cell potential using half-cells that differ only in the concentration of their solutes.

17.5 Batteries and Fuel Cells

Batteries are galvanic cells, or a series of cells, that produce an electric current. When cells are combined into batteries, the potential of the battery is an integer multiple of the potential of a single cell. There are two basic types of batteries: primary and secondary. Primary batteries are "single use" and cannot be recharged. Dry cells and (most) alkaline batteries are examples of primary batteries. The second type is rechargeable and is called a secondary battery. Examples of secondary batteries include nickel-cadmium (NiCd), lead acid, and lithium ion batteries. Fuel cells are similar to batteries in that they generate an electrical current, but require continuous addition of fuel and oxidizer. The hydrogen fuel cell uses hydrogen and oxygen from the air to produce water, and is generally more efficient than internal combustion engines.

17.6 Corrosion

Corrosion is the degradation of a metal caused by an electrochemical process. Large sums of money are spent each year repairing the effects of, or preventing, corrosion. Some metals, such as aluminum and copper, produce a protective layer when they corrode in air. The thin layer that forms on the surface of the metal prevents oxygen from coming into contact with more of the metal atoms and thus "protects" the remaining metal from further corrosion. Iron corrodes (forms rust) when exposed to water and oxygen. The rust that forms on iron metal flakes off, exposing fresh metal, which also corrodes. One way to prevent, or slow, corrosion is by coating the metal. Coating prevents water and oxygen from contacting the metal. Paint or other coatings will slow corrosion, but they are not effective once scratched. Zinc-plated or galvanized iron exploits the fact that zinc is more likely to oxidize than iron. As long as the coating remains, even if scratched, the zinc will oxidize before the iron. Another method for protecting metals is cathodic protection. In this method, an easily oxidized and inexpensive metal, often zinc or magnesium (the sacrificial anode), is electrically connected to the metal that must be protected. The more active metal is the sacrificial anode, and is the anode in a galvanic cell. The "protected" metal is the cathode, and remains unoxidized. One advantage of cathodic protection is that the sacrificial anode can be monitored and replaced if needed.

17.7 Electrolysis

Using electricity to force a nonspontaneous process to occur is electrolysis. Electrolytic cells are electrochemical cells with negative cell potentials (meaning a positive Gibbs free energy), and so are nonspontaneous. Electrolysis can occur in electrolytic cells by introducing a power supply, which supplies the energy to force the electrons to flow in the nonspontaneous direction. Electrolysis is done in solutions, which contain enough ions so current can flow. If the solution contains only one material, like the electrolysis of molten sodium chloride, it is a simple matter to determine what is oxidized and what is reduced. In more complicated systems, like the electrolysis of aqueous sodium chloride, more than one species can be oxidized or reduced and the standard reduction potentials are used to determine the most likely oxidation (the half-reaction with the largest [most positive] standard reduction potential) and reduction (the half-reaction with the smallest [least positive] standard reduction potential). Sometimes unexpected half-reactions occur because of overpotential. Overpotential is the difference between the theoretical half-reaction reduction potential and the actual voltage required. When present, the applied potential must be increased, making it possible for a different reaction to occur in the electrolytic cell. The total charge, Q, that passes through an electrolytic cell can be expressed as the current (I) multiplied by time ($Q = It$) or as the moles of electrons (n) multiplied by Faraday's constant ($Q = nF$). These relationships can be used to determine things like the amount of material used or generated during electrolysis, how long the reaction must proceed, or what value of the current is required.

Exercises

17.1 Balancing Oxidation-Reduction Reactions

1. If a 2.5 A current is run through a circuit for 35 minutes, how many coulombs of charge moved through the circuit?

2. For the scenario in the previous question, how many electrons moved through the circuit?

3. For each of the following balanced half-reactions, determine whether an oxidation or reduction is occurring.

(a) $Fe^{3+} + 3e^- \longrightarrow Fe$

(b) $Cr \longrightarrow Cr^{3+} + 3e^-$

(c) $MnO_4{}^{2-} \longrightarrow MnO_4{}^- + e^-$

(d) $Li^+ + e^- \longrightarrow Li$

4. For each of the following unbalanced half-reactions, determine whether an oxidation or reduction is occurring.

(a) $Cl^- \longrightarrow Cl_2$

(b) $Mn^{2+} \longrightarrow MnO_2$

(c) $H_2 \longrightarrow H^+$

(d) $NO_3{}^- \longrightarrow NO$

5. Given the following pairs of balanced half-reactions, determine the balanced reaction for each pair of half-reactions in an acidic solution.

(a) $Ca \longrightarrow Ca^{2+} + 2e^-, \quad F_2 + 2e^- \longrightarrow 2F^-$

(b) $Li \longrightarrow Li^+ + e^-, \quad Cl_2 + 2e^- \longrightarrow 2Cl^-$

(c) $Fe \longrightarrow Fe^{3+} + 3e^-, \quad Br_2 + 2e^- \longrightarrow 2Br^-$

(d) $Ag \longrightarrow Ag^+ + e^-, \quad MnO_4{}^- + 4H^+ + 3e^- \longrightarrow MnO_2 + 2H_2O$

6. Balance the following in acidic solution:

(a) $H_2O_2 + Sn^{2+} \longrightarrow H_2O + Sn^{4+}$

(b) $PbO_2 + Hg \longrightarrow Hg_2{}^{2+} + Pb^{2+}$

(c) $Al + Cr_2O_7{}^{2-} \longrightarrow Al^{3+} + Cr^{3+}$

7. Identify the species that undergoes oxidation, the species that undergoes reduction, the oxidizing agent, and the reducing agent in each of the reactions of the previous problem.

8. Balance the following in basic solution:

(a) $SO_3{}^{2-}(aq) + Cu(OH)_2(s) \longrightarrow SO_4{}^{2-}(aq) + Cu(OH)(s)$

(b) $O_2(g) + Mn(OH)_2(s) \longrightarrow MnO_2(s)$

(c) $NO_3{}^-(aq) + H_2(g) \longrightarrow NO(g)$

(d) $Al(s) + CrO_4{}^{2-}(aq) \longrightarrow Al(OH)_3(s) + Cr(OH)_4{}^-(aq)$

9. Identify the species that was oxidized, the species that was reduced, the oxidizing agent, and the reducing agent in each of the reactions of the previous problem.

10. Why is it not possible for hydroxide ion (OH^-) to appear in either of the half-reactions or the overall equation when balancing oxidation-reduction reactions in acidic solution?

11. Why is it not possible for hydrogen ion (H^+) to appear in either of the half-reactions or the overall equation when balancing oxidation-reduction reactions in basic solution?

12. Why must the charge balance in oxidation-reduction reactions?

17.2 Galvanic Cells

13. Write the following balanced reactions using cell notation. Use platinum as an inert electrode, if needed.

(a) $Mg(s) + Ni^{2+}(aq) \longrightarrow Mg^{2+}(aq) + Ni(s)$

(b) $2Ag^+(aq) + Cu(s) \longrightarrow Cu^{2+}(aq) + 2Ag(s)$

(c) $Mn(s) + Sn(NO_3)_2(aq) \longrightarrow Mn(NO_3)_2(aq) + Au(s)$

(d) $3CuNO_3(aq) + Au(NO_3)_3(aq) \longrightarrow 3Cu(NO_3)_2(aq) + Au(s)$

14. Given the following cell notations, determine the species oxidized, species reduced, and the oxidizing agent and reducing agent, without writing the balanced reactions.

(a) $Mg(s) \mid Mg^{2+}(aq) \parallel Cu^{2+}(aq) \mid Cu(s)$

(b) $Ni(s) \mid Ni^{2+}(aq) \parallel Ag^+(aq) \mid Ag(s)$

15. For the cell notations in the previous problem, write the corresponding balanced reactions.

16. Balance the following reactions and write the reactions using cell notation. Ignore any inert electrodes, as they are never part of the half-reactions.

(a) $Al(s) + Zr^{4+}(aq) \longrightarrow Al^{3+}(aq) + Zr(s)$

(b) $Ag^+(aq) + NO(g) \longrightarrow Ag(s) + NO_3{}^-(aq)$ (acidic solution)

(c) $SiO_3{}^{2-}(aq) + Mg(s) \longrightarrow Si(s) + Mg(OH)_2(s)$ (basic solution)

(d) $ClO_3{}^-(aq) + MnO_2(s) \longrightarrow Cl^-(aq) + MnO_4{}^-(aq)$ (basic solution)

17. Identify the species oxidized, species reduced, and the oxidizing agent and reducing agent for all the reactions in the previous problem.

18. From the information provided, use cell notation to describe the following systems:

(a) In one half-cell, a solution of $Pt(NO_3)_2$ forms Pt metal, while in the other half-cell, Cu metal goes into a $Cu(NO_3)_2$ solution with all solute concentrations 1 M.

(b) The cathode consists of a gold electrode in a 0.55 M $Au(NO_3)_3$ solution and the anode is a magnesium electrode in 0.75 M $Mg(NO_3)_2$ solution.

(c) One half-cell consists of a silver electrode in a 1 M $AgNO_3$ solution, and in the other half-cell, a copper electrode in 1 M $Cu(NO_3)_2$ is oxidized.

19. Why is a salt bridge necessary in galvanic cells like the one in Figure 17.4?

20. An active (metal) electrode was found to gain mass as the oxidation-reduction reaction was allowed to proceed. Was the electrode part of the anode or cathode? Explain.

21. An active (metal) electrode was found to lose mass as the oxidation-reduction reaction was allowed to proceed. Was the electrode part of the anode or cathode? Explain.

22. The mass of three different metal electrodes, each from a different galvanic cell, were determined before and after the current generated by the oxidation-reduction reaction in each cell was allowed to flow for a few minutes. The first metal electrode, given the label A, was found to have increased in mass; the second metal electrode, given the label B, did not change in mass; and the third metal electrode, given the label C, was found to have lost mass. Make an educated guess as to which electrodes were active and which were inert electrodes, and which were anode(s) and which were the cathode(s).

17.3 Standard Reduction Potentials

23. For each reaction listed, determine its standard cell potential at 25 °C and whether the reaction is spontaneous at standard conditions.

(a) $Mg(s) + Ni^{2+}(aq) \longrightarrow Mg^{2+}(aq) + Ni(s)$

(b) $2Ag^+(aq) + Cu(s) \longrightarrow Cu^{2+}(aq) + 2Ag(s)$

(c) $Mn(s) + Sn(NO_3)_2(aq) \longrightarrow Mn(NO_3)_2(aq) + Sn(s)$

(d) $3Fe(NO_3)_2(aq) + Au(NO_3)_3(aq) \longrightarrow 3Fe(NO_3)_3(aq) + Au(s)$

24. For each reaction listed, determine its standard cell potential at 25 °C and whether the reaction is spontaneous at standard conditions.

(a) $Mn(s) + Ni^{2+}(aq) \longrightarrow Mn^{2+}(aq) + Ni(s)$

(b) $3Cu^{2+}(aq) + 2Al(s) \longrightarrow 2Al^{3+}(aq) + 2Cu(s)$

(c) $Na(s) + LiNO_3(aq) \longrightarrow NaNO_3(aq) + Li(s)$

(d) $Ca(NO_3)_2(aq) + Ba(s) \longrightarrow Ba(NO_3)_2(aq) + Ca(s)$

25. Determine the overall reaction and its standard cell potential at 25 °C for this reaction. Is the reaction spontaneous at standard conditions?

$Cu(s) \mid Cu^{2+}(aq) \parallel Au^{3+}(aq) \mid Au(s)$

26. Determine the overall reaction and its standard cell potential at 25 °C for the reaction involving the galvanic cell made from a half-cell consisting of a silver electrode in 1 M silver nitrate solution and a half-cell consisting of a zinc electrode in 1 M zinc nitrate. Is the reaction spontaneous at standard conditions?

27. Determine the overall reaction and its standard cell potential at 25 °C for the reaction involving the galvanic cell in which cadmium metal is oxidized to 1 M cadmium(II) ion and a half-cell consisting of an aluminum electrode in 1 M aluminum nitrate solution. Is the reaction spontaneous at standard conditions?

28. Determine the overall reaction and its standard cell potential at 25 °C for these reactions. Is the reaction spontaneous at standard conditions? Assume the standard reduction for $Br_2(l)$ is the same as for $Br_2(aq)$.

$Pt(s) \mid H_2(g) \mid H^+(aq) \parallel Br_2(aq) \mid Br^-(aq) \mid Pt(s)$

17.4 The Nernst Equation

29. For the standard cell potentials given here, determine the $\Delta G°$ for the cell in kJ.

(a) 0.000 V, n = 2

(b) +0.434 V, n = 2

(c) −2.439 V, n = 1

30. For the $\Delta G°$ values given here, determine the standard cell potential for the cell.

(a) 12 kJ/mol, n = 3

(b) −45 kJ/mol, n = 1

31. Determine the standard cell potential and the cell potential under the stated conditions for the electrochemical reactions described here. State whether each is spontaneous or nonspontaneous under each set of conditions at 298.15 K.

(a) $Hg(l) + S^{2-}(aq, 0.10\ M) + 2Ag^{+}(aq, 0.25\ M) \longrightarrow 2Ag(s) + HgS(s)$

(b) The galvanic cell made from a half-cell consisting of an aluminum electrode in 0.015 M aluminum nitrate solution and a half-cell consisting of a nickel electrode in 0.25 M nickel(II) nitrate solution.

(c) The cell made of a half-cell in which 1.0 M aqueous bromine is oxidized to 0.11 M bromide ion and a half-cell in which aluminum ion at 0.023 M is reduced to aluminum metal. Assume the standard reduction potential for $Br_2(l)$ is the same as that of $Br_2(aq)$.

32. Determine ΔG and $\Delta G°$ for each of the reactions in the previous problem.

33. Use the data in Appendix L to determine the equilibrium constant for the following reactions. Assume 298.15 K if no temperature is given.

(a) $AgCl(s) \rightleftharpoons Ag^{+}(aq) + Cl^{-}(aq)$

(b) $CdS(s) \rightleftharpoons Cd^{2+}(aq) + S^{2-}(aq)$ at 377 K

(c) $Hg^{2+}(aq) + 4Br^{-}(aq) \rightleftharpoons [HgBr_4]^{2-}(aq)$

(d) $H_2O(l) \rightleftharpoons H^{+}(aq) + OH^{-}(aq)$ at 25 °C

17.5 Batteries and Fuel Cells

34. What are the desirable qualities of an electric battery?

35. List some things that are typically considered when selecting a battery for a new application.

36. Consider a battery made from one half-cell that consists of a copper electrode in 1 M $CuSO_4$ solution and another half-cell that consists of a lead electrode in 1 M $Pb(NO_3)_2$ solution.

(a) What are the reactions at the anode, cathode, and the overall reaction?

(b) What is the standard cell potential for the battery?

(c) Most devices designed to use dry-cell batteries can operate between 1.0 and 1.5 V. Could this cell be used to make a battery that could replace a dry-cell battery? Why or why not.

(d) Suppose sulfuric acid is added to the half-cell with the lead electrode and some $PbSO_4(s)$ forms. Would the cell potential increase, decrease, or remain the same?

37. Consider a battery with the overall reaction: $Cu(s) + 2Ag^{+}(aq) \longrightarrow 2Ag(s) + Cu^{2+}(aq)$.

(a) What is the reaction at the anode and cathode?

(b) A battery is "dead" when it has no cell potential. What is the value of Q when this battery is dead?

(c) If a particular dead battery was found to have $[Cu^{2+}] = 0.11\ M$, what was the concentration of silver ion?

38. An inventor proposes using a SHE (standard hydrogen electrode) in a new battery for smartphones that also removes toxic carbon monoxide from the air:

Anode: $CO(g) + H_2O(l) \longrightarrow CO_2(g) + 2H^{+}(aq) + 2e^{-}$ $E°_{anode} = -0.53\ V$

Cathode: $2H^{+}(aq) + 2e^{-} \longrightarrow H_2(g)$ $E°_{cathode} = 0\ V$

Overall: $CO(g) + H_2O(l) \longrightarrow CO_2(g) + H_2(g)$ $E°_{cell} = +0.53\ V$

Would this make a good battery for smartphones? Why or why not?

39. Why do batteries go dead, but fuel cells do not?

40. Explain what happens to battery voltage as a battery is used, in terms of the Nernst equation.

41. Using the information thus far in this chapter, explain why battery-powered electronics perform poorly in low temperatures.

17.6 Corrosion

42. Which member of each pair of metals is more likely to corrode (oxidize)?

(a) Mg or Ca

(b) Au or Hg

(c) Fe or Zn

(d) Ag or Pt

43. Consider the following metals: Ag, Au, Mg, Ni, and Zn. Which of these metals could be used as a sacrificial anode in the cathodic protection of an underground steel storage tank? Steel is mostly iron, so use −0.447 V as the standard reduction potential for steel.

44. Aluminum $(E^\circ_{Al^{3+}/Al} = -2.07 \text{ V})$ is more easily oxidized than iron $(E^\circ_{Fe^{3+}/Fe} = -0.477 \text{ V})$, and yet when both are exposed to the environment, untreated aluminum has very good corrosion resistance while the corrosion resistance of untreated iron is poor. Explain this observation.

45. If a sample of iron and a sample of zinc come into contact, the zinc corrodes but the iron does not. If a sample of iron comes into contact with a sample of copper, the iron corrodes but the copper does not. Explain this phenomenon.

46. Suppose you have three different metals, A, B, and C. When metals A and B come into contact, B corrodes and A does not corrode. When metals A and C come into contact, A corrodes and C does not corrode. Based on this information, which metal corrodes and which metal does not corrode when B and C come into contact?

47. Why would a sacrificial anode made of lithium metal be a bad choice despite its $E^\circ_{Li^+/Li} = -3.04 \text{ V}$, which appears to be able to protect all the other metals listed in the standard reduction potential table?

17.7 Electrolysis

48. Identify the reaction at the anode, reaction at the cathode, the overall reaction, and the approximate potential required for the electrolysis of the following molten salts. Assume standard states and that the standard reduction potentials in Appendix L are the same as those at each of the melting points. Assume the efficiency is 100%.

(a) $CaCl_2$

(b) LiH

(c) $AlCl_3$

(d) $CrBr_3$

49. What mass of each product is produced in each of the electrolytic cells of the previous problem if a total charge of 3.33×10^5 C passes through each cell? Assume the voltage is sufficient to perform the reduction.

50. How long would it take to reduce 1 mole of each of the following ions using the current indicated? Assume the voltage is sufficient to perform the reduction.

(a) Al^{3+}, 1.234 A

(b) Ca^{2+}, 22.2 A

(c) Cr^{5+}, 37.45 A

(d) Au^{3+}, 3.57 A

51. A current of 2.345 A passes through the cell shown in Figure 17.20 for 45 minutes. What is the volume of the hydrogen collected at room temperature if the pressure is exactly 1 atm? Assume the voltage is sufficient to perform the reduction. (Hint: Is hydrogen the only gas present above the water?)

52. An irregularly shaped metal part made from a particular alloy was galvanized with zinc using a $Zn(NO_3)_2$ solution. When a current of 2.599 A was used, it took exactly 1 hour to deposit a 0.01123-mm layer of zinc on the part. What was the total surface area of the part? The density of zinc is 7.140 g/cm^3. Assume the efficiency is 100%.

Chapter 18

Representative Metals, Metalloids, and Nonmetals

Figure 18.1 Purity is extremely important when preparing silicon wafers. Technicians in a cleanroom prepare silicon without impurities (left). The CEO of VLSI Research, Don Hutcheson, shows off a pure silicon wafer (center). A silicon wafer covered in Pentium chips is an enlarged version of the silicon wafers found in many electronics used today (right). (credit middle: modification of work by "Intel Free Press"/Flickr; credit right: modification of work by Naotake Murayama)

Chapter Outline

Introduction

The development of the periodic table in the mid-1800s came from observations that there was a periodic relationship between the properties of the elements. Chemists, who have an understanding of the variations of these properties, have been able to use this knowledge to solve a wide variety of technical challenges. For example, silicon and other semiconductors form the backbone of modern electronics because of our ability to fine-tune the electrical properties of these materials. This chapter explores important properties of representative metals, metalloids, and nonmetals in the periodic table.

18.1 Periodicity

By the end of this section, you will be able to:

- Classify elements
- Make predictions about the periodicity properties of the representative elements

We begin this section by examining the behaviors of representative metals in relation to their positions in the periodic table. The primary focus of this section will be the application of periodicity to the representative metals.

It is possible to divide elements into groups according to their electron configurations. The **representative elements** are elements where the s and p orbitals are filling. The transition elements are elements where the d orbitals (groups 3–11 on the periodic table) are filling, and the inner transition metals are the elements where the f orbitals are filling. The d orbitals fill with the elements in group 11; therefore, the elements in group 12 qualify as representative elements because the last electron enters an s orbital. Metals among the representative elements are the **representative metals**. Metallic character results from an element's ability to lose its outer valence electrons and results in high thermal and electrical conductivity, among other physical and chemical properties. There are 20 nonradioactive representative metals in groups 1, 2, 3, 12, 13, 14, and 15 of the periodic table (the elements shaded in yellow in Figure 18.2). The radioactive elements copernicium, flerovium, polonium, and livermorium are also metals but are beyond the scope of this chapter.

In addition to the representative metals, some of the representative elements are metalloids. A **metalloid** is an element that has properties that are between those of metals and nonmetals; these elements are typically semiconductors.

The remaining representative elements are nonmetals. Unlike **metals**, which typically form cations and ionic compounds (containing ionic bonds), nonmetals tend to form anions or molecular compounds. In general, the combination of a metal and a nonmetal produces a salt. A salt is an ionic compound consisting of cations and anions.

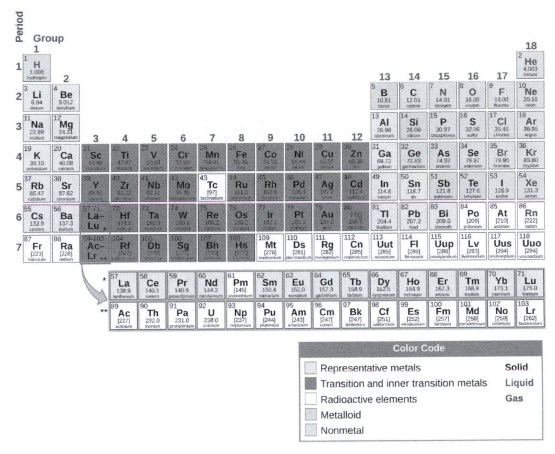

Figure 18.2 The location of the representative metals is shown in the periodic table. Nonmetals are shown in green, metalloids in purple, and the transition metals and inner transition metals in blue.

Most of the representative metals do not occur naturally in an uncombined state because they readily react with water and oxygen in the air. However, it is possible to isolate elemental beryllium, magnesium, zinc, cadmium, mercury, aluminum, tin, and lead from their naturally occurring minerals and use them because they react very slowly with air. Part of the reason why these elements react slowly is that these elements react with air to form a protective coating. The formation of this protective coating is **passivation**. The coating is a nonreactive film of oxide or some other compound. Elemental magnesium, aluminum, zinc, and tin are important in the fabrication of many familiar items, including wire, cookware, foil, and many household and personal objects. Although beryllium, cadmium, mercury, and lead are readily available, there are limitations in their use because of their toxicity.

Group 1: The Alkali Metals

The alkali metals lithium, sodium, potassium, rubidium, cesium, and francium constitute group 1 of the periodic table. Although hydrogen is in group 1 (and also in group 17), it is a nonmetal and deserves separate consideration later in this chapter. The name alkali metal is in reference to the fact that these metals and their oxides react with water to form very basic (alkaline) solutions.

The properties of the alkali metals are similar to each other as expected for elements in the same family. The alkali metals have the largest atomic radii and the lowest first ionization energy in their periods. This combination makes it very easy to remove the single electron in the outermost (valence) shell of each. The easy loss of this valence electron means that these metals readily form stable cations with a charge of 1+. Their reactivity increases with increasing atomic number due to the ease of losing the lone valence electron (decreasing ionization energy). Since oxidation is so easy, the reverse, reduction, is difficult, which explains why it is hard to isolate the elements. The solid alkali metals are very soft; lithium, shown in Figure 18.3, has the lowest density of any metal (0.5 g/cm^3).

The alkali metals all react vigorously with water to form hydrogen gas and a basic solution of the metal hydroxide. This means they are easier to oxidize than is hydrogen. As an example, the reaction of lithium with water is:

$$2\text{Li}(s) + 2\text{H}_2\text{O}(l) \longrightarrow 2\text{LiOH}(aq) + \text{H}_2(g)$$

Figure 18.3   Lithium floats in paraffin oil because its density is less than the density of paraffin oil.

Alkali metals react directly with all the nonmetals (except the noble gases) to yield binary ionic compounds containing 1+ metal ions. These metals are so reactive that it is necessary to avoid contact with both moisture and oxygen in the air. Therefore, they are stored in sealed containers under mineral oil, as shown in Figure 18.4, to prevent contact with air and moisture. The pure metals never exist free (uncombined) in nature due to their high reactivity. In addition, this high reactivity makes it necessary to prepare the metals by electrolysis of alkali metal compounds.

Figure 18.4 To prevent contact with air and water, potassium for laboratory use comes as sticks or beads stored under kerosene or mineral oil, or in sealed containers. (credit: http://images-of-elements.com/potassium.php)

Unlike many other metals, the reactivity and softness of the alkali metals make these metals unsuitable for structural applications. However, there are applications where the reactivity of the alkali metals is an advantage. For example, the production of metals such as titanium and zirconium relies, in part, on the ability of sodium to reduce compounds of these metals. The manufacture of many organic compounds, including certain dyes, drugs, and perfumes, utilizes reduction by lithium or sodium.

Sodium and its compounds impart a bright yellow color to a flame, as seen in Figure 18.5. Passing an electrical discharge through sodium vapor also produces this color. In both cases, this is an example of an emission spectrum as discussed in the chapter on electronic structure. Streetlights sometime employ sodium vapor lights because the sodium vapor penetrates fog better than most other light. This is because the fog does not scatter yellow light as much as it scatters white light. The other alkali metals and their salts also impart color to a flame. Lithium creates a bright, crimson color, whereas the others create a pale, violet color.

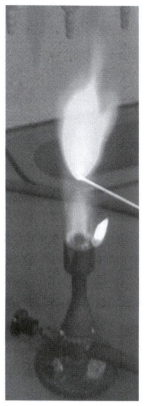

Figure 18.5 Dipping a wire into a solution of a sodium salt and then heating the wire causes emission of a bright yellow light, characteristic of sodium.

Link to Learning

This video (http://openstaxcollege.org/l/16alkalih2o) demonstrates the reactions of the alkali metals with water.

Group 2: The Alkaline Earth Metals

The **alkaline earth metals** (beryllium, magnesium, calcium, strontium, barium, and radium) constitute group 2 of the periodic table. The name alkaline metal comes from the fact that the oxides of the heavier members of the group react with water to form alkaline solutions. The nuclear charge increases when going from group 1 to group 2. Because of this charge increase, the atoms of the alkaline earth metals are smaller and have higher first ionization energies than the alkali metals within the same period. The higher ionization energy makes the alkaline earth metals less reactive than the alkali metals; however, they are still very reactive elements. Their reactivity increases, as expected, with increasing size and decreasing ionization energy. In chemical reactions, these metals readily lose both valence electrons to form compounds in which they exhibit an oxidation state of 2+. Due to their high reactivity, it is common to produce the alkaline earth metals, like the alkali metals, by electrolysis. Even though the ionization energies are low, the two metals with the highest ionization energies (beryllium and magnesium) do form compounds that exhibit some covalent characters. Like the alkali metals, the heavier alkaline earth metals impart color to a flame. As in the case of the alkali metals, this is part of the emission spectrum of these elements. Calcium and strontium produce shades of red, whereas barium produces a green color.

Magnesium is a silver-white metal that is malleable and ductile at high temperatures. Passivation decreases the reactivity of magnesium metal. Upon exposure to air, a tightly adhering layer of magnesium oxycarbonate forms on the surface of the metal and inhibits further reaction. (The carbonate comes from the reaction of carbon dioxide in the atmosphere.) Magnesium is the lightest of the widely used structural metals, which is why most magnesium production is for lightweight alloys.

Magnesium (shown in Figure 18.6), calcium, strontium, and barium react with water and air. At room temperature, barium shows the most vigorous reaction. The products of the reaction with water are hydrogen and the metal hydroxide. The formation of hydrogen gas indicates that the heavier alkaline earth metals are better reducing agents (more easily oxidized) than is hydrogen. As expected, these metals react with both acids and nonmetals to form ionic compounds. Unlike most salts of the alkali metals, many of the common salts of the alkaline earth metals are insoluble in water because of the high lattice energies of these compounds, containing a divalent metal ion.

Figure 18.6 From left to right: Mg(s), warm water at pH 7, and the resulting solution with a pH greater than 7, as indicated by the pink color of the phenolphthalein indicator. (credit: modification of work by Sahar Atwa)

The potent reducing power of hot magnesium is useful in preparing some metals from their oxides. Indeed, magnesium's affinity for oxygen is so great that burning magnesium reacts with carbon dioxide, producing elemental carbon:

$$2Mg(s) + CO_2(g) \longrightarrow 2MgO(s) + C(s)$$

For this reason, a CO_2 fire extinguisher will not extinguish a magnesium fire. Additionally, the brilliant white light emitted by burning magnesium makes it useful in flares and fireworks.

Group 12

The elements in group 12 are transition elements; however, the last electron added is not a *d* electron, but an *s* electron. Since the last electron added is an *s* electron, these elements qualify as representative metals, or post-transition metals. The group 12 elements behave more like the alkaline earth metals than transition metals. Group 12 contains the four elements zinc, cadmium, mercury, and copernicium. Each of these elements has two electrons in its outer shell (ns^2). When atoms of these metals form cations with a charge of 2+, where the two outer electrons are lost, they have pseudo-noble gas electron configurations. Mercury is sometimes an exception because it also exhibits an oxidation state of 1+ in compounds that contain a diatomic Hg_2^{2+} ion. In their elemental forms and in compounds, cadmium and mercury are both toxic.

Zinc is the most reactive in group 12, and mercury is the least reactive. (This is the reverse of the reactivity trend of the metals of groups 1 and 2, in which reactivity increases down a group. The increase in reactivity with increasing atomic number only occurs for the metals in groups 1 and 2.) The decreasing reactivity is due to the formation of ions with a pseudo-noble gas configuration and to other factors that are beyond the scope of this discussion. The chemical behaviors of zinc and cadmium are quite similar to each other but differ from that of mercury.

Zinc and cadmium have lower reduction potentials than hydrogen, and, like the alkali metals and alkaline earth metals, they will produce hydrogen gas when they react with acids. The reaction of zinc with hydrochloric acid, shown in Figure 18.7, is:

$$Zn(s) + 2H_3O^+(aq) + 2Cl^-(aq) \longrightarrow H_2(g) + Zn^{2+}(aq) + 2Cl^-(aq) + 2H_2O(l)$$

Figure 18.7   Zinc is an active transition metal. It dissolves in hydrochloric acid, forming a solution of colorless Zn^{2+} ions, Cl^- ions, and hydrogen gas.

Zinc is a silvery metal that quickly tarnishes to a blue-gray appearance. This change in color is due to an adherent coating of a basic carbonate, $Zn_2(OH)_2CO_3$, which passivates the metal to inhibit further corrosion. Dry cell and alkaline batteries contain a zinc anode. Brass (Cu and Zn) and some bronze (Cu, Sn, and sometimes Zn) are important zinc alloys. About half of zinc production serves to protect iron and other metals from corrosion. This protection may take the form of a sacrificial anode (also known as a galvanic anode, which is a means of providing cathodic protection for various metals) or as a thin coating on the protected metal. Galvanized steel is steel with a protective coating of zinc.

Chemistry in Everyday Life

Sacrificial Anodes

A sacrificial anode, or galvanic anode, is a means of providing cathodic protection of various metals. Cathodic protection refers to the prevention of corrosion by converting the corroding metal into a cathode. As a cathode, the metal resists corrosion, which is an oxidation process. Corrosion occurs at the sacrificial anode instead of at the cathode.

The construction of such a system begins with the attachment of a more active metal (more negative reduction potential) to the metal needing protection. Attachment may be direct or via a wire. To complete the circuit, a *salt bridge* is necessary. This salt bridge is often seawater or ground water. Once the circuit is complete, oxidation (corrosion) occurs at the anode and not the cathode.

The commonly used sacrificial anodes are magnesium, aluminum, and zinc. Magnesium has the most negative reduction potential of the three and serves best when the salt bridge is less efficient due to a low electrolyte concentration such as in freshwater. Zinc and aluminum work better in saltwater than does magnesium. Aluminum is lighter than zinc and has a higher capacity; however, an oxide coating may passivate the aluminum. In special cases, other materials are useful. For example, iron will protect copper.

Mercury is very different from zinc and cadmium. Mercury is the only metal that is liquid at 25 °C. Many metals dissolve in mercury, forming solutions called amalgams (see the feature on Amalgams), which are alloys of mercury with one or more other metals. Mercury, shown in Figure 18.8, is a nonreactive element that is more difficult to oxidize than hydrogen. Thus, it does not displace hydrogen from acids; however, it will react with strong oxidizing acids, such as nitric acid:

$$Hg(l) + HCl(aq) \longrightarrow \text{no reaction}$$
$$3Hg(l) + 8HNO_3(aq) \longrightarrow 3Hg(NO_3)_2(aq) + 4H_2O(l) + 2NO(g)$$

The clear NO initially formed quickly undergoes further oxidation to the reddish brown NO_2.

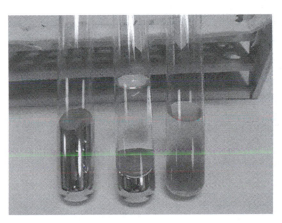

Figure 18.8   From left to right: Hg(*l*), Hg + concentrated HCl, Hg + concentrated HNO₃. (credit: Sahar Atwa)

Most mercury compounds decompose when heated. Most mercury compounds contain mercury with a 2+-oxidation state. When there is a large excess of mercury, it is possible to form compounds containing the Hg_2^{2+} ion. All mercury compounds are toxic, and it is necessary to exercise great care in their synthesis.

Chemistry in Everyday Life

Amalgams

An amalgam is an alloy of mercury with one or more other metals. This is similar to considering steel to be an alloy of iron with other metals. Most metals will form an amalgam with mercury, with the main exceptions being iron, platinum, tungsten, and tantalum.

Due to toxicity issues with mercury, there has been a significant decrease in the use of amalgams. Historically, amalgams were important in electrolytic cells and in the extraction of gold. Amalgams of the alkali metals still find use because they are strong reducing agents and easier to handle than the pure alkali metals.

Prospectors had a problem when they found finely divided gold. They learned that adding mercury to their pans collected the gold into the mercury to form an amalgam for easier collection. Unfortunately, losses of small amounts of mercury over the years left many streams in California polluted with mercury.

Dentists use amalgams containing silver and other metals to fill cavities. There are several reasons to use an amalgam including low cost, ease of manipulation, and longevity compared to alternate materials. Dental amalgams are approximately 50% mercury by weight, which, in recent years, has become a concern due to the toxicity of mercury.

After reviewing the best available data, the Food and Drug Administration (FDA) considers amalgam-based fillings to be safe for adults and children over six years of age. Even with multiple fillings, the mercury levels in the patients remain far below the lowest levels associated with harm. Clinical studies have found no link between dental amalgams and health problems. Health issues may not be the same in cases of children under six or pregnant women. The FDA conclusions are in line with the opinions of the Environmental Protection Agency (EPA) and Centers for Disease Control (CDC). The only health consideration noted is that some people are allergic to the amalgam or one of its components.

Group 13

Group 13 contains the metalloid boron and the metals aluminum, gallium, indium, and thallium. The lightest element, boron, is semiconducting, and its binary compounds tend to be covalent and not ionic. The remaining elements of the group are metals, but their oxides and hydroxides change characters. The oxides and hydroxides of aluminum and gallium exhibit both acidic and basic behaviors. A substance, such as these two, that will react with both acids and bases is amphoteric. This characteristic illustrates the combination of nonmetallic and metallic behaviors of these two elements. Indium and thallium oxides and hydroxides exhibit only basic behavior, in accordance with the clearly metallic character of these two elements. The melting point of gallium is unusually low (about 30 °C) and will melt in your hand.

Aluminum is amphoteric because it will react with both acids and bases. A typical reaction with an acid is:

$$2Al(s) + 6HCl(aq) \longrightarrow 2AlCl_3(aq) + 3H_2(g)$$

The products of the reaction of aluminum with a base depend upon the reaction conditions, with the following being one possibility:

$$2Al(s) + 2NaOH(aq) + 6H_2O(l) \longrightarrow 2Na[Al(OH)_4](aq) + 3H_2(g)$$

With both acids and bases, the reaction with aluminum generates hydrogen gas.

The group 13 elements have a valence shell electron configuration of ns^2np^1. Aluminum normally uses all of its valence electrons when it reacts, giving compounds in which it has an oxidation state of 3+. Although many of these compounds are covalent, others, such as AlF_3 and $Al_2(SO_4)_3$, are ionic. Aqueous solutions of aluminum salts contain the cation $[Al(H_2O)_6]^{3+}$, abbreviated as $Al^{3+}(aq)$. Gallium, indium, and thallium also form ionic compounds containing M^{3+} ions. These three elements exhibit not only the expected oxidation state of 3+ from the three valence electrons but also an oxidation state (in this case, 1+) that is two below the expected value. This phenomenon, the inert pair effect, refers to the formation of a stable ion with an oxidation state two lower than expected for the group. The pair of electrons is the valence s orbital for those elements. In general, the inert pair effect is important for the lower p-block elements. In an aqueous solution, the $Tl^+(aq)$ ion is more stable than is $Tl^{3+}(aq)$. In general, these metals will react with air and water to form 3+ ions; however, thallium reacts to give thallium(I) derivatives. The metals of group 13 all react directly with nonmetals such as sulfur, phosphorus, and the halogens, forming binary compounds.

The metals of group 13 (Al, Ga, In, and Tl) are all reactive. However, passivation occurs as a tough, hard, thin film of the metal oxide forms upon exposure to air. Disruption of this film may counter the passivation, allowing the metal to react. One way to disrupt the film is to expose the passivated metal to mercury. Some of the metal dissolves in the mercury to form an amalgam, which sheds the protective oxide layer to expose the metal to further reaction. The formation of an amalgam allows the metal to react with air and water.

The most important uses of aluminum are in the construction and transportation industries, and in the manufacture of aluminum cans and aluminum foil. These uses depend on the lightness, toughness, and strength of the metal, as well as its resistance to corrosion. Because aluminum is an excellent conductor of heat and resists corrosion, it is useful in the manufacture of cooking utensils.

Aluminum is a very good reducing agent and may replace other reducing agents in the isolation of certain metals from their oxides. Although more expensive than reduction by carbon, aluminum is important in the isolation of Mo, W, and Cr from their oxides.

Group 14

The metallic members of group 14 are tin, lead, and flerovium. Carbon is a typical nonmetal. The remaining elements of the group, silicon and germanium, are examples of semimetals or metalloids. Tin and lead form the stable divalent cations, Sn^{2+} and Pb^{2+}, with oxidation states two below the group oxidation state of 4+. The stability of this oxidation state is a consequence of the inert pair effect. Tin and lead also form covalent compounds with a formal 4+-oxidation state. For example, $SnCl_4$ and $PbCl_4$ are low-boiling covalent liquids.

(a)   (b)

Figure 18.9 (a) Tin(II) chloride is an ionic solid; (b) tin(IV) chloride is a covalent liquid.

Tin reacts readily with nonmetals and acids to form tin(II) compounds (indicating that it is more easily oxidized than hydrogen) and with nonmetals to form either tin(II) or tin(IV) compounds (shown in Figure 18.9), depending on the stoichiometry and reaction conditions. Lead is less reactive. It is only slightly easier to oxidize than hydrogen, and oxidation normally requires a hot concentrated acid.

Many of these elements exist as allotropes. **Allotropes** are two or more forms of the same element in the same physical state with different chemical and physical properties. There are two common allotropes of tin. These allotropes are grey (brittle) tin and white tin. As with other allotropes, the difference between these forms of tin is in the arrangement of the atoms. White tin is stable above 13.2 °C and is malleable like other metals. At low temperatures, gray tin is the more stable form. Gray tin is brittle and tends to break down to a powder. Consequently, articles made of tin will disintegrate in cold weather, particularly if the cold spell is lengthy. The change progresses slowly from the spot of origin, and the gray tin that is first formed catalyzes further change. In a way, this effect is similar to the spread of an infection in a plant or animal body, leading people to call this process tin disease or tin pest.

The principal use of tin is in the coating of steel to form tin plate-sheet iron, which constitutes the tin in tin cans. Important tin alloys are bronze (Cu and Sn) and solder (Sn and Pb). Lead is important in the lead storage batteries in automobiles.

Group 15

Bismuth, the heaviest member of group 15, is a less reactive metal than the other representative metals. It readily gives up three of its five valence electrons to active nonmetals to form the tri-positive ion, Bi^{3+}. It forms compounds with the group oxidation state of 5+ only when treated with strong oxidizing agents. The stability of the 3+-oxidation state is another example of the inert pair effect.

18.2 Occurrence and Preparation of the Representative Metals

By the end of this section, you will be able to:

- Identify natural sources of representative metals
- Describe electrolytic and chemical reduction processes used to prepare these elements from natural sources

Because of their reactivity, we do not find most representative metals as free elements in nature. However, compounds that contain ions of most representative metals are abundant. In this section, we will consider the two common techniques used to isolate the metals from these compounds—electrolysis and chemical reduction.

These metals primarily occur in minerals, with lithium found in silicate or phosphate minerals, and sodium and potassium found in salt deposits from evaporation of ancient seas and in silicates. The alkaline earth metals occur as silicates and, with the exception of beryllium, as carbonates and sulfates. Beryllium occurs as the mineral beryl, $Be_3Al_2Si_6O_{18}$, which, with certain impurities, may be either the gemstone emerald or aquamarine. Magnesium is in seawater and, along with the heavier alkaline earth metals, occurs as silicates, carbonates, and sulfates. Aluminum occurs abundantly in many types of clay and in bauxite, an impure aluminum oxide hydroxide. The principle tin ore is the oxide cassiterite, SnO_2, and the principle lead and thallium ores are the sulfides or the products of weathering of the sulfides. The remaining representative metals occur as impurities in zinc or aluminum ores.

Electrolysis

Ions of metals in of groups 1 and 2, along with aluminum, are very difficult to reduce; therefore, it is necessary to prepare these elements by electrolysis, an important process discussed in the chapter on electrochemistry. Briefly, electrolysis involves using electrical energy to drive unfavorable chemical reactions to completion; it is useful in the isolation of reactive metals in their pure forms. Sodium, aluminum, and magnesium are typical examples.

The Preparation of Sodium

The most important method for the production of sodium is the electrolysis of molten sodium chloride; the set-up is a **Downs cell**, shown in Figure 18.10. The reaction involved in this process is:

$$2NaCl(l) \xrightarrow[\text{600 °C}]{\text{electrolysis}} 2Na(l) + Cl_2(g)$$

The electrolysis cell contains molten sodium chloride (melting point 801 °C), to which calcium chloride has been added to lower the melting point to 600 °C (a colligative effect). The passage of a direct current through the cell causes the sodium ions to migrate to the negatively charged cathode and pick up electrons, reducing the ions to sodium metal. Chloride ions migrate to the positively charged anode, lose electrons, and undergo oxidation to chlorine gas. The overall cell reaction comes from adding the following reactions:

$$\text{at the cathode: } 2Na^+ + 2e^- \longrightarrow 2Na(l)$$
$$\text{at the anode: } 2Cl^- \longrightarrow Cl_2(g) + 2e^-$$
$$\text{overall change: } 2Na^+ + 2Cl^- \longrightarrow 2Na(l) + Cl_2(g)$$

Separation of the molten sodium and chlorine prevents recombination. The liquid sodium, which is less dense than molten sodium chloride, floats to the surface and flows into a collector. The gaseous chlorine goes to storage tanks. Chlorine is also a valuable product.

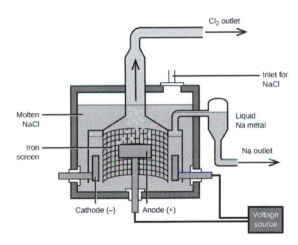

Figure 18.10 Pure sodium metal is isolated by electrolysis of molten sodium chloride using a Downs cell. It is not possible to isolate sodium by electrolysis of aqueous solutions of sodium salts because hydrogen ions are more easily reduced than are sodium ions; as a result, hydrogen gas forms at the cathode instead of the desired sodium metal. The high temperature required to melt NaCl means that liquid sodium metal forms.

The Preparation of Aluminum

The preparation of aluminum utilizes a process invented in 1886 by Charles M. Hall, who began to work on the problem while a student at Oberlin College in Ohio. Paul L. T. Héroult discovered the process independently a month or two later in France. In honor to the two inventors, this electrolysis cell is known as the **Hall–Héroult cell**. The Hall–Héroult cell is an electrolysis cell for the production of aluminum. Figure 18.11 illustrates the Hall–Héroult cell.

The production of aluminum begins with the purification of bauxite, the most common source of aluminum. The reaction of bauxite, AlO(OH), with hot sodium hydroxide forms soluble sodium aluminate, while clay and other impurities remain undissolved:

$$AlO(OH)(s) + NaOH(aq) + H_2O(l) \longrightarrow Na[Al(OH)_4](aq)$$

After the removal of the impurities by filtration, the addition of acid to the aluminate leads to the reprecipitation of aluminum hydroxide:

$$Na[Al(OH)_4](aq) + H_3O^+(aq) \longrightarrow Al(OH)_3(s) + Na^+(aq) + 2H_2O(l)$$

The next step is to remove the precipitated aluminum hydroxide by filtration. Heating the hydroxide produces aluminum oxide, Al_2O_3, which dissolves in a molten mixture of cryolite, Na_3AlF_6, and calcium fluoride, CaF_2. Electrolysis of this solution takes place in a cell like that shown in Figure 18.11. Reduction of aluminum ions to the metal occurs at the cathode, while oxygen, carbon monoxide, and carbon dioxide form at the anode.

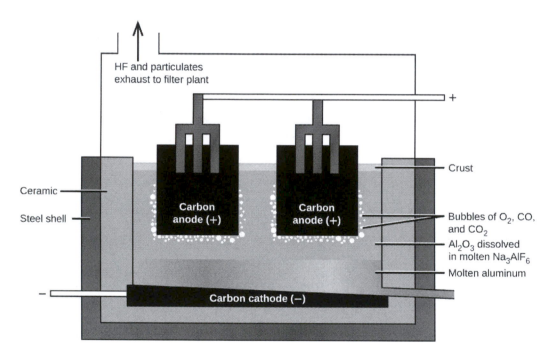

Figure 18.11 An electrolytic cell is used for the production of aluminum. The electrolysis of a solution of cryolite and calcium fluoride results in aluminum metal at the cathode, and oxygen, carbon monoxide, and carbon dioxide at the anode.

The Preparation of Magnesium

Magnesium is the other metal that is isolated in large quantities by electrolysis. Seawater, which contains approximately 0.5% magnesium chloride, serves as the major source of magnesium. Addition of calcium hydroxide to seawater precipitates magnesium hydroxide. The addition of hydrochloric acid to magnesium hydroxide, followed by evaporation of the resultant aqueous solution, leaves pure magnesium chloride. The electrolysis of molten magnesium chloride forms liquid magnesium and chlorine gas:

$$MgCl_2(aq) + Ca(OH)_2(aq) \longrightarrow Mg(OH)_2(s) + CaCl_2(aq)$$
$$Mg(OH)_2(s) + 2HCl(aq) \longrightarrow MgCl_2(aq) + 2H_2O(l)$$
$$MgCl_2(l) \longrightarrow Mg(l) + Cl_2(g)$$

Some production facilities have moved away from electrolysis completely. In the next section, we will see how the Pidgeon process leads to the chemical reduction of magnesium.

Chemical Reduction

It is possible to isolate many of the representative metals by **chemical reduction** using other elements as reducing agents. In general, chemical reduction is much less expensive than electrolysis, and for this reason, chemical reduction is the method of choice for the isolation of these elements. For example, it is possible to produce potassium, rubidium, and cesium by chemical reduction, as it is possible to reduce the molten chlorides of these metals with sodium metal. This may be surprising given that these metals are more reactive than sodium; however, the metals formed are more volatile than sodium and can be distilled for collection. The removal of the metal vapor leads to a shift in the equilibrium to produce more metal (see how reactions can be driven in the discussions of Le Châtelier's principle in the chapter on fundamental equilibrium concepts).

The production of magnesium, zinc, and tin provide additional examples of chemical reduction.

The Preparation of Magnesium

The **Pidgeon process** involves the reaction of magnesium oxide with elemental silicon at high temperatures to form pure magnesium:

$$Si(s) + 2MgO(s) \xrightarrow{\Delta} SiO_2(s) + 2Mg(g)$$

Although this reaction is unfavorable in terms of thermodynamics, the removal of the magnesium vapor produced takes advantage of Le Châtelier's principle to continue the forward progress of the reaction. Over 75% of the world's production of magnesium, primarily in China, comes from this process.

The Preparation of Zinc

Zinc ores usually contain zinc sulfide, zinc oxide, or zinc carbonate. After separation of these compounds from the ores, heating in air converts the ore to zinc oxide by one of the following reactions:

$$2ZnS(s) + 3O_2(g) \xrightarrow{\Delta} 2ZnO(s) + 2SO_2(g)$$
$$ZnCO_3(s) \xrightarrow{\Delta} ZnO(s) + Co_2(g)$$

Carbon, in the form of coal, reduces the zinc oxide to form zinc vapor:

$$ZnO(s) + C(s) \longrightarrow Zn(g) + CO(g)$$

The zinc can be distilled (boiling point 907 °C) and condensed. This zinc contains impurities of cadmium (767 °C), iron (2862 °C), lead (1750 °C), and arsenic (613 °C). Careful redistillation produces pure zinc. Arsenic and cadmium are distilled from the zinc because they have lower boiling points. At higher temperatures, the zinc is distilled from the other impurities, mainly lead and iron.

The Preparation of Tin

The ready reduction of tin(IV) oxide by the hot coals of a campfire accounts for the knowledge of tin in the ancient world. In the modern process, the roasting of tin ores containing SnO_2 removes contaminants such as arsenic and sulfur as volatile oxides. Treatment of the remaining material with hydrochloric acid removes the oxides of other metals. Heating the purified ore with carbon at temperature above 1000 °C produces tin:

$$SnO_2(s) + 2C(s) \xrightarrow{\Delta} Sn(s) + 2CO(g)$$

The molten tin collects at the bottom of the furnace and is drawn off and cast into blocks.

18.3 Structure and General Properties of the Metalloids

By the end of this section, you will be able to:

- Describe the general preparation, properties, and uses of the metalloids
- Describe the preparation, properties, and compounds of boron and silicon

A series of six elements called the metalloids separate the metals from the nonmetals in the periodic table. The metalloids are boron, silicon, germanium, arsenic, antimony, and tellurium. These elements look metallic; however, they do not conduct electricity as well as metals so they are semiconductors. They are semiconductors because their electrons are more tightly bound to their nuclei than are those of metallic conductors. Their chemical behavior falls between that of metals and nonmetals. For example, the pure metalloids form covalent crystals like the nonmetals, but like the metals, they generally do not form monatomic anions. This intermediate behavior is in part due to their intermediate electronegativity values. In this section, we will briefly discuss the chemical behavior of metalloids and deal with two of these elements—boron and silicon—in more detail.

The metalloid boron exhibits many similarities to its neighbor carbon and its diagonal neighbor silicon. All three elements form covalent compounds. However, boron has one distinct difference in that its $2s^2 2p^1$ outer electron structure gives it one less valence electron than it has valence orbitals. Although boron exhibits an oxidation state of 3+ in most of its stable compounds, this electron deficiency provides boron with the ability to form other, sometimes fractional, oxidation states, which occur, for example, in the boron hydrides.

Silicon has the valence shell electron configuration $3s^2 3p^2$, and it commonly forms tetrahedral structures in which it is sp^3 hybridized with a formal oxidation state of 4+. The major differences between the chemistry of carbon and silicon result from the relative strength of the carbon-carbon bond, carbon's ability to form stable bonds to itself, and the presence of the empty $3d$ valence-shell orbitals in silicon. Silicon's empty d orbitals and boron's empty p orbital enable tetrahedral silicon compounds and trigonal planar boron compounds to act as Lewis acids. Carbon, on the other hand, has no available valence shell orbitals; tetrahedral carbon compounds cannot act as Lewis acids. Germanium is very similar to silicon in its chemical behavior.

Arsenic and antimony generally form compounds in which an oxidation state of 3+ or 5+ is exhibited; however, arsenic can form arsenides with an oxidation state of 3−. These elements tarnish only slightly in dry air but readily oxidize when warmed.

Tellurium combines directly with most elements. The most stable tellurium compounds are the tellurides—salts of Te^{2-} formed with active metals and lanthanides—and compounds with oxygen, fluorine, and chlorine, in which tellurium normally exhibits an oxidation state 2+ or 4+. Although tellurium(VI) compounds are known (for example, TeF_6), there is a marked resistance to oxidation to this maximum group oxidation state.

Structures of the Metalloids

Covalent bonding is the key to the crystal structures of the metalloids. In this regard, these elements resemble nonmetals in their behavior.

Elemental silicon, germanium, arsenic, antimony, and tellurium are lustrous, metallic-looking solids. Silicon and germanium crystallize with a diamond structure. Each atom within the crystal has covalent bonds to four neighboring atoms at the corners of a regular tetrahedron. Single crystals of silicon and germanium are giant, three-dimensional molecules. There are several allotropes of arsenic with the most stable being layer like and containing puckered sheets of arsenic atoms. Each arsenic atom forms covalent bonds to three other atoms within the sheet. The crystal structure of antimony is similar to that of arsenic, both shown in Figure 18.12. The structures of arsenic and antimony are similar to the structure of graphite, covered later in this chapter. Tellurium forms crystals that contain infinite spiral chains of tellurium atoms. Each atom in the chain bonds to two other atoms.

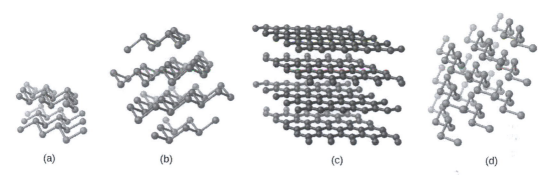

Figure 18.12   (a) Arsenic and (b) antimony have a layered structure similar to that of (c) graphite, except that the layers are puckered rather than planar. (d) Elemental tellurium forms spiral chains.

Pure crystalline boron is transparent. The crystals consist of icosahedra, as shown in Figure 18.13, with a boron atom at each corner. In the most common form of boron, the icosahedra pack together in a manner similar to the cubic closest packing of spheres. All boron-boron bonds within each icosahedron are identical and are approximately 176 pm in length. In the different forms of boron, there are different arrangements and connections between the icosahedra.

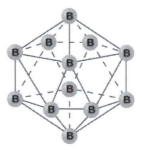

Figure 18.13   An icosahedron is a symmetrical, solid shape with 20 faces, each of which is an equilateral triangle. The faces meet at 12 corners.

The name silicon is derived from the Latin word for flint, *silex*. The metalloid silicon readily forms compounds containing Si-O-Si bonds, which are of prime importance in the mineral world. This bonding capability is in contrast to the nonmetal carbon, whose ability to form carbon-carbon bonds gives it prime importance in the plant and animal worlds.

Occurrence, Preparation, and Compounds of Boron and Silicon

Boron constitutes less than 0.001% by weight of the earth's crust. In nature, it only occurs in compounds with oxygen. Boron is widely distributed in volcanic regions as boric acid, $B(OH)_3$, and in dry lake regions, including the desert areas of California, as borates and salts of boron oxyacids, such as borax, $Na_2B_4O_7 \cdot 10H_2O$.

Elemental boron is chemically inert at room temperature, reacting with only fluorine and oxygen to form boron trifluoride, BF_3, and boric oxide, B_2O_3, respectively. At higher temperatures, boron reacts with all nonmetals, except tellurium and the noble gases, and with nearly all metals; it oxidizes to B_2O_3 when heated with concentrated nitric or sulfuric acid. Boron does not react with nonoxidizing acids. Many boron compounds react readily with water to give boric acid, $B(OH)_3$ (sometimes written as H_3BO_3).

Reduction of boric oxide with magnesium powder forms boron (95–98.5% pure) as a brown, amorphous powder:

$$B_2O_3(s) + 3Mg(s) \longrightarrow 2B(s) + 3MgO(s)$$

An **amorphous** substance is a material that appears to be a solid, but does not have a long-range order like a true solid. Treatment with hydrochloric acid removes the magnesium oxide. Further purification of the boron begins with conversion of the impure boron into boron trichloride. The next step is to heat a mixture of boron trichloride and hydrogen:

$$2BCl_3(g) + 3H_2(g) \xrightarrow{1500\,°C} 2B(s) + 6HCl(g) \qquad \Delta H° = 253.7 \text{ kJ}$$

Silicon makes up nearly one-fourth of the mass of the earth's crust—second in abundance only to oxygen. The crust is composed almost entirely of minerals in which the silicon atoms are at the center of the silicon-oxygen tetrahedron, which connect in a variety of ways to produce, among other things, chains, layers, and three-dimensional frameworks. These minerals constitute the bulk of most common rocks, soil, and clays. In addition, materials such as bricks, ceramics, and glasses contain silicon compounds.

It is possible to produce silicon by the high-temperature reduction of silicon dioxide with strong reducing agents, such as carbon and magnesium:

$$SiO_2(s) + 2C(s) \xrightarrow{\Delta} Si(s) + 2CO(g)$$
$$SiO_2(s) + 2Mg(s) \xrightarrow{\Delta} Si(s) + 2MgO(s)$$

Extremely pure silicon is necessary for the manufacture of semiconductor electronic devices. This process begins with the conversion of impure silicon into silicon tetrahalides, or silane (SiH_4), followed by decomposition at high temperatures. Zone refining, illustrated in Figure 18.14, completes the purification. In this method, a rod of silicon is heated at one end by a heat source that produces a thin cross-section of molten silicon. Slowly lowering the rod through the heat source moves the molten zone from one end of the rod to other. As this thin, molten region moves, impurities in the silicon dissolve in the liquid silicon and move with the molten region. Ultimately, the impurities move to one end of the rod, which is then cut off.

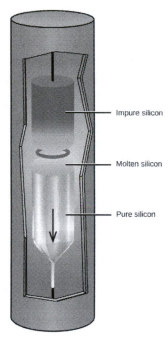

Figure 18.14   A zone-refining apparatus used to purify silicon.

This highly purified silicon, containing no more than one part impurity per million parts of silicon, is the most important element in the computer industry. Pure silicon is necessary in semiconductor electronic devices such as transistors, computer chips, and solar cells.

Like some metals, passivation of silicon occurs due the formation of a very thin film of oxide (primarily silicon dioxide, SiO_2). Silicon dioxide is soluble in hot aqueous base; thus, strong bases destroy the passivation. Removal of the passivation layer allows the base to dissolve the silicon, forming hydrogen gas and silicate anions. For example:

$$Si(s) + 4OH^-(aq) \longrightarrow SiO_4{}^{4-}(aq) + 2H_2(g)$$

Silicon reacts with halogens at high temperatures, forming volatile tetrahalides, such as SiF_4.

Unlike carbon, silicon does not readily form double or triple bonds. Silicon compounds of the general formula SiX_4, where X is a highly electronegative group, can act as Lewis acids to form six-coordinate silicon. For example, silicon tetrafluoride, SiF_4, reacts with sodium fluoride to yield $Na_2[SiF_6]$, which contains the octahedral $\left[SiF_6\right]^{2-}$ ion in which silicon is sp^3d^2 hybridized:

$$2NaF(s) + SiF_4(g) \longrightarrow Na_2SiF_6(s)$$

Antimony reacts readily with stoichiometric amounts of fluorine, chlorine, bromine, or iodine, yielding trihalides or, with excess fluorine or chlorine, forming the pentahalides SbF_5 and $SbCl_5$. Depending on the stoichiometry, it forms antimony(III) sulfide, Sb_2S_3, or antimony(V) sulfide when heated with sulfur. As expected, the metallic nature of the element is greater than that of arsenic, which lies immediately above it in group 15.

Boron and Silicon Halides

Boron trihalides—BF_3, BCl_3, BBr_3, and BI_3—can be prepared by the direct reaction of the elements. These nonpolar molecules contain boron with sp^2 hybridization and a trigonal planar molecular geometry. The fluoride and chloride compounds are colorless gasses, the bromide is a liquid, and the iodide is a white crystalline solid.

Except for boron trifluoride, the boron trihalides readily hydrolyze in water to form boric acid and the corresponding hydrohalic acid. Boron trichloride reacts according to the equation:

$$BCl_3(g) + 3H_2O(l) \longrightarrow B(OH)_3(aq) + 3HCl(aq)$$

Boron trifluoride reacts with hydrofluoric acid, to yield a solution of fluoroboric acid, HBF_4:

$$BF_3(aq) + HF(aq) + H_2O(l) \longrightarrow H_3O^+(aq) + BF_4{}^-(aq)$$

In this reaction, the BF_3 molecule acts as the Lewis acid (electron pair acceptor) and accepts a pair of electrons from a fluoride ion:

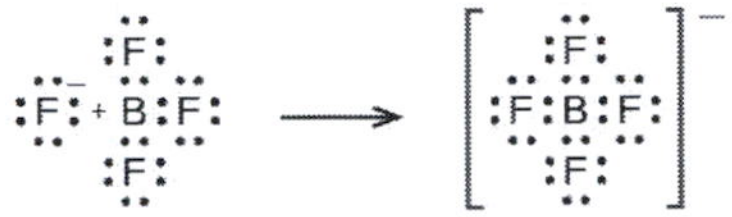

All the tetrahalides of silicon, SiX_4, have been prepared. Silicon tetrachloride can be prepared by direct chlorination at elevated temperatures or by heating silicon dioxide with chlorine and carbon:

$$SiO_2(s) + 2C(s) + 2Cl_2(g) \xrightarrow{\Delta} SiCl_4(g) + 2CO(g)$$

Silicon tetrachloride is a covalent tetrahedral molecule, which is a nonpolar, low-boiling (57 °C), colorless liquid.

It is possible to prepare silicon tetrafluoride by the reaction of silicon dioxide with hydrofluoric acid:

$$SiO_2(s) + 4HF(g) \longrightarrow SiF_4(g) + 2H_2O(l) \qquad \Delta H° = -191.2 \text{ kJ}$$

Hydrofluoric acid is the only common acid that will react with silicon dioxide or silicates. This reaction occurs because the silicon-fluorine bond is the only bond that silicon forms that is stronger than the silicon-oxygen bond. For this reason, it is possible to store all common acids, other than hydrofluoric acid, in glass containers.

Except for silicon tetrafluoride, silicon halides are extremely sensitive to water. Upon exposure to water, $SiCl_4$ reacts rapidly with hydroxide groups, replacing all four chlorine atoms to produce unstable orthosilicic acid, $Si(OH)_4$ or H_4SiO_4, which slowly decomposes into SiO_2.

Boron and Silicon Oxides and Derivatives

Boron burns at 700 °C in oxygen, forming boric oxide, B_2O_3. Boric oxide is necessary for the production of heat-resistant borosilicate glass, like that shown in Figure 18.15 and certain optical glasses. Boric oxide dissolves in hot water to form boric acid, $B(OH)_3$:

$$B_2O_3(s) + 3H_2O(l) \longrightarrow 2B(OH)_3(aq)$$

Figure 18.15 Laboratory glassware, such as Pyrex and Kimax, is made of borosilicate glass because it does not break when heated. The inclusion of borates in the glass helps to mediate the effects of thermal expansion and contraction. This reduces the likelihood of thermal shock, which causes silicate glass to crack upon rapid heating or cooling. (credit: "Tweenk"/Wikimedia Commons)

The boron atom in $B(OH)_3$ is sp^2 hybridized and is located at the center of an equilateral triangle with oxygen atoms at the corners. In solid $B(OH)_3$, hydrogen bonding holds these triangular units together. Boric acid, shown in Figure 18.16, is a very weak acid that does not act as a proton donor but rather as a Lewis acid, accepting an unshared pair of electrons from the Lewis base OH^-:

$$B(OH)_3(aq) + 2H_2O(l) \rightleftharpoons B(OH)_4{}^-(aq) + H_3O^+(aq) \qquad K_a = 5.8 \times 10^{-10}$$

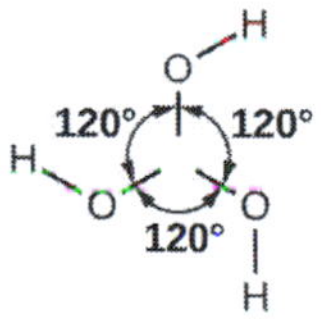

Figure 18.16 Boric acid has a planar structure with three –OH groups spread out equally at 120° angles from each other.

Heating boric acid to 100 °C causes molecules of water to split out between pairs of adjacent –OH groups to form metaboric acid, HBO_2. At about 150 °C, additional B-O-B linkages form, connecting the BO_3 groups together with shared oxygen atoms to form tetraboric acid, $H_2B_4O_7$. Complete water loss, at still higher temperatures, results in boric oxide.

Borates are salts of the oxyacids of boron. Borates result from the reactions of a base with an oxyacid or from the fusion of boric acid or boric oxide with a metal oxide or hydroxide. Borate anions range from the simple trigonal planar $BO_3{}^{3-}$ ion to complex species containing chains and rings of three- and four-coordinated boron atoms. The structures of the anions found in CaB_2O_4, $K[B_5O_6(OH)_4]\cdot2H_2O$ (commonly written $KB_5O_8\cdot4H_2O$) and $Na_2[B_4O_5(OH)_4]\cdot8H_2O$ (commonly written $Na_2B_4O_7\cdot10H_2O$) are shown in Figure 18.17. Commercially, the most important borate is borax, $Na_2[B_4O_5(OH)_4]\cdot8H_2O$, which is an important component of some laundry detergents. Most of the supply of borax comes directly from dry lakes, such as Searles Lake in California, or is prepared from kernite, $Na_2B_4O_7\cdot4H_2O$.

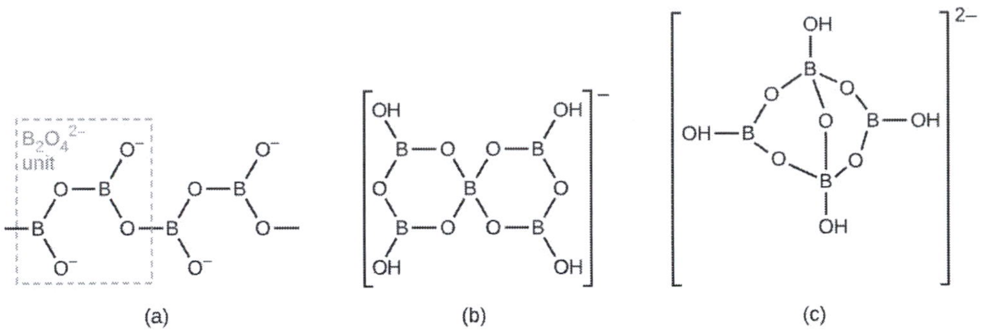

Figure 18.17 The borate anions are (a) CaB_2O_4, (b) $KB_5O_8 \cdot 4H_2O$, and (c) $Na_2B_4O_7 \cdot 10H_2O$. The anion in CaB_2O_4 is an "infinite" chain.

Silicon dioxide, silica, occurs in both crystalline and amorphous forms. The usual crystalline form of silicon dioxide is quartz, a hard, brittle, clear, colorless solid. It is useful in many ways—for architectural decorations, semiprecious jewels, and frequency control in radio transmitters. Silica takes many crystalline forms, or **polymorphs**, in nature. Trace amounts of Fe^{3+} in quartz give amethyst its characteristic purple color. The term *quartz* is also used for articles such as tubing and lenses that are manufactured from amorphous silica. Opal is a naturally occurring form of amorphous silica.

The contrast in structure and physical properties between silicon dioxide and carbon dioxide is interesting, as illustrated in Figure 18.18. Solid carbon dioxide (dry ice) contains single CO_2 molecules with each of the two oxygen atoms attached to the carbon atom by double bonds. Very weak intermolecular forces hold the molecules together in the crystal. The volatility of dry ice reflect these weak forces between molecules. In contrast, silicon dioxide is a covalent network solid. In silicon dioxide, each silicon atom links to four oxygen atoms by single bonds directed toward the corners of a regular tetrahedron, and SiO_4 tetrahedra share oxygen atoms. This arrangement gives a three dimensional, continuous, silicon-oxygen network. A quartz crystal is a macromolecule of silicon dioxide. The difference between these two compounds is the ability of the group 14 elements to form strong π bonds. Second-period elements, such as carbon, form very strong π bonds, which is why carbon dioxide forms small molecules with strong double bonds. Elements below the second period, such as silicon, do not form π bonds as readily as second-period elements, and when they do form, the π bonds are weaker than those formed by second-period elements. For this reason, silicon dioxide does not contain π bonds but only σ bonds.

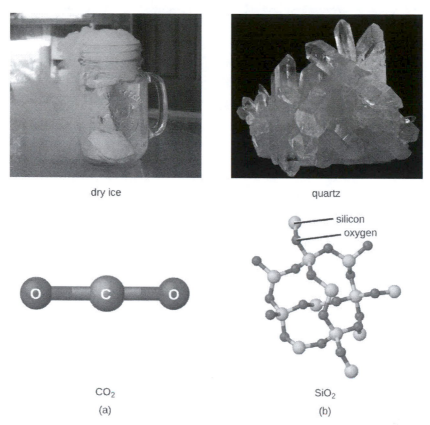

dry ice   quartz

CO$_2$ SiO$_2$

(a)   (b)

Figure 18.18 Because carbon tends to form double and triple bonds and silicon does not, (a) carbon dioxide is a discrete molecule with two C=O double bonds and (b) silicon dioxide is an infinite network of oxygen atoms bridging between silicon atoms with each silicon atom possessing four Si-O single bonds. (credit a photo: modification of work by Erica Gerdes; credit b photo: modification of work by Didier Descouens)

At 1600 °C, quartz melts to yield a viscous liquid. When the liquid cools, it does not crystallize readily but usually supercools and forms a glass, also called silica. The SiO$_4$ tetrahedra in glassy silica have a random arrangement characteristic of supercooled liquids, and the glass has some very useful properties. Silica is highly transparent to both visible and ultraviolet light. For this reason, it is important in the manufacture of lamps that give radiation rich in ultraviolet light and in certain optical instruments that operate with ultraviolet light. The coefficient of expansion of silica glass is very low; therefore, rapid temperature changes do not cause it to fracture. CorningWare and other ceramic cookware contain amorphous silica.

Silicates are salts containing anions composed of silicon and oxygen. In nearly all silicates, sp^3-hybridized silicon atoms occur at the centers of tetrahedra with oxygen at the corners. There is a variation in the silicon-to-oxygen ratio that occurs because silicon-oxygen tetrahedra may exist as discrete, independent units or may share oxygen atoms at corners in a variety of ways. In addition, the presence of a variety of cations gives rise to the large number of silicate minerals.

Many ceramics are composed of silicates. By including small amounts of other compounds, it is possible to modify the physical properties of the silicate materials to produce ceramics with useful characteristics.

18.4 Structure and General Properties of the Nonmetals

By the end of this section, you will be able to:

*   Describe structure and properties of nonmetals

The nonmetals are elements located in the upper right portion of the periodic table. Their properties and behavior are quite different from those of metals on the left side. Under normal conditions, more than half of the nonmetals are gases, one is a liquid, and the rest include some of the softest and hardest of solids. The nonmetals exhibit a rich variety of chemical behaviors. They include the most reactive and least reactive of elements, and they form many different ionic and covalent compounds. This section presents an overview of the properties and chemical behaviors of the nonmetals, as well as the chemistry of specific elements. Many of these nonmetals are important in biological systems.

In many cases, trends in electronegativity enable us to predict the type of bonding and the physical states in compounds involving the nonmetals. We know that electronegativity decreases as we move down a given group and increases as we move from left to right across a period. The nonmetals have higher electronegativities than do metals, and compounds formed between metals and nonmetals are generally ionic in nature because of the large differences in electronegativity between them. The metals form cations, the nonmetals form anions, and the resulting compounds are solids under normal conditions. On the other hand, compounds formed between two or more nonmetals have small differences in electronegativity between the atoms, and covalent bonding—sharing of electrons—results. These substances tend to be molecular in nature and are gases, liquids, or volatile solids at room temperature and pressure.

In normal chemical processes, nonmetals do not form monatomic positive ions (cations) because their ionization energies are too high. All monatomic nonmetal ions are anions; examples include the chloride ion, Cl^-, the nitride ion, N^{3-}, and the selenide ion, Se^{2-}.

The common oxidation states that the nonmetals exhibit in their ionic and covalent compounds are shown in Figure 18.19. Remember that an element exhibits a positive oxidation state when combined with a more electronegative element and that it exhibits a negative oxidation state when combined with a less electronegative element.

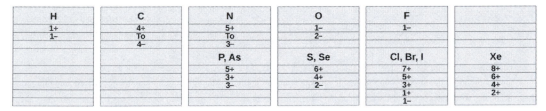

Figure 18.19 Nonmetals exhibit these common oxidation states in ionic and covalent compounds.

The first member of each nonmetal group exhibits different behaviors, in many respects, from the other group members. The reasons for this include smaller size, greater ionization energy, and (most important) the fact that the first member of each group has only four valence orbitals (one 2s and three 2p) available for bonding, whereas other group members have empty d orbitals in their valence shells, making possible five, six, or even more bonds around the central atom. For example, nitrogen forms only NF_3, whereas phosphorus forms both PF_3 and PF_5.

Another difference between the first group member and subsequent members is the greater ability of the first member to form π bonds. This is primarily a function of the smaller size of the first member of each group, which allows better overlap of atomic orbitals. Nonmetals, other than the first member of each group, rarely form π bonds to nonmetals that are the first member of a group. For example, sulfur-oxygen π bonds are well known, whereas sulfur does not normally form stable π bonds to itself.

The variety of oxidation states displayed by most of the nonmetals means that many of their chemical reactions involve changes in oxidation state through oxidation-reduction reactions. There are four general aspects of the oxidation-reduction chemistry:

1. Nonmetals oxidize most metals. The oxidation state of the metal becomes positive as it undergoes oxidation and that of the nonmetal becomes negative as it undergoes reduction. For example:

$$4Fe(s) + 3O_2(g) \longrightarrow 2Fe_2O_3(s)$$
$$0 0 +3-2$$

2. With the exception of nitrogen and carbon, which are poor oxidizing agents, a more electronegative nonmetal oxidizes a less electronegative nonmetal or the anion of the nonmetal:

$$S(s) + O_2(g) \longrightarrow 2SO_2(s)$$
$$0 0 +4-2$$

$$Cl_2(g) + 2I^-(aq) \longrightarrow I_2(s) + 2Cl^-(aq)$$
$$0 0$$

3. Fluorine and oxygen are the strongest oxidizing agents within their respective groups; each oxidizes all the elements that lie below it in the group. Within any period, the strongest oxidizing agent is in group 17. A nonmetal often oxidizes an element that lies to its left in the same period. For example:

$$2As(s) + 3Br_2(l) \longrightarrow 2AsBr_3(s)$$
$$0 0 +3-1$$

4. The stronger a nonmetal is as an oxidizing agent, the more difficult it is to oxidize the anion formed by the nonmetal. This means that the most stable negative ions are formed by elements at the top of the group or in group 17 of the period.

5. Fluorine and oxygen are the strongest oxidizing elements known. Fluorine does not form compounds in which it exhibits positive oxidation states; oxygen exhibits a positive oxidation state only when combined with fluorine. For example:

$$2F_2(g) + 2OH^-(aq) \longrightarrow OF_2(g) + 2F^-(aq) + H_2O(l)$$
$$0 +2 -1$$

With the exception of most of the noble gases, all nonmetals form compounds with oxygen, yielding covalent oxides. Most of these oxides are acidic, that is, they react with water to form oxyacids. Recall from the acid-base chapter that an oxyacid is an acid consisting of hydrogen, oxygen, and some other element. Notable exceptions are carbon monoxide, CO, nitrous oxide, N_2O, and nitric oxide, NO. There are three characteristics of these acidic oxides:

1. Oxides such as SO_2 and N_2O_5, in which the nonmetal exhibits one of its common oxidation states, are **acid anhydrides** and react with water to form acids with no change in oxidation state. The product is an oxyacid. For example:

$$SO_2(g) + H_2O(l) \longrightarrow H_2SO_3(aq)$$

$$N_2O_5(s) + H_2O(l) \longrightarrow 2HNO_3(aq)$$

2. Those oxides such as NO_2 and ClO_2, in which the nonmetal does not exhibit one of its common oxidation states, also react with water. In these reactions, the nonmetal is both oxidized and reduced. For example:

$$3NO_2(g) + H_2O(l) \longrightarrow 2HNO_3(aq) + NO(g)$$
$$+4 +5 +2$$

Reactions in which the same element is both oxidized and reduced are called **disproportionation reactions**.

3. The acid strength increases as the electronegativity of the central atom increases. To learn more, see the discussion in the chapter on acid-base chemistry.

The binary hydrogen compounds of the nonmetals also exhibit an acidic behavior in water, although only HCl, HBr, and HI are strong acids. The acid strength of the nonmetal hydrogen compounds increases from left to right across a period and down a group. For example, ammonia, NH_3, is a weaker acid than is water, H_2O, which is weaker than is hydrogen fluoride, HF. Water, H_2O, is also a weaker acid than is hydrogen sulfide, H_2S, which is weaker than is hydrogen selenide, H_2Se. Weaker acidic character implies greater basic character.

Structures of the Nonmetals

The structures of the nonmetals differ dramatically from those of metals. Metals crystallize in closely packed arrays that do not contain molecules or covalent bonds. Nonmetal structures contain covalent bonds, and many nonmetals consist of individual molecules. The electrons in nonmetals are localized in covalent bonds, whereas in a metal, there is delocalization of the electrons throughout the solid.

The noble gases are all monatomic, whereas the other nonmetal gases—hydrogen, nitrogen, oxygen, fluorine, and chlorine—normally exist as the diatomic molecules H_2, N_2, O_2, F_2, and Cl_2. The other halogens are also diatomic; Br_2 is a liquid and I_2 exists as a solid under normal conditions. The changes in state as one moves down the halogen family offer excellent examples of the increasing strength of intermolecular London forces with increasing molecular mass and increasing polarizability.

Oxygen has two allotropes: O_2, dioxygen, and O_3, ozone. Phosphorus has three common allotropes, commonly referred to by their colors: white, red, and black. Sulfur has several allotropes. There are also many carbon allotropes. Most people know of diamond, graphite, and charcoal, but fewer people know of the recent discovery of fullerenes, carbon nanotubes, and graphene.

Descriptions of the physical properties of three nonmetals that are characteristic of molecular solids follow.

Carbon

Carbon occurs in the uncombined (elemental) state in many forms, such as diamond, graphite, charcoal, coke, carbon black, graphene, and fullerene.

Diamond, shown in Figure 18.20, is a very hard crystalline material that is colorless and transparent when pure. Each atom forms four single bonds to four other atoms at the corners of a tetrahedron (sp^3 hybridization); this makes the diamond a giant molecule. Carbon-carbon single bonds are very strong, and, because they extend throughout the crystal to form a three-dimensional network, the crystals are very hard and have high melting points (~4400 °C).

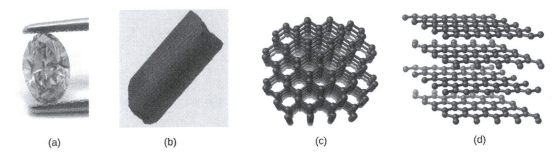

(a)   (b)   (c)   (d)

Figure 18.20   (a) Diamond and (b) graphite are two forms of carbon. (c) In the crystal structure of diamond, the covalent bonds form three-dimensional tetrahedrons. (d) In the crystal structure of graphite, each planar layer is composed of six-membered rings. (credit a: modification of work by "Fancy Diamonds"/Flickr; credit b: modification of work from http://images-of-elements.com/carbon.php)

Graphite, also shown in Figure 18.20, is a soft, slippery, grayish-black solid that conducts electricity. These properties relate to its structure, which consists of layers of carbon atoms, with each atom surrounded by three other carbon atoms in a trigonal planar arrangement. Each carbon atom in graphite forms three σ bonds, one to each of its nearest neighbors, by means of sp^2-hybrid orbitals. The unhybridized p orbital on each carbon atom will overlap unhybridized orbitals on adjacent carbon atoms in the same layer to form π bonds. Many resonance forms are necessary to describe the electronic structure of a graphite layer; Figure 18.21 illustrates two of these forms.

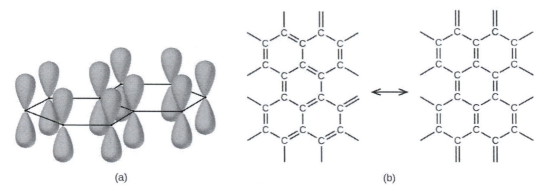

(a)   (b)

Figure 18.21   (a) Carbon atoms in graphite have unhybridized *p* orbitals. Each *p* orbital is perpendicular to the plane of carbon atoms. (b) These are two of the many resonance forms of graphite necessary to describe its electronic structure as a resonance hybrid.

Atoms within a graphite layer are bonded together tightly by the σ and π bonds; however, the forces between layers are weak. London dispersion forces hold the layers together. To learn more, see the discussion of these weak forces in the chapter on liquids and solids. The weak forces between layers give graphite the soft, flaky character that makes it useful as the so-called "lead" in pencils and the slippery character that makes it useful as a lubricant. The loosely held electrons in the resonating π bonds can move throughout the solid and are responsible for the electrical conductivity of graphite.

Other forms of elemental carbon include carbon black, charcoal, and coke. Carbon black is an amorphous form of carbon prepared by the incomplete combustion of natural gas, CH_4. It is possible to produce charcoal and coke by heating wood and coal, respectively, at high temperatures in the absence of air.

Recently, new forms of elemental carbon molecules have been identified in the soot generated by a smoky flame and in the vapor produced when graphite is heated to very high temperatures in a vacuum or in helium. One of these new forms, first isolated by Professor Richard Smalley and coworkers at Rice University, consists of icosahedral (soccer-ball-shaped) molecules that contain 60 carbon atoms, C_{60}. This is buckminsterfullerene (often called bucky balls) after the architect Buckminster Fuller, who designed domed structures, which have a similar appearance (Figure 18.22).

Figure 18.22   The molecular structure of C_{60}, buckminsterfullerene, is icosahedral.

Chemistry in Everyday Life

Nanotubes and Graphene

Graphene and carbon nanotubes are two recently discovered allotropes of carbon. Both of the forms bear some relationship to graphite. Graphene is a single layer of graphite (one atom thick), as illustrated in Figure 18.23, whereas carbon nanotubes roll the layer into a small tube, as illustrated in Figure 18.23.

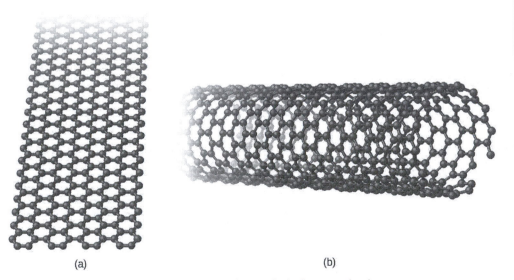

(a)   (b)

Figure 18.23   (a) Graphene and (b) carbon nanotubes are both allotropes of carbon.

Graphene is a very strong, lightweight, and efficient conductor of heat and electricity discovered in 2003. As in graphite, the carbon atoms form a layer of six-membered rings with sp^2-hybridized carbon atoms at the corners. Resonance stabilizes the system and leads to its conductivity. Unlike graphite, there is no stacking of the layers to give a three-dimensional structure. Andre Geim and Kostya Novoselov at the University of Manchester won the 2010 Nobel Prize in Physics for their pioneering work characterizing graphene.

The simplest procedure for preparing graphene is to use a piece of adhesive tape to remove a single layer of graphene from the surface of a piece of graphite. This method works because there are only weak London dispersion forces between the layers in graphite. Alternative methods are to deposit a single layer of carbon atoms on the surface of some other material (ruthenium, iridium, or copper) or to synthesize it at the surface of silicon carbide via the sublimation of silicon.

There currently are no commercial applications of graphene. However, its unusual properties, such as high electron mobility and thermal conductivity, should make it suitable for the manufacture of many advanced electronic devices and for thermal management applications.

Carbon nanotubes are carbon allotropes, which have a cylindrical structure. Like graphite and graphene, nanotubes consist of rings of sp^2-hybridized carbon atoms. Unlike graphite and graphene, which occur in layers, the layers wrap into a tube and bond together to produce a stable structure. The walls of the tube may be one atom or multiple atoms thick.

Carbon nanotubes are extremely strong materials that are harder than diamond. Depending upon the shape of the nanotube, it may be a conductor or semiconductor. For some applications, the conducting form is preferable, whereas other applications utilize the semiconducting form.

The basis for the synthesis of carbon nanotubes is the generation of carbon atoms in a vacuum. It is possible to produce carbon atoms by an electrical discharge through graphite, vaporization of graphite with a laser, and the decomposition of a carbon compound.

The strength of carbon nanotubes will eventually lead to some of their most exciting applications, as a thread produced from several nanotubes will support enormous weight. However, the current applications only employ bulk nanotubes. The addition of nanotubes to polymers improves the mechanical, thermal, and electrical properties of the bulk material. There are currently nanotubes in some bicycle parts, skis, baseball bats, fishing rods, and surfboards.

Phosphorus

The name *phosphorus* comes from the Greek words meaning *light bringing*. When phosphorus was first isolated, scientists noted that it glowed in the dark and burned when exposed to air. Phosphorus is the only member of its group that does not occur in the uncombined state in nature; it exists in many allotropic forms. We will consider two of those forms: white phosphorus and red phosphorus.

White phosphorus is a white, waxy solid that melts at 44.2 °C and boils at 280 °C. It is insoluble in water (in which it is stored—see Figure 18.24), is very soluble in carbon disulfide, and bursts into flame in air. As a solid, as a liquid, as a gas, and in solution, white phosphorus exists as P_4 molecules with four phosphorus atoms at the corners of a regular tetrahedron, as illustrated in Figure 18.24. Each phosphorus atom covalently bonds to the other three atoms in the molecule by single covalent bonds. White phosphorus is the most reactive allotrope and is very toxic.

(a)   (b)   (c)   (d)

Figure 18.24   (a) Because white phosphorus bursts into flame in air, it is stored in water. (b) The structure of white phosphorus consists of P_4 molecules arranged in a tetrahedron. (c) Red phosphorus is much less reactive than is white phosphorus. (d) The structure of red phosphorus consists of networks of P_4 tetrahedra joined by P-P single bonds. (credit a: modification of work from http://images-of-elements.com/phosphorus.php)

Heating white phosphorus to 270–300 °C in the absence of air yields red phosphorus. Red phosphorus (shown in Figure 18.24 and Figure 18.24) is denser, has a higher melting point (~600 °C), is much less reactive, is essentially nontoxic, and is easier and safer to handle than is white phosphorus. Its structure is highly polymeric and appears to contain three-dimensional networks of P_4 tetrahedra joined by P-P single bonds. Red phosphorus is insoluble in solvents that dissolve white phosphorus. When red phosphorus is heated, P_4 molecules sublime from the solid.

Sulfur

The allotropy of sulfur is far greater and more complex than that of any other element. Sulfur is the brimstone referred to in the Bible and other places, and references to sulfur occur throughout recorded history—right up to the relatively recent discovery that it is a component of the atmospheres of Venus and of Io, a moon of Jupiter. The most common and most stable allotrope of sulfur is yellow, rhombic sulfur, so named because of the shape or its crystals. Rhombic sulfur is the form to which all other allotropes revert at room temperature. Crystals of rhombic sulfur melt at 113 °C. Cooling this liquid gives long needles of monoclinic sulfur. This form is stable from 96 °C to the melting point, 119 °C. At room temperature, it gradually reverts to the rhombic form.

Both rhombic sulfur and monoclinic sulfur contain S_8 molecules in which atoms form eight-membered, puckered rings that resemble crowns, as illustrated in Figure 18.25. Each sulfur atom is bonded to each of its two neighbors in the ring by covalent S-S single bonds.

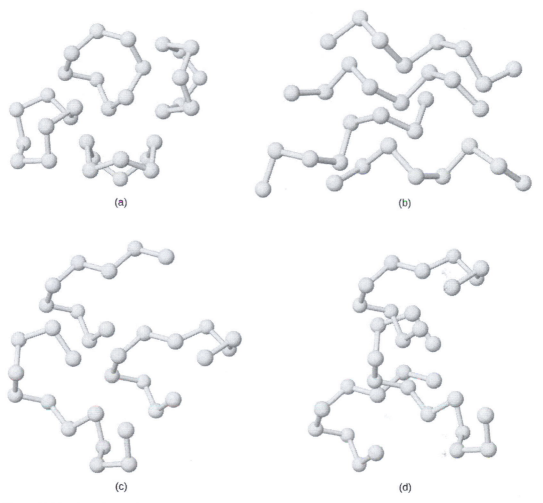

Figure 18.25   These four sulfur allotropes show eight-membered, puckered rings. Each sulfur atom bonds to each of its two neighbors in the ring by covalent S-S single bonds. Here are (a) individual S_8 rings, (b) S_8 chains formed when the rings open, (c) longer chains formed by adding sulfur atoms to S_8 chains, and (d) part of the very long sulfur chains formed at higher temperatures.

When rhombic sulfur melts, the straw-colored liquid is quite mobile; its viscosity is low because S_8 molecules are essentially spherical and offer relatively little resistance as they move past each other. As the temperature rises, S-S bonds in the rings break, and polymeric chains of sulfur atoms result. These chains combine end to end, forming still longer chains that tangle with one another. The liquid gradually darkens in color and becomes so viscous that finally (at about 230 °C) it does not pour easily. The dangling atoms at the ends of the chains of sulfur atoms are responsible for the dark red color because their electronic structure differs from those of sulfur atoms that have bonds to two adjacent sulfur atoms. This causes them to absorb light differently and results in a different visible color. Cooling the liquid rapidly produces a rubberlike amorphous mass, called plastic sulfur.

Sulfur boils at 445 °C and forms a vapor consisting of S_2, S_6, and S_8 molecules; at about 1000 °C, the vapor density corresponds to the formula S_2, which is a paramagnetic molecule like O_2 with a similar electronic structure and a weak sulfur-sulfur double bond.

As seen in this discussion, an important feature of the structural behavior of the nonmetals is that the elements usually occur with eight electrons in their valence shells. If necessary, the elements form enough covalent bonds to supplement the electrons already present to possess an octet. For example, members of group 15 have five valence elements and require only three additional electrons to fill their valence shells. These elements form three covalent bonds in their free state: triple bonds in the N_2 molecule or single bonds to three different atoms in arsenic and phosphorus. The elements of group 16 require only two additional electrons. Oxygen forms a double bond in the O_2 molecule, and sulfur, selenium, and tellurium form two single bonds in various rings and chains. The halogens form diatomic molecules in which each atom is involved in only one bond. This provides the electron required necessary to complete the octet on the halogen atom. The noble gases do not form covalent bonds to other noble gas atoms because they already have a filled outer shell.

18.5 Occurrence, Preparation, and Compounds of Hydrogen

By the end of this section, you will be able to:

- Describe the properties, preparation, and compounds of hydrogen

Hydrogen is the most abundant element in the universe. The sun and other stars are composed largely of hydrogen. Astronomers estimate that 90% of the atoms in the universe are hydrogen atoms. Hydrogen is a component of more compounds than any other element. Water is the most abundant compound of hydrogen found on earth. Hydrogen is an important part of petroleum, many minerals, cellulose and starch, sugar, fats, oils, alcohols, acids, and thousands of other substances.

At ordinary temperatures, hydrogen is a colorless, odorless, tasteless, and nonpoisonous gas consisting of the diatomic molecule H_2. Hydrogen is composed of three isotopes, and unlike other elements, these isotopes have different names and chemical symbols: protium, ^1H, deuterium, ^2H (or "D"), and tritium ^3H (or "T"). In a naturally occurring sample of hydrogen, there is one atom of deuterium for every 7000 H atoms and one atom of radioactive tritium for every 10^{18} H atoms. The chemical properties of the different isotopes are very similar because they have identical electron structures, but they differ in some physical properties because of their differing atomic masses. Elemental deuterium and tritium have lower vapor pressure than ordinary hydrogen. Consequently, when liquid hydrogen evaporates, the heavier isotopes are concentrated in the last portions to evaporate. Electrolysis of heavy water, D_2O, yields deuterium. Most tritium originates from nuclear reactions.

Preparation of Hydrogen

Elemental hydrogen must be prepared from compounds by breaking chemical bonds. The most common methods of preparing hydrogen follow.

From Steam and Carbon or Hydrocarbons

Water is the cheapest and most abundant source of hydrogen. Passing steam over coke (an impure form of elemental carbon) at 1000 °C produces a mixture of carbon monoxide and hydrogen known as water gas:

$$C(s) + H_2O(g) \xrightarrow{1000\,°C} \underset{\text{water gas}}{CO(g) + H_2(g)}$$

Water gas is as an industrial fuel. It is possible to produce additional hydrogen by mixing the water gas with steam in the presence of a catalyst to convert the CO to CO_2. This reaction is the water gas shift reaction.

It is also possible to prepare a mixture of hydrogen and carbon monoxide by passing hydrocarbons from natural gas or petroleum and steam over a nickel-based catalyst. Propane is an example of a hydrocarbon reactant:

$$C_3H_8(g) + 3H_2O(g) \xrightarrow[\text{catalyst}]{900\,°C} 3CO(g) + 7H_2(g)$$

Electrolysis

Hydrogen forms when direct current electricity passes through water containing an electrolyte such as H_2SO_4, as illustrated in Figure 18.26. Bubbles of hydrogen form at the cathode, and oxygen evolves at the anode. The net reaction is:

$$2H_2O(l) + \text{electrical energy} \longrightarrow 2H_2(g) + O_2(g)$$

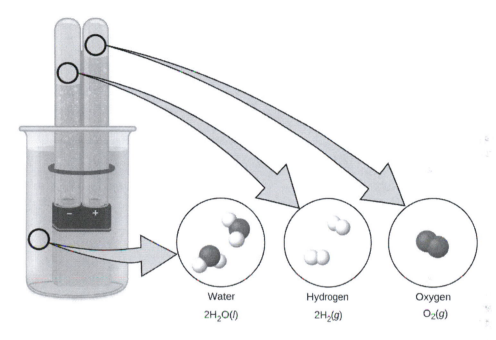

Water   Hydrogen   Oxygen
$2H_2O(l)$ $2H_2(g)$ $O_2(g)$

Figure 18.26   The electrolysis of water produces hydrogen and oxygen. Because there are twice as many hydrogen atoms as oxygen atoms and both elements are diatomic, there is twice the volume of hydrogen produced at the cathode as there is oxygen produced at the anode.

Reaction of Metals with Acids

This is the most convenient laboratory method of producing hydrogen. Metals with lower reduction potentials reduce the hydrogen ion in dilute acids to produce hydrogen gas and metal salts. For example, as shown in Figure 18.27, iron in dilute hydrochloric acid produces hydrogen gas and iron(II) chloride:

$$Fe(s) + 2H_3O^+(aq) + 2Cl^-(aq) \longrightarrow Fe^{2+}(aq) + 2Cl^-(aq) + H_2(g) + 2H_2O(l)$$

Figure 18.27   The reaction of iron with an acid produces hydrogen. Here, iron reacts with hydrochloric acid. (credit: Mark Ott)

Reaction of Ionic Metal Hydrides with Water

It is possible to produce hydrogen from the reaction of hydrides of the active metals, which contain the very strongly basic H^- anion, with water:

$$CaH_2(s) + 2H_2O(l) \longrightarrow Ca^{2+}(aq) + 2OH^-(aq) + 2H_2(g)$$

Metal hydrides are expensive but convenient sources of hydrogen, especially where space and weight are important factors. They are important in the inflation of life jackets, life rafts, and military balloons.

Reactions

Under normal conditions, hydrogen is relatively inactive chemically, but when heated, it enters into many chemical reactions.

Two thirds of the world's hydrogen production is devoted to the manufacture of ammonia, which is a fertilizer and used in the manufacture of nitric acid. Large quantities of hydrogen are also important in the process of **hydrogenation**, discussed in the chapter on organic chemistry.

It is possible to use hydrogen as a nonpolluting fuel. The reaction of hydrogen with oxygen is a very exothermic reaction, releasing 286 kJ of energy per mole of water formed. Hydrogen burns without explosion under controlled conditions. The oxygen-hydrogen torch, because of the high heat of combustion of hydrogen, can achieve temperatures up to 2800 °C. The hot flame of this torch is useful in cutting thick sheets of many metals. Liquid hydrogen is also an important rocket fuel (Figure 18.28).

Figure 18.28 Before the fleet's retirement in 2011, liquid hydrogen and liquid oxygen were used in the three main engines of a space shuttle. Two compartments in the large tank held these liquids until the shuttle was launched. (credit: "reynermedia"/Flickr)

An uncombined hydrogen atom consists of a nucleus and one valence electron in the $1s$ orbital. The $n = 1$ valence shell has a capacity for two electrons, and hydrogen can rightfully occupy two locations in the periodic table. It is possible to consider hydrogen a group 1 element because hydrogen can lose an electron to form the cation, H^+. It is also possible to consider hydrogen to be a group 17 element because it needs only one electron to fill its valence orbital to form a hydride ion, H^-, or it can share an electron to form a single, covalent bond. In reality, hydrogen is a unique element that almost deserves its own location in the periodic table.

Reactions with Elements

When heated, hydrogen reacts with the metals of group 1 and with Ca, Sr, and Ba (the more active metals in group 2). The compounds formed are crystalline, ionic hydrides that contain the hydride anion, H^-, a strong reducing agent and a strong base, which reacts vigorously with water and other acids to form hydrogen gas.

The reactions of hydrogen with nonmetals generally produce *acidic* hydrogen compounds with hydrogen in the 1+ oxidation state. The reactions become more exothermic and vigorous as the electronegativity of the nonmetal increases. Hydrogen reacts with nitrogen and sulfur only when heated, but it reacts explosively with fluorine (forming HF) and, under some conditions, with chlorine (forming HCl). A mixture of hydrogen and oxygen explodes if ignited. Because of the explosive nature of the reaction, it is necessary to exercise caution when handling hydrogen (or any other combustible gas) to avoid the formation of an explosive mixture in a confined space. Although most hydrides of the nonmetals are acidic, ammonia and phosphine (PH_3) are very, very weak acids and generally function as bases. There is a summary of these reactions of hydrogen with the elements in Table 18.1.

Chemical Reactions of Hydrogen with Other Elements

General Equation	Comments
MH or $MH_2 \longrightarrow MOH$ or $M(OH)_2 + H_2$	ionic hydrides with group 1 and Ca, Sr, and Ba
$H_2 + C \longrightarrow$ (no reaction)	
$3H_2 + N_2 \longrightarrow 2NH_3$	requires high pressure and temperature; low yield
$2H_2 + O_2 \longrightarrow 2H_2O$	exothermic and potentially explosive
$H_2 + S \longrightarrow H_2S$	requires heating; low yield

Table 18.1

Chemical Reactions of Hydrogen with Other Elements

General Equation	Comments
$H_2 + X_2 \longrightarrow 2HX$	X = F, Cl, Br, and I; explosive with F_2; low yield with I_2

Table 18.1

Reaction with Compounds

Hydrogen reduces the heated oxides of many metals, with the formation of the metal and water vapor. For example, passing hydrogen over heated CuO forms copper and water.

Hydrogen may also reduce the metal ions in some metal oxides to lower oxidation states:

$$H_2(g) \; + \; MnO_2(s) \xrightarrow{\Delta} MnO(s) + H_2O(g)$$

Hydrogen Compounds

Other than the noble gases, each of the nonmetals forms compounds with hydrogen. For brevity, we will discuss only a few hydrogen compounds of the nonmetals here.

Nitrogen Hydrogen Compounds

Ammonia, NH_3, forms naturally when any nitrogen-containing organic material decomposes in the absence of air. The laboratory preparation of ammonia is by the reaction of an ammonium salt with a strong base such as sodium hydroxide. The acid-base reaction with the weakly acidic ammonium ion gives ammonia, illustrated in Figure 18.29. Ammonia also forms when ionic nitrides react with water. The nitride ion is a much stronger base than the hydroxide ion:

$$Mg_3N_2(s) \; + \; 6H_2O(l) \longrightarrow 3Mg(OH)_2(s) + 2NH_3(g)$$

The commercial production of ammonia is by the direct combination of the elements in the **Haber process**:

$$N_2(g) + 3H_2(g) \overset{\text{catalyst}}{\rightleftharpoons} 2NH_3(g) \qquad\qquad \Delta H° = -92 \text{ kJ}$$

Figure 18.29 The structure of ammonia is shown with a central nitrogen atom and three hydrogen atoms.

Ammonia is a colorless gas with a sharp, pungent odor. Smelling salts utilize this powerful odor. Gaseous ammonia readily liquefies to give a colorless liquid that boils at −33 °C. Due to intermolecular hydrogen bonding, the enthalpy of vaporization of liquid ammonia is higher than that of any other liquid except water, so ammonia is useful as a refrigerant. Ammonia is quite soluble in water (658 L at STP dissolves in 1 L H_2O).

The chemical properties of ammonia are as follows:

1. Ammonia acts as a Brønsted base, as discussed in the chapter on acid-base chemistry. The ammonium ion is similar in size to the potassium ion; compounds of the two ions exhibit many similarities in their structures and solubilities.

2. Ammonia can display acidic behavior, although it is a much weaker acid than water. Like other acids, ammonia reacts with metals, although it is so weak that high temperatures are necessary. Hydrogen and

(depending on the stoichiometry) amides (salts of $NH_2{}^-$), imides (salts of NH^{2-}), or nitrides (salts of N^{3-}) form.

3. The nitrogen atom in ammonia has its lowest possible oxidation state (3−) and thus is not susceptible to reduction. However, it can be oxidized. Ammonia burns in air, giving NO and water. Hot ammonia and the ammonium ion are active reducing agents. Of particular interest are the oxidations of ammonium ion by nitrite ion, $NO_2{}^-$, to yield pure nitrogen and by nitrate ion to yield nitrous oxide, N_2O.

4. There are a number of compounds that we can consider derivatives of ammonia through the replacement of one or more hydrogen atoms with some other atom or group of atoms. Inorganic derivations include chloramine, NH_2Cl, and hydrazine, N_2H_4:

| ammonia | chloramine | hydrazine |
| (a) | (b) | (c) |

Chloramine, NH_2Cl, results from the reaction of sodium hypochlorite, NaOCl, with ammonia in basic solution. In the presence of a large excess of ammonia at low temperature, the chloramine reacts further to produce hydrazine, N_2H_4:

$$NH_3(aq) + OCl^-(aq) \longrightarrow NH_2Cl(aq) + OH^-(aq)$$
$$NH_2Cl(aq) + NH_3(aq) + OH^-(aq) \longrightarrow N_2H_4(aq) + Cl^-(aq) + H_2O(l)$$

Anhydrous hydrazine is relatively stable in spite of its positive free energy of formation:

$$N_2(g) + 2H_2(g) \longrightarrow N_2H_4(l) \qquad \Delta G_f^\circ = 149.2 \text{ kJ mol}^{-1}$$

Hydrazine is a fuming, colorless liquid that has some physical properties remarkably similar to those of H_2O (it melts at 2 °C, boils at 113.5 °C, and has a density at 25 °C of 1.00 g/mL). It burns rapidly and completely in air with substantial evolution of heat:

$$N_2H_4(l) + O_2(g) \longrightarrow N_2(g) + 2H_2O(l) \qquad \Delta H^\circ = -621.5 \text{ kJ mol}^{-1}$$

Like ammonia, hydrazine is both a Brønsted base and a Lewis base, although it is weaker than ammonia. It reacts with strong acids and forms two series of salts that contain the $N_2H_5{}^+$ and $N_2H_6{}^{2+}$ ions, respectively. Some rockets use hydrazine as a fuel.

Phosphorus Hydrogen Compounds

The most important hydride of phosphorus is phosphine, PH_3, a gaseous analog of ammonia in terms of both formula and structure. Unlike ammonia, it is not possible to form phosphine by direct union of the elements. There are two methods for the preparation of phosphine. One method is by the action of an acid on an ionic phosphide. The other method is the disproportionation of white phosphorus with hot concentrated base to produce phosphine and the hydrogen phosphite ion:

$$AlP(s) + 3H_3O^+(aq) \longrightarrow PH_3(g) + Al^{3+}(aq) + 3H_2O(l)$$
$$P_4(s) + 4OH^-(aq) + 2H_2O(l) \longrightarrow 2HPO_3{}^{2-}(aq) + 2PH_3(g)$$

Phosphine is a colorless, very poisonous gas, which has an odor like that of decaying fish. Heat easily decomposes phosphine $(4PH_3 \longrightarrow P_4 + 6H_2)$, and the compound burns in air. The major uses of phosphine are as a fumigant for grains and in semiconductor processing. Like ammonia, gaseous phosphine unites with gaseous hydrogen halides, forming phosphonium compounds like PH_4Cl and PH_4I. Phosphine is a much weaker base than ammonia; therefore, these compounds decompose in water, and the insoluble PH_3 escapes from solution.

Sulfur Hydrogen Compounds

Hydrogen sulfide, H_2S, is a colorless gas that is responsible for the offensive odor of rotten eggs and of many hot springs. Hydrogen sulfide is as toxic as hydrogen cyanide; therefore, it is necessary to exercise great care in handling it. Hydrogen sulfide is particularly deceptive because it paralyzes the olfactory nerves; after a short exposure, one does not smell it.

The production of hydrogen sulfide by the direct reaction of the elements ($H_2 + S$) is unsatisfactory because the yield is low. A more effective preparation method is the reaction of a metal sulfide with a dilute acid. For example:

$$FeS(s) + 2H_3O^+(aq) \longrightarrow Fe^{2+}(aq) + H_2S(g) + 2H_2O(l)$$

It is easy to oxidize the sulfur in metal sulfides and in hydrogen sulfide, making metal sulfides and H_2S good reducing agents. In acidic solutions, hydrogen sulfide reduces Fe^{3+} to Fe^{2+}, MnO_4^- to Mn^{2+}, $Cr_2O_7^{2-}$ to Cr^{3+}, and HNO_3 to NO_2. The sulfur in H_2S usually oxidizes to elemental sulfur, unless a large excess of the oxidizing agent is present. In which case, the sulfide may oxidize to SO_3^{2-} or SO_4^{2-} (or to SO_2 or SO_3 in the absence of water):

$$2H_2S(g) + O_2(g) \longrightarrow 2S(s) + 2H_2O(l)$$

This oxidation process leads to the removal of the hydrogen sulfide found in many sources of natural gas. The deposits of sulfur in volcanic regions may be the result of the oxidation of H_2S present in volcanic gases.

Hydrogen sulfide is a weak diprotic acid that dissolves in water to form hydrosulfuric acid. The acid ionizes in two stages, yielding hydrogen sulfide ions, HS^-, in the first stage and sulfide ions, S^{2-}, in the second. Since hydrogen sulfide is a weak acid, aqueous solutions of soluble sulfides and hydrogen sulfides are basic:

$$S^{2-}(aq) + H_2O(l) \rightleftharpoons HS^-(aq) + OH^-(aq)$$
$$HS^-(aq) + H_2O(l) \rightleftharpoons H_2S(g) + OH^-(aq)$$

Halogen Hydrogen Compounds

Binary compounds containing only hydrogen and a halogen are **hydrogen halides**. At room temperature, the pure hydrogen halides HF, HCl, HBr, and HI are gases.

In general, it is possible to prepare the halides by the general techniques used to prepare other acids. Fluorine, chlorine, and bromine react directly with hydrogen to form the respective hydrogen halide. This is a commercially important reaction for preparing hydrogen chloride and hydrogen bromide.

The acid-base reaction between a nonvolatile strong acid and a metal halide will yield a hydrogen halide. The escape of the gaseous hydrogen halide drives the reaction to completion. For example, the usual method of preparing hydrogen fluoride is by heating a mixture of calcium fluoride, CaF_2, and concentrated sulfuric acid:

$$CaF_2(s) + H_2SO_4(aq) \longrightarrow CaSO_4(s) + 2HF(g)$$

Gaseous hydrogen fluoride is also a by-product in the preparation of phosphate fertilizers by the reaction of fluoroapatite, $Ca_5(PO_4)_3F$, with sulfuric acid. The reaction of concentrated sulfuric acid with a chloride salt produces hydrogen chloride both commercially and in the laboratory.

In most cases, sodium chloride is the chloride of choice because it is the least expensive chloride. Hydrogen bromide and hydrogen iodide cannot be prepared using sulfuric acid because this acid is an oxidizing agent capable of oxidizing both bromide and iodide. However, it is possible to prepare both hydrogen bromide and hydrogen iodide using an acid such as phosphoric acid because it is a weaker oxidizing agent. For example:

$$H_3PO_4(l) + Br^-(aq) \longrightarrow HBr(g) + H_2PO_4^-(aq)$$

All of the hydrogen halides are very soluble in water, forming hydrohalic acids. With the exception of hydrogen fluoride, which has a strong hydrogen-fluoride bond, they are strong acids. Reactions of hydrohalic acids with metals, metal hydroxides, oxides, or carbonates produce salts of the halides. Most chloride salts are soluble in water. AgCl, $PbCl_2$, and Hg_2Cl_2 are the commonly encountered exceptions.

The halide ions give the substances the properties associated with $X^-(aq)$. The heavier halide ions (Cl^-, Br^-, and I^-) can act as reducing agents, and the lighter halogens or other oxidizing agents will oxidize them:

$$Cl_2(aq) + 2e^- \longrightarrow 2Cl^-(aq) \qquad E° = 1.36 \text{ V}$$
$$Br_2(aq) + 2e^- \longrightarrow 2Br^-(aq) \qquad E° = 1.09 \text{ V}$$
$$I_2(aq) + 2e^- \longrightarrow 2I^-(aq) \qquad E° = 0.54 \text{ V}$$

For example, bromine oxidizes iodine:

$$Br_2(aq) + 2HI(aq) \longrightarrow 2HBr(aq) + I_2(aq) \qquad E° = 0.55 \text{ V}$$

Hydrofluoric acid is unique in its reactions with sand (silicon dioxide) and with glass, which is a mixture of silicates:

$$SiO_2(s) + 4HF(aq) \longrightarrow SiF_4(g) + 2H_2O(l)$$
$$CaSiO_3(s) + 6HF(aq) \longrightarrow CaF_2(s) + SiF_4(g) + 3H_2O(l)$$

The volatile silicon tetrafluoride escapes from these reactions. Because hydrogen fluoride attacks glass, it can frost or etch glass and is used to etch markings on thermometers, burets, and other glassware.

The largest use for hydrogen fluoride is in production of hydrochlorofluorocarbons for refrigerants, in plastics, and in propellants. The second largest use is in the manufacture of cryolite, Na_3AlF_6, which is important in the production of aluminum. The acid is also important in the production of other inorganic fluorides (such as BF_3), which serve as catalysts in the industrial synthesis of certain organic compounds.

Hydrochloric acid is relatively inexpensive. It is an important and versatile acid in industry and is important for the manufacture of metal chlorides, dyes, glue, glucose, and various other chemicals. A considerable amount is also important for the activation of oil wells and as pickle liquor—an acid used to remove oxide coating from iron or steel that is to be galvanized, tinned, or enameled. The amounts of hydrobromic acid and hydroiodic acid used commercially are insignificant by comparison.

18.6 Occurrence, Preparation, and Properties of Carbonates

By the end of this section, you will be able to:

- Describe the preparation, properties, and uses of some representative metal carbonates

The chemistry of carbon is extensive; however, most of this chemistry is not relevant to this chapter. The other aspects of the chemistry of carbon will appear in the chapter covering organic chemistry. In this chapter, we will focus on the carbonate ion and related substances. The metals of groups 1 and 2, as well as zinc, cadmium, mercury, and lead(II), form ionic **carbonates**—compounds that contain the carbonate anions, $CO_3{}^{2-}$. The metals of group 1, magnesium, calcium, strontium, and barium also form **hydrogen carbonates**—compounds that contain the hydrogen carbonate anion, $HCO_3{}^-$, also known as the **bicarbonate anion**.

With the exception of magnesium carbonate, it is possible to prepare carbonates of the metals of groups 1 and 2 by the reaction of carbon dioxide with the respective oxide or hydroxide. Examples of such reactions include:

$$Na_2O(s) + CO_2(g) \longrightarrow Na_2CO_3(s)$$
$$Ca(OH)_2(s) + CO_2(g) \longrightarrow CaCO_3(s) + H_2O(l)$$

The carbonates of the alkaline earth metals of group 12 and lead(II) are not soluble. These carbonates precipitate upon mixing a solution of soluble alkali metal carbonate with a solution of soluble salts of these metals. Examples of net ionic equations for the reactions are:

$$Ca^{2+}(aq) + CO_3{}^{2-}(aq) \longrightarrow CaCO_3(s)$$
$$Pb^{2+}(aq) + CO_3{}^{2-}(aq) \longrightarrow PbCO_3(s)$$

Pearls and the shells of most mollusks are calcium carbonate. Tin(II) or one of the trivalent or tetravalent ions such as Al^{3+} or Sn^{4+} behave differently in this reaction as carbon dioxide and the corresponding oxide form instead of the carbonate.

Alkali metal hydrogen carbonates such as $NaHCO_3$ and $CsHCO_3$ form by saturating a solution of the hydroxides with carbon dioxide. The net ionic reaction involves hydroxide ion and carbon dioxide:

$$OH^-(aq) + CO_2(aq) \longrightarrow HCO_3^{\ -}(aq)$$

It is possible to isolate the solids by evaporation of the water from the solution.

Although they are insoluble in pure water, alkaline earth carbonates dissolve readily in water containing carbon dioxide because hydrogen carbonate salts form. For example, caves and sinkholes form in limestone when $CaCO_3$ dissolves in water containing dissolved carbon dioxide:

$$CaCO_3(s) + CO_2(aq) + H_2O(l) \longrightarrow Ca^{2+}(aq) + 2HCO_3^{\ -}(aq)$$

Hydrogen carbonates of the alkaline earth metals remain stable only in solution; evaporation of the solution produces the carbonate. Stalactites and stalagmites, like those shown in Figure 18.30, form in caves when drops of water containing dissolved calcium hydrogen carbonate evaporate to leave a deposit of calcium carbonate.

(a)   (b)

Figure 18.30   (a) Stalactites and (b) stalagmites are cave formations of calcium carbonate. (credit a: modification of work by Arvind Govindaraj; credit b: modification of work by the National Park Service.)

The two carbonates used commercially in the largest quantities are sodium carbonate and calcium carbonate. In the United States, sodium carbonate is extracted from the mineral trona, $Na_3(CO_3)(HCO_3)(H_2O)_2$. Following recrystallization to remove clay and other impurities, heating the recrystallized trona produces Na_2CO_3:

$$2Na_3(CO_3)(HCO_3)(H_2O)_2(s) \longrightarrow 3Na_2CO_3(s) + 5H_2O(l) + CO_2(g)$$

Carbonates are moderately strong bases. Aqueous solutions are basic because the carbonate ion accepts hydrogen ion from water in this reversible reaction:

$$CO_3^{\ 2-}(aq) + H_2O(l) \rightleftharpoons HCO_3^{\ -}(aq) + OH^-(aq)$$

Carbonates react with acids to form salts of the metal, gaseous carbon dioxide, and water. The reaction of calcium carbonate, the active ingredient of the antacid Tums, with hydrochloric acid (stomach acid), as shown in Figure 18.31, illustrates the reaction:

$$CaCO_3(s) + 2HCl(aq) \longrightarrow CaCl_2(aq) + CO_2(g) + H_2O(l)$$

Figure 18.31 The reaction of calcium carbonate with hydrochloric acid is shown. (credit: Mark Ott)

Other applications of carbonates include glass making—where carbonate ions serve as a source of oxide ions—and synthesis of oxides.

Hydrogen carbonates are amphoteric because they act as both weak acids and weak bases. Hydrogen carbonate ions act as acids and react with solutions of soluble hydroxides to form a carbonate and water:

$$KHCO_3(aq) + KOH(aq) \longrightarrow K_2CO_3(aq) + H_2O(l)$$

With acids, hydrogen carbonates form a salt, carbon dioxide, and water. Baking soda (bicarbonate of soda or sodium bicarbonate) is sodium hydrogen carbonate. Baking powder contains baking soda and a solid acid such as potassium hydrogen tartrate (cream of tartar), $KHC_4H_4O_6$. As long as the powder is dry, no reaction occurs; immediately after the addition of water, the acid reacts with the hydrogen carbonate ions to form carbon dioxide:

$$HC_4H_4O_6{}^-(aq) + HCO_3{}^-(aq) \longrightarrow C_4H_4O_6{}^{2-}(aq) + CO_2(g) + H_2O(l)$$

Dough will trap the carbon dioxide, causing it to expand during baking, producing the characteristic texture of baked goods.

18.7 Occurrence, Preparation, and Properties of Nitrogen

By the end of this section, you will be able to:

- Describe the properties, preparation, and uses of nitrogen

Most pure nitrogen comes from the fractional distillation of liquid air. The atmosphere consists of 78% nitrogen by volume. This means there are more than 20 million tons of nitrogen over every square mile of the earth's surface. Nitrogen is a component of proteins and of the genetic material (DNA/RNA) of all plants and animals.

Under ordinary conditions, nitrogen is a colorless, odorless, and tasteless gas. It boils at 77 K and freezes at 63 K. Liquid nitrogen is a useful coolant because it is inexpensive and has a low boiling point. Nitrogen is very unreactive because of the very strong triple bond between the nitrogen atoms. The only common reactions at room temperature occur with lithium to form Li_3N, with certain transition metal complexes, and with hydrogen or oxygen in nitrogen-fixing bacteria. The general lack of reactivity of nitrogen makes the remarkable ability of some bacteria to synthesize nitrogen compounds using atmospheric nitrogen gas as the source one of the most exciting chemical events on our planet. This process is one type of **nitrogen fixation**. In this case, nitrogen fixation is the process where organisms convert atmospheric nitrogen into biologically useful chemicals. Nitrogen fixation also occurs when lightning passes through air, causing molecular nitrogen to react with oxygen to form nitrogen oxides, which are then carried down to the soil.

Chemistry in Everyday Life

Nitrogen Fixation

All living organisms require nitrogen compounds for survival. Unfortunately, most of these organisms cannot absorb nitrogen from its most abundant source—the atmosphere. Atmospheric nitrogen consists of N_2 molecules, which are very unreactive due to the strong nitrogen-nitrogen triple bond. However, a few organisms can overcome this problem through a process known as nitrogen fixation, illustrated in Figure 18.32.

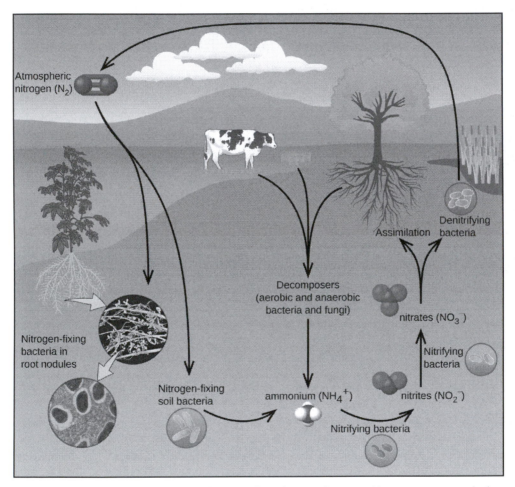

Figure 18.32 All living organisms require nitrogen. A few microorganisms are able to process atmospheric nitrogen using nitrogen fixation. (credit "roots": modification of work by the United States Department of Agriculture; credit "root nodules": modification of work by Louisa Howard)

Nitrogen fixation is the process where organisms convert atmospheric nitrogen into biologically useful chemicals. To date, the only known kind of biological organisms capable of nitrogen fixation are microorganisms. These organisms employ enzymes called nitrogenases, which contain iron and molybdenum.

> Many of these microorganisms live in a symbiotic relationship with plants, with the best-known example being the presence of rhizobia in the root nodules of legumes.

Large volumes of atmospheric nitrogen are necessary for making ammonia—the principal starting material used for preparation of large quantities of other nitrogen-containing compounds. Most other uses for elemental nitrogen depend on its inactivity. It is helpful when a chemical process requires an inert atmosphere. Canned foods and luncheon meats cannot oxidize in a pure nitrogen atmosphere, so they retain a better flavor and color, and spoil less rapidly, when sealed in nitrogen instead of air. This technology allows fresh produce to be available year-round, regardless of growing season.

There are compounds with nitrogen in all of its oxidation states from 3− to 5+. Much of the chemistry of nitrogen involves oxidation-reduction reactions. Some active metals (such as alkali metals and alkaline earth metals) can reduce nitrogen to form metal nitrides. In the remainder of this section, we will examine nitrogen-oxygen chemistry.

There are well-characterized nitrogen oxides in which nitrogen exhibits each of its positive oxidation numbers from 1+ to 5+. When ammonium nitrate is carefully heated, nitrous oxide (dinitrogen oxide) and water vapor form. Stronger heating generates nitrogen gas, oxygen gas, and water vapor. No one should ever attempt this reaction—it can be very explosive. In 1947, there was a major ammonium nitrate explosion in Texas City, Texas, and, in 2013, there was another major explosion in West, Texas. In the last 100 years, there were nearly 30 similar disasters worldwide, resulting in the loss of numerous lives. In this oxidation-reduction reaction, the nitrogen in the nitrate ion oxidizes the nitrogen in the ammonium ion. Nitrous oxide, shown in Figure 18.33, is a colorless gas possessing a mild, pleasing odor and a sweet taste. It finds application as an anesthetic for minor operations, especially in dentistry, under the name "laughing gas."

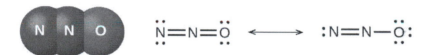

Figure 18.33   Nitrous oxide, N_2O, is an anesthetic that has these molecular (left) and resonance (right) structures.

Low yields of nitric oxide, NO, form when heating nitrogen and oxygen together. NO also forms when lightning passes through air during thunderstorms. Burning ammonia is the commercial method of preparing nitric oxide. In the laboratory, the reduction of nitric acid is the best method for preparing nitric oxide. When copper reacts with dilute nitric acid, nitric oxide is the principal reduction product:

$$3Cu(s) + 8HNO_3(aq) \longrightarrow 2NO(g) + 3Cu(NO_3)_2(aq) + 4H_2O(l)$$

Gaseous nitric oxide is the most thermally stable of the nitrogen oxides and is the simplest known thermally stable molecule with an unpaired electron. It is one of the air pollutants generated by internal combustion engines, resulting from the reaction of atmospheric nitrogen and oxygen during the combustion process.

At room temperature, nitric oxide is a colorless gas consisting of diatomic molecules. As is often the case with molecules that contain an unpaired electron, two molecules combine to form a dimer by pairing their unpaired electrons to form a bond. Liquid and solid NO both contain N_2O_2 dimers, like that shown in Figure 18.34. Most substances with unpaired electrons exhibit color by absorbing visible light; however, NO is colorless because the absorption of light is not in the visible region of the spectrum.

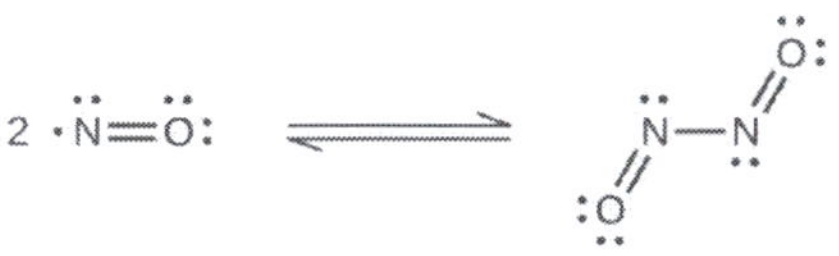

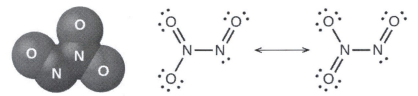

Figure 18.34   This shows the equilibrium between NO and N_2O_2. The molecule, N_2O_2, absorbs light.

Cooling a mixture of equal parts nitric oxide and nitrogen dioxide to −21 °C produces dinitrogen trioxide, a blue liquid consisting of N_2O_3 molecules (shown in Figure 18.35). Dinitrogen trioxide exists only in the liquid and solid states. When heated, it reverts to a mixture of NO and NO_2.

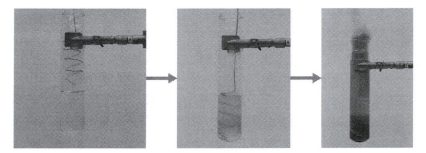

Figure 18.35   Dinitrogen trioxide, N_2O_3, only exists in liquid or solid states and has these molecular (left) and resonance (right) structures.

It is possible to prepare nitrogen dioxide in the laboratory by heating the nitrate of a heavy metal, or by the reduction of concentrated nitric acid with copper metal, as shown in Figure 18.36. Commercially, it is possible to prepare nitrogen dioxide by oxidizing nitric oxide with air.

Figure 18.36   The reaction of copper metal with concentrated HNO_3 produces a solution of $Cu(NO_3)_2$ and brown fumes of NO_2. (credit: modification of work by Mark Ott)

The nitrogen dioxide molecule (illustrated in Figure 18.37) contains an unpaired electron, which is responsible for its color and paramagnetism. It is also responsible for the dimerization of NO_2. At low pressures or at high temperatures, nitrogen dioxide has a deep brown color that is due to the presence of the NO_2 molecule. At low temperatures, the color almost entirely disappears as dinitrogen tetraoxide, N_2O_4, forms. At room temperature, an equilibrium exists:

$$2NO_2(g) \;\rightleftharpoons\; N_2O_4(g) \qquad\qquad K_P = 6.86$$

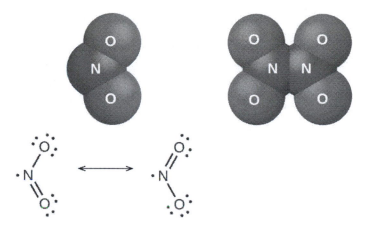

Figure 18.37 The molecular and resonance structures for nitrogen dioxide (NO_2, left) and dinitrogen tetraoxide (N_2O_4, right) are shown.

Dinitrogen pentaoxide, N_2O_5 (illustrated in Figure 18.38), is a white solid that is formed by the dehydration of nitric acid by phosphorus(V) oxide (tetraphosphorus decoxide):

$$P_4O_{10}(s) + 4HNO_3(l) \longrightarrow 4HPO_3(s) + 2N_2O_5(s)$$

It is unstable above room temperature, decomposing to N_2O_4 and O_2.

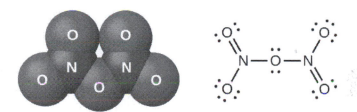

Figure 18.38 This image shows the molecular structure and one resonance structure of a molecule of dinitrogen pentaoxide, N_2O_5.

The oxides of nitrogen(III), nitrogen(IV), and nitrogen(V) react with water and form nitrogen-containing oxyacids. Nitrogen(III) oxide, N_2O_3, is the anhydride of nitrous acid; HNO_2 forms when N_2O_3 reacts with water. There are no stable oxyacids containing nitrogen with an oxidation state of 4+; therefore, nitrogen(IV) oxide, NO_2, disproportionates in one of two ways when it reacts with water. In cold water, a mixture of HNO_2 and HNO_3 forms. At higher temperatures, HNO_3 and NO will form. Nitrogen(V) oxide, N_2O_5, is the anhydride of nitric acid; HNO_3 is produced when N_2O_5 reacts with water:

$$N_2O_5(s) + H_2O(l) \longrightarrow 2HNO_3(aq)$$

The nitrogen oxides exhibit extensive oxidation-reduction behavior. Nitrous oxide resembles oxygen in its behavior when heated with combustible substances. N_2O is a strong oxidizing agent that decomposes when heated to form nitrogen and oxygen. Because one-third of the gas liberated is oxygen, nitrous oxide supports combustion better than air (one-fifth oxygen). A glowing splinter bursts into flame when thrust into a bottle of this gas. Nitric oxide acts both as an oxidizing agent and as a reducing agent. For example:

$$\text{oxidizing agent: } P_4(s) + 6NO(g) \longrightarrow P_4O_6(s) + 3N_2(g)$$

$$\text{reducing agent: } Cl_2(g) + 2NO(g) \longrightarrow 2ClNO(g)$$

Nitrogen dioxide (or dinitrogen tetraoxide) is a good oxidizing agent. For example:

$$NO_2(g) + CO(g) \longrightarrow NO(g) + CO_2(g)$$
$$NO_2(g) + 2HCl(aq) \longrightarrow NO(g) + Cl_2(g) + H_2O(l)$$

18.8 Occurrence, Preparation, and Properties of Phosphorus

By the end of this section, you will be able to:

- Describe the properties, preparation, and uses of phosphorus

The industrial preparation of phosphorus is by heating calcium phosphate, obtained from phosphate rock, with sand and coke:

$$2Ca_3(PO_4)_2(s) + 6SiO_2(s) + 10C(s) \xrightarrow{\Delta} 6CaSiO_3(l) + 10CO(g) + P_4(g)$$

The phosphorus distills out of the furnace and is condensed into a solid or burned to form P_4O_{10}. The preparation of many other phosphorus compounds begins with P_4O_{10}. The acids and phosphates are useful as fertilizers and in the chemical industry. Other uses are in the manufacture of special alloys such as ferrophosphorus and phosphor bronze. Phosphorus is important in making pesticides, matches, and some plastics. Phosphorus is an active nonmetal. In compounds, phosphorus usually occurs in oxidation states of 3−, 3+, and 5+. Phosphorus exhibits oxidation numbers that are unusual for a group 15 element in compounds that contain phosphorus-phosphorus bonds; examples include diphosphorus tetrahydride, H_2P-PH_2, and tetraphosphorus trisulfide, P_4S_3, illustrated in Figure 18.39.

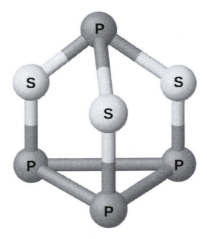

Figure 18.39 P_4S_3 is a component of the heads of strike-anywhere matches.

Phosphorus Oxygen Compounds

Phosphorus forms two common oxides, phosphorus(III) oxide (or tetraphosphorus hexaoxide), P_4O_6, and phosphorus(V) oxide (or tetraphosphorus decaoxide), P_4O_{10}, both shown in Figure 18.40. Phosphorus(III) oxide is a white crystalline solid with a garlic-like odor. Its vapor is very poisonous. It oxidizes slowly in air and inflames when heated to 70 °C, forming P_4O_{10}. Phosphorus(III) oxide dissolves slowly in cold water to form phosphorous acid, H_3PO_3.

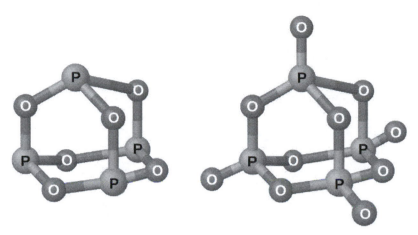

Figure 18.40   This image shows the molecular structures of P_4O_6 (left) and P_4O_{10} (right).

Phosphorus(V) oxide, P_4O_{10}, is a white powder that is prepared by burning phosphorus in excess oxygen. Its enthalpy of formation is very high (−2984 kJ), and it is quite stable and a very poor oxidizing agent. Dropping P_4O_{10} into water produces a hissing sound, heat, and orthophosphoric acid:

$$P_4O_{10}(s) + 6H_2O(l) \longrightarrow 4H_3PO_4(aq)$$

Because of its great affinity for water, phosphorus(V) oxide is an excellent drying agent for gases and solvents, and for removing water from many compounds.

Phosphorus Halogen Compounds

Phosphorus will react directly with the halogens, forming trihalides, PX_3, and pentahalides, PX_5. The trihalides are much more stable than the corresponding nitrogen trihalides; nitrogen pentahalides do not form because of nitrogen's inability to form more than four bonds.

The chlorides PCl_3 and PCl_5, both shown in Figure 18.41, are the most important halides of phosphorus. Phosphorus trichloride is a colorless liquid that is prepared by passing chlorine over molten phosphorus. Phosphorus pentachloride is an off-white solid that is prepared by oxidizing the trichloride with excess chlorine. The pentachloride sublimes when warmed and forms an equilibrium with the trichloride and chlorine when heated.

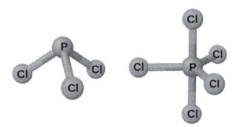

Figure 18.41   This image shows the molecular structure of PCl_3 (left) and PCl_5 (right) in the gas phase.

Like most other nonmetal halides, both phosphorus chlorides react with an excess of water and yield hydrogen chloride and an oxyacid: PCl_3 yields phosphorous acid H_3PO_3 and PCl_5 yields phosphoric acid, H_3PO_4.

The pentahalides of phosphorus are Lewis acids because of the empty valence d orbitals of phosphorus. These compounds readily react with halide ions (Lewis bases) to give the anion $PX_6{}^-$. Whereas phosphorus pentafluoride is a molecular compound in all states, X-ray studies show that solid phosphorus pentachloride is an ionic compound, $[PCl_4{}^+][PCl_6{}^-]$, as are phosphorus pentabromide, $[PBr_4{}^+][Br^-]$, and phosphorus pentaiodide, $[PI_4{}^+][I^-]$.

18.9 Occurrence, Preparation, and Compounds of Oxygen

By the end of this section, you will be able to:

- Describe the properties, preparation, and compounds of oxygen
- Describe the preparation, properties, and uses of some representative metal oxides, peroxides, and hydroxides

Oxygen is the most abundant element on the earth's crust. The earth's surface is composed of the crust, atmosphere, and hydrosphere. About 50% of the mass of the earth's crust consists of oxygen (combined with other elements, principally silicon). Oxygen occurs as O_2 molecules and, to a limited extent, as O_3 (ozone) molecules in air. It forms about 20% of the mass of the air. About 89% of water by mass consists of combined oxygen. In combination with carbon, hydrogen, and nitrogen, oxygen is a large part of plants and animals.

Oxygen is a colorless, odorless, and tasteless gas at ordinary temperatures. It is slightly denser than air. Although it is only slightly soluble in water (49 mL of gas dissolves in 1 L at STP), oxygen's solubility is very important to aquatic life.

Most of the oxygen isolated commercially comes from air and the remainder from the electrolysis of water. The separation of oxygen from air begins with cooling and compressing the air until it liquefies. As liquid air warms, oxygen with its higher boiling point (90 K) separates from nitrogen, which has a lower boiling point (77 K). It is possible to separate the other components of air at the same time based on differences in their boiling points.

Oxygen is essential in combustion processes such as the burning of fuels. Plants and animals use the oxygen from the air in respiration. The administration of oxygen-enriched air is an important medical practice when a patient is receiving an inadequate supply of oxygen because of shock, pneumonia, or some other illness.

The chemical industry employs oxygen for oxidizing many substances. A significant amount of oxygen produced commercially is important in the removal of carbon from iron during steel production. Large quantities of pure oxygen are also necessary in metal fabrication and in the cutting and welding of metals with oxyhydrogen and oxyacetylene torches.

Liquid oxygen is important to the space industry. It is an oxidizing agent in rocket engines. It is also the source of gaseous oxygen for life support in space.

As we know, oxygen is very important to life. The energy required for the maintenance of normal body functions in human beings and in other organisms comes from the slow oxidation of chemical compounds. Oxygen is the final oxidizing agent in these reactions. In humans, oxygen passes from the lungs into the blood, where it combines with hemoglobin, producing oxyhemoglobin. In this form, blood transports the oxygen to tissues, where it is transferred to the tissues. The ultimate products are carbon dioxide and water. The blood carries the carbon dioxide through the veins to the lungs, where the blood releases the carbon dioxide and collects another supply of oxygen. Digestion and assimilation of food regenerate the materials consumed by oxidation in the body; the energy liberated is the same as if the food burned outside the body.

Green plants continually replenish the oxygen in the atmosphere by a process called **photosynthesis**. The products of photosynthesis may vary, but, in general, the process converts carbon dioxide and water into glucose (a sugar) and oxygen using the energy of light:

$$6CO_2(g) + 6H_2O(l) \xrightarrow[\text{light}]{\text{chlorophyll}} C_6H_{12}O_6(aq) + 6O_2(g)$$

carbon   water   glucose   oxygen

dioxide

Thus, the oxygen that became carbon dioxide and water by the metabolic processes in plants and animals returns to the atmosphere by photosynthesis.

When dry oxygen is passed between two electrically charged plates, **ozone** (O_3, illustrated in Figure 18.42), an allotrope of oxygen possessing a distinctive odor, forms. The formation of ozone from oxygen is an endothermic reaction, in which the energy comes from an electrical discharge, heat, or ultraviolet light:

$$3O_2(g) \xrightarrow{\text{electric discharge}} 2O_3(g) \qquad\qquad \Delta H° = 287 \text{ kJ}$$

The sharp odor associated with sparking electrical equipment is due, in part, to ozone.

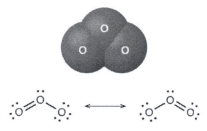

Figure 18.42   The image shows the bent ozone (O_3) molecule and the resonance structures necessary to describe its bonding.

Ozone forms naturally in the upper atmosphere by the action of ultraviolet light from the sun on the oxygen there. Most atmospheric ozone occurs in the stratosphere, a layer of the atmosphere extending from about 10 to 50 kilometers above the earth's surface. This ozone acts as a barrier to harmful ultraviolet light from the sun by absorbing it via a chemical decomposition reaction:

$$O_3(g) \xrightarrow{\text{ultraviolet light}} O(g) + O_2(g)$$

The reactive oxygen atoms recombine with molecular oxygen to complete the ozone cycle. The presence of stratospheric ozone decreases the frequency of skin cancer and other damaging effects of ultraviolet radiation. It has been clearly demonstrated that chlorofluorocarbons, CFCs (known commercially as Freons), which were present as aerosol propellants in spray cans and as refrigerants, caused depletion of ozone in the stratosphere. This occurred because ultraviolet light also causes CFCs to decompose, producing atomic chlorine. The chlorine atoms react with ozone molecules, resulting in a net removal of O_3 molecules from stratosphere. This process is explored in detail in our coverage of chemical kinetics. There is a worldwide effort to reduce the amount of CFCs used commercially, and the ozone hole is already beginning to decrease in size as atmospheric concentrations of atomic chlorine decrease. While ozone in the stratosphere helps protect us, ozone in the troposphere is a problem. This ozone is a toxic component of photochemical smog.

The uses of ozone depend on its reactivity with other substances. It can be used as a bleaching agent for oils, waxes, fabrics, and starch: It oxidizes the colored compounds in these substances to colorless compounds. It is an alternative to chlorine as a disinfectant for water.

Reactions

Elemental oxygen is a strong oxidizing agent. It reacts with most other elements and many compounds.

Reaction with Elements

Oxygen reacts directly at room temperature or at elevated temperatures with all other elements except the noble gases, the halogens, and few second- and third-row transition metals of low reactivity (those with higher reduction potentials than copper). Rust is an example of the reaction of oxygen with iron. The more active metals form peroxides or superoxides. Less active metals and the nonmetals give oxides. Two examples of these reactions are:

$$2Mg(s) + O_2(g) \longrightarrow 2MgO(s)$$
$$P_4(s) + 5O_2(g) \longrightarrow P_4O_{10}(s)$$

The oxides of halogens, at least one of the noble gases, and metals with higher reduction potentials than copper do not form by the direct action of the elements with oxygen.

Reaction with Compounds

Elemental oxygen also reacts with some compounds. If it is possible to oxidize any of the elements in a given compound, further oxidation by oxygen can occur. For example, hydrogen sulfide, H_2S, contains sulfur with an oxidation state of $2-$. Because the sulfur does not exhibit its maximum oxidation state, we would expect H_2S to react with oxygen. It does, yielding water and sulfur dioxide. The reaction is:

$$2H_2S(g) + 3O_2(g) \longrightarrow 2H_2O(l) + 2SO_2(g)$$

It is also possible to oxidize oxides such as CO and P_4O_6 that contain an element with a lower oxidation state. The ease with which elemental oxygen picks up electrons is mirrored by the difficulty of removing electrons from oxygen in most oxides. Of the elements, only the very reactive fluorine can oxidize oxides to form oxygen gas.

Oxides, Peroxides, and Hydroxides

Compounds of the representative metals with oxygen fall into three categories: (1) **oxides**, containing oxide ions, O^{2-}; (2) **peroxides**, containing peroxides ions, $O_2{}^{2-}$, with oxygen-oxygen covalent single bonds and a very limited number of **superoxides**, containing superoxide ions, $O_2{}^-$, with oxygen-oxygen covalent bonds that have a bond order of $1\frac{1}{2}$, In addition, there are (3) **hydroxides**, containing hydroxide ions, OH^-. All representative metals form oxides. Some of the metals of group 2 also form peroxides, MO_2, and the metals of group 1 also form peroxides, M_2O_2, and superoxides, MO_2.

Oxides

It is possible to produce the oxides of most representative metals by heating the corresponding hydroxides (forming the oxide and gaseous water) or carbonates (forming the oxide and gaseous CO_2). Equations for example reactions are:

$$2Al(OH)_3(s) \xrightarrow{\Delta} Al_2O_3(s) + 3H_2O(g)$$
$$CaCO_3(s) \xrightarrow{\Delta} CaO(s) + CO_2(g)$$

However, alkali metal salts generally are very stable and do not decompose easily when heated. Alkali metal oxides result from the oxidation-reduction reactions created by heating nitrates or hydroxides with the metals. Equations for sample reactions are:

$$2KNO_3(s) + 10K(s) \xrightarrow{\Delta} 6K_2O(s) + N_2(g)$$
$$2LiOH(s) + 2Li(s) \xrightarrow{\Delta} 2Li_2O(s) + H_2(g)$$

With the exception of mercury(II) oxide, it is possible to produce the oxides of the metals of groups 2–15 by burning the corresponding metal in air. The heaviest member of each group, the member for which the inert pair effect is most pronounced, forms an oxide in which the oxidation state of the metal ion is two less than the group oxidation state (inert pair effect). Thus, Tl_2O, PbO, and Bi_2O_3 form when burning thallium, lead, and bismuth, respectively.

The oxides of the lighter members of each group exhibit the group oxidation state. For example, SnO_2 forms from burning tin. Mercury(II) oxide, HgO, forms slowly when mercury is warmed below 500 °C; it decomposes at higher temperatures.

Burning the members of groups 1 and 2 in air is not a suitable way to form the oxides of these elements. These metals are reactive enough to combine with nitrogen in the air, so they form mixtures of oxides and ionic nitrides. Several also form peroxides or superoxides when heated in air.

Ionic oxides all contain the oxide ion, a very powerful hydrogen ion acceptor. With the exception of the very insoluble aluminum oxide, Al_2O_3, tin(IV), SnO_2, and lead(IV), PbO_2, the oxides of the representative metals react with acids to form salts. Some equations for these reactions are:

$$Na_2O + 2HNO_3(aq) \longrightarrow 2NaNO_3(aq) + H_2O(l)$$
$$CaO(s) + 2HCL(aq) \longrightarrow CaCl_2(aq) + H_2O(l)$$
$$SnO(s) + 2HClO_4(aq) \longrightarrow Sn(ClO_4)_2(aq) + H_2O(l)$$

The oxides of the metals of groups 1 and 2 and of thallium(I) oxide react with water and form hydroxides. Examples of such reactions are:

$$Na_2O(s) + H_2O(l) \longrightarrow NaOH(aq)$$
$$CaO(s) + H_2O(l) \longrightarrow Ca(OH)_2(aq)$$
$$Tl_2O(s) + H_2O(aq) \longrightarrow 2TlOH(aq)$$

The oxides of the alkali metals have little industrial utility, unlike magnesium oxide, calcium oxide, and aluminum oxide. Magnesium oxide is important in making firebrick, crucibles, furnace linings, and thermal insulation—applications that require chemical and thermal stability. Calcium oxide, sometimes called *quicklime* or lime in the industrial market, is very reactive, and its principal uses reflect its reactivity. Pure calcium oxide emits an intense white light when heated to a high temperature (as illustrated in Figure 18.43). Blocks of calcium oxide heated by gas flames were the stage lights in theaters before electricity was available. This is the source of the phrase "in the limelight."

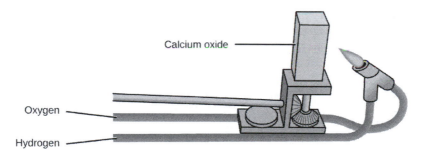

Figure 18.43 Calcium oxide has many industrial uses. When it is heated at high temperatures, it emits an intense white light.

Calcium oxide and calcium hydroxide are inexpensive bases used extensively in chemical processing, although most of the useful products prepared from them do not contain calcium. Calcium oxide, CaO, is made by heating calcium carbonate, $CaCO_3$, which is widely and inexpensively available as limestone or oyster shells:

$$CaCO_3(s) \longrightarrow CaO(s) + CO_2(g)$$

Although this decomposition reaction is reversible, it is possible to obtain a 100% yield of CaO by allowing the CO_2 to escape. It is possible to prepare calcium hydroxide by the familiar acid-base reaction of a soluble metal oxide with water:

$$CaO(s) + H_2O(l) \longrightarrow Ca(OH)_2(s)$$

Both CaO and Ca(OH)$_2$ are useful as bases; they accept protons and neutralize acids.

Alumina (Al$_2$O$_3$) occurs in nature as the mineral corundum, a very hard substance used as an abrasive for grinding and polishing. Corundum is important to the jewelry trade as ruby and sapphire. The color of ruby is due to the presence of a small amount of chromium; other impurities produce the wide variety of colors possible for sapphires. Artificial rubies and sapphires are now manufactured by melting aluminum oxide (melting point = 2050 °C) with small amounts of oxides to produce the desired colors and cooling the melt in such a way as to produce large crystals. Ruby lasers use synthetic ruby crystals.

Zinc oxide, ZnO, was a useful white paint pigment; however, pollutants tend to discolor the compound. The compound is also important in the manufacture of automobile tires and other rubber goods, and in the preparation of medicinal ointments. For example, zinc-oxide-based sunscreens, as shown in Figure 18.44, help prevent sunburn. The zinc oxide in these sunscreens is present in the form of very small grains known as nanoparticles. Lead dioxide is a constituent of charged lead storage batteries. Lead(IV) tends to revert to the more stable lead(II) ion by gaining two electrons, so lead dioxide is a powerful oxidizing agent.

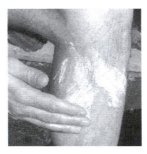

Figure 18.44   Zinc oxide protects exposed skin from sunburn. (credit: modification of work by "osseous"/Flickr)

Peroxides and Superoxides

Peroxides and superoxides are strong oxidizers and are important in chemical processes. Hydrogen peroxide, H$_2$O$_2$, prepared from metal peroxides, is an important bleach and disinfectant. Peroxides and superoxides form when the metal or metal oxides of groups 1 and 2 react with pure oxygen at elevated temperatures. Sodium peroxide and the peroxides of calcium, strontium, and barium form by heating the corresponding metal or metal oxide in pure oxygen:

$$2Na(s) + O_2(g) \xrightarrow{\Delta} Na_2O_2(s)$$
$$2Na_2O(s) + O_2(g) \xrightarrow{\Delta} 2Na_2O_2(s)$$
$$2SrO(s) + O_2(g) \xrightarrow{\Delta} 2SrO_2(s)$$

The peroxides of potassium, rubidium, and cesium can be prepared by heating the metal or its oxide in a carefully controlled amount of oxygen:

$$2K(s) + O_2(g) \longrightarrow K_2O_2(s) \qquad \text{(2 mol K per mol O}_2\text{)}$$

With an excess of oxygen, the superoxides KO$_2$, RbO$_2$, and CsO$_2$ form. For example:

$$K(s) + O_2(g) \longrightarrow KO_2(s) \qquad \text{(1 mol K per mol O}_2\text{)}$$

The stability of the peroxides and superoxides of the alkali metals increases as the size of the cation increases.

Hydroxides

Hydroxides are compounds that contain the OH$^-$ ion. It is possible to prepare these compounds by two general types of reactions. Soluble metal hydroxides can be produced by the reaction of the metal or metal oxide with water.

Insoluble metal hydroxides form when a solution of a soluble salt of the metal combines with a solution containing hydroxide ions.

With the exception of beryllium and magnesium, the metals of groups 1 and 2 react with water to form hydroxides and hydrogen gas. Examples of such reactions include:

$$2Li(s) + 2H_2O(l) \longrightarrow 2LiOH(aq) + H_2(g)$$
$$Ca(s) + 2H_2O(l) \longrightarrow Ca(OH)_2(aq) + H_2(g)$$

However, these reactions can be violent and dangerous; therefore, it is preferable to produce soluble metal hydroxides by the reaction of the respective oxide with water:

$$Li_2O(s) + H_2O(l) \longrightarrow 2LiOH(aq)$$
$$CaO(s) + H_2O(l) \longrightarrow Ca(OH)_2(aq)$$

Most metal oxides are **base anhydrides**. This is obvious for the soluble oxides because they form metal hydroxides. Most other metal oxides are insoluble and do not form hydroxides in water; however, they are still base anhydrides because they will react with acids.

It is possible to prepare the insoluble hydroxides of beryllium, magnesium, and other representative metals by the addition of sodium hydroxide to a solution of a salt of the respective metal. The net ionic equations for the reactions involving a magnesium salt, an aluminum salt, and a zinc salt are:

$$Mg^{2+}(aq) + 2OH^-(aq) \longrightarrow Mg(OH)_2(s)$$
$$Al^{3+}(aq) + 3OH^-(aq) \longrightarrow Al(OH)_3(s)$$
$$Zn^{2+}(aq) + 2OH^-(aq) \longrightarrow Zn(OH)_2(s)$$

An excess of hydroxide must be avoided when preparing aluminum, gallium, zinc, and tin(II) hydroxides, or the hydroxides will dissolve with the formation of the corresponding complex ions: $Al(OH)_4{}^-$, $Ga(OH)_4{}^-$, $Zn(OH)_4{}^{2-}$, and $Sn(OH)_3{}^-$ (see Figure 18.45). The important aspect of complex ions for this chapter is that they form by a Lewis acid-base reaction with the metal being the Lewis acid.

 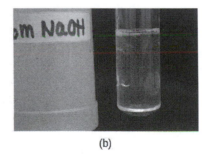

(a)   (b)

Figure 18.45   (a) Mixing solutions of NaOH and $Zn(NO_3)_2$ produces a white precipitate of $Zn(OH)_2$. (b) Addition of an excess of NaOH results in dissolution of the precipitate. (credit: modification of work by Mark Ott)

Industry uses large quantities of sodium hydroxide as a cheap, strong base. Sodium chloride is the starting material for the production of NaOH because NaCl is a less expensive starting material than the oxide. Sodium hydroxide is among the top 10 chemicals in production in the United States, and this production was almost entirely by electrolysis of solutions of sodium chloride. This process is the **chlor-alkali process**, and it is the primary method for producing chlorine.

Sodium hydroxide is an ionic compound and melts without decomposition. It is very soluble in water, giving off a great deal of heat and forming very basic solutions: 40 grams of sodium hydroxide dissolves in only 60 grams of water at 25 °C. Sodium hydroxide is employed in the production of other sodium compounds and is used to neutralize acidic solutions during the production of other chemicals such as petrochemicals and polymers.

Many of the applications of hydroxides are for the neutralization of acids (such as the antacid shown in Figure 18.46) and for the preparation of oxides by thermal decomposition. An aqueous suspension of magnesium hydroxide constitutes the antacid milk of magnesia. Because of its ready availability (from the reaction of water with calcium oxide prepared by the decomposition of limestone, $CaCO_3$), low cost, and activity, calcium hydroxide is used extensively in commercial applications needing a cheap, strong base. The reaction of hydroxides with appropriate acids is also used to prepare salts.

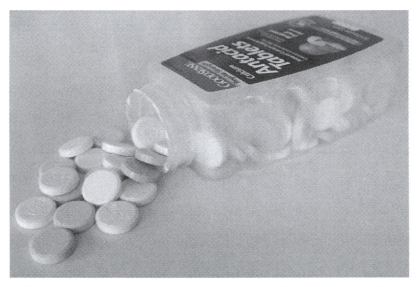

Figure 18.46   Calcium carbonate, $CaCO_3$, can be consumed in the form of an antacid to neutralize the effects of acid in your stomach. (credit: "Midnightcomm"/Wikimedia Commons)

Chemistry in Everyday Life

The Chlor-Alkali Process

Although they are very different chemically, there is a link between chlorine and sodium hydroxide because there is an important electrochemical process that produces the two chemicals simultaneously. The process known as the chlor-alkali process, utilizes sodium chloride, which occurs in large deposits in many parts of the world. This is an electrochemical process to oxidize chloride ion to chlorine and generate sodium hydroxide.

Passing a direct current of electricity through a solution of NaCl causes the chloride ions to migrate to the positive electrode where oxidation to gaseous chlorine occurs when the ion gives up an electron to the electrode:

$$2Cl^-(aq) \longrightarrow Cl_2(g) + 2e^- \qquad \text{(at the positive electrode)}$$

The electrons produced travel through the outside electrical circuit to the negative electrode. Although the positive sodium ions migrate toward this negative electrode, metallic sodium does not form because sodium ions are too difficult to reduce under the conditions used. (Recall that metallic sodium is active enough to react with water and hence, even if produced, would immediately react with water to produce sodium ions again.)

Instead, water molecules pick up electrons from the electrode and undergo reduction to form hydrogen gas and hydroxide ions:

$$2H_2O(l) + 2e^- \text{ (from the negative electrode)} \longrightarrow H_2(g) + 2OH^-(aq)$$

The overall result is the conversion of the aqueous solution of NaCl to an aqueous solution of NaOH, gaseous Cl_2, and gaseous H_2:

$$2Na^+(aq) + 2Cl^-(aq) + 2H_2O(l) \xrightarrow{\text{electrolysis}} 2Na^+(aq) + 2OH^-(aq) + Cl_2(g) + H_2(g)$$

Nonmetal Oxygen Compounds

Most nonmetals react with oxygen to form nonmetal oxides. Depending on the available oxidation states for the element, a variety of oxides might form. Fluorine will combine with oxygen to form fluorides such as OF_2, where the oxygen has a 2+-oxidation state.

Sulfur Oxygen Compounds

The two common oxides of sulfur are sulfur dioxide, SO_2, and sulfur trioxide, SO_3. The odor of burning sulfur comes from sulfur dioxide. Sulfur dioxide, shown in Figure 18.47, occurs in volcanic gases and in the atmosphere near industrial plants that burn fuel containing sulfur compounds.

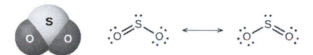

Figure 18.47 This image shows the molecular structure (left) and resonance forms (right) of sulfur dioxide.

Commercial production of sulfur dioxide is from either burning sulfur or roasting sulfide ores such as ZnS, FeS_2, and Cu_2S in air. (Roasting, which forms the metal oxide, is the first step in the separation of many metals from their ores.) A convenient method for preparing sulfur dioxide in the laboratory is by the action of a strong acid on either sulfite salts containing the $SO_3{}^{2-}$ ion or hydrogen sulfite salts containing $HSO_3{}^-$. Sulfurous acid, H_2SO_3, forms first, but quickly decomposes into sulfur dioxide and water. Sulfur dioxide also forms when many reducing agents react with hot, concentrated sulfuric acid. Sulfur trioxide forms slowly when heating sulfur dioxide and oxygen together, and the reaction is exothermic:

$$2SO_2(g) + O_2(g) \longrightarrow 2SO_3(g) \qquad \Delta H° = -197.8 \text{ kJ}$$

Sulfur dioxide is a gas at room temperature, and the SO_2 molecule is bent. Sulfur trioxide melts at 17 °C and boils at 43 °C. In the vapor state, its molecules are single SO_3 units (shown in Figure 18.48), but in the solid state, SO_3 exists in several polymeric forms.

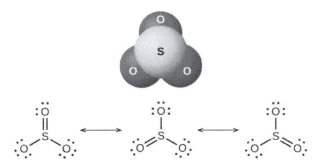

Figure 18.48   This image shows the structure (top) of sulfur trioxide in the gas phase and its resonance forms (bottom).

The sulfur oxides react as Lewis acids with many oxides and hydroxides in Lewis acid-base reactions, with the formation of **sulfites** or **hydrogen sulfites**, and **sulfates** or **hydrogen sulfates**, respectively.

Halogen Oxygen Compounds

The halogens do not react directly with oxygen, but it is possible to prepare binary oxygen-halogen compounds by the reactions of the halogens with oxygen-containing compounds. Oxygen compounds with chlorine, bromine, and iodine are oxides because oxygen is the more electronegative element in these compounds. On the other hand, fluorine compounds with oxygen are fluorides because fluorine is the more electronegative element.

As a class, the oxides are extremely reactive and unstable, and their chemistry has little practical importance. Dichlorine oxide, formally called dichlorine monoxide, and chlorine dioxide, both shown in Figure 18.49, are the only commercially important compounds. They are important as bleaching agents (for use with pulp and flour) and for water treatment.

(a)   (b)

Figure 18.49   This image shows the structures of the (a) Cl_2O and (b) ClO_2 molecules.

Nonmetal Oxyacids and Their Salts

Nonmetal oxides form acids when allowed to react with water; these are acid anhydrides. The resulting oxyanions can form salts with various metal ions.

Nitrogen Oxyacids and Salts

Nitrogen pentaoxide, N_2O_5, and NO_2 react with water to form nitric acid, HNO_3. Alchemists, as early as the eighth century, knew nitric acid (shown in Figure 18.50) as *aqua fortis* (meaning "strong water"). The acid was useful in the separation of gold from silver because it dissolves silver but not gold. Traces of nitric acid occur in the atmosphere after thunderstorms, and its salts are widely distributed in nature. There are tremendous deposits of Chile saltpeter,

$NaNO_3$, in the desert region near the boundary of Chile and Peru. Bengal saltpeter, KNO_3, occurs in India and in other countries of the Far East.

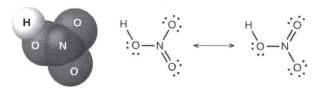

Figure 18.50 This image shows the molecular structure (left) of nitric acid, HNO_3 and its resonance forms (right).

In the laboratory, it is possible to produce nitric acid by heating a nitrate salt (such as sodium or potassium nitrate) with concentrated sulfuric acid:

$$NaNO_3(s) + H_2SO_4(l) \xrightarrow{\Delta} NaHSO_4(s) + HNO_3(g)$$

The **Ostwald process** is the commercial method for producing nitric acid. This process involves the oxidation of ammonia to nitric oxide, NO; oxidation of nitric oxide to nitrogen dioxide, NO_2; and further oxidation and hydration of nitrogen dioxide to form nitric acid:

$$4NH_3(g) + 5O_2(g) \longrightarrow 4NO(g) + 6H_2O(g)$$
$$2NO(g) + O_2(g) \longrightarrow 2NO_2(g)$$
$$3NO_2(g) + H_2O(l) \longrightarrow 2HNO_3(aq) + NO(g)$$

Or

$$4NO_2(g) + O_2(g) + 2H_2O(g) \longrightarrow 4HNO_3(l)$$

Pure nitric acid is a colorless liquid. However, it is often yellow or brown in color because NO_2 forms as the acid decomposes. Nitric acid is stable in aqueous solution; solutions containing 68% of the acid are commercially available concentrated nitric acid. It is both a strong oxidizing agent and a strong acid.

The action of nitric acid on a metal rarely produces H_2 (by reduction of H^+) in more than small amounts. Instead, the reduction of nitrogen occurs. The products formed depend on the concentration of the acid, the activity of the metal, and the temperature. Normally, a mixture of nitrates, nitrogen oxides, and various reduction products form. Less active metals such as copper, silver, and lead reduce concentrated nitric acid primarily to nitrogen dioxide. The reaction of dilute nitric acid with copper produces NO. In each case, the nitrate salts of the metals crystallize upon evaporation of the resultant solutions.

Nonmetallic elements, such as sulfur, carbon, iodine, and phosphorus, undergo oxidation by concentrated nitric acid to their oxides or oxyacids, with the formation of NO_2:

$$S(s) + 6HNO_3(aq) \longrightarrow H_2SO_4(aq) + 6NO_2(g) + 2H_2O(l)$$
$$C(s) + 4HNO_3(aq) \longrightarrow CO_2(g) + 4NO_2(g) + 2H_2O(l)$$

Nitric acid oxidizes many compounds; for example, concentrated nitric acid readily oxidizes hydrochloric acid to chlorine and chlorine dioxide. A mixture of one part concentrated nitric acid and three parts concentrated hydrochloric acid (called *aqua regia*, which means royal water) reacts vigorously with metals. This mixture is particularly useful in dissolving gold, platinum, and other metals that are more difficult to oxidize than hydrogen. A simplified equation to represent the action of *aqua regia* on gold is:

$$Au(s) + 4HCl(aq) + 3HNO_3(aq) \longrightarrow HAuCl_4(aq) + 3NO_2(g) + 3H_2O(l)$$

Nitrates, salts of nitric acid, form when metals, oxides, hydroxides, or carbonates react with nitric acid. Most nitrates are soluble in water; indeed, one of the significant uses of nitric acid is to prepare soluble metal nitrates.

Nitric acid finds extensive use in the laboratory and in chemical industries as a strong acid and strong oxidizing agent. It is important in the manufacture of explosives, dyes, plastics, and drugs. Salts of nitric acid (nitrates) are valuable as fertilizers. Gunpowder is a mixture of potassium nitrate, sulfur, and charcoal.

The reaction of N_2O_3 with water gives a pale blue solution of nitrous acid, HNO_2. However, HNO_2 (shown in Figure 18.51) is easier to prepare by the addition of an acid to a solution of nitrite; nitrous acid is a weak acid, so the nitrite ion is basic in aqueous solution:

$$NO_2^-(aq) + H_3O^+(aq) \longrightarrow HNO_2(aq) + H_2O(l)$$

Nitrous acid is very unstable and exists only in solution. It disproportionates slowly at room temperature (rapidly when heated) into nitric acid and nitric oxide. Nitrous acid is an active oxidizing agent with strong reducing agents, and strong oxidizing agents oxidize it to nitric acid.

Figure 18.51   This image shows the molecular structure of a molecule of nitrous acid, HNO_2.

Sodium nitrite, $NaNO_2$, is an additive to meats such as hot dogs and cold cuts. The nitrite ion has two functions. It limits the growth of bacteria that can cause food poisoning, and it prolongs the meat's retention of its red color. The addition of sodium nitrite to meat products is controversial because nitrous acid reacts with certain organic compounds to form a class of compounds known as nitrosamines. Nitrosamines produce cancer in laboratory animals. This has prompted the FDA to limit the amount of $NaNO_2$ in foods.

The nitrites are much more stable than the acid, but nitrites, like nitrates, can explode. Nitrites, like nitrates, are also soluble in water ($AgNO_2$ is only slightly soluble).

Phosphorus Oxyacids and Salts

Pure orthophosphoric acid, H_3PO_4 (shown in Figure 18.52), forms colorless, deliquescent crystals that melt at 42 °C. The common name of this compound is phosphoric acid, and is commercially available as a viscous 82% solution known as syrupy phosphoric acid. One use of phosphoric acid is as an additive to many soft drinks.

One commercial method of preparing orthophosphoric acid is to treat calcium phosphate rock with concentrated sulfuric acid:

$$Ca_3(PO_4)_2(s) + 3H_2SO_4(aq) \longrightarrow 2H_3PO_4(aq) + 3CaSO_4(s)$$

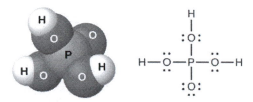

Figure 18.52   Orthophosphoric acid, H_3PO_4, is colorless when pure and has this molecular (left) and Lewis structure (right).

Dilution of the products with water, followed by filtration to remove calcium sulfate, gives a dilute acid solution contaminated with calcium dihydrogen phosphate, $Ca(H_2PO_4)_2$, and other compounds associated with calcium phosphate rock. It is possible to prepare pure orthophosphoric acid by dissolving P_4O_{10} in water.

The action of water on P_4O_6, PCl_3, PBr_3, or PI_3 forms phosphorous acid, H_3PO_3 (shown in Figure 18.53). The best method for preparing pure phosphorous acid is by hydrolyzing phosphorus trichloride:

$$PCl_3(l) + 3H_2O(l) \longrightarrow H_3PO_3(aq) + 3HCl(g)$$

Heating the resulting solution expels the hydrogen chloride and leads to the evaporation of water. When sufficient water evaporates, white crystals of phosphorous acid will appear upon cooling. The crystals are deliquescent, very soluble in water, and have an odor like that of garlic. The solid melts at 70.1 °C and decomposes at about 200 °C by disproportionation into phosphine and orthophosphoric acid:

$$4H_3PO_3(l) \longrightarrow PH_3(g) + 3H_3PO_4(l)$$

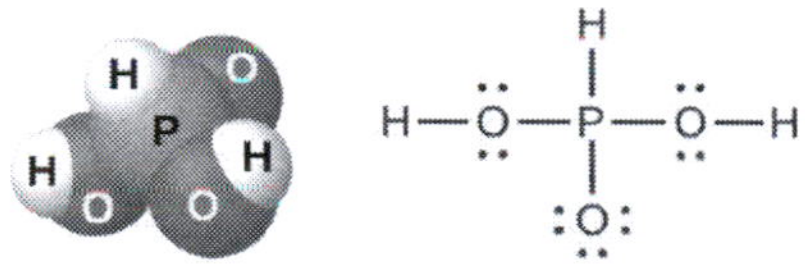

Figure 18.53   In a molecule of phosphorous acid, H_3PO_3, only the two hydrogen atoms bonded to an oxygen atom are acidic.

Phosphorous acid forms only two series of salts, which contain the dihydrogen phosphite ion, $H_2PO_3^{-}$, or the hydrogen phosphate ion, HPO_3^{2-}, respectively. It is not possible to replace the third atom of hydrogen because it is not very acidic, as it is not easy to ionize the P-H bond.

Sulfur Oxyacids and Salts

The preparation of sulfuric acid, H_2SO_4 (shown in Figure 18.54), begins with the oxidation of sulfur to sulfur trioxide and then converting the trioxide to sulfuric acid. Pure sulfuric acid is a colorless, oily liquid that freezes at 10.5 °C. It fumes when heated because the acid decomposes to water and sulfur trioxide. The heating process causes the loss of more sulfur trioxide than water, until reaching a concentration of 98.33% acid. Acid of this concentration boils at 338 °C without further change in concentration (a constant boiling solution) and is commercially concentrated H_2SO_4. The amount of sulfuric acid used in industry exceeds that of any other manufactured compound.

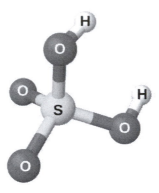

Figure 18.54 Sulfuric acid has a tetrahedral molecular structure.

The strong affinity of concentrated sulfuric acid for water makes it a good dehydrating agent. It is possible to dry gases and immiscible liquids that do not react with the acid by passing them through the acid.

Sulfuric acid is a strong diprotic acid that ionizes in two stages. In aqueous solution, the first stage is essentially complete. The secondary ionization is not nearly so complete, and HSO_4^- is a moderately strong acid (about 25% ionized in solution of a HSO_4^- salt: $K_a = 1.2 \times 10^{-2}$).

Being a diprotic acid, sulfuric acid forms both sulfates, such as Na_2SO_4, and hydrogen sulfates, such as $NaHSO_4$. Most sulfates are soluble in water; however, the sulfates of barium, strontium, calcium, and lead are only slightly soluble in water.

Among the important sulfates are $Na_2SO_4 \cdot 10H_2O$ and Epsom salts, $MgSO_4 \cdot 7H_2O$. Because the HSO_4^- ion is an acid, hydrogen sulfates, such as $NaHSO_4$, exhibit acidic behavior, and this compound is the primary ingredient in some household cleansers.

Hot, concentrated sulfuric acid is an oxidizing agent. Depending on its concentration, the temperature, and the strength of the reducing agent, sulfuric acid oxidizes many compounds and, in the process, undergoes reduction to SO_2, HSO_3^-, SO_3^{2-}, S, H_2S, or S^{2-}.

Sulfur dioxide dissolves in water to form a solution of sulfurous acid, as expected for the oxide of a nonmetal. Sulfurous acid is unstable, and it is not possible to isolate anhydrous H_2SO_3. Heating a solution of sulfurous acid expels the sulfur dioxide. Like other diprotic acids, sulfurous acid ionizes in two steps: The hydrogen sulfite ion, HSO_3^-, and the sulfite ion, SO_3^{2-}, form. Sulfurous acid is a moderately strong acid. Ionization is about 25% in the first stage, but it is much less in the second ($K_{a1} = 1.2 \times 10^{-2}$ and $K_{a2} = 6.2 \times 10^{-8}$).

In order to prepare solid sulfite and hydrogen sulfite salts, it is necessary to add a stoichiometric amount of a base to a sulfurous acid solution and then evaporate the water. These salts also form from the reaction of SO_2 with oxides and hydroxides. Heating solid sodium hydrogen sulfite forms sodium sulfite, sulfur dioxide, and water:

$$2NaHSO_3(s) \xrightarrow{\Delta} Na_2SO_3(s) + SO_2(g) + H_2O(l)$$

Strong oxidizing agents can oxidize sulfurous acid. Oxygen in the air oxidizes it slowly to the more stable sulfuric acid:

$$2H_2SO_3(aq) + O_2(g) + 2H_2O(l) \xrightarrow{\Delta} 2H_3O^+(aq) + 2HSO_4^-(aq)$$

Solutions of sulfites are also very susceptible to air oxidation to produce sulfates. Thus, solutions of sulfites always contain sulfates after exposure to air.

Halogen Oxyacids and Their Salts

The compounds HXO, HXO_2, HXO_3, and HXO_4, where X represents Cl, Br, or I, are the hypohalous, halous, halic, and perhalic acids, respectively. The strengths of these acids increase from the hypohalous acids, which are very weak acids, to the perhalic acids, which are very strong. Table 18.2 lists the known acids, and, where known, their pK_a values are given in parentheses.

Oxyacids of the Halogens

Name	Fluorine	Chlorine	Bromine	Iodine
hypohalous	HOF	HOCl (7.5)	HOBr (8.7)	HOI (11)
halous		$HClO_2$ (2.0)		
halic		$HClO_3$	$HBrO_3$	HIO_3 (0.8)
perhalic		$HClO_4$	$HBrO_4$	HIO_4 (1.6)
paraperhalic				H_5IO_6 (1.6)

Table 18.2

The only known oxyacid of fluorine is the very unstable hypofluorous acid, HOF, which is prepared by the reaction of gaseous fluorine with ice:

$$F_2(g) + H_2O(s) \longrightarrow HOF(g) + HF(g)$$

The compound is very unstable and decomposes above −40 °C. This compound does not ionize in water, and there are no known salts. It is uncertain whether the name hypofluorous acid is even appropriate for HOF; a more appropriate name might be hydrogen hypofluorite.

The reactions of chlorine and bromine with water are analogous to that of fluorine with ice, but these reactions do not go to completion, and mixtures of the halogen and the respective hypohalous and hydrohalic acids result. Other than HOF, the hypohalous acids only exist in solution. The hypohalous acids are all very weak acids; however, HOCl is a stronger acid than HOBr, which, in turn, is stronger than HOI.

The addition of base to solutions of the hypohalous acids produces solutions of salts containing the basic hypohalite ions, OX^-. It is possible to isolate these salts as solids. All of the hypohalites are unstable with respect to disproportionation in solution, but the reaction is slow for hypochlorite. Hypobromite and hypoiodite disproportionate rapidly, even in the cold:

$$3XO^-(aq) \longrightarrow 2X^-(aq) + XO_3{}^-(aq)$$

Sodium hypochlorite is an inexpensive bleach (Clorox) and germicide. The commercial preparation involves the electrolysis of cold, dilute, aqueous sodium chloride solutions under conditions where the resulting chlorine and hydroxide ion can react. The net reaction is:

$$Cl^-(aq) + H_2O(l) \xrightarrow{\text{electrical energy}} ClO^-(aq) + H_2(g)$$

The only definitely known halous acid is chlorous acid, $HClO_2$, obtained by the reaction of barium chlorite with dilute sulfuric acid:

$$Ba(ClO_2)_2(aq) + H_2SO_4(aq) \longrightarrow BaSO_4(s) + 2HClO_2(aq)$$

Filtering the insoluble barium sulfate leaves a solution of $HClO_2$. Chlorous acid is not stable; it slowly decomposes in solution to yield chlorine dioxide, hydrochloric acid, and water. Chlorous acid reacts with bases to give salts containing the chlorite ion (shown in Figure 18.55). Sodium chlorite finds an extensive application in the bleaching of paper because it is a strong oxidizing agent and does not damage the paper.

Figure 18.55 Chlorite ions, ClO_2^-, are produced when chlorous acid reacts with bases.

Chloric acid, $HClO_3$, and bromic acid, $HBrO_3$, are stable only in solution. The reaction of iodine with concentrated nitric acid produces stable white iodic acid, HIO_3:

$$I_2(s) + 10HNO_3(aq) \longrightarrow 2HIO_3(s) + 10NO_2(g) + 4H_2O(l)$$

It is possible to obtain the lighter halic acids from their barium salts by reaction with dilute sulfuric acid. The reaction is analogous to that used to prepare chlorous acid. All of the halic acids are strong acids and very active oxidizing agents. The acids react with bases to form salts containing chlorate ions (shown in Figure 18.56). Another preparative method is the electrochemical oxidation of a hot solution of a metal halide to form the appropriate metal chlorates. Sodium chlorate is a weed killer; potassium chlorate is used as an oxidizing agent.

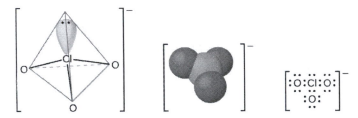

Figure 18.56 Chlorate ions, ClO_3^-, are produced when halic acids react with bases.

Perchloric acid, $HClO_4$, forms when treating a perchlorate, such as potassium perchlorate, with sulfuric acid under reduced pressure. The $HClO_4$ can be distilled from the mixture:

$$KClO_4(s) + H_2SO_4(aq) \longrightarrow HClO_4(g) + KHSO_4(s)$$

Dilute aqueous solutions of perchloric acid are quite stable thermally, but concentrations above 60% are unstable and dangerous. Perchloric acid and its salts are powerful oxidizing agents, as the very electronegative chlorine is more stable in a lower oxidation state than 7+. Serious explosions have occurred when heating concentrated solutions with easily oxidized substances. However, its reactions as an oxidizing agent are slow when perchloric acid is cold and dilute. The acid is among the strongest of all acids. Most salts containing the perchlorate ion (shown in Figure 18.57) are soluble. It is possible to prepare them from reactions of bases with perchloric acid and, commercially, by the electrolysis of hot solutions of their chlorides.

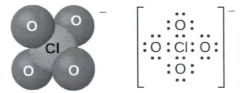

Figure 18.57   Perchlorate ions, ClO_4^-, can be produced when perchloric acid reacts with a base or by electrolysis of hot solutions of their chlorides.

Perbromate salts are difficult to prepare, and the best syntheses currently involve the oxidation of bromates in basic solution with fluorine gas followed by acidification. There are few, if any, commercial uses of this acid or its salts.

There are several different acids containing iodine in the 7+-oxidation state; they include metaperiodic acid, HIO_4, and paraperiodic acid, H_5IO_6. These acids are strong oxidizing agents and react with bases to form the appropriate salts.

18.10 Occurrence, Preparation, and Properties of Sulfur

By the end of this section, you will be able to:

- Describe the properties, preparation, and uses of sulfur

Sulfur exists in nature as elemental deposits as well as sulfides of iron, zinc, lead, and copper, and sulfates of sodium, calcium, barium, and magnesium. Hydrogen sulfide is often a component of natural gas and occurs in many volcanic gases, like those shown in Figure 18.58. Sulfur is a constituent of many proteins and is essential for life.

Figure 18.58   Volcanic gases contain hydrogen sulfide. (credit: Daniel Julie/Wikimedia Commons)

The **Frasch process**, illustrated in Figure 18.59, is important in the mining of free sulfur from enormous underground deposits in Texas and Louisiana. Superheated water (170 °C and 10 atm pressure) is forced down the outermost of three concentric pipes to the underground deposit. The hot water melts the sulfur. The innermost pipe conducts compressed air into the liquid sulfur. The air forces the liquid sulfur, mixed with air, to flow up through the outlet pipe. Transferring the mixture to large settling vats allows the solid sulfur to separate upon cooling. This sulfur is 99.5% to 99.9% pure and requires no purification for most uses.

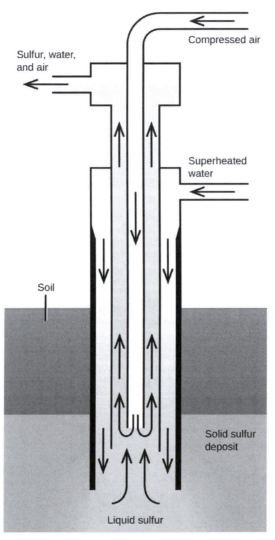

Figure 18.59 The Frasch process is used to mine sulfur from underground deposits.

Larger amounts of sulfur also come from hydrogen sulfide recovered during the purification of natural gas.

Sulfur exists in several allotropic forms. The stable form at room temperature contains eight-membered rings, and so the true formula is S_8. However, chemists commonly use S to simplify the coefficients in chemical equations; we will follow this practice in this book.

Like oxygen, which is also a member of group 16, sulfur exhibits a distinctly nonmetallic behavior. It oxidizes metals, giving a variety of binary sulfides in which sulfur exhibits a negative oxidation state (2−). Elemental sulfur oxidizes less electronegative nonmetals, and more electronegative nonmetals, such as oxygen and the halogens, will oxidize it. Other strong oxidizing agents also oxidize sulfur. For example, concentrated nitric acid oxidizes sulfur to the sulfate ion, with the concurrent formation of nitrogen(IV) oxide:

$$S(s) + 6HNO_3(aq) \longrightarrow 2H_3O^+(aq) + SO_4{}^{2-}(aq) + 6NO_2(g)$$

The chemistry of sulfur with an oxidation state of 2− is similar to that of oxygen. Unlike oxygen, however, sulfur forms many compounds in which it exhibits positive oxidation states.

18.11 Occurrence, Preparation, and Properties of Halogens

By the end of this section, you will be able to:

- Describe the preparation, properties, and uses of halogens
- Describe the properties, preparation, and uses of halogen compounds

The elements in group 17 are the halogens. These are the elements fluorine, chlorine, bromine, iodine, and astatine. These elements are too reactive to occur freely in nature, but their compounds are widely distributed. Chlorides are the most abundant; although fluorides, bromides, and iodides are less common, they are reasonably available. In this section, we will examine the occurrence, preparation, and properties of halogens. Next, we will examine halogen compounds with the representative metals followed by an examination of the interhalogens. This section will conclude with some applications of halogens.

Occurrence and Preparation

All of the halogens occur in seawater as halide ions. The concentration of the chloride ion is 0.54 M; that of the other halides is less than 10^{-4} M. Fluoride also occurs in minerals such as CaF_2, $Ca(PO_4)_3F$, and Na_3AlF_6. Chloride also occurs in the Great Salt Lake and the Dead Sea, and in extensive salt beds that contain NaCl, KCl, or $MgCl_2$. Part of the chlorine in your body is present as hydrochloric acid, which is a component of stomach acid. Bromine compounds occur in the Dead Sea and underground brines. Iodine compounds are found in small quantities in Chile saltpeter, underground brines, and sea kelp. Iodine is essential to the function of the thyroid gland.

The best sources of halogens (except iodine) are halide salts. It is possible to oxidize the halide ions to free diatomic halogen molecules by various methods, depending on the ease of oxidation of the halide ion. Fluoride is the most difficult to oxidize, whereas iodide is the easiest.

The major method for preparing fluorine is electrolytic oxidation. The most common electrolysis procedure is to use a molten mixture of potassium hydrogen fluoride, KHF_2, and anhydrous hydrogen fluoride. Electrolysis causes HF to decompose, forming fluorine gas at the anode and hydrogen at the cathode. It is necessary to keep the two gases separated to prevent their explosive recombination to reform hydrogen fluoride.

Most commercial chlorine comes from the electrolysis of the chloride ion in aqueous solutions of sodium chloride; this is the chlor-alkali process discussed previously. Chlorine is also a product of the electrolytic production of metals such as sodium, calcium, and magnesium from their fused chlorides. It is also possible to prepare chlorine by the chemical oxidation of the chloride ion in acid solution with strong oxidizing agents such as manganese dioxide (MnO_2) or sodium dichromate ($Na_2Cr_2O_7$). The reaction with manganese dioxide is:

$$MnO_2(s) + 2Cl^-(aq) + 4H_3O^+(aq) \longrightarrow Mn^{2+}(aq) + Cl_2(g) + 6H_2O(l)$$

The commercial preparation of bromine involves the oxidation of bromide ion by chlorine:

$$2Br^-(aq) + Cl_2(g) \longrightarrow Br_2(l) + 2Cl^-(aq)$$

Chlorine is a stronger oxidizing agent than bromine. This method is important for the production of essentially all domestic bromine.

Some iodine comes from the oxidation of iodine chloride, ICl, or iodic acid, HlO_3. The commercial preparation of iodine utilizes the reduction of sodium iodate, $NaIO_3$, an impurity in deposits of Chile saltpeter, with sodium hydrogen sulfite:

$$2IO_3^-(aq) + 5HSO_3^-(aq) \longrightarrow 3HSO_4^-(aq) + 2SO_4^{2-}(aq) + H_2O(l) + I_2(s)$$

Properties of the Halogens

Fluorine is a pale yellow gas, chlorine is a greenish-yellow gas, bromine is a deep reddish-brown liquid, and iodine is a grayish-black crystalline solid. Liquid bromine has a high vapor pressure, and the reddish vapor is readily visible in Figure 18.60. Iodine crystals have a noticeable vapor pressure. When gently heated, these crystals sublime and form a beautiful deep violet vapor.

Figure 18.60   Chlorine is a pale yellow-green gas (left), gaseous bromine is deep orange (center), and gaseous iodine is purple (right). (Fluorine is so reactive that it is too dangerous to handle.) (credit: Sahar Atwa)

Bromine is only slightly soluble in water, but it is miscible in all proportions in less polar (or nonpolar) solvents such as chloroform, carbon tetrachloride, and carbon disulfide, forming solutions that vary from yellow to reddish-brown, depending on the concentration.

Iodine is soluble in chloroform, carbon tetrachloride, carbon disulfide, and many hydrocarbons, giving violet solutions of I_2 molecules. Iodine dissolves only slightly in water, giving brown solutions. It is quite soluble in aqueous solutions of iodides, with which it forms brown solutions. These brown solutions result because iodine molecules have empty valence d orbitals and can act as weak Lewis acids towards the iodide ion. The equation for the reversible reaction of iodine (Lewis acid) with the iodide ion (Lewis base) to form triiodide ion, I_3^-, is:

$$I_2(s) + I^-(aq) \longrightarrow I_3^-(aq)$$

The easier it is to oxidize the halide ion, the more difficult it is for the halogen to act as an oxidizing agent. Fluorine generally oxidizes an element to its highest oxidation state, whereas the heavier halogens may not. For example, when excess fluorine reacts with sulfur, SF_6 forms. Chlorine gives SCl_2 and bromine, S_2Br_2. Iodine does not react with sulfur.

Fluorine is the most powerful oxidizing agent of the known elements. It spontaneously oxidizes most other elements; therefore, the reverse reaction, the oxidation of fluorides, is very difficult to accomplish. Fluorine reacts directly and forms binary fluorides with all of the elements except the lighter noble gases (He, Ne, and Ar). Fluorine is such a strong oxidizing agent that many substances ignite on contact with it. Drops of water inflame in fluorine and form O_2, OF_2, H_2O_2, O_3, and HF. Wood and asbestos ignite and burn in fluorine gas. Most hot metals burn vigorously in fluorine. However, it is possible to handle fluorine in copper, iron, or nickel containers because an adherent film of the fluoride salt passivates their surfaces. Fluorine is the only element that reacts directly with the noble gas xenon.

Although it is a strong oxidizing agent, chlorine is less active than fluorine. Mixing chlorine and hydrogen in the dark makes the reaction between them to be imperceptibly slow. Exposure of the mixture to light causes the two to react explosively. Chlorine is also less active towards metals than fluorine, and oxidation reactions usually require higher temperatures. Molten sodium ignites in chlorine. Chlorine attacks most nonmetals (C, N_2, and O_2 are notable exceptions), forming covalent molecular compounds. Chlorine generally reacts with compounds that contain only carbon and hydrogen (hydrocarbons) by adding to multiple bonds or by substitution.

In cold water, chlorine undergoes a disproportionation reaction:

$$Cl_2(aq) + 2H_2O(l) \longrightarrow HOCl(aq) + H_3O^+(aq) + Cl^-(aq)$$

Half the chlorine atoms oxidize to the 1+ oxidation state (hypochlorous acid), and the other half reduce to the 1− oxidation state (chloride ion). This disproportionation is incomplete, so chlorine water is an equilibrium mixture of chlorine molecules, hypochlorous acid molecules, hydronium ions, and chloride ions. When exposed to light, this solution undergoes a photochemical decomposition:

$$2HOCl(aq) + 2H_2O(l) \xrightarrow{\text{sunlight}} 2H_3O^+(aq) + 2Cl^-(aq) + O_2(g)$$

The nonmetal chlorine is more electronegative than any other element except fluorine, oxygen, and nitrogen. In general, very electronegative elements are good oxidizing agents; therefore, we would expect elemental chlorine to oxidize all of the other elements except for these three (and the nonreactive noble gases). Its oxidizing property, in fact, is responsible for its principal use. For example, phosphorus(V) chloride, an important intermediate in the preparation of insecticides and chemical weapons, is manufactured by oxidizing the phosphorus with chlorine:

$$P_4(s) + 10Cl_2(g) \longrightarrow 4PCl_5(l)$$

A great deal of chlorine is also used to oxidize, and thus to destroy, organic or biological materials in water purification and in bleaching.

The chemical properties of bromine are similar to those of chlorine, although bromine is the weaker oxidizing agent and its reactivity is less than that of chlorine.

Iodine is the least reactive of the halogens. It is the weakest oxidizing agent, and the iodide ion is the most easily oxidized halide ion. Iodine reacts with metals, but heating is often required. It does not oxidize other halide ions.

Compared with the other halogens, iodine reacts only slightly with water. Traces of iodine in water react with a mixture of starch and iodide ion, forming a deep blue color. This reaction is a very sensitive test for the presence of iodine in water.

Halides of the Representative Metals

Thousands of salts of the representative metals have been prepared. The binary halides are an important subclass of salts. A salt is an ionic compound composed of cations and anions, other than hydroxide or oxide ions. In general, it is possible to prepare these salts from the metals or from oxides, hydroxides, or carbonates. We will illustrate the general types of reactions for preparing salts through reactions used to prepare binary halides.

The binary compounds of a metal with the halogens are the **halides**. Most binary halides are ionic. However, mercury, the elements of group 13 with oxidation states of 3+, tin(IV), and lead(IV) form covalent binary halides.

The direct reaction of a metal and a halogen produce the halide of the metal. Examples of these oxidation-reduction reactions include:

$$Cd(s) + Cl_2(g) \longrightarrow CdCl_2(s)$$
$$2Ga(l) + 3Br_2(l) \longrightarrow 2GaBr_3(s)$$

Link to Learning

Reactions of the alkali metals with elemental halogens are very exothermic and often quite violent. Under controlled conditions, they provide exciting demonstrations for budding students of chemistry. You can view the initial heating (http://openstaxcollege.org/l/16sodium) of the sodium that removes the coating of sodium hydroxide, sodium peroxide, and residual mineral oil to expose the reactive surface. The reaction with chlorine gas then proceeds very nicely.

If a metal can exhibit two oxidation states, it may be necessary to control the stoichiometry in order to obtain the halide with the lower oxidation state. For example, preparation of tin(II) chloride requires a 1:1 ratio of Sn to Cl_2, whereas preparation of tin(IV) chloride requires a 1:2 ratio:

$$Sn(s) + Cl_2(g) \longrightarrow SnCl_2(s)$$
$$Sn(s) + 2Cl_2(g) \longrightarrow SnCl_4(l)$$

The active representative metals—those that are easier to oxidize than hydrogen—react with gaseous hydrogen halides to produce metal halides and hydrogen. The reaction of zinc with hydrogen fluoride is:

$$Zn(s) + 2HF(g) \longrightarrow ZnF_2(s) + H_2(g)$$

The active representative metals also react with solutions of hydrogen halides to form hydrogen and solutions of the corresponding halides. Examples of such reactions include:

$$Cd(s) + 2HBr(aq) \longrightarrow CdBr_2(aq) + H_2(g)$$
$$Sn(s) + 2HI(aq) \longrightarrow SnI_2(aq) + H_2(g)$$

Hydroxides, carbonates, and some oxides react with solutions of the hydrogen halides to form solutions of halide salts. It is possible to prepare additional salts by the reaction of these hydroxides, carbonates, and oxides with aqueous solution of other acids:

$$CaCo_3(s) + 2HCl(aq) \longrightarrow CaCl_2(aq) + CO_2(g) + H_2O(l)$$
$$TlOH(aq) + HF(aq) \longrightarrow TlF(aq) + H_2O(l)$$

A few halides and many of the other salts of the representative metals are insoluble. It is possible to prepare these soluble salts by metathesis reactions that occur when solutions of soluble salts are mixed (see Figure 18.61). Metathesis reactions are examined in the chapter on the stoichiometry of chemical reactions.

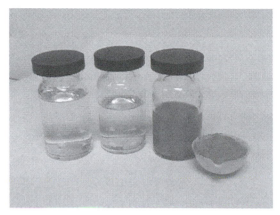

Figure 18.61   Solid HgI_2 forms when solutions of KI and $Hg(NO_3)_2$ are mixed. (credit: Sahar Atwa)

Several halides occur in large quantities in nature. The ocean and underground brines contain many halides. For example, magnesium chloride in the ocean is the source of magnesium ions used in the production of magnesium. Large underground deposits of sodium chloride, like the salt mine shown in Figure 18.62, occur in many parts of the world. These deposits serve as the source of sodium and chlorine in almost all other compounds containing these elements. The chlor-alkali process is one example.

Figure 18.62   Underground deposits of sodium chloride are found throughout the world and are often mined. This is a tunnel in the Kłodawa salt mine in Poland. (credit: Jarek Zok)

Interhalogens

Compounds formed from two or more different halogens are **interhalogens**. Interhalogen molecules consist of one atom of the heavier halogen bonded by single bonds to an odd number of atoms of the lighter halogen. The structures of IF$_3$, IF$_5$, and IF$_7$ are illustrated in Figure 18.63. Formulas for other interhalogens, each of which comes from the reaction of the respective halogens, are in Table 18.3.

IF$_3$ IF$_5$ IF$_7$

Figure 18.63   The structure of IF$_3$ is T-shaped (left), IF$_5$ is square pyramidal (center), and IF$_7$ is pentagonal bipyramidal (right).

Note from Table 18.3 that fluorine is able to oxidize iodine to its maximum oxidation state, 7+, whereas bromine and chlorine, which are more difficult to oxidize, achieve only the 5+-oxidation state. A 7+-oxidation state is the limit for the halogens. Because smaller halogens are grouped about a larger one, the maximum number of smaller atoms possible increases as the radius of the larger atom increases. Many of these compounds are unstable, and most are extremely reactive. The interhalogens react like their component halides; halogen fluorides, for example, are stronger oxidizing agents than are halogen chlorides.

The ionic polyhalides of the alkali metals, such as KI$_3$, KICl$_2$, KICl$_4$, CsIBr$_2$, and CsBrCl$_2$, which contain an anion composed of at least three halogen atoms, are closely related to the interhalogens. As seen previously, the formation of the polyhalide anion I$_3^-$ is responsible for the solubility of iodine in aqueous solutions containing an iodide ion.

Interhalogens

YX	YX$_3$	YX$_5$	YX$_7$
ClF(g)	ClF$_3$(g)	ClF$_5$(g)	
BrF(g)	BrF$_3$(l)	BrF$_5$(l)	
BrCl(g)			
IF(s)	IF$_3$(s)	IF$_5$(l)	IF$_7$(g)
ICl(l)	ICl$_3$(s)		
IBr(s)			

Table 18.3

Applications

The fluoride ion and fluorine compounds have many important uses. Compounds of carbon, hydrogen, and fluorine are replacing Freons (compounds of carbon, chlorine, and fluorine) as refrigerants. Teflon is a polymer composed of $-CF_2CF_2-$ units. Fluoride ion is added to water supplies and to some toothpastes as SnF$_2$ or NaF to fight tooth decay. Fluoride partially converts teeth from Ca$_5$(PO$_4$)$_3$(OH) into Ca$_5$(PO$_4$)$_3$F.

Chlorine is important to bleach wood pulp and cotton cloth. The chlorine reacts with water to form hypochlorous acid, which oxidizes colored substances to colorless ones. Large quantities of chlorine are important in chlorinating hydrocarbons (replacing hydrogen with chlorine) to produce compounds such as tetrachloride (CCl_4), chloroform ($CHCl_3$), and ethyl chloride (C_2H_5Cl), and in the production of polyvinyl chloride (PVC) and other polymers. Chlorine is also important to kill the bacteria in community water supplies.

Bromine is important in the production of certain dyes, and sodium and potassium bromides are used as sedatives. At one time, light-sensitive silver bromide was a component of photographic film.

Iodine in alcohol solution with potassium iodide is an antiseptic (tincture of iodine). Iodide salts are essential for the proper functioning of the thyroid gland; an iodine deficiency may lead to the development of a goiter. Iodized table salt contains 0.023% potassium iodide. Silver iodide is useful in the seeding of clouds to induce rain; it was important in the production of photographic film and iodoform, CHI_3, is an antiseptic.

18.12 Occurrence, Preparation, and Properties of the Noble Gases

By the end of this section, you will be able to:

- Describe the properties, preparation, and uses of the noble gases

The elements in group 18 are the noble gases (helium, neon, argon, krypton, xenon, and radon). They earned the name "noble" because they were assumed to be nonreactive since they have filled valence shells. In 1962, Dr. Neil Bartlett at the University of British Columbia proved this assumption to be false.

These elements are present in the atmosphere in small amounts. Some natural gas contains 1–2% helium by mass. Helium is isolated from natural gas by liquefying the condensable components, leaving only helium as a gas. The United States possesses most of the world's commercial supply of this element in its helium-bearing gas fields. Argon, neon, krypton, and xenon come from the fractional distillation of liquid air. Radon comes from other radioactive elements. More recently, it was observed that this radioactive gas is present in very small amounts in soils and minerals. Its accumulation in well-insulated, tightly sealed buildings, however, constitutes a health hazard, primarily lung cancer.

The boiling points and melting points of the noble gases are extremely low relative to those of other substances of comparable atomic or molecular masses. This is because only weak London dispersion forces are present, and these forces can hold the atoms together only when molecular motion is very slight, as it is at very low temperatures. Helium is the only substance known that does not solidify on cooling at normal pressure. It remains liquid close to absolute zero (0.001 K) at ordinary pressures, but it solidifies under elevated pressure.

Helium is used for filling balloons and lighter-than-air craft because it does not burn, making it safer to use than hydrogen. Helium at high pressures is not a narcotic like nitrogen. Thus, mixtures of oxygen and helium are important for divers working under high pressures. Using a helium-oxygen mixture avoids the disoriented mental state known as nitrogen narcosis, the so-called rapture of the deep. Helium is important as an inert atmosphere for the melting and welding of easily oxidizable metals and for many chemical processes that are sensitive to air.

Liquid helium (boiling point, 4.2 K) is an important coolant to reach the low temperatures necessary for cryogenic research, and it is essential for achieving the low temperatures necessary to produce superconduction in traditional superconducting materials used in powerful magnets and other devices. This cooling ability is necessary for the magnets used for magnetic resonance imaging, a common medical diagnostic procedure. The other common coolant is liquid nitrogen (boiling point, 77 K), which is significantly cheaper.

Neon is a component of neon lamps and signs. Passing an electric spark through a tube containing neon at low pressure generates the familiar red glow of neon. It is possible to change the color of the light by mixing argon or mercury vapor with the neon or by utilizing glass tubes of a special color.

Argon was useful in the manufacture of gas-filled electric light bulbs, where its lower heat conductivity and chemical inertness made it preferable to nitrogen for inhibiting the vaporization of the tungsten filament and prolonging the life of the bulb. Fluorescent tubes commonly contain a mixture of argon and mercury vapor. Argon is the third most abundant gas in dry air.

Krypton-xenon flash tubes are used to take high-speed photographs. An electric discharge through such a tube gives a very intense light that lasts only $\frac{1}{50,000}$ of a second. Krypton forms a difluoride, KrF_2, which is thermally unstable at room temperature.

Stable compounds of xenon form when xenon reacts with fluorine. Xenon difluoride, XeF_2, forms after heating an excess of xenon gas with fluorine gas and then cooling. The material forms colorless crystals, which are stable at room temperature in a dry atmosphere. Xenon tetrafluoride, XeF_4, and xenon hexafluoride, XeF_6, are prepared in an analogous manner, with a stoichiometric amount of fluorine and an excess of fluorine, respectively. Compounds with oxygen are prepared by replacing fluorine atoms in the xenon fluorides with oxygen.

When XeF_6 reacts with water, a solution of XeO_3 results and the xenon remains in the 6+-oxidation state:

$$XeF_6(s) + 3H_2O(l) \longrightarrow XeO_3(aq) + 6HF(aq)$$

Dry, solid xenon trioxide, XeO_3, is extremely explosive—it will spontaneously detonate. Both XeF_6 and XeO_3 disproportionate in basic solution, producing xenon, oxygen, and salts of the perxenate ion, XeO_6^{4-}, in which xenon reaches its maximum oxidation sate of 8+.

Radon apparently forms RnF_2—evidence of this compound comes from radiochemical tracer techniques.

Unstable compounds of argon form at low temperatures, but stable compounds of helium and neon are not known.

Key Terms

acid anhydride compound that reacts with water to form an acid or acidic solution

alkaline earth metal any of the metals (beryllium, magnesium, calcium, strontium, barium, and radium) occupying group 2 of the periodic table; they are reactive, divalent metals that form basic oxides

allotropes two or more forms of the same element, in the same physical state, with different chemical structures

amorphous solid material such as a glass that does not have a regular repeating component to its three-dimensional structure; a solid but not a crystal

base anhydride metal oxide that behaves as a base towards acids

bicarbonate anion salt of the hydrogen carbonate ion, HCO_3^-

bismuth heaviest member of group 15; a less reactive metal than other representative metals

borate compound containing boron-oxygen bonds, typically with clusters or chains as a part of the chemical structure

carbonate salt of the anion CO_3^{2-}; often formed by the reaction of carbon dioxide with bases

chemical reduction method of preparing a representative metal using a reducing agent

chlor-alkali process electrolysis process for the synthesis of chlorine and sodium hydroxide

disproportionation reaction chemical reaction where a single reactant is simultaneously reduced and oxidized; it is both the reducing agent and the oxidizing agent

Downs cell electrochemical cell used for the commercial preparation of metallic sodium (and chlorine) from molten sodium chloride

Frasch process important in the mining of free sulfur from enormous underground deposits

Haber process main industrial process used to produce ammonia from nitrogen and hydrogen; involves the use of an iron catalyst and elevated temperatures and pressures

halide compound containing an anion of a group 17 element in the 1− oxidation state (fluoride, F^-; chloride, Cl^-; bromide, Br^-; and iodide, I^-)

Hall–Héroult cell electrolysis apparatus used to isolate pure aluminum metal from a solution of alumina in molten cryolite

hydrogen carbonate salt of carbonic acid, H_2CO_3 (containing the anion HCO_3^-) in which one hydrogen atom has been replaced; an acid carbonate; also known as *bicarbonate ion*

hydrogen halide binary compound formed between hydrogen and the halogens: HF, HCl, HBr, and HI

hydrogen sulfate HSO_4^- ion

hydrogen sulfite HSO_3^- ion

hydrogenation addition of hydrogen (H_2) to reduce a compound

hydroxide compound of a metal with the hydroxide ion OH^- or the group $-OH$

interhalogen compound formed from two or more different halogens

metal atoms of the metallic elements of groups 1, 2, 12, 13, 14, 15, and 16, which form ionic compounds by losing electrons from their outer s or p orbitals

metalloid element that has properties that are between those of metals and nonmetals; these elements are typically semiconductors

nitrate NO_3^- ion; salt of nitric acid

nitrogen fixation formation of nitrogen compounds from molecular nitrogen

Ostwald process industrial process used to convert ammonia into nitric acid

oxide binary compound of oxygen with another element or group, typically containing O^{2-} ions or the group –O– or =O

ozone allotrope of oxygen; O_3

passivation metals with a protective nonreactive film of oxide or other compound that creates a barrier for chemical reactions; physical or chemical removal of the passivating film allows the metals to demonstrate their expected chemical reactivity

peroxide molecule containing two oxygen atoms bonded together or as the anion, O_2^{2-}

photosynthesis process whereby light energy promotes the reaction of water and carbon dioxide to form carbohydrates and oxygen; this allows photosynthetic organisms to store energy

Pidgeon process chemical reduction process used to produce magnesium through the thermal reaction of magnesium oxide with silicon

polymorph variation in crystalline structure that results in different physical properties for the resulting compound

representative element element where the s and p orbitals are filling

representative metal metal among the representative elements

silicate compound containing silicon-oxygen bonds, with silicate tetrahedra connected in rings, sheets, or three-dimensional networks, depending on the other elements involved in the formation of the compounds

sulfate SO_4^{2-} ion

sulfite SO_3^{2-} ion

superoxide oxide containing the anion O_2^-

Summary

18.1 Periodicity
This section focuses on the periodicity of the representative elements. These are the elements where the electrons are entering the s and p orbitals. The representative elements occur in groups 1, 2, and 12–18. These elements are representative metals, metalloids, and nonmetals. The alkali metals (group 1) are very reactive, readily form ions with a charge of 1+ to form ionic compounds that are usually soluble in water, and react vigorously with water to form hydrogen gas and a basic solution of the metal hydroxide. The outermost electrons of the alkaline earth metals (group 2) are more difficult to remove than the outer electron of the alkali metals, leading to the group 2 metals being less

reactive than those in group 1. These elements easily form compounds in which the metals exhibit an oxidation state of 2+. Zinc, cadmium, and mercury (group 12) commonly exhibit the group oxidation state of 2+ (although mercury also exhibits an oxidation state of 1+ in compounds that contain Hg_2^{2+}). Aluminum, gallium, indium, and thallium (group 13) are easier to oxidize than is hydrogen. Aluminum, gallium, and indium occur with an oxidation state 3+ (however, thallium also commonly occurs as the Tl^+ ion). Tin and lead form stable divalent cations and covalent compounds in which the metals exhibit the 4+-oxidation state.

18.2 Occurrence and Preparation of the Representative Metals

Because of their chemical reactivity, it is necessary to produce the representative metals in their pure forms by reduction from naturally occurring compounds. Electrolysis is important in the production of sodium, potassium, and aluminum. Chemical reduction is the primary method for the isolation of magnesium, zinc, and tin. Similar procedures are important for the other representative metals.

18.3 Structure and General Properties of the Metalloids

The elements boron, silicon, germanium, arsenic, antimony, and tellurium separate the metals from the nonmetals in the periodic table. These elements, called metalloids or sometimes semimetals, exhibit properties characteristic of both metals and nonmetals. The structures of these elements are similar in many ways to those of nonmetals, but the elements are electrical semiconductors.

18.4 Structure and General Properties of the Nonmetals

Nonmetals have structures that are very different from those of the metals, primarily because they have greater electronegativity and electrons that are more tightly bound to individual atoms. Most nonmetal oxides are acid anhydrides, meaning that they react with water to form acidic solutions. Molecular structures are common for most of the nonmetals, and several have multiple allotropes with varying physical properties.

18.5 Occurrence, Preparation, and Compounds of Hydrogen

Hydrogen is the most abundant element in the universe and its chemistry is truly unique. Although it has some chemical reactivity that is similar to that of the alkali metals, hydrogen has many of the same chemical properties of a nonmetal with a relatively low electronegativity. It forms ionic hydrides with active metals, covalent compounds in which it has an oxidation state of 1− with less electronegative elements, and covalent compounds in which it has an oxidation state of 1+ with more electronegative nonmetals. It reacts explosively with oxygen, fluorine, and chlorine, less readily with bromine, and much less readily with iodine, sulfur, and nitrogen. Hydrogen reduces the oxides of metals with lower reduction potentials than chromium to form the metal and water. The hydrogen halides are all acidic when dissolved in water.

18.6 Occurrence, Preparation, and Properties of Carbonates

The usual method for the preparation of the carbonates of the alkali and alkaline earth metals is by reaction of an oxide or hydroxide with carbon dioxide. Other carbonates form by precipitation. Metal carbonates or hydrogen carbonates such as limestone ($CaCO_3$), the antacid Tums ($CaCO_3$), and baking soda ($NaHCO_3$) are common examples. Carbonates and hydrogen carbonates decompose in the presence of acids and most decompose on heating.

18.7 Occurrence, Preparation, and Properties of Nitrogen

Nitrogen exhibits oxidation states ranging from 3− to 5+. Because of the stability of the N≡N triple bond, it requires a great deal of energy to make compounds from molecular nitrogen. Active metals such as the alkali metals and alkaline earth metals can reduce nitrogen to form metal nitrides. Nitrogen oxides and nitrogen hydrides are also important substances.

18.8 Occurrence, Preparation, and Properties of Phosphorus

Phosphorus (group 15) commonly exhibits oxidation states of 3− with active metals and of 3+ and 5+ with more electronegative nonmetals. The halogens and oxygen will oxidize phosphorus. The oxides are phosphorus(V) oxide, P_4O_{10}, and phosphorus(III) oxide, P_4O_6. The two common methods for preparing orthophosphoric acid, H_3PO_4, are either the reaction of a phosphate with sulfuric acid or the reaction of water with phosphorus(V) oxide. Orthophosphoric acid is a triprotic acid that forms three types of salts.

18.9 Occurrence, Preparation, and Compounds of Oxygen

Oxygen is one of the most reactive elements. This reactivity, coupled with its abundance, makes the chemistry of oxygen very rich and well understood.

Compounds of the representative metals with oxygen exist in three categories (1) oxides, (2) peroxides and superoxides, and (3) hydroxides. Heating the corresponding hydroxides, nitrates, or carbonates is the most common method for producing oxides. Heating the metal or metal oxide in oxygen may lead to the formation of peroxides and superoxides. The soluble oxides dissolve in water to form solutions of hydroxides. Most metals oxides are base anhydrides and react with acids. The hydroxides of the representative metals react with acids in acid-base reactions to form salts and water. The hydroxides have many commercial uses.

All nonmetals except fluorine form multiple oxides. Nearly all of the nonmetal oxides are acid anhydrides. The acidity of oxyacids requires that the hydrogen atoms bond to the oxygen atoms in the molecule rather than to the other nonmetal atom. Generally, the strength of the oxyacid increases with the number of oxygen atoms bonded to the nonmetal atom and not to a hydrogen.

18.10 Occurrence, Preparation, and Properties of Sulfur

Sulfur (group 16) reacts with almost all metals and readily forms the sulfide ion, S^{2-}, in which it has as oxidation state of 2−. Sulfur reacts with most nonmetals.

18.11 Occurrence, Preparation, and Properties of Halogens

The halogens form halides with less electronegative elements. Halides of the metals vary from ionic to covalent; halides of nonmetals are covalent. Interhalogens form by the combination of two or more different halogens.

All of the representative metals react directly with elemental halogens or with solutions of the hydrohalic acids (HF, HCl, HBr, and HI) to produce representative metal halides. Other laboratory preparations involve the addition of aqueous hydrohalic acids to compounds that contain such basic anions, such as hydroxides, oxides, or carbonates.

18.12 Occurrence, Preparation, and Properties of the Noble Gases

The most significant property of the noble gases (group 18) is their inactivity. They occur in low concentrations in the atmosphere. They find uses as inert atmospheres, neon signs, and as coolants. The three heaviest noble gases react with fluorine to form fluorides. The xenon fluorides are the best characterized as the starting materials for a few other noble gas compounds.

Exercises

18.1 Periodicity

1. How do alkali metals differ from alkaline earth metals in atomic structure and general properties?

2. Why does the reactivity of the alkali metals decrease from cesium to lithium?

3. Predict the formulas for the nine compounds that may form when each species in column 1 of Table 18.3 reacts with each species in column 2.

1	2
Na	I
Sr	Se
Al	O

4. Predict the best choice in each of the following. You may wish to review the chapter on electronic structure for relevant examples.

(a) the most metallic of the elements Al, Be, and Ba

(b) the most covalent of the compounds NaCl, $CaCl_2$, and $BeCl_2$

(c) the lowest first ionization energy among the elements Rb, K, and Li

(d) the smallest among Al, Al^+, and Al^{3+}

(e) the largest among Cs^+, Ba^{2+}, and Xe

5. Sodium chloride and strontium chloride are both white solids. How could you distinguish one from the other?

6. The reaction of quicklime, CaO, with water produces slaked lime, $Ca(OH)_2$, which is widely used in the construction industry to make mortar and plaster. The reaction of quicklime and water is highly exothermic:

$$CaO(s) + H_2O(l) \longrightarrow Ca(OH)_2(s) \qquad \Delta H = -350 \text{ kJ mol}^{-1}$$

(a) What is the enthalpy of reaction per gram of quicklime that reacts?

(b) How much heat, in kilojoules, is associated with the production of 1 ton of slaked lime?

7. Write a balanced equation for the reaction of elemental strontium with each of the following:

(a) oxygen

(b) hydrogen bromide

(c) hydrogen

(d) phosphorus

(e) water

8. How many moles of ionic species are present in 1.0 L of a solution marked 1.0 M mercury(I) nitrate?

9. What is the mass of fish, in kilograms, that one would have to consume to obtain a fatal dose of mercury, if the fish contains 30 parts per million of mercury by weight? (Assume that all the mercury from the fish ends up as mercury(II) chloride in the body and that a fatal dose is 0.20 g of $HgCl_2$.) How many pounds of fish is this?

10. The elements sodium, aluminum, and chlorine are in the same period.

(a) Which has the greatest electronegativity?

(b) Which of the atoms is smallest?

(c) Write the Lewis structure for the simplest covalent compound that can form between aluminum and chlorine.

(d) Will the oxide of each element be acidic, basic, or amphoteric?

11. Does metallic tin react with HCl?

12. What is tin pest, also known as tin disease?

13. Compare the nature of the bonds in $PbCl_2$ to that of the bonds in $PbCl_4$.

14. Is the reaction of rubidium with water more or less vigorous than that of sodium? How does the rate of reaction of magnesium compare?

18.2 Occurrence and Preparation of the Representative Metals

15. Write an equation for the reduction of cesium chloride by elemental calcium at high temperature.

16. Why is it necessary to keep the chlorine and sodium, resulting from the electrolysis of sodium chloride, separate during the production of sodium metal?

17. Give balanced equations for the overall reaction in the electrolysis of molten lithium chloride and for the reactions occurring at the electrodes. You may wish to review the chapter on electrochemistry for relevant examples.

18. The electrolysis of molten sodium chloride or of aqueous sodium chloride produces chlorine.

Calculate the mass of chlorine produced from 3.00 kg sodium chloride in each case. You may wish to review the chapter on electrochemistry for relevant examples.

19. What mass, in grams, of hydrogen gas forms during the complete reaction of 10.01 g of calcium with water?

20. How many grams of oxygen gas are necessary to react completely with 3.01×10^{21} atoms of magnesium to yield magnesium oxide?

21. Magnesium is an active metal; it burns in the form of powder, ribbons, and filaments to provide flashes of brilliant light. Why is it possible to use magnesium in construction?

22. Why is it possible for an active metal like aluminum to be useful as a structural metal?

23. Describe the production of metallic aluminum by electrolytic reduction.

24. What is the common ore of tin and how is tin separated from it?

25. A chemist dissolves a 1.497-g sample of a type of metal (an alloy of Sn, Pb, Sb, and Cu) in nitric acid, and metastannic acid, H_2SnO_3, is precipitated. She heats the precipitate to drive off the water, which leaves 0.4909 g of tin(IV) oxide. What was the percentage of tin in the original sample?

26. Consider the production of 100 kg of sodium metal using a current of 50,000 A, assuming a 100% yield.

(a) How long will it take to produce the 100 kg of sodium metal?

(b) What volume of chlorine at 25 °C and 1.00 atm forms?

27. What mass of magnesium forms when 100,000 A is passed through a $MgCl_2$ melt for 1.00 h if the yield of magnesium is 85% of the theoretical yield?

18.3 Structure and General Properties of the Metalloids

28. Give the hybridization of the metalloid and the molecular geometry for each of the following compounds or ions. You may wish to review the chapters on chemical bonding and advanced covalent bonding for relevant examples.

(a) GeH_4

(b) SbF_3

(c) $Te(OH)_6$

(d) H_2Te

(e) GeF_2

(f) $TeCl_4$

(g) $SiF_6{}^{2-}$

(h) $SbCl_5$

(i) TeF_6

29. Write a Lewis structure for each of the following molecules or ions. You may wish to review the chapter on chemical bonding.

(a) H_3BPH_3

(b) $BF_4{}^{-}$

(c) BBr_3

(d) $B(CH_3)_3$

(e) $B(OH)_3$

30. Describe the hybridization of boron and the molecular structure about the boron in each of the following:

(a) H_3BPH_3

(b) $BF_4{}^{-}$

(c) BBr_3

(d) $B(CH_3)_3$

(e) $B(OH)_3$

31. Using only the periodic table, write the complete electron configuration for silicon, including any empty orbitals in the valence shell. You may wish to review the chapter on electronic structure.

32. Write a Lewis structure for each of the following molecules and ions:

(a) $(CH_3)_3SiH$

(b) $SiO_4{}^{4-}$

(c) Si_2H_6

(d) $Si(OH)_4$

(e) $SiF_6{}^{2-}$

33. Describe the hybridization of silicon and the molecular structure of the following molecules and ions:

(a) $(CH_3)_3SiH$

(b) $SiO_4{}^{4-}$

(c) Si_2H_6

(d) $Si(OH)_4$

(e) $SiF_6{}^{2-}$

34. Describe the hybridization and the bonding of a silicon atom in elemental silicon.

35. Classify each of the following molecules as polar or nonpolar. You may wish to review the chapter on chemical bonding.

(a) SiH_4

(b) Si_2H_6

(c) $SiCl_3H$

(d) SiF_4

(e) $SiCl_2F_2$

36. Silicon reacts with sulfur at elevated temperatures. If 0.0923 g of silicon reacts with sulfur to give 0.3030 g of silicon sulfide, determine the empirical formula of silicon sulfide.

37. Name each of the following compounds:

(a) TeO_2

(b) Sb_2S_3

(c) GeF_4

(d) SiH_4

(e) GeH_4

38. Write a balanced equation for the reaction of elemental boron with each of the following (most of these reactions require high temperature):

(a) F_2

(b) O_2

(c) S

(d) Se

(e) Br_2

39. Why is boron limited to a maximum coordination number of four in its compounds?

40. Write a formula for each of the following compounds:

(a) silicon dioxide

(b) silicon tetraiodide

(c) silane

(d) silicon carbide

(e) magnesium silicide

41. From the data given in Appendix I , determine the standard enthalpy change and the standard free energy change for each of the following reactions:

(a) $BF_3(g) + 3H_2O(l) \longrightarrow B(OH)_3(s) + 3HF(g)$

(b) $BCl_3(g) + 3H_2O(l) \longrightarrow B(OH)_3(s) + 3HCl(g)$

(c) $B_2H_6(g) + 6H_2O(l) \longrightarrow 2B(OH)_3(s) + 6H_2(g)$

42. A hydride of silicon prepared by the reaction of Mg_2Si with acid exerted a pressure of 306 torr at 26 °C in a bulb with a volume of 57.0 mL. If the mass of the hydride was 0.0861 g, what is its molecular mass? What is the molecular formula for the hydride?

43. Suppose you discovered a diamond completely encased in a silicate rock. How would you chemically free the diamond without harming it?

18.4 Structure and General Properties of the Nonmetals

44. Carbon forms a number of allotropes, two of which are graphite and diamond. Silicon has a diamond structure. Why is there no allotrope of silicon with a graphite structure?

45. Nitrogen in the atmosphere exists as very stable diatomic molecules. Why does phosphorus form less stable P_4 molecules instead of P_2 molecules?

46. Write balanced chemical equations for the reaction of the following acid anhydrides with water:

(a) SO_3

(b) N_2O_3

(c) Cl_2O_7

(d) P_4O_{10}

(e) NO_2

47. Determine the oxidation number of each element in each of the following compounds:

(a) HCN

(b) OF_2

(c) $AsCl_3$

48. Determine the oxidation state of sulfur in each of the following:

(a) SO_3

(b) SO_2

(c) $SO_3{}^{2-}$

49. Arrange the following in order of increasing electronegativity: F; Cl; O; and S.

50. Why does white phosphorus consist of tetrahedral P_4 molecules while nitrogen consists of diatomic N_2 molecules?

18.5 Occurrence, Preparation, and Compounds of Hydrogen

51. Why does hydrogen not exhibit an oxidation state of 1− when bonded to nonmetals?

52. The reaction of calcium hydride, CaH_2, with water can be characterized as a Lewis acid-base reaction:
$$CaH_2(s) + 2H_2O(l) \longrightarrow Ca(OH)_2(aq) + 2H_2(g)$$

Identify the Lewis acid and the Lewis base among the reactants. The reaction is also an oxidation-reduction reaction. Identify the oxidizing agent, the reducing agent, and the changes in oxidation number that occur in the reaction.

53. In drawing Lewis structures, we learn that a hydrogen atom forms only one bond in a covalent compound. Why?

54. What mass of CaH_2 is necessary to react with water to provide enough hydrogen gas to fill a balloon at 20 °C and 0.8 atm pressure with a volume of 4.5 L? The balanced equation is:
$$CaH_2(s) + 2H_2O(l) \longrightarrow Ca(OH)_2(aq) + 2H_2(g)$$

55. What mass of hydrogen gas results from the reaction of 8.5 g of KH with water?
$$KH + H_2O \longrightarrow KOH + H_2$$

18.6 Occurrence, Preparation, and Properties of Carbonates

56. Carbon forms the $CO_3{}^{2-}$ ion, yet silicon does not form an analogous $SiO_3{}^{2-}$ ion. Why?

57. Complete and balance the following chemical equations:

(a) hardening of plaster containing slaked lime
$$Ca(OH)_2 + CO_2 \longrightarrow$$

(b) removal of sulfur dioxide from the flue gas of power plants
$$CaO + SO_2 \longrightarrow$$

(c) the reaction of baking powder that produces carbon dioxide gas and causes bread to rise
$$NaHCO_3 + NaH_2PO_4 \longrightarrow$$

58. Heating a sample of $Na_2CO_3 \cdot xH_2O$ weighing 4.640 g until the removal of the water of hydration leaves 1.720 g of anhydrous Na_2CO_3. What is the formula of the hydrated compound?

18.7 Occurrence, Preparation, and Properties of Nitrogen

59. Write the Lewis structures for each of the following:

(a) NH^{2-}

(b) N_2F_4

(c) $NH_2{}^{-}$

(d) NF_3

(e) $N_3{}^{-}$

60. For each of the following, indicate the hybridization of the nitrogen atom (for $N_3{}^{-}$, the central nitrogen).

(a) N_2F_4

(b) $NH_2{}^{-}$

(c) NF_3

(d) $N_3{}^{-}$

61. Explain how ammonia can function both as a Brønsted base and as a Lewis base.

62. Determine the oxidation state of nitrogen in each of the following. You may wish to review the chapter on chemical bonding for relevant examples.

(a) NCl_3

(b) $ClNO$

(c) N_2O_5

(d) N_2O_3

(e) $NO_2{}^-$

(f) N_2O_4

(g) N_2O

(h) $NO_3{}^-$

(i) HNO_2

(j) HNO_3

63. For each of the following, draw the Lewis structure, predict the ONO bond angle, and give the hybridization of the nitrogen. You may wish to review the chapters on chemical bonding and advanced theories of covalent bonding for relevant examples.

(a) NO_2

(b) $NO_2{}^-$

(c) $NO_2{}^+$

64. How many grams of gaseous ammonia will the reaction of 3.0 g hydrogen gas and 3.0 g of nitrogen gas produce?

65. Although PF_5 and AsF_5 are stable, nitrogen does not form NF_5 molecules. Explain this difference among members of the same group.

66. The equivalence point for the titration of a 25.00-mL sample of CsOH solution with 0.1062 M HNO_3 is at 35.27 mL. What is the concentration of the CsOH solution?

18.8 Occurrence, Preparation, and Properties of Phosphorus

67. Write the Lewis structure for each of the following. You may wish to review the chapter on chemical bonding and molecular geometry.

(a) PH_3

(b) $PH_4{}^+$

(c) P_2H_4

(d) $PO_4{}^{3-}$

(e) PF_5

68. Describe the molecular structure of each of the following molecules or ions listed. You may wish to review the chapter on chemical bonding and molecular geometry.

(a) PH_3

(b) $PH_4{}^+$

(c) P_2H_4

(d) $PO_4{}^{3-}$

69. Complete and balance each of the following chemical equations. (In some cases, there may be more than one correct answer.)

(a) $P_4 + Al \longrightarrow$

(b) $P_4 + Na \longrightarrow$

(c) $P_4 + F_2 \longrightarrow$

(d) $P_4 + Cl_2 \longrightarrow$

(e) $P_4 + O_2 \longrightarrow$

(f) $P_4O_6 + O_2 \longrightarrow$

70. Describe the hybridization of phosphorus in each of the following compounds: P_4O_{10}, P_4O_6, PH_4I (an ionic compound), PBr_3, H_3PO_4, H_3PO_3, PH_3, and P_2H_4. You may wish to review the chapter on advanced theories of covalent bonding.

71. What volume of 0.200 M NaOH is necessary to neutralize the solution produced by dissolving 2.00 g of PCl_3 is an excess of water? Note that when H_3PO_3 is titrated under these conditions, only one proton of the acid molecule reacts.

72. How much $POCl_3$ can form from 25.0 g of PCl_5 and the appropriate amount of H_2O?

73. How many tons of $Ca_3(PO_4)_2$ are necessary to prepare 5.0 tons of phosphorus if the yield is 90%?

74. Write equations showing the stepwise ionization of phosphorous acid.

75. Draw the Lewis structures and describe the geometry for the following:

(a) PF_4^+

(b) PF_5

(c) PF_6^-

(d) POF_3

76. Why does phosphorous acid form only two series of salts, even though the molecule contains three hydrogen atoms?

77. Assign an oxidation state to phosphorus in each of the following:

(a) NaH_2PO_3

(b) PF_5

(c) P_4O_6

(d) K_3PO_4

(e) Na_3P

(f) $Na_4P_2O_7$

78. Phosphoric acid, one of the acids used in some cola drinks, is produced by the reaction of phosphorus(V) oxide, an acidic oxide, with water. Phosphorus(V) oxide is prepared by the combustion of phosphorus.

(a) Write the empirical formula of phosphorus(V) oxide.

(b) What is the molecular formula of phosphorus(V) oxide if the molar mass is about 280.

(c) Write balanced equations for the production of phosphorus(V) oxide and phosphoric acid.

(d) Determine the mass of phosphorus required to make 1.00×10^4 kg of phosphoric acid, assuming a yield of 98.85%.

18.9 Occurrence, Preparation, and Compounds of Oxygen

79. Predict the product of burning francium in air.

80. Using equations, describe the reaction of water with potassium and with potassium oxide.

81. Write balanced chemical equations for the following reactions:

(a) zinc metal heated in a stream of oxygen gas

(b) zinc carbonate heated until loss of mass stops

(c) zinc carbonate added to a solution of acetic acid, CH_3CO_2H

(d) zinc added to a solution of hydrobromic acid

82. Write balanced chemical equations for the following reactions:

(a) cadmium burned in air

(b) elemental cadmium added to a solution of hydrochloric acid

(c) cadmium hydroxide added to a solution of acetic acid, CH_3CO_2H

83. Illustrate the amphoteric nature of aluminum hydroxide by citing suitable equations.

84. Write balanced chemical equations for the following reactions:

(a) metallic aluminum burned in air

(b) elemental aluminum heated in an atmosphere of chlorine

(c) aluminum heated in hydrogen bromide gas

(d) aluminum hydroxide added to a solution of nitric acid

85. Write balanced chemical equations for the following reactions:

(a) sodium oxide added to water

(b) cesium carbonate added to an excess of an aqueous solution of HF

(c) aluminum oxide added to an aqueous solution of $HClO_4$

(d) a solution of sodium carbonate added to solution of barium nitrate

(e) titanium metal produced from the reaction of titanium tetrachloride with elemental sodium

86. What volume of 0.250 M H_2SO_4 solution is required to neutralize a solution that contains 5.00 g of $CaCO_3$?

87. Which is the stronger acid, $HClO_4$ or $HBrO_4$? Why?

88. Write a balanced chemical equation for the reaction of an excess of oxygen with each of the following. Remember that oxygen is a strong oxidizing agent and tends to oxidize an element to its maximum oxidation state.

(a) Mg

(b) Rb

(c) Ga

(d) C_2H_2

(e) CO

89. Which is the stronger acid, H_2SO_4 or H_2SeO_4? Why? You may wish to review the chapter on acid-base equilibria.

18.10 Occurrence, Preparation, and Properties of Sulfur

90. Explain why hydrogen sulfide is a gas at room temperature, whereas water, which has a lower molecular mass, is a liquid.

91. Give the hybridization and oxidation state for sulfur in SO_2, in SO_3, and in H_2SO_4.

92. Which is the stronger acid, $NaHSO_3$ or $NaHSO_4$?

93. Determine the oxidation state of sulfur in SF_6, SO_2F_2, and KHS.

94. Which is a stronger acid, sulfurous acid or sulfuric acid? Why?

95. Oxygen forms double bonds in O_2, but sulfur forms single bonds in S_8. Why?

96. Give the Lewis structure of each of the following:

(a) SF_4

(b) K_2SO_4

(c) SO_2Cl_2

(d) H_2SO_3

(e) SO_3

97. Write two balanced chemical equations in which sulfuric acid acts as an oxidizing agent.

98. Explain why sulfuric acid, H_2SO_4, which is a covalent molecule, dissolves in water and produces a solution that contains ions.

99. How many grams of Epsom salts ($MgSO_4 \cdot 7H_2O$) will form from 5.0 kg of magnesium?

18.11 Occurrence, Preparation, and Properties of Halogens

100. What does it mean to say that mercury(II) halides are weak electrolytes?

101. Why is $SnCl_4$ not classified as a salt?

102. The following reactions are all similar to those of the industrial chemicals. Complete and balance the equations for these reactions:

(a) reaction of a weak base and a strong acid

$NH_3 + HClO_4 \longrightarrow$

(b) preparation of a soluble silver salt for silver plating

$Ag_2CO_3 + HNO_3 \longrightarrow$

(c) preparation of strontium hydroxide by electrolysis of a solution of strontium chloride

$SrCl_2(aq) + H_2O(l) \xrightarrow{\text{electrolysis}}$

103. Which is the stronger acid, $HClO_3$ or $HBrO_3$? Why?

104. What is the hybridization of iodine in IF_3 and IF_5?

105. Predict the molecular geometries and draw Lewis structures for each of the following. You may wish to review the chapter on chemical bonding and molecular geometry.

(a) IF_5

(b) I_3^-

(c) PCl_5

(d) SeF_4

(e) ClF_3

106. Which halogen has the highest ionization energy? Is this what you would predict based on what you have learned about periodic properties?

107. Name each of the following compounds:

(a) BrF_3

(b) $NaBrO_3$

(c) PBr_5

(d) $NaClO_4$

(e) $KClO$

108. Explain why, at room temperature, fluorine and chlorine are gases, bromine is a liquid, and iodine is a solid.

109. What is the oxidation state of the halogen in each of the following?

(a) H_5IO_6

(b) IO_4^-

(c) ClO_2

(d) ICl_3

(e) F_2

110. Physiological saline concentration—that is, the sodium chloride concentration in our bodies—is approximately 0.16 M. A saline solution for contact lenses is prepared to match the physiological concentration. If you purchase 25 mL of contact lens saline solution, how many grams of sodium chloride have you bought?

18.12 Occurrence, Preparation, and Properties of the Noble Gases

111. Give the hybridization of xenon in each of the following. You may wish to review the chapter on the advanced theories of covalent bonding.

(a) XeF_2

(b) XeF_4

(c) XeO_3

(d) XeO_4

(e) $XeOF_4$

112. What is the molecular structure of each of the following molecules? You may wish to review the chapter on chemical bonding and molecular geometry.

(a) XeF_2

(b) XeF_4

(c) XeO_3

(d) XeO_4

(e) $XeOF_4$

113. Indicate whether each of the following molecules is polar or nonpolar. You may wish to review the chapter on chemical bonding and molecular geometry.

(a) XeF_2

(b) XeF_4

(c) XeO_3

(d) XeO_4

(e) $XeOF_4$

114. What is the oxidation state of the noble gas in each of the following? You may wish to review the chapter on chemical bonding and molecular geometry.

(a) XeO_2F_2

(b) KrF_2

(c) $XeF_3{}^+$

(d) $XeO_6{}^{4-}$

(e) XeO_3

115. A mixture of xenon and fluorine was heated. A sample of the white solid that formed reacted with hydrogen to yield 81 mL of xenon (at STP) and hydrogen fluoride, which was collected in water, giving a solution of hydrofluoric acid. The hydrofluoric acid solution was titrated, and 68.43 mL of 0.3172 M sodium hydroxide was required to reach the equivalence point. Determine the empirical formula for the white solid and write balanced chemical equations for the reactions involving xenon.

116. Basic solutions of Na_4XeO_6 are powerful oxidants. What mass of $Mn(NO_3)_2 \cdot 6H_2O$ reacts with 125.0 mL of a 0.1717 *M* basic solution of Na_4XeO_6 that contains an excess of sodium hydroxide if the products include Xe and solution of sodium permanganate?

Chapter 19

Transition Metals and Coordination Chemistry

Figure 19.1 Transition metals often form vibrantly colored complexes. The minerals malachite (green), azurite (blue), and proustite (red) are some examples. (credit left: modification of work by James St. John; credit middle: modification of work by Stephanie Clifford; credit right: modification of work by Terry Wallace)

Chapter Outline

19.1 Occurrence, Preparation, and Properties of Transition Metals and Their Compounds

19.2 Coordination Chemistry of Transition Metals

19.3 Spectroscopic and Magnetic Properties of Coordination Compounds

Introduction

We have daily contact with many transition metals. Iron occurs everywhere—from the rings in your spiral notebook and the cutlery in your kitchen to automobiles, ships, buildings, and in the hemoglobin in your blood. Titanium is useful in the manufacture of lightweight, durable products such as bicycle frames, artificial hips, and jewelry. Chromium is useful as a protective plating on plumbing fixtures and automotive detailing.

In addition to being used in their pure elemental forms, many compounds containing transition metals have numerous other applications. Silver nitrate is used to create mirrors, zirconium silicate provides friction in automotive brakes, and many important cancer-fighting agents, like the drug cisplatin and related species, are platinum compounds.

The variety of properties exhibited by transition metals is due to their complex valence shells. Unlike most main group metals where one oxidation state is normally observed, the valence shell structure of transition metals means that they usually occur in several different stable oxidation states. In addition, electron transitions in these elements can correspond with absorption of photons in the visible electromagnetic spectrum, leading to colored compounds. Because of these behaviors, transition metals exhibit a rich and fascinating chemistry.

19.1 Occurrence, Preparation, and Properties of Transition Metals and Their Compounds

By the end of this section, you will be able to:

- Outline the general approach for the isolation of transition metals from natural sources
- Describe typical physical and chemical properties of the transition metals
- Identify simple compound classes for transition metals and describe their chemical properties

Transition metals are defined as those elements that have (or readily form) partially filled *d* orbitals. As shown in Figure 19.2, the **d-block elements** in groups 3–11 are transition elements. The **f-block elements**, also called *inner transition metals* (the lanthanides and actinides), also meet this criterion because the *d* orbital is partially occupied before the *f* orbitals. The *d* orbitals fill with the copper family (group 11); for this reason, the next family (group 12) are technically not transition elements. However, the group 12 elements do display some of the same chemical properties and are commonly included in discussions of transition metals. Some chemists do treat the group 12 elements as transition metals.

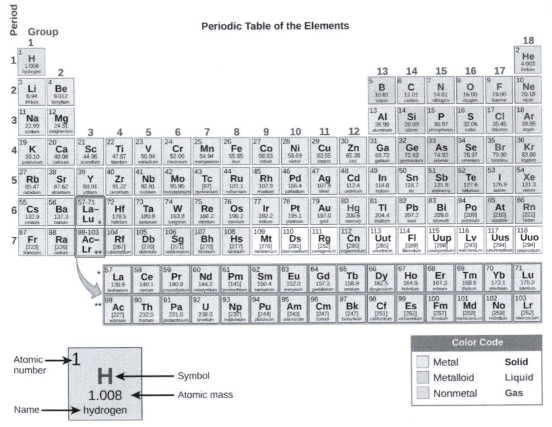

Figure 19.2 The transition metals are located in groups 3–11 of the periodic table. The inner transition metals are in the two rows below the body of the table.

The *d*-block elements are divided into the **first transition series** (the elements Sc through Cu), the **second transition series** (the elements Y through Ag), and the **third transition series** (the element La and the elements Hf through Au). Actinium, Ac, is the first member of the **fourth transition series**, which also includes Rf through Rg.

The *f*-block elements are the elements Ce through Lu, which constitute the **lanthanide series** (or **lanthanoid series**), and the elements Th through Lr, which constitute the **actinide series** (or **actinoid series**). Because lanthanum behaves very much like the lanthanide elements, it is considered a lanthanide element, even though its electron configuration makes it the first member of the third transition series. Similarly, the behavior of actinium means it is part of the actinide series, although its electron configuration makes it the first member of the fourth transition series.

Example 19.1

Valence Electrons in Transition Metals

Review how to write electron configurations, covered in the chapter on electronic structure and periodic properties of elements. Recall that for the transition and inner transition metals, it is necessary to remove the *s* electrons before the *d* or *f* electrons. Then, for each ion, give the electron configuration:

(a) cerium(III)

(b) lead(II)

(c) Ti^{2+}

(d) Am^{3+}

(e) Pd^{2+}

For the examples that are transition metals, determine to which series they belong.

Solution

For ions, the *s*-valence electrons are lost prior to the *d* or *f* electrons.

(a) $Ce^{3+}[Xe]4f^1$; Ce^{3+} is an inner transition element in the lanthanide series.

(b) $Pb^{2+}[Xe]6s^25d^{10}4f^{14}$; the electrons are lost from the *p* orbital. This is a main group element.
(c) titanium(II) $[Ar]3d^2$; first transition series

(d) americium(III) $[Rn]5f^6$; actinide

(e) palladium(II) $[Kr]4d^8$; second transition series

Check Your Learning

Give an example of an ion from the first transition series with no *d* electrons.

Answer: V^{5+} is one possibility. Other examples include Sc^{3+}, Ti^{4+}, Cr^{6+}, and Mn^{7+}.

Chemistry in Everyday Life

Uses of Lanthanides in Devices

Lanthanides (elements 57–71) are fairly abundant in the earth's crust, despite their historic characterization as **rare earth elements**. Thulium, the rarest naturally occurring lanthanoid, is more common in the earth's crust than silver (4.5×10^{-5}% versus 0.79×10^{-5}% by mass). There are 17 rare earth elements, consisting of the 15 lanthanoids plus scandium and yttrium. They are called rare because they were once difficult to extract economically, so it was rare to have a pure sample; due to similar chemical properties, it is difficult to separate any one lanthanide from the others. However, newer separation methods, such as ion exchange resins similar to those found in home water softeners, make the separation of these elements easier and more economical. Most ores that contain these elements have low concentrations of all the rare earth elements mixed together.

The commercial applications of lanthanides are growing rapidly. For example, europium is important in flat screen displays found in computer monitors, cell phones, and televisions. Neodymium is useful in laptop hard drives and in the processes that convert crude oil into gasoline (Figure 19.3). Holmium is found in dental and medical equipment. In addition, many alternative energy technologies rely heavily on lanthanoids. Neodymium and dysprosium are key components of hybrid vehicle engines and the magnets used in wind turbines.

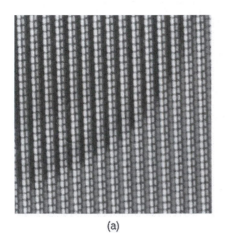
(a)

(b)

Figure 19.3   (a) Europium is used in display screens for televisions, computer monitors, and cell phones. (b) Neodymium magnets are commonly found in computer hard drives. (credit b: modification of work by "KUERT Datenrettung"/Flickr)

As the demand for lanthanide materials has increased faster than supply, prices have also increased. In 2008, dysprosium cost $110/kg; by 2014, the price had increased to $470/kg. Increasing the supply of lanthanoid elements is one of the most significant challenges facing the industries that rely on the optical and magnetic properties of these materials.

The transition elements have many properties in common with other metals. They are almost all hard, high-melting solids that conduct heat and electricity well. They readily form alloys and lose electrons to form stable cations. In addition, transition metals form a wide variety of stable **coordination compounds**, in which the central metal atom or ion acts as a Lewis acid and accepts one or more pairs of electrons. Many different molecules and ions can donate lone pairs to the metal center, serving as Lewis bases. In this chapter, we shall focus primarily on the chemical behavior of the elements of the first transition series.

Properties of the Transition Elements

Transition metals demonstrate a wide range of chemical behaviors. As can be seen from their reduction potentials (see Appendix H), some transition metals are strong reducing agents, whereas others have very low reactivity. For example, the lanthanides all form stable 3+ aqueous cations. The driving force for such oxidations is similar to that of alkaline earth metals such as Be or Mg, forming Be^{2+} and Mg^{2+}. On the other hand, materials like platinum and gold have much higher reduction potentials. Their ability to resist oxidation makes them useful materials for constructing circuits and jewelry.

Ions of the lighter d-block elements, such as Cr^{3+}, Fe^{3+}, and Co^{2+}, form colorful hydrated ions that are stable in water. However, ions in the period just below these (Mo^{3+}, Ru^{3+}, and Ir^{2+}) are unstable and react readily with oxygen from the air. The majority of simple, water-stable ions formed by the heavier d-block elements are oxyanions such as $MoO_4{}^{2-}$ and $ReO_4{}^{-}$.

Ruthenium, osmium, rhodium, iridium, palladium, and platinum are the **platinum metals**. With difficulty, they form simple cations that are stable in water, and, unlike the earlier elements in the second and third transition series, they do not form stable oxyanions.

Both the d- and f-block elements react with nonmetals to form binary compounds; heating is often required. These elements react with halogens to form a variety of halides ranging in oxidation state from 1+ to 6+. On heating, oxygen reacts with all of the transition elements except palladium, platinum, silver, and gold. The oxides of these latter metals can be formed using other reactants, but they decompose upon heating. The f-block elements, the elements of group 3, and the elements of the first transition series except copper react with aqueous solutions of acids, forming hydrogen gas and solutions of the corresponding salts.

Transition metals can form compounds with a wide range of oxidation states. Some of the observed oxidation states of the elements of the first transition series are shown in Figure 19.4. As we move from left to right across the first transition series, we see that the number of common oxidation states increases at first to a maximum towards the middle of the table, then decreases. The values in the table are typical values; there are other known values, and it is possible to synthesize new additions. For example, in 2014, researchers were successful in synthesizing a new oxidation state of iridium (9+).

21 Sc	22 Ti	23 V	24 Cr	25 Mn	26 Fe	27 Co	28 Ni	29 Cu	30 Zn
								1+	
		2+	2+	2+	2+	2+	2+	2+	2+
3+	3+	3+	3+	3+	3+	3+	3+	3+	
	4+	4+	4+	4+					
		5+							
			6+	6+	6+				
				7+					

Figure 19.4 Transition metals of the first transition series can form compounds with varying oxidation states.

For the elements scandium through manganese (the first half of the first transition series), the highest oxidation state corresponds to the loss of all of the electrons in both the s and d orbitals of their valence shells. The titanium(IV) ion, for example, is formed when the titanium atom loses its two $3d$ and two $4s$ electrons. These highest oxidation states are the most stable forms of scandium, titanium, and vanadium. However, it is not possible to continue to remove all of the valence electrons from metals as we continue through the series. Iron is known to form oxidation states from 2+ to 6+, with iron(II) and iron(III) being the most common. Most of the elements of the first transition series form ions with a charge of 2+ or 3+ that are stable in water, although those of the early members of the series can be readily oxidized by air.

The elements of the second and third transition series generally are more stable in higher oxidation states than are the elements of the first series. In general, the atomic radius increases down a group, which leads to the ions of the second and third series being larger than are those in the first series. Removing electrons from orbitals that are located farther from the nucleus is easier than removing electrons close to the nucleus. For example, molybdenum and tungsten, members of group 6, are limited mostly to an oxidation state of 6+ in aqueous solution. Chromium, the lightest member of the group, forms stable Cr^{3+} ions in water and, in the absence of air, less stable Cr^{2+} ions. The sulfide with the highest oxidation state for chromium is Cr_2S_3, which contains the Cr^{3+} ion. Molybdenum and tungsten form sulfides in which the metals exhibit oxidation states of 4+ and 6+.

Example 19.2

Activity of the Transition Metals

Which is the strongest oxidizing agent in acidic solution: dichromate ion, which contains chromium(VI), permanganate ion, which contains manganese(VII), or titanium dioxide, which contains titanium(IV)?

Solution

First, we need to look up the reduction half reactions (in Appendix L) for each oxide in the specified oxidation state:

$$Cr_2O_7{}^{2-} + 14H^+ + 6e^- \longrightarrow 2Cr^{3+} + 7H_2O \qquad +1.33 \text{ V}$$
$$MnO_4{}^- + 8H^+ + 5e^- \longrightarrow Mn^{2+} + H_2O \qquad +1.51 \text{ V}$$
$$TiO_2 + 4H^+ + 2e^- \longrightarrow Ti^{2+} + 2H_2O \qquad -0.50 \text{ V}$$

A larger reduction potential means that it is easier to reduce the reactant. Permanganate, with the largest reduction potential, is the strongest oxidizer under these conditions. Dichromate is next, followed by titanium dioxide as the weakest oxidizing agent (the hardest to reduce) of this set.

Check Your Learning

Predict what reaction (if any) will occur between HCl and Co(s), and between HBr and Pt(s). You will need to use the standard reduction potentials from Appendix L.

Answer: $Co(s) + 2HCl \longrightarrow H_2 + CoCl_2(aq)$; no reaction because Pt(s) will not be oxidized by H^+

Preparation of the Transition Elements

Ancient civilizations knew about iron, copper, silver, and gold. The time periods in human history known as the Bronze Age and Iron Age mark the advancements in which societies learned to isolate certain metals and use them to make tools and goods. Naturally occurring ores of copper, silver, and gold can contain high concentrations of these metals in elemental form (Figure 19.5). Iron, on the other hand, occurs on earth almost exclusively in oxidized forms, such as rust (Fe_2O_3). The earliest known iron implements were made from iron meteorites. Surviving iron artifacts dating from approximately 4000 to 2500 BC are rare, but all known examples contain specific alloys of iron and nickel that occur only in extraterrestrial objects, not on earth. It took thousands of years of technological advances before civilizations developed iron **smelting**, the ability to extract a pure element from its naturally occurring ores and for iron tools to become common.

(a)

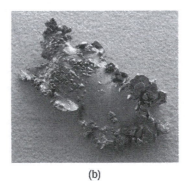
(b)

(c)

Figure 19.5 Transition metals occur in nature in various forms. Examples include (a) a nugget of copper, (b) a deposit of gold, and (c) an ore containing oxidized iron. (credit a: modification of work by http://images-of-elements.com/copper-2.jpg; credit c: modification of work by http://images-of-elements.com/iron-ore.jpg)

Generally, the transition elements are extracted from minerals found in a variety of ores. However, the ease of their recovery varies widely, depending on the concentration of the element in the ore, the identity of the other elements present, and the difficulty of reducing the element to the free metal.

In general, it is not difficult to reduce ions of the *d*-block elements to the free element. Carbon is a sufficiently strong reducing agent in most cases. However, like the ions of the more active main group metals, ions of the *f*-block elements must be isolated by electrolysis or by reduction with an active metal such as calcium.

We shall discuss the processes used for the isolation of iron, copper, and silver because these three processes illustrate the principal means of isolating most of the *d*-block metals. In general, each of these processes involves three principal steps: preliminary treatment, smelting, and refining.

1. Preliminary treatment. In general, there is an initial treatment of the ores to make them suitable for the extraction of the metals. This usually involves crushing or grinding the ore, concentrating the metal-bearing components, and sometimes treating these substances chemically to convert them into compounds that are easier to reduce to the metal.

2. Smelting. The next step is the extraction of the metal in the molten state, a process called smelting, which includes reduction of the metallic compound to the metal. Impurities may be removed by the addition of a compound that forms a slag—a substance with a low melting point that can be readily separated from the molten metal.

3. Refining. The final step in the recovery of a metal is refining the metal. Low boiling metals such as zinc and mercury can be refined by distillation. When fused on an inclined table, low melting metals like tin flow away from higher-melting impurities. Electrolysis is another common method for refining metals.

Isolation of Iron

The early application of iron to the manufacture of tools and weapons was possible because of the wide distribution of iron ores and the ease with which iron compounds in the ores could be reduced by carbon. For a long time, charcoal was the form of carbon used in the reduction process. The production and use of iron became much more widespread about 1620, when coke was introduced as the reducing agent. Coke is a form of carbon formed by heating coal in the absence of air to remove impurities.

The first step in the metallurgy of iron is usually roasting the ore (heating the ore in air) to remove water, decomposing carbonates into oxides, and converting sulfides into oxides. The oxides are then reduced in a blast furnace that is 80–100 feet high and about 25 feet in diameter (Figure 19.6) in which the roasted ore, coke, and limestone (impure $CaCO_3$) are introduced continuously into the top. Molten iron and slag are withdrawn at the bottom. The entire stock in a furnace may weigh several hundred tons.

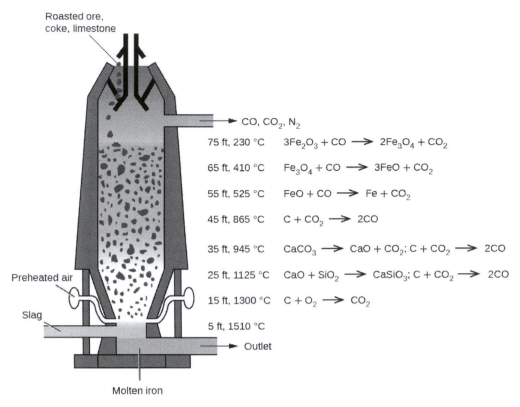

Figure 19.6 Within a blast furnace, different reactions occur in different temperature zones. Carbon monoxide is generated in the hotter bottom regions and rises upward to reduce the iron oxides to pure iron through a series of reactions that take place in the upper regions.

Near the bottom of a furnace are nozzles through which preheated air is blown into the furnace. As soon as the air enters, the coke in the region of the nozzles is oxidized to carbon dioxide with the liberation of a great deal of heat. The hot carbon dioxide passes upward through the overlying layer of white-hot coke, where it is reduced to carbon monoxide:

$$CO_2(g) + C(s) \longrightarrow 2CO(g)$$

The carbon monoxide serves as the reducing agent in the upper regions of the furnace. The individual reactions are indicated in Figure 19.6.

The iron oxides are reduced in the upper region of the furnace. In the middle region, limestone (calcium carbonate) decomposes, and the resulting calcium oxide combines with silica and silicates in the ore to form slag. The slag is mostly calcium silicate and contains most of the commercially unimportant components of the ore:

$$CaO(s) + SiO_2(s) \longrightarrow CaSiO_3(l)$$

Just below the middle of the furnace, the temperature is high enough to melt both the iron and the slag. They collect in layers at the bottom of the furnace; the less dense slag floats on the iron and protects it from oxidation. Several times a day, the slag and molten iron are withdrawn from the furnace. The iron is transferred to casting machines or to a steelmaking plant (Figure 19.7).

Figure 19.7   Molten iron is shown being cast as steel. (credit: Clint Budd)

Much of the iron produced is refined and converted into steel. **Steel** is made from iron by removing impurities and adding substances such as manganese, chromium, nickel, tungsten, molybdenum, and vanadium to produce alloys with properties that make the material suitable for specific uses. Most steels also contain small but definite percentages of carbon (0.04%–2.5%). However, a large part of the carbon contained in iron must be removed in the manufacture of steel; otherwise, the excess carbon would make the iron brittle.

Link to Learning

You can watch an animation of steelmaking (http://openstaxcollege.org/l/16steelmaking) that walks you through the process.

Isolation of Copper

The most important ores of copper contain copper sulfides (such as covellite, CuS), although copper oxides (such as tenorite, CuO) and copper hydroxycarbonates [such as malachite, $Cu_2(OH)_2CO_3$] are sometimes found. In the production of copper metal, the concentrated sulfide ore is roasted to remove part of the sulfur as sulfur dioxide. The remaining mixture, which consists of Cu_2S, FeS, FeO, and SiO_2, is mixed with limestone, which serves as a flux (a material that aids in the removal of impurities), and heated. Molten slag forms as the iron and silica are removed by Lewis acid-base reactions:

$$CaCO_3(s) + SiO_2(s) \longrightarrow CaSiO_3(l) + CO_2(g)$$
$$FeO(s) + SiO_2(s) \longrightarrow FeSiO_3(l)$$

In these reactions, the silicon dioxide behaves as a Lewis acid, which accepts a pair of electrons from the Lewis base (the oxide ion).

Reduction of the Cu_2S that remains after smelting is accomplished by blowing air through the molten material. The air converts part of the Cu_2S into Cu_2O. As soon as copper(I) oxide is formed, it is reduced by the remaining copper(I) sulfide to metallic copper:

$$2Cu_2S(l) + 3O_2(g) \longrightarrow 2Cu_2O(l) + 2SO_2(g)$$
$$2Cu_2O(l) + Cu_2S(l) \longrightarrow 6Cu(l) + SO_2(g)$$

The copper obtained in this way is called blister copper because of its characteristic appearance, which is due to the air blisters it contains (Figure 19.8). This impure copper is cast into large plates, which are used as anodes in the electrolytic refining of the metal (which is described in the chapter on electrochemistry).

Figure 19.8 Blister copper is obtained during the conversion of copper-containing ore into pure copper. (credit: "Tortie tude"/Wikimedia Commons)

Isolation of Silver

Silver sometimes occurs in large nuggets (Figure 19.9) but more frequently in veins and related deposits. At one time, panning was an effective method of isolating both silver and gold nuggets. Due to their low reactivity, these metals, and a few others, occur in deposits as nuggets. The discovery of platinum was due to Spanish explorers in Central America mistaking platinum nuggets for silver. When the metal is not in the form of nuggets, it often useful to employ a process called **hydrometallurgy** to separate silver from its ores. Hydrology involves the separation of a metal from a mixture by first converting it into soluble ions and then extracting and reducing them to precipitate the pure metal. In the presence of air, alkali metal cyanides readily form the soluble dicyanoargentate(I) ion, $[Ag(CN)_2]^-$, from silver metal or silver-containing compounds such as Ag_2S and $AgCl$. Representative equations are:

$$4Ag(s) + 8CN^-(aq) + O_2(g) + 2H_2O(l) \longrightarrow 4[Ag(CN)_2]^-(aq) + 4OH^-(aq)$$
$$2Ag_2S(s) + 8CN^-(aq) + O_2(g) + 2H_2O(l) \longrightarrow 4[Ag(CN)_2]^-(aq) + 2S(s) + 4OH^-(aq)$$
$$AgCl(s) + 2CN^-(aq) \longrightarrow [Ag(CN)_2]^-(aq) + Cl^-(aq)$$

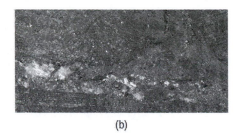

(a)   (b)

Figure 19.9   Naturally occurring free silver may be found as nuggets (a) or in veins (b). (credit a: modification of work by "Teravolt"/Wikimedia Commons; credit b: modification of work by James St. John)

The silver is precipitated from the cyanide solution by the addition of either zinc or iron(II) ions, which serves as the reducing agent:

$$2[Ag(CN)_2]^-(aq) + Zn(s) \longrightarrow 2Ag(s) + [Zn(CN)_4]^{2-}(aq)$$

Example 19.3

Refining Redox

One of the steps for refining silver involves converting silver into dicyanoargenate(I) ions:

$$4Ag(s) + 8CN^-(aq) + O_2(g) + 2H_2O(l) \longrightarrow 4[Ag(CN)_2]^-(aq) + 4OH^-(aq)$$

Explain why oxygen must be present to carry out the reaction. Why does the reaction not occur as:

$$4Ag(s) + 8CN^-(aq) \longrightarrow 4[Ag(CN)_2]^-(aq)?$$

Solution

The charges, as well as the atoms, must balance in reactions. The silver atom is being oxidized from the 0 oxidation state to the 1+ state. Whenever something loses electrons, something must also gain electrons (be reduced) to balance the equation. Oxygen is a good oxidizing agent for these reactions because it can gain electrons to go from the 0 oxidation state to the 2− state.

Check Your Learning

During the refining of iron, carbon must be present in the blast furnace. Why is carbon necessary to convert iron oxide into iron?

Answer: The carbon is converted into CO, which is the reducing agent that accepts electrons so that iron(III) can be reduced to iron(0).

Transition Metal Compounds

The bonding in the simple compounds of the transition elements ranges from ionic to covalent. In their lower oxidation states, the transition elements form ionic compounds; in their higher oxidation states, they form covalent compounds or polyatomic ions. The variation in oxidation states exhibited by the transition elements gives these compounds a metal-based, oxidation-reduction chemistry. The chemistry of several classes of compounds containing elements of the transition series follows.

Halides

Anhydrous halides of each of the transition elements can be prepared by the direct reaction of the metal with halogens. For example:

$$2Fe(s) + 3Cl_2(g) \longrightarrow 2FeCl_3(s)$$

Heating a metal halide with additional metal can be used to form a halide of the metal with a lower oxidation state:

$$Fe(s) + 2FeCl_3(s) \longrightarrow 3FeCl_2(s)$$

The stoichiometry of the metal halide that results from the reaction of the metal with a halogen is determined by the relative amounts of metal and halogen and by the strength of the halogen as an oxidizing agent. Generally, fluorine forms fluoride-containing metals in their highest oxidation states. The other halogens may not form analogous compounds.

In general, the preparation of stable water solutions of the halides of the metals of the first transition series is by the addition of a hydrohalic acid to carbonates, hydroxides, oxides, or other compounds that contain basic anions. Sample reactions are:

$$NiCO_3(s) + 2HF(aq) \longrightarrow NiF_2(aq) + H_2O(l) + CO_2(g)$$
$$Co(OH)_2(s) + 2HBr(aq) \longrightarrow CoBr_2(aq) + 2H_2O(l)$$

Most of the first transition series metals also dissolve in acids, forming a solution of the salt and hydrogen gas. For example:

$$Cr(s) + 2HCl(aq) \longrightarrow CrCl_2(aq) + H_2(g)$$

The polarity of bonds with transition metals varies based not only upon the electronegativities of the atoms involved but also upon the oxidation state of the transition metal. Remember that bond polarity is a continuous spectrum with electrons being shared evenly (covalent bonds) at one extreme and electrons being transferred completely (ionic bonds) at the other. No bond is ever 100% ionic, and the degree to which the electrons are evenly distributed determines many properties of the compound. Transition metal halides with low oxidation numbers form more ionic bonds. For example, titanium(II) chloride and titanium(III) chloride ($TiCl_2$ and $TiCl_3$) have high melting points that are characteristic of ionic compounds, but titanium(IV) chloride ($TiCl_4$) is a volatile liquid, consistent with having covalent titanium-chlorine bonds. All halides of the heavier d-block elements have significant covalent characteristics.

The covalent behavior of the transition metals with higher oxidation states is exemplified by the reaction of the metal tetrahalides with water. Like covalent silicon tetrachloride, both the titanium and vanadium tetrahalides react with water to give solutions containing the corresponding hydrohalic acids and the metal oxides:

$$SiCl_4(l) + 2H_2O(l) \longrightarrow SiO_2(s) + 4HCl(aq)$$
$$TiCl_4(l) + 2H_2O(l) \longrightarrow TiO_2(s) + 4HCl(aq)$$

Oxides

As with the halides, the nature of bonding in oxides of the transition elements is determined by the oxidation state of the metal. Oxides with low oxidation states tend to be more ionic, whereas those with higher oxidation states are more covalent. These variations in bonding are because the electronegativities of the elements are not fixed values. The electronegativity of an element increases with increasing oxidation state. Transition metals in low oxidation states have lower electronegativity values than oxygen; therefore, these metal oxides are ionic. Transition metals in very high oxidation states have electronegativity values close to that of oxygen, which leads to these oxides being covalent.

The oxides of the first transition series can be prepared by heating the metals in air. These oxides are Sc_2O_3, TiO_2, V_2O_5, Cr_2O_3, Mn_3O_4, Fe_3O_4, Co_3O_4, NiO, and CuO.

Alternatively, these oxides and other oxides (with the metals in different oxidation states) can be produced by heating the corresponding hydroxides, carbonates, or oxalates in an inert atmosphere. Iron(II) oxide can be prepared by heating iron(II) oxalate, and cobalt(II) oxide is produced by heating cobalt(II) hydroxide:

$$FeC_2O_4(s) \longrightarrow FeO(s) + CO(g) + CO_2(g)$$
$$Co(OH)_2(s) \longrightarrow CoO(s) + H_2O(g)$$

With the exception of CrO_3 and Mn_2O_7, transition metal oxides are not soluble in water. They can react with acids and, in a few cases, with bases. Overall, oxides of transition metals with the lowest oxidation states are basic (and react with acids), the intermediate ones are amphoteric, and the highest oxidation states are primarily acidic. Basic metal oxides at a low oxidation state react with aqueous acids to form solutions of salts and water. Examples include the reaction of cobalt(II) oxide accepting protons from nitric acid, and scandium(III) oxide accepting protons from hydrochloric acid:

$$CoO(s) + 2HNO_3(aq) \longrightarrow Co(NO_3)_2(aq) + H_2O(l)$$
$$Sc_2O_3(s) + 6HCl(aq) \longrightarrow 2ScCl_3(aq) + 3H_2O(l)$$

The oxides of metals with oxidation states of 4+ are amphoteric, and most are not soluble in either acids or bases. Vanadium(V) oxide, chromium(VI) oxide, and manganese(VII) oxide are acidic. They react with solutions of hydroxides to form salts of the oxyanions VO_4^{3-}, CrO_4^{2-}, and MnO_4^-. For example, the complete ionic equation for the reaction of chromium(VI) oxide with a strong base is given by:

$$CrO_3(s) + 2Na^+(aq) + 2OH^-(aq) \longrightarrow 2Na^+(aq) + CrO_4^{2-}(aq) + H_2O(l)$$

Chromium(VI) oxide and manganese(VII) oxide react with water to form the acids H_2CrO_4 and $HMnO_4$, respectively.

Hydroxides

When a soluble hydroxide is added to an aqueous solution of a salt of a transition metal of the first transition series, a gelatinous precipitate forms. For example, adding a solution of sodium hydroxide to a solution of cobalt sulfate produces a gelatinous pink or blue precipitate of cobalt(II) hydroxide. The net ionic equation is:

$$Co^{2+}(aq) + 2OH^-(aq) \longrightarrow Co(OH)_2(s)$$

In this and many other cases, these precipitates are hydroxides containing the transition metal ion, hydroxide ions, and water coordinated to the transition metal. In other cases, the precipitates are hydrated oxides composed of the metal ion, oxide ions, and water of hydration:

$$4Fe^{3+}(aq) + 6OH^-(aq) + n\,H_2O(l) \longrightarrow 2Fe_2O_3 \cdot (n+3)H_2O(s)$$

These substances do not contain hydroxide ions. However, both the hydroxides and the hydrated oxides react with acids to form salts and water. When precipitating a metal from solution, it is necessary to avoid an excess of hydroxide ion, as this may lead to complex ion formation as discussed later in this chapter. The precipitated metal hydroxides can be separated for further processing or for waste disposal.

Carbonates

Many of the elements of the first transition series form insoluble carbonates. It is possible to prepare these carbonates by the addition of a soluble carbonate salt to a solution of a transition metal salt. For example, nickel carbonate can be prepared from solutions of nickel nitrate and sodium carbonate according to the following net ionic equation:

$$Ni^{2+}(aq) + CO_3^{2-} \longrightarrow NiCO_3(s)$$

The reactions of the transition metal carbonates are similar to those of the active metal carbonates. They react with acids to form metals salts, carbon dioxide, and water. Upon heating, they decompose, forming the transition metal oxides.

Other Salts

In many respects, the chemical behavior of the elements of the first transition series is very similar to that of the main group metals. In particular, the same types of reactions that are used to prepare salts of the main group metals can be used to prepare simple ionic salts of these elements.

A variety of salts can be prepared from metals that are more active than hydrogen by reaction with the corresponding acids: Scandium metal reacts with hydrobromic acid to form a solution of scandium bromide:

$$2Sc(s) + 6HBr(aq) \longrightarrow 2ScBr_3(aq) + 3H_2(g)$$

The common compounds that we have just discussed can also be used to prepare salts. The reactions involved include the reactions of oxides, hydroxides, or carbonates with acids. For example:

$$Ni(OH)_2(s) + 2H_3O^+(aq) + 2ClO_4{}^-(aq) \longrightarrow Ni^{2+}(aq) + 2ClO_4{}^-(aq) + 4H_2O(l)$$

Substitution reactions involving soluble salts may be used to prepare insoluble salts. For example:

$$Ba^{2+}(aq) + 2Cl^-(aq) + 2K^+(aq) + CrO_4{}^{2-}(aq) \longrightarrow BaCrO_4(s) + 2K^+(aq) + 2Cl^-(aq)$$

In our discussion of oxides in this section, we have seen that reactions of the covalent oxides of the transition elements with hydroxides form salts that contain oxyanions of the transition elements.

How Sciences Interconnect

High Temperature Superconductors

A **superconductor** is a substance that conducts electricity with no resistance. This lack of resistance means that there is no energy loss during the transmission of electricity. This would lead to a significant reduction in the cost of electricity.

Most currently used, commercial superconducting materials, such as NbTi and Nb$_3$Sn, do not become superconducting until they are cooled below 23 K (−250 °C). This requires the use of liquid helium, which has a boiling temperature of 4 K and is expensive and difficult to handle. The cost of liquid helium has deterred the widespread application of superconductors.

One of the most exciting scientific discoveries of the 1980s was the characterization of compounds that exhibit superconductivity at temperatures above 90 K. (Compared to liquid helium, 90 K is a high temperature.) Typical among the high-temperature superconducting materials are oxides containing yttrium (or one of several rare earth elements), barium, and copper in a 1:2:3 ratio. The formula of the ionic yttrium compound is YBa$_2$Cu$_3$O$_7$.

The new materials become superconducting at temperatures close to 90 K (Figure 19.10), temperatures that can be reached by cooling with liquid nitrogen (boiling temperature of 77 K). Not only are liquid nitrogen-cooled materials easier to handle, but the cooling costs are also about 1000 times lower than for liquid helium.

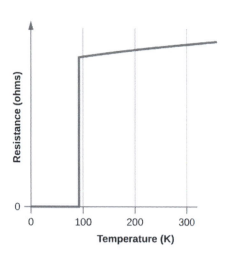

Figure 19.10   The resistance of the high-temperature superconductor $YBa_2Cu_3O_7$ varies with temperature. Note how the resistance falls to zero below 92 K, when the substance becomes superconducting.

Although the brittle, fragile nature of these materials presently hampers their commercial applications, they have tremendous potential that researchers are hard at work improving their processes to help realize. Superconducting transmission lines would carry current for hundreds of miles with no loss of power due to resistance in the wires. This could allow generating stations to be located in areas remote from population centers and near the natural resources necessary for power production. The first project demonstrating the viability of high-temperature superconductor power transmission was established in New York in 2008.

Researchers are also working on using this technology to develop other applications, such as smaller and more powerful microchips. In addition, high-temperature superconductors can be used to generate magnetic fields for applications such as medical devices, magnetic levitation trains, and containment fields for nuclear fusion reactors (Figure 19.11).

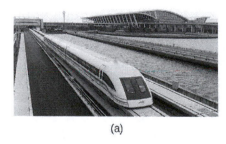

(a)   (b)

Figure 19.11   (a) This magnetic levitation train (or maglev) uses superconductor technology to move along its tracks. (b) A magnet can be levitated using a dish like this as a superconductor. (credit a: modification of work by Alex Needham; credit b: modification of work by Kevin Jarrett)

19.2 Coordination Chemistry of Transition Metals

By the end of this section, you will be able to:

- List the defining traits of coordination compounds
- Describe the structures of complexes containing monodentate and polydentate ligands
- Use standard nomenclature rules to name coordination compounds
- Explain and provide examples of geometric and optical isomerism
- Identify several natural and technological occurrences of coordination compounds

The hemoglobin in your blood, the chlorophyll in green plants, vitamin B-12, and the catalyst used in the manufacture of polyethylene all contain coordination compounds. Ions of the metals, especially the transition metals, are likely to form complexes. Many of these compounds are highly colored (Figure 19.12). In the remainder of this chapter, we will consider the structure and bonding of these remarkable compounds.

Figure 19.12 Metal ions that contain partially filled d subshell usually form colored complex ions; ions with empty d subshell (d^0) or with filled d subshells (d^{10}) usually form colorless complexes. This figure shows, from left to right, solutions containing $[M(H_2O)_6]^{n+}$ ions with $M = Sc^{3+}(d^0)$, $Cr^{3+}(d^3)$, $Co^{2+}(d^7)$, $Ni^{2+}(d^8)$, $Cu^{2+}(d^9)$, and $Zn^{2+}(d^{10})$. (credit: Sahar Atwa)

Remember that in most main group element compounds, the valence electrons of the isolated atoms combine to form chemical bonds that satisfy the octet rule. For instance, the four valence electrons of carbon overlap with electrons from four hydrogen atoms to form CH_4. The one valence electron leaves sodium and adds to the seven valence electrons of chlorine to form the ionic formula unit NaCl (Figure 19.13). Transition metals do not normally bond in this fashion. They primarily form coordinate covalent bonds, a form of the Lewis acid-base interaction in which both of the electrons in the bond are contributed by a donor (Lewis base) to an electron acceptor (Lewis acid). The Lewis acid in coordination complexes, often called a **central metal** ion (or atom), is often a transition metal or inner transition metal, although main group elements can also form **coordination compounds**. The Lewis base donors, called **ligands**, can be a wide variety of chemicals—atoms, molecules, or ions. The only requirement is that they have one or more electron pairs, which can be donated to the central metal. Most often, this involves a **donor atom** with a lone pair of electrons that can form a coordinate bond to the metal.

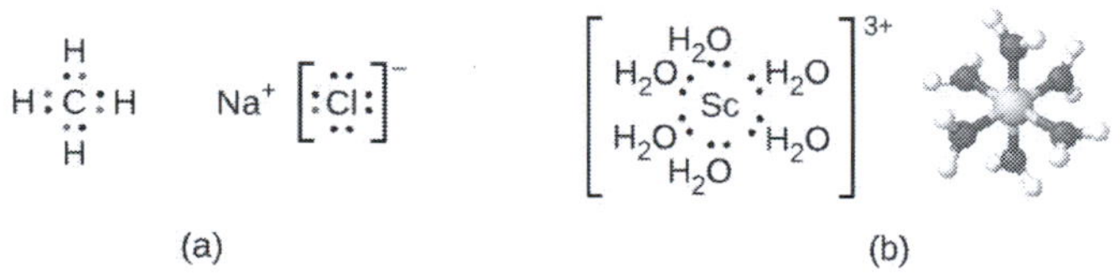

(a)   (b)

Figure 19.13 (a) Covalent bonds involve the sharing of electrons, and ionic bonds involve the transferring of electrons associated with each bonding atom, as indicated by the colored electrons. (b) However, coordinate covalent bonds involve electrons from a Lewis base being donated to a metal center. The lone pairs from six water molecules form bonds to the scandium ion to form an octahedral complex. (Only the donated pairs are shown.)

The **coordination sphere** consists of the central metal ion or atom plus its attached ligands. Brackets in a formula enclose the coordination sphere; species outside the brackets are not part of the coordination sphere. The **coordination number** of the central metal ion or atom is the number of donor atoms bonded to it. The coordination number for the silver ion in $[Ag(NH_3)_2]^+$ is two (Figure 19.14). For the copper(II) ion in $[CuCl_4]^{2-}$, the coordination number is four, whereas for the cobalt(II) ion in $[Co(H_2O)_6]^{2+}$ the coordination number is six. Each of these ligands is **monodentate**, from the Greek for "one toothed," meaning that they connect with the central metal through only one atom. In this case, the number of ligands and the coordination number are equal.

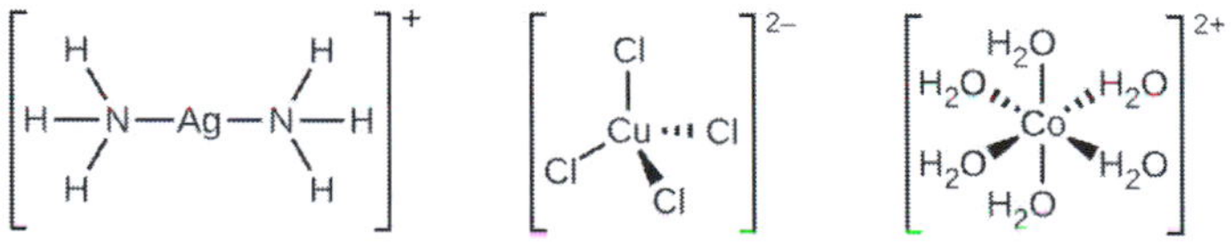

Figure 19.14 The complexes (a) $[Ag(NH_3)_2]^+$, (b) $[Cu(Cl)_4]^{2-}$, and (c) $[Co(H_2O)_6]^{2+}$ have coordination numbers of two, four, and six, respectively. The geometries of these complexes are the same as we have seen with VSEPR theory for main group elements: linear, tetrahedral, and octahedral.

Many other ligands coordinate to the metal in more complex fashions. **Bidentate ligands** are those in which two atoms coordinate to the metal center. For example, ethylenediamine (en, $H_2NCH_2CH_2NH_2$) contains two nitrogen atoms, each of which has a lone pair and can serve as a Lewis base (Figure 19.15). Both of the atoms can coordinate to a single metal center. In the complex $[Co(en)_3]^{3+}$, there are three bidentate en ligands, and the coordination number of the cobalt(III) ion is six. The most common coordination numbers are two, four, and six, but examples of all coordination numbers from 1 to 15 are known.

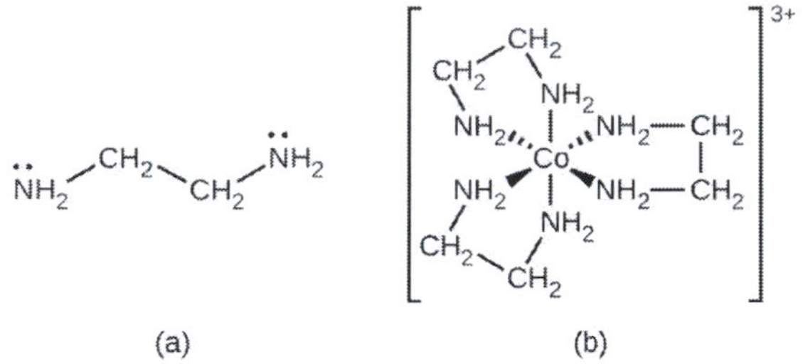

Figure 19.15 (a) The ethylenediamine (en) ligand contains two atoms with lone pairs that can coordinate to the metal center. (b) The cobalt(III) complex $[Co(en)_3]^{3+}$ contains three of these ligands, each forming two bonds to the cobalt ion.

Any ligand that bonds to a central metal ion by more than one donor atom is a **polydentate ligand** (or "many teeth") because it can bite into the metal center with more than one bond. The term **chelate** (pronounced "KEY-late") from the Greek for "claw" is also used to describe this type of interaction. Many polydentate ligands are **chelating ligands**, and a complex consisting of one or more of these ligands and a central metal is a chelate. A chelating ligand is also known as a chelating agent. A chelating ligand holds the metal ion rather like a crab's claw would hold a marble. Figure 19.15 showed one example of a chelate. The heme complex in hemoglobin is another important example (Figure 19.16). It contains a polydentate ligand with four donor atoms that coordinate to iron.

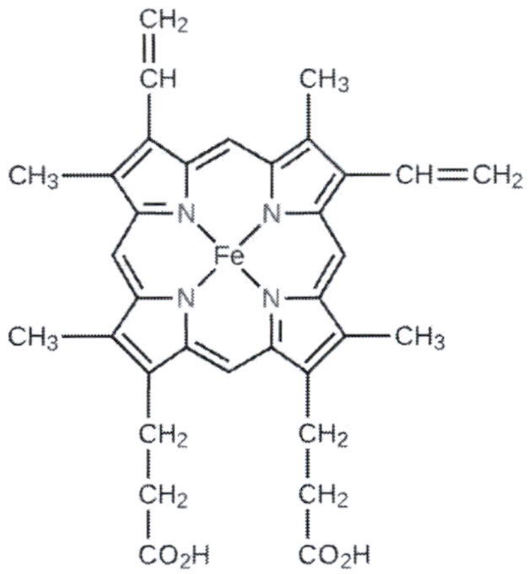

Figure 19.16 The single ligand heme contains four nitrogen atoms that coordinate to iron in hemoglobin to form a chelate.

Polydentate ligands are sometimes identified with prefixes that indicate the number of donor atoms in the ligand. As we have seen, ligands with one donor atom, such as NH_3, Cl^-, and H_2O, are monodentate ligands. Ligands with two donor groups are bidentate ligands. Ethylenediamine, $H_2NCH_2CH_2NH_2$, and the anion of the acid glycine, $NH_2CH_2CO_2^-$ (Figure 19.17) are examples of bidentate ligands. Tridentate ligands, tetradentate ligands, pentadentate ligands, and hexadentate ligands contain three, four, five, and six donor atoms, respectively. The ligand in heme (Figure 19.16) is a tetradentate ligand.

                              Download for free at http://cnx.org/content/col11760/latest/

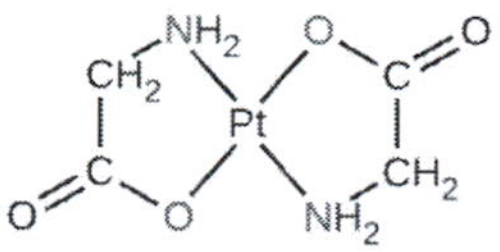

Figure 19.17 Each of the anionic ligands shown attaches in a bidentate fashion to platinum(II), with both a nitrogen and oxygen atom coordinating to the metal.

The Naming of Complexes

The nomenclature of the complexes is patterned after a system suggested by Alfred Werner, a Swiss chemist and Nobel laureate, whose outstanding work more than 100 years ago laid the foundation for a clearer understanding of these compounds. The following five rules are used for naming complexes:

1. If a coordination compound is ionic, name the cation first and the anion second, in accordance with the usual nomenclature.

2. Name the ligands first, followed by the central metal. Name the ligands alphabetically. Negative ligands (anions) have names formed by adding -o to the stem name of the group. For examples, see Table 19.1. For most neutral ligands, the name of the molecule is used. The four common exceptions are *aqua* (H_2O), *amine* (NH_3), *carbonyl* (CO), and *nitrosyl* (NO). For example, name $[Pt(NH_3)_2Cl_4]$ as diaminetetrachloroplatinum(IV).

Examples of Anionic Ligands

Anionic Ligand	Name
F^-	fluoro
Cl^-	chloro
Br^-	bromo
I^-	iodo
CN^-	cyano
NO_3^-	nitrato
OH^-	hydroxo
O^{2-}	oxo
$C_2O_4^{2-}$	oxalato
CO_2^{2-}	carbonato

Table 19.1

3. If more than one ligand of a given type is present, the number is indicated by the prefixes *di-* (for two), *tri-* (for three), *tetra-* (for four), *penta-* (for five), and *hexa-* (for six). Sometimes, the prefixes *bis-* (for two), *tris-* (for three), and *tetrakis-* (for four) are used when the name of the ligand already includes *di-*, *tri-*, or *tetra-*, or when the ligand name begins with a vowel. For example, the ion bis(bipyridyl)osmium(II) uses bis- to signify that there are two ligands attached to Os, and each bipyridyl ligand contains two pyridine groups (C_5H_4N).

When the complex is either a cation or a neutral molecule, the name of the central metal atom is spelled exactly like the name of the element and is followed by a Roman numeral in parentheses to indicate its oxidation state (Table 19.2 and Table 19.3). When the complex is an anion, the suffix -ate is added to the stem of the name of the metal, followed by the Roman numeral designation of its oxidation state (Table 19.4). Sometimes, the Latin name of the metal is used when the English name is clumsy. For example, *ferrate* is used instead of *ironate*, *plumbate* instead *leadate*, and *stannate* instead of *tinate*. The oxidation state of the metal is determined based on the charges of each ligand and the overall charge of the coordination compound. For example, in $[Cr(H_2O)_4Cl_2]Br$, the coordination sphere (in brackets) has a charge of 1+ to balance the bromide ion. The water ligands are neutral, and the chloride ligands are anionic with a charge of 1− each. To determine the oxidation state of the metal, we set the overall charge equal to the sum of the ligands and the metal: $+1 = -2 + x$, so the oxidation state (x) is equal to 3+.

Examples in Which the Complex Is a Cation

$[Co(NH_3)_6]Cl_3$	hexaaminecobalt(III) chloride
$[Pt(NH_3)_4Cl_2]^{2+}$	tetraaminedichloroplatinum(IV) ion
$[Ag(NH_3)_2]^+$	diaminesilver(I) ion
$[Cr(H_2O)_4Cl_2]Cl$	tetraaquadichlorochromium(III) chloride
$[Co(H_2NCH_2CH_2NH_2)_3]_2(SO_4)_3$	tris(ethylenediamine)cobalt(III) sulfate

Table 19.2

Examples in Which the Complex Is Neutral

$[Pt(NH_3)_2Cl_4]$	diaminetetrachloroplatinum(IV)
$[Ni(H_2NCH_2CH_2NH_2)_2Cl_2]$	dichlorobis(ethylenediamine)nickel(II)

Table 19.3

Examples in Which the Complex Is an Anion

$[PtCl_6]^{2-}$	hexachloroplatinate(IV) ion
$Na_2[SnCl_6]$	sodium hexachlorostannate(IV)

Table 19.4

Link to Learning

Do you think you understand naming coordination complexes? You can look over more examples and test yourself with online quizzes (http://openstaxcollege.org/l/16namingcomps) at the University of Sydney's site.

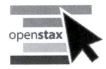

Example 19.4

Coordination Numbers and Oxidation States

Determine the name of the following complexes and give the coordination number of the central metal atom.

(a) $Na_2[PtCl_6]$

(b) $K_3[Fe(C_2O_4)_3]$

(c) $[Co(NH_3)_5Cl]Cl_2$

Solution

(a) There are two Na^+ ions, so the coordination sphere has a negative two charge: $[PtCl_6]^{2-}$. There are six anionic chloride ligands, so $-2 = -6 + x$, and the oxidation state of the platinum is 4+. The name of the complex is sodium hexachloroplatinate(IV), and the coordination number is six. (b) The coordination sphere has a charge of 3− (based on the potassium) and the oxalate ligands each have a charge of 2−, so the metal oxidation state is given by $-3 = -6 + x$, and this is an iron(III) complex. The name is potassium trisoxalatoferrate(III) (note that tris is used instead of tri because the ligand name starts with a vowel). Because oxalate is a bidentate ligand, this complex has a coordination number of six. (c) In this example, the coordination sphere has a cationic charge of 2+. The NH_3 ligand is neutral, but the chloro ligand has a charge of 1−. The oxidation state is found by $+2 = -1 + x$ and is 3+, so the complex is pentaaminechlorocobalt(III) chloride and the coordination number is six.

Check Your Learning

The complex potassium dicyanoargenate(I) is used to make antiseptic compounds. Give the formula and coordination number.

Answer: $K[Ag(CN)_2]$; coordination number two

The Structures of Complexes

The most common structures of the complexes in coordination compounds are octahedral, tetrahedral, and square planar (see **Figure 19.18**). For transition metal complexes, the coordination number determines the geometry around the central metal ion. **Table 19.5** compares coordination numbers to the molecular geometry:

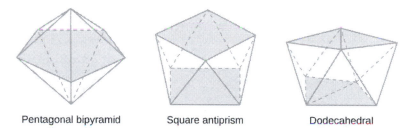

| Pentagonal bipyramid | Square antiprism | Dodecahedral |

Figure 19.18 These are geometries of some complexes with coordination numbers of seven and eight.

Coordination Numbers and Molecular Geometry

Coordination Number	Molecular Geometry	Example
2	linear	$[Ag(NH_3)_2]^+$
3	trigonal planar	$[Cu(CN)_3]^{2-}$
4	tetrahedral(d^0 or d^{10}), low oxidation states for M	$[Ni(CO)_4]$
4	square planar (d^8)	$[NiCl_4]^{2-}$

Table 19.5

Coordination Numbers and Molecular Geometry

Coordination Number	Molecular Geometry	Example
5	trigonal bipyramidal	$[CoCl_5]^{2-}$
5	square pyramidal	$[VO(CN)_4]^{2-}$
6	octahedral	$[CoCl_6]^{3-}$
7	pentagonal bipyramid	$[ZrF_7]^{3-}$
8	square antiprism	$[ReF_8]^{2-}$
8	dodecahedron	$[Mo(CN)_8]^{4-}$
9 and above	more complicated structures	$[ReH_9]^{2-}$

Table 19.5

Unlike main group atoms in which both the bonding and nonbonding electrons determine the molecular shape, the nonbonding *d*-electrons do not change the arrangement of the ligands. Octahedral complexes have a coordination number of six, and the six donor atoms are arranged at the corners of an octahedron around the central metal ion. Examples are shown in Figure 19.19. The chloride and nitrate anions in $[Co(H_2O)_6]Cl_2$ and $[Cr(en)_3](NO_3)_3$, and the potassium cations in $K_2[PtCl_6]$, are outside the brackets and are not bonded to the metal ion.

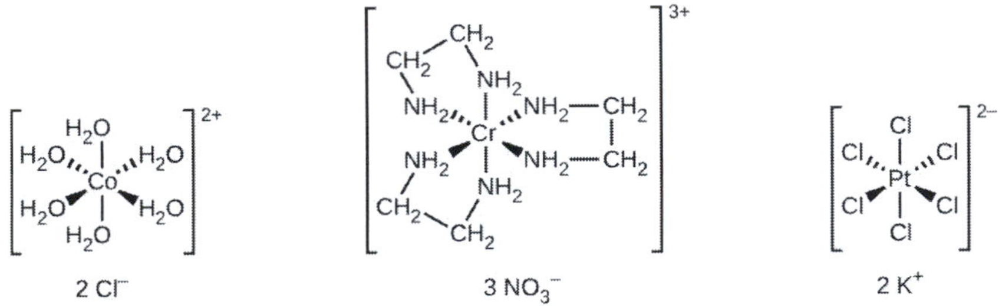

Figure 19.19 Many transition metal complexes adopt octahedral geometries, with six donor atoms forming bond angles of 90° about the central atom with adjacent ligands. Note that only ligands within the coordination sphere affect the geometry around the metal center.

For transition metals with a coordination number of four, two different geometries are possible: tetrahedral or square planar. Unlike main group elements, where these geometries can be predicted from VSEPR theory, a more detailed discussion of transition metal orbitals (discussed in the section on Crystal Field Theory) is required to predict which complexes will be tetrahedral and which will be square planar. In tetrahedral complexes such as $[Zn(CN)_4]^{2-}$ (Figure 19.20), each of the ligand pairs forms an angle of 109.5°. In square planar complexes, such as $[Pt(NH_3)_2Cl_2]$, each ligand has two other ligands at 90° angles (called the *cis* positions) and one additional ligand at an 180° angle, in the *trans* position.

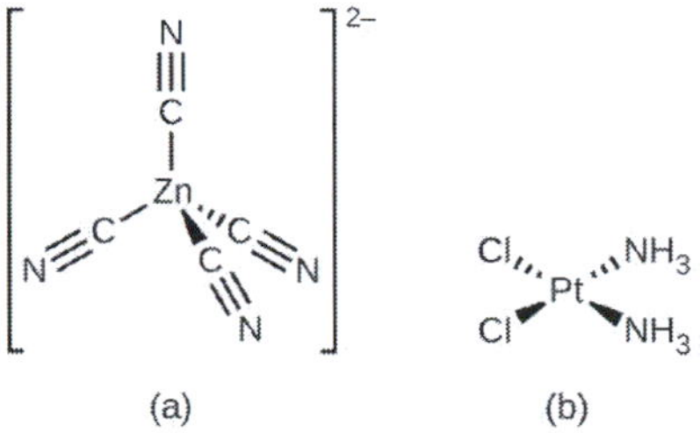

Figure 19.20 Transition metals with a coordination number of four can adopt a tetrahedral geometry (a) as in $K_2[Zn(CN)_4]$ or a square planar geometry (b) as shown in $[Pt(NH_3)_2Cl_2]$.

Isomerism in Complexes

Isomers are different chemical species that have the same chemical formula. Transition metals often form **geometric isomers**, in which the same atoms are connected through the same types of bonds but with differences in their orientation in space. Coordination complexes with two different ligands in the *cis* and *trans* positions from a ligand of interest form isomers. For example, the octahedral $[Co(NH_3)_4Cl_2]^+$ ion has two isomers. In the *cis* **configuration**, the two chloride ligands are adjacent to each other (Figure 19.21). The other isomer, the *trans* **configuration**, has the two chloride ligands directly across from one another.

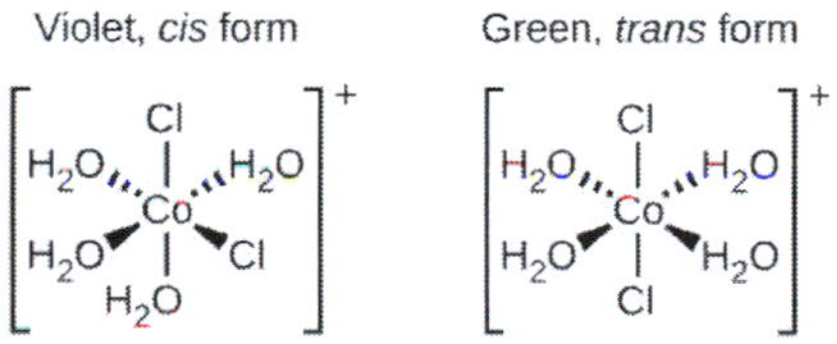

Figure 19.21 The *cis* and *trans* isomers of $[Co(H_2O)_4Cl_2]^+$ contain the same ligands attached to the same metal ion, but the spatial arrangement causes these two compounds to have very different properties.

Different geometric isomers of a substance are different chemical compounds. They exhibit different properties, even though they have the same formula. For example, the two isomers of $[Co(NH_3)_4Cl_2]NO_3$ differ in color; the *cis* form is violet, and the *trans* form is green. Furthermore, these isomers have different dipole moments, solubilities, and reactivities. As an example of how the arrangement in space can influence the molecular properties, consider the polarity of the two $[Co(NH_3)_4Cl_2]NO_3$ isomers. Remember that the polarity of a molecule or ion is determined by the bond dipoles (which are due to the difference in electronegativity of the bonding atoms) and their arrangement in space. In one isomer, *cis* chloride ligands cause more electron density on one side of the molecule than on the other, making it polar. For the *trans* isomer, each ligand is directly across from an identical ligand, so the bond dipoles cancel out, and the molecule is nonpolar.

Example 19.5

Geometric Isomers

Identify which geometric isomer of $[Pt(NH_3)_2Cl_2]$ is shown in Figure 19.20. Draw the other geometric isomer and give its full name.

Solution

In the Figure 19.20, the two chlorine ligands occupy *cis* positions. The other form is shown in Figure 19.22. When naming specific isomers, the descriptor is listed in front of the name. Therefore, this complex is *trans*-diaminedichloroplatinum(II).

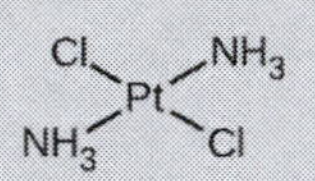

Figure 19.22   The *trans* isomer of $[Pt(NH_3)_2Cl_2]$ has each ligand directly across from an adjacent ligand.

Check Your Learning

Draw the ion *trans*-diaqua-*trans*-dibromo-*trans*-dichlorocobalt(II).

Answer:

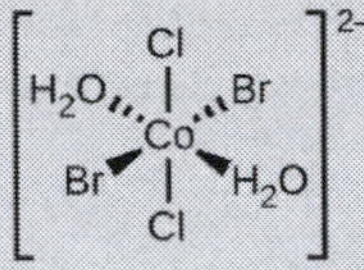

Another important type of isomers are **optical isomers**, or **enantiomers**, in which two objects are exact mirror images of each other but cannot be lined up so that all parts match. This means that optical isomers are nonsuperimposable mirror images. A classic example of this is a pair of hands, in which the right and left hand are mirror images of one another but cannot be superimposed. Optical isomers are very important in organic and biochemistry because living systems often incorporate one specific optical isomer and not the other. Unlike geometric isomers, pairs of optical isomers have identical properties (boiling point, polarity, solubility, etc.). Optical isomers differ only in the way they affect polarized light and how they react with other optical isomers. For coordination complexes, many coordination compounds such as $[M(en)_3]^{n+}$ [in which M^{n+} is a central metal ion such as iron(III) or cobalt(II)] form enantiomers, as shown in Figure 19.23. These two isomers will react differently with other optical isomers. For example, DNA helices are optical isomers, and the form that occurs in nature (right-handed DNA) will bind to only one isomer of $[M(en)_3]^{n+}$ and not the other.

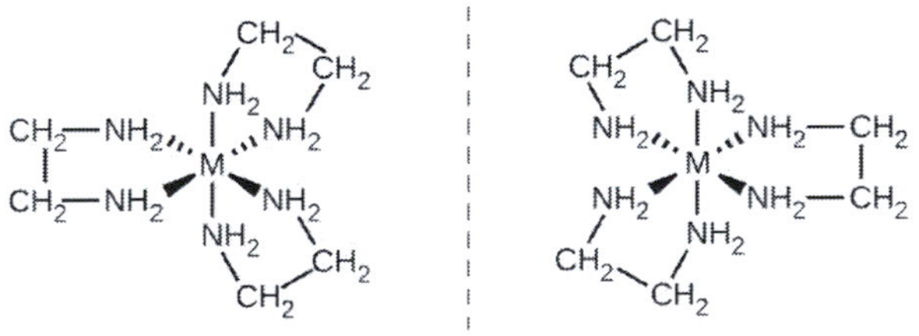

Figure 19.23   The complex $[M(en)_3]^{n+}$ (M^{n+} = a metal ion, en = ethylenediamine) has a nonsuperimposable mirror image.

The $[Co(en)_2Cl_2]^+$ ion exhibits geometric isomerism (*cis/trans*), and its *cis* isomer exists as a pair of optical isomers (Figure 19.24).

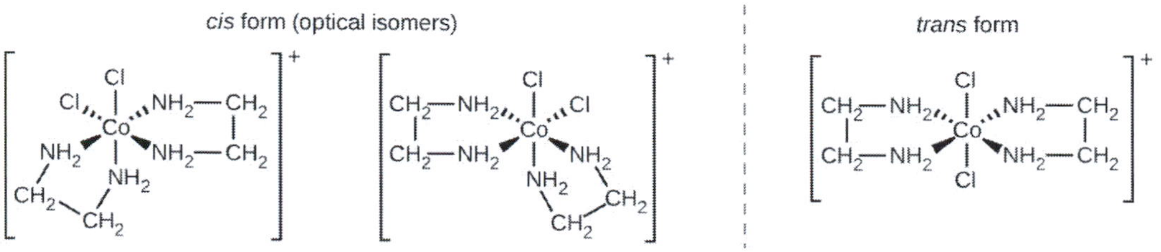

cis form (optical isomers) | *trans* form

Figure 19.24 Three isomeric forms of $[Co(en)_2Cl_2]^+$ exist. The *trans* isomer, formed when the chlorines are positioned at a 180° angle, has very different properties from the *cis* isomers. The mirror images of the *cis* isomer form a pair of optical isomers, which have identical behavior except when reacting with other enantiomers.

Linkage isomers occur when the coordination compound contains a ligand that can bind to the transition metal center through two different atoms. For example, the CN ligand can bind through the carbon atom (cyano) or through the nitrogen atom (isocyano). Similarly, SCN− can be bound through the sulfur or nitrogen atom, affording two distinct compounds ($[Co(NH_3)_5SCN]^{2+}$ or $[Co(NH_3)_5NCS]^{2+}$).

Ionization isomers (or **coordination isomers**) occur when one anionic ligand in the inner coordination sphere is replaced with the counter ion from the outer coordination sphere. A simple example of two ionization isomers are $[CoCl_6][Br]$ and $[CoCl_5Br][Cl]$.

Coordination Complexes in Nature and Technology

Chlorophyll, the green pigment in plants, is a complex that contains magnesium (**Figure 19.25**). This is an example of a main group element in a coordination complex. Plants appear green because chlorophyll absorbs red and purple light; the reflected light consequently appears green. The energy resulting from the absorption of light is used in photosynthesis.

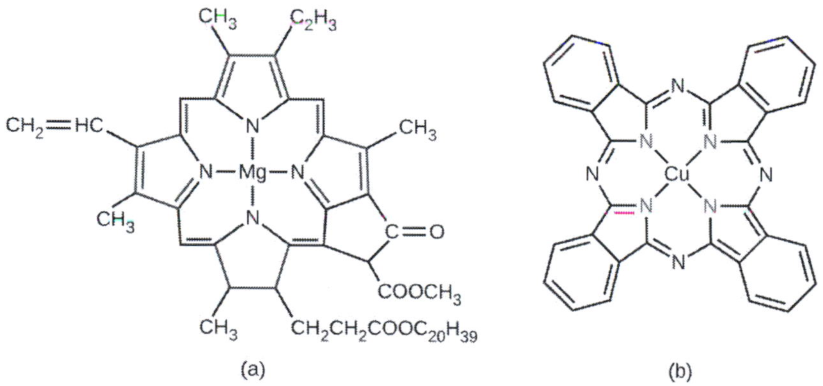

(a)   (b)

Figure 19.25 (a) Chlorophyll comes in several different forms, which all have the same basic structure around the magnesium center. (b) Copper phthalocyanine blue, a square planar copper complex, is present in some blue dyes.

Chemistry in Everyday Life

Transition Metal Catalysts

One of the most important applications of transition metals is as industrial catalysts. As you recall from the chapter on kinetics, a catalyst increases the rate of reaction by lowering the activation energy and is regenerated in the catalytic cycle. Over 90% of all manufactured products are made with the aid of one or more catalysts. The ability to bind ligands and change oxidation states makes transition metal catalysts well suited for catalytic applications. Vanadium oxide is used to produce 230,000,000 tons of sulfuric acid worldwide each year, which in turn is used to make everything from fertilizers to cans for food. Plastics are made with the aid of transition metal catalysts, along with detergents, fertilizers, paints, and more (see Figure 19.26). Very complicated pharmaceuticals are manufactured with catalysts that are selective, reacting with one specific bond out of a large number of possibilities. Catalysts allow processes to be more economical and more environmentally friendly. Developing new catalysts and better understanding of existing systems are important areas of current research.

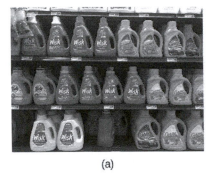
(a)

(b)

(c)

Figure 19.26   (a) Detergents, (b) paints, and (c) fertilizers are all made using transition metal catalysts. (credit a: modification of work by "Mr. Brian"/Flickr; credit b: modification of work by Ewen Roberts; credit c: modification of work by "osseous"/Flickr)

Portrait of a Chemist

Deanna D'Alessandro

Dr. Deanna D'Alessandro develops new metal-containing materials that demonstrate unique electronic, optical, and magnetic properties. Her research combines the fields of fundamental inorganic and physical chemistry with materials engineering. She is working on many different projects that rely on transition metals. For example, one type of compound she is developing captures carbon dioxide waste from power plants and catalytically converts it into useful products (see Figure 19.27).

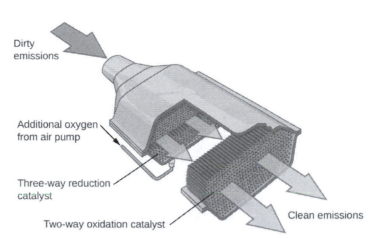

Figure 19.27 Catalytic converters change carbon dioxide emissions from power plants into useful products, and, like the one shown here, are also found in cars.

Another project involves the development of porous, sponge-like materials that are "photoactive." The absorption of light causes the pores of the sponge to change size, allowing gas diffusion to be controlled. This has many potential useful applications, from powering cars with hydrogen fuel cells to making better electronics components. Although not a complex, self-darkening sunglasses are an example of a photoactive substance.

Watch this video (http://openstaxcollege.org/l/16DeannaD) to learn more about this research and listen to Dr. D'Alessandro (shown in Figure 19.28) describe what it is like being a research chemist.

Figure 19.28 Dr. Deanna D'Alessandro is a functional materials researcher. Her work combines the inorganic and physical chemistry fields with engineering, working with transition metals to create new systems to power cars and convert energy (credit: image courtesy of Deanna D'Alessandro).

Many other coordination complexes are also brightly colored. The square planar copper(II) complex phthalocyanine blue (from Figure 19.25) is one of many complexes used as pigments or dyes. This complex is used in blue ink, blue jeans, and certain blue paints.

The structure of heme (Figure 19.29), the iron-containing complex in hemoglobin, is very similar to that in chlorophyll. In hemoglobin, the red heme complex is bonded to a large protein molecule (globin) by the attachment of the protein to the heme ligand. Oxygen molecules are transported by hemoglobin in the blood by being bound to the iron center. When the hemoglobin loses its oxygen, the color changes to a bluish red. Hemoglobin will only transport oxygen if the iron is Fe^{2+}; oxidation of the iron to Fe^{3+} prevents oxygen transport.

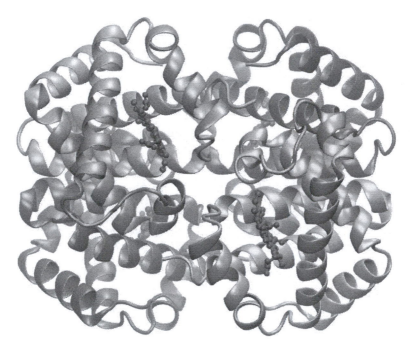

Figure 19.29 Hemoglobin contains four protein subunits, each of which has an iron center attached to a heme ligand (shown in red), which is coordinated to a globin protein. Each subunit is shown in a different color.

Complexing agents often are used for water softening because they tie up such ions as Ca^{2+}, Mg^{2+}, and Fe^{2+}, which make water hard. Many metal ions are also undesirable in food products because these ions can catalyze reactions that change the color of food. Coordination complexes are useful as preservatives. For example, the ligand EDTA, $(HO_2CCH_2)_2NCH_2CH_2N(CH_2CO_2H)_2$, coordinates to metal ions through six donor atoms and prevents the metals from reacting (Figure 19.30). This ligand also is used to sequester metal ions in paper production, textiles, and detergents, and has pharmaceutical uses.

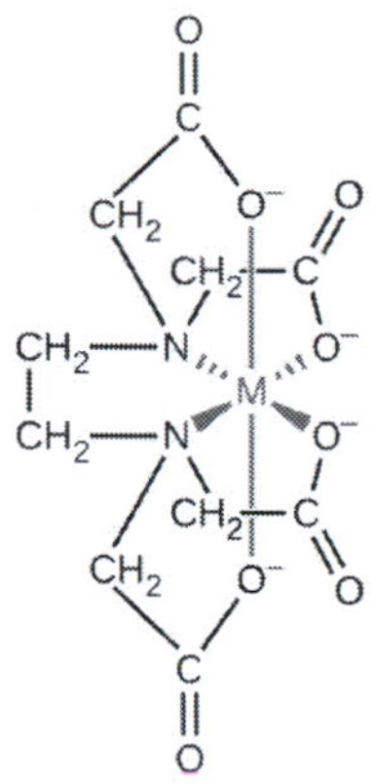

Figure 19.30   The ligand EDTA binds tightly to a variety of metal ions by forming hexadentate complexes.

Complexing agents that tie up metal ions are also used as drugs. British Anti-Lewisite (BAL), $HSCH_2CH(SH)CH_2OH$, is a drug developed during World War I as an antidote for the arsenic-based war gas Lewisite. BAL is now used to treat poisoning by heavy metals, such as arsenic, mercury, thallium, and chromium. The drug is a ligand and functions by making a water-soluble chelate of the metal; the kidneys eliminate this metal chelate (Figure 19.31). Another polydentate ligand, enterobactin, which is isolated from certain bacteria, is used to form complexes of iron and thereby to control the severe iron buildup found in patients suffering from blood diseases such as Cooley's anemia, who require frequent transfusions. As the transfused blood breaks down, the usual metabolic processes that remove iron are overloaded, and excess iron can build up to fatal levels. Enterobactin forms a water-soluble complex with excess iron, and the body can safely eliminate this complex.

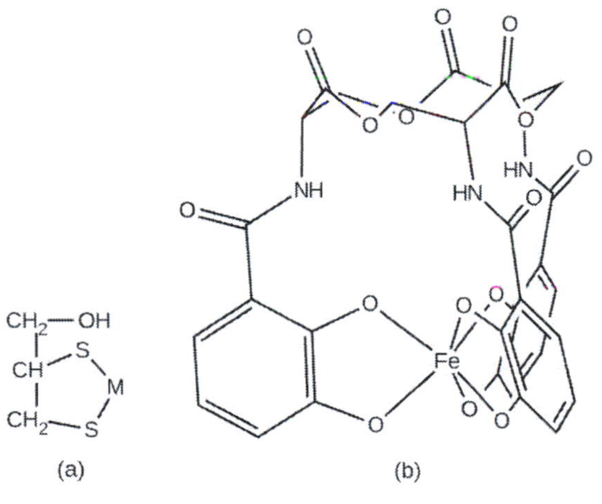

(a)   (b)

Figure 19.31   Coordination complexes are used as drugs. (a) British Anti-Lewisite is used to treat heavy metal poisoning by coordinating metals (M), and enterobactin (b) allows excess iron in the blood to be removed.

Example 19.6

Chelation Therapy

Ligands like BAL and enterobactin are important in medical treatments for heavy metal poisoning. However, chelation therapies can disrupt the normal concentration of ions in the body, leading to serious side effects, so researchers are searching for new chelation drugs. One drug that has been developed is dimercaptosuccinic acid (DMSA), shown in Figure 19.32. Identify which atoms in this molecule could act as donor atoms.

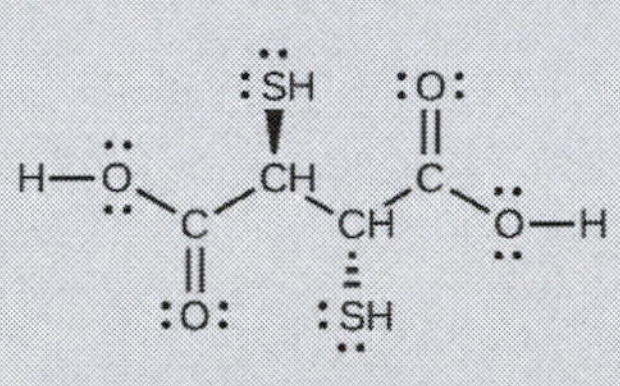

Figure 19.32   Dimercaptosuccinic acid is used to treat heavy metal poisoning.

Solution

All of the oxygen and sulfur atoms have lone pairs of electrons that can be used to coordinate to a metal center, so there are six possible donor atoms. Geometrically, only two of these atoms can be coordinated to a metal at once. The most common binding mode involves the coordination of one sulfur atom and one oxygen atom, forming a five-member ring with the metal.

Check Your Learning

Some alternative medicine practitioners recommend chelation treatments for ailments that are not clearly related to heavy metals, such as cancer and autism, although the practice is discouraged by many scientific organizations.[1] Identify at least two biologically important metals that could be disrupted by chelation therapy.

Answer: Ca, Fe, Zn, and Cu

Ligands are also used in the electroplating industry. When metal ions are reduced to produce thin metal coatings, metals can clump together to form clusters and nanoparticles. When metal coordination complexes are used, the ligands keep the metal atoms isolated from each other. It has been found that many metals plate out as a smoother, more uniform, better-looking, and more adherent surface when plated from a bath containing the metal as a complex ion. Thus, complexes such as $[Ag(CN)_2]^-$ and $[Au(CN)_2]^-$ are used extensively in the electroplating industry.

In 1965, scientists at Michigan State University discovered that there was a platinum complex that inhibited cell division in certain microorganisms. Later work showed that the complex was *cis*-diaminedichloroplatinum(II), $[Pt(NH_3)_2(Cl)_2]$, and that the *trans* isomer was not effective. The inhibition of cell division indicated that this square planar compound could be an anticancer agent. In 1978, the US Food and Drug Administration approved this compound, known as cisplatin, for use in the treatment of certain forms of cancer. Since that time, many similar platinum compounds have been developed for the treatment of cancer. In all cases, these are the *cis* isomers and never the *trans* isomers. The diamine $(NH_3)_2$ portion is retained with other groups, replacing the dichloro $[(Cl)_2]$ portion. The newer drugs include carboplatin, oxaliplatin, and satraplatin.

---

1. National Council against Health Fraud, *NCAHF Policy Statement on Chelation Therapy*, (Peabody, MA, 2002).

19.3 Spectroscopic and Magnetic Properties of Coordination Compounds

By the end of this section, you will be able to:

- Outline the basic premise of crystal field theory (CFT)
- Identify molecular geometries associated with various d-orbital splitting patterns
- Predict electron configurations of split d orbitals for selected transition metal atoms or ions
- Explain spectral and magnetic properties in terms of CFT concepts

The behavior of coordination compounds cannot be adequately explained by the same theories used for main group element chemistry. The observed geometries of coordination complexes are not consistent with hybridized orbitals on the central metal overlapping with ligand orbitals, as would be predicted by valence bond theory. The observed colors indicate that the d orbitals often occur at different energy levels rather than all being degenerate, that is, of equal energy, as are the three p orbitals. To explain the stabilities, structures, colors, and magnetic properties of transition metal complexes, a different bonding model has been developed. Just as valence bond theory explains many aspects of bonding in main group chemistry, crystal field theory is useful in understanding and predicting the behavior of transition metal complexes.

Crystal Field Theory

To explain the observed behavior of transition metal complexes (such as how colors arise), a model involving electrostatic interactions between the electrons from the ligands and the electrons in the unhybridized d orbitals of the central metal atom has been developed. This electrostatic model is **crystal field theory** (CFT). It allows us to understand, interpret, and predict the colors, magnetic behavior, and some structures of coordination compounds of transition metals.

CFT focuses on the nonbonding electrons on the central metal ion in coordination complexes not on the metal-ligand bonds. Like valence bond theory, CFT tells only part of the story of the behavior of complexes. However, it tells the part that valence bond theory does not. In its pure form, CFT ignores any covalent bonding between ligands and metal ions. Both the ligand and the metal are treated as infinitesimally small point charges.

All electrons are negative, so the electrons donated from the ligands will repel the electrons of the central metal. Let us consider the behavior of the electrons in the unhybridized d orbitals in an octahedral complex. The five d orbitals consist of lobe-shaped regions and are arranged in space, as shown in Figure 19.33. In an octahedral complex, the six ligands coordinate along the axes.

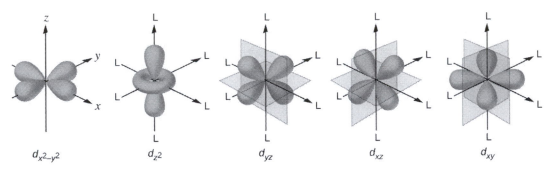

$d_{x^2-y^2}$　　　d_{z^2}　　　d_{yz}　　　d_{xz}　　　d_{xy}

Figure 19.33　The directional characteristics of the five d orbitals are shown here. The shaded portions indicate the phase of the orbitals. The ligands (L) coordinate along the axes. For clarity, the ligands have been omitted from the $d_{x^2-y^2}$ orbital so that the axis labels could be shown.

In an uncomplexed metal ion in the gas phase, the electrons are distributed among the five d orbitals in accord with Hund's rule because the orbitals all have the same energy. However, when ligands coordinate to a metal ion, the energies of the d orbitals are no longer the same.

In octahedral complexes, the lobes in two of the five d orbitals, the d_{z^2} and $d_{x^2-y^2}$ orbitals, point toward the ligands (Figure 19.33). These two orbitals are called the **e_g orbitals** (the symbol actually refers to the symmetry of the orbitals, but we will use it as a convenient name for these two orbitals in an octahedral complex). The other three orbitals, the d_{xy}, d_{xz}, and d_{yz} orbitals, have lobes that point between the ligands and are called the **t_{2g} orbitals** (again, the symbol really refers to the symmetry of the orbitals). As six ligands approach the metal ion along the axes of the octahedron, their point charges repel the electrons in the d orbitals of the metal ion. However, the repulsions between the electrons in the e_g orbitals (the d_{z^2} and $d_{x^2-y^2}$ orbitals) and the ligands are greater than the repulsions between the electrons in the t_{2g} orbitals (the d_{zy}, d_{xz}, and d_{yz} orbitals) and the ligands. This is because the lobes of the e_g orbitals point directly at the ligands, whereas the lobes of the t_{2g} orbitals point between them. Thus, electrons in the e_g orbitals of the metal ion in an octahedral complex have higher potential energies than those of electrons in the t_{2g} orbitals. The difference in energy may be represented as shown in Figure 19.34.

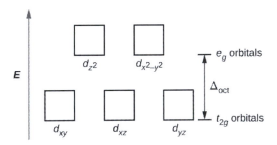

Figure 19.34　In octahedral complexes, the e_g orbitals are destabilized (higher in energy) compared to the t_{2g} orbitals because the ligands interact more strongly with the d orbitals at which they are pointed directly.

The difference in energy between the e_g and the t_{2g} orbitals is called the **crystal field splitting** and is symbolized by **Δoct**, where oct stands for octahedral.

The magnitude of Δ_{oct} depends on many factors, including the nature of the six ligands located around the central metal ion, the charge on the metal, and whether the metal is using $3d$, $4d$, or $5d$ orbitals. Different ligands produce different crystal field splittings. The increasing crystal field splitting produced by ligands is expressed in the **spectrochemical series**, a short version of which is given here:

$$I^- < Br^- < Cl^- < F^- < H_2O < C_2O_4{}^{2-} < NH_3 < en < NO_2{}^- < CN^-$$

a few ligands of the spectrochemical series, in order of increasing field strength of the ligand

In this series, ligands on the left cause small crystal field splittings and are **weak-field ligands**, whereas those on the right cause larger splittings and are **strong-field ligands**. Thus, the Δ_{oct} value for an octahedral complex with iodide ligands (I^-) is much smaller than the Δ_{oct} value for the same metal with cyanide ligands (CN^-).

Electrons in the d orbitals follow the aufbau ("filling up") principle, which says that the orbitals will be filled to give the lowest total energy, just as in main group chemistry. When two electrons occupy the same orbital, the like charges repel each other. The energy needed to pair up two electrons in a single orbital is called the **pairing energy (P)**. Electrons will always singly occupy each orbital in a degenerate set before pairing. P is similar in magnitude to Δ_{oct}. When electrons fill the d orbitals, the relative magnitudes of Δ_{oct} and P determine which orbitals will be occupied.

In $[Fe(CN)_6]^{4-}$, the strong field of six cyanide ligands produces a large Δ_{oct}. Under these conditions, the electrons require less energy to pair than they require to be excited to the e_g orbitals ($\Delta_{oct} > P$). The six $3d$ electrons of the Fe^{2+} ion pair in the three t_{2g} orbitals (Figure 19.35). Complexes in which the electrons are paired because of the large crystal field splitting are called **low-spin complexes** because the number of unpaired electrons (spins) is minimized.

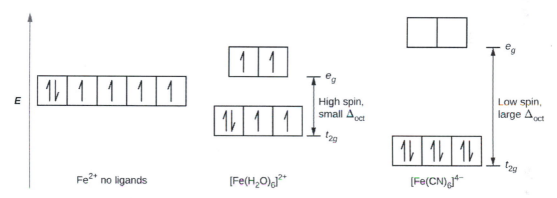

Figure 19.35 Iron(II) complexes have six electrons in the $5d$ orbitals. In the absence of a crystal field, the orbitals are degenerate. For coordination complexes with strong-field ligands such as $[Fe(CN)_6]^{4-}$, Δ_{oct} is greater than P, and the electrons pair in the lower energy t_{2g} orbitals before occupying the eg orbitals. With weak-field ligands such as H_2O, the ligand field splitting is less than the pairing energy, Δ_{oct} less than P, so the electrons occupy all d orbitals singly before any pairing occurs.

In $[Fe(H_2O)_6]^{2+}$, on the other hand, the weak field of the water molecules produces only a small crystal field splitting ($\Delta_{oct} < P$). Because it requires less energy for the electrons to occupy the e_g orbitals than to pair together, there will be an electron in each of the five $3d$ orbitals before pairing occurs. For the six d electrons on the iron(II) center in $[Fe(H_2O)_6]^{2+}$, there will be one pair of electrons and four unpaired electrons (Figure 19.35). Complexes such as the $[Fe(H_2O)_6]^{2+}$ ion, in which the electrons are unpaired because the crystal field splitting is not large enough to cause them to pair, are called **high-spin complexes** because the number of unpaired electrons (spins) is maximized.

A similar line of reasoning shows why the $[Fe(CN)_6]^{3-}$ ion is a low-spin complex with only one unpaired electron, whereas both the $[Fe(H_2O)_6]^{3+}$ and $[FeF_6]^{3-}$ ions are high-spin complexes with five unpaired electrons.

Example 19.7

High- and Low-Spin Complexes

Predict the number of unpaired electrons.

(a) $K_3[CrI_6]$

(b) $[Cu(en)_2(H_2O)_2]Cl_2$

(c) $Na_3[Co(NO_2)_6]$

Solution

The complexes are octahedral.

(a) Cr^{3+} has a d^3 configuration. These electrons will all be unpaired.

(b) Cu^{2+} is d^9, so there will be one unpaired electron.

(c) Co^{3+} has d^6 valence electrons, so the crystal field splitting will determine how many are paired. Nitrite is a strong-field ligand, so the complex will be low spin. Six electrons will go in the t_{2g} orbitals, leaving 0 unpaired.

Check Your Learning

The size of the crystal field splitting only influences the arrangement of electrons when there is a choice between pairing electrons and filling the higher-energy orbitals. For which d-electron configurations will there be a difference between high- and low-spin configurations in octahedral complexes?

Answer: d^4, d^5, d^6, and d^7

Example 19.8

CFT for Other Geometries

CFT is applicable to molecules in geometries other than octahedral. In octahedral complexes, remember that the lobes of the e_g set point directly at the ligands. For tetrahedral complexes, the d orbitals remain in place, but now we have only four ligands located between the axes (Figure 19.36). None of the orbitals points directly at the tetrahedral ligands. However, the e_g set (along the Cartesian axes) overlaps with the ligands less than does the t_{2g} set. By analogy with the octahedral case, predict the energy diagram for the d orbitals in a tetrahedral crystal field. To avoid confusion, the octahedral e_g set becomes a tetrahedral e set, and the octahedral t_{2g} set becomes a t_2 set.

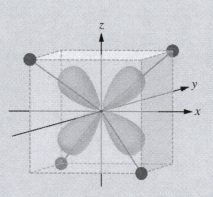

Figure 19.36 This diagram shows the orientation of the tetrahedral ligands with respect to the axis system for the orbitals.

Solution

Since CFT is based on electrostatic repulsion, the orbitals closer to the ligands will be destabilized and raised in energy relative to the other set of orbitals. The splitting is less than for octahedral complexes because the overlap is less, so Δ_{tet} is usually small $\left(\Delta_{tet} = \frac{4}{9}\Delta_{oct}\right)$:

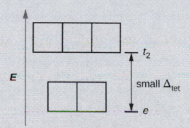

Check Your Learning

Explain how many unpaired electrons a tetrahedral d^4 ion will have.

Answer: 4; because Δ_{tet} is small, all tetrahedral complexes are high spin and the electrons go into the t_2 orbitals before pairing

The other common geometry is square planar. It is possible to consider a square planar geometry as an octahedral structure with a pair of *trans* ligands removed. The removed ligands are assumed to be on the z-axis. This changes the distribution of the d orbitals, as orbitals on or near the z-axis become more stable, and those on or near the x- or y-axes become less stable. This results in the octahedral t_{2g} and the e_g sets splitting and gives a more complicated pattern with no simple Δ_{oct}. The basic pattern is:

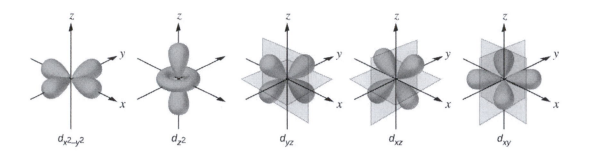

Magnetic Moments of Molecules and Ions

Experimental evidence of magnetic measurements supports the theory of high- and low-spin complexes. Remember that molecules such as O_2 that contain unpaired electrons are paramagnetic. Paramagnetic substances are attracted to magnetic fields. Many transition metal complexes have unpaired electrons and hence are paramagnetic. Molecules such as N_2 and ions such as Na^+ and $[Fe(CN)_6]^{4-}$ that contain no unpaired electrons are diamagnetic. Diamagnetic substances have a slight tendency to be repelled by magnetic fields.

When an electron in an atom or ion is unpaired, the magnetic moment due to its spin makes the entire atom or ion paramagnetic. The size of the magnetic moment of a system containing unpaired electrons is related directly to the number of such electrons: the greater the number of unpaired electrons, the larger the magnetic moment. Therefore, the observed magnetic moment is used to determine the number of unpaired electrons present. The measured magnetic moment of low-spin d^6 $[Fe(CN)_6]^{4-}$ confirms that iron is diamagnetic, whereas high-spin d^6 $[Fe(H_2O)_6]^{2+}$ has four unpaired electrons with a magnetic moment that confirms this arrangement.

Colors of Transition Metal Complexes

When atoms or molecules absorb light at the proper frequency, their electrons are excited to higher-energy orbitals. For many main group atoms and molecules, the absorbed photons are in the ultraviolet range of the electromagnetic spectrum, which cannot be detected by the human eye. For coordination compounds, the energy difference between the d orbitals often allows photons in the visible range to be absorbed.

The human eye perceives a mixture of all the colors, in the proportions present in sunlight, as white light. Complementary colors, those located across from each other on a color wheel, are also used in color vision. The eye perceives a mixture of two complementary colors, in the proper proportions, as white light. Likewise, when a color is missing from white light, the eye sees its complement. For example, when red photons are absorbed from white light, the eyes see the color green. When violet photons are removed from white light, the eyes see lemon yellow. The blue color of the $[Cu(NH_3)_4]^{2+}$ ion results because this ion absorbs orange and red light, leaving the complementary colors of blue and green (Figure 19.37).

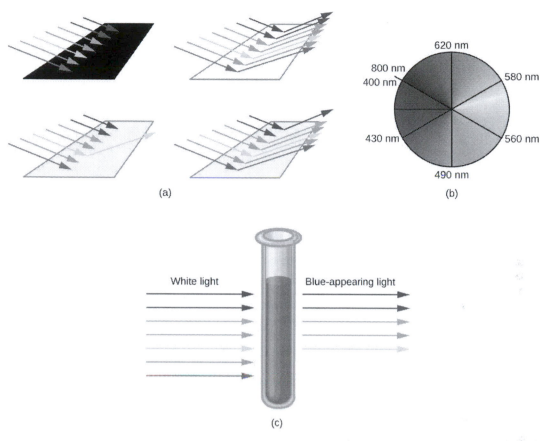

(a)

(b)

(c)

Figure 19.37 (a) An object is black if it absorbs all colors of light. If it reflects all colors of light, it is white. An object has a color if it absorbs all colors except one, such as this yellow strip. The strip also appears yellow if it absorbs the complementary color from white light (in this case, indigo). (b) Complementary colors are located directly across from one another on the color wheel. (c) A solution of $[Cu(NH_3)_4]^{2+}$ ions absorbs red and orange light, so the transmitted light appears as the complementary color, blue.

Example 19.9

Colors of Complexes

The octahedral complex $[Ti(H_2O)_6]^{3+}$ has a single d electron. To excite this electron from the ground state t_{2g} orbital to the e_g orbital, this complex absorbs light from 450 to 600 nm. The maximum absorbance corresponds to Δ_{oct} and occurs at 499 nm. Calculate the value of Δ_{oct} in Joules and predict what color the solution will appear.

Solution

Using Planck's equation (refer to the section on electromagnetic energy), we calculate:

$$v = \frac{c}{\lambda} \text{ so } \frac{3.00 \times 10^8 \text{ m/s}}{\frac{499 \text{ nm} \times 1 \text{ m}}{10^9 \text{ nm}}} = 6.01 \times 10^{14} \text{ Hz}$$

$$E = hnu \text{ so } 6.63 \times 10^{-34} \text{ J·s} \times 6.01 \times 10^{14} \text{ Hz} = 3.99 \times 10^{-19} \text{ Joules/ion}$$

Because the complex absorbs 600 nm (orange) through 450 (blue), the indigo, violet, and red wavelengths will be transmitted, and the complex will appear purple.

Check Your Learning

A complex that appears green, absorbs photons of what wavelengths?

Answer: red, 620–800 nm

Small changes in the relative energies of the orbitals that electrons are transitioning between can lead to drastic shifts in the color of light absorbed. Therefore, the colors of coordination compounds depend on many factors. As shown in Figure 19.38, different aqueous metal ions can have different colors. In addition, different oxidation states of one metal can produce different colors, as shown for the vanadium complexes in the link below.

Figure 19.38 The partially filled d orbitals of the stable ions $Cr^{3+}(aq)$, $Fe^{3+}(aq)$, and $Co^{2+}(aq)$ (left, center and right, respectively) give rise to various colors. (credit: Sahar Atwa)

The specific ligands coordinated to the metal center also influence the color of coordination complexes. For example, the iron(II) complex $[Fe(H_2O)_6]SO_4$ appears blue-green because the high-spin complex absorbs photons in the red wavelengths (Figure 19.39). In contrast, the low-spin iron(II) complex $K_4[Fe(CN)_6]$ appears pale yellow because it absorbs higher-energy violet photons.

(a)   (b)

Figure 19.39 Both (a) hexaaquairon(II) sulfate and (b) potassium hexacyanoferrate(II) contain d^6 iron(II) octahedral metal centers, but they absorb photons in different ranges of the visible spectrum.

In general, strong-field ligands cause a large split in the energies of d orbitals of the central metal atom (large Δ_{oct}). Transition metal coordination compounds with these ligands are yellow, orange, or red because they absorb higher-energy violet or blue light. On the other hand, coordination compounds of transition metals with weak-field ligands are often blue-green, blue, or indigo because they absorb lower-energy yellow, orange, or red light.

A coordination compound of the Cu^+ ion has a d^{10} configuration, and all the e_g orbitals are filled. To excite an electron to a higher level, such as the $4p$ orbital, photons of very high energy are necessary. This energy corresponds to very short wavelengths in the ultraviolet region of the spectrum. No visible light is absorbed, so the eye sees no change, and the compound appears white or colorless. A solution containing $[Cu(CN)_2]^-$, for example, is colorless. On the other hand, octahedral Cu^{2+} complexes have a vacancy in the e_g orbitals, and electrons can be excited to this level. The wavelength (energy) of the light absorbed corresponds to the visible part of the spectrum, and Cu^{2+} complexes are almost always colored—blue, blue-green violet, or yellow (Figure 19.40). Although CFT successfully describes many properties of coordination complexes, molecular orbital explanations (beyond the introductory scope provided here) are required to understand fully the behavior of coordination complexes.

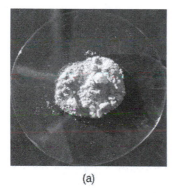

(a)   (b)

Figure 19.40   (a) Copper(I) complexes with d^{10} configurations such as CuI tend to be colorless, whereas (b) d^9 copper(II) complexes such as $Cu(NO_3)_2 \cdot 5H_2O$ are brightly colored.

Key Terms

actinide series (also, actinoid series) actinium and the elements in the second row or the *f*-block, atomic numbers 89–103

bidentate ligand ligand that coordinates to one central metal through coordinate bonds from two different atoms

central metal ion or atom to which one or more ligands is attached through coordinate covalent bonds

chelate complex formed from a polydentate ligand attached to a central metal

chelating ligand ligand that attaches to a central metal ion by bonds from two or more donor atoms

***cis* configuration** configuration of a geometrical isomer in which two similar groups are on the same side of an imaginary reference line on the molecule

coordination compound stable compound in which the central metal atom or ion acts as a Lewis acid and accepts one or more pairs of electrons

coordination compound substance consisting of atoms, molecules, or ions attached to a central atom through Lewis acid-base interactions

coordination number number of coordinate covalent bonds to the central metal atom in a complex or the number of closest contacts to an atom in a crystalline form

coordination sphere central metal atom or ion plus the attached ligands of a complex

crystal field splitting (Δ_{oct}) difference in energy between the t_{2g} and e_g sets or t and e sets of orbitals

crystal field theory model that explains the energies of the orbitals in transition metals in terms of electrostatic interactions with the ligands but does not include metal ligand bonding

***d*-block element** one of the elements in groups 3–11 with valence electrons in *d* orbitals

donor atom atom in a ligand with a lone pair of electrons that forms a coordinate covalent bond to a central metal

e_g orbitals set of two *d* orbitals that are oriented on the Cartesian axes for coordination complexes; in octahedral complexes, they are higher in energy than the t_{2g} orbitals

***f*-block element** (also, inner transition element) one of the elements with atomic numbers 58–71 or 90–103 that have valence electrons in *f* orbitals; they are frequently shown offset below the periodic table

first transition series transition elements in the fourth period of the periodic table (first row of the *d*-block), atomic numbers 21–29

fourth transition series transition elements in the seventh period of the periodic table (fourth row of the *d*-block), atomic numbers 89 and 104–111

geometric isomers isomers that differ in the way in which atoms are oriented in space relative to each other, leading to different physical and chemical properties

high-spin complex complex in which the electrons maximize the total electron spin by singly populating all of the orbitals before pairing two electrons into the lower-energy orbitals

hydrometallurgy process in which a metal is separated from a mixture by first converting it into soluble ions, extracting the ions, and then reducing the ions to precipitate the pure metal

ionization isomer (or coordination isomer) isomer in which an anionic ligand is replaced by the counter ion in the inner coordination sphere

lanthanide series (also, lanthanoid series) lanthanum and the elements in the first row or the *f*-block, atomic numbers 57–71

ligand ion or neutral molecule attached to the central metal ion in a coordination compound

linkage isomer coordination compound that possesses a ligand that can bind to the transition metal in two different ways (CN^- vs. NC^-)

low-spin complex complex in which the electrons minimize the total electron spin by pairing in the lower-energy orbitals before populating the higher-energy orbitals

monodentate ligand that attaches to a central metal through just one coordinate covalent bond

optical isomer (also, enantiomer) molecule that is a nonsuperimposable mirror image with identical chemical and physical properties, except when it reacts with other optical isomers

pairing energy (P) energy required to place two electrons with opposite spins into a single orbital

platinum metals group of six transition metals consisting of ruthenium, osmium, rhodium, iridium, palladium, and platinum that tend to occur in the same minerals and demonstrate similar chemical properties

polydentate ligand ligand that is attached to a central metal ion by bonds from two or more donor atoms, named with prefixes specifying how many donors are present (e.g., hexadentate = six coordinate bonds formed)

rare earth element collection of 17 elements including the lanthanides, scandium, and yttrium that often occur together and have similar chemical properties, making separation difficult

second transition series transition elements in the fifth period of the periodic table (second row of the *d*-block), atomic numbers 39–47

smelting process of extracting a pure metal from a molten ore

spectrochemical series ranking of ligands according to the magnitude of the crystal field splitting they induce

steel material made from iron by removing impurities in the iron and adding substances that produce alloys with properties suitable for specific uses

strong-field ligand ligand that causes larger crystal field splittings

superconductor material that conducts electricity with no resistance

t_{2g} **orbitals** set of three *d* orbitals aligned between the Cartesian axes for coordination complexes; in octahedral complexes, they are lowered in energy compared to the e_g orbitals according to CFT

third transition series transition elements in the sixth period of the periodic table (third row of the *d*-block), atomic numbers 57 and 72–79

trans **configuration** configuration of a geometrical isomer in which two similar groups are on opposite sides of an imaginary reference line on the molecule

weak-field ligand ligand that causes small crystal field splittings

Summary

19.1 Occurrence, Preparation, and Properties of Transition Metals and Their Compounds

The transition metals are elements with partially filled d orbitals, located in the d-block of the periodic table. The reactivity of the transition elements varies widely from very active metals such as scandium and iron to almost inert elements, such as the platinum metals. The type of chemistry used in the isolation of the elements from their ores depends upon the concentration of the element in its ore and the difficulty of reducing ions of the elements to the metals. Metals that are more active are more difficult to reduce.

Transition metals exhibit chemical behavior typical of metals. For example, they oxidize in air upon heating and react with elemental halogens to form halides. Those elements that lie above hydrogen in the activity series react with acids, producing salts and hydrogen gas. Oxides, hydroxides, and carbonates of transition metal compounds in low oxidation states are basic. Halides and other salts are generally stable in water, although oxygen must be excluded in some cases. Most transition metals form a variety of stable oxidation states, allowing them to demonstrate a wide range of chemical reactivity.

19.2 Coordination Chemistry of Transition Metals

The transition elements and main group elements can form coordination compounds, or complexes, in which a central metal atom or ion is bonded to one or more ligands by coordinate covalent bonds. Ligands with more than one donor atom are called polydentate ligands and form chelates. The common geometries found in complexes are tetrahedral and square planar (both with a coordination number of four) and octahedral (with a coordination number of six). *Cis* and *trans* configurations are possible in some octahedral and square planar complexes. In addition to these geometrical isomers, optical isomers (molecules or ions that are mirror images but not superimposable) are possible in certain octahedral complexes. Coordination complexes have a wide variety of uses including oxygen transport in blood, water purification, and pharmaceutical use.

19.3 Spectroscopic and Magnetic Properties of Coordination Compounds

Crystal field theory treats interactions between the electrons on the metal and the ligands as a simple electrostatic effect. The presence of the ligands near the metal ion changes the energies of the metal d orbitals relative to their energies in the free ion. Both the color and the magnetic properties of a complex can be attributed to this crystal field splitting. The magnitude of the splitting (Δ_{oct}) depends on the nature of the ligands bonded to the metal. Strong-field ligands produce large splitting and favor low-spin complexes, in which the t_{2g} orbitals are completely filled before any electrons occupy the e_g orbitals. Weak-field ligands favor formation of high-spin complexes. The t_{2g} and the e_g orbitals are singly occupied before any are doubly occupied.

Exercises

19.1 Occurrence, Preparation, and Properties of Transition Metals and Their Compounds

1. Write the electron configurations for each of the following elements:

(a) Sc

(b) Ti

(c) Cr

(d) Fe

(e) Ru

2. Write the electron configurations for each of the following elements and its ions:

(a) Ti

(b) Ti^{2+}

(c) Ti^{3+}

(d) Ti^{4+}

3. Write the electron configurations for each of the following elements and its 3+ ions:

(a) La

(b) Sm

(c) Lu

4. Why are the lanthanoid elements not found in nature in their elemental forms?

5. Which of the following elements is most likely to be used to prepare La by the reduction of La_2O_3: Al, C, or Fe? Why?

6. Which of the following is the strongest oxidizing agent: $VO_4{}^{3}$, $CrO_4{}^{2-}$, or $MnO_4{}^{-}$?

7. Which of the following elements is most likely to form an oxide with the formula MO_3: Zr, Nb, or Mo?

8. The following reactions all occur in a blast furnace. Which of these are redox reactions?

(a) $3Fe_2O_3(s) + CO(g) \longrightarrow 2Fe_3O_4(s) + CO_2(g)$

(b) $Fe_3O_4(s) + CO(g) \longrightarrow 3FeO(s) + CO_2(g)$

(c) $FeO(s) + CO(g) \longrightarrow Fe(l) + CO_2(g)$

(d) $C(s) + O_2(g) \longrightarrow CO_2(g)$

(e) $C(s) + CO_2(g) \longrightarrow 2CO(g)$

(f) $CaCO_3(s) \longrightarrow CaO(s) + CO_2(g)$

(g) $CaO(s) + SiO_2(s) \longrightarrow CaSiO_3(l)$

9. Why is the formation of slag useful during the smelting of iron?

10. Would you expect an aqueous manganese(VII) oxide solution to have a pH greater or less than 7.0? Justify your answer.

11. Iron(II) can be oxidized to iron(III) by dichromate ion, which is reduced to chromium(III) in acid solution. A 2.5000-g sample of iron ore is dissolved and the iron converted into iron(II). Exactly 19.17 mL of 0.0100 M $Na_2Cr_2O_7$ is required in the titration. What percentage of the ore sample was iron?

12. How many cubic feet of air at a pressure of 760 torr and 0 °C is required per ton of Fe_2O_3 to convert that Fe_2O_3 into iron in a blast furnace? For this exercise, assume air is 19% oxygen by volume.

13. Find the potentials of the following electrochemical cell:

Cd | Cd^{2+}, $M = 0.10$ ‖ Ni^{2+}, $M = 0.50$ | Ni

14. A 2.5624-g sample of a pure solid alkali metal chloride is dissolved in water and treated with excess silver nitrate. The resulting precipitate, filtered and dried, weighs 3.03707 g. What was the percent by mass of chloride ion in the original compound? What is the identity of the salt?

15. The standard reduction potential for the reaction $[Co(H_2O)_6]^{3+}(aq) + e^- \longrightarrow [Co(H_2O)_6]^{2+}(aq)$ is about 1.8 V. The reduction potential for the reaction $[Co(NH_3)_6]^{3+}(aq) + e^- \longrightarrow [Co(NH_3)_6]^{2+}(aq)$ is +0.1 V. Calculate the cell potentials to show whether the complex ions, $[Co(H_2O)_6]^{2+}$ and/or $[Co(NH_3)_6]^{2+}$, can be oxidized to the corresponding cobalt(III) complex by oxygen.

16. Predict the products of each of the following reactions. (Note: In addition to using the information in this chapter, also use the knowledge you have accumulated at this stage of your study, including information on the prediction of reaction products.)

(a) $MnCO_3(s) + HI(aq) \longrightarrow$

(b) $CoO(s) + O_2(g) \longrightarrow$

(c) $La(s) + O_2(g) \longrightarrow$

(d) $V(s) + VCl_4(s) \longrightarrow$

(e) $Co(s) + xsF_2(g) \longrightarrow$

(f) $CrO_3(s) + CsOH(aq) \longrightarrow$

17. Predict the products of each of the following reactions. (Note: In addition to using the information in this chapter, also use the knowledge you have accumulated at this stage of your study, including information on the prediction of reaction products.)

(a) $Fe(s) + H_2SO_4(aq) \longrightarrow$

(b) $FeCl_3(aq) + NaOH(aq) \longrightarrow$

(c) $Mn(OH)_2(s) + HBr(aq) \longrightarrow$

(d) $Cr(s) + O_2(g) \longrightarrow$

(e) $Mn_2O_3(s) + HCl(aq) \longrightarrow$

(f) $Ti(s) + xsF_2(g) \longrightarrow$

18. Describe the electrolytic process for refining copper.

19. Predict the products of the following reactions and balance the equations.

(a) Zn is added to a solution of $Cr_2(SO_4)_3$ in acid.

(b) $FeCl_2$ is added to a solution containing an excess of $Cr_2O_7{}^{2-}$ in hydrochloric acid.

(c) Cr^{2+} is added to $Cr_2O_7{}^{2-}$ in acid solution.

(d) Mn is heated with CrO_3.

(e) CrO is added to $2HNO_3$ in water.

(f) $FeCl_3$ is added to an aqueous solution of NaOH.

20. What is the gas produced when iron(II) sulfide is treated with a nonoxidizing acid?

21. Predict the products of each of the following reactions and then balance the chemical equations.

(a) Fe is heated in an atmosphere of steam.

(b) NaOH is added to a solution of $Fe(NO_3)_3$.

(c) $FeSO_4$ is added to an acidic solution of $KMnO_4$.

(d) Fe is added to a dilute solution of H_2SO_4.

(e) A solution of $Fe(NO_3)_2$ and HNO_3 is allowed to stand in air.

(f) $FeCO_3$ is added to a solution of $HClO_4$.

(g) Fe is heated in air.

22. Balance the following equations by oxidation-reduction methods; note that *three* elements change oxidation state.
$Co(NO_3)_2(s) \longrightarrow Co_2O_3(s) + NO_2(g) + O_2(g)$

23. Dilute sodium cyanide solution is slowly dripped into a slowly stirred silver nitrate solution. A white precipitate forms temporarily but dissolves as the addition of sodium cyanide continues. Use chemical equations to explain this observation. Silver cyanide is similar to silver chloride in its solubility.

24. Predict which will be more stable, $[CrO_4]^{2-}$ or $[WO_4]^{2-}$, and explain.

25. Give the oxidation state of the metal for each of the following oxides of the first transition series. (Hint: Oxides of formula M_3O_4 are examples of *mixed valence compounds* in which the metal ion is present in more than one oxidation state. It is possible to write these compound formulas in the equivalent format $MO \cdot M_2O_3$, to permit estimation of the metal's two oxidation states.)

(a) Sc_2O_3

(b) TiO_2

(c) V_2O_5

(d) CrO_3

(e) MnO_2

(f) Fe_3O_4

(g) Co_3O_4

(h) NiO

(i) Cu_2O

19.2 Coordination Chemistry of Transition Metals

26. Indicate the coordination number for the central metal atom in each of the following coordination compounds:

(a) $[Pt(H_2O)_2Br_2]$

(b) $[Pt(NH_3)(py)(Cl)(Br)]$ (py = pyridine, C_5H_5N)

(c) $[Zn(NH_3)_2Cl_2]$

(d) $[Zn(NH_3)(py)(Cl)(Br)]$

(e) $[Ni(H_2O)_4Cl_2]$

(f) $[Fe(en)_2(CN)_2]^+$ (en = ethylenediamine, $C_2H_8N_2$)

27. Give the coordination numbers and write the formulas for each of the following, including all isomers where appropriate:

(a) tetrahydroxozincate(II) ion (tetrahedral)

(b) hexacyanopalladate(IV) ion

(c) dichloroaurate(I) ion (note that *aurum* is Latin for "gold")

(d) diaminedichloroplatinum(II)

(e) potassium diaminetetrachlorochromate(III)

(f) hexaaminecobalt(III) hexacyanochromate(III)

(g) dibromobis(ethylenediamine) cobalt(III) nitrate

28. Give the coordination number for each metal ion in the following compounds:

(a) $[Co(CO_3)_3]^{3-}$ (note that CO_3^{2-} is bidentate in this complex)

(b) $[Cu(NH_3)_4]^{2+}$

(c) $[Co(NH_3)_4Br_2]_2(SO_4)_3$

(d) $[Pt(NH_3)_4][PtCl_4]$

(e) $[Cr(en)_3](NO_3)_3$

(f) $[Pd(NH_3)_2Br_2]$ (square planar)

(g) $K_3[Cu(Cl)_5]$

(h) $[Zn(NH_3)_2Cl_2]$

29. Sketch the structures of the following complexes. Indicate any *cis*, *trans*, and optical isomers.

(a) $[Pt(H_2O)_2Br_2]$ (square planar)

(b) $[Pt(NH_3)(py)(Cl)(Br)]$ (square planar, py = pyridine, C_5H_5N)

(c) $[Zn(NH_3)_3Cl]^+$ (tetrahedral)

(d) $[Pt(NH_3)_3Cl]^+$ (square planar)

(e) $[Ni(H_2O)_4Cl_2]$

(f) $[Co(C_2O_4)_2Cl_2]^{3-}$ (note that $C_2O_4{}^{2-}$ is the bidentate oxalate ion, $-O_2CCO_2{}^-$)

30. Draw diagrams for any *cis*, *trans*, and optical isomers that could exist for the following (en is ethylenediamine):

(a) $[Co(en)_2(NO_2)Cl]^+$

(b) $[Co(en)_2Cl_2]^+$

(c) $[Pt(NH_3)_2Cl_4]$

(d) $[Cr(en)_3]^{3+}$

(e) $[Pt(NH_3)_2Cl_2]$

31. Name each of the compounds or ions given in Exercise 19.28, including the oxidation state of the metal.

32. Name each of the compounds or ions given in Exercise 19.30.

33. Specify whether the following complexes have isomers.

(a) tetrahedral $[Ni(CO)_2(Cl)_2]$

(b) trigonal bipyramidal $[Mn(CO)_4NO]$

(c) $[Pt(en)_2Cl_2]Cl_2$

34. Predict whether the carbonate ligand $CO_3{}^{2-}$ will coordinate to a metal center as a monodentate, bidentate, or tridentate ligand.

35. Draw the geometric, linkage, and ionization isomers for $[CoCl_5CN][CN]$.

19.3 Spectroscopic and Magnetic Properties of Coordination Compounds

36. Determine the number of unpaired electrons expected for $[Fe(NO_2)_6]^{3-}$ and for $[FeF_6]^{3-}$ in terms of crystal field theory.

37. Draw the crystal field diagrams for $[Fe(NO_2)_6]^{4-}$ and $[FeF_6]^{3-}$. State whether each complex is high spin or low spin, paramagnetic or diamagnetic, and compare Δ_{oct} to P for each complex.

38. Give the oxidation state of the metal, number of *d* electrons, and the number of unpaired electrons predicted for $[Co(NH_3)_6]Cl_3$.

39. The solid anhydrous solid $CoCl_2$ is blue in color. Because it readily absorbs water from the air, it is used as a humidity indicator to monitor if equipment (such as a cell phone) has been exposed to excessive levels of moisture. Predict what product is formed by this reaction, and how many unpaired electrons this complex will have.

40. Is it possible for a complex of a metal in the transition series to have six unpaired electrons? Explain.

41. How many unpaired electrons are present in each of the following?

(a) $[CoF_6]^{3-}$ (high spin)

(b) $[Mn(CN)_6]^{3-}$ (low spin)

(c) $[Mn(CN)_6]^{4-}$ (low spin)

(d) $[MnCl_6]^{4-}$ (high spin)

(e) $[RhCl_6]^{3-}$ (low spin)

42. Explain how the diphosphate ion, $[O_3P-O-PO_3]^{4-}$, can function as a water softener that prevents the precipitation of Fe^{2+} as an insoluble iron salt.

43. For complexes of the same metal ion with no change in oxidation number, the stability increases as the number of electrons in the t_{2g} orbitals increases. Which complex in each of the following pairs of complexes is more stable?

(a) $[Fe(H_2O)_6]^{2+}$ or $[Fe(CN)_6]^{4-}$

(b) $[Co(NH_3)_6]^{3+}$ or $[CoF_6]^{3-}$

(c) $[Mn(CN)_6]^{4-}$ or $[MnCl_6]^{4-}$

44. Trimethylphosphine, $P(CH_3)_3$, can act as a ligand by donating the lone pair of electrons on the phosphorus atom. If trimethylphosphine is added to a solution of nickel(II) chloride in acetone, a blue compound that has a molecular mass of approximately 270 g and contains 21.5% Ni, 26.0% Cl, and 52.5% $P(CH_3)_3$ can be isolated. This blue compound does not have any isomeric forms. What are the geometry and molecular formula of the blue compound?

45. Would you expect the complex $[Co(en)_3]Cl_3$ to have any unpaired electrons? Any isomers?

46. Would you expect the $Mg_3[Cr(CN)_6]_2$ to be diamagnetic or paramagnetic? Explain your reasoning.

47. Would you expect salts of the gold(I) ion, Au^+, to be colored? Explain.

48. $[CuCl_4]^{2-}$ is green. $[Cu(H_2O)_6]^{2+}$ is blue. Which absorbs higher-energy photons? Which is predicted to have a larger crystal field splitting?

Chapter 20

Organic Chemistry

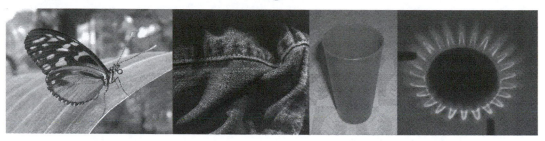

Figure 20.1 All organic compounds contain carbon and most are formed by living things, although they are also formed by geological and artificial processes. (credit left: modification of work by Jon Sullivan; credit left middle: modification of work by Deb Tremper; credit right middle: modification of work by "annszyp"/Wikimedia Commons; credit right: modification of work by George Shuklin)

Chapter Outline

20.1 Hydrocarbons

20.2 Alcohols and Ethers

20.3 Aldehydes, Ketones, Carboxylic Acids, and Esters

20.4 Amines and Amides

Introduction

All living things on earth are formed mostly of carbon compounds. The prevalence of carbon compounds in living things has led to the epithet "carbon-based" life. The truth is we know of no other kind of life. Early chemists regarded substances isolated from *organisms* (plants and animals) as a different type of matter that could not be synthesized artificially, and these substances were thus known as *organic compounds*. The widespread belief called vitalism held that organic compounds were formed by a vital force present only in living organisms. The German chemist Friedrich Wohler was one of the early chemists to refute this aspect of vitalism, when, in 1828, he reported the synthesis of urea, a component of many body fluids, from nonliving materials. Since then, it has been recognized that organic molecules obey the same natural laws as inorganic substances, and the category of organic compounds has evolved to include both natural and synthetic compounds that contain carbon. Some carbon-containing compounds are *not* classified as organic, for example, carbonates and cyanides, and simple oxides, such as CO and CO_2. Although a single, precise definition has yet to be identified by the chemistry community, most agree that a defining trait of organic molecules is the presence of carbon as the principal element, bonded to hydrogen and other carbon atoms.

Today, organic compounds are key components of plastics, soaps, perfumes, sweeteners, fabrics, pharmaceuticals, and many other substances that we use every day. The value to us of organic compounds ensures that organic chemistry is an important discipline within the general field of chemistry. In this chapter, we discuss why the element carbon gives rise to a vast number and variety of compounds, how those compounds are classified, and the role of organic compounds in representative biological and industrial settings.

20.1 Hydrocarbons

By the end of this section, you will be able to:

- Explain the importance of hydrocarbons and the reason for their diversity
- Name saturated and unsaturated hydrocarbons, and molecules derived from them
- Describe the reactions characteristic of saturated and unsaturated hydrocarbons
- Identify structural and geometric isomers of hydrocarbons

The largest database[1] of organic compounds lists about 10 million substances, which include compounds originating from living organisms and those synthesized by chemists. The number of potential organic compounds has been estimated[2] at 10^{60}—an astronomically high number. The existence of so many organic molecules is a consequence of the ability of carbon atoms to form up to four strong bonds to other carbon atoms, resulting in chains and rings of many different sizes, shapes, and complexities.

The simplest **organic compounds** contain only the elements carbon and hydrogen, and are called hydrocarbons. Even though they are composed of only two types of atoms, there is a wide variety of hydrocarbons because they may consist of varying lengths of chains, branched chains, and rings of carbon atoms, or combinations of these structures. In addition, hydrocarbons may differ in the types of carbon-carbon bonds present in their molecules. Many hydrocarbons are found in plants, animals, and their fossils; other hydrocarbons have been prepared in the laboratory. We use hydrocarbons every day, mainly as fuels, such as natural gas, acetylene, propane, butane, and the principal components of gasoline, diesel fuel, and heating oil. The familiar plastics polyethylene, polypropylene, and polystyrene are also hydrocarbons. We can distinguish several types of hydrocarbons by differences in the bonding between carbon atoms. This leads to differences in geometries and in the hybridization of the carbon orbitals.

Alkanes

Alkanes, or **saturated hydrocarbons**, contain only single covalent bonds between carbon atoms. Each of the carbon atoms in an alkane has sp^3 hybrid orbitals and is bonded to four other atoms, each of which is either carbon or hydrogen. The Lewis structures and models of methane, ethane, and pentane are illustrated in Figure 20.2. Carbon chains are usually drawn as straight lines in Lewis structures, but one has to remember that Lewis structures are not intended to indicate the geometry of molecules. Notice that the carbon atoms in the structural models (the ball-and-stick and space-filling models) of the pentane molecule do not lie in a straight line. Because of the sp^3 hybridization, the bond angles in carbon chains are close to 109.5°, giving such chains in an alkane a zigzag shape.

The structures of alkanes and other organic molecules may also be represented in a less detailed manner by condensed structural formulas (or simply, *condensed formulas*). Instead of the usual format for chemical formulas in which each element symbol appears just once, a condensed formula is written to suggest the bonding in the molecule. These formulas have the appearance of a Lewis structure from which most or all of the bond symbols have been removed. Condensed structural formulas for ethane and pentane are shown at the bottom of Figure 20.2, and several additional examples are provided in the exercises at the end of this chapter.

---

1. This is the Beilstein database, now available through the Reaxys site (www.elsevier.com/online-tools/reaxys).
2. Peplow, Mark. "Organic Synthesis: The Robo-Chemist," *Nature* 512 (2014): 20–2.

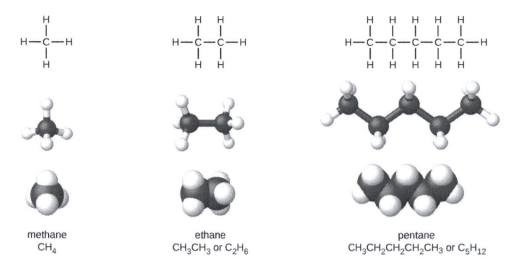

Figure 20.2 Pictured are the Lewis structures, ball-and-stick models, and space-filling models for molecules of methane, ethane, and pentane.

A common method used by organic chemists to simplify the drawings of larger molecules is to use a **skeletal structure** (also called a line-angle structure). In this type of structure, carbon atoms are not symbolized with a C, but represented by each end of a line or bend in a line. Hydrogen atoms are not drawn if they are attached to a carbon. Other atoms besides carbon and hydrogen are represented by their elemental symbols. Figure 20.3 shows three different ways to draw the same structure.

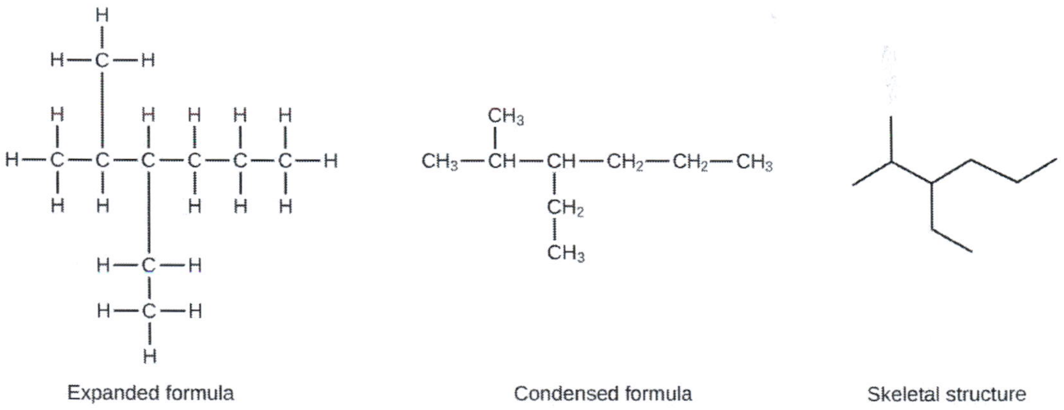

Figure 20.3 The same structure can be represented three different ways: an expanded formula, a condensed formula, and a skeletal structure.

Example 20.1

Drawing Skeletal Structures

Draw the skeletal structures for these two molecules:

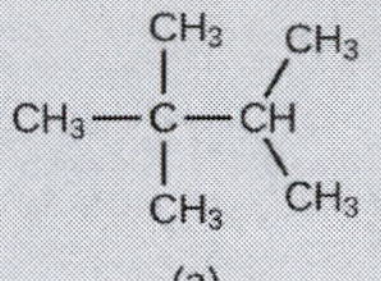

$CH_3CH_2CH_2CH_2CH_2CH_2CH_3$

(a) (b)

Solution

Each carbon atom is converted into the end of a line or the place where lines intersect. All hydrogen atoms attached to the carbon atoms are left out of the structure (although we still need to recognize they are there):

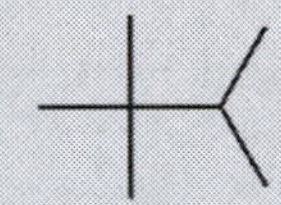

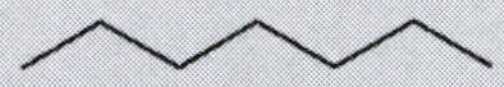

(a) (b)

Check Your Learning

Draw the skeletal structures for these two molecules:

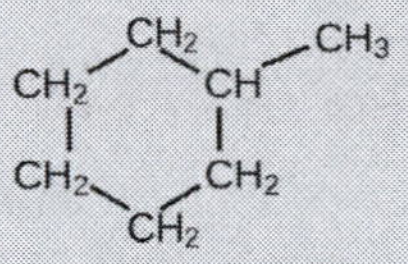

$CH_3CH_2CH_2CH_2CH_3$

(a) (b)

Answer:

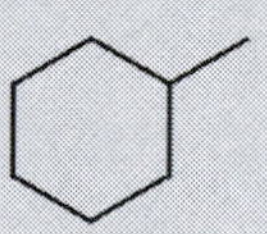

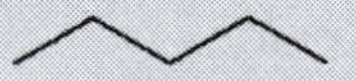

(a) (b)

Example 20.2

Interpreting Skeletal Structures

Identify the chemical formula of the molecule represented here:

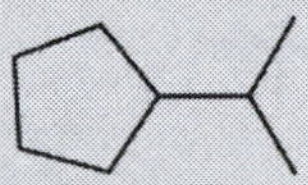

Solution

There are eight places where lines intersect or end, meaning that there are eight carbon atoms in the molecule. Since we know that carbon atoms tend to make four bonds, each carbon atom will have the number of hydrogen atoms that are required for four bonds. This compound contains 16 hydrogen atoms for a molecular formula of C_8H_{16}.

Location of the hydrogen atoms:

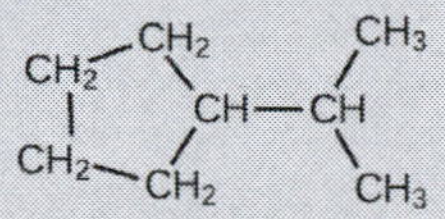

Check Your Learning

Identify the chemical formula of the molecule represented here:

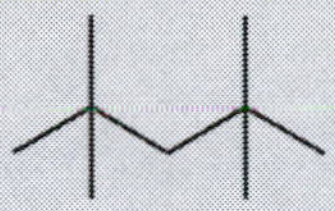

Answer: C_9H_{20}

All alkanes are composed of carbon and hydrogen atoms, and have similar bonds, structures, and formulas; noncyclic alkanes all have a formula of C_nH_{2n+2}. The number of carbon atoms present in an alkane has no limit. Greater numbers of atoms in the molecules will lead to stronger intermolecular attractions (dispersion forces) and correspondingly different physical properties of the molecules. Properties such as melting point and boiling point (Table 20.1) usually change smoothly and predictably as the number of carbon and hydrogen atoms in the molecules change.

Properties of Some Alkanes

Alkane	Molecular Formula	Melting Point (°C)	Boiling Point (°C)	Phase at STP[4]	Number of Structural Isomers
methane	CH_4	−182.5	−161.5	gas	1
ethane	C_2H_6	−183.3	−88.6	gas	1
propane	C_3H_8	−187.7	−42.1	gas	1
butane	C_4H_{10}	−138.3	−0.5	gas	2
pentane	C_5H_{12}	−129.7	36.1	liquid	3
hexane	C_6H_{14}	−95.3	68.7	liquid	5
heptane	C_7H_{16}	−90.6	98.4	liquid	9
octane	C_8H_{18}	−56.8	125.7	liquid	18
nonane	C_9H_{20}	−53.6	150.8	liquid	35
decane	$C_{10}H_{22}$	−29.7	174.0	liquid	75
tetradecane	$C_{14}H_{30}$	5.9	253.5	solid	1858
octadecane	$C_{18}H_{38}$	28.2	316.1	solid	60,523

Table 20.1

---

3. Physical properties for C_4H_{10} and heavier molecules are those of the *normal isomer*, *n*-butane, *n*-pentane, etc.
4. STP indicates a temperature of 0 °C and a pressure of 1 atm.

Hydrocarbons with the same formula, including alkanes, can have different structures. For example, two alkanes have the formula C_4H_{10}: They are called *n*-butane and 2-methylpropane (or isobutane), and have the following Lewis structures:

n-butane   2-methylpropane

The compounds *n*-butane and 2-methylpropane are structural isomers (the term constitutional isomers is also commonly used). Constitutional isomers have the same molecular formula but different spatial arrangements of the atoms in their molecules. The *n*-butane molecule contains an *unbranched chain*, meaning that no carbon atom is bonded to more than two other carbon atoms. We use the term *normal*, or the prefix *n*, to refer to a chain of carbon atoms without branching. The compound 2–methylpropane has a branched chain (the carbon atom in the center of the Lewis structure is bonded to three other carbon atoms)

Identifying isomers from Lewis structures is not as easy as it looks. Lewis structures that look different may actually represent the same isomers. For example, the three structures in Figure 20.4 all represent the same molecule, *n*-butane, and hence are not different isomers. They are identical because each contains an unbranched chain of four carbon atoms.

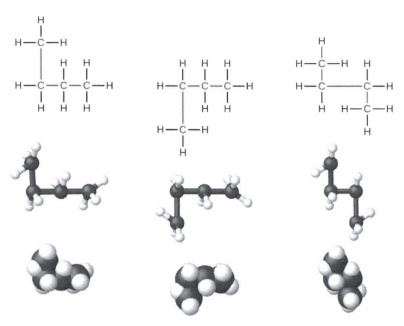

Figure 20.4   These three representations of the structure of n-butane are not isomers because they all contain the same arrangement of atoms and bonds.

The Basics of Organic Nomenclature: Naming Alkanes

The International Union of Pure and Applied Chemistry (IUPAC) has devised a system of nomenclature that begins with the names of the alkanes and can be adjusted from there to account for more complicated structures. The nomenclature for alkanes is based on two rules:

1. To name an alkane, first identify the longest chain of carbon atoms in its structure. A two-carbon chain is called ethane; a three-carbon chain, propane; and a four-carbon chain, butane. Longer chains are named as follows: pentane (five-carbon chain), hexane (6), heptane (7), octane (8), nonane (9), and decane (10). These prefixes can be seen in the names of the alkanes described in Table 20.1.

2. Add prefixes to the name of the longest chain to indicate the positions and names of **substituents**. Substituents are branches or functional groups that replace hydrogen atoms on a chain. The position of a substituent or branch is identified by the number of the carbon atom it is bonded to in the chain. We number the carbon atoms in the chain by counting from the end of the chain nearest the substituents. Multiple substituents are named individually and placed in alphabetical order at the front of the name.

$\overset{1}{C}H_3\overset{2}{C}H_2\overset{3}{C}H_3$	$\overset{1}{C}H_3\overset{2}{C}H\overset{3}{C}H_3$ | Cl	$\overset{3}{C}H_3\overset{2}{C}H\overset{1}{C}H_3$ | CH_3	$\overset{6}{C}H_3\overset{5}{C}H_2\overset{4}{C}H\overset{3}{C}H_2\overset{2}{C}H\overset{1}{C}H_3$ | | F F	$\overset{1}{C}H_2\overset{2}{C}H_2\overset{3}{C}H\overset{4}{C}H_2\overset{5}{C}H_2\overset{6}{C}H_3$ | | Br Cl
propane	2-chloropropane	2-methylpropane	2,4-difluorohexane	1-bromo-3-chlorohexane

When more than one substituent is present, either on the same carbon atom or on different carbon atoms, the substituents are listed alphabetically. Because the carbon atom numbering begins at the end closest to a substituent, the longest chain of carbon atoms is numbered in such a way as to produce the lowest number for the substituents. The ending *-o* replaces *-ide* at the end of the name of an electronegative substituent (in ionic compounds, the negatively charged ion ends with *-ide* like chloride; in organic compounds, such atoms are treated as substituents and the *-o* ending is used). The number of substituents of the same type is indicated by the prefixes *di-* (two), *tri-* (three), *tetra-* (four), and so on (for example, *difluoro-* indicates two fluoride substituents).

Example 20.3

Naming Halogen-substituted Alkanes

Name the molecule whose structure is shown here:

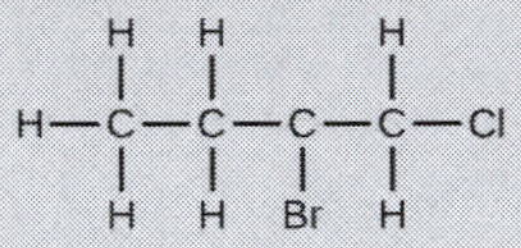

Solution

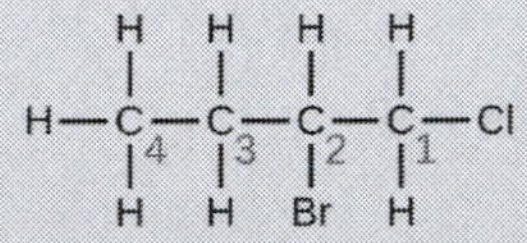

The four-carbon chain is numbered from the end with the chlorine atom. This puts the substituents on positions 1 and 2 (numbering from the other end would put the substituents on positions 3 and 4). Four carbon atoms means that the base name of this compound will be butane. The bromine at position 2 will be described by adding 2-bromo-; this will come at the beginning of the name, since bromo- comes before chloro- alphabetically. The chlorine at position 1 will be described by adding 1-chloro-, resulting in the name of the molecule being 2-bromo-1-chlorobutane.

Check Your Learning

Name the following molecule:

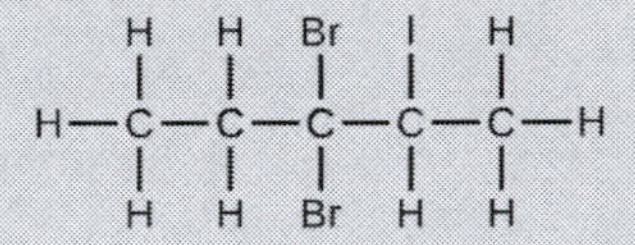

Answer:　3,3-dibromo-2-iodopentane

We call a substituent that contains one less hydrogen than the corresponding alkane an alkyl group. The name of an **alkyl group** is obtained by dropping the suffix *-ane* of the alkane name and adding *-yl*:

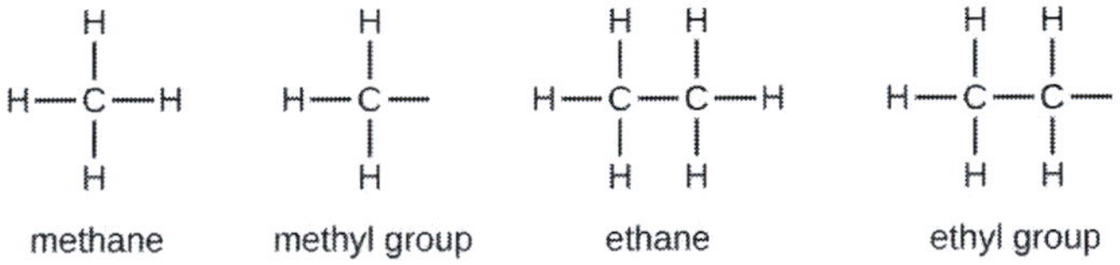

The open bonds in the methyl and ethyl groups indicate that these alkyl groups are bonded to another atom.

Example 20.4

Naming Substituted Alkanes

Name the molecule whose structure is shown here:

Solution

The longest carbon chain runs horizontally across the page and contains six carbon atoms (this makes the base of the name hexane, but we will also need to incorporate the name of the branch). In this case, we want to number from right to left (as shown by the red numbers) so the branch is connected to carbon 3 (imagine the numbers from left to right—this would put the branch on carbon 4, violating our rules). The branch attached to position 3 of our chain contains two carbon atoms (numbered in blue)—so we take our name for two carbons *eth-* and attach *-yl* at the end to signify we are describing a branch. Putting all the pieces together, this molecule is 3-ethylhexane.

Check Your Learning

Name the following molecule:

Answer: 4-propyloctane

Some hydrocarbons can form more than one type of alkyl group when the hydrogen atoms that would be removed have different "environments" in the molecule. This diversity of possible alkyl groups can be identified in the following way: The four hydrogen atoms in a methane molecule are equivalent; they all have the same environment. They are equivalent because each is bonded to a carbon atom (the same carbon atom) that is bonded to three hydrogen atoms. (It may be easier to see the equivalency in the ball and stick models in Figure 20.2. Removal of any one of the four hydrogen atoms from methane forms a methyl group. Likewise, the six hydrogen atoms in ethane are equivalent (Figure 20.2) and removing any one of these hydrogen atoms produces an ethyl group. Each of the six hydrogen atoms is bonded to a carbon atom that is bonded to two other hydrogen atoms and a carbon atom. However, in both propane and 2–methylpropane, there are hydrogen atoms in two different environments, distinguished by the adjacent atoms or groups of atoms:

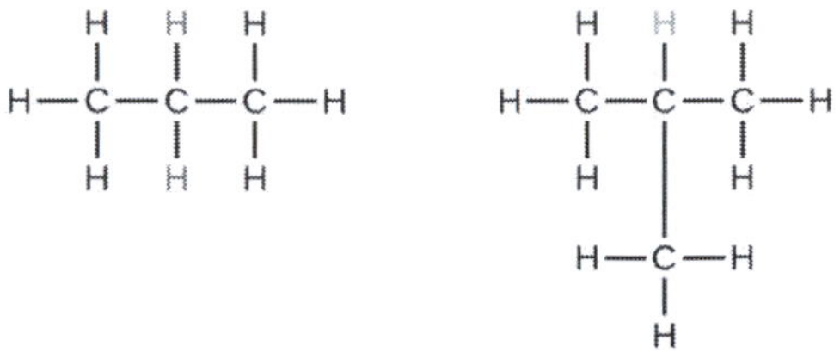

propane   2-methylpropane

Each of the six equivalent hydrogen atoms of the first type in propane and each of the nine equivalent hydrogen atoms of that type in 2-methylpropane (all shown in black) are bonded to a carbon atom that is bonded to only one other carbon atom. The two purple hydrogen atoms in propane are of a second type. They differ from the six hydrogen atoms of the first type in that they are bonded to a carbon atom bonded to two other carbon atoms. The green hydrogen atom in 2-methylpropane differs from the other nine hydrogen atoms in that molecule and from the purple hydrogen atoms in propane. The green hydrogen atom in 2-methylpropane is bonded to a carbon atom bonded to three other carbon atoms. Two different alkyl groups can be formed from each of these molecules, depending on which hydrogen atom is removed. The names and structures of these and several other alkyl groups are listed in Figure 20.5.

Alkyl Group	Structure
methyl	CH_3-
ethyl	CH_3CH_2-
n-propyl	$CH_3CH_2CH_2-$
isopropyl	$CH_3\overset{\mid}{C}HCH_3$
n-butyl	$CH_3CH_2CH_2CH_2-$
sec-butyl	$CH_3CH_2\overset{\mid}{C}HCH_3$
isobutyl	CH_3CHCH_2- $\overset{\mid}{C}H_3$
tert-butyl	$CH_3\overset{\mid}{\underset{\mid}{C}}CH_3$ CH_3

Figure 20.5   This listing gives the names and formulas for various alkyl groups formed by the removal of hydrogen atoms from different locations.

Note that alkyl groups do not exist as stable independent entities. They are always a part of some larger molecule. The location of an alkyl group on a hydrocarbon chain is indicated in the same way as any other substituent:

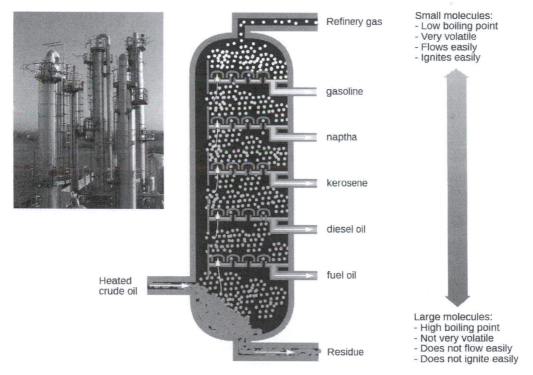

$$3\text{-ethylheptane} \qquad 2,2,4\text{-trimethylpentane} \qquad 4\text{-isopropylheptane}$$

Alkanes are relatively stable molecules, but heat or light will activate reactions that involve the breaking of C–H or C–C single bonds. Combustion is one such reaction:

$$CH_4(g) + 2O_2(g) \longrightarrow CO_2(g) + 2H_2O(g)$$

Alkanes burn in the presence of oxygen, a highly exothermic oxidation-reduction reaction that produces carbon dioxide and water. As a consequence, alkanes are excellent fuels. For example, methane, CH_4, is the principal component of natural gas. Butane, C_4H_{10}, used in camping stoves and lighters is an alkane. Gasoline is a liquid mixture of continuous- and branched-chain alkanes, each containing from five to nine carbon atoms, plus various additives to improve its performance as a fuel. Kerosene, diesel oil, and fuel oil are primarily mixtures of alkanes with higher molecular masses. The main source of these liquid alkane fuels is crude oil, a complex mixture that is separated by fractional distillation. Fractional distillation takes advantage of differences in the boiling points of the components of the mixture (see Figure 20.6). You may recall that boiling point is a function of intermolecular interactions, which was discussed in the chapter on solutions and colloids.

Figure 20.6 In a column for the fractional distillation of crude oil, oil heated to about 425 °C in the furnace vaporizes when it enters the base of the tower. The vapors rise through bubble caps in a series of trays in the tower. As the vapors gradually cool, fractions of higher, then of lower, boiling points condense to liquids and are drawn off. (credit left: modification of work by Luigi Chiesa)

In a **substitution reaction**, another typical reaction of alkanes, one or more of the alkane's hydrogen atoms is replaced with a different atom or group of atoms. No carbon-carbon bonds are broken in these reactions, and the hybridization of the carbon atoms does not change. For example, the reaction between ethane and molecular chlorine depicted here is a substitution reaction:

$$
\underset{\text{ethane}}{\text{H—C—C—H}} + \text{Cl—Cl} \xrightarrow{\text{Heat or light}} \underset{\text{chloroethane}}{\text{H—C—C—Cl:}} + \text{H—Cl:}
$$

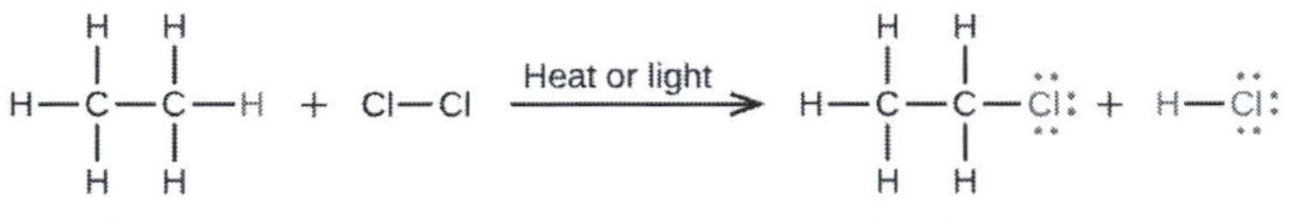

The C–Cl portion of the chloroethane molecule is an example of a **functional group**, the part or moiety of a molecule that imparts a specific chemical reactivity. The types of functional groups present in an organic molecule are major determinants of its chemical properties and are used as a means of classifying organic compounds as detailed in the remaining sections of this chapter.

Link to Learning

Want more practice naming alkanes? Watch this brief video tutorial (http://openstaxcollege.org/l/16alkanes) to review the nomenclature process.

Alkenes

Organic compounds that contain one or more double or triple bonds between carbon atoms are described as unsaturated. You have likely heard of unsaturated fats. These are complex organic molecules with long chains of carbon atoms, which contain at least one double bond between carbon atoms. Unsaturated hydrocarbon molecules that contain one or more double bonds are called **alkenes**. Carbon atoms linked by a double bond are bound together by two bonds, one σ bond and one π bond. Double and triple bonds give rise to a different geometry around the carbon atom that participates in them, leading to important differences in molecular shape and properties. The differing geometries are responsible for the different properties of unsaturated versus saturated fats.

Ethene, C_2H_4, is the simplest alkene. Each carbon atom in ethene, commonly called ethylene, has a trigonal planar structure. The second member of the series is propene (propylene) (Figure 20.7); the butene isomers follow in the series. Four carbon atoms in the chain of butene allows for the formation of isomers based on the position of the double bond, as well as a new form of isomerism.

Download for free at http://cnx.org/content/col11760/latest/

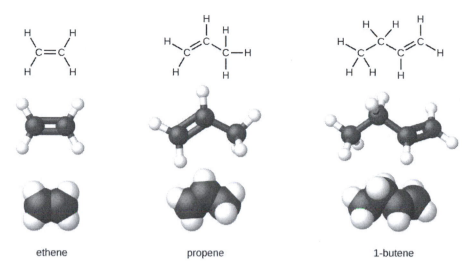

ethene   propene   1-butene

Figure 20.7 Expanded structures, ball-and-stick structures, and space-filling models for the alkenes ethene, propene, and 1-butene are shown.

Ethylene (the common industrial name for ethene) is a basic raw material in the production of polyethylene and other important compounds. Over 135 million tons of ethylene were produced worldwide in 2010 for use in the polymer, petrochemical, and plastic industries. Ethylene is produced industrially in a process called cracking, in which the long hydrocarbon chains in a petroleum mixture are broken into smaller molecules.

Chemistry in Everyday Life

Recycling Plastics

Polymers (from Greek words *poly* meaning "many" and *mer* meaning "parts") are large molecules made up of repeating units, referred to as monomers. Polymers can be natural (starch is a polymer of sugar residues and proteins are polymers of amino acids) or synthetic [like polyethylene, polyvinyl chloride (PVC), and polystyrene]. The variety of structures of polymers translates into a broad range of properties and uses that make them integral parts of our everyday lives. Adding functional groups to the structure of a polymer can result in significantly different properties (see the discussion about Kevlar later in this chapter).

An example of a polymerization reaction is shown in Figure 20.8. The monomer ethylene (C_2H_4) is a gas at room temperature, but when polymerized, using a transition metal catalyst, it is transformed into a solid material made up of long chains of $-CH_2-$ units called polyethylene. Polyethylene is a commodity plastic used primarily for packaging (bags and films).

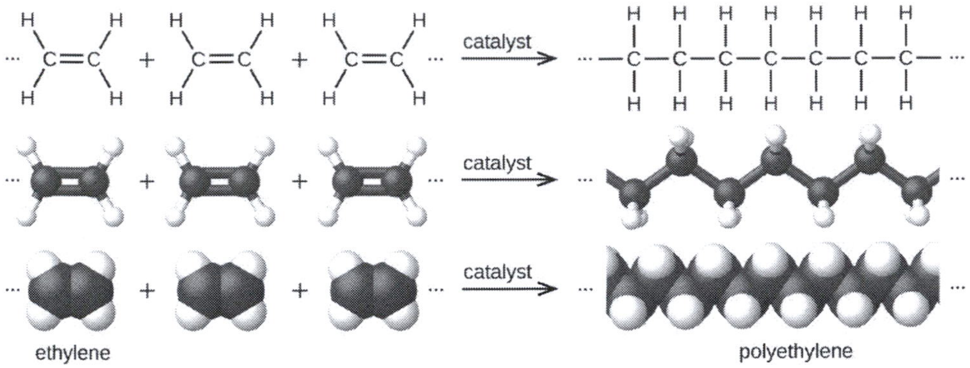

Figure 20.8   The reaction for the polymerization of ethylene to polyethylene is shown.

Polyethylene is a member of one subset of synthetic polymers classified as plastics. Plastics are synthetic organic solids that can be molded; they are typically organic polymers with high molecular masses. Most of the monomers that go into common plastics (ethylene, propylene, vinyl chloride, styrene, and ethylene terephthalate) are derived from petrochemicals and are not very biodegradable, making them candidate materials for recycling. Recycling plastics helps minimize the need for using more of the petrochemical supplies and also minimizes the environmental damage caused by throwing away these nonbiodegradable materials.

Plastic recycling is the process of recovering waste, scrap, or used plastics, and reprocessing the material into useful products. For example, polyethylene terephthalate (soft drink bottles) can be melted down and used for plastic furniture, in carpets, or for other applications. Other plastics, like polyethylene (bags) and polypropylene (cups, plastic food containers), can be recycled or reprocessed to be used again. Many areas of the country have recycling programs that focus on one or more of the commodity plastics that have been assigned a recycling code (see Figure 20.9). These operations have been in effect since the 1970s and have made the production of some plastics among the most efficient industrial operations today.

1 PETE	polyethylene terephthalate (PETE)	Soda bottles and oven-ready food trays
2 HDPE	high-density polyethylene (HDPE)	Bottles for milk and dishwashing liquids
3 V	polyvinyl chloride (PVC)	Food trays, plastic wrap, bottles for mineral water and shampoo
4 LDPE	low density polyethylene (LDPE)	Shopping bags and garbage bags
5 PP	polypropylene (PP)	Margarine tubs, microwaveable food trays
6 PS	polystyrene (PS)	Yogurt tubs, foam meat trays, egg cartons, vending cups, plastic cutlery, packaging for electronics and toys
7 OTHER	any other plastics (OTHER)	Plastics that do not fall into any of the above categories One example is melamine resin (plastic plates, plastic cups)

Figure 20.9 Each type of recyclable plastic is imprinted with a code for easy identification.

The name of an alkene is derived from the name of the alkane with the same number of carbon atoms. The presence of the double bond is signified by replacing the suffix -ane with the suffix -ene. The location of the double bond is identified by naming the smaller of the numbers of the carbon atoms participating in the double bond:

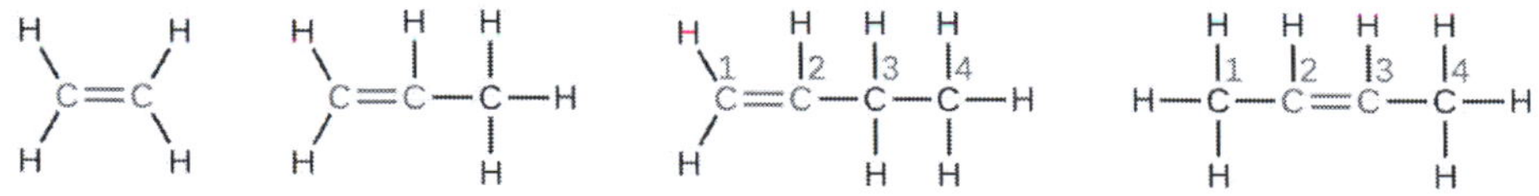

| ethene (ethylene) | propene (propylene) | 1-butene | 2-butene |

Isomers of Alkenes

Molecules of 1-butene and 2-butene are structural isomers; the arrangement of the atoms in these two molecules differs. As an example of arrangement differences, the first carbon atom in 1-butene is bonded to two hydrogen atoms; the first carbon atom in 2-butene is bonded to three hydrogen atoms.

The compound 2-butene and some other alkenes also form a second type of isomer called a geometric isomer. In a set of geometric isomers, the same types of atoms are attached to each other in the same order, but the geometries of the two molecules differ. Geometric isomers of alkenes differ in the orientation of the groups on either side of a $C = C$ bond.

Carbon atoms are free to rotate around a single bond but not around a double bond; a double bond is rigid. This makes it possible to have two isomers of 2-butene, one with both methyl groups on the same side of the double bond and one with the methyl groups on opposite sides. When structures of butene are drawn with 120° bond angles around the sp^2-hybridized carbon atoms participating in the double bond, the isomers are apparent. The 2-butene isomer in which the two methyl groups are on the same side is called a *cis*-isomer; the one in which the two methyl groups are on opposite sides is called a *trans*-isomer (Figure 20.10). The different geometries produce different physical properties, such as boiling point, that may make separation of the isomers possible:

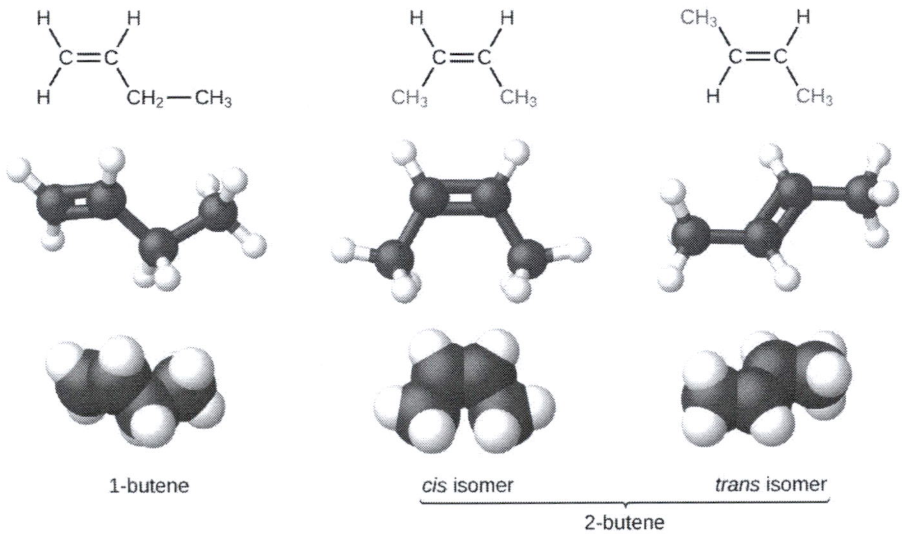

Figure 20.10   These molecular models show the structural and geometric isomers of butene.

Alkenes are much more reactive than alkanes because the $C = C$ moiety is a reactive functional group. A π bond, being a weaker bond, is disrupted much more easily than a σ bond. Thus, alkenes undergo a characteristic reaction in which the π bond is broken and replaced by two σ bonds. This reaction is called an addition reaction. The hybridization of the carbon atoms in the double bond in an alkene changes from sp^2 to sp^3 during an addition reaction. For example, halogens add to the double bond in an alkene instead of replacing hydrogen, as occurs in an alkane:

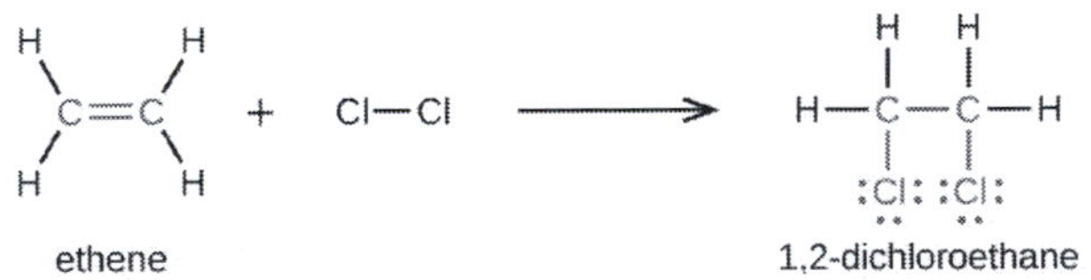

Example 20.5

Alkene Reactivity and Naming

Provide the IUPAC names for the reactant and product of the halogenation reaction shown here:

$$\underset{CH=CH}{\overset{CH_3 \qquad CH_2-CH_3}{\diagdown \qquad \diagup}} \quad + \quad Cl-Cl \quad \longrightarrow$$

Solution

The reactant is a five-carbon chain that contains a carbon-carbon double bond, so the base name will be pentene. We begin counting at the end of the chain closest to the double bond—in this case, from the left—the double bond spans carbons 2 and 3, so the name becomes 2-pentene. Since there are two carbon-containing groups attached to the two carbon atoms in the double bond—and they are on the same side of the double bond—this molecule is the *cis*-isomer, making the name of the starting alkene *cis*-2-pentene. The product of the halogenation reaction will have two chlorine atoms attached to the carbon atoms that were a part of the carbon-carbon double bond:

$$\underset{\underset{Cl \qquad Cl}{\diagup \qquad \diagdown}}{\overset{CH_3 \qquad \quad CH_2-CH_3}{\underset{CH-CH}{\diagdown \qquad \diagup}}}$$

This molecule is now a substituted alkane and will be named as such. The base of the name will be pentane. We will count from the end that numbers the carbon atoms where the chlorine atoms are attached as 2 and 3, making the name of the product 2,3-dichloropentane.

Check Your Learning

Provide names for the reactant and product of the reaction shown:

$$\underset{\underset{H \quad H \qquad\qquad H \quad H}{\big|\quad\big|\qquad\qquad\big|\quad\big|}}{\overset{\overset{H \quad H \quad H \quad H \quad H \quad H}{\big|\quad\big|\quad\big|\quad\big|\quad\big|\quad\big|}}{H-C-C-C=C-C-C-H}} \quad + \quad Cl-Cl \quad \longrightarrow$$

Answer: reactant: trans-3-hexene, product: 3,4-dichlorohexane

Alkynes

Hydrocarbon molecules with one or more triple bonds are called **alkynes**; they make up another series of unsaturated hydrocarbons. Two carbon atoms joined by a triple bond are bound together by one σ bond and two π bonds. The *sp*-hybridized carbons involved in the triple bond have bond angles of 180°, giving these types of bonds a linear, rod-like shape.

The simplest member of the alkyne series is ethyne, C_2H_2, commonly called acetylene. The Lewis structure for ethyne, a linear molecule, is:

$$H-C\equiv C-H$$

ethyne (acetylene)

The IUPAC nomenclature for alkynes is similar to that for alkenes except that the suffix *-yne* is used to indicate a triple bond in the chain. For example, $CH_3CH_2C \equiv CH$ is called 1-butyne.

Example 20.6

Structure of Alkynes

Describe the geometry and hybridization of the carbon atoms in the following molecule:

$$\overset{1}{CH_3}-\overset{2}{C}\equiv\overset{3}{C}-\overset{4}{CH_3}$$

Solution

Carbon atoms 1 and 4 have four single bonds and are thus tetrahedral with sp^3 hybridization. Carbon atoms 2 and 3 are involved in the triple bond, so they have linear geometries and would be classified as sp hybrids.

Check Your Learning

Identify the hybridization and bond angles at the carbon atoms in the molecule shown:

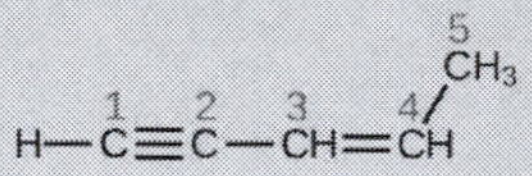

$$\overset{5}{\underset{\qquad}{CH_3}}$$
$$H-\overset{1}{C}\equiv\overset{2}{C}-\overset{3}{C}H=\overset{4}{C}H$$

Answer: carbon 1: sp, 180°; carbon 2: sp, 180°; carbon 3: sp^2, 120°; carbon 4: sp^2, 120°; carbon 5: sp^3, 109.5°

Chemically, the alkynes are similar to the alkenes. Since the $C \equiv C$ functional group has two π bonds, alkynes typically react even more readily, and react with twice as much reagent in addition reactions. The reaction of acetylene with bromine is a typical example:

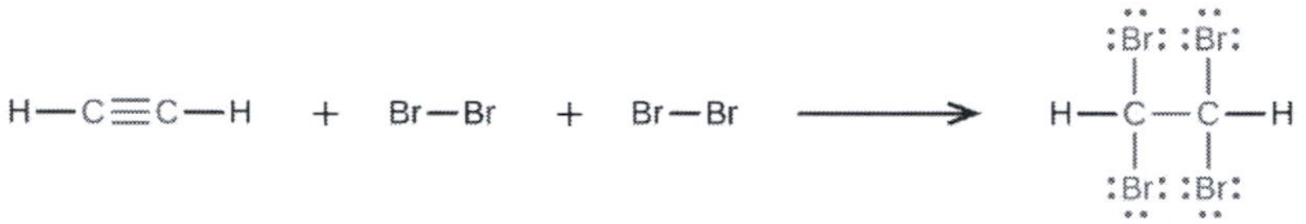

$$H-C\equiv C-H \quad + \quad Br-Br \quad + \quad Br-Br \quad \longrightarrow \quad H-\overset{\displaystyle :\overset{..}{Br}:}{\underset{\displaystyle :\overset{..}{Br}:}{C}}-\overset{\displaystyle :\overset{..}{Br}:}{\underset{\displaystyle :\overset{..}{Br}:}{C}}-H$$

1,1,2,2-tetrabromoethane

Acetylene and the other alkynes also burn readily. An acetylene torch takes advantage of the high heat of combustion for acetylene.

Aromatic Hydrocarbons

Benzene, C_6H_6, is the simplest member of a large family of hydrocarbons, called **aromatic hydrocarbons**. These compounds contain ring structures and exhibit bonding that must be described using the resonance hybrid concept of valence bond theory or the delocalization concept of molecular orbital theory. (To review these concepts, refer to the earlier chapters on chemical bonding). The resonance structures for benzene, C_6H_6, are:

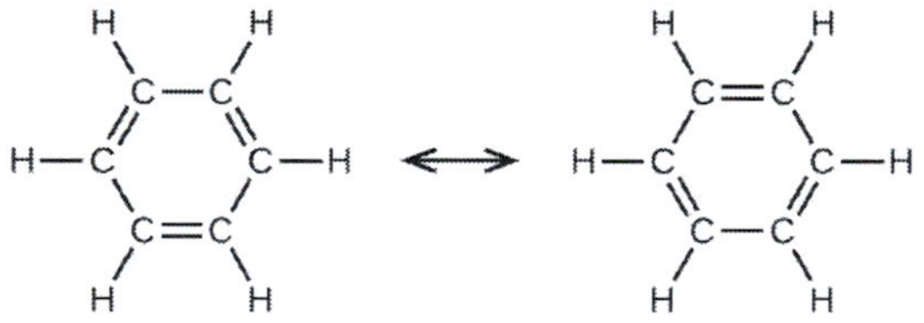

Valence bond theory describes the benzene molecule and other planar aromatic hydrocarbon molecules as hexagonal rings of sp^2-hybridized carbon atoms with the unhybridized p orbital of each carbon atom perpendicular to the plane of the ring. Three valence electrons in the sp^2 hybrid orbitals of each carbon atom and the valence electron of each hydrogen atom form the framework of σ bonds in the benzene molecule. The fourth valence electron of each carbon atom is shared with an adjacent carbon atom in their unhybridized p orbitals to yield the π bonds. Benzene does not, however, exhibit the characteristics typical of an alkene. Each of the six bonds between its carbon atoms is equivalent and exhibits properties that are intermediate between those of a C–C single bond and a $C = C$ double bond. To represent this unique bonding, structural formulas for benzene and its derivatives are typically drawn with single bonds between the carbon atoms and a circle within the ring as shown in **Figure 20.11**.

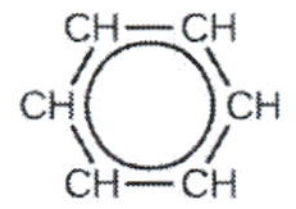

Figure 20.11 This condensed formula shows the unique bonding structure of benzene.

There are many derivatives of benzene. The hydrogen atoms can be replaced by many different substituents. Aromatic compounds more readily undergo substitution reactions than addition reactions; replacement of one of the hydrogen atoms with another substituent will leave the delocalized double bonds intact. The following are typical examples of substituted benzene derivatives:

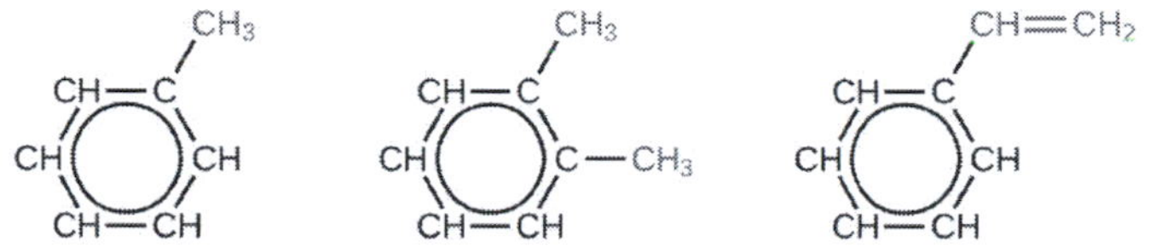

toluene xylene styrene

Toluene and xylene are important solvents and raw materials in the chemical industry. Styrene is used to produce the polymer polystyrene.

Example 20.7

Structure of Aromatic Hydrocarbons

One possible isomer created by a substitution reaction that replaces a hydrogen atom attached to the aromatic ring of toluene with a chlorine atom is shown here. Draw two other possible isomers in which the chlorine atom replaces a different hydrogen atom attached to the aromatic ring:

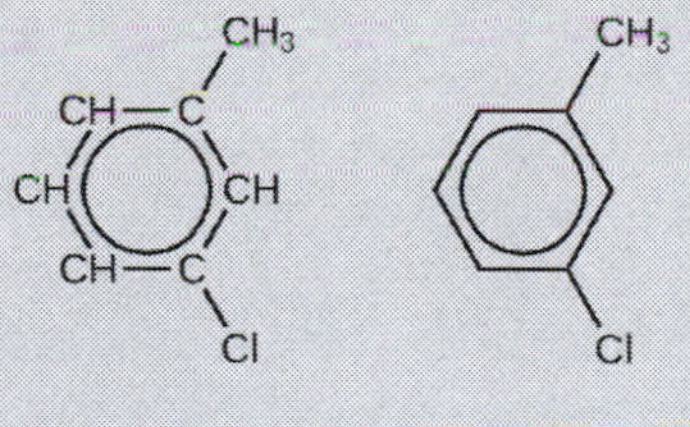

Solution

Since the six-carbon ring with alternating double bonds is necessary for the molecule to be classified as aromatic, appropriate isomers can be produced only by changing the positions of the chloro-substituent relative to the methyl-substituent:

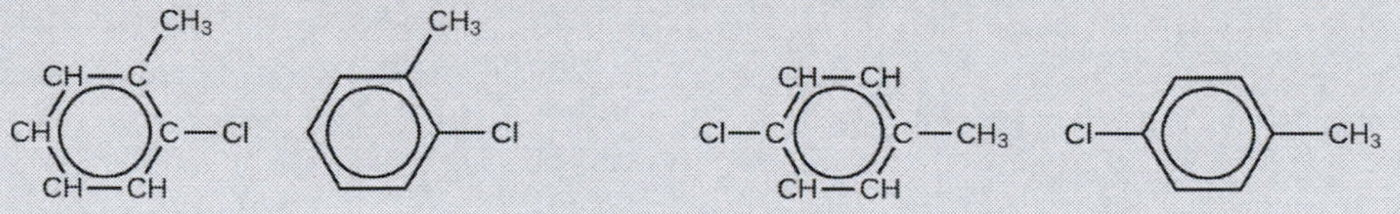

Check Your Learning

Draw three isomers of a six-membered aromatic ring compound substituted with two bromines.

Answer:

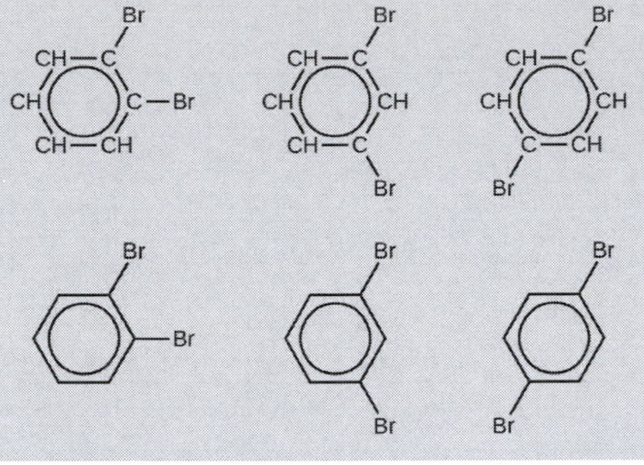

20.2 Alcohols and Ethers

By the end of this section, you will be able to:

*   Describe the structure and properties of alcohols
*   Describe the structure and properties of ethers
*   Name and draw structures for alcohols and ethers

In this section, we will learn about alcohols and ethers.

Alcohols

Incorporation of an oxygen atom into carbon- and hydrogen-containing molecules leads to new functional groups and new families of compounds. When the oxygen atom is attached by single bonds, the molecule is either an alcohol or ether.

Alcohols are derivatives of hydrocarbons in which an –OH group has replaced a hydrogen atom. Although all alcohols have one or more hydroxyl (–OH) functional groups, they do not behave like bases such as NaOH and KOH. NaOH and KOH are ionic compounds that contain OH^- ions. Alcohols are covalent molecules; the –OH group in an alcohol molecule is attached to a carbon atom by a covalent bond.

Ethanol, CH_3CH_2OH, also called ethyl alcohol, is a particularly important alcohol for human use. Ethanol is the alcohol produced by some species of yeast that is found in wine, beer, and distilled drinks. It has long been prepared by humans harnessing the metabolic efforts of yeasts in fermenting various sugars:

$$C_6H_{12}O_6(aq) \xrightarrow{\text{Yeast}} 2C_2H_5OH(aq) + 2CO_2(g)$$

glucose   ethanol

Large quantities of ethanol are synthesized from the addition reaction of water with ethylene using an acid as a catalyst:

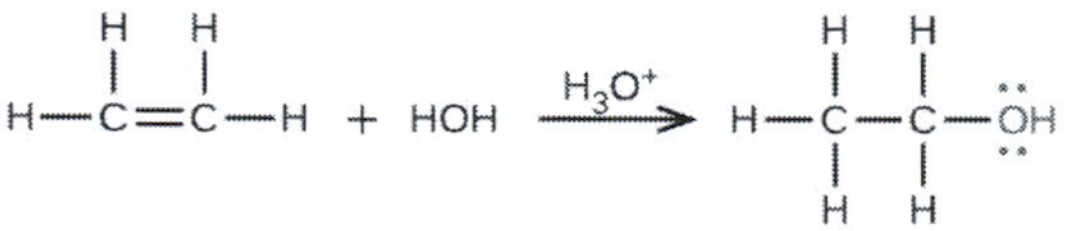

Alcohols containing two or more hydroxyl groups can be made. Examples include 1,2-ethanediol (ethylene glycol, used in antifreeze) and 1,2,3-propanetriol (glycerine, used as a solvent for cosmetics and medicines):

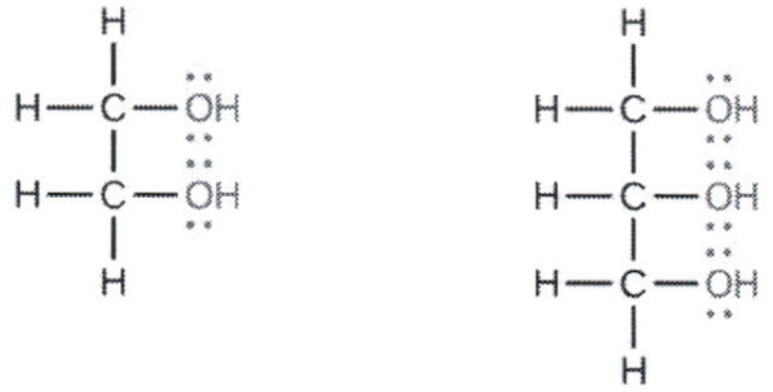

1,2-ethanediol   1,2,3-propanetriol

Naming Alcohols

The name of an alcohol comes from the hydrocarbon from which it was derived. The final *-e* in the name of the hydrocarbon is replaced by *-ol*, and the carbon atom to which the –OH group is bonded is indicated by a number placed before the name.[5]

Example 20.8

Naming Alcohols

Consider the following example. How should it be named?

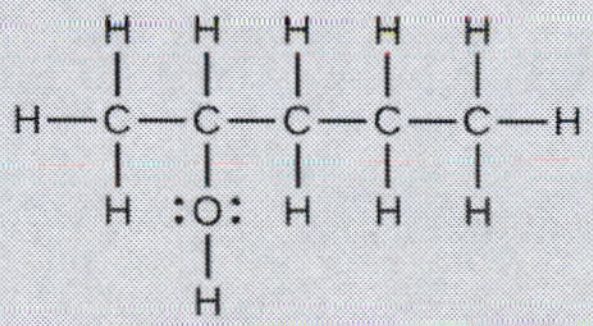

Solution

The carbon chain contains five carbon atoms. If the hydroxyl group was not present, we would have named this molecule pentane. To address the fact that the hydroxyl group is present, we change the ending of the name to *-ol*. In this case, since the –OH is attached to carbon 2 in the chain, we would name this molecule 2-pentanol.

Check Your Learning

Name the following molecule:

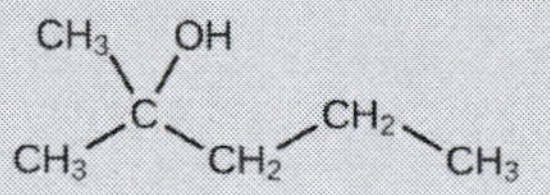

Answer: 2-methyl-2-pentanol

---

5. The IUPAC adopted new nomenclature guidelines in 2013 that require this number to be placed as an "infix" rather than a prefix. For example, the new name for 2-propanol would be propan-2-ol. Widespread adoption of this new nomenclature will take some time, and students are encouraged to be familiar with both the old and new naming protocols.

Ethers

Ethers are compounds that contain the functional group –O–. Ethers do not have a designated suffix like the other types of molecules we have named so far. In the IUPAC system, the oxygen atom and the smaller carbon branch are named as an alkoxy substituent and the remainder of the molecule as the base chain, as in alkanes. As shown in the following compound, the red symbols represent the smaller alkyl group and the oxygen atom, which would be named "methoxy." The larger carbon branch would be ethane, making the molecule methoxyethane. Many ethers are referred to with common names instead of the IUPAC system names. For common names, the two branches connected to the oxygen atom are named separately and followed by "ether." The common name for the compound shown in Example 20.9 is ethylmethyl ether:

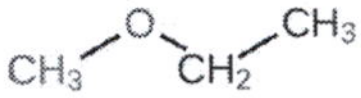
CH3
O
CH2
CH3

Example 20.9

Naming Ethers

Provide the IUPAC and common name for the ether shown here:

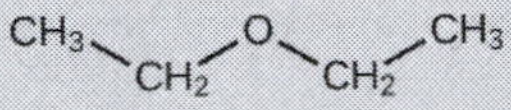
CH3
CH2
O
CH2
CH3

Solution

IUPAC: The molecule is made up of an ethoxy group attached to an ethane chain, so the IUPAC name would be ethoxyethane.

Common: The groups attached to the oxygen atom are both ethyl groups, so the common name would be diethyl ether.

Check Your Learning

Provide the IUPAC and common name for the ether shown:

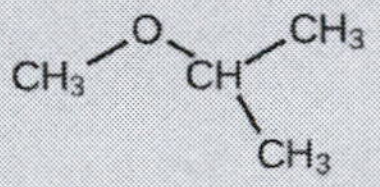
CH3
O
CH
CH3
CH3

Answer: IUPAC: 2-methoxypropane; common: isopropylmethyl ether

Ethers can be obtained from alcohols by the elimination of a molecule of water from two molecules of the alcohol. For example, when ethanol is treated with a limited amount of sulfuric acid and heated to 140 °C, diethyl ether and water are formed:

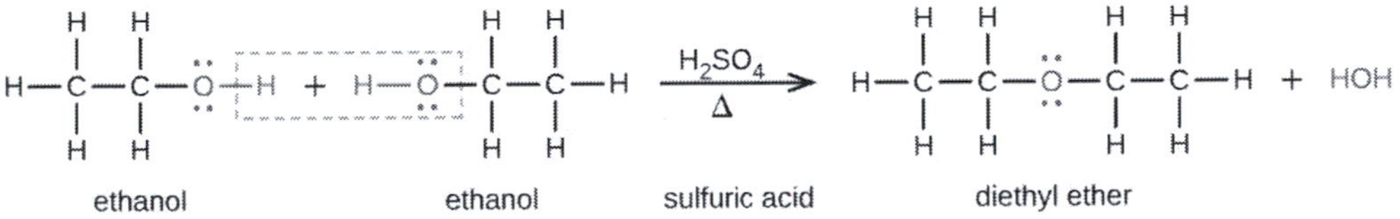
H2SO4
Δ
HOH
ethanol
ethanol
sulfuric acid
diethyl ether

In the general formula for ethers, R—O—R, the hydrocarbon groups (R) may be the same or different. Diethyl ether, the most widely used compound of this class, is a colorless, volatile liquid that is highly flammable. It was first used in 1846 as an anesthetic, but better anesthetics have now largely taken its place. Diethyl ether and other ethers are presently used primarily as solvents for gums, fats, waxes, and resins. *Tertiary*-butyl methyl ether, $C_4H_9OCH_3$ (abbreviated MTBE—italicized portions of names are not counted when ranking the groups alphabetically—so butyl comes before methyl in the common name), is used as an additive for gasoline. MTBE belongs to a group of chemicals known as oxygenates due to their capacity to increase the oxygen content of gasoline.

Link to Learning

Want more practice naming ethers? This brief video review (http://openstaxcollege.org/l/16ethers) summarizes the nomenclature for ethers.

Chemistry in Everyday Life

Carbohydrates and Diabetes

Carbohydrates are large biomolecules made up of carbon, hydrogen, and oxygen. The dietary forms of carbohydrates are foods rich in these types of molecules, like pastas, bread, and candy. The name "carbohydrate" comes from the formula of the molecules, which can be described by the general formula $C_m(H_2O)_n$, which shows that they are in a sense "carbon and water" or "hydrates of carbon." In many cases, m and n have the same value, but they can be different. The smaller carbohydrates are generally referred to as "sugars," the biochemical term for this group of molecules is "saccharide" from the Greek word for sugar (Figure 20.12). Depending on the number of sugar units joined together, they may be classified as monosaccharides (one sugar unit), disaccharides (two sugar units), oligosaccharides (a few sugars), or polysaccharides (the polymeric version of sugars—polymers were described in the feature box earlier in this chapter on recycling plastics). The scientific names of sugars can be recognized by the suffix -ose at the end of the name (for instance, fruit sugar is a monosaccharide called "fructose" and milk sugar is a disaccharide called lactose composed of two monosaccharides, glucose and galactose, connected together). Sugars contain some of the functional groups we have discussed: Note the alcohol groups present in the structures and how monosaccharide units are linked to form a disaccharide by formation of an ether.

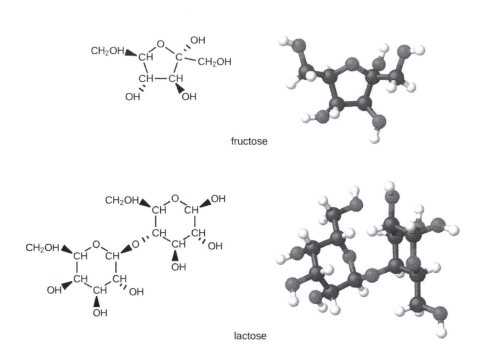

fructose

lactose

Figure 20.12 The illustrations show the molecular structures of fructose, a five-carbon monosaccharide, and of lactose, a disaccharide composed of two isomeric, six-carbon sugars.

Organisms use carbohydrates for a variety of functions. Carbohydrates can store energy, such as the polysaccharides glycogen in animals or starch in plants. They also provide structural support, such as the polysaccharide cellulose in plants and the modified polysaccharide chitin in fungi and animals. The sugars ribose and deoxyribose are components of the backbones of RNA and DNA, respectively. Other sugars play key roles in the function of the immune system, in cell-cell recognition, and in many other biological roles.

Diabetes is a group of metabolic diseases in which a person has a high sugar concentration in their blood (Figure 20.13). Diabetes may be caused by insufficient insulin production by the pancreas or by the body's cells not responding properly to the insulin that is produced. In a healthy person, insulin is produced when it is needed and functions to transport glucose from the blood into the cells where it can be used for energy. The long-term complications of diabetes can include loss of eyesight, heart disease, and kidney failure.

In 2013, it was estimated that approximately 3.3% of the world's population (~380 million people) suffered from diabetes, resulting in over a million deaths annually. Prevention involves eating a healthy diet, getting plenty of exercise, and maintaining a normal body weight. Treatment involves all of these lifestyle practices and may require injections of insulin.

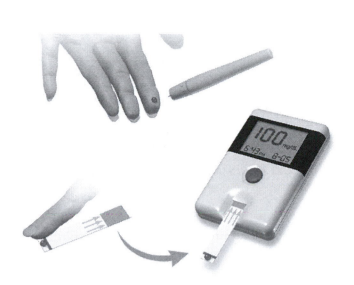

Figure 20.13 Diabetes is a disease characterized by high concentrations of glucose in the blood. Treating diabetes involves making lifestyle changes, monitoring blood-sugar levels, and sometimes insulin injections. (credit: "Blausen Medical Communications"/Wikimedia Commons)

20.3 Aldehydes, Ketones, Carboxylic Acids, and Esters

By the end of this section, you will be able to:

- Describe the structure and properties of aldehydes, ketones, carboxylic acids and esters

Another class of organic molecules contains a carbon atom connected to an oxygen atom by a double bond, commonly called a carbonyl group. The trigonal planar carbon in the carbonyl group can attach to two other substituents leading to several subfamilies (aldehydes, ketones, carboxylic acids and esters) described in this section.

Aldehydes and Ketones

Both **aldehydes** and **ketones** contain a **carbonyl group**, a functional group with a carbon-oxygen double bond. The names for aldehyde and ketone compounds are derived using similar nomenclature rules as for alkanes and alcohols, and include the class-identifying suffixes -*al* and -*one*, respectively:

$$\overset{O}{\underset{}{\underset{\|}{-C-}}}$$

In an aldehyde, the carbonyl group is bonded to at least one hydrogen atom. In a ketone, the carbonyl group is bonded to two carbon atoms:

$$\overset{O}{\underset{R-C-H}{\|}} \qquad \overset{O}{\underset{R-C-R}{\|}}$$

Functional group of an aldehyde   Functional group of a ketone

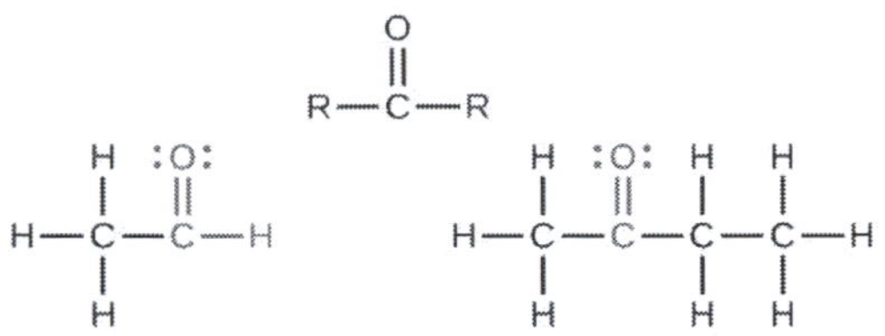

CH_3CHO
An aldehyde
ethanal (acetaldehyde)

$CH_3COCH_2CH_3$
A ketone
butanone

As text, an aldehyde group is represented as –CHO; a ketone is represented as –C(O)– or –CO–.

In both aldehydes and ketones, the geometry around the carbon atom in the carbonyl group is trigonal planar; the carbon atom exhibits sp^2 hybridization. Two of the sp^2 orbitals on the carbon atom in the carbonyl group are used to form σ bonds to the other carbon or hydrogen atoms in a molecule. The remaining sp^2 hybrid orbital forms a σ bond to the oxygen atom. The unhybridized *p* orbital on the carbon atom in the carbonyl group overlaps a *p* orbital on the oxygen atom to form the π bond in the double bond.

Like the $C=O$ bond in carbon dioxide, the $C=O$ bond of a carbonyl group is polar (recall that oxygen is significantly more electronegative than carbon, and the shared electrons are pulled toward the oxygen atom and away from the carbon atom). Many of the reactions of aldehydes and ketones start with the reaction between a Lewis base and the carbon atom at the positive end of the polar $C=O$ bond to yield an unstable intermediate that subsequently undergoes one or more structural rearrangements to form the final product (Figure 20.14).

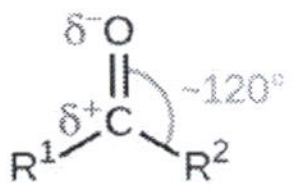

Figure 20.14 The carbonyl group is polar, and the geometry of the bonds around the central carbon is trigonal planar.

The importance of molecular structure in the reactivity of organic compounds is illustrated by the reactions that produce aldehydes and ketones. We can prepare a carbonyl group by oxidation of an alcohol—for organic molecules, oxidation of a carbon atom is said to occur when a carbon-hydrogen bond is replaced by a carbon-oxygen bond. The reverse reaction—replacing a carbon-oxygen bond by a carbon-hydrogen bond—is a reduction of that carbon atom. Recall that oxygen is generally assigned a –2 oxidation number unless it is elemental or attached to a fluorine. Hydrogen is generally assigned an oxidation number of +1 unless it is attached to a metal. Since carbon does not have a specific rule, its oxidation number is determined algebraically by factoring the atoms it is attached to and the overall charge of the molecule or ion. In general, a carbon atom attached to an oxygen atom will have a more positive oxidation number and a carbon atom attached to a hydrogen atom will have a more negative oxidation number. This should fit nicely with your understanding of the polarity of C–O and C–H bonds. The other reagents and possible products of these reactions are beyond the scope of this chapter, so we will focus only on the changes to the carbon atoms:

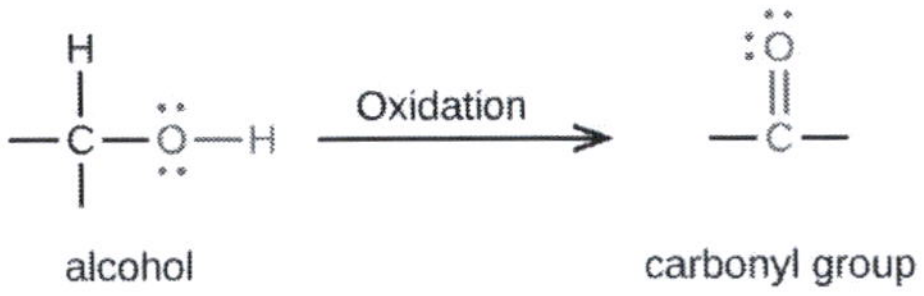

Example 20.10

Oxidation and Reduction in Organic Chemistry

Methane represents the completely reduced form of an organic molecule that contains one carbon atom. Sequentially replacing each of the carbon-hydrogen bonds with a carbon-oxygen bond would lead to an alcohol, then an aldehyde, then a carboxylic acid (discussed later), and, finally, carbon dioxide:

$$CH_4 \longrightarrow CH_3OH \longrightarrow CH_2O \longrightarrow HCO_2H \longrightarrow CO_2$$

What are the oxidation numbers for the carbon atoms in the molecules shown here?

Solution

In this example, we can calculate the oxidation number (review the chapter on oxidation-reduction reactions if necessary) for the carbon atom in each case (note how this would become difficult for larger molecules with additional carbon atoms and hydrogen atoms, which is why organic chemists use the definition dealing with replacing C–H bonds with C–O bonds described). For CH_4, the carbon atom carries a –4 oxidation number (the hydrogen atoms are assigned oxidation numbers of +1 and the carbon atom balances that by having an oxidation number of –4). For the alcohol (in this case, methanol), the carbon atom has an oxidation number of –2 (the oxygen atom is assigned –2, the four hydrogen atoms each are assigned +1, and the carbon atom balances the sum by having an oxidation number of –2; note that compared to the carbon atom in CH_4, this carbon atom has lost two electrons so it was oxidized); for the aldehyde, the carbon atom's oxidation number is 0 (–2 for the oxygen atom and +1 for each hydrogen atom already balances to 0, so the oxidation number for the carbon atom is 0); for the carboxylic acid, the carbon atom's oxidation number is +2 (two oxygen atoms each at –2 and two hydrogen atoms at +1); and for carbon dioxide, the carbon atom's oxidation number is +4 (here, the carbon atom needs to balance the –4 sum from the two oxygen atoms).

Check Your Learning

Indicate whether the marked carbon atoms in the three molecules here are oxidized or reduced relative to the marked carbon atom in ethanol:

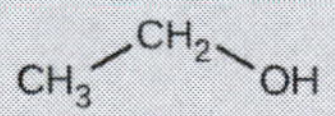

$$CH_3 \diagup^{CH_2} \diagdown OH$$

There is no need to calculate oxidation states in this case; instead, just compare the types of atoms bonded to the marked carbon atoms:

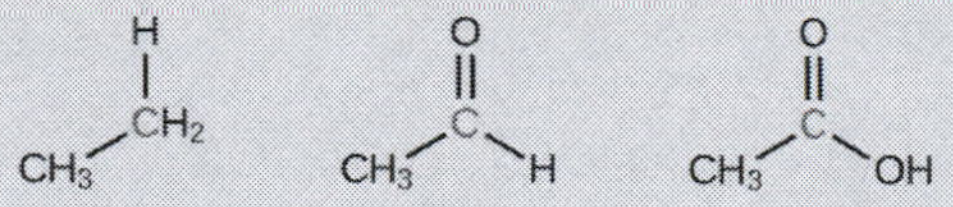

(a)　　　　(b)　　　　(c)

Answer: (a) reduced (bond to oxygen atom replaced by bond to hydrogen atom); (b) oxidized (one bond to hydrogen atom replaced by one bond to oxygen atom); (c) oxidized (2 bonds to hydrogen atoms have been replaced by bonds to an oxygen atom)

Aldehydes are commonly prepared by the oxidation of alcohols whose –OH functional group is located on the carbon atom at the end of the chain of carbon atoms in the alcohol:

$$CH_3CH_2CH_2OH \longrightarrow CH_3CH_2CHO$$

alcohol　　　　　　　　　　　　aldehyde

Alcohols that have their –OH groups in the middle of the chain are necessary to synthesize a ketone, which requires the carbonyl group to be bonded to two other carbon atoms:

$$CH_3CH(OH)CH_3 \longrightarrow CH_3COCH_3$$

alcohol   ketone

An alcohol with its –OH group bonded to a carbon atom that is bonded to no or one other carbon atom will form an aldehyde. An alcohol with its –OH group attached to two other carbon atoms will form a ketone. If three carbons are attached to the carbon bonded to the –OH, the molecule will not have a C–H bond to be replaced, so it will not be susceptible to oxidation.

Formaldehyde, an aldehyde with the formula HCHO, is a colorless gas with a pungent and irritating odor. It is sold in an aqueous solution called formalin, which contains about 37% formaldehyde by weight. Formaldehyde causes coagulation of proteins, so it kills bacteria (and any other living organism) and stops many of the biological processes that cause tissue to decay. Thus, formaldehyde is used for preserving tissue specimens and embalming bodies. It is also used to sterilize soil or other materials. Formaldehyde is used in the manufacture of Bakelite, a hard plastic having high chemical and electrical resistance.

Dimethyl ketone, CH_3COCH_3, commonly called acetone, is the simplest ketone. It is made commercially by fermenting corn or molasses, or by oxidation of 2-propanol. Acetone is a colorless liquid. Among its many uses are as a solvent for lacquer (including fingernail polish), cellulose acetate, cellulose nitrate, acetylene, plastics, and varnishes; as a paint and varnish remover; and as a solvent in the manufacture of pharmaceuticals and chemicals.

Carboxylic Acids and Esters

The odor of vinegar is caused by the presence of acetic acid, a carboxylic acid, in the vinegar. The odor of ripe bananas and many other fruits is due to the presence of esters, compounds that can be prepared by the reaction of a carboxylic acid with an alcohol. Because esters do not have hydrogen bonds between molecules, they have lower vapor pressures than the alcohols and carboxylic acids from which they are derived (see Figure 20.15).

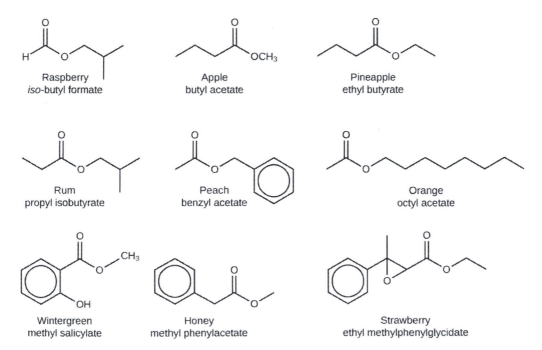

Figure 20.15   Esters are responsible for the odors associated with various plants and their fruits.

Both **carboxylic acids** and **esters** contain a carbonyl group with a second oxygen atom bonded to the carbon atom in the carbonyl group by a single bond. In a carboxylic acid, the second oxygen atom also bonds to a hydrogen atom. In an ester, the second oxygen atom bonds to another carbon atom. The names for carboxylic acids and esters include prefixes that denote the lengths of the carbon chains in the molecules and are derived following nomenclature rules similar to those for inorganic acids and salts (see these examples):

ethanoic acid
(acetic acid)

methyl ethanoate
(methyl acetate)

The functional groups for an acid and for an ester are shown in red in these formulas.

The hydrogen atom in the functional group of a carboxylic acid will react with a base to form an ionic salt:

propionic acid

proprionate ion

Carboxylic acids are weak acids (see the chapter on acids and bases), meaning they are not 100% ionized in water. Generally only about 1% of the molecules of a carboxylic acid dissolved in water are ionized at any given time. The remaining molecules are undissociated in solution.

We prepare carboxylic acids by the oxidation of aldehydes or alcohols whose –OH functional group is located on the carbon atom at the end of the chain of carbon atoms in the alcohol:

alcohol

aldehyde

carboxylic acid

Esters are produced by the reaction of acids with alcohols. For example, the ester ethyl acetate, $CH_3CO_2CH_2CH_3$, is formed when acetic acid reacts with ethanol:

acetic acid

ethanol

ethyl acetate

The simplest carboxylic acid is formic acid, HCO_2H, known since 1670. Its name comes from the Latin word *formicus*, which means "ant"; it was first isolated by the distillation of red ants. It is partially responsible for the pain and irritation of ant and wasp stings, and is responsible for a characteristic odor of ants that can be sometimes detected in their nests.

Acetic acid, CH_3CO_2H, constitutes 3–6% vinegar. Cider vinegar is produced by allowing apple juice to ferment without oxygen present. Yeast cells present in the juice carry out the fermentation reactions. The fermentation reactions change the sugar present in the juice to ethanol, then to acetic acid. Pure acetic acid has a penetrating odor and produces painful burns. It is an excellent solvent for many organic and some inorganic compounds, and it is essential in the production of cellulose acetate, a component of many synthetic fibers such as rayon.

The distinctive and attractive odors and flavors of many flowers, perfumes, and ripe fruits are due to the presence of one or more esters (Figure 20.16). Among the most important of the natural esters are fats (such as lard, tallow, and butter) and oils (such as linseed, cottonseed, and olive oils), which are esters of the trihydroxyl alcohol glycerine, $C_3H_5(OH)_3$, with large carboxylic acids, such as palmitic acid, $CH_3(CH_2)_{14}CO_2H$, stearic acid, $CH_3(CH_2)_{16}CO_2H$, and oleic acid, $CH_3(CH_2)_7 CH = CH(CH_2)_7 CO_2 H$. Oleic acid is an unsaturated acid; it contains a $C = C$ double bond. Palmitic and stearic acids are saturated acids that contain no double or triple bonds.

Figure 20.16 Over 350 different volatile molecules (many members of the ester family) have been identified in strawberries. (credit: Rebecca Siegel)

20.4 Amines and Amides

By the end of this section, you will be able to:

- Describe the structure and properties of an amine
- Describe the structure and properties of an amide

Amines are molecules that contain carbon-nitrogen bonds. The nitrogen atom in an amine has a lone pair of electrons and three bonds to other atoms, either carbon or hydrogen. Various nomenclatures are used to derive names for amines, but all involve the class-identifying suffix *–ine* as illustrated here for a few simple examples:

$$CH_3 - \overset{..}{N} - H \qquad CH_3 - \overset{..}{N} - CH_3 \qquad CH_3 - \overset{..}{N} - CH_3$$
$$\quad\quad | \qquad\qquad\qquad | \qquad\qquad\qquad\quad |$$
$$\quad\quad H \qquad\qquad\qquad H \qquad\qquad\qquad\quad CH_3$$

methyl amine   dimethyl amine   trimethyl amine

In some amines, the nitrogen atom replaces a carbon atom in an aromatic hydrocarbon. Pyridine (Figure 20.17) is one such heterocyclic amine. A heterocyclic compound contains atoms of two or more different elements in its ring structure.

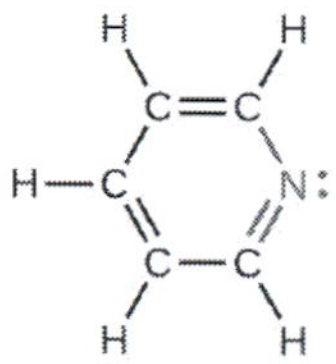

Figure 20.17   The illustration shows one of the resonance structures of pyridine.

How Sciences Interconnect

DNA in Forensics and Paternity

The genetic material for all living things is a polymer of four different molecules, which are themselves a combination of three subunits. The genetic information, the code for developing an organism, is contained in the specific sequence of the four molecules, similar to the way the letters of the alphabet can be sequenced to form words that convey information. The information in a DNA sequence is used to form two other types of polymers, one of which are proteins. The proteins interact to form a specific type of organism with individual characteristics.

A genetic molecule is called DNA, which stands for deoxyribonucleic acid. The four molecules that make up DNA are called nucleotides. Each nucleotide consists of a single- or double-ringed molecule containing nitrogen, carbon, oxygen, and hydrogen called a nitrogenous base. Each base is bonded to a five-carbon sugar called deoxyribose. The sugar is in turn bonded to a phosphate group $(-PO_4{}^{3-})$ When new DNA is made, a polymerization reaction occurs that binds the phosphate group of one nucleotide to the sugar group of a second nucleotide. The nitrogenous bases of each nucleotide stick out from this sugar-phosphate backbone. DNA is actually formed from two such polymers coiled around each other and held together by hydrogen bonds between the nitrogenous bases. Thus, the two backbones are on the outside of the coiled pair of strands, and the bases are on the inside. The shape of the two strands wound around each other is called a double helix (see Figure 20.18).

It probably makes sense that the sequence of nucleotides in the DNA of a cat differs from those of a dog. But it is also true that the sequences of the DNA in the cells of two individual pugs differ. Likewise, the sequences of DNA in you and a sibling differ (unless your sibling is an identical twin), as do those between you and an unrelated individual. However, the DNA sequences of two related individuals are more similar than the sequences of two unrelated individuals, and these similarities in sequence can be observed in various ways. This is the principle behind DNA fingerprinting, which is a method used to determine whether two DNA samples came from related (or the same) individuals or unrelated individuals.

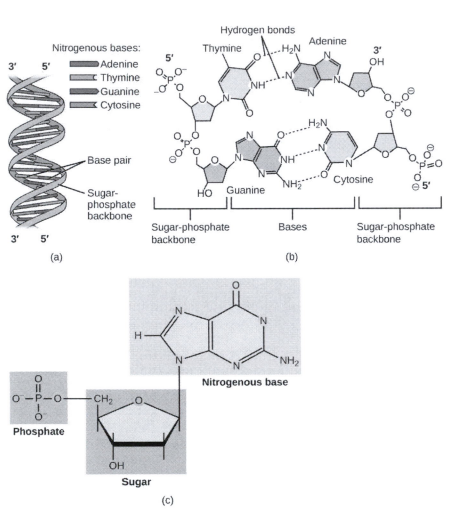

Figure 20.18 DNA is an organic molecule and the genetic material for all living organisms. (a) DNA is a double helix consisting of two single DNA strands hydrogen bonded together at each nitrogenous base. (b) This detail shows the hydrogen bonding (dotted lines) between nitrogenous bases on each DNA strand and the way in which each nucleotide is joined to the next, forming a backbone of sugars and phosphate groups along each strand. (c) This detail shows the structure of one of the four nucleotides that makes up the DNA polymer. Each nucleotide consists of a nitrogenous base (a double-ring molecule, in this case), a five-carbon sugar (deoxyribose), and a phosphate group.

Using similarities in sequences, technicians can determine whether a man is the father of a child (the identity of the mother is rarely in doubt, except in the case of an adopted child and a potential birth mother). Likewise, forensic geneticists can determine whether a crime scene sample of human tissue, such as blood or skin cells, contains DNA that matches exactly the DNA of a suspect.

Like ammonia, amines are weak bases due to the lone pair of electrons on their nitrogen atoms:

$$H{-}\overset{\cdot\cdot}{\underset{\underset{H}{|}}{N}}{-}H \;+\; H^+ \longrightarrow \left[H{-}\overset{\overset{H}{|}}{\underset{\underset{H}{|}}{N}}{-}H \right]^+$$

ammonia   ammonium ion

$$CH_3{-}\overset{\cdot\cdot}{\underset{\underset{H}{|}}{N}}{-}H \;+\; H^+ \longrightarrow \left[CH_3{-}\overset{\overset{H}{|}}{\underset{\underset{H}{|}}{N}}{-}H \right]^+$$

methyl amine   methyl ammonium ion

The basicity of an amine's nitrogen atom plays an important role in much of the compound's chemistry. Amine functional groups are found in a wide variety of compounds, including natural and synthetic dyes, polymers, vitamins, and medications such as penicillin and codeine. They are also found in many molecules essential to life, such as amino acids, hormones, neurotransmitters, and DNA.

How Sciences Interconnect

Addictive Alkaloids

Since ancient times, plants have been used for medicinal purposes. One class of substances, called *alkaloids*, found in many of these plants has been isolated and found to contain cyclic molecules with an amine functional group. These amines are bases. They can react with H_3O^+ in a dilute acid to form an ammonium salt, and this property is used to extract them from the plant:

$$R_3N + H_3O^+ + Cl^- \longrightarrow \left[R_3NH^+\right]Cl^- + H_2O$$

The name alkaloid means "like an alkali." Thus, an alkaloid reacts with acid. The free compound can be recovered after extraction by reaction with a base:

$$\left[R_3NH^+\right]Cl^- + OH^- \longrightarrow R_3N + H_2O + Cl^-$$

The structures of many naturally occurring alkaloids have profound physiological and psychotropic effects in humans. Examples of these drugs include nicotine, morphine, codeine, and heroin. The plant produces these substances, collectively called secondary plant compounds, as chemical defenses against the numerous pests that attempt to feed on the plant:

In these diagrams, as is common in representing structures of large organic compounds, carbon atoms in the rings and the hydrogen atoms bonded to them have been omitted for clarity. The solid wedges indicate bonds that extend out of the page. The dashed wedges indicate bonds that extend into the page. Notice that small changes to a part of the molecule change the properties of morphine, codeine, and heroin. Morphine, a strong narcotic used to relieve pain, contains two hydroxyl functional groups, located at the bottom of the molecule in this structural formula. Changing one of these hydroxyl groups to a methyl ether group forms codeine, a less potent drug used as a local anesthetic. If both hydroxyl groups are converted to esters of acetic acid, the powerfully addictive drug heroin results (Figure 20.19).

Figure 20.19 Poppies can be used in the production of opium, a plant latex that contains morphine from which other opiates, such as heroin, can be synthesized. (credit: Karen Roe)

Amides are molecules that contain nitrogen atoms connected to the carbon atom of a carbonyl group. Like amines, various nomenclature rules may be used to name amides, but all include use of the class-specific suffix *-amide*:

acetamide   hexanamide

Amides can be produced when carboxylic acids react with amines or ammonia in a process called amidation. A water molecule is eliminated from the reaction, and the amide is formed from the remaining pieces of the carboxylic acid and the amine (note the similarity to formation of an ester from a carboxylic acid and an alcohol discussed in the previous section):

carboxylic acid   amine   amide   water

The reaction between amines and carboxylic acids to form amides is biologically important. It is through this reaction that amino acids (molecules containing both amine and carboxylic acid substituents) link together in a polymer to form proteins.

How Sciences Interconnect

Proteins and Enzymes

Proteins are large biological molecules made up of long chains of smaller molecules called amino acids. Organisms rely on proteins for a variety of functions—proteins transport molecules across cell membranes, replicate DNA, and catalyze metabolic reactions, to name only a few of their functions. The properties of proteins are functions of the combination of amino acids that compose them and can vary greatly. Interactions between amino acid sequences in the chains of proteins result in the folding of the chain into specific, three-dimensional structures that determine the protein's activity.

Amino acids are organic molecules that contain an amine functional group ($-NH_2$), a carboxylic acid functional group ($-COOH$), and a side chain (that is specific to each individual amino acid). Most living things build proteins from the same 20 different amino acids. Amino acids connect by the formation of a peptide bond, which is a covalent bond formed between two amino acids when the carboxylic acid group of one amino acid reacts with the amine group of the other amino acid. The formation of the bond results in the production of a molecule of water (in general, reactions that result in the production of water when two other molecules combine are referred to as condensation reactions). The resulting bond—between the carbonyl group carbon atom and the amine nitrogen atom is called a peptide link or peptide bond. Since each of the original amino acids has an unreacted group (one has an unreacted amine and the other an unreacted carboxylic acid), more peptide bonds can form to other amino acids, extending the structure. (Figure 20.20) A chain of connected amino acids is called a polypeptide. Proteins contain at least one long polypeptide chain.

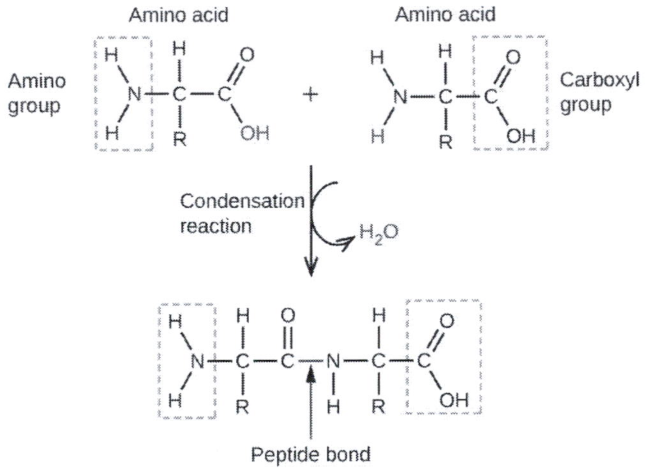

Figure 20.20 This condensation reaction forms a dipeptide from two amino acids and leads to the formation of water.

Enzymes are large biological molecules, mostly composed of proteins, which are responsible for the thousands of metabolic processes that occur in living organisms. Enzymes are highly specific catalysts; they speed up the rates of certain reactions. Enzymes function by lowering the activation energy of the reaction they are catalyzing, which can dramatically increase the rate of the reaction. Most reactions catalyzed by enzymes have rates that are millions of times faster than the noncatalyzed version. Like all catalysts, enzymes are not consumed during the reactions that they catalyze. Enzymes do differ from other catalysts in how specific they are for their substrates (the molecules that an enzyme will convert into a different product). Each enzyme is only capable of speeding up one or a few very specific reactions or types of reactions. Since the function of enzymes is so specific, the lack or malfunctioning of an enzyme can lead to serious health consequences. One disease that is the result of an enzyme malfunction is phenylketonuria. In this disease, the enzyme that catalyzes the first step in the degradation of the amino acid phenylalanine is not functional (Figure 20.21). Untreated, this can lead to an accumulation of phenylalanine, which can lead to intellectual disabilities.

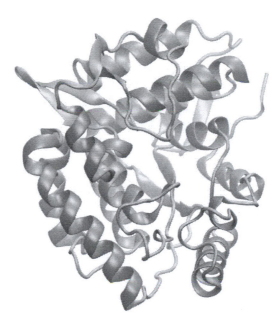

Figure 20.21 A computer rendering shows the three-dimensional structure of the enzyme phenylalanine hydroxylase. In the disease phenylketonuria, a defect in the shape of phenylalanine hydroxylase causes it to lose its function in breaking down phenylalanine.

Chemistry in Everyday Life

Kevlar

Kevlar (Figure 20.22) is a synthetic polymer made from two monomers 1,4-phenylene-diamine and terephthaloyl chloride (Kevlar is a registered trademark of DuPont). Kevlar's first commercial use was as a replacement for steel in racing tires. Kevlar is typically spun into ropes or fibers. The material has a high tensile strength-to-weight ratio (it is about 5 times stronger than an equal weight of steel), making it useful for many applications from bicycle tires to sails to body armor.

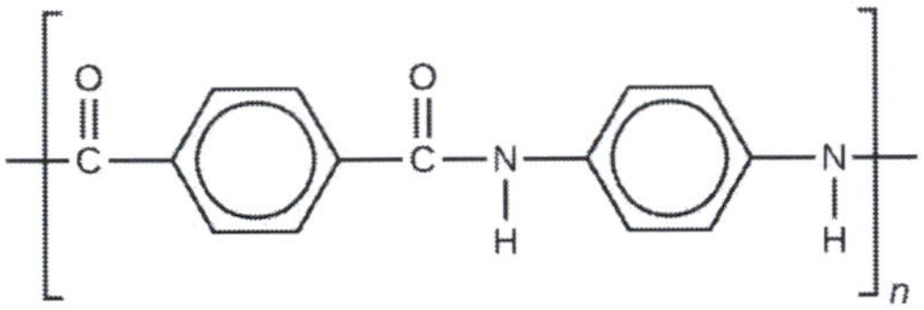

Figure 20.22 This illustration shows the formula for polymeric Kevlar.

The material owes much of its strength to hydrogen bonds between polymer chains (refer back to the chapter on intermolecular interactions). These bonds form between the carbonyl group oxygen atom (which has a partial negative charge due to oxygen's electronegativity) on one monomer and the partially positively charged hydrogen atom in the N–H bond of an adjacent monomer in the polymer structure (see dashed line in Figure 20.23). There is additional strength derived from the interaction between the unhybridized *p* orbitals in the six-membered rings, called aromatic stacking.

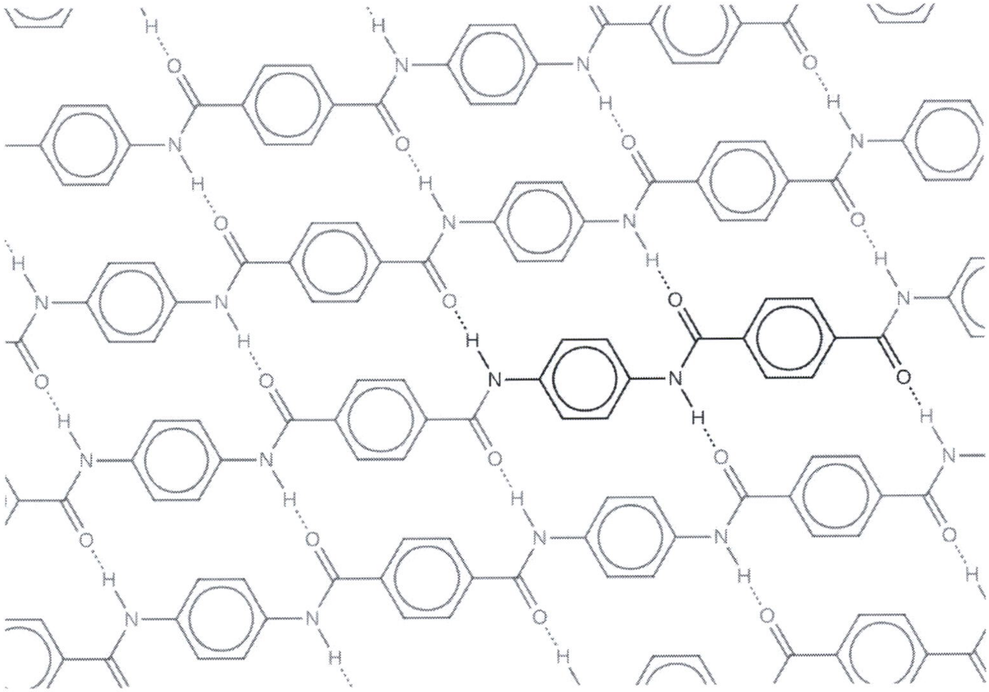

Figure 20.23 The diagram shows the polymer structure of Kevlar, with hydrogen bonds between polymer chains represented by dotted lines.

Kevlar may be best known as a component of body armor, combat helmets, and face masks. Since the 1980s, the US military has used Kevlar as a component of the PASGT (personal armor system for ground troops) helmet and vest. Kevlar is also used to protect armored fighting vehicles and aircraft carriers. Civilian applications include protective gear for emergency service personnel such as body armor for police officers and heat-resistant clothing for fire fighters. Kevlar based clothing is considerably lighter and thinner than equivalent gear made from other materials (Figure 20.24).

(a)

(b)

(c)

Figure 20.24 (a) These soldiers are sorting through pieces of a Kevlar helmet that helped absorb a grenade blast. Kevlar is also used to make (b) canoes and (c) marine mooring lines. (credit a: modification of work by "Cla68"/Wikimedia Commons; credit b: modification of work by "OakleyOriginals"/Flickr; credit c: modification of work by Casey H. Kyhl)

In addition to its better-known uses, Kevlar is also often used in cryogenics for its very low thermal conductivity (along with its high strength). Kevlar maintains its high strength when cooled to the temperature of liquid nitrogen (–196 °C).

The table here summarizes the structures discussed in this chapter:

Compound Name	Structure of Compound and Functional Group (red)	Example	
		Formula	Name
alkene	$C=C$	C_2H_4	ethene
alkyne	$C\equiv C$	C_2H_2	ethyne
alcohol	$R-\overset{..}{\underset{..}{O}}-H$	CH_3CH_2OH	ethanol
ether	$R-\overset{..}{\underset{..}{O}}-R'$	$(C_2H_5)_2O$	diethyl ether
aldehyde	$R-\overset{:O:}{\overset{\|}{C}}-H$	CH_3CHO	ethanal
ketone	$R-\overset{:O:}{\overset{\|}{C}}-R'$	$CH_3COCH_2CH_3$	methyl ethyl ketone
carboxylic acid	$R-\overset{:O:}{\overset{\|}{C}}-\overset{..}{\underset{..}{O}}-H$	CH_3COOH	acetic acid
ester	$R-\overset{:O:}{\overset{\|}{C}}-\overset{..}{\underset{..}{O}}-R'$	$CH_3CO_2CH_2CH_3$	ethyl acetate
amine	$R-\overset{..}{\underset{\|}{N}}-H \quad R-\overset{..}{\underset{\|}{N}}-H \quad R-\overset{..}{\underset{\|}{N}}-R''$ $\qquad\quad H \qquad\qquad R' \qquad\qquad R'$	$C_2H_5NH_2$	ethylamine
amide	$R-\overset{:O:}{\overset{\|}{C}}-\overset{..}{\underset{\|}{N}}-R'$ $\qquad\qquad\quad H$	CH_3CONH_2	acetamide

Key Terms

alcohol organic compound with a hydroxyl group (–OH) bonded to a carbon atom

aldehyde organic compound containing a carbonyl group bonded to two hydrogen atoms or a hydrogen atom and a carbon substituent

alkane molecule consisting of only carbon and hydrogen atoms connected by single (σ) bonds

alkene molecule consisting of carbon and hydrogen containing at least one carbon-carbon double bond

alkyl group substituent, consisting of an alkane missing one hydrogen atom, attached to a larger structure

alkyne molecule consisting of carbon and hydrogen containing at least one carbon-carbon triple bond

amide organic molecule that features a nitrogen atom connected to the carbon atom in a carbonyl group

amine organic molecule in which a nitrogen atom is bonded to one or more alkyl group

aromatic hydrocarbon cyclic molecule consisting of carbon and hydrogen with delocalized alternating carbon-carbon single and double bonds, resulting in enhanced stability

carbonyl group carbon atom double bonded to an oxygen atom

carboxylic acid organic compound containing a carbonyl group with an attached hydroxyl group

ester organic compound containing a carbonyl group with an attached oxygen atom that is bonded to a carbon substituent

ether organic compound with an oxygen atom that is bonded to two carbon atoms

functional group part of an organic molecule that imparts a specific chemical reactivity to the molecule

ketone organic compound containing a carbonyl group with two carbon substituents attached to it

organic compound natural or synthetic compound that contains carbon

saturated hydrocarbon molecule containing carbon and hydrogen that has only single bonds between carbon atoms

skeletal structure shorthand method of drawing organic molecules in which carbon atoms are represented by the ends of lines and bends in between lines, and hydrogen atoms attached to the carbon atoms are not shown (but are understood to be present by the context of the structure)

substituent branch or functional group that replaces hydrogen atoms in a larger hydrocarbon chain

substitution reaction reaction in which one atom replaces another in a molecule

Summary

20.1 Hydrocarbons

Strong, stable bonds between carbon atoms produce complex molecules containing chains, branches, and rings. The chemistry of these compounds is called organic chemistry. Hydrocarbons are organic compounds composed of only carbon and hydrogen. The alkanes are saturated hydrocarbons—that is, hydrocarbons that contain only single bonds. Alkenes contain one or more carbon-carbon double bonds. Alkynes contain one or more carbon-carbon triple bonds. Aromatic hydrocarbons contain ring structures with delocalized π electron systems.

20.2 Alcohols and Ethers

Many organic compounds that are not hydrocarbons can be thought of as derivatives of hydrocarbons. A hydrocarbon derivative can be formed by replacing one or more hydrogen atoms of a hydrocarbon by a functional group, which contains at least one atom of an element other than carbon or hydrogen. The properties of hydrocarbon derivatives are determined largely by the functional group. The –OH group is the functional group of an alcohol. The –R–O–R– group is the functional group of an ether.

20.3 Aldehydes, Ketones, Carboxylic Acids, and Esters

Functional groups related to the carbonyl group include the –CHO group of an aldehyde, the –CO– group of a ketone, the –CO$_2$H group of a carboxylic acid, and the –CO$_2$R group of an ester. The carbonyl group, a carbon-oxygen double bond, is the key structure in these classes of organic molecules: Aldehydes contain at least one hydrogen atom attached to the carbonyl carbon atom, ketones contain two carbon groups attached to the carbonyl carbon atom, carboxylic acids contain a hydroxyl group attached to the carbonyl carbon atom, and esters contain an oxygen atom attached to another carbon group connected to the carbonyl carbon atom. All of these compounds contain oxidized carbon atoms relative to the carbon atom of an alcohol group.

20.4 Amines and Amides

The addition of nitrogen into an organic framework leads to two families of molecules. Compounds containing a nitrogen atom bonded in a hydrocarbon framework are classified as amines. Compounds that have a nitrogen atom bonded to one side of a carbonyl group are classified as amides. Amines are a basic functional group. Amines and carboxylic acids can combine in a condensation reaction to form amides.

Exercises

20.1 Hydrocarbons

1. Write the chemical formula and Lewis structure of the following, each of which contains five carbon atoms:

(a) an alkane

(b) an alkene

(c) an alkyne

2. What is the difference between the hybridization of carbon atoms' valence orbitals in saturated and unsaturated hydrocarbons?

3. On a microscopic level, how does the reaction of bromine with a saturated hydrocarbon differ from its reaction with an unsaturated hydrocarbon? How are they similar?

4. On a microscopic level, how does the reaction of bromine with an alkene differ from its reaction with an alkyne? How are they similar?

5. Explain why unbranched alkenes can form geometric isomers while unbranched alkanes cannot. Does this explanation involve the macroscopic domain or the microscopic domain?

6. Explain why these two molecules are not isomers:

Download for free at http://cnx.org/content/col11760/latest/

7. Explain why these two molecules are not isomers:

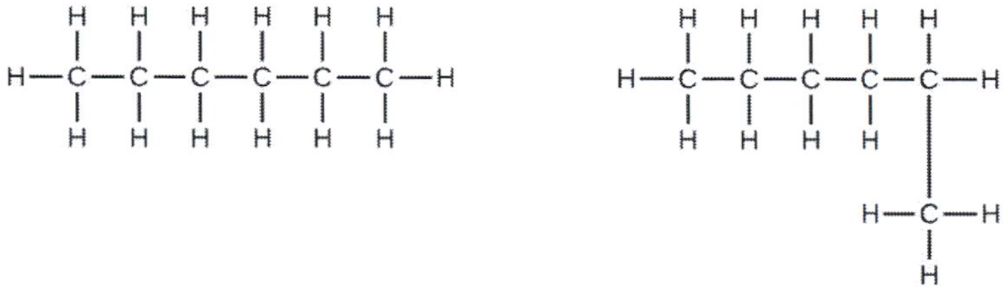

8. How does the carbon-atom hybridization change when polyethylene is prepared from ethylene?

9. Write the Lewis structure and molecular formula for each of the following hydrocarbons:

(a) hexane

(b) 3-methylpentane

(c) *cis*-3-hexene

(d) 4-methyl-1-pentene

(e) 3-hexyne

(f) 4-methyl-2-pentyne

10. Write the chemical formula, condensed formula, and Lewis structure for each of the following hydrocarbons:

(a) heptane

(b) 3-methylhexane

(c) *trans*-3-heptene

(d) 4-methyl-1-hexene

(e) 2-heptyne

(f) 3,4-dimethyl-1-pentyne

11. Give the complete IUPAC name for each of the following compounds:

(a) $CH_3CH_2CBr_2CH_3$

(b) $(CH_3)_3CCl$

(c)

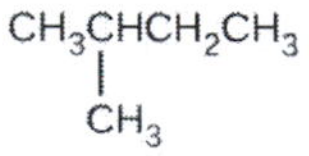

$$CH_3CHCH_2CH_3$$
$$|$$
$$CH_3$$

(d) $CH_3CH_2C \equiv CH\ CH_3CH_2C \equiv CH$

(e)

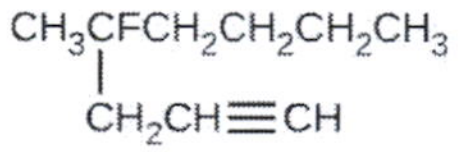

$$CH_3CFCH_2CH_2CH_2CH_3$$
$$|$$
$$CH_2CH \equiv CH$$

(f)

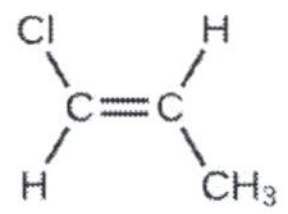

(g) $(CH_3)_2 CHCH_2 CH = CH_2$

12. Give the complete IUPAC name for each of the following compounds:

(a) $(CH_3)_2CHF$

(b) $CH_3CHClCHClCH_3$

(c)

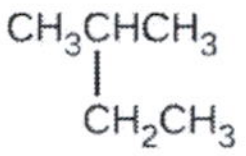

(d) $CH_3 CH_2 CH = CHCH_3$

(e)

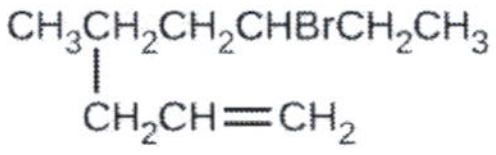

(f) $(CH_3)_3 CCH_2 C \equiv CH$

13. Butane is used as a fuel in disposable lighters. Write the Lewis structure for each isomer of butane.

14. Write Lewis structures and name the five structural isomers of hexane.

15. Write Lewis structures for the *cis–trans* isomers of $CH_3 CH = CHCl$.

16. Write structures for the three isomers of the aromatic hydrocarbon xylene, $C_6H_4(CH_3)_2$.

17. Isooctane is the common name of the isomer of C_8H_{18} used as the standard of 100 for the gasoline octane rating:

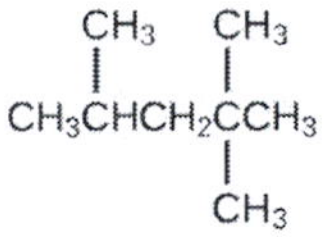

(a) What is the IUPAC name for the compound?

(b) Name the other isomers that contain a five-carbon chain with three methyl substituents.

18. Write Lewis structures and IUPAC names for the alkyne isomers of C_4H_6.

19. Write Lewis structures and IUPAC names for all isomers of C_4H_9Cl.

20. Name and write the structures of all isomers of the propyl and butyl alkyl groups.

21. Write the structures for all the isomers of the $–C_5H_{11}$ alkyl group.

22. Write Lewis structures and describe the molecular geometry at each carbon atom in the following compounds:

(a) *cis*-3-hexene

(b) *cis*-1-chloro-2-bromoethene

(c) 2-pentyne

(d) *trans*-6-ethyl-7-methyl-2-octene

23.   Benzene is one of the compounds used as an octane enhancer in unleaded gasoline. It is manufactured by the catalytic conversion of acetylene to benzene:

$$3C_2H_2 \longrightarrow C_6H_6$$

Draw Lewis structures for these compounds, with resonance structures as appropriate, and determine the hybridization of the carbon atoms in each.

24.   Teflon is prepared by the polymerization of tetrafluoroethylene. Write the equation that describes the polymerization using Lewis symbols.

25.   Write two complete, balanced equations for each of the following reactions, one using condensed formulas and one using Lewis structures.

(a) 1 mol of 1-butyne reacts with 2 mol of iodine.

(b) Pentane is burned in air.

26.   Write two complete, balanced equations for each of the following reactions, one using condensed formulas and one using Lewis structures.

(a) 2-butene reacts with chlorine.

(b) benzene burns in air.

27.   What mass of 2-bromopropane could be prepared from 25.5 g of propene? Assume a 100% yield of product.

28.   Acetylene is a very weak acid; however, it will react with moist silver(I) oxide and form water and a compound composed of silver and carbon. Addition of a solution of HCl to a 0.2352-g sample of the compound of silver and carbon produced acetylene and 0.2822 g of AgCl.

(a) What is the empirical formula of the compound of silver and carbon?

(b) The production of acetylene on addition of HCl to the compound of silver and carbon suggests that the carbon is present as the acetylide ion, $C_2{}^{2-}$. Write the formula of the compound showing the acetylide ion.

29.   Ethylene can be produced by the pyrolysis of ethane:

$$C_2H_6 \longrightarrow C_2H_4 + H_2$$

How many kilograms of ethylene is produced by the pyrolysis of 1.000×10^3 kg of ethane, assuming a 100.0% yield?

20.2 Alcohols and Ethers

30.   Why do the compounds hexane, hexanol, and hexene have such similar names?

31.   Write condensed formulas and provide IUPAC names for the following compounds:

(a) ethyl alcohol (in beverages)

(b) methyl alcohol (used as a solvent, for example, in shellac)

(c) ethylene glycol (antifreeze)

(d) isopropyl alcohol (used in rubbing alcohol)

(e) glycerine

32.   Give the complete IUPAC name for each of the following compounds:

(a)

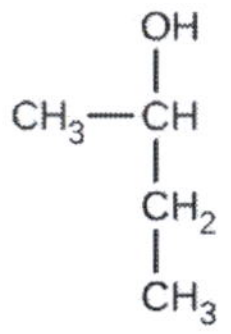

(b)

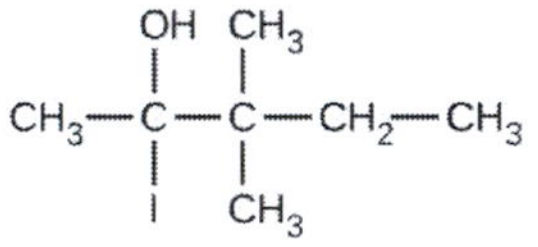

(c)

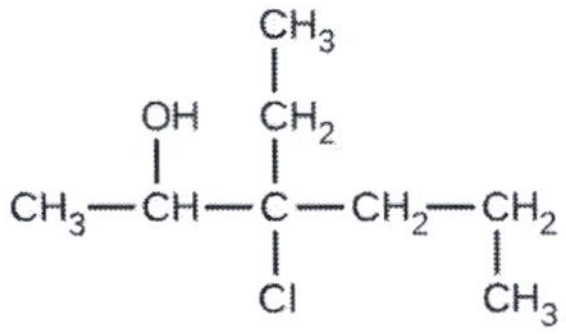

33. Give the complete IUPAC name and the common name for each of the following compounds:

(a)

$$CH_3—CH_2—O—CH_2—CH_2—CH_2—CH_3$$

(b)

$$CH_3—CH_2—O—CH_2—CH_2—CH_3$$

(c)

$$CH_3—O—CH_2—CH_2—CH_3$$

34. Write the condensed structures of both isomers with the formula C_2H_6O. Label the functional group of each isomer.

35. Write the condensed structures of all isomers with the formula $C_2H_6O_2$. Label the functional group (or groups) of each isomer.

36. Draw the condensed formulas for each of the following compounds:

(a) dipropyl ether

(b) 2,2-dimethyl-3-hexanol

(c) 2-ethoxybutane

37. MTBE, Methyl *tert*-butyl ether, $CH_3OC(CH_3)_3$, is used as an oxygen source in oxygenated gasolines. MTBE is manufactured by reacting 2-methylpropene with methanol.

(a) Using Lewis structures, write the chemical equation representing the reaction.

(b) What volume of methanol, density 0.7915 g/mL, is required to produce exactly 1000 kg of MTBE, assuming a 100% yield?

38. Write two complete balanced equations for each of the following reactions, one using condensed formulas and one using Lewis structures.

(a) propanol is converted to dipropyl ether

(b) propene is treated with water in dilute acid.

39. Write two complete balanced equations for each of the following reactions, one using condensed formulas and one using Lewis structures.

(a) 2-butene is treated with water in dilute acid

(b) ethanol is dehydrated to yield ethene

20.3 Aldehydes, Ketones, Carboxylic Acids, and Esters

40. Order the following molecules from least to most oxidized, based on the marked carbon atom:

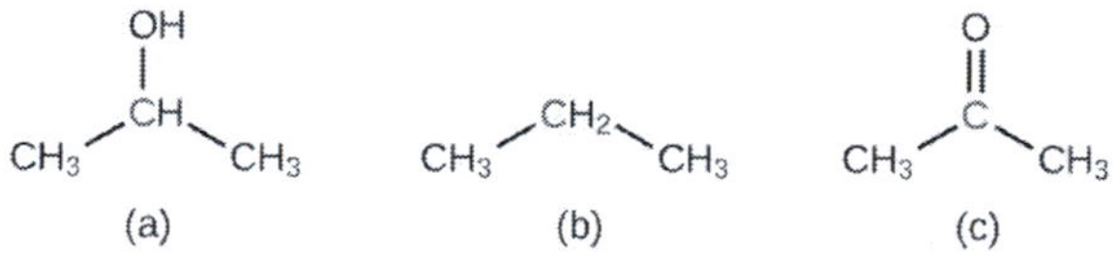

41. Predict the products of oxidizing the molecules shown in this problem. In each case, identify the product that will result from the minimal increase in oxidation state for the highlighted carbon atom:

(a)

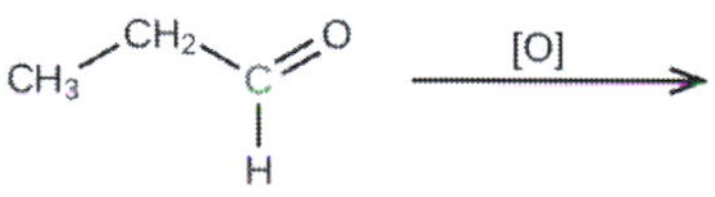

(b)

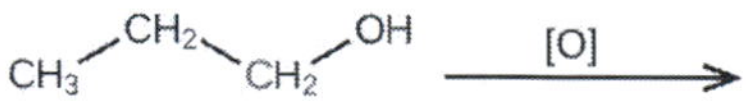

(c)

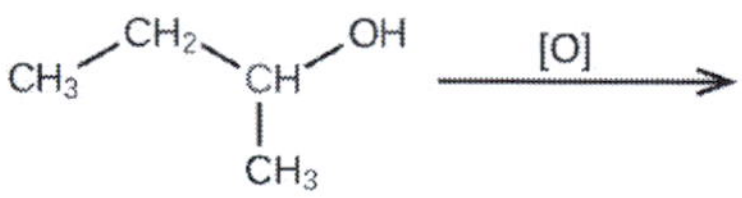

42. Predict the products of reducing the following molecules. In each case, identify the product that will result from the minimal decrease in oxidation state for the highlighted carbon atom:

(a)

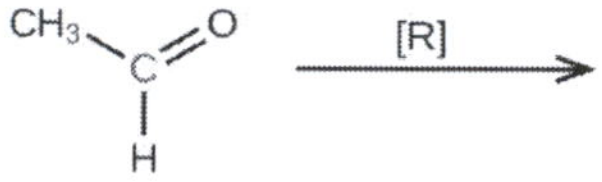

(b)

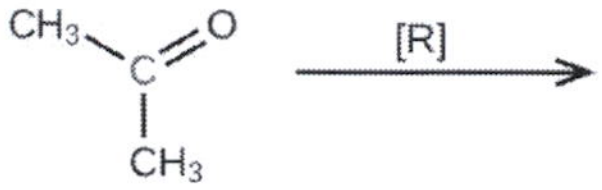

(c)

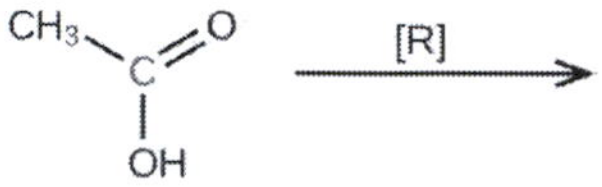

43. Explain why it is not possible to prepare a ketone that contains only two carbon atoms.

44. How does hybridization of the substituted carbon atom change when an alcohol is converted into an aldehyde? An aldehyde to a carboxylic acid?

45. Fatty acids are carboxylic acids that have long hydrocarbon chains attached to a carboxylate group. How does a saturated fatty acid differ from an unsaturated fatty acid? How are they similar?

46. Write a condensed structural formula, such as CH_3CH_3, and describe the molecular geometry at each carbon atom.

(a) propene

(b) 1-butanol

(c) ethyl propyl ether

(d) *cis*-4-bromo-2-heptene

(e) 2,2,3-trimethylhexane

(f) formaldehyde

47. Write a condensed structural formula, such as CH_3CH_3, and describe the molecular geometry at each carbon atom.

(a) 2-propanol

(b) acetone

(c) dimethyl ether

(d) acetic acid

(e) 3-methyl-1-hexene

48. The foul odor of rancid butter is caused by butyric acid, $CH_3CH_2CH_2CO_2H$.

(a) Draw the Lewis structure and determine the oxidation number and hybridization for each carbon atom in the molecule.

(b) The esters formed from butyric acid are pleasant-smelling compounds found in fruits and used in perfumes. Draw the Lewis structure for the ester formed from the reaction of butyric acid with 2-propanol.

49. Write the two-resonance structures for the acetate ion.

50. Write two complete, balanced equations for each of the following reactions, one using condensed formulas and one using Lewis structures:

(a) ethanol reacts with propionic acid

(b) benzoic acid, $C_6H_5CO_2H$, is added to a solution of sodium hydroxide

51. Write two complete balanced equations for each of the following reactions, one using condensed formulas and one using Lewis structures.

(a) 1-butanol reacts with acetic acid

(b) propionic acid is poured onto solid calcium carbonate

52. Yields in organic reactions are sometimes low. What is the percent yield of a process that produces 13.0 g of ethyl acetate from 10.0 g of CH_3CO_2H?

53. Alcohols A, B, and C all have the composition $C_4H_{10}O$. Molecules of alcohol A contain a branched carbon chain and can be oxidized to an aldehyde; molecules of alcohol B contain a linear carbon chain and can be oxidized to a ketone; and molecules of alcohol C can be oxidized to neither an aldehyde nor a ketone. Write the Lewis structures of these molecules.

20.4 Amines and Amides

54. Write the Lewis structures of both isomers with the formula C_2H_7N.

55. What is the molecular structure about the nitrogen atom in trimethyl amine and in the trimethyl ammonium ion, $(CH_3)_3NH^+$? What is the hybridization of the nitrogen atom in trimethyl amine and in the trimethyl ammonium ion?

56. Write the two resonance structures for the pyridinium ion, $C_5H_5NH^+$.

57. Draw Lewis structures for pyridine and its conjugate acid, the pyridinium ion, $C_5H_5NH^+$. What are the geometries and hybridizations about the nitrogen atoms in pyridine and in the pyridinium ion?

58. Write the Lewis structures of all isomers with the formula C_3H_7ON that contain an amide linkage.

59. Write two complete balanced equations for the following reaction, one using condensed formulas and one using Lewis structures.

Methyl amine is added to a solution of HCl.

60. Write two complete, balanced equations for each of the following reactions, one using condensed formulas and one using Lewis structures.

Ethylammonium chloride is added to a solution of sodium hydroxide.

61. Identify any carbon atoms that change hybridization and the change in hybridization during the reactions in Exercise 20.26.

62. Identify any carbon atoms that change hybridization and the change in hybridization during the reactions in Exercise 20.39.

63. Identify any carbon atoms that change hybridization and the change in hybridization during the reactions in Exercise 20.51.

Chapter 21

Nuclear Chemistry

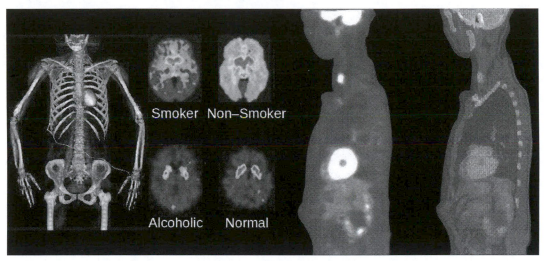

Smoker   Non–Smoker

Alcoholic   Normal

Figure 21.1 Nuclear chemistry provides the basis for many useful diagnostic and therapeutic methods in medicine, such as these positron emission tomography (PET) scans. The PET/computed tomography scan on the left shows muscle activity. The brain scans in the center show chemical differences in dopamine signaling in the brains of addicts and nonaddicts. The images on the right show an oncological application of PET scans to identify lymph node metastasis.

Chapter Outline

21.1 Nuclear Structure and Stability

21.2 Nuclear Equations

21.3 Radioactive Decay

21.4 Transmutation and Nuclear Energy

21.5 Uses of Radioisotopes

21.6 Biological Effects of Radiation

Introduction

The chemical reactions that we have considered in previous chapters involve changes in the *electronic* structure of the species involved, that is, the arrangement of the electrons around atoms, ions, or molecules. *Nuclear* structure, the numbers of protons and neutrons within the nuclei of the atoms involved, remains unchanged during chemical reactions.

This chapter will introduce the topic of nuclear chemistry, which began with the discovery of radioactivity in 1896 by French physicist Antoine Becquerel and has become increasingly important during the twentieth and twenty-first centuries, providing the basis for various technologies related to energy, medicine, geology, and many other areas.

21.1 Nuclear Structure and Stability

By the end of this section, you will be able to:

- Describe nuclear structure in terms of protons, neutrons, and electrons
- Calculate mass defect and binding energy for nuclei
- Explain trends in the relative stability of nuclei

Nuclear chemistry is the study of reactions that involve changes in nuclear structure. The chapter on atoms, molecules, and ions introduced the basic idea of nuclear structure, that the nucleus of an atom is composed of protons and, with the exception of ^1_1H, neutrons. Recall that the number of protons in the nucleus is called the atomic number (Z) of the element, and the sum of the number of protons and the number of neutrons is the mass number (A). Atoms with the same atomic number but different mass numbers are isotopes of the same element. When referring to a single type of nucleus, we often use the term **nuclide** and identify it by the notation ^A_ZX, where X is the symbol for the element, A is the mass number, and Z is the atomic number (for example, $^{14}_6\text{C}$). Often a nuclide is referenced by the name of the element followed by a hyphen and the mass number. For example, $^{14}_6\text{C}$ is called "carbon-14."

Protons and neutrons, collectively called **nucleons**, are packed together tightly in a nucleus. With a radius of about 10^{-15} meters, a nucleus is quite small compared to the radius of the entire atom, which is about 10^{-10} meters. Nuclei are extremely dense compared to bulk matter, averaging 1.8×10^{14} grams per cubic centimeter. For example, water has a density of 1 gram per cubic centimeter, and iridium, one of the densest elements known, has a density of 22.6 g/cm^3. If the earth's density were equal to the average nuclear density, the earth's radius would be only about 200 meters (earth's actual radius is approximately 6.4×10^6 meters, 30,000 times larger). Example 21.1 demonstrates just how great nuclear densities can be in the natural world.

Example 21.1

Density of a Neutron Star

Neutron stars form when the core of a very massive star undergoes gravitational collapse, causing the star's outer layers to explode in a supernova. Composed almost completely of neutrons, they are the densest-known stars in the universe, with densities comparable to the average density of an atomic nucleus. A neutron star in a faraway galaxy has a mass equal to 2.4 solar masses (1 solar mass = $M_\odot$ = mass of the sun = 1.99×10^{30} kg) and a diameter of 26 km.

(a) What is the density of this neutron star?

(b) How does this neutron star's density compare to the density of a uranium nucleus, which has a diameter of about 15 fm (1 fm = 10^{-15} m)?

Solution

We can treat both the neutron star and the U-235 nucleus as spheres. Then the density for both is given by:

$$d = \frac{m}{V} \qquad \text{with} \qquad V = \frac{4}{3}\pi r^3$$

(a) The radius of the neutron star is $\frac{1}{2} \times 26$ km = $\frac{1}{2} \times 2.6 \times 10^4$ m = 1.3×10^4 m, so the density of the neutron star is:

$$d = \frac{m}{V} = \frac{m}{\frac{4}{3}\pi r^3} = \frac{2.4(1.99 \times 10^{30} \text{ kg})}{\frac{4}{3}\pi(1.3 \times 10^4 \text{ m})^3} = 5.2 \times 10^{17} \text{ kg/m}^3$$

(b) The radius of the U-235 nucleus is $\frac{1}{2} \times 15 \times 10^{-15}$ m $= 7.5 \times 10^{-15}$ m, so the density of the U-235 nucleus is:

$$d = \frac{m}{V} = \frac{m}{\frac{4}{3}\pi r^3} = \frac{235 \text{ amu}\left(\frac{1.66 \times 10^{-27} \text{kg}}{1 \text{ amu}}\right)}{\frac{4}{3}\pi\left(7.5 \times 10^{-15} \text{m}\right)^3} = 2.2 \times 10^{17} \text{ kg/m}^3$$

These values are fairly similar (same order of magnitude), but the nucleus is more than twice as dense as the neutron star.

Check Your Learning

Find the density of a neutron star with a mass of 1.97 solar masses and a diameter of 13 km, and compare it to the density of a hydrogen nucleus, which has a diameter of 1.75 fm (1 fm = 1 $\times$ 10^{-15} m).

Answer: The density of the neutron star is 3.4×10^{18} kg/m^3. The density of a hydrogen nucleus is 6.0×10^{17} kg/m^3. The neutron star is 5.7 times denser than the hydrogen nucleus.

To hold positively charged protons together in the very small volume of a nucleus requires very strong attractive forces because the positively charged protons repel one another strongly at such short distances. The force of attraction that holds the nucleus together is the **strong nuclear force**. (The strong force is one of the four fundamental forces that are known to exist. The others are the electromagnetic force, the gravitational force, and the nuclear weak force.) This force acts between protons, between neutrons, and between protons and neutrons. It is very different from the electrostatic force that holds negatively charged electrons around a positively charged nucleus (the attraction between opposite charges). Over distances less than 10^{-15} meters and within the nucleus, the strong nuclear force is much stronger than electrostatic repulsions between protons; over larger distances and outside the nucleus, it is essentially nonexistent.

Link to Learning

Visit this website (http://openstaxcollege.org/l/16fourfund) for more information about the four fundamental forces.

Nuclear Binding Energy

As a simple example of the energy associated with the strong nuclear force, consider the helium atom composed of two protons, two neutrons, and two electrons. The total mass of these six subatomic particles may be calculated as:

$$(2 \times 1.0073 \text{ amu}) + (2 \times 1.0087 \text{ amu}) + (2 \times 0.00055 \text{ amu}) = 4.0331 \text{ amu}$$
$$\text{protons} \qquad\qquad \text{neutrons} \qquad\qquad \text{electrons}$$

However, mass spectrometric measurements reveal that the mass of an ^4_2He atom is 4.0026 amu, less than the combined masses of its six constituent subatomic particles. This difference between the calculated and experimentally measured masses is known as the **mass defect** of the atom. In the case of helium, the mass defect indicates a "loss" in mass of 4.0331 amu − 4.0026 amu = 0.0305 amu. The loss in mass accompanying the formation of an atom from protons, neutrons, and electrons is due to the conversion of that mass into energy that is evolved as the atom forms. The **nuclear binding energy** is the energy produced when the atoms' nucleons are bound together; this is also the energy needed to break a nucleus into its constituent protons and neutrons. In comparison to chemical bond energies, nuclear binding energies are *vastly* greater, as we will learn in this section. Consequently, the energy changes associated with nuclear reactions are vastly greater than are those for chemical reactions.

The conversion between mass and energy is most identifiably represented by the **mass-energy equivalence equation** as stated by Albert Einstein:

$$E = mc^2$$

where E is energy, m is mass of the matter being converted, and c is the speed of light in a vacuum. This equation can be used to find the amount of energy that results when matter is converted into energy. Using this mass-energy equivalence equation, the nuclear binding energy of a nucleus may be calculated from its mass defect, as demonstrated in Example 21.2. A variety of units are commonly used for nuclear binding energies, including **electron volts (eV)**, with 1 eV equaling the amount of energy necessary to the move the charge of an electron across an electric potential difference of 1 volt, making 1 eV = 1.602×10^{-19} J.

Example 21.2

Calculation of Nuclear Binding Energy

Determine the binding energy for the nuclide ^4_2He in:

(a) joules per mole of nuclei

(b) joules per nucleus

(c) MeV per nucleus

Solution

The mass defect for a ^4_2He nucleus is 0.0305 amu, as shown previously. Determine the binding energy in joules per nuclide using the mass-energy equivalence equation. To accommodate the requested energy units, the mass defect must be expressed in kilograms (recall that 1 J = 1 kg m^2/s^2).

(a) First, express the mass defect in g/mol. This is easily done considering the *numerical equivalence* of atomic mass (amu) and molar mass (g/mol) that results from the definitions of the amu and mole units (refer to the previous discussion in the chapter on atoms, molecules, and ions if needed). The mass defect is therefore 0.0305 g/mol. To accommodate the units of the other terms in the mass-energy equation, the mass must be expressed in kg, since 1 J = 1 kg m^2/s^2. Converting grams into kilograms yields a mass defect of 3.05×10^{-5} kg/mol. Substituting this quantity into the mass-energy equivalence equation yields:

$$E = mc^2 = \frac{3.05 \times 10^{-5} \text{ kg}}{\text{mol}} \times \left(\frac{2.998 \times 10^8 \text{ m}}{\text{s}}\right)^2 = 2.74 \times 10^{12} \text{ kg m}^2\text{s}^{-2}\text{mol}^{-1}$$

$$= 2.74 \times 10^{12} \text{ J mol}^{-1} = 2.74 \text{ TJ mol}^{-1}$$

Note that this tremendous amount of energy is associated with the conversion of a very small amount of matter (about 30 mg, roughly the mass of typical drop of water).

(b) The binding energy for a single nucleus is computed from the molar binding energy using Avogadro's number:

$$E = 2.74 \times 10^{12} \text{ J mol}^{-1} \times \frac{1 \text{ mol}}{6.022 \times 10^{23} \text{ nuclei}} = 4.55 \times 10^{-12} \text{ J} = 4.55 \text{ pJ}$$

(c) Recall that 1 eV = 1.602×10^{-19} J. Using the binding energy computed in part (b):

$$E = 4.55 \times 10^{-12} \text{ J} \times \frac{1 \text{ eV}}{1.602 \times 10^{-19} \text{ J}} = 2.84 \times 10^7 \text{ eV} = 28.4 \text{ MeV}$$

Check Your Learning

What is the binding energy for the nuclide $^{19}_9\text{F}$ (atomic mass: 18.9984 amu) in MeV per nucleus?

Answer: 148.4 MeV

Because the energy changes for breaking and forming bonds are so small compared to the energy changes for breaking or forming nuclei, the changes in mass during all ordinary chemical reactions are virtually undetectable. As described in the chapter on thermochemistry, the most energetic chemical reactions exhibit enthalpies on the order of *thousands* of kJ/mol, which is equivalent to mass differences in the nanogram range (10^{-9} g). On the other hand, nuclear binding energies are typically on the order of *billions* of kJ/mol, corresponding to mass differences in the milligram range (10^{-3} g).

Nuclear Stability

A nucleus is stable if it cannot be transformed into another configuration without adding energy from the outside. Of the thousands of nuclides that exist, about 250 are stable. A plot of the number of neutrons versus the number of protons for stable nuclei reveals that the stable isotopes fall into a narrow band. This region is known as the **band of stability** (also called the belt, zone, or valley of stability). The straight line in Figure 21.2 represents nuclei that have a 1:1 ratio of protons to neutrons (n:p ratio). Note that the lighter stable nuclei, in general, have equal numbers of protons and neutrons. For example, nitrogen-14 has seven protons and seven neutrons. Heavier stable nuclei, however, have increasingly more neutrons than protons. For example: iron-56 has 30 neutrons and 26 protons, an n:p ratio of 1.15, whereas the stable nuclide lead-207 has 125 neutrons and 82 protons, an n:p ratio equal to 1.52. This is because larger nuclei have more proton-proton repulsions, and require larger numbers of neutrons to provide compensating strong forces to overcome these electrostatic repulsions and hold the nucleus together.

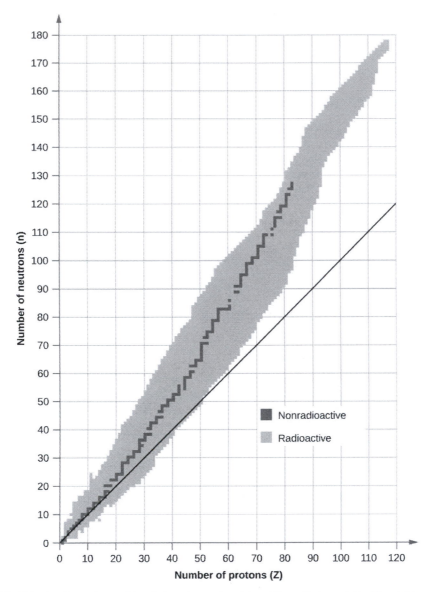

Figure 21.2 This plot shows the nuclides that are known to exist and those that are stable. The stable nuclides are indicated in blue, and the unstable nuclides are indicated in green. Note that all isotopes of elements with atomic numbers greater than 83 are unstable. The solid line is the line where n = Z.

The nuclei that are to the left or to the right of the band of stability are unstable and exhibit **radioactivity**. They change spontaneously (decay) into other nuclei that are either in, or closer to, the band of stability. These nuclear decay reactions convert one unstable isotope (or **radioisotope**) into another, more stable, isotope. We will discuss the nature and products of this radioactive decay in subsequent sections of this chapter.

Several observations may be made regarding the relationship between the stability of a nucleus and its structure. Nuclei with even numbers of protons, neutrons, or both are more likely to be stable (see Table 21.1). Nuclei with certain numbers of nucleons, known as **magic numbers**, are stable against nuclear decay. These numbers of protons or neutrons (2, 8, 20, 28, 50, 82, and 126) make complete shells in the nucleus. These are similar in concept to the stable electron shells observed for the noble gases. Nuclei that have magic numbers of both protons and neutrons, such as $_2^4\text{He}$, $_8^{16}\text{O}$, $_{20}^{40}\text{Ca}$, and $_{82}^{208}\text{Pb}$, are called "double magic" and are particularly stable. These trends in nuclear stability may be rationalized by considering a quantum mechanical model of nuclear energy states analogous to that used to describe electronic states earlier in this textbook. The details of this model are beyond the scope of this chapter.

Stable Nuclear Isotopes

Number of Stable Isotopes	Proton Number	Neutron Number
157	even	even
53	even	odd
50	odd	even
5	odd	odd

Table 21.1

The relative stability of a nucleus is correlated with its **binding energy per nucleon**, the total binding energy for the nucleus divided by the number or nucleons in the nucleus. For instance, we saw in Example 21.2 that the binding energy for a $_2^4\text{He}$ nucleus is 28.4 MeV. The binding energy *per nucleon* for a $_2^4\text{He}$ nucleus is therefore:

$$\frac{28.4 \text{ MeV}}{4 \text{ nucleons}} = 7.10 \text{ MeV/nucleon}$$

In Example 21.3, we learn how to calculate the binding energy per nucleon of a nuclide on the curve shown in Figure 21.3.

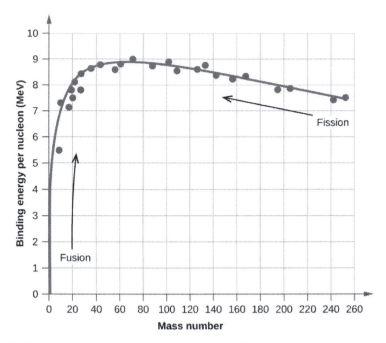

Figure 21.3 The binding energy per nucleon is largest for nuclides with mass number of approximately 56.

Example 21.3

Calculation of Binding Energy per Nucleon

The iron nuclide $^{56}_{26}\text{Fe}$ lies near the top of the binding energy curve (**Figure 21.3**) and is one of the most stable nuclides. What is the binding energy per nucleon (in MeV) for the nuclide $^{56}_{26}\text{Fe}$ (atomic mass of 55.9349 amu)?

Solution

As in **Example 21.2**, we first determine the mass defect of the nuclide, which is the difference between the mass of 26 protons, 30 neutrons, and 26 electrons, and the observed mass of an $^{56}_{26}\text{Fe}$ atom:

$$\text{Mass defect} = [(26 \times 1.0073 \text{ amu}) + (30 \times 1.0087 \text{ amu}) + (26 \times 0.00055 \text{ amu})] - 55.9349 \text{ amu}$$
$$= 56.4651 \text{ amu} - 55.9349 \text{ amu}$$
$$= 0.5302 \text{ amu}$$

We next calculate the binding energy for one nucleus from the mass defect using the mass-energy equivalence equation:

$$E = mc^2 = 0.5302 \text{ amu} \times \frac{1.6605 \times 10^{-27} \text{ kg}}{1 \text{ amu}} \times \left(2.998 \times 10^8 \text{ m/s}\right)^2$$

$$= 7.913 \times 10^{-11} \text{ kg·m/s}^2$$

$$= 7.913 \times 10^{-11} \text{ J}$$

We then convert the binding energy in joules per nucleus into units of MeV per nuclide:

$$7.913 \times 10^{-11} \text{ J} \times \frac{1 \text{ MeV}}{1.602 \times 10^{-13} \text{ J}} = 493.9 \text{ MeV}$$

Finally, we determine the binding energy per nucleon by dividing the total nuclear binding energy by the number of nucleons in the atom:

$$\text{Binding energy per nucleon} = \frac{493.9 \text{ MeV}}{56} = 8.820 \text{ MeV/nucleon}$$

Note that this is almost 25% larger than the binding energy per nucleon for $^{4}_{2}\text{He}$.

(Note also that this is the same process as in Example 21.1, but with the additional step of dividing the total nuclear binding energy by the number of nucleons.)

Check Your Learning

What is the binding energy per nucleon in $^{19}_{9}\text{F}$ (atomic mass, 18.9984 amu)?

Answer: 7.810 MeV/nucleon

21.2 Nuclear Equations

By the end of this section, you will be able to:

- Identify common particles and energies involved in nuclear reactions
- Write and balance nuclear equations

Changes of nuclei that result in changes in their atomic numbers, mass numbers, or energy states are **nuclear reactions**. To describe a nuclear reaction, we use an equation that identifies the nuclides involved in the reaction, their mass numbers and atomic numbers, and the other particles involved in the reaction.

Types of Particles in Nuclear Reactions

Many entities can be involved in nuclear reactions. The most common are protons, neutrons, alpha particles, beta particles, positrons, and gamma rays, as shown in Figure 21.4. Protons $\left(^{1}_{1}\text{p}, \text{ also represented by the symbol } ^{1}_{1}\text{H}\right)$ and neutrons $\left(^{1}_{0}\text{n}\right)$ are the constituents of atomic nuclei, and have been described previously. **Alpha particles** $\left(^{4}_{2}\text{He}, \text{ also represented by the symbol } ^{4}_{2}\alpha\right)$ are high-energy helium nuclei. **Beta particles** $\left(^{0}_{-1}\beta, \text{ also represented by the symbol } ^{0}_{-1}\text{e}\right)$ are high-energy electrons, and gamma rays are photons of very high-energy electromagnetic radiation. **Positrons** $\left(^{0}_{+1}\text{e}, \text{ also represented by the symbol } ^{0}_{+1}\beta\right)$ are positively charged electrons ("anti-electrons"). The subscripts and superscripts are necessary for balancing nuclear equations, but are usually optional in other circumstances. For example, an alpha particle is a helium nucleus (He) with a charge of +2 and a mass number of 4, so it is symbolized $^{4}_{2}\text{He}$. This works because, in general, the ion charge is not important in the balancing of nuclear equations.

Name	Symbol(s)	Representation	Description
Alpha particle	^4_2He or $^4_2\alpha$		(High-energy) helium nuclei consisting of two protons and two neutrons
Beta particle	^0_1e or $^{\;\;0}_{-1}\beta$		(High-energy) electrons
Positron	$^{\;\;0}_{+1}\text{e}$ or $^{\;\;0}_{+1}\beta$		Particles with the same mass as an electron but with 1 unit of positive charge
Proton	^1_1H or ^1_1p		Nuclei of hydrogen atoms
Neutron	^1_0n		Particles with a mass approximately equal to that of a proton but with no charge
Gamma ray	γ	 $\rightsquigarrow \gamma$	Very high-energy electromagnetic radiation

Figure 21.4 Although many species are encountered in nuclear reactions, this table summarizes the names, symbols, representations, and descriptions of the most common of these.

Note that positrons are exactly like electrons, except they have the opposite charge. They are the most common example of **antimatter**, particles with the same mass but the opposite state of another property (for example, charge) than ordinary matter. When antimatter encounters ordinary matter, both are annihilated and their mass is converted into energy in the form of **gamma rays (γ)**—and other much smaller subnuclear particles, which are beyond the scope of this chapter—according to the mass-energy equivalence equation $E = mc^2$, seen in the preceding section. For example, when a positron and an electron collide, both are annihilated and two gamma ray photons are created:

$$^{\;\;0}_{-1}\text{e} + ^{\;\;0}_{+1}\text{e} \longrightarrow \gamma + \gamma$$

As seen in the chapter discussing light and electromagnetic radiation, gamma rays compose short wavelength, high-energy electromagnetic radiation and are (much) more energetic than better-known X-rays that can behave as particles in the wave-particle duality sense. Gamma rays are produced when a nucleus undergoes a transition from a higher to a lower energy state, similar to how a photon is produced by an electronic transition from a higher to a lower energy level. Due to the much larger energy differences between nuclear energy shells, gamma rays emanating from a nucleus have energies that are typically millions of times larger than electromagnetic radiation emanating from electronic transitions.

Balancing Nuclear Reactions

A balanced chemical reaction equation reflects the fact that during a chemical reaction, bonds break and form, and atoms are rearranged, but the total numbers of atoms of each element are conserved and do not change. A balanced nuclear reaction equation indicates that there is a rearrangement during a nuclear reaction, but of subatomic particles rather than atoms. Nuclear reactions also follow conservation laws, and they are balanced in two ways:

1. The sum of the mass numbers of the reactants equals the sum of the mass numbers of the products.

2. The sum of the charges of the reactants equals the sum of the charges of the products.

If the atomic number and the mass number of all but one of the particles in a nuclear reaction are known, we can identify the particle by balancing the reaction. For instance, we could determine that $^{17}_{8}O$ is a product of the nuclear reaction of $^{14}_{7}N$ and $^{4}_{2}He$ if we knew that a proton, $^{1}_{1}H$, was one of the two products. Example 21.4 shows how we can identify a nuclide by balancing the nuclear reaction.

Example 21.4

Balancing Equations for Nuclear Reactions

The reaction of an α particle with magnesium-25 ($^{25}_{12}Mg$) produces a proton and a nuclide of another element. Identify the new nuclide produced.

Solution

The nuclear reaction can be written as:

$$^{25}_{12}Mg + ^{4}_{2}He \longrightarrow ^{1}_{1}H + ^{A}_{Z}X$$

where A is the mass number and Z is the atomic number of the new nuclide, X. Because the sum of the mass numbers of the reactants must equal the sum of the mass numbers of the products:

$$25 + 4 = A + 1, \text{ or } A = 28$$

Similarly, the charges must balance, so:

$$12 + 2 = Z + 1, \text{ and } Z = 13$$

Check the periodic table: The element with nuclear charge = +13 is aluminum. Thus, the product is $^{28}_{13}Al$.

Check Your Learning

The nuclide $^{125}_{53}I$ combines with an electron and produces a new nucleus and no other massive particles. What is the equation for this reaction?

Answer: $^{125}_{53}I + ^{0}_{-1}e \longrightarrow ^{125}_{52}Te$

Following are the equations of several nuclear reactions that have important roles in the history of nuclear chemistry:

- The first naturally occurring unstable element that was isolated, polonium, was discovered by the Polish scientist Marie Curie and her husband Pierre in 1898. It decays, emitting α particles:

$$^{212}_{84}Po \longrightarrow ^{208}_{82}Pb + ^{4}_{2}He$$

- The first nuclide to be prepared by artificial means was an isotope of oxygen, ^{17}O. It was made by Ernest Rutherford in 1919 by bombarding nitrogen atoms with α particles:

$$^{14}_{7}N + ^{4}_{2}\alpha \longrightarrow ^{17}_{8}O + ^{1}_{1}H$$

- James Chadwick discovered the neutron in 1932, as a previously unknown neutral particle produced along with ^{12}C by the nuclear reaction between ^{9}Be and ^{4}He:

$$^{9}_{4}Be + ^{4}_{2}He \longrightarrow ^{12}_{6}C + ^{1}_{0}n$$

- The first element to be prepared that does not occur naturally on the earth, technetium, was created by bombardment of molybdenum by deuterons (heavy hydrogen, $^{2}_{1}H$), by Emilio Segre and Carlo Perrier in 1937:

$$^{2}_{1}H + ^{97}_{42}Mo \longrightarrow 2^{1}_{0}n + ^{97}_{43}Tc$$

- The first controlled nuclear chain reaction was carried out in a reactor at the University of Chicago in 1942. One of the many reactions involved was:

$$^{235}_{92}U + ^{1}_{0}n \longrightarrow ^{87}_{35}Br + ^{146}_{57}La + 3^{1}_{0}n$$

21.3 Radioactive Decay

By the end of this section, you will be able to:

- Recognize common modes of radioactive decay
- Identify common particles and energies involved in nuclear decay reactions
- Write and balance nuclear decay equations
- Calculate kinetic parameters for decay processes, including half-life
- Describe common radiometric dating techniques

Following the somewhat serendipitous discovery of radioactivity by Becquerel, many prominent scientists began to investigate this new, intriguing phenomenon. Among them were Marie Curie (the first woman to win a Nobel Prize, and the only person to win two Nobel Prizes in different sciences—chemistry and physics), who was the first to coin the term "radioactivity," and Ernest Rutherford (of gold foil experiment fame), who investigated and named three of the most common types of radiation. During the beginning of the twentieth century, many radioactive substances were discovered, the properties of radiation were investigated and quantified, and a solid understanding of radiation and nuclear decay was developed.

The spontaneous change of an unstable nuclide into another is **radioactive decay**. The unstable nuclide is called the **parent nuclide**; the nuclide that results from the decay is known as the **daughter nuclide**. The daughter nuclide may be stable, or it may decay itself. The radiation produced during radioactive decay is such that the daughter nuclide lies closer to the band of stability than the parent nuclide, so the location of a nuclide relative to the band of stability can serve as a guide to the kind of decay it will undergo (Figure 21.5).

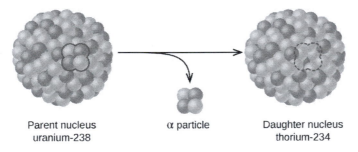

| Parent nucleus | α particle | Daughter nucleus |
| uranium-238 | | thorium-234 |

Figure 21.5   A nucleus of uranium-238 (the parent nuclide) undergoes α decay to form thorium-234 (the daughter nuclide). The alpha particle removes two protons (green) and two neutrons (gray) from the uranium-238 nucleus.

Link to Learning

Although the radioactive decay of a nucleus is too small to see with the naked eye, we can indirectly view radioactive decay in an environment called a cloud chamber. Click here (http://openstaxcollege.org/l/16cloudchamb) to learn about cloud chambers and to view an interesting Cloud Chamber Demonstration from the Jefferson Lab.

Types of Radioactive Decay

Ernest Rutherford's experiments involving the interaction of radiation with a magnetic or electric field (Figure 21.6) helped him determine that one type of radiation consisted of positively charged and relatively massive α particles; a second type was made up of negatively charged and much less massive β particles; and a third was uncharged electromagnetic waves, γ rays. We now know that α particles are high-energy helium nuclei, β particles are high-energy electrons, and γ radiation compose high-energy electromagnetic radiation. We classify different types of radioactive decay by the radiation produced.

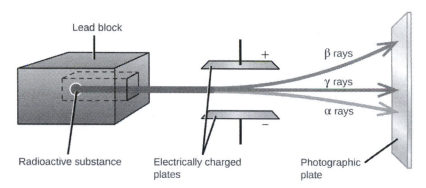

Figure 21.6 Alpha particles, which are attracted to the negative plate and deflected by a relatively small amount, must be positively charged and relatively massive. Beta particles, which are attracted to the positive plate and deflected a relatively large amount, must be negatively charged and relatively light. Gamma rays, which are unaffected by the electric field, must be uncharged.

Alpha (α) decay is the emission of an α particle from the nucleus. For example, polonium-210 undergoes α decay:

$$^{210}_{84}\text{Po} \longrightarrow {}^{4}_{2}\text{He} + {}^{206}_{82}\text{Pb} \qquad \text{or} \qquad {}^{210}_{84}\text{Po} \longrightarrow {}^{4}_{2}\alpha + {}^{206}_{82}\text{Pb}$$

Alpha decay occurs primarily in heavy nuclei (A > 200, Z > 83). Because the loss of an α particle gives a daughter nuclide with a mass number four units smaller and an atomic number two units smaller than those of the parent nuclide, the daughter nuclide has a larger n:p ratio than the parent nuclide. If the parent nuclide undergoing α decay lies below the band of stability (refer to Figure 21.2), the daughter nuclide will lie closer to the band.

Beta (β) decay is the emission of an electron from a nucleus. Iodine-131 is an example of a nuclide that undergoes β decay:

$$^{131}_{53}\text{I} \longrightarrow {}^{0}_{-1}\text{e} + {}^{131}_{54}\text{X} \qquad \text{or} \qquad {}^{131}_{53}\text{I} \longrightarrow {}^{0}_{-1}\beta + {}^{131}_{54}\text{Xe}$$

Beta decay, which can be thought of as the conversion of a neutron into a proton and a β particle, is observed in nuclides with a large n:p ratio. The beta particle (electron) emitted is from the atomic nucleus and is not one of the electrons surrounding the nucleus. Such nuclei lie above the band of stability. Emission of an electron does not change the mass number of the nuclide but does increase the number of its protons and decrease the number of its neutrons. Consequently, the n:p ratio is decreased, and the daughter nuclide lies closer to the band of stability than did the parent nuclide.

Gamma emission (γ emission) is observed when a nuclide is formed in an excited state and then decays to its ground state with the emission of a γ ray, a quantum of high-energy electromagnetic radiation. The presence of a nucleus in an excited state is often indicated by an asterisk (*). Cobalt-60 emits γ radiation and is used in many applications including cancer treatment:

$$^{60}_{27}\text{Co*} \longrightarrow {}^{0}_{0}\gamma + {}^{60}_{27}\text{Co}$$

There is no change in mass number or atomic number during the emission of a γ ray unless the γ emission accompanies one of the other modes of decay.

Positron emission (β⁺ decay) is the emission of a positron from the nucleus. Oxygen-15 is an example of a nuclide that undergoes positron emission:

$$^{15}_{8}\text{O} \longrightarrow {}^{0}_{+1}\text{e} + {}^{15}_{7}\text{N} \qquad \text{or} \qquad ^{15}_{8}\text{O} \longrightarrow {}^{0}_{+1}\beta + {}^{15}_{7}\text{N}$$

Positron emission is observed for nuclides in which the n:p ratio is low. These nuclides lie below the band of stability. Positron decay is the conversion of a proton into a neutron with the emission of a positron. The n:p ratio increases, and the daughter nuclide lies closer to the band of stability than did the parent nuclide.

Electron capture occurs when one of the inner electrons in an atom is captured by the atom's nucleus. For example, potassium-40 undergoes electron capture:

$$^{40}_{19}\text{K} + {}^{0}_{-1}\text{e} \longrightarrow {}^{40}_{18}\text{Ar}$$

Electron capture occurs when an inner shell electron combines with a proton and is converted into a neutron. The loss of an inner shell electron leaves a vacancy that will be filled by one of the outer electrons. As the outer electron drops into the vacancy, it will emit energy. In most cases, the energy emitted will be in the form of an X-ray. Like positron emission, electron capture occurs for "proton-rich" nuclei that lie below the band of stability. Electron capture has the same effect on the nucleus as does positron emission: The atomic number is decreased by one and the mass number does not change. This increases the n:p ratio, and the daughter nuclide lies closer to the band of stability than did the parent nuclide. Whether electron capture or positron emission occurs is difficult to predict. The choice is primarily due to kinetic factors, with the one requiring the smaller activation energy being the one more likely to occur.

Figure 21.7 summarizes these types of decay, along with their equations and changes in atomic and mass numbers.

Type	Nuclear equation	Representation	Change in mass/atomic numbers
Alpha decay	$^{A}_{Z}\text{X} \quad {}^{4}_{2}\text{He} + {}^{A-4}_{Z-2}\text{Y}$		A: decrease by 4 Z: decrease by 2
Beta decay	$^{A}_{Z}\text{X} \quad {}^{0}_{-1}\text{e} + {}^{A}_{Z+1}\text{Y}$	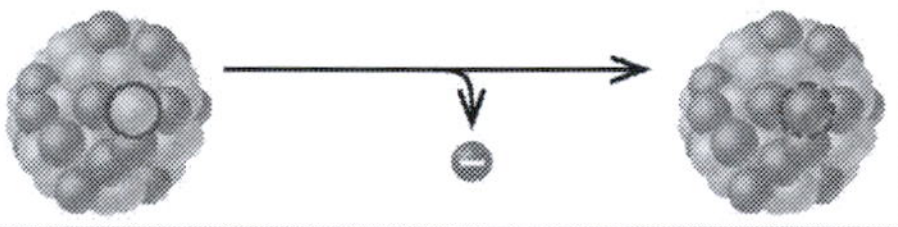	A: unchanged Z: increase by 1
Gamma decay	$^{A}_{Z}\text{X} \quad {}^{0}_{0}\gamma + {}^{A}_{Z}\text{Y}$	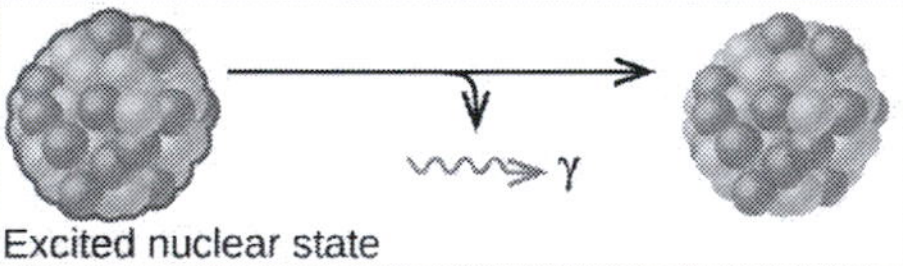 Excited nuclear state	A: unchanged Z: unchanged
Positron emission	$^{A}_{Z}\text{X} \quad {}^{0}_{+1}\text{e} + {}^{A}_{Y-1}\text{Y}$	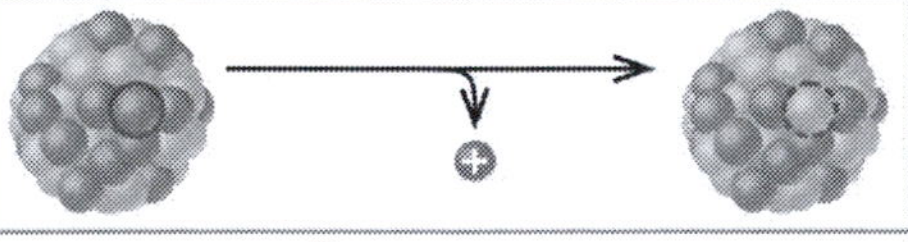	A: unchanged Z: decrease by 1
Electron capture	$^{A}_{Z}\text{X} \quad {}^{0}_{-1}\text{e} + {}^{A}_{Y-1}\text{Y}$	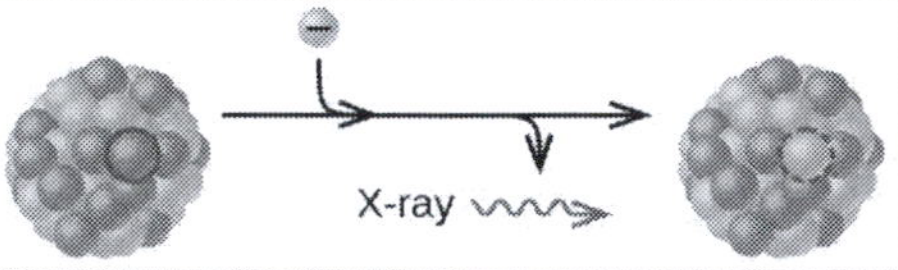 X-ray	A: unchanged Z: decrease by 1

Figure 21.7   This table summarizes the type, nuclear equation, representation, and any changes in the mass or atomic numbers for various types of decay.

PET Scan

Positron emission tomography (PET) scans use radiation to diagnose and track health conditions and monitor medical treatments by revealing how parts of a patient's body function (Figure 21.8). To perform a PET scan, a positron-emitting radioisotope is produced in a cyclotron and then attached to a substance that is used by the part of the body being investigated. This "tagged" compound, or radiotracer, is then put into the patient (injected via IV or breathed in as a gas), and how it is used by the tissue reveals how that organ or other area of the body functions.

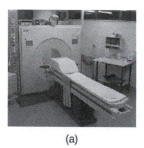
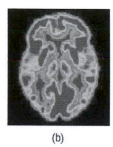
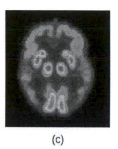

(a)   (b)   (c)

Figure 21.8   A PET scanner (a) uses radiation to provide an image of how part of a patient's body functions. The scans it produces can be used to image a healthy brain (b) or can be used for diagnosing medical conditions such as Alzheimer's disease (c). (credit a: modification of work by Jens Maus)

For example, F-18 is produced by proton bombardment of ^{18}O ($^{18}_{8}O + ^{1}_{1}p \longrightarrow ^{18}_{9}F + ^{1}_{0}n$) and incorporated into a glucose analog called fludeoxyglucose (FDG). How FDG is used by the body provides critical diagnostic information; for example, since cancers use glucose differently than normal tissues, FDG can reveal cancers. The ^{18}F emits positrons that interact with nearby electrons, producing a burst of gamma radiation. This energy is detected by the scanner and converted into a detailed, three-dimensional, color image that shows how that part of the patient's body functions. Different levels of gamma radiation produce different amounts of brightness and colors in the image, which can then be interpreted by a radiologist to reveal what is going on. PET scans can detect heart damage and heart disease, help diagnose Alzheimer's disease, indicate the part of a brain that is affected by epilepsy, reveal cancer, show what stage it is, and how much it has spread, and whether treatments are effective. Unlike magnetic resonance imaging and X-rays, which only show how something looks, the big advantage of PET scans is that they show how something functions. PET scans are now usually performed in conjunction with a computed tomography scan.

Radioactive Decay Series

The naturally occurring radioactive isotopes of the heaviest elements fall into chains of successive disintegrations, or decays, and all the species in one chain constitute a radioactive family, or **radioactive decay series**. Three of these series include most of the naturally radioactive elements of the periodic table. They are the uranium series, the actinide series, and the thorium series. The neptunium series is a fourth series, which is no longer significant on the earth because of the short half-lives of the species involved. Each series is characterized by a parent (first member) that has a long half-life and a series of daughter nuclides that ultimately lead to a stable end-product—that is, a nuclide on the band of stability (Figure 21.9). In all three series, the end-product is a stable isotope of lead. The neptunium series, previously thought to terminate with bismuth-209, terminates with thallium-205.

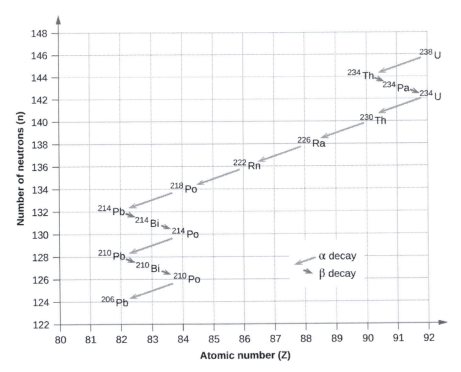

Figure 21.9 Uranium-238 undergoes a radioactive decay series consisting of 14 separate steps before producing stable lead-206. This series consists of eight α decays and six β decays.

Radioactive Half-Lives

Radioactive decay follows first-order kinetics. Since first-order reactions have already been covered in detail in the kinetics chapter, we will now apply those concepts to nuclear decay reactions. Each radioactive nuclide has a characteristic, constant **half-life** ($t_{1/2}$), the time required for half of the atoms in a sample to decay. An isotope's half-life allows us to determine how long a sample of a useful isotope will be available, and how long a sample of an undesirable or dangerous isotope must be stored before it decays to a low-enough radiation level that is no longer a problem.

For example, cobalt-60, an isotope that emits gamma rays used to treat cancer, has a half-life of 5.27 years (Figure 21.10). In a given cobalt-60 source, since half of the $_{27}^{60}\text{Co}$ nuclei decay every 5.27 years, both the amount of material and the intensity of the radiation emitted is cut in half every 5.27 years. (Note that for a given substance, the intensity of radiation that it produces is directly proportional to the rate of decay of the substance and the amount of the substance.) This is as expected for a process following first-order kinetics. Thus, a cobalt-60 source that is used for cancer treatment must be replaced regularly to continue to be effective.

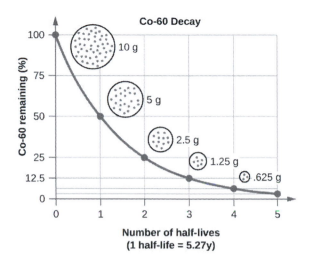

Figure 21.10 For cobalt-60, which has a half-life of 5.27 years, 50% remains after 5.27 years (one half-life), 25% remains after 10.54 years (two half-lives), 12.5% remains after 15.81 years (three half-lives), and so on.

Since nuclear decay follows first-order kinetics, we can adapt the mathematical relationships used for first-order chemical reactions. We generally substitute the number of nuclei, N, for the concentration. If the rate is stated in nuclear decays per second, we refer to it as the activity of the radioactive sample. The rate for radioactive decay is:

decay rate = λN with λ = the decay constant for the particular radioisotope

The decay constant, λ, which is the same as a rate constant discussed in the kinetics chapter. It is possible to express the decay constant in terms of the half-life, $t_{1/2}$:

$$\lambda = \frac{\ln 2}{t_{1/2}} = \frac{0.693}{t_{1/2}} \qquad \text{or} \qquad t_{1/2} = \frac{\ln 2}{\lambda} = \frac{0.693}{\lambda}$$

The first-order equations relating amount, N, and time are:

$$N_t = N_0 e^{-kt} \qquad \text{or} \qquad t = -\frac{1}{\lambda} \ln\left(\frac{N_t}{N_0}\right)$$

where N_0 is the initial number of nuclei or moles of the isotope, and N_t is the number of nuclei/moles remaining at time t. Example 21.5 applies these calculations to find the rates of radioactive decay for specific nuclides.

Example 21.5

Rates of Radioactive Decay

$^{60}_{27}\text{Co}$ decays with a half-life of 5.27 years to produce $^{60}_{28}\text{Ni}$.

(a) What is the decay constant for the radioactive disintegration of cobalt-60?

(b) Calculate the fraction of a sample of the $^{60}_{27}\text{Co}$ isotope that will remain after 15 years.

(c) How long does it take for a sample of $^{60}_{27}\text{Co}$ to disintegrate to the extent that only 2.0% of the original amount remains?

Solution

(a) The value of the rate constant is given by:

$$\lambda = \frac{\ln 2}{t_{1/2}} = \frac{0.693}{5.27 \text{ y}} = 0.132 \text{ y}^{-1}$$

(b) The fraction of $^{60}_{27}\text{Co}$ that is left after time t is given by $\frac{N_t}{N_0}$. Rearranging the first-order relationship $N_t = N_0 e^{-\lambda t}$ to solve for this ratio yields:

$$\frac{N_t}{N_0} = e^{-\lambda t} = e^{-(0.132/\text{y})(15.0/\text{y})} = 0.138$$

The fraction of $^{60}_{27}\text{Co}$ that will remain after 15.0 years is 0.138. Or put another way, 13.8% of the $^{60}_{27}\text{Co}$ originally present will remain after 15 years.

(c) 2.00% of the original amount of $^{60}_{27}\text{Co}$ is equal to $0.0200 \times N_0$. Substituting this into the equation for time for first-order kinetics, we have:

$$t = -\frac{1}{\lambda} \ln\left(\frac{N_t}{N_0}\right) = -\frac{1}{0.132 \text{ y}^{-1}} \ln\left(\frac{0.0200 \times N_0}{N_0}\right) = 29.6 \text{ y}$$

Check Your Learning

Radon-222, $^{222}_{86}\text{Rn}$, has a half-life of 3.823 days. How long will it take a sample of radon-222 with a mass of 0.750 g to decay into other elements, leaving only 0.100 g of radon-222?

Answer: 11.1 days

Because each nuclide has a specific number of nucleons, a particular balance of repulsion and attraction, and its own degree of stability, the half-lives of radioactive nuclides vary widely. For example: the half-life of $^{209}_{83}\text{Bi}$ is 1.9 $\times$ 10^{19} years; $^{239}_{94}\text{Ra}$ is 24,000 years; $^{222}_{86}\text{Rn}$ is 3.82 days; and element-111 (Rg for roentgenium) is 1.5 $\times$ 10^{-3} seconds. The half-lives of a number of radioactive isotopes important to medicine are shown in Table 21.2, and others are listed in Appendix M.

Half-lives of Radioactive Isotopes Important to Medicine

Type[1]	Decay Mode	Half-Life	Uses
F-18	β^+ decay	110. minutes	PET scans
Co-60	β decay, γ decay	5.27 years	cancer treatment
Tc-99m	γ decay	8.01 hours	scans of brain, lung, heart, bone
I-131	β decay	8.02 days	thyroid scans and treatment
Tl-201	electron capture	73 hours	heart and arteries scans; cardiac stress tests

Table 21.2

Radiometric Dating

Several radioisotopes have half-lives and other properties that make them useful for purposes of "dating" the origin of objects such as archaeological artifacts, formerly living organisms, or geological formations. This process is **radiometric dating** and has been responsible for many breakthrough scientific discoveries about the geological

---

1. The "m" in Tc-99m stands for "metastable," indicating that this is an unstable, high-energy state of Tc-99. Metastable isotopes emit γ radiation to rid themselves of excess energy and become (more) stable.

history of the earth, the evolution of life, and the history of human civilization. We will explore some of the most common types of radioactive dating and how the particular isotopes work for each type.

Radioactive Dating Using Carbon-14

The radioactivity of carbon-14 provides a method for dating objects that were a part of a living organism. This method of radiometric dating, which is also called **radiocarbon dating** or carbon-14 dating, is accurate for dating carbon-containing substances that are up to about 30,000 years old, and can provide reasonably accurate dates up to a maximum of about 50,000 years old.

Naturally occurring carbon consists of three isotopes: $^{12}_{6}C$, which constitutes about 99% of the carbon on earth; $^{13}_{6}C$, about 1% of the total; and trace amounts of $^{14}_{6}C$. Carbon-14 forms in the upper atmosphere by the reaction of nitrogen atoms with neutrons from cosmic rays in space:

$$^{14}_{7}N + ^{1}_{0}n \longrightarrow ^{14}_{6}C + ^{1}_{1}H$$

All isotopes of carbon react with oxygen to produce CO_2 molecules. The ratio of $^{14}_{6}CO_2$ to $^{12}_{6}CO_2$ depends on the ratio of $^{14}_{6}CO$ to $^{12}_{6}CO$ in the atmosphere. The natural abundance of $^{14}_{6}CO$ in the atmosphere is approximately 1 part per trillion; until recently, this has generally been constant over time, as seen is gas samples found trapped in ice. The incorporation of $^{14}_{6}C^{14}_{6}CO_2$ and $^{12}_{6}CO_2$ into plants is a regular part of the photosynthesis process, which means that the $^{14}_{6}C$: $^{12}_{6}C$ ratio found in a living plant is the same as the $^{14}_{6}C$: $^{12}_{6}C$ ratio in the atmosphere. But when the plant dies, it no longer traps carbon through photosynthesis. Because $^{12}_{6}C$ is a stable isotope and does not undergo radioactive decay, its concentration in the plant does not change. However, carbon-14 decays by β emission with a half-life of 5730 years:

$$^{14}_{6}C \longrightarrow ^{12}_{7}N + ^{0}_{-1}e$$

Thus, the $^{14}_{6}C$: $^{12}_{6}C$ ratio gradually decreases after the plant dies. The decrease in the ratio with time provides a measure of the time that has elapsed since the death of the plant (or other organism that ate the plant). Figure 21.11 visually depicts this process.

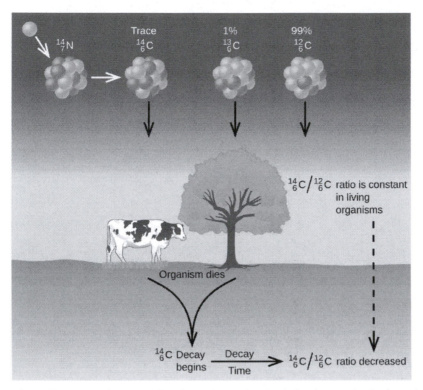

Figure 21.11 Along with stable carbon-12, radioactive carbon-14 is taken in by plants and animals, and remains at a constant level within them while they are alive. After death, the C-14 decays and the C-14:C-12 ratio in the remains decreases. Comparing this ratio to the C-14:C-12 ratio in living organisms allows us to determine how long ago the organism lived (and died).

For example, with the half-life of $^{14}_{6}C$ being 5730 years, if the $^{14}_{6}C$: $^{12}_{6}C$ ratio in a wooden object found in an archaeological dig is half what it is in a living tree, this indicates that the wooden object is 5730 years old. Highly accurate determinations of $^{14}_{6}C$: $^{12}_{6}C$ ratios can be obtained from very small samples (as little as a milligram) by the use of a mass spectrometer.

Link to Learning

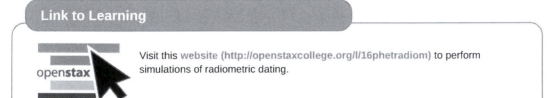

Visit this **website (http://openstaxcollege.org/l/16phetradiom)** to perform simulations of radiometric dating.

Example 21.6

Radiocarbon Dating

A tiny piece of paper (produced from formerly living plant matter) taken from the Dead Sea Scrolls has an activity of 10.8 disintegrations per minute per gram of carbon. If the initial C-14 activity was 13.6 disintegrations/min/g of C, estimate the age of the Dead Sea Scrolls.

Solution

The rate of decay (number of disintegrations/minute/gram of carbon) is proportional to the amount of radioactive C-14 left in the paper, so we can substitute the rates for the amounts, N, in the relationship:

$$t = -\frac{1}{\lambda}\ln\left(\frac{N_t}{N_0}\right) \longrightarrow t = -\frac{1}{\lambda}\ln\left(\frac{\text{Rate}_t}{\text{Rate}_0}\right)$$

where the subscript 0 represents the time when the plants were cut to make the paper, and the subscript t represents the current time.

The decay constant can be determined from the half-life of C-14, 5730 years:

$$\lambda = \frac{\ln 2}{t_{1/2}} = \frac{0.693}{5730 \text{ y}} = 1.21 \times 10^{-4} \text{ y}^{-1}$$

Substituting and solving, we have:

$$t = -\frac{1}{\lambda}\ln\left(\frac{\text{Rate}_t}{\text{Rate}_0}\right) = -\frac{1}{1.21 \times 10^{-4} \text{ y}^{-1}}\ln\left(\frac{10.8 \text{ dis/min/g C}}{13.6 \text{ dis/min/g C}}\right) = 1910 \text{ y}$$

Therefore, the Dead Sea Scrolls are approximately 1900 years old (Figure 21.12).

Figure 21.12 Carbon-14 dating has shown that these pages from the Dead Sea Scrolls were written or copied on paper made from plants that died between 100 BC and AD 50.

Check Your Learning

More accurate dates of the reigns of ancient Egyptian pharaohs have been determined recently using plants that were preserved in their tombs. Samples of seeds and plant matter from King Tutankhamun's tomb have a C-14 decay rate of 9.07 disintegrations/min/g of C. How long ago did King Tut's reign come to an end?

Answer: about 3350 years ago, or approximately 1340 BC

There have been some significant, well-documented changes to the $^{14}_{6}\text{C}$: $^{12}_{6}\text{C}$ ratio. The accuracy of a straightforward application of this technique depends on the $^{14}_{6}\text{C}$: $^{12}_{6}\text{C}$ ratio in a living plant being the same now as it was in an earlier era, but this is not always valid. Due to the increasing accumulation of CO_2 molecules (largely $^{12}_{6}\text{CO}_2$) in the atmosphere caused by combustion of fossil fuels (in which essentially all of the $^{14}_{6}\text{C}$ has decayed), the ratio of $^{14}_{6}\text{C}$: $^{12}_{6}\text{C}$ in the atmosphere may be changing. This manmade increase in $^{12}_{6}\text{CO}_2$ in the atmosphere causes the $^{14}_{6}\text{C}$: $^{12}_{6}\text{C}$ ratio to decrease, and this in turn affects the ratio in currently living organisms on the earth.

Fortunately, however, we can use other data, such as tree dating via examination of annual growth rings, to calculate correction factors. With these correction factors, accurate dates can be determined. In general, radioactive dating only works for about 10 half-lives; therefore, the limit for carbon-14 dating is about 57,000 years.

Radioactive Dating Using Nuclides Other than Carbon-14

Radioactive dating can also use other radioactive nuclides with longer half-lives to date older events. For example, uranium-238 (which decays in a series of steps into lead-206) can be used for establishing the age of rocks (and the approximate age of the oldest rocks on earth). Since U-238 has a half-life of 4.5 billion years, it takes that amount of time for half of the original U-238 to decay into Pb-206. In a sample of rock that does not contain appreciable amounts of Pb-208, the most abundant isotope of lead, we can assume that lead was not present when the rock was formed. Therefore, by measuring and analyzing the ratio of U-238:Pb-206, we can determine the age of the rock. This assumes that all of the lead-206 present came from the decay of uranium-238. If there is additional lead-206 present, which is indicated by the presence of other lead isotopes in the sample, it is necessary to make an adjustment. Potassium-argon dating uses a similar method. K-40 decays by positron emission and electron capture to form Ar-40 with a half-life of 1.25 billion years. If a rock sample is crushed and the amount of Ar-40 gas that escapes is measured, determination of the Ar-40:K-40 ratio yields the age of the rock. Other methods, such as rubidium-strontium dating (Rb-87 decays into Sr-87 with a half-life of 48.8 billion years), operate on the same principle. To estimate the lower limit for the earth's age, scientists determine the age of various rocks and minerals, making the assumption that the earth is older than the oldest rocks and minerals in its crust. As of 2014, the oldest known rocks on earth are the Jack Hills zircons from Australia, found by uranium-lead dating to be almost 4.4 billion years old.

Example 21.7

Radioactive Dating of Rocks

An igneous rock contains 9.58×10^{-5} g of U-238 and 2.51×10^{-5} g of Pb-206, and much, much smaller amounts of Pb-208. Determine the approximate time at which the rock formed.

Solution

The sample of rock contains very little Pb-208, the most common isotope of lead, so we can safely assume that all the Pb-206 in the rock was produced by the radioactive decay of U-238. When the rock formed, it contained all of the U-238 currently in it, plus some U-238 that has since undergone radioactive decay.

The amount of U-238 currently in the rock is:

$$9.58 \times 10^{-5} \; \cancel{g \, U} \times \left(\frac{1 \; \text{mol U}}{238 \; \cancel{g \, U}} \right) = 4.03 \times 10^{-7} \; \text{mol U}$$

Because when one mole of U-238 decays, it produces one mole of Pb-206, the amount of U-238 that has undergone radioactive decay since the rock was formed is:

$$2.51 \times 10^{-5} \; \cancel{g \, Pb} \times \left(\frac{1 \; \cancel{\text{mol Pb}}}{206 \; \cancel{g \, Pb}} \right) \times \left(\frac{1 \; \text{mol U}}{1 \; \cancel{\text{mol Pb}}} \right) = 1.22 \times 10^{-7} \; \text{mol U}$$

The total amount of U-238 originally present in the rock is therefore:

$$4.03 \times 10^{-7} \; \text{mol} + 1.22 \times 10^{-7} \; \text{mol} = 5.25 \times 10^{-7} \; \text{mol U}$$

The amount of time that has passed since the formation of the rock is given by:

$$t = -\frac{1}{\lambda} \ln \left(\frac{N_t}{N_0} \right)$$

with N_0 representing the original amount of U-238 and N_t representing the present amount of U-238.

U-238 decays into Pb-206 with a half-life of 4.5×10^9 y, so the decay constant λ is:

$$\lambda = \frac{\ln 2}{t_{1/2}} = \frac{0.693}{4.5 \times 10^9 \text{ y}} = 1.54 \times 10^{-10} \text{ y}^{-1}$$

Substituting and solving, we have:

$$t = -\frac{1}{1.54 \times 10^{-10} \text{ y}^{-1}} \ln\left(\frac{4.03 \times 10^{-7} \text{ mol U}}{5.25 \times 10^{-7} \text{ mol U}}\right) = 1.7 \times 10^9 \text{ y}$$

Therefore, the rock is approximately 1.7 billion years old.

Check Your Learning

A sample of rock contains 6.14×10^{-4} g of Rb-87 and 3.51×10^{-5} g of Sr-87. Calculate the age of the rock. (The half-life of the β decay of Rb-87 is 4.7×10^{10} y.)

Answer: 3.7×10^9 y

21.4 Transmutation and Nuclear Energy

By the end of this section, you will be able to:

- Describe the synthesis of transuranium nuclides
- Explain nuclear fission and fusion processes
- Relate the concepts of critical mass and nuclear chain reactions
- Summarize basic requirements for nuclear fission and fusion reactors

After the discovery of radioactivity, the field of nuclear chemistry was created and developed rapidly during the early twentieth century. A slew of new discoveries in the 1930s and 1940s, along with World War II, combined to usher in the Nuclear Age in the mid-twentieth century. Science learned how to create new substances, and certain isotopes of certain elements were found to possess the capacity to produce unprecedented amounts of energy, with the potential to cause tremendous damage during war, as well as produce enormous amounts of power for society's needs during peace.

Synthesis of Nuclides

Nuclear transmutation is the conversion of one nuclide into another. It can occur by the radioactive decay of a nucleus, or the reaction of a nucleus with another particle. The first manmade nucleus was produced in Ernest Rutherford's laboratory in 1919 by a **transmutation** reaction, the bombardment of one type of nuclei with other nuclei or with neutrons. Rutherford bombarded nitrogen atoms with high-speed α particles from a natural radioactive isotope of radium and observed protons resulting from the reaction:

$$^{14}_{7}\text{N} + ^{4}_{2}\text{He} \longrightarrow ^{17}_{8}\text{O} + ^{1}_{1}\text{H}$$

The $^{17}_{8}\text{O}$ and $^{1}_{1}\text{H}$ nuclei that are produced are stable, so no further (nuclear) changes occur.

To reach the kinetic energies necessary to produce transmutation reactions, devices called **particle accelerators** are used. These devices use magnetic and electric fields to increase the speeds of nuclear particles. In all accelerators, the particles move in a vacuum to avoid collisions with gas molecules. When neutrons are required for transmutation reactions, they are usually obtained from radioactive decay reactions or from various nuclear reactions occurring in nuclear reactors. The Chemistry in Everyday Life feature that follows discusses a famous particle accelerator that made worldwide news.

Chemistry in Everyday Life

CERN Particle Accelerator

Located near Geneva, the CERN ("Conseil Européen pour la Recherche Nucléaire," or European Council for Nuclear Research) Laboratory is the world's premier center for the investigations of the fundamental particles that make up matter. It contains the 27-kilometer (17 mile) long, circular Large Hadron Collider (LHC), the largest particle accelerator in the world (Figure 21.13). In the LHC, particles are boosted to high energies and are then made to collide with each other or with stationary targets at nearly the speed of light. Superconducting electromagnets are used to produce a strong magnetic field that guides the particles around the ring. Specialized, purpose-built detectors observe and record the results of these collisions, which are then analyzed by CERN scientists using powerful computers.

Figure 21.13   A small section of the LHC is shown with workers traveling along it. (credit: Christophe Delaere)

In 2012, CERN announced that experiments at the LHC showed the first observations of the Higgs boson, an elementary particle that helps explain the origin of mass in fundamental particles. This long-anticipated discovery made worldwide news and resulted in the awarding of the 2103 Nobel Prize in Physics to François Englert and Peter Higgs, who had predicted the existence of this particle almost 50 years previously.

Link to Learning

Famous physicist Brian Cox talks about his work on the Large Hadron Collider at CERN, providing an entertaining and engaging tour (http://openstaxcollege.org/l/16tedCERN) of this massive project and the physics behind it.

View a short video (http://openstaxcollege.org/l/16CERNvideo) from CERN, describing the basics of how its particle accelerators work.

Prior to 1940, the heaviest-known element was uranium, whose atomic number is 92. Now, many artificial elements have been synthesized and isolated, including several on such a large scale that they have had a profound effect on society. One of these—element 93, neptunium (Np)—was first made in 1940 by McMillan and Abelson by bombarding uranium-238 with neutrons. The reaction creates unstable uranium-239, with a half-life of 23.5 minutes, which then decays into neptunium-239. Neptunium-239 is also radioactive, with a half-life of 2.36 days, and it decays into plutonium-239. The nuclear reactions are:

$$^{238}_{92}U + ^{1}_{0}n \longrightarrow ^{239}_{92}U$$

$$^{239}_{92}U \longrightarrow ^{239}_{93}Np + ^{0}_{-1}e \quad t_{1/2} \qquad \text{half-life} = 23.5 \text{ min}$$

$$^{239}_{93}Np \longrightarrow ^{239}_{94}Pu + ^{0}_{-1}e \quad t_{1/2} \qquad \text{half-life} = 2.36 \text{ days}$$

Plutonium is now mostly formed in nuclear reactors as a byproduct during the decay of uranium. Some of the neutrons that are released during U-235 decay combine with U-238 nuclei to form uranium-239; this undergoes β decay to form neptunium-239, which in turn undergoes β decay to form plutonium-239 as illustrated in the preceding three equations. It is possible to summarize these equations as:

$$^{238}_{92}U + ^{1}_{0}n \longrightarrow ^{239}_{92}U \xrightarrow{\beta^-} ^{239}_{93}Np \xrightarrow{\beta^-} ^{239}_{94}Pu$$

Heavier isotopes of plutonium—Pu-240, Pu-241, and Pu-242—are also produced when lighter plutonium nuclei capture neutrons. Some of this highly radioactive plutonium is used to produce military weapons, and the rest presents a serious storage problem because they have half-lives from thousands to hundreds of thousands of years.

Although they have not been prepared in the same quantity as plutonium, many other synthetic nuclei have been produced. Nuclear medicine has developed from the ability to convert atoms of one type into other types of atoms. Radioactive isotopes of several dozen elements are currently used for medical applications. The radiation produced by their decay is used to image or treat various organs or portions of the body, among other uses.

The elements beyond element 92 (uranium) are called **transuranium elements**. As of this writing, 22 transuranium elements have been produced and officially recognized by IUPAC; several other elements have formation claims that are waiting for approval. Some of these elements are shown in Table 21.3.

Preparation of Some of the Transuranium Elements

Name	Symbol	Atomic Number	Reaction
americium	Am	95	$^{239}_{94}Pu + ^{1}_{0}n \longrightarrow ^{240}_{95}Am + ^{0}_{-1}e$
curium	Cm	96	$^{239}_{94}Pu + ^{4}_{2}He \longrightarrow ^{242}_{96}Cm + ^{1}_{0}n$
californium	Cf	98	$^{242}_{96}Cm + ^{4}_{2}He \longrightarrow ^{243}_{97}Bk + 2^{1}_{0}n$
einsteinium	Es	99	$^{238}_{92}U + 15^{1}_{0}n \longrightarrow ^{253}_{99}Es + 7^{0}_{-1}e$
mendelevium	Md	101	$^{253}_{99}Es + ^{4}_{2}He \longrightarrow ^{256}_{101}Md + ^{1}_{0}n$
nobelium	No	102	$^{246}_{96}Cm + ^{12}_{6}C \longrightarrow ^{254}_{102}No + 4^{1}_{0}n$
rutherfordium	Rf	104	$^{249}_{98}Cf + ^{12}_{6}C \longrightarrow ^{257}_{104}Rf + 4^{1}_{0}n$
seaborgium	Sg	106	$^{206}_{82}Pb + ^{54}_{24}Cr \longrightarrow ^{257}_{106}Sg + 3^{1}_{0}n$ $^{249}_{98}Cf + ^{18}_{8}O \longrightarrow ^{263}_{106}Sg + 4^{1}_{0}n$
meitnerium	Mt	107	$^{209}_{83}Bi + ^{58}_{26}Fe \longrightarrow ^{266}_{109}Mt + ^{1}_{0}n$

Table 21.3

Nuclear Fission

Many heavier elements with smaller binding energies per nucleon can decompose into more stable elements that have intermediate mass numbers and larger binding energies per nucleon—that is, mass numbers and binding energies per nucleon that are closer to the "peak" of the binding energy graph near 56 (see Figure 21.3). Sometimes neutrons are also produced. This decomposition is called **fission**, the breaking of a large nucleus into smaller pieces. The breaking is rather random with the formation of a large number of different products. Fission usually does not occur naturally, but is induced by bombardment with neutrons. The first reported nuclear fission occurred in 1939 when three German scientists, Lise Meitner, Otto Hahn, and Fritz Strassman, bombarded uranium-235 atoms with slow-moving neutrons that split the U-238 nuclei into smaller fragments that consisted of several neutrons and elements near the middle of the periodic table. Since then, fission has been observed in many other isotopes, including most actinide isotopes that have an odd number of neutrons. A typical nuclear fission reaction is shown in Figure 21.14.

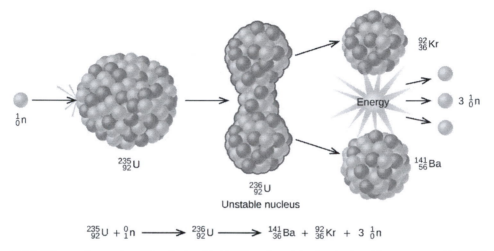

$$^{235}_{92}U + ^{0}_{1}n \longrightarrow ^{236}_{92}U \longrightarrow ^{141}_{36}Ba + ^{92}_{36}Kr + 3\ ^{1}_{0}n$$

Figure 21.14   When a slow neutron hits a fissionable U-235 nucleus, it is absorbed and forms an unstable U-236 nucleus. The U-236 nucleus then rapidly breaks apart into two smaller nuclei (in this case, Ba-141 and Kr-92) along with several neutrons (usually two or three), and releases a very large amount of energy.

Among the products of Meitner, Hahn, and Strassman's fission reaction were barium, krypton, lanthanum, and cerium, all of which have nuclei that are more stable than uranium-235. Since then, hundreds of different isotopes have been observed among the products of fissionable substances. A few of the many reactions that occur for U-235, and a graph showing the distribution of its fission products and their yields, are shown in Figure 21.15. Similar fission reactions have been observed with other uranium isotopes, as well as with a variety of other isotopes such as those of plutonium.

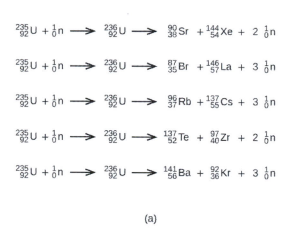

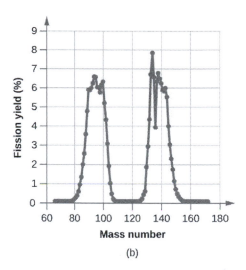

$$^{235}_{92}U + ^{1}_{0}n \longrightarrow ^{236}_{92}U \longrightarrow ^{90}_{38}Sr + ^{144}_{54}Xe + 2\,^{1}_{0}n$$

$$^{235}_{92}U + ^{1}_{0}n \longrightarrow ^{236}_{92}U \longrightarrow ^{87}_{35}Br + ^{146}_{57}La + 3\,^{1}_{0}n$$

$$^{235}_{92}U + ^{1}_{0}n \longrightarrow ^{236}_{92}U \longrightarrow ^{96}_{37}Rb + ^{137}_{55}Cs + 3\,^{1}_{0}n$$

$$^{235}_{92}U + ^{1}_{0}n \longrightarrow ^{236}_{92}U \longrightarrow ^{137}_{52}Te + ^{97}_{40}Zr + 2\,^{1}_{0}n$$

$$^{235}_{92}U + ^{1}_{0}n \longrightarrow ^{236}_{92}U \longrightarrow ^{141}_{56}Ba + ^{92}_{36}Kr + 3\,^{1}_{0}n$$

(a)

(b)

Figure 21.15 (a) Nuclear fission of U-235 produces a range of fission products. (b) The larger fission products of U-235 are typically one isotope with a mass number around 85–105, and another isotope with a mass number that is about 50% larger, that is, about 130–150.

Link to Learning

View this link (http://openstaxcollege.org/l/16fission) to see a simulation of nuclear fission.

A tremendous amount of energy is produced by the fission of heavy elements. For instance, when one mole of U-235 undergoes fission, the products weigh about 0.2 grams less than the reactants; this "lost" mass is converted into a very large amount of energy, about 1.8×10^{10} kJ per mole of U-235. Nuclear fission reactions produce incredibly large amounts of energy compared to chemical reactions. The fission of 1 kilogram of uranium-235, for example, produces about 2.5 million times as much energy as is produced by burning 1 kilogram of coal.

As described earlier, when undergoing fission U-235 produces two "medium-sized" nuclei, and two or three neutrons. These neutrons may then cause the fission of other uranium-235 atoms, which in turn provide more neutrons that can cause fission of even more nuclei, and so on. If this occurs, we have a nuclear **chain reaction** (see Figure 21.16). On the other hand, if too many neutrons escape the bulk material without interacting with a nucleus, then no chain reaction will occur.

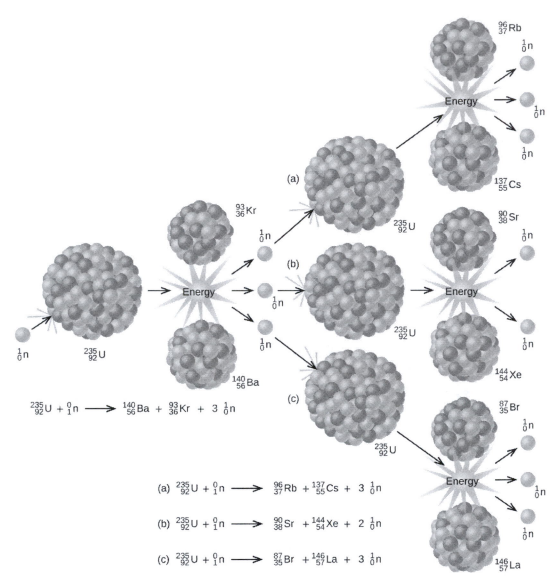

$$^{235}_{92}U + ^{0}_{1}n \longrightarrow ^{140}_{56}Ba + ^{93}_{36}Kr + 3\,^{1}_{0}n$$

(a) $^{235}_{92}U + ^{0}_{1}n \longrightarrow ^{96}_{37}Rb + ^{137}_{55}Cs + 3\,^{1}_{0}n$

(b) $^{235}_{92}U + ^{0}_{1}n \longrightarrow ^{90}_{38}Sr + ^{144}_{54}Xe + 2\,^{1}_{0}n$

(c) $^{235}_{92}U + ^{0}_{1}n \longrightarrow ^{87}_{35}Br + ^{146}_{57}La + 3\,^{1}_{0}n$

Figure 21.16 The fission of a large nucleus, such as U-235, produces two or three neutrons, each of which is capable of causing fission of another nucleus by the reactions shown. If this process continues, a nuclear chain reaction occurs.

Material that can sustain a nuclear fission chain reaction is said to be **fissile** or **fissionable**. (Technically, fissile material can undergo fission with neutrons of any energy, whereas fissionable material requires high-energy neutrons.) Nuclear fission becomes self-sustaining when the number of neutrons produced by fission equals or exceeds the number of neutrons absorbed by splitting nuclei plus the number that escape into the surroundings. The amount of a fissionable material that will support a self-sustaining chain reaction is a **critical mass**. An amount of fissionable material that cannot sustain a chain reaction is a **subcritical mass**. An amount of material in which there is an increasing rate of fission is known as a **supercritical mass**. The critical mass depends on the type of material: its purity, the temperature, the shape of the sample, and how the neutron reactions are controlled (Figure 21.17).

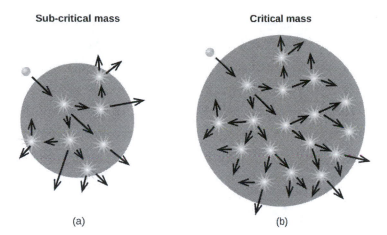

Sub-critical mass   **Critical mass**

(a)   (b)

Figure 21.17   (a) In a subcritical mass, the fissile material is too small and allows too many neutrons to escape the material, so a chain reaction does not occur. (b) In a critical mass, a large enough number of neutrons in the fissile material induce fission to create a chain reaction.

An atomic bomb (Figure 21.18) contains several pounds of fissionable material, $^{235}_{92}U$ or $^{239}_{94}Pu$, a source of neutrons, and an explosive device for compressing it quickly into a small volume. When fissionable material is in small pieces, the proportion of neutrons that escape through the relatively large surface area is great, and a chain reaction does not take place. When the small pieces of fissionable material are brought together quickly to form a body with a mass larger than the critical mass, the relative number of escaping neutrons decreases, and a chain reaction and explosion result.

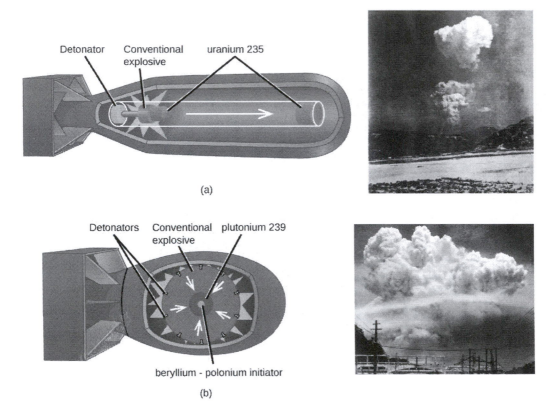

Figure 21.18 (a) The nuclear fission bomb that destroyed Hiroshima on August 6, 1945, consisted of two subcritical masses of U-235, where conventional explosives were used to fire one of the subcritical masses into the other, creating the critical mass for the nuclear explosion. (b) The plutonium bomb that destroyed Nagasaki on August 12, 1945, consisted of a hollow sphere of plutonium that was rapidly compressed by conventional explosives. This led to a concentration of plutonium in the center that was greater than the critical mass necessary for the nuclear explosion.

Fission Reactors

Chain reactions of fissionable materials can be controlled and sustained without an explosion in a **nuclear reactor** (Figure 21.19). Any nuclear reactor that produces power via the fission of uranium or plutonium by bombardment with neutrons must have at least five components: nuclear fuel consisting of fissionable material, a nuclear moderator, reactor coolant, control rods, and a shield and containment system. We will discuss these components in greater detail later in the section. The reactor works by separating the fissionable nuclear material such that a critical mass cannot be formed, controlling both the flux and absorption of neutrons to allow shutting down the fission reactions. In a nuclear reactor used for the production of electricity, the energy released by fission reactions is trapped as thermal energy and used to boil water and produce steam. The steam is used to turn a turbine, which powers a generator for the production of electricity.

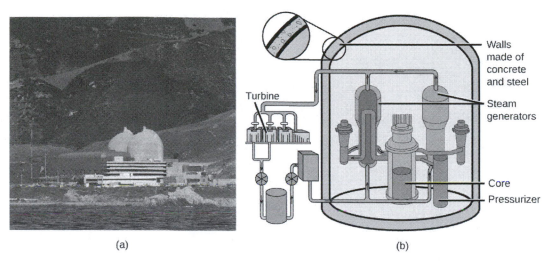

Figure 21.19   (a) The Diablo Canyon Nuclear Power Plant near San Luis Obispo is the only nuclear power plant currently in operation in California. The domes are the containment structures for the nuclear reactors, and the brown building houses the turbine where electricity is generated. Ocean water is used for cooling. (b) The Diablo Canyon uses a pressurized water reactor, one of a few different fission reactor designs in use around the world, to produce electricity. Energy from the nuclear fission reactions in the core heats water in a closed, pressurized system. Heat from this system produces steam that drives a turbine, which in turn produces electricity. (credit a: modification of work by "Mike" Michael L. Baird; credit b: modification of work by the Nuclear Regulatory Commission)

Nuclear Fuels

Nuclear fuel consists of a fissionable isotope, such as uranium-235, which must be present in sufficient quantity to provide a self-sustaining chain reaction. In the United States, uranium ores contain from 0.05–0.3% of the uranium oxide U_3O_8; the uranium in the ore is about 99.3% nonfissionable U-238 with only 0.7% fissionable U-235. Nuclear reactors require a fuel with a higher concentration of U-235 than is found in nature; it is normally enriched to have about 5% of uranium mass as U-235. At this concentration, it is not possible to achieve the supercritical mass necessary for a nuclear explosion. Uranium can be enriched by gaseous diffusion (the only method currently used in the US), using a gas centrifuge, or by laser separation.

In the gaseous diffusion enrichment plant where U-235 fuel is prepared, UF_6 (uranium hexafluoride) gas at low pressure moves through barriers that have holes just barely large enough for UF_6 to pass through. The slightly lighter $^{235}UF_6$ molecules diffuse through the barrier slightly faster than the heavier $^{238}UF_6$ molecules. This process is repeated through hundreds of barriers, gradually increasing the concentration of $^{235}UF_6$ to the level needed by the nuclear reactor. The basis for this process, Graham's law, is described in the chapter on gases. The enriched UF_6 gas is collected, cooled until it solidifies, and then taken to a fabrication facility where it is made into fuel assemblies. Each fuel assembly consists of fuel rods that contain many thimble-sized, ceramic-encased, enriched uranium (usually UO_2) fuel pellets. Modern nuclear reactors may contain as many as 10 million fuel pellets. The amount of energy in each of these pellets is equal to that in almost a ton of coal or 150 gallons of oil.

Nuclear Moderators

Neutrons produced by nuclear reactions move too fast to cause fission (refer back to Figure 21.17). They must first be slowed to be absorbed by the fuel and produce additional nuclear reactions. A **nuclear moderator** is a substance that slows the neutrons to a speed that is low enough to cause fission. Early reactors used high-purity graphite as a moderator. Modern reactors in the US exclusively use heavy water $(_1^2 H_2 O)$ or light water (ordinary H_2O), whereas some reactors in other countries use other materials, such as carbon dioxide, beryllium, or graphite.

Reactor Coolants

A nuclear **reactor coolant** is used to carry the heat produced by the fission reaction to an external boiler and turbine, where it is transformed into electricity. Two overlapping coolant loops are often used; this counteracts the transfer of radioactivity from the reactor to the primary coolant loop. All nuclear power plants in the US use water as a coolant. Other coolants include molten sodium, lead, a lead-bismuth mixture, or molten salts.

Control Rods

Nuclear reactors use **control rods** (Figure 21.20) to control the fission rate of the nuclear fuel by adjusting the number of slow neutrons present to keep the rate of the chain reaction at a safe level. Control rods are made of boron, cadmium, hafnium, or other elements that are able to absorb neutrons. Boron-10, for example, absorbs neutrons by a reaction that produces lithium-7 and alpha particles:

$$_5^{10} B + _0^1 n \longrightarrow _3^7 Li + _2^4 He$$

When control rod assemblies are inserted into the fuel element in the reactor core, they absorb a larger fraction of the slow neutrons, thereby slowing the rate of the fission reaction and decreasing the power produced. Conversely, if the control rods are removed, fewer neutrons are absorbed, and the fission rate and energy production increase. In an emergency, the chain reaction can be shut down by fully inserting all of the control rods into the nuclear core between the fuel rods.

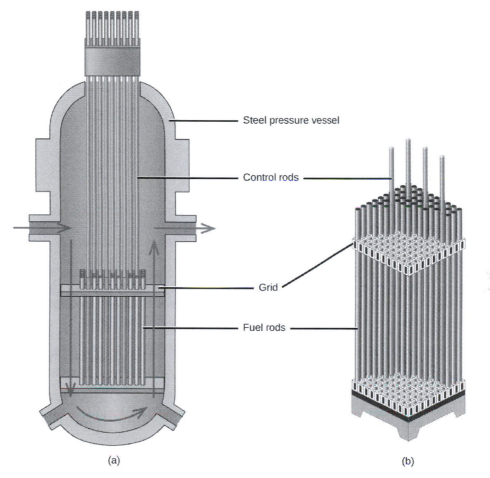

Figure 21.20 The nuclear reactor core shown in (a) contains the fuel and control rod assembly shown in (b). (credit: modification of work by E. Generalic, http://glossary.periodni.com/glossary.php?en=control+rod)

Shield and Containment System

During its operation, a nuclear reactor produces neutrons and other radiation. Even when shut down, the decay products are radioactive. In addition, an operating reactor is thermally very hot, and high pressures result from the circulation of water or another coolant through it. Thus, a reactor must withstand high temperatures and pressures, and must protect operating personnel from the radiation. Reactors are equipped with a **containment system** (or shield) that consists of three parts:

1. The reactor vessel, a steel shell that is 3–20-centimeters thick and, with the moderator, absorbs much of the radiation produced by the reactor

2. A main shield of 1–3 meters of high-density concrete

3. A personnel shield of lighter materials that protects operators from γ rays and X-rays

In addition, reactors are often covered with a steel or concrete dome that is designed to contain any radioactive materials might be released by a reactor accident.

Link to Learning

Click here to watch a 3-minute video (http://openstaxcollege.org/l/
16nucreactors) from the Nuclear Energy Institute on how nuclear reactors work.

Nuclear power plants are designed in such a way that they cannot form a supercritical mass of fissionable material and therefore cannot create a nuclear explosion. But as history has shown, failures of systems and safeguards can cause catastrophic accidents, including chemical explosions and nuclear meltdowns (damage to the reactor core from overheating). The following Chemistry in Everyday Life feature explores three infamous meltdown incidents.

Chemistry in Everyday Life

Nuclear Accidents

The importance of cooling and containment are amply illustrated by three major accidents that occurred with the nuclear reactors at nuclear power generating stations in the United States (Three Mile Island), the former Soviet Union (Chernobyl), and Japan (Fukushima).

In March 1979, the cooling system of the Unit 2 reactor at Three Mile Island Nuclear Generating Station in Pennsylvania failed, and the cooling water spilled from the reactor onto the floor of the containment building. After the pumps stopped, the reactors overheated due to the high radioactive decay heat produced in the first few days after the nuclear reactor shut down. The temperature of the core climbed to at least 2200 °C, and the upper portion of the core began to melt. In addition, the zirconium alloy cladding of the fuel rods began to react with steam and produced hydrogen:

$$Zr(s) + 2H_2O(g) \longrightarrow ZrO_2(s) + 2H_2(g)$$

The hydrogen accumulated in the confinement building, and it was feared that there was danger of an explosion of the mixture of hydrogen and air in the building. Consequently, hydrogen gas and radioactive gases (primarily krypton and xenon) were vented from the building. Within a week, cooling water circulation was restored and the core began to cool. The plant was closed for nearly 10 years during the cleanup process.

Although zero discharge of radioactive material is desirable, the discharge of radioactive krypton and xenon, such as occurred at the Three Mile Island plant, is among the most tolerable. These gases readily disperse in the atmosphere and thus do not produce highly radioactive areas. Moreover, they are noble gases and are not incorporated into plant and animal matter in the food chain. Effectively none of the heavy elements of the core of the reactor were released into the environment, and no cleanup of the area outside of the containment building was necessary (Figure 21.21).

(a)   (b)

Figure 21.21 (a) In this 2010 photo of Three Mile Island, the remaining structures from the damaged Unit 2 reactor are seen on the left, whereas the separate Unit 1 reactor, unaffected by the accident, continues generating power to this day (right). (b) President Jimmy Carter visited the Unit 2 control room a few days after the accident in 1979.

Another major nuclear accident involving a reactor occurred in April 1986, at the Chernobyl Nuclear Power Plant in Ukraine, which was still a part of the former Soviet Union. While operating at low power during an unauthorized experiment with some of its safety devices shut off, one of the reactors at the plant became unstable. Its chain reaction became uncontrollable and increased to a level far beyond what the reactor was designed for. The steam pressure in the reactor rose to between 100 and 500 times the full power pressure and ruptured the reactor. Because the reactor was not enclosed in a containment building, a large amount of radioactive material spewed out, and additional fission products were released, as the graphite (carbon) moderator of the core ignited and burned. The fire was controlled, but over 200 plant workers and firefighters developed acute radiation sickness and at least 32 soon died from the effects of the radiation. It is predicted that about 4000 more deaths will occur among emergency workers and former Chernobyl residents from radiation-induced cancer and leukemia. The reactor has since been encapsulated in steel and concrete, a now-decaying structure known as the sarcophagus. Almost 30 years later, significant radiation problems still persist in the area, and Chernobyl largely remains a wasteland.

In 2011, the Fukushima Daiichi Nuclear Power Plant in Japan was badly damaged by a 9.0-magnitude earthquake and resulting tsunami. Three reactors up and running at the time were shut down automatically, and emergency generators came online to power electronics and coolant systems. However, the tsunami quickly flooded the emergency generators and cut power to the pumps that circulated coolant water through the reactors. High-temperature steam in the reactors reacted with zirconium alloy to produce hydrogen gas. The gas escaped into the containment building, and the mixture of hydrogen and air exploded. Radioactive material was released from the containment vessels as the result of deliberate venting to reduce the hydrogen pressure, deliberate discharge of coolant water into the sea, and accidental or uncontrolled events.

An evacuation zone around the damaged plant extended over 12.4 miles away, and an estimated 200,000 people were evacuated from the area. All 48 of Japan's nuclear power plants were subsequently shut down, remaining shuttered as of December 2014. Since the disaster, public opinion has shifted from largely favoring to largely opposing increasing the use of nuclear power plants, and a restart of Japan's atomic energy program is still stalled (Figure 21.22).

 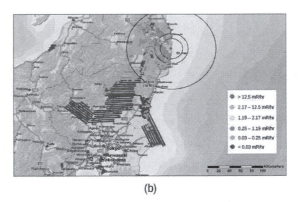

(a)   (b)

Figure 21.22 (a) After the accident, contaminated waste had to be removed, and (b) an evacuation zone was set up around the plant in areas that received heavy doses of radioactive fallout. (credit a: modification of work by "Live Action Hero"/Flickr)

The energy produced by a reactor fueled with enriched uranium results from the fission of uranium as well as from the fission of plutonium produced as the reactor operates. As discussed previously, the plutonium forms from the combination of neutrons and the uranium in the fuel. In any nuclear reactor, only about 0.1% of the mass of the fuel is converted into energy. The other 99.9% remains in the fuel rods as fission products and unused fuel. All of the fission products absorb neutrons, and after a period of several months to a few years, depending on the reactor, the fission products must be removed by changing the fuel rods. Otherwise, the concentration of these fission products would increase and absorb more neutrons until the reactor could no longer operate.

Spent fuel rods contain a variety of products, consisting of unstable nuclei ranging in atomic number from 25 to 60, some transuranium elements, including plutonium and americium, and unreacted uranium isotopes. The unstable nuclei and the transuranium isotopes give the spent fuel a dangerously high level of radioactivity. The long-lived isotopes require thousands of years to decay to a safe level. The ultimate fate of the nuclear reactor as a significant source of energy in the United States probably rests on whether or not a politically and scientifically satisfactory technique for processing and storing the components of spent fuel rods can be developed.

Link to Learning

Explore the information in this link (http://openstaxcollege.org/l/16wastemgmt) to learn about the approaches to nuclear waste management.

Nuclear Fusion and Fusion Reactors

The process of converting very light nuclei into heavier nuclei is also accompanied by the conversion of mass into large amounts of energy, a process called **fusion**. The principal source of energy in the sun is a net fusion reaction in which four hydrogen nuclei fuse and produce one helium nucleus and two positrons. This is a net reaction of a more complicated series of events:

$$4\,^1_1\text{H} \longrightarrow \,^4_2\text{He} + 2\,^0_{+1}$$

A helium nucleus has a mass that is 0.7% less than that of four hydrogen nuclei; this lost mass is converted into energy during the fusion. This reaction produces about 3.6×10^{11} kJ of energy per mole of $_2^4\text{He}$ produced. This is somewhat larger than the energy produced by the nuclear fission of one mole of U-235 (1.8×10^{10} kJ), and over 3 million times larger than the energy produced by the (chemical) combustion of one mole of octane (5471 kJ).

It has been determined that the nuclei of the heavy isotopes of hydrogen, a deuteron, $_1^2\text{H}$ and a triton, $_1^3\text{H}$, undergo fusion at extremely high temperatures (thermonuclear fusion). They form a helium nucleus and a neutron:

$$_1^2\text{H} + _1^3\text{H} \longrightarrow _2^4\text{He} + 2_0^1\text{n}$$

This change proceeds with a mass loss of 0.0188 amu, corresponding to the release of 1.69×10^9 kilojoules per mole of $_2^4\text{He}$ formed. The very high temperature is necessary to give the nuclei enough kinetic energy to overcome the very strong repulsive forces resulting from the positive charges on their nuclei so they can collide.

Useful fusion reactions require very high temperatures for their initiation—about 15,000,000 K or more. At these temperatures, all molecules dissociate into atoms, and the atoms ionize, forming plasma. These conditions occur in an extremely large number of locations throughout the universe—stars are powered by fusion. Humans have already figured out how to create temperatures high enough to achieve fusion on a large scale in thermonuclear weapons. A thermonuclear weapon such as a hydrogen bomb contains a nuclear fission bomb that, when exploded, gives off enough energy to produce the extremely high temperatures necessary for fusion to occur.

Another much more beneficial way to create fusion reactions is in a **fusion reactor**, a nuclear reactor in which fusion reactions of light nuclei are controlled. Because no solid materials are stable at such high temperatures, mechanical devices cannot contain the plasma in which fusion reactions occur. Two techniques to contain plasma at the density and temperature necessary for a fusion reaction are currently the focus of intensive research efforts: containment by a magnetic field and by the use of focused laser beams (Figure 21.23). A number of large projects are working to attain one of the biggest goals in science: getting hydrogen fuel to ignite and produce more energy than the amount supplied to achieve the extremely high temperatures and pressures that are required for fusion. At the time of this writing, there are no self-sustaining fusion reactors operating in the world, although small-scale controlled fusion reactions have been run for very brief periods.

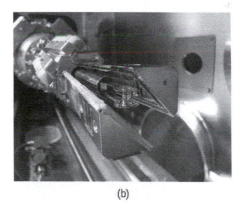

(a)   (b)

Figure 21.23   (a) This model is of the International Thermonuclear Experimental Reactor (ITER) reactor. Currently under construction in the south of France with an expected completion date of 2027, the ITER will be the world's largest experimental Tokamak nuclear fusion reactor with a goal of achieving large-scale sustained energy production. (b) In 2012, the National Ignition Facility at Lawrence Livermore National Laboratory briefly produced over 500,000,000,000 watts (500 terawatts, or 500 TW) of peak power and delivered 1,850,000 joules (1.85 MJ) of energy, the largest laser energy ever produced and 1000 times the power usage of the entire United States in any given moment. Although lasting only a few billionths of a second, the 192 lasers attained the conditions needed for nuclear fusion ignition. This image shows the target prior to the laser shot. (credit a: modification of work by Stephan Mosel)

21.5 Uses of Radioisotopes

By the end of this section, you will be able to:

- List common applications of radioactive isotopes

Radioactive isotopes have the same chemical properties as stable isotopes of the same element, but they emit radiation, which can be detected. If we replace one (or more) atom(s) with radioisotope(s) in a compound, we can track them by monitoring their radioactive emissions. This type of compound is called a **radioactive tracer** (or **radioactive label**). Radioisotopes are used to follow the paths of biochemical reactions or to determine how a substance is distributed within an organism. Radioactive tracers are also used in many medical applications, including both diagnosis and treatment. They are used to measure engine wear, analyze the geological formation around oil wells, and much more.

Radioisotopes have revolutionized medical practice (see Appendix M), where they are used extensively. Over 10 million nuclear medicine procedures and more than 100 million nuclear medicine tests are performed annually in the United States. Four typical examples of radioactive tracers used in medicine are technetium-99 $\left(^{99}_{43}\text{Tc}\right)$, thallium-201 $\left(^{201}_{81}\text{Tl}\right)$, iodine-131 $\left(^{131}_{53}\text{I}\right)$, and sodium-24 $\left(^{24}_{11}\text{Na}\right)$. Damaged tissues in the heart, liver, and lungs absorb certain compounds of technetium-99 preferentially. After it is injected, the location of the technetium compound, and hence the damaged tissue, can be determined by detecting the γ rays emitted by the Tc-99 isotope. Thallium-201 (Figure 21.24) becomes concentrated in healthy heart tissue, so the two isotopes, Tc-99 and Tl-201, are used together to study heart tissue. Iodine-131 concentrates in the thyroid gland, the liver, and some parts of the brain. It can therefore be used to monitor goiter and treat thyroid conditions, such as Grave's disease, as well as liver and brain tumors. Salt solutions containing compounds of sodium-24 are injected into the bloodstream to help locate obstructions to the flow of blood.

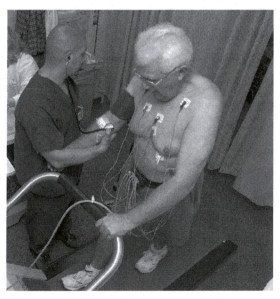

Figure 21.24   Administering thallium-201 to a patient and subsequently performing a stress test offer medical professionals an opportunity to visually analyze heart function and blood flow. (credit: modification of work by "BlueOctane"/Wikimedia Commons)

Radioisotopes used in medicine typically have short half-lives—for example, the ubiquitous Tc-99m has a half-life of 6.01 hours. This makes Tc-99m essentially impossible to store and prohibitively expensive to transport, so it is made on-site instead. Hospitals and other medical facilities use Mo-99 (which is primarily extracted from U-235 fission products) to generate Tc-99. Mo-99 undergoes β decay with a half-life of 66 hours, and the Tc-99 is then chemically extracted (Figure 21.25). The parent nuclide Mo-99 is part of a molybdate ion, $MoO_4{}^{2-}$; when it decays, it forms the pertechnetate ion, $TcO_4{}^-$. These two water-soluble ions are separated by column chromatography, with the higher charge molybdate ion adsorbing onto the alumina in the column, and the lower charge pertechnetate ion passing through the column in the solution. A few micrograms of Mo-99 can produce enough Tc-99 to perform as many as 10,000 tests.

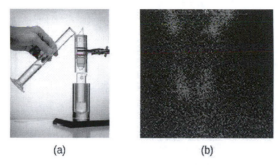

(a)   (b)

Figure 21.25   (a) The first Tc-99m generator (circa 1958) is used to separate Tc-99 from Mo-99. The $MoO_4{}^{2-}$ is retained by the matrix in the column, whereas the $TcO_4{}^-$ passes through and is collected. (b) Tc-99 was used in this scan of the neck of a patient with Grave's disease. The scan shows the location of high concentrations of Tc-99. (credit a: modification of work by the Department of Energy; credit b: modification of work by "MBq"/Wikimedia Commons)

Radioisotopes can also be used, typically in higher doses than as a tracer, as treatment. **Radiation therapy** is the use of high-energy radiation to damage the DNA of cancer cells, which kills them or keeps them from dividing (Figure 21.26). A cancer patient may receive **external beam radiation therapy** delivered by a machine outside the body, or **internal radiation therapy (brachytherapy)** from a radioactive substance that has been introduced into the body. Note that **chemotherapy** is similar to internal radiation therapy in that the cancer treatment is injected into the body, but differs in that chemotherapy uses chemical rather than radioactive substances to kill the cancer cells.

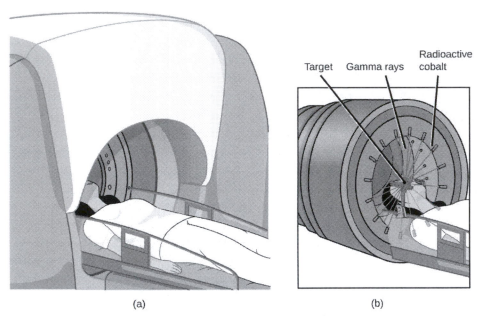

Figure 21.26   The cartoon in (a) shows a cobalt-60 machine used in the treatment of cancer. The diagram in (b) shows how the gantry of the Co-60 machine swings through an arc, focusing radiation on the targeted region (tumor) and minimizing the amount of radiation that passes through nearby regions.

Cobalt-60 is a synthetic radioisotope produced by the neutron activation of Co-59, which then undergoes β decay to form Ni-60, along with the emission of γ radiation. The overall process is:

$$\ce{^{59}_{27}Co} + \ce{^{1}_{0}n} \longrightarrow \ce{^{60}_{27}Co} \longrightarrow \ce{^{60}_{28}Ni} + \ce{^{0}_{-1}\beta} + 2\ce{^{0}_{0}\gamma}$$

The overall decay scheme for this is shown graphically in Figure 21.27.

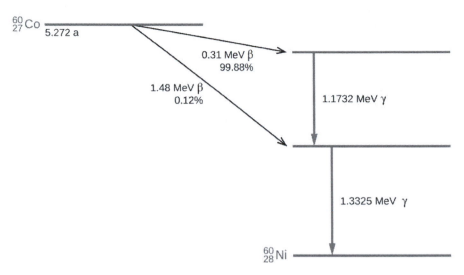

Figure 21.27 Co-60 undergoes a series of radioactive decays. The γ emissions are used for radiation therapy.

Radioisotopes are used in diverse ways to study the mechanisms of chemical reactions in plants and animals. These include labeling fertilizers in studies of nutrient uptake by plants and crop growth, investigations of digestive and milk-producing processes in cows, and studies on the growth and metabolism of animals and plants.

For example, the radioisotope C-14 was used to elucidate the details of how photosynthesis occurs. The overall reaction is:

$$6CO_2(g) + 6H_2O(l) \longrightarrow C_6H_{12}O_6(s) + 6O_2(g),$$

but the process is much more complex, proceeding through a series of steps in which various organic compounds are produced. In studies of the pathway of this reaction, plants were exposed to CO_2 containing a high concentration of $^{14}_{6}C$. At regular intervals, the plants were analyzed to determine which organic compounds contained carbon-14 and how much of each compound was present. From the time sequence in which the compounds appeared and the amount of each present at given time intervals, scientists learned more about the pathway of the reaction.

Commercial applications of radioactive materials are equally diverse (Figure 21.28). They include determining the thickness of films and thin metal sheets by exploiting the penetration power of various types of radiation. Flaws in metals used for structural purposes can be detected using high-energy gamma rays from cobalt-60 in a fashion similar to the way X-rays are used to examine the human body. In one form of pest control, flies are controlled by sterilizing male flies with γ radiation so that females breeding with them do not produce offspring. Many foods are preserved by radiation that kills microorganisms that cause the foods to spoil.

(a)   (b)

Figure 21.28   Common commercial uses of radiation include (a) X-ray examination of luggage at an airport and (b) preservation of food. (credit a: modification of work by the Department of the Navy; credit b: modification of work by the US Department of Agriculture)

Americium-241, an α emitter with a half-life of 458 years, is used in tiny amounts in ionization-type smoke detectors (Figure 21.29). The α emissions from Am-241 ionize the air between two electrode plates in the ionizing chamber. A battery supplies a potential that causes movement of the ions, thus creating a small electric current. When smoke enters the chamber, the movement of the ions is impeded, reducing the conductivity of the air. This causes a marked drop in the current, triggering an alarm.

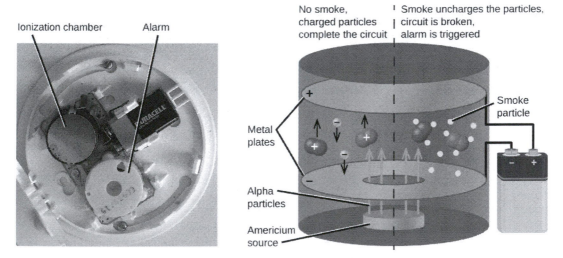

Figure 21.29   Inside a smoke detector, Am-241 emits α particles that ionize the air, creating a small electric current. During a fire, smoke particles impede the flow of ions, reducing the current and triggering an alarm. (credit a: modification of work by "Muffet"/Wikimedia Commons)

21.6 Biological Effects of Radiation

By the end of this section, you will be able to:

- Describe the biological impact of ionizing radiation
- Define units for measuring radiation exposure
- Explain the operation of common tools for detecting radioactivity
- List common sources of radiation exposure in the US

The increased use of radioisotopes has led to increased concerns over the effects of these materials on biological systems (such as humans). All radioactive nuclides emit high-energy particles or electromagnetic waves. When this radiation encounters living cells, it can cause heating, break chemical bonds, or ionize molecules. The most serious biological damage results when these radioactive emissions fragment or ionize molecules. For example, alpha and beta particles emitted from nuclear decay reactions possess much higher energies than ordinary chemical bond energies. When these particles strike and penetrate matter, they produce ions and molecular fragments that are extremely reactive. The damage this does to biomolecules in living organisms can cause serious malfunctions in normal cell processes, taxing the organism's repair mechanisms and possibly causing illness or even death (Figure 21.30).

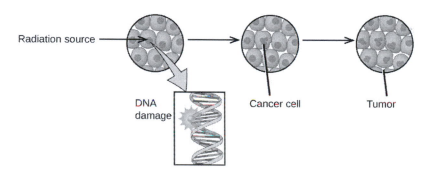

Figure 21.30 Radiation can harm biological systems by damaging the DNA of cells. If this damage is not properly repaired, the cells may divide in an uncontrolled manner and cause cancer.

Ionizing and Nonionizing Radiation

There is a large difference in the magnitude of the biological effects of **nonionizing radiation** (for example, light and microwaves) and **ionizing radiation**, emissions energetic enough to knock electrons out of molecules (for example, α and β particles, γ rays, X-rays, and high-energy ultraviolet radiation) (Figure 21.31).

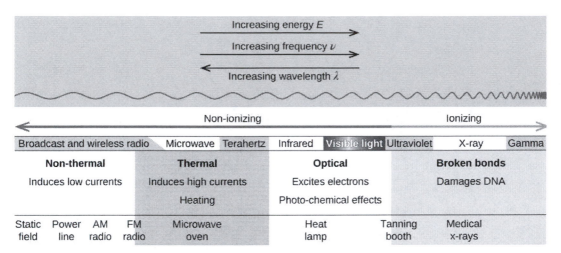

Figure 21.31   Lower frequency, lower-energy electromagnetic radiation is nonionizing, and higher frequency, higher-energy electromagnetic radiation is ionizing.

Energy absorbed from nonionizing radiation speeds up the movement of atoms and molecules, which is equivalent to heating the sample. Although biological systems are sensitive to heat (as we might know from touching a hot stove or spending a day at the beach in the sun), a large amount of nonionizing radiation is necessary before dangerous levels are reached. Ionizing radiation, however, may cause much more severe damage by breaking bonds or removing electrons in biological molecules, disrupting their structure and function. The damage can also be done indirectly, by first ionizing H_2O (the most abundant molecule in living organisms), which forms a H_2O^+ ion that reacts with water, forming a hydronium ion and a hydroxyl radical:

$$H_2O + radiation \xrightarrow{\quad e^- \quad} H_2O^+ + H_2O \longrightarrow H_3O^+ + OH^-$$

Because the hydroxyl radical has an unpaired electron, it is highly reactive. (This is true of any substance with unpaired electrons, known as a free radical.) This hydroxyl radical can react with all kinds of biological molecules (DNA, proteins, enzymes, and so on), causing damage to the molecules and disrupting physiological processes. Examples of direct and indirect damage are shown in Figure 21.32.

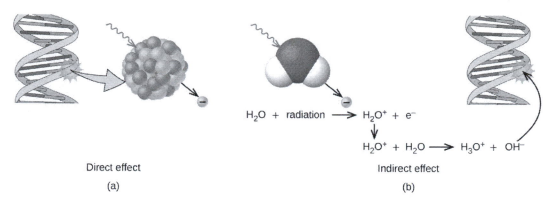

$$H_2O + radiation \longrightarrow H_2O^+ + e^-$$

$$H_2O^+ + H_2O \longrightarrow H_3O^+ + OH^-$$

Direct effect

(a)

Indirect effect

(b)

Figure 21.32 Ionizing radiation can (a) directly damage a biomolecule by ionizing it or breaking its bonds, or (b) create an H_2O^+ ion, which reacts with H_2O to form a hydroxyl radical, which in turn reacts with the biomolecule, causing damage indirectly.

Biological Effects of Exposure to Radiation

Radiation can harm either the whole body (somatic damage) or eggs and sperm (genetic damage). Its effects are more pronounced in cells that reproduce rapidly, such as the stomach lining, hair follicles, bone marrow, and embryos. This is why patients undergoing radiation therapy often feel nauseous or sick to their stomach, lose hair, have bone aches, and so on, and why particular care must be taken when undergoing radiation therapy during pregnancy.

Different types of radiation have differing abilities to pass through material (Figure 21.33). A very thin barrier, such as a sheet or two of paper, or the top layer of skin cells, usually stops alpha particles. Because of this, alpha particle sources are usually not dangerous if outside the body, but are quite hazardous if ingested or inhaled (see the Chemistry in Everyday Life feature on Radon Exposure). Beta particles will pass through a hand, or a thin layer of material like paper or wood, but are stopped by a thin layer of metal. Gamma radiation is very penetrating and can pass through a thick layer of most materials. Some high-energy gamma radiation is able to pass through a few feet of concrete. Certain dense, high atomic number elements (such as lead) can effectively attenuate gamma radiation with thinner material and are used for shielding. The ability of various kinds of emissions to cause ionization varies greatly, and some particles have almost no tendency to produce ionization. Alpha particles have about twice the ionizing power of fast-moving neutrons, about 10 times that of β particles, and about 20 times that of γ rays and X-rays.

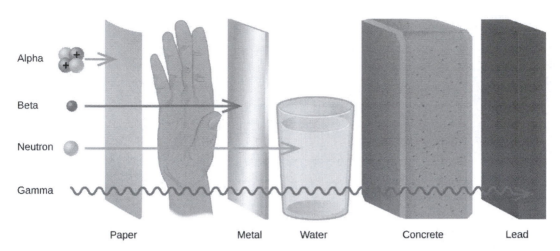

Figure 21.33 The ability of different types of radiation to pass through material is shown. From least to most penetrating, they are alpha < beta < neutron < gamma.

Radon Exposure

For many people, one of the largest sources of exposure to radiation is from radon gas (Rn-222). Radon-222 is an α emitter with a half–life of 3.82 days. It is one of the products of the radioactive decay series of U-238 (Figure 21.9), which is found in trace amounts in soil and rocks. The radon gas that is produced slowly escapes from the ground and gradually seeps into homes and other structures above. Since it is about eight times more dense than air, radon gas accumulates in basements and lower floors, and slowly diffuses throughout buildings (Figure 21.34).

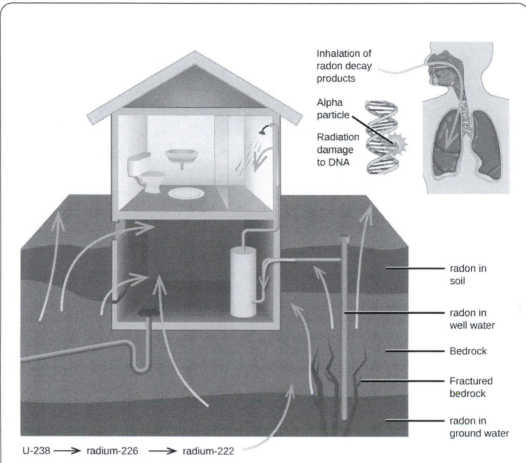

Inhalation of
radon decay
products

Alpha
particle

Radiation
damage
to DNA

radon in
soil

radon in
well water

Bedrock

Fractured
bedrock

radon in
ground water

U-238 ⟶ radium-226 ⟶ radium-222

Figure 21.34 Radon-222 seeps into houses and other buildings from rocks that contain uranium-238, a radon emitter. The radon enters through cracks in concrete foundations and basement floors, stone or porous cinderblock foundations, and openings for water and gas pipes.

Radon is found in buildings across the country, with amounts depending on where you live. The average concentration of radon inside houses in the US (1.25 pCi/L) is about three times the levels found in outside air, and about one in six houses have radon levels high enough that remediation efforts to reduce the radon concentration are recommended. Exposure to radon increases one's risk of getting cancer (especially lung cancer), and high radon levels can be as bad for health as smoking a carton of cigarettes a day. Radon is the number one cause of lung cancer in nonsmokers and the second leading cause of lung cancer overall. Radon exposure is believed to cause over 20,000 deaths in the US per year.

Measuring Radiation Exposure

Several different devices are used to detect and measure radiation, including Geiger counters, scintillation counters (scintillators), and radiation dosimeters (Figure 21.35). Probably the best-known radiation instrument, the **Geiger counter** (also called the Geiger-Müller counter) detects and measures radiation. Radiation causes the ionization of the gas in a Geiger-Müller tube. The rate of ionization is proportional to the amount of radiation. A **scintillation counter** contains a scintillator—a material that emits light (luminesces) when excited by ionizing radiation—and a sensor that converts the light into an electric signal. **Radiation dosimeters** also measure ionizing radiation and are often used to determine personal radiation exposure. Commonly used types are electronic, film badge, thermoluminescent, and quartz fiber dosimeters.

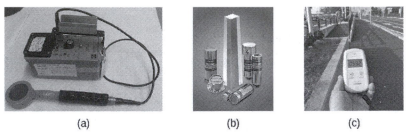

Figure 21.35 Devices such as (a) Geiger counters, (b) scintillators, and (c) dosimeters can be used to measure radiation. (credit c: modification of work by "osaMu"/Wikimedia commons)

A variety of units are used to measure various aspects of radiation (Figure 21.36). The SI unit for rate of radioactive decay is the **becquerel (Bq)**, with 1 Bq = 1 disintegration per second. The **curie (Ci)** and **millicurie (mCi)** are much larger units and are frequently used in medicine (1 curie = 1 Ci = 3.7×10^{10} disintegrations per second). The SI unit for measuring radiation dose is the **gray (Gy)**, with 1 Gy = 1 J of energy absorbed per kilogram of tissue. In medical applications, the **radiation absorbed dose (rad)** is more often used (1 rad = 0.01 Gy; 1 rad results in the absorption of 0.01 J/kg of tissue). The SI unit measuring tissue damage caused by radiation is the **sievert (Sv)**. This takes into account both the energy and the biological effects of the type of radiation involved in the radiation dose. The **roentgen equivalent for man (rem)** is the unit for radiation damage that is used most frequently in medicine (1 rem = 1 Sv). Note that the tissue damage units (rem or Sv) includes the energy of the radiation dose (rad or Gy) along with a biological factor referred to as the **RBE** (for **relative biological effectiveness**) that is an approximate measure of the relative damage done by the radiation. These are related by:

$$\text{number of rems} = \text{RBE} \times \text{number of rads}$$

with RBE approximately 10 for α radiation, 2(+) for protons and neutrons, and 1 for β and γ radiation.

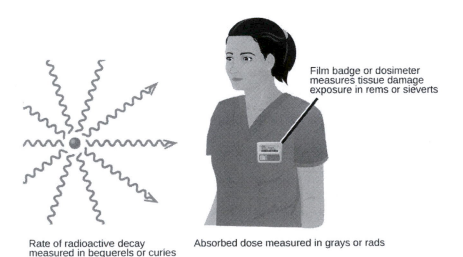

Film badge or dosimeter measures tissue damage exposure in rems or sieverts

Rate of radioactive decay measured in bequerels or curies

Absorbed dose measured in grays or rads

Figure 21.36 Different units are used to measure the rate of emission from a radioactive source, the energy that is absorbed from the source, and the amount of damage the absorbed radiation does.

Units of Radiation Measurement

Table 21.4 summarizes the units used for measuring radiation.

Units Used for Measuring Radiation

Measurement Purpose	Unit	Quantity Measured	Description
activity of source	becquerel (Bq)	radioactive decays or emissions	amount of sample that undergoes 1 decay/second
	curie (Ci)		amount of sample that undergoes 3.7×10^{10} decays/second
absorbed dose	gray (Gy)	energy absorbed per kg of tissue	1 Gy = 1 J/kg tissue
	radiation absorbed dose (rad)		1 rad = 0.01 J/kg tissue
biologically effective dose	sievert (Sv)	tissue damage	Sv = RBE $\times$ Gy
	roentgen equivalent for man (rem)		Rem = RBE $\times$ rad

Table 21.4

Example 21.8

Amount of Radiation

Cobalt-60 ($t_{1/2}$ = 5.26 y) is used in cancer therapy since the γ rays it emits can be focused in small areas where the cancer is located. A 5.00-g sample of Co-60 is available for cancer treatment.

(a) What is its activity in Bq?

(b) What is its activity in Ci?

Solution

The activity is given by:

$$\text{Activity} = \lambda N = \left(\frac{\ln 2}{t_{1/2}}\right)N = \left(\frac{\ln 2}{5.26 \text{ y}}\right) \times 5.00 \text{ g} = 0.659 \frac{\text{g}}{\text{y}} \text{ of Co–60 that decay}$$

And to convert this to decays per second:

$$0.659 \frac{\text{g}}{\text{y}} \times \frac{1 \text{ y}}{365 \text{ d}} \times \frac{1 \text{ d}}{24 \text{ h}} \times \frac{1 \text{ h}}{3600 \text{ s}} \times \frac{1 \text{ mol}}{59.9 \text{ g}} \times \frac{6.02 \times 10^{23} \text{ atoms}}{1 \text{ mol}} \times \frac{1 \text{ decay}}{1 \text{ atom}}$$

$$= 2.10 \times 10^{14} \frac{\text{decay}}{\text{s}}$$

(a) Since 1 Bq = $\frac{1 \text{ decay}}{\text{s}}$, the activity in Becquerel (Bq) is:

$$2.10 \times 10^{14} \frac{\text{decay}}{\text{s}} \times \left(\frac{1 \text{ Bq}}{1 \frac{\text{decay}}{\text{s}}}\right) = 2.10 \times 10^{14} \text{ Bq}$$

(b) Since 1 Ci = $\frac{3.7 \times 10^{11} \text{ decay}}{\text{s}}$, the activity in curie (Ci) is:

$$2.10 \times 10^{14} \frac{\text{decay}}{\text{s}} \times \left(\frac{1 \text{ Ci}}{3.7 \times 10^{11} \frac{\text{decay}}{\text{s}}} \right) = 5.7 \times 10^2 \text{ Ci}$$

Check Your Learning

Tritium is a radioactive isotope of hydrogen ($t_{1/2}$ = 12.32 y) that has several uses, including self-powered lighting, in which electrons emitted in tritium radioactive decay cause phosphorus to glow. Its nucleus contains one proton and two neutrons, and the atomic mass of tritium is 3.016 amu. What is the activity of a sample containing 1.00mg of tritium (a) in Bq and (b) in Ci?

Answer: (a) 3.56 $\times$ 10^{11} Bq; (b) 0.962 Ci

Effects of Long-term Radiation Exposure on the Human Body

The effects of radiation depend on the type, energy, and location of the radiation source, and the length of exposure. As shown in Figure 21.37, the average person is exposed to background radiation, including cosmic rays from the sun and radon from uranium in the ground (see the Chemistry in Everyday Life feature on Radon Exposure); radiation from medical exposure, including CAT scans, radioisotope tests, X-rays, and so on; and small amounts of radiation from other human activities, such as airplane flights (which are bombarded by increased numbers of cosmic rays in the upper atmosphere), radioactivity from consumer products, and a variety of radionuclides that enter our bodies when we breathe (for example, carbon-14) or through the food chain (for example, potassium-40, strontium-90, and iodine-131).

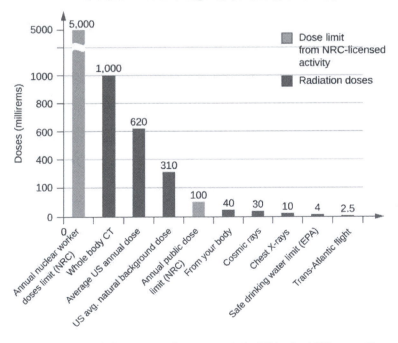

Figure 21.37 The total annual radiation exposure for a person in the US is about 620 mrem. The various sources and their relative amounts are shown in this bar graph. (source: U.S. Nuclear Regulatory Commission)

A short-term, sudden dose of a large amount of radiation can cause a wide range of health effects, from changes in blood chemistry to death. Short-term exposure to tens of rems of radiation will likely cause very noticeable symptoms or illness; a dose of about 500 rems is estimated to have a 50% probability of causing the death of the victim within 30 days of exposure. Exposure to radioactive emissions has a cumulative effect on the body during a person's lifetime, which is another reason why it is important to avoid any unnecessary exposure to radiation. Health effects of short-term exposure to radiation are shown in Table 21.5.

Health Effects of Radiation

Exposure (rem)	Health Effect	Time to Onset (without treatment)
5–10	changes in blood chemistry	—
50	nausea	hours
55	fatigue	—
70	vomiting	—
75	hair loss	2–3 weeks
90	diarrhea	—
100	hemorrhage	—
400	possible death	within 2 months
1000	destruction of intestinal lining	—
	internal bleeding	—
	death	1–2 weeks
2000	damage to central nervous system	—
	loss of consciousness;	minutes
	death	hours to days

Table 21.5

It is impossible to avoid some exposure to ionizing radiation. We are constantly exposed to background radiation from a variety of natural sources, including cosmic radiation, rocks, medical procedures, consumer products, and even our own atoms. We can minimize our exposure by blocking or shielding the radiation, moving farther from the source, and limiting the time of exposure.

2. Source: US Environmental Protection Agency

Key Terms

alpha (α) decay loss of an alpha particle during radioactive decay

alpha particle $\left(\alpha \text{ or } {}^{4}_{2}\text{He} \text{ or } {}^{4}_{2}\alpha\right)$ high-energy helium nucleus; a helium atom that has lost two electrons and contains two protons and two neutrons

antimatter particles with the same mass but opposite properties (such as charge) of ordinary particles

band of stability (also, belt of stability, zone of stability, or valley of stability) region of graph of number of protons versus number of neutrons containing stable (nonradioactive) nuclides

becquerel (Bq) SI unit for rate of radioactive decay; 1 Bq = 1 disintegration/s

beta (β) decay breakdown of a neutron into a proton, which remains in the nucleus, and an electron, which is emitted as a beta particle

beta particle $\left(\beta \text{ or } {}^{0}_{-1}\text{e} \text{ or } {}^{0}_{-1}\beta\right)$ high-energy electron

binding energy per nucleon total binding energy for the nucleus divided by the number of nucleons in the nucleus

chain reaction repeated fission caused when the neutrons released in fission bombard other atoms

chemotherapy similar to internal radiation therapy, but chemical rather than radioactive substances are introduced into the body to kill cancer cells

containment system (also, shield) a three-part structure of materials that protects the exterior of a nuclear fission reactor and operating personnel from the high temperatures, pressures, and radiation levels inside the reactor

control rod material inserted into the fuel assembly that absorbs neutrons and can be raised or lowered to adjust the rate of a fission reaction

critical mass amount of fissionable material that will support a self-sustaining (nuclear fission) chain reaction

curie (Ci) larger unit for rate of radioactive decay frequently used in medicine; $1 \text{ Ci} = 3.7 \times 10^{10}$ disintegrations/s

daughter nuclide nuclide produced by the radioactive decay of another nuclide; may be stable or may decay further

electron capture combination of a core electron with a proton to yield a neutron within the nucleus

electron volt (eV) measurement unit of nuclear binding energies, with 1 eV equaling the amount energy due to the moving an electron across an electric potential difference of 1 volt

external beam radiation therapy radiation delivered by a machine outside the body

fissile (or fissionable) when a material is capable of sustaining a nuclear fission reaction

fission splitting of a heavier nucleus into two or more lighter nuclei, usually accompanied by the conversion of mass into large amounts of energy

fusion combination of very light nuclei into heavier nuclei, accompanied by the conversion of mass into large amounts of energy

fusion reactor nuclear reactor in which fusion reactions of light nuclei are controlled

gamma (γ) emission decay of an excited-state nuclide accompanied by emission of a gamma ray

gamma ray $(\gamma \text{ or } {}^{0}_{0}\gamma)$ short wavelength, high-energy electromagnetic radiation that exhibits wave-particle duality

Geiger counter instrument that detects and measures radiation via the ionization produced in a Geiger-Müller tube

gray (Gy) SI unit for measuring radiation dose; 1 Gy = 1 J absorbed/kg tissue

half-life ($t_{1/2}$) time required for half of the atoms in a radioactive sample to decay

internal radiation therapy (also, brachytherapy) radiation from a radioactive substance introduced into the body to kill cancer cells

ionizing radiation radiation that can cause a molecule to lose an electron and form an ion

magic number nuclei with specific numbers of nucleons that are within the band of stability

mass defect difference between the mass of an atom and the summed mass of its constituent subatomic particles (or the mass "lost" when nucleons are brought together to form a nucleus)

mass-energy equivalence equation Albert Einstein's relationship showing that mass and energy are equivalent

millicurie (mCi) larger unit for rate of radioactive decay frequently used in medicine; 1 Ci = 3.7 $\times$ 10^{10} disintegrations/s

nonionizing radiation radiation that speeds up the movement of atoms and molecules; it is equivalent to heating a sample, but is not energetic enough to cause the ionization of molecules

nuclear binding energy energy lost when an atom's nucleons are bound together (or the energy needed to break a nucleus into its constituent protons and neutrons)

nuclear chemistry study of the structure of atomic nuclei and processes that change nuclear structure

nuclear fuel fissionable isotope present in sufficient quantities to provide a self-sustaining chain reaction in a nuclear reactor

nuclear moderator substance that slows neutrons to a speed low enough to cause fission

nuclear reaction change to a nucleus resulting in changes in the atomic number, mass number, or energy state

nuclear reactor environment that produces energy via nuclear fission in which the chain reaction is controlled and sustained without explosion

nuclear transmutation conversion of one nuclide into another nuclide

nucleon collective term for protons and neutrons in a nucleus

nuclide nucleus of a particular isotope

parent nuclide unstable nuclide that changes spontaneously into another (daughter) nuclide

particle accelerator device that uses electric and magnetic fields to increase the kinetic energy of nuclei used in transmutation reactions

positron $({}^{0}_{+1}\beta \text{ or } {}^{0}_{+1}e)$ antiparticle to the electron; it has identical properties to an electron, except for having the opposite (positive) charge

positron emission (also, β^+ decay) conversion of a proton into a neutron, which remains in the nucleus, and a positron, which is emitted

radiation absorbed dose (rad) SI unit for measuring radiation dose, frequently used in medical applications; 1 rad = 0.01 Gy

radiation dosimeter device that measures ionizing radiation and is used to determine personal radiation exposure

radiation therapy use of high-energy radiation to damage the DNA of cancer cells, which kills them or keeps them from dividing

radioactive decay spontaneous decay of an unstable nuclide into another nuclide

radioactive decay series chains of successive disintegrations (radioactive decays) that ultimately lead to a stable end-product

radioactive tracer (also, radioactive label) radioisotope used to track or follow a substance by monitoring its radioactive emissions

radioactivity phenomenon exhibited by an unstable nucleon that spontaneously undergoes change into a nucleon that is more stable; an unstable nucleon is said to be radioactive

radiocarbon dating highly accurate means of dating objects 30,000–50,000 years old that were derived from once-living matter; achieved by calculating the ratio of $^{14}_{6}C$: $^{12}_{6}C$ in the object vs. the ratio of $^{14}_{6}C$: $^{12}_{6}C$ in the present-day atmosphere

radioisotope isotope that is unstable and undergoes conversion into a different, more stable isotope

radiometric dating use of radioisotopes and their properties to date the formation of objects such as archeological artifacts, formerly living organisms, or geological formations

reactor coolant assembly used to carry the heat produced by fission in a reactor to an external boiler and turbine where it is transformed into electricity

relative biological effectiveness (RBE) measure of the relative damage done by radiation

roentgen equivalent man (rem) unit for radiation damage, frequently used in medicine; 1 rem = 1 Sv

scintillation counter instrument that uses a scintillator—a material that emits light when excited by ionizing radiation—to detect and measure radiation

sievert (Sv) SI unit measuring tissue damage caused by radiation; takes into account energy and biological effects of radiation

strong nuclear force force of attraction between nucleons that holds a nucleus together

subcritical mass amount of fissionable material that cannot sustain a chain reaction; less than a critical mass

supercritical mass amount of material in which there is an increasing rate of fission

transmutation reaction bombardment of one type of nuclei with other nuclei or neutrons

transuranium element element with an atomic number greater than 92; these elements do not occur in nature

Key Equations

- $E = mc^2$
- decay rate $= \lambda N$

- $t_{1/2} = \dfrac{\ln 2}{\lambda} = \dfrac{0.693}{\lambda}$

- rem = RBE $\times$ rad

- Sv = RBE $\times$ Gy

Summary

21.1 Nuclear Structure and Stability

An atomic nucleus consists of protons and neutrons, collectively called nucleons. Although protons repel each other, the nucleus is held tightly together by a short-range, but very strong, force called the strong nuclear force. A nucleus has less mass than the total mass of its constituent nucleons. This "missing" mass is the mass defect, which has been converted into the binding energy that holds the nucleus together according to Einstein's mass-energy equivalence equation, $E = mc^2$. Of the many nuclides that exist, only a small number are stable. Nuclides with even numbers of protons or neutrons, or those with magic numbers of nucleons, are especially likely to be stable. These stable nuclides occupy a narrow band of stability on a graph of number of protons versus number of neutrons. The binding energy per nucleon is largest for the elements with mass numbers near 56; these are the most stable nuclei.

21.2 Nuclear Equations

Nuclei can undergo reactions that change their number of protons, number of neutrons, or energy state. Many different particles can be involved in nuclear reactions. The most common are protons, neutrons, positrons (which are positively charged electrons), alpha (α) particles (which are high-energy helium nuclei), beta (β) particles (which are high-energy electrons), and gamma (γ) rays (which compose high-energy electromagnetic radiation). As with chemical reactions, nuclear reactions are always balanced. When a nuclear reaction occurs, the total mass (number) and the total charge remain unchanged.

21.3 Radioactive Decay

Nuclei that have unstable n:p ratios undergo spontaneous radioactive decay. The most common types of radioactivity are α decay, β decay, γ emission, positron emission, and electron capture. Nuclear reactions also often involve γ rays, and some nuclei decay by electron capture. Each of these modes of decay leads to the formation of a new nucleus with a more stable n:p ratio. Some substances undergo radioactive decay series, proceeding through multiple decays before ending in a stable isotope. All nuclear decay processes follow first-order kinetics, and each radioisotope has its own characteristic half-life, the time that is required for half of its atoms to decay. Because of the large differences in stability among nuclides, there is a very wide range of half-lives of radioactive substances. Many of these substances have found useful applications in medical diagnosis and treatment, determining the age of archaeological and geological objects, and more.

21.4 Transmutation and Nuclear Energy

It is possible to produce new atoms by bombarding other atoms with nuclei or high-speed particles. The products of these transmutation reactions can be stable or radioactive. A number of artificial elements, including technetium, astatine, and the transuranium elements, have been produced in this way.

Nuclear power as well as nuclear weapon detonations can be generated through fission (reactions in which a heavy nucleus is split into two or more lighter nuclei and several neutrons). Because the neutrons may induce additional fission reactions when they combine with other heavy nuclei, a chain reaction can result. Useful power is obtained if the fission process is carried out in a nuclear reactor. The conversion of light nuclei into heavier nuclei (fusion) also produces energy. At present, this energy has not been contained adequately and is too expensive to be feasible for commercial energy production.

21.5 Uses of Radioisotopes

Compounds known as radioactive tracers can be used to follow reactions, track the distribution of a substance, diagnose and treat medical conditions, and much more. Other radioactive substances are helpful for controlling pests, visualizing structures, providing fire warnings, and for many other applications. Hundreds of millions of nuclear medicine tests and procedures, using a wide variety of radioisotopes with relatively short half-lives, are performed every year in the US. Most of these radioisotopes have relatively short half-lives; some are short enough that the radioisotope must be made on-site at medical facilities. Radiation therapy uses high-energy radiation to kill cancer cells by damaging their DNA. The radiation used for this treatment may be delivered externally or internally.

21.6 Biological Effects of Radiation

We are constantly exposed to radiation from a variety of naturally occurring and human-produced sources. This radiation can affect living organisms. Ionizing radiation is the most harmful because it can ionize molecules or break chemical bonds, which damages the molecule and causes malfunctions in cell processes. It can also create reactive hydroxyl radicals that damage biological molecules and disrupt physiological processes. Radiation can cause somatic or genetic damage, and is most harmful to rapidly reproducing cells. Types of radiation differ in their ability to penetrate material and damage tissue, with alpha particles the least penetrating but potentially most damaging and gamma rays the most penetrating.

Various devices, including Geiger counters, scintillators, and dosimeters, are used to detect and measure radiation, and monitor radiation exposure. We use several units to measure radiation: becquerels or curies for rates of radioactive decay; gray or rads for energy absorbed; and rems or sieverts for biological effects of radiation. Exposure to radiation can cause a wide range of health effects, from minor to severe, and including death. We can minimize the effects of radiation by shielding with dense materials such as lead, moving away from the source, and limiting time of exposure.

Exercises

21.1 Nuclear Structure and Stability

1. Write the following isotopes in hyphenated form (e.g., "carbon-14")

(a) $^{24}_{11}\text{Na}$

(b) $^{29}_{13}\text{Al}$

(c) $^{73}_{36}\text{Kr}$

(d) $^{194}_{77}\text{Ir}$

2. Write the following isotopes in nuclide notation (e.g., $"^{14}_{6}\text{C}"$)

(a) oxygen-14

(b) copper-70

(c) tantalum-175

(d) francium-217

3. For the following isotopes that have missing information, fill in the missing information to complete the notation

(a) $^{34}_{14}\text{X}$

(b) $^{36}_{\text{X}}\text{P}$

(c) $^{57}_{\text{X}}\text{Mn}$

(d) $^{121}_{56}\text{X}$

4. For each of the isotopes in Exercise 21.1, determine the numbers of protons, neutrons, and electrons in a neutral atom of the isotope.

5. Write the nuclide notation, including charge if applicable, for atoms with the following characteristics:

(a) 25 protons, 20 neutrons, 24 electrons

(b) 45 protons, 24 neutrons, 43 electrons

(c) 53 protons, 89 neutrons, 54 electrons

(d) 97 protons, 146 neutrons, 97 electrons

6. Calculate the density of the $^{24}_{12}Mg$ nucleus in g/mL, assuming that it has the typical nuclear diameter of 1×10^{-13} cm and is spherical in shape.

7. What are the two principal differences between nuclear reactions and ordinary chemical changes?

8. The mass of the atom $^{23}_{11}Na$ is 22.9898 amu.

(a) Calculate its binding energy per atom in millions of electron volts.

(b) Calculate its binding energy per nucleon.

9. Which of the following nuclei lie within the band of stability shown in Figure 21.2?

(a) chlorine-37

(b) calcium-40

(c) ^{204}Bi

(d) ^{56}Fe

(e) ^{206}Pb

(f) ^{211}Pb

(g) ^{222}Rn

(h) carbon-14

10. Which of the following nuclei lie within the band of stability shown in Figure 21.2?

(a) argon-40

(b) oxygen-16

(c) ^{122}Ba

(d) ^{58}Ni

(e) ^{205}Tl

(f) ^{210}Tl

(g) ^{226}Ra

(h) magnesium-24

21.2 Nuclear Equations

11. Write a brief description or definition of each of the following:

(a) nucleon

(b) α particle

(c) β particle

(d) positron

(e) γ ray

(f) nuclide

(g) mass number

(h) atomic number

12. Which of the various particles (α particles, β particles, and so on) that may be produced in a nuclear reaction are actually nuclei?

13. Complete each of the following equations by adding the missing species:

(a) $^{27}_{13}\text{Al} + ^4_2\text{He} \longrightarrow \text{?} + ^1_0\text{n}$

(b) $^{239}_{94}\text{Pu} + \text{?} \longrightarrow ^{242}_{96}\text{Cm} + ^1_0\text{n}$

(c) $^{14}_7\text{N} + ^4_2\text{He} \longrightarrow \text{?} + ^1_1\text{H}$

(d) $^{235}_{92}\text{U} \longrightarrow \text{?} + ^{135}_{55}\text{Cs} + 4^1_0\text{n}$

14. Complete each of the following equations:

(a) $^7_3\text{Li} + \text{?} \longrightarrow 2^4_2\text{He}$

(b) $^{14}_6\text{C} \longrightarrow ^{14}_7\text{N} + \text{?}$

(c) $^{27}_{13}\text{Al} + ^4_2\text{He} \longrightarrow \text{?} + ^1_0\text{n}$

(d) $^{250}_{96}\text{Cm} \longrightarrow \text{?} + ^{98}_{38}\text{Sr} + 4^1_0\text{n}$

15. Write a balanced equation for each of the following nuclear reactions:

(a) the production of ^{17}O from ^{14}N by α particle bombardment

(b) the production of ^{14}C from ^{14}N by neutron bombardment

(c) the production of ^{233}Th from ^{232}Th by neutron bombardment

(d) the production of ^{239}U from ^{238}U by ^2_1H bombardment

16. Technetium-99 is prepared from ^{98}Mo. Molybdenum-98 combines with a neutron to give molybdenum-99, an unstable isotope that emits a β particle to yield an excited form of technetium-99, represented as $^{99}\text{Tc}^*$. This excited nucleus relaxes to the ground state, represented as ^{99}Tc, by emitting a γ ray. The ground state of ^{99}Tc then emits a β particle. Write the equations for each of these nuclear reactions.

17. The mass of the atom $^{19}_9\text{F}$ is 18.99840 amu.

(a) Calculate its binding energy per atom in millions of electron volts.

(b) Calculate its binding energy per nucleon.

18. For the reaction $^{14}_6\text{C} \longrightarrow ^{14}_7\text{N} + \text{?}$, if 100.0 g of carbon reacts, what volume of nitrogen gas (N_2) is produced at 273K and 1 atm?

21.3 Radioactive Decay

19. What are the types of radiation emitted by the nuclei of radioactive elements?

20. What changes occur to the atomic number and mass of a nucleus during each of the following decay scenarios?

(a) an α particle is emitted

(b) a β particle is emitted

(c) γ radiation is emitted

(d) a positron is emitted

(e) an electron is captured

21. What is the change in the nucleus that results from the following decay scenarios?

(a) emission of a β particle

(b) emission of a β^+ particle

(c) capture of an electron

22. Many nuclides with atomic numbers greater than 83 decay by processes such as electron emission. Explain the observation that the emissions from these unstable nuclides also normally include α particles.

23. Why is electron capture accompanied by the emission of an X-ray?

24. Explain, in terms of Figure 21.2, how unstable heavy nuclides (atomic number > 83) may decompose to form nuclides of greater stability (a) if they are below the band of stability and (b) if they are above the band of stability.

25. Which of the following nuclei is most likely to decay by positron emission? Explain your choice.

(a) chromium-53

(b) manganese-51

(c) iron-59

26. The following nuclei do not lie in the band of stability. How would they be expected to decay? Explain your answer.

(a) $^{34}_{15}P$

(b) $^{239}_{92}U$

(c) $^{38}_{20}Ca$

(d) $^{3}_{1}H$

(e) $^{245}_{94}Pu$

27. The following nuclei do not lie in the band of stability. How would they be expected to decay?

(a) $^{28}_{15}P$

(b) $^{235}_{92}U$

(c) $^{37}_{20}Ca$

(d) $^{9}_{3}Li$

(e) $^{245}_{96}Cm$

28. Predict by what mode(s) of spontaneous radioactive decay each of the following unstable isotopes might proceed:

(a) $^{6}_{2}He$

(b) $^{60}_{30}Zn$

(c) $^{235}_{91}Pa$

(d) $^{241}_{94}Np$

(e) ^{18}F

(f) ^{129}Ba

(g) ^{237}Pu

29. Write a nuclear reaction for each step in the formation of $^{218}_{84}Po$ from $^{238}_{98}U$, which proceeds by a series of decay reactions involving the step-wise emission of α, β, β, α, α, α particles, in that order.

30. Write a nuclear reaction for each step in the formation of $^{208}_{82}Pb$ from $^{228}_{90}Th$, which proceeds by a series of decay reactions involving the step-wise emission of α, α, α, α, β, β, α particles, in that order.

31. Define the term half-life and illustrate it with an example.

32. A 1.00×10^{-6}-g sample of nobelium, $^{254}_{102}\text{No}$, has a half-life of 55 seconds after it is formed. What is the percentage of $^{254}_{102}\text{No}$ remaining at the following times?

(a) 5.0 min after it forms

(b) 1.0 h after it forms

33. ^{239}Pu is a nuclear waste byproduct with a half-life of 24,000 y. What fraction of the ^{239}Pu present today will be present in 1000 y?

34. The isotope ^{208}Tl undergoes β decay with a half-life of 3.1 min.

(a) What isotope is produced by the decay?

(b) How long will it take for 99.0% of a sample of pure ^{208}Tl to decay?

(c) What percentage of a sample of pure ^{208}Tl remains un-decayed after 1.0 h?

35. If 1.000 g of $^{226}_{88}\text{Ra}$ produces 0.0001 mL of the gas $^{222}_{86}\text{Rn}$ at STP (standard temperature and pressure) in 24 h, what is the half-life of ^{226}Ra in years?

36. The isotope $^{90}_{38}\text{Sr}$ is one of the extremely hazardous species in the residues from nuclear power generation. The strontium in a 0.500-g sample diminishes to 0.393 g in 10.0 y. Calculate the half-life.

37. Technetium-99 is often used for assessing heart, liver, and lung damage because certain technetium compounds are absorbed by damaged tissues. It has a half-life of 6.0 h. Calculate the rate constant for the decay of $^{99}_{43}\text{Tc}$.

38. What is the age of mummified primate skin that contains 8.25% of the original quantity of ^{14}C?

39. A sample of rock was found to contain 8.23 mg of rubidium-87 and 0.47 mg of strontium-87.

(a) Calculate the age of the rock if the half-life of the decay of rubidium by β emission is 4.7×10^{10} y.

(b) If some $^{87}_{38}\text{Sr}$ was initially present in the rock, would the rock be younger, older, or the same age as the age calculated in (a)? Explain your answer.

40. A laboratory investigation shows that a sample of uranium ore contains 5.37 mg of $^{238}_{92}\text{U}$ and 2.52 mg of $^{206}_{82}\text{Pb}$. Calculate the age of the ore. The half-life of $^{238}_{92}\text{U}$ is 4.5×10^{9} yr.

41. Plutonium was detected in trace amounts in natural uranium deposits by Glenn Seaborg and his associates in 1941. They proposed that the source of this ^{239}Pu was the capture of neutrons by ^{238}U nuclei. Why is this plutonium not likely to have been trapped at the time the solar system formed 4.7×10^{9} years ago?

42. A $^{7}_{4}\text{Be}$ atom (mass = 7.0169 amu) decays into a $^{7}_{3}\text{Li}$ atom (mass = 7.0160 amu) by electron capture. How much energy (in millions of electron volts, MeV) is produced by this reaction?

43. A $^{8}_{5}\text{B}$ atom (mass = 8.0246 amu) decays into a $^{8}_{4}\text{B}$ atom (mass = 8.0053 amu) by loss of a $β^{+}$ particle (mass = 0.00055 amu) or by electron capture. How much energy (in millions of electron volts) is produced by this reaction?

44. Isotopes such as ^{26}Al (half-life: 7.2×10^{5} years) are believed to have been present in our solar system as it formed, but have since decayed and are now called extinct nuclides.

(a) ^{26}Al decays by $β^{+}$ emission or electron capture. Write the equations for these two nuclear transformations.

(b) The earth was formed about 4.7×10^{9} (4.7 billion) years ago. How old was the earth when 99.999999% of the ^{26}Al originally present had decayed?

45. Write a balanced equation for each of the following nuclear reactions:

(a) bismuth-212 decays into polonium-212

(b) beryllium-8 and a positron are produced by the decay of an unstable nucleus

(c) neptunium-239 forms from the reaction of uranium-238 with a neutron and then spontaneously converts into plutonium-239

(d) strontium-90 decays into yttrium-90

46. Write a balanced equation for each of the following nuclear reactions:

(a) mercury-180 decays into platinum-176

(b) zirconium-90 and an electron are produced by the decay of an unstable nucleus

(c) thorium-232 decays and produces an alpha particle and a radium-228 nucleus, which decays into actinium-228 by beta decay

(d) neon-19 decays into fluorine-19

21.4 Transmutation and Nuclear Energy

47. Write the balanced nuclear equation for the production of the following transuranium elements:

(a) berkelium-244, made by the reaction of Am-241 and He-4

(b) fermium-254, made by the reaction of Pu-239 with a large number of neutrons

(c) lawrencium-257, made by the reaction of Cf-250 and B-11

(d) dubnium-260, made by the reaction of Cf-249 and N-15

48. How does nuclear fission differ from nuclear fusion? Why are both of these processes exothermic?

49. Both fusion and fission are nuclear reactions. Why is a very high temperature required for fusion, but not for fission?

50. Cite the conditions necessary for a nuclear chain reaction to take place. Explain how it can be controlled to produce energy, but not produce an explosion.

51. Describe the components of a nuclear reactor.

52. In usual practice, both a moderator and control rods are necessary to operate a nuclear chain reaction safely for the purpose of energy production. Cite the function of each and explain why both are necessary.

53. Describe how the potential energy of uranium is converted into electrical energy in a nuclear power plant.

54. The mass of a hydrogen atom $(_1^1 H)$ is 1.007825 amu; that of a tritium atom $(_1^3 H)$ is 3.01605 amu; and that of an α particle is 4.00150 amu. How much energy in kilojoules per mole of $_2^4 He$ produced is released by the following fusion reaction: $_1^1 H + _1^3 H \longrightarrow _2^4 He$.

21.5 Uses of Radioisotopes

55. How can a radioactive nuclide be used to show that the equilibrium:
$$AgCl(s) \rightleftharpoons Ag^+(aq) + Cl^-(aq)$$
is a dynamic equilibrium?

56. Technetium-99m has a half-life of 6.01 hours. If a patient injected with technetium-99m is safe to leave the hospital once 75% of the dose has decayed, when is the patient allowed to leave?

57. Iodine that enters the body is stored in the thyroid gland from which it is released to control growth and metabolism. The thyroid can be imaged if iodine-131 is injected into the body. In larger doses, I-133 is also used as a means of treating cancer of the thyroid. I-131 has a half-life of 8.70 days and decays by $β^-$ emission.

(a) Write an equation for the decay.

(b) How long will it take for 95.0% of a dose of I-131 to decay?

21.6 Biological Effects of Radiation

58. If a hospital were storing radioisotopes, what is the minimum containment needed to protect against:

(a) cobalt-60 (a strong γ emitter used for irradiation)

(b) molybdenum-99 (a beta emitter used to produce technetium-99 for imaging)

59. Based on what is known about Radon-222's primary decay method, why is inhalation so dangerous?

60. Given specimens uranium-232 ($t_{1/2}$ = 68.9 y) and uranium-233 ($t_{1/2}$ = 159,200 y) of equal mass, which one would have greater activity and why?

61. A scientist is studying a 2.234 g sample of thorium-229 ($t_{1/2}$ = 7340 y) in a laboratory.

(a) What is its activity in Bq?

(b) What is its activity in Ci?

62. Given specimens neon-24 ($t_{1/2}$ = 3.38 min) and bismuth-211 ($t_{1/2}$ = 2.14 min) of equal mass, which one would have greater activity and why?

Appendix A

The Periodic Table

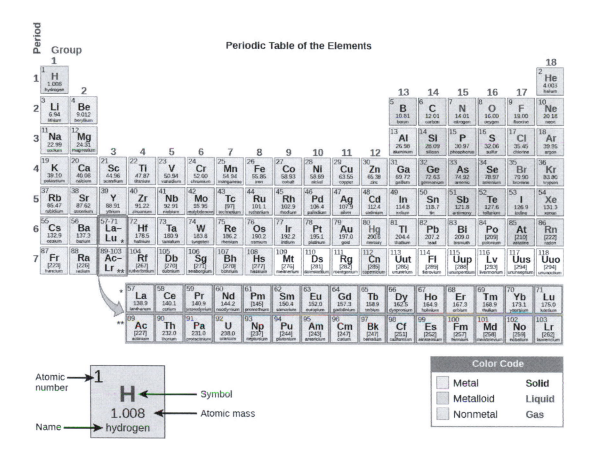

Periodic Table of the Elements

Appendix B

Essential Mathematics

Exponential Arithmetic

Exponential notation is used to express very large and very small numbers as a product of two numbers. The first number of the product, the *digit term*, is usually a number not less than 1 and not greater than 10. The second number of the product, the *exponential term*, is written as 10 with an exponent. Some examples of exponential notation are:

$$
\begin{align*}
1000 &= 1 \times 10^3 \\
100 &= 1 \times 10^2 \\
10 &= 1 \times 10^1 \\
1 &= 1 \times 10^0 \\
0.1 &= 1 \times 10^{-1} \\
0.001 &= 1 \times 10^{-3} \\
2386 &= 2.386 \times 1000 = 2.386 \times 10^3 \\
0.123 &= 1.23 \times 0.1 = 1.23 \times 10^{-1}
\end{align*}
$$

The power (exponent) of 10 is equal to the number of places the decimal is shifted to give the digit number. The exponential method is particularly useful notation for every large and very small numbers. For example, $1,230,000,000 = 1.23 \times 10^9$, and $0.00000000036 \times 10^{-10}$.

Addition of Exponentials

Convert all numbers to the same power of 10, add the digit terms of the numbers, and if appropriate, convert the digit term back to a number between 1 and 10 by adjusting the exponential term.

Example B1

Adding Exponentials

Add 5.00×10^{-5} and 3.00×10^{-3}.

Solution

$$
\begin{align*}
3.00 \times 10^{-3} &= 300 \times 10^{-5} \\
(5.00 \times 10^{-5}) + (300 \times 10^{-5}) &= 305 \times 10^{-5} = 3.05 \times 10^{-3}
\end{align*}
$$

Subtraction of Exponentials

Convert all numbers to the same power of 10, take the difference of the digit terms, and if appropriate, convert the digit term back to a number between 1 and 10 by adjusting the exponential term.

Example B2

Subtracting Exponentials

Subtract 4.0×10^{-7} from 5.0×10^{-6}.

Solution

$$4.0 \times 10^{-7} = 0.40 \times 10^{-6}$$
$$(5.0 \times 10^{-6}) - (0.40 \times 10^{-6}) = 4.6 \times 10^{-6}$$

Multiplication of Exponentials

Multiply the digit terms in the usual way and add the exponents of the exponential terms.

Example B3

Multiplying Exponentials

Multiply 4.2×10^{-8} by 2.0×10^{3}.

Solution

$$(4.2 \times 10^{-8}) \times (2.0 \times 10^{3}) = (4.2 \times 2.0) \times 10^{(-8)+(+3)} = 8.4 \times 10^{-5}$$

Division of Exponentials

Divide the digit term of the numerator by the digit term of the denominator and subtract the exponents of the exponential terms.

Example B4

Dividing Exponentials

Divide 3.6×10^{5} by 6.0×10^{-4}.

Solution

$$\frac{3.6 \times 10^{-5}}{6.0 \times 10^{-4}} = \left(\frac{3.6}{6.0}\right) \times 10^{(-5)-(-4)} = 0.60 \times 10^{-1} = 6.0 \times 10^{-2}$$

Squaring of Exponentials

Square the digit term in the usual way and multiply the exponent of the exponential term by 2.

Example B5

Squaring Exponentials

Square the number 4.0×10^{-6}.

Solution

$$(4.0 \times 10^{-6})^2 = 4 \times 4 \times 10^{2 \times (-6)} = 16 \times 10^{-12} = 1.6 \times 10^{-11}$$

Cubing of Exponentials

Cube the digit term in the usual way and multiply the exponent of the exponential term by 3.

Example B6

Cubing Exponentials

Cube the number 2×10^4.

Solution

$$(2 \times 10^4)^3 = 2 \times 2 \times 2 \times 10^{3 \times 4} = 8 \times 10^{12}$$

Taking Square Roots of Exponentials

If necessary, decrease or increase the exponential term so that the power of 10 is evenly divisible by 2. Extract the square root of the digit term and divide the exponential term by 2.

Example B7

Finding the Square Root of Exponentials

Find the square root of 1.6×10^{-7}.

Solution

$$1.6 \times 10^{-7} = 16 \times 10^{-8}$$
$$\sqrt{16 \times 10^{-8}} = \sqrt{16} \times \sqrt{10^{-8}} = \sqrt{16} \times 10^{-\frac{8}{2}} = 4.0 \times 10^{-4}$$

Significant Figures

A beekeeper reports that he has 525,341 bees. The last three figures of the number are obviously inaccurate, for during the time the keeper was counting the bees, some of them died and others hatched; this makes it quite difficult to determine the exact number of bees. It would have been more accurate if the beekeeper had reported the number 525,000. In other words, the last three figures are not significant, except to set the position of the decimal point. Their exact values have no meaning useful in this situation. In reporting any information as numbers, use only as many significant figures as the accuracy of the measurement warrants.

The importance of significant figures lies in their application to fundamental computation. In addition and subtraction, the sum or difference should contain as many digits to the right of the decimal as that in the least certain of the numbers used in the computation (indicated by underscoring in the following example).

Example B8

Addition and Subtraction with Significant Figures

Add 4.383 g and 0.0023 g.

Solution

$$\begin{array}{r} 4.38\underline{3} \text{ g} \\ 0.002\underline{3} \text{ g} \\ \hline 4.38\underline{5} \text{ g} \end{array}$$

In multiplication and division, the product or quotient should contain no more digits than that in the factor containing the least number of significant figures.

Example B9

Multiplication and Division with Significant Figures

Multiply 0.6238 by 6.6.

Solution

$$0.623\underline{8} \times 6.\underline{6} = 4.\underline{1}$$

When rounding numbers, increase the retained digit by 1 if it is followed by a number larger than 5 ("round up"). Do not change the retained digit if the digits that follow are less than 5 ("round down"). If the retained digit is followed by 5, round up if the retained digit is odd, or round down if it is even (after rounding, the retained digit will thus always be even).

The Use of Logarithms and Exponential Numbers

The common logarithm of a number (log) is the power to which 10 must be raised to equal that number. For example, the common logarithm of 100 is 2, because 10 must be raised to the second power to equal 100. Additional examples follow.

Logarithms and Exponential Numbers

Number	Number Expressed Exponentially	Common Logarithm
1000	10^3	3
10	10^1	1
1	10^0	0
0.1	10^{-1}	−1
0.001	10^{-3}	−3

Table B1

What is the common logarithm of 60? Because 60 lies between 10 and 100, which have logarithms of 1 and 2, respectively, the logarithm of 60 is 1.7782; that is,

$$60 = 10^{1.7782}$$

The common logarithm of a number less than 1 has a negative value. The logarithm of 0.03918 is −1.4069, or

$$0.03918 = 10^{-1.4069} = \frac{1}{10^{1.4069}}$$

To obtain the common logarithm of a number, use the *log* button on your calculator. To calculate a number from its logarithm, take the inverse log of the logarithm, or calculate 10^x (where x is the logarithm of the number).

The natural logarithm of a number (ln) is the power to which e must be raised to equal the number; e is the constant 2.7182818. For example, the natural logarithm of 10 is 2.303; that is,

$$10 = e^{2.303} = 2.7182818^{2.303}$$

To obtain the natural logarithm of a number, use the *ln* button on your calculator. To calculate a number from its natural logarithm, enter the natural logarithm and take the inverse ln of the natural logarithm, or calculate e^x (where x is the natural logarithm of the number).

Logarithms are exponents; thus, operations involving logarithms follow the same rules as operations involving exponents.

1. The logarithm of a product of two numbers is the sum of the logarithms of the two numbers.

$$\log xy = \log x + \log y, \text{ and } \ln xy = \ln x + \ln y$$

2. The logarithm of the number resulting from the division of two numbers is the difference between the logarithms of the two numbers.

$$\log \frac{x}{y} = \log x - \log y, \text{ and } \ln \frac{x}{y} = \ln x - \ln y$$

3. The logarithm of a number raised to an exponent is the product of the exponent and the logarithm of the number.

$$\log x^n = n\log x \text{ and } \ln x^n = n\ln x$$

The Solution of Quadratic Equations

Mathematical functions of this form are known as second-order polynomials or, more commonly, quadratic functions.

$$ax^2 + bx + c = 0$$

The solution or roots for any quadratic equation can be calculated using the following formula:

$$x = \frac{-b \pm \sqrt{b^2 - 4ac}}{2a}$$

Example B10

Solving Quadratic Equations

Solve the quadratic equation $3x^2 + 13x - 10 = 0$.

Solution

Substituting the values $a = 3$, $b = 13$, $c = -10$ in the formula, we obtain

$$x = \frac{-13 \pm \sqrt{(13)^2 - 4 \times 3 \times (-10)}}{2 \times 3}$$

$$x = \frac{-13 \pm \sqrt{169 + 120}}{6} = \frac{-13 \pm \sqrt{289}}{6} = \frac{-13 \pm 17}{6}$$

The two roots are therefore

$$x = \frac{-13 + 17}{6} = \frac{2}{3} \text{ and } x = \frac{-13 - 17}{6} = -5$$

Quadratic equations constructed on physical data always have real roots, and of these real roots, often only those having positive values are of any significance.

Two-Dimensional (x-y) Graphing

The relationship between any two properties of a system can be represented graphically by a two-dimensional data plot. Such a graph has two axes: a horizontal one corresponding to the independent variable, or the variable whose value is being controlled (x), and a vertical axis corresponding to the dependent variable, or the variable whose value is being observed or measured (y).

When the value of y is changing as a function of x (that is, different values of x correspond to different values of y), a graph of this change can be plotted or sketched. The graph can be produced by using specific values for (x,y) data pairs.

Example B11

Graphing the Dependence of *y* on *x*

x	y
1	5
2	10
3	7
4	14

This table contains the following points: (1,5), (2,10), (3,7), and (4,14). Each of these points can be plotted on a graph and connected to produce a graphical representation of the dependence of *y* on *x*.

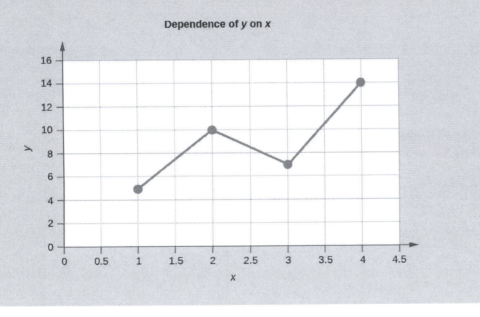

If the function that describes the dependence of *y* on *x* is known, it may be used to compute x,y data pairs that may subsequently be plotted.

Example B12

Plotting Data Pairs

If we know that $y = x^2 + 2$, we can produce a table of a few (x,y) values and then plot the line based on the data shown here.

x	$y = x^2 + 2$
1	3
2	6
3	11
4	18

$$y = x^2 + 2$$

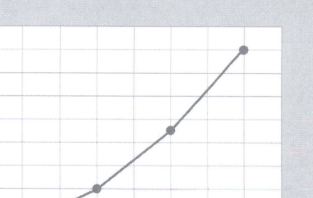

Appendix C

Units And Conversion Factors

Units of Length

meter (m)	= 39.37 inches (in.) = 1.094 yards (yd)	angstrom (Å)	= 10^{-8} cm (exact, definition) = 10^{-10} m (exact, definition)
centimeter (cm)	= 0.01 m (exact, definition)	yard (yd)	= 0.9144 m
millimeter (mm)	= 0.001 m (exact, definition)	inch (in.)	= 2.54 cm (exact, definition)
kilometer (km)	= 1000 m (exact, definition)	mile (US)	= 1.60934 km

Table C1

Units of Volume

liter (L)	= 0.001 m^3 (exact, definition) = 1000 cm^3 (exact, definition) = 1.057 (US) quarts	liquid quart (US)	= 32 (US) liquid ounces (exact, definition) = 0.25 (US) gallon (exact, definition) = 0.9463 L
milliliter (mL)	= 0.001 L (exact, definition) = 1 cm^3 (exact, definition)	dry quart	= 1.1012 L
microliter (μL)	= 10^{-6} L (exact, definition) = 10^{-3} cm^3 (exact, definition)	cubic foot (US)	= 28.316 L

Table C2

Units of Mass

gram (g)	= 0.001 kg (exact, definition)	ounce (oz) (avoirdupois)	= 28.35 g
milligram (mg)	= 0.001 g (exact, definition)	pound (lb) (avoirdupois)	= 0.4535924 kg
kilogram (kg)	= 1000 g (exact, definition) = 2.205 lb	ton (short)	=2000 lb (exact, definition) = 907.185 kg
ton (metric)	=1000 kg (exact, definition) = 2204.62 lb	ton (long)	= 2240 lb (exact, definition) = 1.016 metric ton

Table C3

Units of Energy

4.184 joule (J)	= 1 thermochemical calorie (cal)
1 thermochemical calorie (cal)	= 4.184 × 10^7 erg
erg	= 10^{-7} J (exact, definition)

Table C4

Units of Energy

electron-volt (eV)	$= 1.60218 \times 10^{-19}$ J $= 23.061$ kcal mol^{-1}
liter·atmosphere	$= 24.217$ cal $= 101.325$ J (exact, definition)
nutritional calorie (Cal)	$= 1000$ cal (exact, definition) $= 4184$ J
British thermal unit (BTU)	$= 1054.804$ J[1]

Table C4

Units of Pressure

torr	$= 1$ mm Hg (exact, definition)
pascal (Pa)	$= $ N m^{-2} (exact, definition) $= $ kg m^{-1} s^{-2} (exact, definition)
atmosphere (atm)	$= 760$ mm Hg (exact, definition) $= 760$ torr (exact, definition) $= 101,325$ N m^{-2} (exact, definition) $= 101,325$ Pa (exact, definition)
bar	$= 10^5$ Pa (exact, definition) $= 10^5$ kg m^{-1} s^{-2} (exact, definition)

Table C5

1. BTU is the amount of energy needed to heat one pound of water by one degree Fahrenheit. Therefore, the exact relationship of BTU to joules and other energy units depends on the temperature at which BTU is measured. 59 °F (15 °C) is the most widely used reference temperature for BTU definition in the United States. At this temperature, the conversion factor is the one provided in this table.

Appendix D

Fundamental Physical Constants

Fundamental Physical Constants

Name and Symbol	Value
atomic mass unit (amu)	$1.6605402 \times 10^{-27}$ kg
Avogadro's number	6.0221367×10^{23} mol^{-1}
Boltzmann's constant (k)	1.380658×10^{-23} J K^{-1}
charge-to-mass ratio for electron (e/m_e)	$1.75881962 \times 10^{11}$ C kg^{-1}
electron charge (e)	$1.60217733 \times 10^{-19}$ C
electron rest mass (m_e)	$9.1093897 \times 10^{-31}$ kg
Faraday's constant (F)	9.6485309×10^4 C mol^{-1}
gas constant (R)	8.205784×10^{-2} L atm mol^{-1} K^{-1} = 8.314510 J mol^{-1} K^{-1}
molar volume of an ideal gas, 1 atm, 0 °C	22.41409 L mol^{-1}
molar volume of an ideal gas, 1 bar, 0 °C	22.71108 L mol^{-1}
neutron rest mass (m_n)	$1.6749274 \times 10^{-27}$ kg
Planck's constant (h)	$6.6260755 \times 10^{-34}$ J s
proton rest mass (m_p)	$1.6726231 \times 10^{-27}$ kg
Rydberg constant (R)	1.0973731534×10^7 m^{-1} = 2.1798736 $\times 10^{-18}$ J
speed of light (in vacuum) (c)	2.99792458×10^8 m s^{-1}

Table D1

Appendix E

Water Properties

Water Density (kg/m³) at Different Temperatures (°C)

Temperature[1]	Density
0	999.8395
4	999.9720 (density maximum)
10	999.7026
15	999.1026
20	998.2071
22	997.7735
25	997.0479
30	995.6502
40	992.2
60	983.2
80	971.8
100	958.4

Table E1

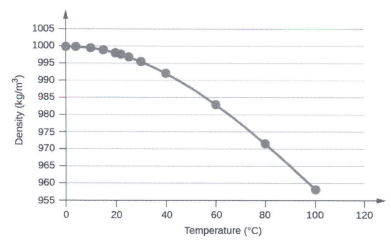

1. Data for t < 0 °C are for supercooled water

Water Vapor Pressure at Different Temperatures (°C)

Temperature	Vapor Pressure (torr)	Vapor Pressure (Pa)
0	4.6	613.2812
4	6.1	813.2642
10	9.2	1226.562
15	12.8	1706.522
20	17.5	2333.135
22	19.8	2639.776
25	23.8	3173.064
30	31.8	4239.64
35	42.2	5626.188
40	55.3	7372.707
45	71.9	9585.852
50	92.5	12332.29
55	118.0	15732
60	149.4	19918.31
65	187.5	24997.88
70	233.7	31157.35
75	289.1	38543.39
80	355.1	47342.64
85	433.6	57808.42
90	525.8	70100.71
95	633.9	84512.82
100	760.0	101324.7

Table E2

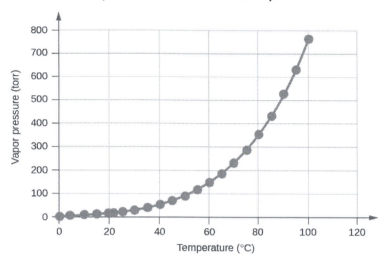

Vapor Pressure as a Function of Temperature

Water K_w and pK_w at Different Temperatures (°C)

Temperature	K_w 10^{-14}	pK_w[2]
0	0.112	14.95
5	0.182	14.74
10	0.288	14.54
15	0.465	14.33
20	0.671	14.17
25	0.991	14.00
30	1.432	13.84
35	2.042	13.69
40	2.851	13.55
45	3.917	13.41
50	5.297	13.28
55	7.080	13.15
60	9.311	13.03
75	19.95	12.70
100	56.23	12.25

Table E3

2. $pK_w = -\log_{10}(K_w)$

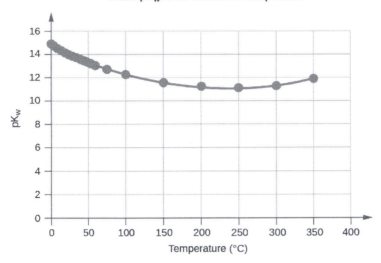

Water pK$_w$ as a Function of Temperature

Specific Heat Capacity for Water

$C°(H_2O(l)) = 4184 \text{ J·K}^{-1}\text{·kg}^{-1} = 4.184 \text{ J·g}^{-1}\text{·°C}^{-1}$
$C°(H_2O(s)) = 1864 \text{ J·K}^{-1}\text{·kg}^{-1}$
$C°(H_2O(g)) = 2093 \text{ J·K}^{-1}\text{·kg}^{-1}$

Table E4

Standard Water Melting and Boiling Temperatures and Enthalpies of the Transitions

	Temperature (K)	ΔH (kJ/mol)
melting	273.15	6.088
boiling	373.15	40.656 (44.016 at 298 K)

Table E5

Water Cryoscopic (Freezing Point Depression) and Ebullioscopic (Boiling Point Elevation) Constants

$K_f = 1.86\text{°C·kg·mol}^{-1}$ (cryoscopic constant)
$K_b = 0.51\text{°C·kg·mol}^{-1}$ (ebullioscopic constant)

Table E6

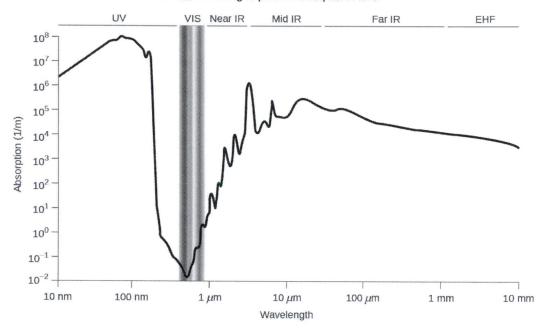

Figure E1 Water full-range spectral absorption curve. This curve shows the full-range spectral absorption for water. The *y*-axis signifies the absorption in 1/cm. If we divide 1 by this value, we will obtain the length of the path (in cm) after which the intensity of a light beam passing through water decays by a factor of the base of the natural logarithm e (e = 2.718281828).

Appendix F
Composition Of Commercial Acids And Bases

Composition of Commercial Acids and Bases

Acid or Base[1]	Density (g/mL)[2]	Percentage by Mass	Molarity
acetic acid, glacial	1.05	99.5%	17.4
aqueous ammonia[3]	0.90	28%	14.8
hydrochloric acid	1.18	36%	11.6
nitric acid	1.42	71%	16.0
perchloric acid	1.67	70%	11.65
phosphoric acid	1.70	85%	14.7
sodium hydroxide	1.53	50%	19.1
sulfuric acid	1.84	96%	18.0

Table F1

1. Acids and bases are commercially available as aqueous solutions. This table lists properties (densities and concentrations) of common acid and base solutions. Nominal values are provided in cases where the manufacturer cites a range of concentrations and densities.
2. This column contains specific gravity data. In the case of this table, specific gravity is the ratio of density of a substance to the density of pure water at the same conditions. Specific gravity is often cited on commercial labels.
3. This solution is sometimes called "ammonium hydroxide," although this term is not chemically accurate.

Appendix G
Standard Thermodynamic Properties For Selected Substances

Standard Thermodynamic Properties for Selected Substances

Substance	ΔH_f° (kJ mol⁻)	ΔG_f° (kJ mol⁻¹)	S_{298}° (J K⁻¹ mol⁻¹)
aluminum			
Al(s)	0	0	28.3
Al(g)	324.4	285.7	164.54
Al³⁺(aq)	−531	−485	−321.7
Al₂O₃(s)	−1676	−1582	50.92
AlF₃(s)	−1510.4	−1425	66.5
AlCl₃(s)	−704.2	−628.8	110.67
AlCl₃·6H₂O(s)	−2691.57	−2269.40	376.56
Al₂S₃(s)	−724.0	−492.4	116.9
Al₂(SO₄)₃(s)	−3445.06	−3506.61	239.32
antimony			
Sb(s)	0	0	45.69
Sb(g)	262.34	222.17	180.16
Sb₄O₆(s)	−1440.55	−1268.17	220.92
SbCl₃(g)	−313.8	−301.2	337.80
SbCl₅(g)	−394.34	−334.29	401.94
Sb₂S₃(s)	−174.89	−173.64	182.00
SbCl₃(s)	−382.17	−323.72	184.10
SbOCl(s)	−374.0	—	—
arsenic			
As(s)	0	0	35.1
As(g)	302.5	261.0	174.21
As₄(g)	143.9	92.4	314
As₄O₆(s)	−1313.94	−1152.52	214.22

Table G1

Standard Thermodynamic Properties for Selected Substances

Substance	ΔH_f° (kJ mol⁻)	ΔG_f° (kJ mol⁻¹)	S_{298}° (J K⁻¹ mol⁻¹)
As₂O₅(s)	−924.87	−782.41	105.44
AsCl₃(g)	−261.50	−248.95	327.06
As₂S₃(s)	−169.03	−168.62	163.59
AsH₃(g)	66.44	68.93	222.78
H₃AsO₄(s)	−906.3	—	—
barium			
Ba(s)	0	0	62.5
Ba(g)	180	146	170.24
Ba²⁺(aq)	−537.6	−560.8	9.6
BaO(s)	−548.0	−520.3	72.1
BaCl₂(s)	−855.0	−806.7	123.7
BaSO₄(s)	−1473.2	−1362.3	132.2
beryllium			
Be(s)	0	0	9.50
Be(g)	324.3	286.6	136.27
BeO(s)	−609.4	−580.1	13.8
bismuth			
Bi(s)	0	0	56.74
Bi(g)	207.1	168.2	187.00
Bi₂O₃(s)	−573.88	−493.7	151.5
BiCl₃(s)	−379.07	−315.06	176.98
Bi₂S₃(s)	−143.1	−140.6	200.4
boron			
B(s)	0	0	5.86
B(g)	565.0	521.0	153.4
B₂O₃(s)	−1273.5	−1194.3	53.97
B₂H₆(g)	36.4	87.6	232.1
H₃BO₃(s)	−1094.33	−968.92	88.83
BF₃(g)	−1136.0	−1119.4	254.4
BCl₃(g)	−403.8	−388.7	290.1
B₃N₃H₆(l)	−540.99	−392.79	199.58

Table G1

Standard Thermodynamic Properties for Selected Substances

Substance	ΔH_f° (kJ mol^{-1})	ΔG_f° (kJ mol^{-1})	S_{298}° (J K^{-1} mol^{-1})
HBO$_2$(s)	−794.25	−723.41	37.66
bromine			
Br$_2$(l)	0	0	152.23
Br$_2$(g)	30.91	3.142	245.5
Br(g)	111.88	82.429	175.0
Br$^-$(aq)	−120.9	−102.82	80.71
BrF$_3$(g)	−255.60	−229.45	292.42
HBr(g)	−36.3	−53.43	198.7
cadmium			
Cd(s)	0	0	51.76
Cd(g)	112.01	77.41	167.75
Cd^{2+}(aq)	−75.90	−77.61	−73.2
CdO(s)	−258.2	−228.4	54.8
CdCl$_2$(s)	−391.5	−343.9	115.3
CdSO$_4$(s)	−933.3	−822.7	123.0
CdS(s)	−161.9	−156.5	64.9
calcium			
Ca(s)	0	0	41.6
Ca(g)	178.2	144.3	154.88
Ca^{2+}(aq)	−542.96	−553.04	−55.2
CaO(s)	−634.9	−603.3	38.1
Ca(OH)$_2$(s)	−985.2	−897.5	83.4
CaSO$_4$(s)	−1434.5	−1322.0	106.5
CaSO$_4$·2H$_2$O(s)	−2022.63	−1797.45	194.14
CaCO$_3$(s) (calcite)	−1220.0	−1081.4	110.0
CaSO$_3$·H$_2$O(s)	−1752.68	−1555.19	184.10
carbon			
C(s) (graphite)	0	0	5.740
C(s) (diamond)	1.89	2.90	2.38
C(g)	716.681	671.2	158.1
CO(g)	−110.52	−137.15	197.7

Table G1

Standard Thermodynamic Properties for Selected Substances

Substance	ΔH_f° (kJ mol⁻)	ΔG_f° (kJ mol⁻¹)	S_{298}° (J K⁻¹ mol⁻¹)
$CO_2(g)$	−393.51	−394.36	213.8
$CO_3{}^{2-}(aq)$	−677.1	−527.8	−56.9
$CH_4(g)$	−74.6	−50.5	186.3
$CH_3OH(l)$	−239.2	−166.6	126.8
$CH_3OH(g)$	−201.0	−162.3	239.9
$CCl_4(l)$	−128.2	−62.5	214.4
$CCl_4(g)$	−95.7	−58.2	309.7
$CHCl_3(l)$	−134.1	−73.7	201.7
$CHCl_3(g)$	−103.14	−70.34	295.71
$CS_2(l)$	89.70	65.27	151.34
$CS_2(g)$	116.9	66.8	238.0
$C_2H_2(g)$	227.4	209.2	200.9
$C_2H_4(g)$	52.4	68.4	219.3
$C_2H_6(g)$	−84.0	−32.0	229.2
$CH_3CO_2H(l)$	−484.3	−389.9	159.8
$CH_3CO_2H(g)$	−434.84	−376.69	282.50
$C_2H_5OH(l)$	−277.6	−174.8	160.7
$C_2H_5OH(g)$	−234.8	−167.9	281.6
$HCO_3{}^-(aq)$	−691.11	−587.06	95
$C_3H_8(g)$	−103.8	−23.4	270.3
$C_6H_6(g)$	82.927	129.66	269.2
$C_6H_6(l)$	49.1	124.50	173.4
$CH_2Cl_2(l)$	−124.2	−63.2	177.8
$CH_2Cl_2(g)$	−95.4	−65.90	270.2
$CH_3Cl(g)$	−81.9	−60.2	234.6
$C_2H_5Cl(l)$	−136.52	−59.31	190.79
$C_2H_5Cl(g)$	−112.17	−60.39	276.00
$C_2N_2(g)$	308.98	297.36	241.90
$HCN(l)$	108.9	125.0	112.8
$HCN(g)$	135.5	124.7	201.8
cesium			

Table G1

Standard Thermodynamic Properties for Selected Substances

Substance	ΔH_f° (kJ mol^{-1})	ΔG_f° (kJ mol^{-1})	S_{298}° (J K^{-1} mol^{-1})
$Cs^+(aq)$	−248	−282.0	133
chlorine			
$Cl_2(g)$	0	0	223.1
$Cl(g)$	121.3	105.70	165.2
$Cl^-(aq)$	−167.2	−131.2	56.5
$ClF(g)$	−54.48	−55.94	217.78
$ClF_3(g)$	−158.99	−118.83	281.50
$Cl_2O(g)$	80.3	97.9	266.2
$Cl_2O_7(l)$	238.1	—	—
$Cl_2O_7(g)$	272.0	—	—
$HCl(g)$	−92.307	−95.299	186.9
$HClO_4(l)$	−40.58	—	—
chromium			
$Cr(s)$	0	0	23.77
$Cr(g)$	396.6	351.8	174.50
$CrO_4{}^{2-}(aq)$	−881.2	−727.8	50.21
$Cr_2O_7{}^{2-}(aq)$	−1490.3	−1301.1	261.9
$Cr_2O_3(s)$	−1139.7	−1058.1	81.2
$CrO_3(s)$	−589.5	—	—
$(NH_4)_2Cr_2O_7(s)$	−1806.7	—	—
cobalt			
$Co(s)$	0	0	30.0
$Co^{2+}(aq)$	−67.4	−51.5	−155
$Co^{3+}(aq)$	92	134	−305.0
$CoO(s)$	−237.9	−214.2	52.97
$Co_3O_4(s)$	−910.02	−794.98	114.22
$Co(NO_3)_2(s)$	−420.5	—	—
copper			
$Cu(s)$	0	0	33.15
$Cu(g)$	338.32	298.58	166.38
$Cu^+(aq)$	51.9	50.2	−26

Table G1

Standard Thermodynamic Properties for Selected Substances

Substance	ΔH_f° (kJ mol⁻¹)	ΔG_f° (kJ mol⁻¹)	S_{298}° (J K⁻¹ mol⁻¹)
$Cu^{2+}(aq)$	64.77	65.49	−99.6
$CuO(s)$	−157.3	−129.7	42.63
$Cu_2O(s)$	−168.6	−146.0	93.14
$CuS(s)$	−53.1	−53.6	66.5
$Cu_2S(s)$	−79.5	−86.2	120.9
$CuSO_4(s)$	−771.36	−662.2	109.2
$Cu(NO_3)_2(s)$	−302.9	—	—
fluorine			
$F_2(g)$	0	0	202.8
$F(g)$	79.4	62.3	158.8
$F^-(aq)$	−332.6	−278.8	−13.8
$F_2O(g)$	24.7	41.9	247.43
$HF(g)$	−273.3	−275.4	173.8
hydrogen			
$H_2(g)$	0	0	130.7
$H(g)$	217.97	203.26	114.7
$H^+(aq)$	0	0	0
$OH^-(aq)$	−230.0	−157.2	−10.75
$H_3O^+(aq)$	−285.8		69.91
$H_2O(l)$	−285.83	−237.1	70.0
$H_2O(g)$	−241.82	−228.59	188.8
$H_2O_2(l)$	−187.78	−120.35	109.6
$H_2O_2(g)$	−136.3	−105.6	232.7
$HF(g)$	−273.3	−275.4	173.8
$HCl(g)$	−92.307	−95.299	186.9
$HBr(g)$	−36.3	−53.43	198.7
$HI(g)$	26.48	1.70	206.59
$H_2S(g)$	−20.6	−33.4	205.8
$H_2Se(g)$	29.7	15.9	219.0
iodine			
$I_2(s)$	0	0	116.14

Table G1

Standard Thermodynamic Properties for Selected Substances

Substance	ΔH_f° (kJ mol^{-1})	ΔG_f° (kJ mol^{-1})	S_{298}° (J K^{-1} mol^{-1})
$I_2(g)$	62.438	19.3	260.7
$I(g)$	106.84	70.2	180.8
$I^-(aq)$	−55.19	−51.57	11.13
$IF(g)$	95.65	−118.49	236.06
$ICl(g)$	17.78	−5.44	247.44
$IBr(g)$	40.84	3.72	258.66
$IF_7(g)$	−943.91	−818.39	346.44
$HI(g)$	26.48	1.70	206.59
iron			
$Fe(s)$	0	0	27.3
$Fe(g)$	416.3	370.7	180.5
$Fe^{2+}(aq)$	−89.1	−78.90	−137.7
$Fe^{3+}(aq)$	−48.5	−4.7	−315.9
$Fe_2O_3(s)$	−824.2	−742.2	87.40
$Fe_3O_4(s)$	−1118.4	−1015.4	146.4
$Fe(CO)_5(l)$	−774.04	−705.42	338.07
$Fe(CO)_5(g)$	−733.87	−697.26	445.18
$FeCl_2(s)$	−341.79	−302.30	117.95
$FeCl_3(s)$	−399.49	−334.00	142.3
$FeO(s)$	−272.0	−255.2	60.75
$Fe(OH)_2(s)$	−569.0	−486.5	88.
$Fe(OH)_3(s)$	−823.0	−696.5	106.7
$FeS(s)$	−100.0	−100.4	60.29
$Fe_3C(s)$	25.10	20.08	104.60
lead			
$Pb(s)$	0	0	64.81
$Pb(g)$	195.2	162.	175.4
$Pb^{2+}(aq)$	−1.7	−24.43	10.5
$PbO(s)$ (yellow)	−217.32	−187.89	68.70
$PbO(s)$ (red)	−218.99	−188.93	66.5
$Pb(OH)_2(s)$	−515.9	—	—

Table G1

Standard Thermodynamic Properties for Selected Substances

Substance	ΔH_f° (kJ mol⁻)	ΔG_f° (kJ mol⁻¹)	S_{298}° (J K⁻¹ mol⁻¹)
PbS(s)	−100.4	−98.7	91.2
Pb(NO₃)₂(s)	−451.9	—	—
PbO₂(s)	−277.4	−217.3	68.6
PbCl₂(s)	−359.4	−314.1	136.0
lithium			
Li(s)	0	0	29.1
Li(g)	159.3	126.6	138.8
Li⁺(aq)	−278.5	−293.3	13.4
LiH(s)	−90.5	−68.3	20.0
Li(OH)(s)	−487.5	−441.5	42.8
LiF(s)	−616.0	−587.5	35.7
Li₂CO₃(s)	−1216.04	−1132.19	90.17
magnesium			
Mg²⁺(aq)	−466.9	−454.8	−138.1
manganese			
Mn(s)	0	0	32.0
Mn(g)	280.7	238.5	173.7
Mn²⁺(aq)	−220.8	−228.1	−73.6
MnO(s)	−385.2	−362.9	59.71
MnO₂(s)	−520.03	−465.1	53.05
Mn₂O₃(s)	−958.97	−881.15	110.46
Mn₃O₄(s)	−1378.83	−1283.23	155.64
MnO₄⁻(aq)	−541.4	−447.2	191.2
MnO₄²⁻(aq)	−653.0	−500.7	59
mercury			
Hg(l)	0	0	75.9
Hg(g)	61.4	31.8	175.0
Hg²⁺(aq)		164.8	
Hg²⁺(aq)	172.4	153.9	84.5
HgO(s) (red)	−90.83	−58.5	70.29
HgO(s) (yellow)	−90.46	−58.43	71.13

Table G1

Standard Thermodynamic Properties for Selected Substances

Substance	ΔH_f° (kJ mol⁻)	ΔG_f° (kJ mol⁻¹)	S_{298}° (J K⁻¹ mol⁻¹)
$HgCl_2(s)$	−224.3	−178.6	146.0
$Hg_2Cl_2(s)$	−265.4	−210.7	191.6
$HgS(s)$ (red)	−58.16	−50.6	82.4
$HgS(s)$ (black)	−53.56	−47.70	88.28
$HgSO_4(s)$	−707.51	−594.13	0.00
nickel			
$Ni^{2+}(aq)$	−64.0	−46.4	−159
nitrogen			
$N_2(g)$	0	0	191.6
$N(g)$	472.704	455.5	153.3
$NO(g)$	90.25	87.6	210.8
$NO_2(g)$	33.2	51.30	240.1
$N_2O(g)$	81.6	103.7	220.0
$N_2O_3(g)$	83.72	139.41	312.17
$NO_3^-(aq)$	−205.0	−108.7	146.4
$N_2O_4(g)$	11.1	99.8	304.4
$N_2O_5(g)$	11.3	115.1	355.7
$NH_3(g)$	−45.9	−16.5	192.8
$NH_4^+(aq)$	−132.5	−79.31	113.4
$N_2H_4(l)$	50.63	149.43	121.21
$N_2H_4(g)$	95.4	159.4	238.5
$NH_4NO_3(s)$	−365.56	−183.87	151.08
$NH_4Cl(s)$	−314.43	−202.87	94.6
$NH_4Br(s)$	−270.8	−175.2	113.0
$NH_4I(s)$	−201.4	−112.5	117.0
$NH_4NO_2(s)$	−256.5	—	—
$HNO_3(l)$	−174.1	−80.7	155.6
$HNO_3(g)$	−133.9	−73.5	266.9
oxygen			
$O_2(g)$	0	0	205.2
$O(g)$	249.17	231.7	161.1

Table G1

Standard Thermodynamic Properties for Selected Substances

Substance	ΔH_f° (kJ mol$^-$)	ΔG_f° (kJ mol^{-1})	S_{298}° (J K^{-1} mol^{-1})
$O_3(g)$	142.7	163.2	238.9
phosphorus			
$P_4(s)$	0	0	164.4
$P_4(g)$	58.91	24.4	280.0
$P(g)$	314.64	278.25	163.19
$PH_3(g)$	5.4	13.5	210.2
$PCl_3(g)$	−287.0	−267.8	311.78
$PCl_5(g)$	−374.9	−305.0	364.4
$P_4O_6(s)$	−1640.1	—	—
$P_4O_{10}(s)$	−2984.0	−2697.0	228.86
$PO_4{}^{3-}(aq)$	−1277	−1019	−222
$HPO_3(s)$	−948.5	—	—
$HPO_4{}^{2-}(aq)$	−1292.1	−1089.3	−33
$H_2PO_4{}^{2-}(aq)$	−1296.3	−1130.4	90.4
$H_3PO_2(s)$	−604.6	—	—
$H_3PO_3(s)$	−964.4	—	—
$H_3PO_4(s)$	−1279.0	−1119.1	110.50
$H_3PO_4(l)$	−1266.9	−1124.3	110.5
$H_4P_2O_7(s)$	−2241.0	—	—
$POCl_3(l)$	−597.1	−520.8	222.5
$POCl_3(g)$	−558.5	−512.9	325.5
potassium			
$K(s)$	0	0	64.7
$K(g)$	89.0	60.5	160.3
$K^+(aq)$	−252.4	−283.3	102.5
$KF(s)$	−576.27	−537.75	66.57
$KCl(s)$	−436.5	−408.5	82.6
rubidium			
$Rb^+(aq)$	−246	−282.2	124
silicon			
$Si(s)$	0	0	18.8

Table G1

Standard Thermodynamic Properties for Selected Substances

Substance	ΔH_f° (kJ mol$^-$)	ΔG_f° (kJ mol^{-1})	S_{298}° (J K^{-1} mol^{-1})
Si(g)	450.0	405.5	168.0
SiO$_2$(s)	−910.7	−856.3	41.5
SiH$_4$(g)	34.3	56.9	204.6
H$_2$SiO$_3$(s)	−1188.67	−1092.44	133.89
H$_4$SiO$_4$(s)	−1481.14	−1333.02	192.46
SiF$_4$(g)	−1615.0	−1572.8	282.8
SiCl$_4$(l)	−687.0	−619.8	239.7
SiCl$_4$(g)	−662.75	−622.58	330.62
SiC(s, beta cubic)	−73.22	−70.71	16.61
SiC(s, alpha hexagonal)	−71.55	−69.04	16.48
silver			
Ag(s)	0	0	42.55
Ag(g)	284.9	246.0	172.89
Ag$^+$(aq)	105.6	77.11	72.68
Ag$_2$O(s)	−31.05	−11.20	121.3
AgCl(s)	−127.0	−109.8	96.3
Ag$_2$S(s)	−32.6	−40.7	144.0
sodium			
Na(s)	0	0	51.3
Na(g)	107.5	77.0	153.7
Na$^+$(aq)	−240.1	−261.9	59
Na$_2$O(s)	−414.2	−375.5	75.1
NaCl(s)	−411.2	−384.1	72.1
strontium			
Sr^{2+}(aq)	−545.8	−557.3	−32.6
sulfur			
S$_8$(s) (rhombic)	0	0	256.8
S(g)	278.81	238.25	167.82
S^{2-}(aq)	41.8	83.7	22
SO$_2$(g)	−296.83	−300.1	248.2
SO$_3$(g)	−395.72	−371.06	256.76

Table G1

Standard Thermodynamic Properties for Selected Substances

Substance	ΔH_f° (kJ mol⁻)	ΔG_f° (kJ mol⁻¹)	S_{298}° (J K⁻¹ mol⁻¹)
$SO_4{}^{2-}(aq)$	−909.3	−744.5	20.1
$S_2O_3{}^{2-}(aq)$	−648.5	−522.5	67
$H_2S(g)$	−20.6	−33.4	205.8
$HS^-(aq)$	−17.7	12.6	61.1
$H_2SO_4(l)$	−813.989	690.00	156.90
$HSO_4{}^{2-}(aq)$	−885.75	−752.87	126.9
$H_2S_2O_7(s)$	−1273.6	—	—
$SF_4(g)$	−728.43	−684.84	291.12
$SF_6(g)$	−1220.5	−1116.5	291.5
$SCl_2(l)$	−50	—	—
$SCl_2(g)$	−19.7	—	—
$S_2Cl_2(l)$	−59.4	—	—
$S_2Cl_2(g)$	−19.50	−29.25	319.45
$SOCl_2(g)$	−212.55	−198.32	309.66
$SOCl_2(l)$	−245.6	—	—
$SO_2Cl_2(l)$	−394.1	—	—
$SO_2Cl_2(g)$	−354.80	−310.45	311.83
tin			
$Sn(s)$	0	0	51.2
$Sn(g)$	301.2	266.2	168.5
$SnO(s)$	−285.8	−256.9	56.5
$SnO_2(s)$	−577.6	−515.8	49.0
$SnCl_4(l)$	−511.3	−440.1	258.6
$SnCl_4(g)$	−471.5	−432.2	365.8
titanium			
$Ti(s)$	0	0	30.7
$Ti(g)$	473.0	428.4	180.3
$TiO_2(s)$	−944.0	−888.8	50.6
$TiCl_4(l)$	−804.2	−737.2	252.4
$TiCl_4(g)$	−763.2	−726.3	353.2
tungsten			

Table G1

Standard Thermodynamic Properties for Selected Substances

Substance	ΔH_f° (kJ mol^{-})	ΔG_f° (kJ mol^{-1})	S_{298}° (J K^{-1} mol^{-1})
W(s)	0	0	32.6
W(g)	849.4	807.1	174.0
WO$_3$(s)	−842.9	−764.0	75.9
zinc			
Zn(s)	0	0	41.6
Zn(g)	130.73	95.14	160.98
Zn^{2+}(aq)	−153.9	−147.1	−112.1
ZnO(s)	−350.5	−320.5	43.7
ZnCl$_2$(s)	−415.1	−369.43	111.5
ZnS(s)	−206.0	−201.3	57.7
ZnSO$_4$(s)	−982.8	−871.5	110.5
ZnCO$_3$(s)	−812.78	−731.57	82.42
complexes			
[Co(NH$_3$)$_4$(NO$_2$)$_2$]NO$_3$, cis	−898.7	—	—
[Co(NH$_3$)$_4$(NO$_2$)$_2$]NO$_3$, $trans$	−896.2	—	—
NH$_4$[Co(NH$_3$)$_2$(NO$_2$)$_4$]	−837.6	—	—
[Co(NH$_3$)$_6$][Co(NH$_3$)$_2$(NO$_2$)$_4$]$_3$	−2733.0	—	—
[Co(NH$_3$)$_4$Cl$_2$]Cl, cis	−874.9	—	—
[Co(NH$_3$)$_4$Cl$_2$]Cl, $trans$	−877.4	—	—
[Co(en)$_2$(NO$_2$)$_2$]NO$_3$, cis	−689.5	—	—
[Co(en)$_2$Cl$_2$]Cl, cis	−681.2	—	—
[Co(en)$_2$Cl$_2$]Cl, $trans$	−677.4	—	—
[Co(en)$_3$](ClO$_4$)$_3$	−762.7	—	—
[Co(en)$_3$]Br$_2$	−595.8	—	—
[Co(en)$_3$]I$_2$	−475.3	—	—
[Co(en)$_3$]I$_3$	−519.2	—	—
[Co(NH$_3$)$_6$](ClO$_4$)$_3$	−1034.7	−221.1	615
[Co(NH$_3$)$_5$NO$_2$](NO$_3$)$_2$	−1088.7	−412.9	331
[Co(NH$_3$)$_6$](NO$_3$)$_3$	−1282.0	−524.5	448
[Co(NH$_3$)$_5$Cl]Cl$_2$	−1017.1	−582.5	366.1
[Pt(NH$_3$)$_4$]Cl$_2$	−725.5	—	—

Table G1

Standard Thermodynamic Properties for Selected Substances

Substance	ΔH_f° (kJ mol$^-$)	ΔG_f° (kJ mol^{-1})	S_{298}° (J K^{-1} mol^{-1})
[Ni(NH$_3$)$_6$]Cl$_2$	−994.1	—	—
[Ni(NH$_3$)$_6$]Br$_2$	−923.8	—	—
[Ni(NH$_3$)$_6$]I$_2$	−808.3	—	—

Table G1

Appendix H

Ionization Constants Of Weak Acids

Ionization Constants of Weak Acids

Acid	Formula	K_a at 25 °C	Lewis Structure
acetic	CH_3CO_2H	1.8×10^{-5}	
arsenic	H_3AsO_4	5.5×10^{-3}	
	$H_2AsO_4^{-}$	1.7×10^{-7}	
	$HAsO_4^{2-}$	5.1×10^{-12}	
arsenous	H_3AsO_3	5.1×10^{-10}	
boric	H_3BO_3	5.4×10^{-10}	
carbonic	H_2CO_3	4.3×10^{-7}	
	HCO_3^{-}	5.6×10^{-11}	

Table H1

Ionization Constants of Weak Acids

Acid	Formula	K_a at 25 °C	Lewis Structure
cyanic	HCNO	2×10^{-4}	
formic	HCO_2H	1.8×10^{-4}	
hydrazoic	HN_3	2.5×10^{-5}	
hydrocyanic	HCN	4.9×10^{-10}	
hydrofluoric	HF	3.5×10^{-4}	
hydrogen peroxide	H_2O_2	2.4×10^{-12}	
hydrogen selenide	H_2Se	1.29×10^{-4}	
	HSe^-	1×10^{-12}	
hydrogen sulfate ion	HSO_4^-	1.2×10^{-2}	
hydrogen sulfide	H_2S	8.9×10^{-8}	
	HS^-	1.0×10^{-19}	
hydrogen telluride	H_2Te	2.3×10^{-3}	
	HTe^-	1.6×10^{-11}	
hypobromous	HBrO	2.8×10^{-9}	
hypochlorous	HClO	2.9×10^{-8}	
nitrous	HNO_2	4.6×10^{-4}	

Table H1

Ionization Constants of Weak Acids

Acid	Formula	K_a at 25 °C	Lewis Structure
oxalic	$H_2C_2O_4$	6.0×10^{-2}	
	$HC_2O_4^-$	6.1×10^{-5}	
phosphoric	H_3PO_4	7.5×10^{-3}	
	$H_2PO_4^-$	6.2×10^{-8}	
	HPO_4^{2-}	4.2×10^{-13}	
phosphorous	H_3PO_3	5×10^{-2}	
	$H_2PO_3^-$	2.0×10^{-7}	
sulfurous	H_2SO_3	1.6×10^{-2}	
	HSO_3^-	6.4×10^{-8}	

Table H1

Appendix I
Ionization Constants Of Weak Bases

Ionization Constants of Weak Bases

Base	Lewis Structure	K_b at 25 °C
ammonia		1.8×10^{-5}
dimethylamine		5.9×10^{-4}
methylamine		4.4×10^{-4}
phenylamine (aniline)		4.3×10^{-10}
trimethylamine		6.3×10^{-5}

Table I1

Appendix J

Solubility Products

Solubility Products

Substance	K_{sp} at 25 °C
aluminum	
$Al(OH)_3$	2×10^{-32}
barium	
$BaCO_3$	1.6×10^{-9}
$BaC_2O_4 \cdot 2H_2O$	1.1×10^{-7}
$BaSO_4$	2.3×10^{-8}
$BaCrO_4$	8.5×10^{-11}
BaF_2	2.4×10^{-5}
$Ba(OH)_2 \cdot 8H_2O$	5.0×10^{-3}
$Ba_3(PO_4)_2$	6×10^{-39}
$Ba_3(AsO_4)_2$	1.1×10^{-13}
bismuth	
$BiO(OH)$	4×10^{-10}
$BiOCl$	1.8×10^{-31}
Bi_2S_3	1×10^{-97}
cadmium	
$Cd(OH)_2$	5.9×10^{-15}
CdS	1.0×10^{-28}
$CdCO_3$	5.2×10^{-12}
calcium	
$Ca(OH)_2$	1.3×10^{-6}
$CaCO_3$	8.7×10^{-9}
$CaSO4 \cdot 2H_2O$	6.1×10^{-5}
$CaC_2O_4 \cdot H_2O$	1.96×10^{-8}
$Ca_3(PO_4)_2$	1.3×10^{-32}
$CaHPO_4$	7×10^{-7}
CaF_2	4.0×10^{-11}
chromium	

Table J1

Solubility Products

Substance	K_{sp} at 25 °C
$Cr(OH)_3$	6.7×10^{-31}
cobalt	
$Co(OH)_2$	2.5×10^{-16}
$CoS(\alpha)$	5×10^{-22}
$CoS(\beta)$	3×10^{-26}
$CoCO_3$	1.4×10^{-13}
$Co(OH)_3$	2.5×10^{-43}
copper	
$CuCl$	1.2×10^{-6}
$CuBr$	6.27×10^{-9}
CuI	1.27×10^{-12}
$CuSCN$	1.6×10^{-11}
Cu_2S	2.5×10^{-48}
$Cu(OH)_2$	2.2×10^{-20}
CuS	8.5×10^{-45}
$CuCO_3$	2.5×10^{-10}
iron	
$Fe(OH)_2$	1.8×10^{-15}
$FeCO_3$	2.1×10^{-11}
FeS	3.7×10^{-19}
$Fe(OH)_3$	4×10^{-38}
lead	
$Pb(OH)_2$	1.2×10^{-15}
PbF_2	4×10^{-8}
$PbCl_2$	1.6×10^{-5}
$PbBr_2$	4.6×10^{-6}
PbI_2	1.4×10^{-8}
$PbCO_3$	1.5×10^{-15}
PbS	7×10^{-29}
$PbCrO_4$	2×10^{-16}
$PbSO_4$	1.3×10^{-8}

Table J1

Solubility Products

Substance	K_{sp} at 25 °C
$Pb_3(PO_4)_2$	1×10^{-54}
magnesium	
$Mg(OH)_2$	8.9×10^{-12}
$MgCO_3 \cdot 3H_2O$	$ca\ 1 \times 10^{-5}$
$MgNH_4PO_4$	3×10^{-13}
MgF_2	6.4×10^{-9}
MgC_2O_4	7×10^{-7}
manganese	
$Mn(OH)_2$	2×10^{-13}
$MnCO_3$	8.8×10^{-11}
MnS	2.3×10^{-13}
mercury	
$Hg_2O \cdot H_2O$	3.6×10^{-26}
Hg_2Cl_2	1.1×10^{-18}
Hg_2Br_2	1.3×10^{-22}
Hg_2I_2	4.5×10^{-29}
Hg_2CO_3	9×10^{-15}
Hg_2SO_4	7.4×10^{-7}
Hg_2S	1.0×10^{-47}
Hg_2CrO_4	2×10^{-9}
HgS	1.6×10^{-54}
nickel	
$Ni(OH)_2$	1.6×10^{-16}
$NiCO_3$	1.4×10^{-7}
$NiS(\alpha)$	4×10^{-20}
$NiS(\beta)$	1.3×10^{-25}
potassium	
$KClO_4$	1.05×10^{-2}
K_2PtCl_6	7.48×10^{-6}
$KHC_4H_4O_6$	3×10^{-4}
silver	

Table J1

Solubility Products

Substance	K_{sp} at 25 °C
$\frac{1}{2}Ag_2O(Ag^+ + OH^-)$	2×10^{-8}
AgCl	1.6×10^{-10}
AgBr	5.0×10^{-13}
AgI	1.5×10^{-16}
AgCN	1.2×10^{-16}
AgSCN	1.0×10^{-12}
Ag_2S	1.6×10^{-49}
Ag_2CO_3	8.1×10^{-12}
Ag_2CrO_4	9.0×10^{-12}
$Ag_4Fe(CN)_6$	1.55×10^{-41}
Ag_2SO_4	1.2×10^{-5}
Ag_3PO_4	1.8×10^{-18}
strontium	
$Sr(OH)_2 \cdot 8H_2O$	3.2×10^{-4}
$SrCO_3$	7×10^{-10}
$SrCrO_4$	3.6×10^{-5}
$SrSO_4$	3.2×10^{-7}
$SrC_2O_4 \cdot H_2O$	4×10^{-7}
thallium	
TlCl	1.7×10^{-4}
TlSCN	1.6×10^{-4}
Tl_2S	6×10^{-22}
$Tl(OH)_3$	6.3×10^{-46}
tin	
$Sn(OH)_2$	3×10^{-27}
SnS	1×10^{-26}
$Sn(OH)_4$	1.0×10^{-57}
zinc	
$ZnCO_3$	2×10^{-10}

Table J1

Appendix K
Formation Constants For Complex Ions

Formation Constants for Complex Ions

Equilibrium	K_f
$Al^{3+} + 6F^- \rightleftharpoons [AlF_6]^{3-}$	7×10^{19}
$Cd^{2+} + 4NH_3 \rightleftharpoons [Cd(NH_3)_4]^{2+}$	1.3×10^7
$Cd^{2+} + 4CN^- \rightleftharpoons [Cd(CN)_4]^{2-}$	3×10^{18}
$Co^{2+} + 6NH_3 \rightleftharpoons [Co(NH_3)_6]^{2+}$	1.3×10^5
$Co^{3+} + 6NH_3 \rightleftharpoons [Co(NH_3)_6]^{3+}$	2.3×10^{33}
$Cu^+ + 2CN \rightleftharpoons [Cu(CN)_2]^-$	1.0×10^{16}
$Cu^{2+} + 4NH_3 \rightleftharpoons [Cu(NH_3)_4]^{2+}$	1.7×10^{13}
$Fe^{2+} + 6CN^- \rightleftharpoons [Fe(CN)_6]^{4-}$	1.5×10^{35}
$Fe^{3+} + 6CN^- \rightleftharpoons [Fe(CN)_6]^{3-}$	2×10^{43}
$Fe^{3+} + 6SCN^- \rightleftharpoons [Fe(SCN)_6]^{3-}$	3.2×10^3
$Hg^{2+} + 4Cl^- \rightleftharpoons [HgCl_4]^{2-}$	1.1×10^{16}
$Ni^{2+} + 6NH_3 \rightleftharpoons [Ni(NH_3)_6]^{2+}$	2.0×10^8
$Ag^+ + 2Cl^- \rightleftharpoons [AgCl_2]^-$	1.8×10^5
$Ag^+ + 2CN^- \rightleftharpoons [Ag(CN)_2]^-$	1×10^{21}
$Ag^+ + 2NH_3 \rightleftharpoons [Ag(NH_3)_2]^+$	1.7×10^7
$Zn^{2+} + 4CN^- \rightleftharpoons [Zn(CN)_4]^{2-}$	2.1×10^{19}
$Zn^{2+} + 4OH^- \rightleftharpoons [Zn(OH)_4]^{2-}$	2×10^{15}
$Fe^{3+} + SCN^- \rightleftharpoons [Fe(SCN)]^{2+}$	8.9×10^2
$Ag^+ + 4SCN^- \rightleftharpoons [Ag(SCN)_4]^{3-}$	1.2×10^{10}
$Pb^{2+} + 4I^- \rightleftharpoons [PbI_4]^{2-}$	3.0×10^4

Table K1

Formation Constants for Complex Ions

Equilibrium	K_f
$Pt^{2+} + 4Cl^- \rightleftharpoons [PtCl_4]^{2-}$	1×10^{16}
$Cu^{2+} + 4CN \rightleftharpoons [Cu(CN)_4]^{2-}$	1.0×10^{25}
$Co^{2+} + 4SCN^- \rightleftharpoons [Co(SCN)_4]^{2-}$	1×10^3

Table K1

Appendix L
Standard Electrode (Half-Cell) Potentials

Standard Electrode (Half-Cell) Potentials

Half-Reaction	$E°$ (V)
$Ag^+ + e^- \longrightarrow Ag$	+0.7996
$AgCl + e^- \longrightarrow Ag + Cl^-$	+0.22233
$[Ag(CN)_2]^- + e^- \longrightarrow Ag + 2CN^-$	−0.31
$Ag_2CrO_4 + 2e^- \longrightarrow 2Ag + CrO_4{}^{2-}$	+0.45
$[Ag(NH_3)_2]^+ + e^- \longrightarrow Ag + 2NH_3$	+0.373
$[Ag(S_2O_3)_2]^{3+} + e^- \longrightarrow Ag + 2S_2O_3{}^{2-}$	+0.017
$[AlF_6]^{3-} + 3e^- \longrightarrow Al + 6F^-$	−2.07
$Al^{3+} + 3e^- \longrightarrow Al$	−1.662
$Am^{3+} + 3e^- \longrightarrow Am$	−2.048
$Au^{3+} + 3e^- \longrightarrow Au$	+1.498
$Au^+ + e^- \longrightarrow Au$	+1.692
$Ba^{2+} + 2e^- \longrightarrow Ba$	−2.912
$Be^{2+} + 2e^- \longrightarrow Be$	−1.847
$Br_2(aq) + 2e^- \longrightarrow 2Br^-$	+1.0873
$Ca^{2+} + 2e^- \longrightarrow Ca$	−2.868
$Ce^3 + 3e^- \longrightarrow Ce$	−2.483
$Ce^{4+} + e^- \longrightarrow Ce^{3+}$	+1.61
$Cd^{2+} + 2e^- \longrightarrow Cd$	−0.4030
$[Cd(CN)_4]^{2-} + 2e^- \longrightarrow Cd + 4CN^-$	−1.09
$[Cd(NH_3)_4]^{2+} + 2e^- \longrightarrow Cd + 4NH_3$	−0.61
$CdS + 2e^- \longrightarrow Cd + S^{2-}$	−1.17
$Cl_2 + 2e^- \longrightarrow 2Cl^-$	+1.35827

Table L1

Standard Electrode (Half-Cell) Potentials

Half-Reaction	$E°$ (V)
$ClO_4^- + H_2O + 2e^- \longrightarrow ClO_3^- + 2OH^-$	+0.36
$ClO_3^- + H_2O + 2e^- \longrightarrow ClO_2^- + 2OH^-$	+0.33
$ClO_2^- + H_2O + 2e^- \longrightarrow ClO^- + 2OH^-$	+0.66
$ClO^- + H_2O + 2e^- \longrightarrow Cl^- + 2OH^-$	+0.89
$ClO_4^- + 2H_3O^+ + 2e^- \longrightarrow ClO_3^- + 3H_2O$	+1.189
$ClO_3^- + 3H_3O^+ + 2e^- \longrightarrow HClO_2 + 4H_2O$	+1.21
$HClO + H_3O^+ + 2e^- \longrightarrow Cl^- + 2H_2O$	+1.482
$HClO + H_3O^+ + e^- \longrightarrow \frac{1}{2}Cl_2 + 2H_2O$	+1.611
$HClO_2 + 2H_3O^+ + 2e^- \longrightarrow HClO + 3H_2O$	+1.628
$Co^{3+} + e^- \longrightarrow Co^{2+}$ (2 mol // H_2SO_4)	+1.83
$Co^{2+} + 2e^- \longrightarrow Co$	−0.28
$[Co(NH_3)_6]^{3+} + e^- \longrightarrow [Co(NH_3)_6]^{2+}$	+0.1
$Co(OH)_3 + e^- \longrightarrow Co(OH)_2 + OH^-$	+0.17
$Cr^3 + 3e^- \longrightarrow Cr$	−0.744
$Cr^{3+} + e^- \longrightarrow Cr^{2+}$	−0.407
$Cr^{2+} + 2e^- \longrightarrow Cr$	−0.913
$[Cu(CN)_2]^- + e^- \longrightarrow Cu + 2CN^-$	−0.43
$CrO_4^{2-} + 4H_2O + 3e^- \longrightarrow Cr(OH)_3 + 5OH^-$	−0.13
$Cr_2O_7^{2-} + 14H_3O^+ + 6e^- \longrightarrow 2Cr^{3+} + 21H_2O$	+1.232
$[Cr(OH)_4]^- + 3e^- \longrightarrow Cr + 4OH^-$	−1.2
$Cr(OH)_3 + 3e^- \longrightarrow Cr + 3OH^-$	−1.48
$Cu^{2+} + e^- \longrightarrow Cu^+$	+0.153
$Cu^{2+} + 2e^- \longrightarrow Cu$	+0.34
$Cu^+ + e^- \longrightarrow Cu$	+0.521
$F_2 + 2e^- \longrightarrow 2F^-$	+2.866
$Fe^{2+} + 2e^- \longrightarrow Fe$	−0.447

Table L1

Standard Electrode (Half-Cell) Potentials

Half-Reaction	$E°$ (V)
$Fe^{3+} + e^- \longrightarrow Fe^{2+}$	+0.771
$[Fe(CN)_6]^{3-} + e^- \longrightarrow [Fe(CN)_6]^{4-}$	+0.36
$Fe(OH)_2 + 2e^- \longrightarrow Fe + 2OH^-$	−0.88
$FeS + 2e^- \longrightarrow Fe + S^{2-}$	−1.01
$Ga^{3+} + 3e^- \longrightarrow Ga$	−0.549
$Gd^{3+} + 3e^- \longrightarrow Gd$	−2.279
$\frac{1}{2}H_2 + e^- \longrightarrow H^-$	−2.23
$2H_2O + 2e^- \longrightarrow H_2 + 2OH^-$	−0.8277
$H_2O_2 + 2H_3O^+ + 2e^- \longrightarrow 4H_2O$	+1.776
$2H_3O^+ + 2e^- \longrightarrow H_2 + 2H_2O$	0.00
$HO_2^- + H_2O + 2e^- \longrightarrow 3OH^-$	+0.878
$Hf^{4+} + 4e^- \longrightarrow Hf$	−1.55
$Hg^{2+} + 2e^- \longrightarrow Hg$	+0.851
$2Hg^{2+} + 2e^- \longrightarrow Hg_2^{2+}$	+0.92
$Hg_2^{2+} + 2e^- \longrightarrow 2Hg$	+0.7973
$[HgBr_4]^{2-} + 2e^- \longrightarrow Hg + 4Br^-$	+0.21
$Hg_2Cl_2 + 2e^- \longrightarrow 2Hg + 2Cl^-$	+0.26808
$[Hg(CN)_4]^{2-} + 2e^- \longrightarrow Hg + 4CN^-$	−0.37
$[HgI_4]^{2-} + 2e^- \longrightarrow Hg + 4I^-$	−0.04
$HgS + 2e^- \longrightarrow Hg + S^{2-}$	−0.70
$I_2 + 2e^- \longrightarrow 2I^-$	+0.5355
$In^{3+} + 3e^- \longrightarrow In$	−0.3382
$K^+ + e^- \longrightarrow K$	−2.931
$La^{3+} + 3e^- \longrightarrow La$	−2.52
$Li^+ + e^- \longrightarrow Li$	−3.04
$Lu^{3+} + 3e^- \longrightarrow Lu$	−2.28

Table L1

Standard Electrode (Half-Cell) Potentials

Half-Reaction	$E°$ (V)
$Mg^{2+} + 2e^- \longrightarrow Mg$	−2.372
$Mn^{2+} + 2e^- \longrightarrow Mn$	−1.185
$MnO_2 + 2H_2O + 2e^- \longrightarrow Mn(OH)_2 + 2OH^-$	−0.05
$MnO_4^- + 2H_2O + 3e^- \longrightarrow MnO_2 + 4OH^-$	+0.558
$MnO_2 + 4H^+ + 2e^- \longrightarrow Mn^{2+} + 2H_2O$	+1.23
$MnO_4^- + 8H^+ + 5e^- \longrightarrow Mn^{2+} + 4H_2O$	+1.507
$Na^+ + e^- \longrightarrow Na$	−2.71
$Nd^{3+} + 3e^- \longrightarrow Nd$	−2.323
$Ni^{2+} + 2e^- \longrightarrow Ni$	−0.257
$[Ni(NH_3)_6]^{2+} + 2e^- \longrightarrow Ni + 6NH_3$	−0.49
$NiO_2 + 4H^+ + 2e^- \longrightarrow Ni^{2+} + 2H_2O$	+1.593
$NiO_2 + 2H_2O + 2e^- \longrightarrow Ni(OH)_2 + 2OH^-$	+0.49
$NiS + 2e^- \longrightarrow Ni + S^{2-}$	+0.76
$NO_3^- + 4H^+ + 3e^- \longrightarrow NO + 2H_2O$	+0.957
$NO_3^- + 3H^+ + 2e^- \longrightarrow HNO_2 + H_2O$	+0.92
$NO_3^- + H_2O + 2e^- \longrightarrow NO_2^- + 2OH^-$	+0.10
$Np^{3+} + 3e^- \longrightarrow Np$	−1.856
$O_2 + 2H_2O + 4e^- \longrightarrow 4OH^-$	+0.401
$O_2 + 2H^+ + 2e^- \longrightarrow H_2O_2$	+0.695
$O_2 + 4H^+ + 4e^- \longrightarrow 2H_2O$	+1.229
$Pb^{2+} + 2e^- \longrightarrow Pb$	−0.1262
$PbO_2 + SO_4^{2-} + 4H^+ + 2e^- \longrightarrow PbSO_4 + 2H_2O$	+1.69
$PbS + 2e^- \longrightarrow Pb + S^{2-}$	−0.95
$PbSO_4 + 2e^- \longrightarrow Pb + SO_4^{2-}$	−0.3505
$Pd^{2+} + 2e^- \longrightarrow Pd$	+0.987
$[PdCl_4]^{2-} + 2e^- \longrightarrow Pd + 4Cl^-$	+0.591

Table L1

Standard Electrode (Half-Cell) Potentials

Half-Reaction	$E°$ (V)
$Pt^{2+} + 2e^- \longrightarrow Pt$	+1.20
$[PtBr_4]^{2-} + 2e^- \longrightarrow Pt + 4Br^-$	+0.58
$[PtCl_4]^{2-} + 2e^- \longrightarrow Pt + 4Cl^-$	+0.755
$[PtCl_6]^{2-} + 2e^- \longrightarrow [PtCl_4]^{2-} + 2Cl^-$	+0.68
$Pu^3 + 3e^- \longrightarrow Pu$	−2.03
$Ra^{2+} + 2e^- \longrightarrow Ra$	−2.92
$Rb^+ + e^- \longrightarrow Rb$	−2.98
$[RhCl_6]^{3-} + 3e^- \longrightarrow Rh + 6Cl^-$	+0.44
$S + 2e^- \longrightarrow S^{2-}$	−0.47627
$S + 2H^+ + 2e^- \longrightarrow H_2S$	+0.142
$Sc^{3+} + 3e^- \longrightarrow Sc$	−2.09
$Se + 2H^+ + 2e^- \longrightarrow H_2Se$	−0.399
$[SiF_6]^{2-} + 4e^- \longrightarrow Si + 6F^-$	−1.2
$SiO_3{}^{2-} + 3H_2O + 4e^- \longrightarrow Si + 6OH^-$	−1.697
$SiO_2 + 4H^+ + 4e^- \longrightarrow Si + 2H_2O$	−0.86
$Sm^{3+} + 3e^- \longrightarrow Sm$	−2.304
$Sn^{4+} + 2e^- \longrightarrow Sn^{2+}$	+0.151
$Sn^{2+} + 2e^- \longrightarrow Sn$	−0.1375
$[SnF_6]^{2-} + 4e^- \longrightarrow Sn + 6F^-$	−0.25
$SnS + 2e^- \longrightarrow Sn + S^{2-}$	−0.94
$Sr^{2+} + 2e^- \longrightarrow Sr$	−2.89
$TeO_2 + 4H^+ + 4e^- \longrightarrow Te + 2H_2O$	+0.593
$Th^{4+} + 4e^- \longrightarrow Th$	−1.90
$Ti^{2+} + 2e^- \longrightarrow Ti$	−1.630
$U^{3+} + 3e^- \longrightarrow U$	−1.79
$V^{2+} + 2e^- \longrightarrow V$	−1.19

Table L1

Standard Electrode (Half-Cell) Potentials

Half-Reaction	$E°$ (V)
$Y^{3+} + 3e^- \longrightarrow Y$	−2.37
$Zn^{2+} + 2e^- \longrightarrow Zn$	−0.7618
$[Zn(CN)_4]^{2-} + 2e^- \longrightarrow Zn + 4CN^-$	−1.26
$[Zn(NH_3)_4]^{2+} + 2e^- \longrightarrow Zn + 4NH_3$	−1.04
$Zn(OH)_2 + 2e^- \longrightarrow Zn + 2OH^-$	−1.245
$[Zn(OH)_4]^2 + 2e^- \longrightarrow Zn + 4OH^-$	−1.199
$ZnS + 2e^- \longrightarrow Zn + S^{2-}$	−1.40
$Zr^4 + 4e^- \longrightarrow Zr$	−1.539

Table L1

Appendix M

Half-Lives For Several Radioactive Isotopes

Half-Lives for Several Radioactive Isotopes

Isotope	Half-Life[1]	Type of Emission[2]	Isotope	Half-Life[3]	Type of Emission[4]
$^{14}_{6}C$	5730 y	(β^-)	$^{210}_{83}Bi$	5.01 d	(β^-)
$^{13}_{7}N$	9.97 m	(β^+)	$^{212}_{83}Bi$	60.55 m	$(\alpha \text{ or } \beta^-)$
$^{15}_{9}F$	4.1×10^{-22} s	(p)	$^{210}_{84}Po$	138.4 d	(α)
$^{24}_{11}Na$	15.00 h	(β^-)	$^{212}_{84}Po$	3×10^{-7} s	(α)
$^{32}_{15}P$	14.29 d	(β^-)	$^{216}_{84}Po$	0.15 s	(α)
$^{40}_{19}K$	1.27×10^9 y	$(\beta \text{ or } E.C.)$	$^{218}_{84}Po$	3.05 m	(α)
$^{49}_{26}Fe$	0.08 s	(β^+)	$^{215}_{85}At$	1.0×10^{-4} s	(α)
$^{60}_{26}Fe$	2.6×10^6 y	(β^-)	$^{218}_{85}At$	1.6 s	(α)
$^{60}_{27}Co$	5.27 y	(β^-)	$^{220}_{86}Rn$	55.6 s	(α)
$^{87}_{37}Rb$	4.7×10^{10} y	(β^-)	$^{222}_{86}Rn$	3.82 d	(α)
$^{90}_{38}Sr$	29 y	(β^-)	$^{224}_{88}Ra$	3.66 d	(α)
$^{115}_{49}In$	5.1×10^{15} y	(β^-)	$^{226}_{88}Ra$	1600 y	(α)
$^{131}_{53}I$	8.040 d	(β^-)	$^{228}_{88}Ra$	5.75 y	(β^-)
$^{142}_{58}Ce$	5×10^{15} y	(α)	$^{228}_{89}Ac$	6.13 h	(β^-)
$^{208}_{81}Tl$	3.07 m	(β^-)	$^{228}_{90}Th$	1.913 y	(α)
$^{210}_{82}Pb$	22.3 y	(β^-)	$^{232}_{90}Th$	1.4×10^{10} y	(α)
$^{212}_{82}Pb$	10.6 h	(β^-)	$^{233}_{90}Th$	22 m	(β^-)
$^{214}_{82}Pb$	26.8 m	(β^-)	$^{234}_{90}Th$	24.10 d	(β^-)

Table M1

1. y = years, d = days, h = hours, m = minutes, s = seconds
2. *E.C.* = electron capture, *S.F.* = Spontaneous fission
3. y = years, d = days, h = hours, m = minutes, s = seconds
4. *E.C.* = electron capture, *S.F.* = Spontaneous fission

Half-Lives for Several Radioactive Isotopes

Isotope	Half-Life[5]	Type of Emission[6]	Isotope	Half-Life[7]	Type of Emission[8]
$^{206}_{83}Bi$	6.243 d	$(E.C.)$	$^{233}_{91}Pa$	27 d	(β^-)
$^{233}_{92}U$	1.59×10^5 y	(α)	$^{242}_{96}Cm$	162.8 d	(α)
$^{234}_{92}U$	2.45×10^5 y	(α)	$^{243}_{97}Bk$	4.5 h	$(\alpha \text{ or } E.C.)$
$^{235}_{92}U$	7.03×10^8 y	(α)	$^{253}_{99}Es$	20.47 d	(α)
$^{238}_{92}U$	4.47×10^9 y	(α)	$^{254}_{100}Fm$	3.24 h	$(\alpha \text{ or } S.F.)$
$^{239}_{92}U$	23.54 m	(β^-)	$^{255}_{100}Fm$	20.1 h	(α)
$^{239}_{93}Np$	2.3 d	(β^-)	$^{256}_{101}Md$	76 m	$(\alpha \text{ or } E.C.)$
$^{239}_{94}Pu$	2.407×10^4 y	(α)	$^{254}_{102}No$	55 s	(α)
$^{239}_{94}Pu$	6.54×10^3 y	(α)	$^{257}_{103}Lr$	0.65 s	(α)
$^{241}_{94}Pu$	14.4 y	$(\alpha \text{ or } \beta^-)$	$^{260}_{105}Ha$	1.5 s	$(\alpha \text{ or } S.F.)$
$^{241}_{95}Am$	432.2 y	(α)	$^{263}_{106}Sg$	0.8 s	$(\alpha \text{ or } S.F.)$

Table M1

---

5. y = years, d = days, h = hours, m = minutes, s = seconds
6. *E.C.* = electron capture, *S.F.* = Spontaneous fission
7. y = years, d = days, h = hours, m = minutes, s = seconds
8. *E.C.* = electron capture, *S.F.* = Spontaneous fission

Answer Key

Chapter 12

1. The instantaneous rate is the rate of a reaction at any particular point in time, a period of time that is so short that the concentrations of reactants and products change by a negligible amount. The initial rate is the instantaneous rate of reaction as it starts (as product just begins to form). Average rate is the average of the instantaneous rates over a time period.

3. rate $= +\dfrac{1}{2}\dfrac{\Delta[CIF_3]}{\Delta t} = -\dfrac{\Delta[Cl_2]}{\Delta t} = -\dfrac{1}{3}\dfrac{\Delta[F_2]}{\Delta t}$

5. (a) average rate, $0 - 10$ s $= 0.0375$ mol L^{-1} s^{-1}; average rate, $12 - 18$ s $= 0.0225$ mol L^{-1} s^{-1}; (b) instantaneous rate, 15 s $= 0.0500$ mol L^{-1} s^{-1}; (c) average rate for B formation $= 0.0188$ mol L^{-1} s^{-1}; instantaneous rate for B formation $= 0.0250$ mol L^{-1} s^{-1}

7. Higher molarity increases the rate of the reaction. Higher temperature increases the rate of the reaction. Smaller pieces of magnesium metal will react more rapidly than larger pieces because more reactive surface exists.

9. (a) Depending on the angle selected, the atom may take a long time to collide with the molecule and, when a collision does occur, it may not result in the breaking of the bond and the forming of the other. (b) Particles of reactant must come into contact with each other before they can react.

11. (a) very slow; (b) As the temperature is increased, the reaction proceeds at a faster rate. The amount of reactants decreases, and the amount of products increases. After a while, there is a roughly equal amount of *BC*, *AB*, and *C* in the mixture and a slight excess of *A*.

13. (a) 2; (b) 1

15. (a) The process reduces the rate by a factor of 4. (b) Since CO does not appear in the rate law, the rate is not affected.

17. 4.3×10^{-5} mol/L/s

19. 7.9×10^{-13} mol/L/year

21. rate $= k$; $k = 2.0 \times 10^{-2}$ mol/L/h (about 0.9 g/L/h for the average male); The reaction is zero order.

23. rate $= k[NOC]^2$; $k = 8.0 \times 10^{-8}$ L/mol/s; second order

25. rate $= k[NO]^2[Cl]_2$; $k = 9.12$ L^2 mol^{-2} h^{-1}; second order in NO; first order in Cl_2

27. (a) The rate equation is second order in A and is written as rate $= k[A]^2$. (b) $k = 7.88 \times 10^{-13}$ L mol^{-1} s^{-1}

29. (a) 2.5×10^{-4} mol/L/min

31. rate $= k[I^-][OCl^{-1}]$; $k = 6.1 \times 10^{-2}$ L mol $^{-1}$ s^{-1}

33. Plotting a graph of $\ln[SO_2Cl_2]$ versus t reveals a linear trend; therefore we know this is a first-order reaction:

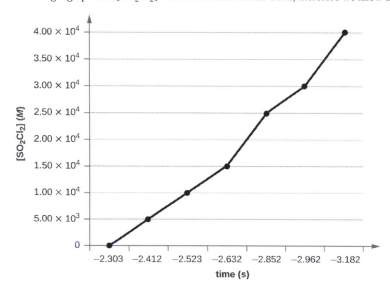

$k = -2.20 \times 10^5$ s^{-1}

35.

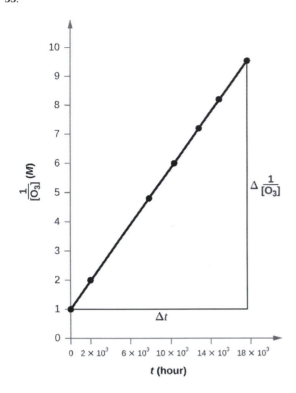

The plot is nicely linear, so the reaction is second order.
$k = 50.1 \text{ L mol}^{-1} \text{ h}^{-1}$

37. 14.3 d

39. 8.3×10^7 s

41. 0.826 s

43. The reaction is first order.
$k = 1.0 \times 10^7 \text{ mol}^{-1} \text{ min}^{-1}$

45. 4.98; 20% remains

47. 252 days

49.

$[A]_0$ (M)	$k \times 10^3$ (s^{-1})
4.88	2.45
3.52	2.51
2.29	2.54
1.81	2.58
5.33	2.35
4.05	2.44

$[A]_0$ (M)	$k \times 10^3$ (s^{-1})
2.95	2.47
1.72	2.43

51. The reactants either may be moving too slowly to have enough kinetic energy to exceed the activation energy for the reaction, or the orientation of the molecules when they collide may prevent the reaction from occurring.

53. The activation energy is the minimum amount of energy necessary to form the activated complex in a reaction. It is usually expressed as the energy necessary to form one mole of activated complex.

55. After finding k at several different temperatures, a plot of ln k versus $\dfrac{-E_a}{R}$ gives a straight line with the slope $\dfrac{1}{T}$, from which E_a may be determined.

57. (a) 4-times faster (b) 128-times faster

59. 3.9×10^{15} s^{-1}

61. 43.0 kJ/mol

63. 177 kJ/mol

65. $E_a = 108$ kJ
$A = 2.0 \times 10^8$ s^{-1}
$k = 3.2 \times 10^{-10}$ s^{-1}
(b) 1.81×10^8 h or 7.6×10^6 day. (c) Assuming that the reaction is irreversible simplifies the calculation because we do not have to account for any reactant that, having been converted to product, returns to the original state.

67. The A atom has enough energy to react with BC; however, the different angles at which it bounces off of BC without reacting indicate that the orientation of the molecule is an important part of the reaction kinetics and determines whether a reaction will occur.

69. No. In general, for the overall reaction, we cannot predict the effect of changing the concentration without knowing the rate equation. Yes. If the reaction is an elementary reaction, then doubling the concentration of A doubles the rate.

71. Rate = $k[A][B]^2$; Rate = $k[A]^3$

73. (a) Rate$_1$ = $k[O_3]$; (b) Rate$_2$ = $k[O_3][Cl]$; (c) Rate$_3$ = $k[ClO][O]$; (d) Rate$_2$ = $k[O_3][NO]$; (e) Rate$_3$ = $k[NO_2][O]$

75. (a) Doubling $[H_2]$ doubles the rate. $[H_2]$ must enter the rate equation to the first power. Doubling [NO] increases the rate by a factor of 4. [NO] must enter the rate law to the second power. (b) Rate = k [NO]^2[H$_2$]; (c) $k = 5.0 \times 10^3$ mol^{-2} L^{-2} min^{-1}; (d) 0.0050 mol/L; (e) Step II is the rate-determining step. If step I gives N_2O_2 in adequate amount, steps 1 and 2 combine to give $2NO + H_2 \longrightarrow H_2O + N_2O$. This reaction corresponds to the observed rate law. Combine steps 1 and 2 with step 3, which occurs by supposition in a rapid fashion, to give the appropriate stoichiometry.

77. The general mode of action for a catalyst is to provide a mechanism by which the reactants can unite more readily by taking a path with a lower reaction energy. The rates of both the forward and the reverse reactions are increased, leading to a faster achievement of equilibrium.

79. (a) Chlorine atoms are a catalyst because they react in the second step but are regenerated in the third step. Thus, they are not used up, which is a characteristic of catalysts. (b) NO is a catalyst for the same reason as in part (a).

81. The lowering of the transition state energy indicates the effect of a catalyst. (a) A; (b) B

83. The energy needed to go from the initial state to the transition state is (a) 10 kJ; (b) 10 kJ

85. Both have the same activation energy, so they also have the same rate.

Chapter 13

1. The reaction can proceed in both the forward and reverse directions.

3. When a system has reached equilibrium, no further changes in the reactant and product concentrations occur; the reactions continue to occur, but at equivalent rates.

5. The concept of equilibrium does not imply equal concentrations, though it is possible.

7. Equilibrium cannot be established between the liquid and the gas phase if the top is removed from the bottle because the system is not closed; one of the components of the equilibrium, the Br_2 vapor, would escape from the bottle until all liquid disappeared. Thus, more liquid would evaporate than can condense back from the gas phase to the liquid phase.

9. (a) $K_c = [Ag^+][Cl^-] < 1$. AgCl is insoluble; thus, the concentrations of ions are much less than 1 M; (b) $K_c = \dfrac{1}{[Pb^{2+}][Cl^-]^2} > 1$ because $PbCl_2$ is insoluble and formation of the solid will reduce the concentration of ions to a low level (<1 M).

11. Since $K_c = \dfrac{[C_6H_6]}{[C_2H_2]^3}$, a value of $K_c \approx 10$ means that C_6H_6 predominates over C_2H_2. In such a case, the reaction would be commercially feasible if the rate to equilibrium is suitable.

13. $K_c > 1$

15. (a) $Q_c = \dfrac{[CH_3Cl][HCl]}{[CH_4][Cl_2]}$; (b) $Q_c = \dfrac{[NO]^2}{[N_2][O_2]}$; (c) $Q_c = \dfrac{[SO_3]^2}{[SO_2]^2[O_2]}$; (d) $Q_c = [SO_2]$; (e) $Q_c = \dfrac{1}{[P_4][O_2]^5}$; (f) $Q_c = \dfrac{[Br]^2}{[Br_2]}$; (g) $Q_c = \dfrac{[CO_2]}{[CH_4][O_2]^2}$; (h) $Q_c = [H_2O]^5$

17. (a) Q_c 25 proceeds left; (b) Q_P 0.22 proceeds right; (c) Q_c undefined proceeds left; (d) Q_P 1.00 proceeds right; (e) Q_P 0 proceeds right; (f) Q_c 4 proceeds left

19. The system will shift toward the reactants to reach equilibrium.

21. (a) homogenous; (b) homogenous; (c) homogenous; (d) heterogeneous; (e) heterogeneous; (f) homogenous; (g) heterogeneous; (h) heterogeneous

23. This situation occurs in (a) and (b).

25. (a) $K_P = 1.6 \times 10^{-4}$; (b) $K_P = 50.2$; (c) $K_c = 5.31 \times 10^{-39}$; (d) $K_c = 4.60 \times 10^{-3}$

27. $K_P = P_{H_2O} = 0.042$.

29. $Q_c = \dfrac{[NH_4^+][OH^-]}{[HN_3]}$

31. The amount of $CaCO_3$ must be so small that P_{CO_2} is less than K_P when the $CaCO_3$ has completely decomposed. In other words, the starting amount of $CaCO_3$ cannot completely generate the full P_{CO_2} required for equilibrium.

33. The change in enthalpy may be used. If the reaction is exothermic, the heat produced can be thought of as a product. If the reaction is endothermic the heat added can be thought of as a reactant. Additional heat would shift an exothermic reaction back to the reactants but would shift an endothermic reaction to the products. Cooling an exothermic reaction causes the reaction to shift toward the product side; cooling an endothermic reaction would cause it to shift to the reactants' side.

35. No, it is not at equilibrium. Because the system is not confined, products continuously escape from the region of the flame; reactants are also added continuously from the burner and surrounding atmosphere.

37. Add N_2; add H_2; decrease the container volume; heat the mixture.

39. (a) ΔT increase = shift right, ΔP increase = shift left; (b) ΔT increase = shift right, ΔP increase = no effect; (c) ΔT increase = shift left, ΔP increase = shift left; (d) ΔT increase = shift left, ΔP increase = shift right.

41. (a) $K_c = \dfrac{[CH_3OH]}{[H_2]^2[CO]}$; (b) [H$_2$] increases, [CO] decreases, [CH$_3$OH] increases; (c), [H$_2$] increases, [CO] decreases, [CH$_3$OH] decreases; (d), [H$_2$] increases, [CO] increases, [CH$_3$OH] increases; (e), [H$_2$] increases, [CO] increases, [CH$_3$OH] decreases; (f), no changes.

43. (a) $K_c = \dfrac{[CO][H_2]}{[H_2O]}$; (b) [H$_2$O] no change, [CO] no change, [H$_2$] no change; (c) [H$_2$O] decreases, [CO] decreases, [H$_2$] decreases; (d) [H$_2$O] increases, [CO] increases, [H$_2$] decreases; (f) [H$_2$O] decreases, [CO] increases, [H$_2$] increases. In (b), (c), (d), and (e), the mass of carbon will change, but its concentration (activity) will not change.

45. Only (b)

47. Add NaCl or some other salt that produces Cl$^-$ to the solution. Cooling the solution forces the equilibrium to the right, precipitating more AgCl(s).

49. (a)

51. $K_c = \dfrac{[C]^2}{[A][B]^2}$. [A] = 0.1 M, [B] = 0.1 M, [C] = 1 M; and [A] = 0.01, [B] = 0.250, [C] = 0.791.

53. $K_c = 6.00 \times 10^{-2}$

55. $K_c = 0.50$

57. The equilibrium equation is
$K_P = 1.9 \times 10^3$

59. $K_P = 3.06$

61. (a) $-2x$, $2x$, $-0.250\ M$, $0.250\ M$; (b) $4x$, $-2x$, $-6x$, $0.32\ M$, $-0.16\ M$, $-0.48\ M$; (c) $-2x$, $3x$, -50 torr, 75 torr; (d) x, $-x$, $-3x$, 5 atm, -5 atm, -15 atm; (e) x, $1.03 \times 10^{-4}\ M$; (f) x, 0.1 atm.

63. Activities of pure crystalline solids equal 1 and are constant; however, the mass of Ni does change.

65. [NH$_3$] = $9.1 \times 10^{-2}\ M$

67. $P_{BrCl} = 4.9 \times 10^{-2}$ atm

69. [CO] = $2.0 \times 10^{-4}\ M$

71. $P_{H_2O} = 3.64 \times 10^{-3}$ atm

73. Calculate Q based on the calculated concentrations and see if it is equal to K_c. Because Q does equal 4.32, the system must be at equilibrium.

75. (a) [NO$_2$] = $1.17 \times 10^{-3}\ M$
[N$_2$O$_4$] = 0.128 M
(b) Percent error $= \dfrac{5.87 \times 10^{-4}}{0.129} \times 100\% = 0.455\%$. The change in concentration of N$_2$O$_4$ is far less than the 5% maximum allowed.

77. (a) [H$_2$S] = 0.810 atm
[H$_2$] = 0.014 atm
[S$_2$] = 0.0072 atm
(b) The $2x$ is dropped from the equilibrium calculation because 0.014 is negligible when subtracted from 0.824. The percent error associated with ignoring $2x$ is $\dfrac{0.014}{0.824} \times 100\% = 1.7\%$, which is less than allowed by the "5% test."
The error is, indeed, negligible.

79. [PCl$_3$] = 1.80 M; [PC$_3$] = 0.195 M; [PCl$_3$] = 0.195 M.

81. [NO$_2$] = 0.19 M
[NO] = 0.0070 M
[O$_2$] = 0.0035 M

83. $P_{O_3} = 4.9 \times 10^{-26}$ atm

85. 507 g

87. 330 g

89. (a) Both gases must increase in pressure.
(b) $P_{N_2O_4} = 8.0$ atm and $P_{NO_2} = 1.0$ atm

91. (a) 0.33 mol.
(b) $[CO]^2 = 0.50$ M Added H_2 forms some water to compensate for the removal of water vapor and as a result of a shift to the left after H_2 is added.

93. $P_{H_2} = 8.64 \times 10^{-11}$ atm
$P_{O_2} = 0.250$ atm
$P_{H_2O} = 0.500$ atm

95. (a) $K_c = \dfrac{[NH_3]^4 [O_2]^7}{[NO_2]^4 [H_2O]^6}$. (b) $[NH_3]$ must increase for Q_c to reach K_c. (c) That decrease in pressure would decrease $[NO_2]$. (d) $P_{O_2} = 49$ torr

97. [fructose] = 0.15 M

99. $P_{N_2O_3} = 1.90$ atm and $P_{NO} = P_{NO_2} = 1.90$ atm

101. (a) HB ionizes to a greater degree and has the larger K_c.
(b) $K_c(HA) = 5 \times 10^{-4}$
$K_c(HB) = 3 \times 10^{-3}$

Chapter 14

1. One example for NH_3 as a conjugate acid: $NH_2{}^- + H^+ \longrightarrow NH_3$; as a conjugate base:
$NH_4{}^+(aq) + OH^-(aq) \longrightarrow NH_3(aq) + H_2O(l)$

3. (a) $H_3O^+(aq) \longrightarrow H^+(aq) + H_2O(l)$; (b) $HCl(l) \longrightarrow H^+(aq) + Cl^-(aq)$; (c) $NH_3(aq) \longrightarrow H^+(aq) + NH_2{}^-(aq)$; (d) $CH_3CO_2H(aq) \longrightarrow H^+(aq) + CH_3CO_2{}^-(aq)$; (e) $NH_4{}^+(aq) \longrightarrow H^+(aq) + NH_3(aq)$; (f) $HSO_4{}^-(aq) \longrightarrow H^+(aq) + SO_4{}^{2-}(aq)$

5. (a) $H_2O(l) + H^+(aq) \longrightarrow H_3O^+(aq)$; (b) $OH^-(aq) + H^+(aq) \longrightarrow H_2O(l)$; (c) $NH_3(aq) + H^+(aq) \longrightarrow NH_4{}^+(aq)$; (d) $CN^-(aq) + H^+(aq) \longrightarrow HCN(aq)$; (e) $S^{2-}(aq) + H^+(aq) \longrightarrow HS^-(aq)$; (f) $H_2PO_4{}^-(aq) + H^+(aq) \longrightarrow H_3PO_4(aq)$

7. (a) H_2O, O^{2-}; (b) H_3O^+, OH^-; (c) H_2CO_3, $CO_3{}^{2-}$; (d) $NH_4{}^+$, $NH_2{}^-$; (e) H_2SO_4, $SO_4{}^{2-}$; (f) $H_3O_2{}^+$, $HO_2{}^-$; (g) H_2S; S^{2-}; (h) $H_6N_2{}^{2+}$, H_4N_2

9. The labels are Brønsted-Lowry acid = BA; its conjugate base = CB; Brønsted-Lowry base = BB; its conjugate acid = CA. (a) $HNO_3(BA)$, $H_2O(BB)$, $H_3O^+(CA)$, $NO_3{}^-(CB)$; (b) $CN^-(BB)$, $H_2O(BA)$, $HCN(CA)$, $OH^-(CB)$; (c) $H_2SO_4(BA)$, $Cl^-(BB)$, $HCl(CA)$, $HSO_4{}^-(CB)$; (d) $HSO_4{}^-(BA)$, $OH^-(BB)$, $SO_4{}^{2-}(CB)$, $H_2O(CA)$; (e) $O^{2-}(BB)$, $H_2O(BA)$ $OH^-(CB$ and $CA)$; (f) $[Cu(H_2O)_3(OH)]^+(BB)$, $[Al(H_2O)_6]^{3+}(BA)$, $[Cu(H_2O)_4]^{2+}(CA)$, $[Al(H_2O)_5(OH)]^{2+}(CB)$; (g) $H_2S(BA)$, $NH_2{}^-(BB)$, $HS^-(CB)$, $NH_3(CA)$

11. Amphiprotic species may either gain or lose a proton in a chemical reaction, thus acting as a base or an acid. An example is H_2O. As an acid:
$H_2O(aq) + NH_3(aq) \rightleftharpoons NH_4{}^+(aq) + OH^-(aq)$. As a base: $H_2O(aq) + HCl(aq) \rightleftharpoons H_3O^+(aq) + Cl^-(aq)$

13. amphiprotic: (a) $NH_3 + H_3O^+ \longrightarrow NH_4OH + H_2O$, $NH_3 + OCH_3^- \longrightarrow NH_2^- + CH_3OH$; (b) $HPO_4^{2-} + OH^- \longrightarrow PO_4^{3-} + H_2O$, $HPO_4^{2-} + HClO_4 \longrightarrow H_2PO_4^- + ClO_4^-$; not amphiprotic: (c) Br^-; (d) NH_4^+; (e) AsO_4^{3-}

15. In a neutral solution $[H_3O^+] = [OH^-]$. At 40 °C,
$[H_3O^+] = [OH^-] = (2.910^{-14})^{1/2} = 1.7 \times 10^{-7}$.

17. $x = 3.101 \times 10^{-7} M = [H_3O^+] = [OH^-]$
$pH = -\log 3.101 \times 10^{-7} = -(-6.5085) = 6.5085$
$pOH = pH = 6.5085$

19. (a) pH = 3.587; pOH = 10.413; (b) pH = 0.68; pOH = 13.32; (c) pOH = 3.85; pH = 10.15; (d) pH = −0.40; pOH = 14.4

21. $[H_3O^+] = 3.0 \times 10^{-7} M$; $[OH^-] = 3.3 \times 10^{-8} M$

23. $[H_3O^+] = 1 \times 10^{-2} M$; $[OH^-] = 1 \times 10^{-12} M$

25. $[OH^-] = 3.1 \times 10^{-12} M$

27. The salt ionizes in solution, but the anion slightly reacts with water to form the weak acid. This reaction also forms OH^-, which causes the solution to be basic.

29. $[H_2O] > [CH_3CO_2H] > [H_3O^+] \approx [CH_3CO_2^-] > [OH^-]$

31. The oxidation state of the sulfur in H_2SO_4 is greater than the oxidation state of the sulfur in H_2SO_3.

33. $Mg(OH)_2(s) + HCl(aq) \longrightarrow Mg^{2+}(aq) + 2Cl^-(aq) + 2H_2O(l)$
    BB   BA   CB   CA

35. $K_a = 2.3 \times 10^{-11}$

37. The strongest base or strongest acid is the one with the larger K_b or K_a, respectively. In these two examples, they are $(CH_3)_2NH$ and $CH_3NH_3^+$.

39. triethylamine.

41. (a) HSO_4^-; higher electronegativity of the central ion. (b) H_2O; NH_3 is a base and water is neutral, or decide on the basis of K_a values. (c) HI; PH_3 is weaker than HCl; HCl is weaker than HI. Thus, PH_3 is weaker than HI. (d) PH_3; in binary compounds of hydrogen with nonmetals, the acidity increases for the element lower in a group. (e) HBr; in a period, the acidity increases from left to right; in a group, it increases from top to bottom. Br is to the left and below S, so HBr is the stronger acid.

43. (a) $NaHSeO_3 < NaHSO_3 < NaHSO_4$; in polyoxy acids, the more electronegative central element—S, in this case—forms the stronger acid. The larger number of oxygen atoms on the central atom (giving it a higher oxidation state) also creates a greater release of hydrogen atoms, resulting in a stronger acid. As a salt, the acidity increases in the same manner. (b) $ClO_2^- < BrO_2^- < IO_2^-$; the basicity of the anions in a series of acids will be the opposite of the acidity in their oxyacids. The acidity increases as the electronegativity of the central atom increases. Cl is more electronegative than Br, and I is the least electronegative of the three. (c) $HOI < HOBr < HOCl$; in a series of the same form of oxyacids, the acidity increases as the electronegativity of the central atom increases. Cl is more electronegative than Br, and I is the least electronegative of the three. (d) $HOCl < HOClO < HOClO_2 < HOClO_3$; in a series of oxyacids of the same central element, the acidity increases as the number of oxygen atoms increases (or as the oxidation state of the central atom increases). (e) $HTe^- < HS^- << PH_2^- < NH_2^-$; PH_2^- and NH_2^- are anions of weak bases, so they act as strong bases toward H^+. HTe^- and HS^- are anions of weak acids, so they have less basic character. In a periodic group, the more electronegative element has the more basic anion. (f) $BrO_4^- < BrO_3^- < BrO_2^- < BrO^-$; with a larger number of oxygen atoms (that is, as the oxidation state of the central ion increases), the corresponding acid becomes more acidic and the anion consequently less basic.

45. $[H_2O] > [C_6H_4OH(CO_2H)] > [H^+]0 > [C_6H_4OH(CO_2)^-] \gg [C_6H_4O(CO_2H)^-] > [OH^-]$

47. Strong electrolytes are 100% ionized, and, as long as the component ions are neither weak acids nor weak bases, the ionic species present result from the dissociation of the strong electrolyte. Equilibrium calculations are necessary when one (or more) of the ions is a weak acid or a weak base.

49. 1. Assume that the change in initial concentration of the acid as the equilibrium is established can be neglected, so this concentration can be assumed constant and equal to the initial value of the total acid concentration. 2. Assume we can neglect the contribution of water to the equilibrium concentration of H_3O^+.

51. (b) The addition of HCl

53. (a) Adding HCl will add H_3O^+ ions, which will then react with the OH^- ions, lowering their concentration. The equilibrium will shift to the right, increasing the concentration of HNO_2, and decreasing the concentration of NO_2^- ions. (b) Adding HNO_2 increases the concentration of HNO_2 and shifts the equilibrium to the left, increasing the concentration of NO_2^- ions and decreasing the concentration of OH^- ions. (c) Adding NaOH adds OH^- ions, which shifts the equilibrium to the left, increasing the concentration of NO_2^- ions and decreasing the concentrations of HNO_2. (d) Adding NaCl has no effect on the concentrations of the ions. (e) Adding KNO_2 adds NO_2^- ions and shifts the equilibrium to the right, increasing the HNO_2 and OH^- ion concentrations.

55. This is a case in which the solution contains a mixture of acids of different ionization strengths. In solution, the HCO_2H exists primarily as HCO_2H molecules because the ionization of the weak acid is suppressed by the strong acid. Therefore, the HCO_2H contributes a negligible amount of hydronium ions to the solution. The stronger acid, HCl, is the dominant producer of hydronium ions because it is completely ionized. In such a solution, the stronger acid determines the concentration of hydronium ions, and the ionization of the weaker acid is fixed by the $[H_3O^+]$ produced by the stronger acid.

57. (a) $K_b = 1.8 \times 10^{-5}$;

(b) $K_a = 4.5 \times 10^{-4}$;

(c) $K_b = 7.4 \times 10^{-5}$;

(d) $K_a = 5.6 \times 10^{-10}$

59. $K_a = 1.2 \times 10^{-2}$

61. (a) $K_b = 4.3 \times 10^{-12}$;

(b) $K_a = 1.4 \times 10^{-10}$;

(c) $K_b = 1 \times 10^{-7}$;

(d) $K_b = 4.2 \times 10^{-3}$;

(e) $K_b = 4.2 \times 10^{-3}$;

(f) $K_b = 8.3 \times 10^{-13}$

63. (a) is the correct statement.

65. $[H_3O^+] = 7.5 \times 10^{-3} M$
$[HNO_2] = 0.126$
$[OH^-] = 1.3 \times 10^{-12} M$
$[BrO^-] = 3.2 \times 10^{-8} M$
$[HBrO] = 0.120 M$

67. $[OH^-] = [NO_4^+] = 0.0014 M$
$[NH_3] = 0.144 M$
$[H_3O^+] = 6.9 \times 10^{-12} M$
$[C_6H_5NH_3^+] = 3.9 \times 10^{-8} M$
$[C_6H_5NH_2] = 0.100 M$

69. (a) $\dfrac{[H_3O^+][ClO^-]}{[HClO]} = \dfrac{(x)(x)}{(0.0092 - x)} \approx \dfrac{(x)(x)}{0.0092} = 3.5 \times 10^{-8}$

Solving for x gives $1.79 \times 10^{-5} M$. This value is less than 5% of 0.0092, so the assumption that it can be neglected is valid. Thus, the concentrations of solute species at equilibrium are:

$[H_3O^+] = [ClO] = 1.8 \times 10^{-5}\ M$
$[HClO] = 0.00092\ M$
$[OH^-] = 5.6 \times 10^{-10}\ M$;

(b) $\dfrac{[C_6H_5NH_3^+][OH^-]}{[C_6H_5NH_2]} = \dfrac{(x)(x)}{(0.0784 - x)} \approx \dfrac{(x)(x)}{0.0784} = 4.6 \times 10^{-10}$

Solving for x gives $6.01 \times 10^{-6}\ M$. This value is less than 5% of 0.0784, so the assumption that it can be neglected is valid. Thus, the concentrations of solute species at equilibrium are:

$[CH_3CO_2^-] = [OH^-] = 6.0 \times 10^{-6}\ M$
$[C_6H_5NH_2] = 0.00784$
$[H_3O^+] = 1.7 \times 10^{-9}\ M$;

(c) $\dfrac{[H_3O^+][CN^-]}{[HCN]} = \dfrac{(x)(x)}{(0.0810 - x)} \approx \dfrac{(x)(x)}{0.0810} = 4 \times 10^{-10}$

Solving for x gives $5.69 \times 10^{-6}\ M$. This value is less than 5% of 0.0810, so the assumption that it can be neglected is valid. Thus, the concentrations of solute species at equilibrium are:

$[H_3O^+] = [CN^-] = 5.7 \times 10^{-6}\ M$
$[HCN] = 0.0810\ M$
$[OH^-] = 1.8 \times 10^{-9}\ M$;

(d) $\dfrac{[(CH_3)_3NH^+][OH^-]}{[(CH_3)_3N]} = \dfrac{(x)(x)}{(0.11 - x)} \approx \dfrac{(x)(x)}{0.11} = 7.4 \times 10^{-5}$

Solving for x gives $2.85 \times 10^{-3}\ M$. This value is less than 5% of 0.11, so the assumption that it can be neglected is valid. Thus, the concentrations of solute species at equilibrium are:

$[(CH_3)_3NH^+] = [OH^-] = 2.9 \times 10^{-3}\ M$
$[(CH_3)_3N] = 0.11\ M$
$[H_3O^+] = 3.5 \times 10^{-12}\ M$;

(e) $\dfrac{[Fe(H_2O)_5(OH)^+][H_3O^+]}{[Fe(H_2O)_6^{2+}]} = \dfrac{(x)(x)}{(0.120 - x)} \approx \dfrac{(x)(x)}{0.120} = 1.6 \times 10^{-7}$

Solving for x gives $1.39 \times 10^{-4}\ M$. This value is less than 5% of 0.120, so the assumption that it can be neglected is valid. Thus, the concentrations of solute species at equilibrium are:

$[Fe(H_2O)_5(OH)^+] = [H_3O^+] = 1.4 \times 10^{-4}\ M$
$[Fe(H_2O)_6^{2+}] = 0.120\ M$
$[OH^-] = 7.2 \times 10^{-11}\ M$

71. $pH = 2.41$

73. $[C_{10}H_{14}N_2] = 0.049\ M$
$[C_{10}H_{14}N_2H^+] = 1.9 \times 10^{-4}\ M$
$[C_{10}H_{14}N_2H_2^{2+}] = 1.4 \times 10^{-11}\ M$
$[OH^-] = 1.9 \times 10^{-4}\ M$
$[H_3O^+] = 5.3 \times 10^{-11}\ M$

75. $K_a = 1.2 \times 10^{-2}$

77. $K_b = 1.77 \times 10^{-5}$

79. (a) acidic; (b) basic; (c) acidic; (d) neutral

81. $[H_3O^+]$ and $[HCO_3^-]$ are equal, H_3O^+ and HCO_3^- are practically equal

83. $[C_6H_4(CO_2H)_2]\ 7.2 \times 10^{-3}\ M$, $[C_6H_4(CO_2H)(CO_2)^-] = [H_3O^+]\ 2.8 \times 10^{-3}\ M$, $[C_6H_4(CO_2)_2^{2-}]\ 3.9 \times 10^{-6}\ M$, $[OH^-]\ 3.6 \times 10^{-12}\ M$

85. (a) $K_{a2} = 1 \times 10^{-5}$;
(b) $K_b = 4.3 \times 10^{-12}$;

(c) $\dfrac{[Te^{2-}][H_3O^+]}{[HTe^-]} = \dfrac{(x)(0.0141 + x)}{(0.0141 - x)} \approx \dfrac{(x)(0.0141)}{0.0141} = 1 \times 10^{-5}$

Solving for x gives $1 \times 10^{-5}\ M$. Therefore, compared with 0.014 M, this value is negligible (0.071%).

87. Excess H_3O^+ is removed primarily by the reaction:
$$H_3O^+(aq) + H_2PO_4^-(aq) \longrightarrow H_3PO_4(aq) + H_2O(l)$$
Excess base is removed by the reaction:
$$OH^-(aq) + H_3PO_4(aq) \longrightarrow H_2PO_4^-(aq) + H_2O(l)$$

89. $[H_3O^+] = 1.5 \times 10^{-4} M$

91. $[OH^-] = 4.2 \times 10^{-4} M$

93. $[NH_4NO_3] = 0.36 M$

95. (a) The added HCl will increase the concentration of H_3O^+ slightly, which will react with $CH_3CO_2^-$ and produce CH_3CO_2H in the process. Thus, $[CH_3CO_2^-]$ decreases and $[CH_3CO_2H]$ increases.
(b) The added KCH_3CO_2 will increase the concentration of $[CH_3CO_2^-]$ which will react with H_3O^+ and produce CH_3CO_2H in the process. Thus, $[H_3O^+]$ decreases slightly and $[CH_3CO_2H]$ increases.
(c) The added NaCl will have no effect on the concentration of the ions.
(d) The added KOH will produce OH^- ions, which will react with the H_3O^+, thus reducing $[H_3O^+]$. Some additional CH_3CO_2H will dissociate, producing $[CH_3CO_2^-]$ ions in the process. Thus, $[CH_3CO_2H]$ decreases slightly and $[CH_3CO_2^-]$ increases.
(e) The added CH_3CO_2H will increase its concentration, causing more of it to dissociate and producing more $[CH_3CO_2^-]$ and H_3O^+ in the process. Thus, $[H_3O^+]$ increases slightly and $[CH_3CO_2^-]$ increases.

97. pH = 8.95

99. 37 g (0.27 mol)

101. (a) pH = 5.222;
(b) The solution is acidic.
(c) pH = 5.221

103. To prepare the best buffer for a weak acid HA and its salt, the ratio $\dfrac{[H_3O^+]}{K_a}$ should be as close to 1 as possible for effective buffer action. The $[H_3O^+]$ concentration in a buffer of pH 3.1 is $[H_3O^+] = 10^{-3.1} = 7.94 \times 10^{-4} M$
We can now solve for K_a of the best acid as follows:

$$\frac{[H_3O^+]}{K_a} = 1$$

$$K_a = \frac{[H_3O^+]}{1} = 7.94 \times 10^{-4}$$

In Table 14.2, the acid with the closest K_a to 7.94×10^{-4} is HF, with a K_a of 7.2×10^{-4}.

105. For buffers with pHs > 7, you should use a weak base and its salt. The most effective buffer will have a ratio $\dfrac{[OH^-]}{K_b}$ that is as close to 1 as possible. The pOH of the buffer is $14.00 - 10.65 = 3.35$. Therefore, $[OH^-]$ is $[OH^-] = 10^{-pOH} = 10^{-3.35} = 4.467 \times 10^{-4} M$.
We can now solve for K_b of the best base as follows:
$$\frac{[OH^-]}{K_b} = 1$$
$K_b = [OH^-] = 4.47 \times 10^{-4}$
In Table 14.3, the base with the closest K_b to 4.47×10^{-4} is CH_3NH_2, with a $K_b = 4.4 \times 10^{-4}$.

107. The molar mass of sodium saccharide is 205.169 g/mol. Using the abbreviations HA for saccharin and NaA for sodium saccharide the number of moles of NaA in the solution is:
9.75×10^{-6} mol
This ionizes initially to form saccharin ions, A^-, with:
$[A^-] = 3.9 \times 10^{-5} M$

109. At the equivalence point in the titration of a weak base with a strong acid, the resulting solution is slightly acidic due to the presence of the conjugate acid. Thus, pick an indicator that changes color in the acidic range and brackets the pH at the equivalence point. Methyl orange is a good example.

111. In an acid solution, the only source of OH^- ions is water. We use K_w to calculate the concentration. If the contribution from water was neglected, the concentration of OH^- would be zero.

113.

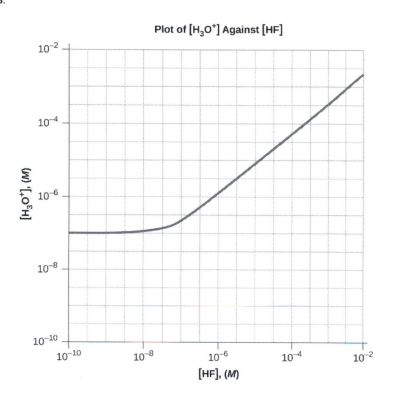

115. (a) pH = 2.50;
(b) pH = 4.01;
(c) pH = 5.60;
(d) pH = 8.35;
(e) pH = 11.08

Chapter 15

1. (a) $AgI(s) \rightleftharpoons Ag^+(aq) + I^-(aq)$
$$ x x$$

(b) $CaCO_3(s) \rightleftharpoons Ca^{2+}(aq) + CO_3{}^{2-}(aq)$
$$ \underline{x} \phantom{Ca^{2+}(aq) +} x$$

(c) $Mg(OH)_2(s) \rightleftharpoons Mg^{2+}(aq) + 2OH^-(aq)$
$$ x \phantom{Mg^{2+}(aq) +} \underline{2x}$$

(d) $Mg_3(PO_4)_2(s) \rightleftharpoons 3Mg^{2+}(aq) + 2PO_4^{3-}(aq)$

$$3x \qquad 2x$$

(e) $Ca_5(PO_4)_3OH(s) \rightleftharpoons 5Ca^{2+}(aq) + 3PO_4^{3-}(aq) + OH^-(aq)$

$$5x \qquad 3x \qquad x$$

3. There is no change. A solid has an activity of 1 whether there is a little or a lot.

5. The solubility of silver bromide at the new temperature must be known. Normally the solubility increases and some of the solid silver bromide will dissolve.

7. CaF_2, $MnCO_3$, and ZnS

9. (a) $LaF_3(s) \rightleftharpoons La^{3+}(aq) + 3F^-(aq) \qquad K_{sp} = [La^{3+}][F^-]^3$;

(b) $CaCO_3(s) \rightleftharpoons Ca^{2+}(aq) + CO_3^{2-}(aq) \qquad K_{sp} = [Ca^{2+}][CO_3^{2-}]$;

(c) $Ag_2SO_4(s) \rightleftharpoons 2Ag^+(aq) + SO_4^{2-}(aq) \qquad K_{sp} = [Ag^+]^2[SO_4^{2-}]$;

(d) $Pb(OH)_2(s) \rightleftharpoons Pb^{2+}(aq) + 2OH^-(aq) \qquad K_{sp} = [Pb^{2+}][OH^-]^2$

11. (a)1.77×10^{-7}; (b) 1.6×10^{-6}; (c) 2.2×10^{-9}; (d) 7.91×10^{-22}

13. (a) 2×10^{-2} M; (b) 1.5×10^{-3} M; (c) 2.27×10^{-9} M; (d) 2.2×10^{-10} M

15. (a) 7.2×10^{-9} M = $[Ag^+]$, $[Cl^-] = 0.025$ M

Check: $\frac{6.4 \times 10^{-9}\ M}{0.025\ M} \times 100\% = 2.9 \times 10^{-5}\%$, an insignificant change;

(b) 2.2×10^{-5} M = $[Ca^{2+}]$, $[F^-] = 0.0013$ M

Check: $\frac{2.25 \times 10^{-5}\ M}{0.00133\ M} \times 100\% = 1.69\%$. This value is less than 5% and can be ignored.

(c) 0.2238 M = $[SO_4^{2-}]$; $[Ag^+] = 2.30 \times 10^{-9}$ M

Check: $\frac{1.15 \times 10^{-9}}{0.2238} \times 100\% = 5.14 \times 10^{-7}$; the condition is satisfied.

(d) $[OH^-] = 2.8 \times 10^{-3}$ M; 5.7×10^{-12} M = $[Zn^{2+}]$

Check: $\frac{5.7 \times 10^{-12}}{2.8 \times 10^{-3}} \times 100\% = 2.0 \times 10^{-7}\%$; x is less than 5% of $[OH^-]$ and is, therefore, negligible.

17. (a) $[Cl^-] = 7.6 \times 10^{-3}$ M

Check: $\frac{7.6 \times 10^{-3}}{0.025} \times 100\% = 30\%$

This value is too large to drop x. Therefore solve by using the quadratic equation:

$[Ti^+] = 3.1 \times 10^{-2}$ M

$[Cl^-] = 6.1 \times 10^{-3}$

(b) $[Ba^{2+}] = 1.7 \times 10^{-3}$ M

Check: $\frac{1.7 \times 10^{-3}}{0.0313} \times 100\% = 5.5\%$

This value is too large to drop x, and the entire equation must be solved.

$[Ba^{2+}] = 1.6 \times 10^{-3}$ M

$[F^-] = 0.0329$ M;

(c) $Mg(NO_3)_2 = 0.02444$ M

$[C_2O_4^{2-}] = 3.5 \times 10^{-3}$

Check: $\frac{3.5 \times 10^{-3}}{0.02444} \times 100\% = 14\%$

This value is greater than 5%, so the quadratic equation must be used:

$[C_2O_4^{2-}] = 3.5 \times 10^{-3}$ M

$[Mg^{2+}] = 0.0275$ M

(d) $[OH^-] = 0.0501$ M

$[Ca^{2+}] = 3.15 \times 10^{-3}$

Check: $\dfrac{3.15 \times 10^{-3}}{0.050} \times 100\% = 6.28\%$

This value is greater than 5%, so a more exact method, such as successive approximations, must be used.
$[Ca^{2+}] = 2.8 \times 10^{-3} M$
$[OH^-] = 0.053 \times 10^{-2} M$

19. The changes in concentration are greater than 5% and thus exceed the maximum value for disregarding the change.

21. $CaSO_4 \cdot 2H_2O$ is the most soluble Ca salt in mol/L, and it is also the most soluble Ca salt in g/L.

23. $4.9 \times 10^{-3} M = [SO_4{}^{2-}] = [Ca^{2+}]$; Since this concentration is higher than $2.60 \times 10^{-3} M$, "gyp" water does not meet the standards.

25. Mass $(CaSO_4 \cdot 2H_2O) = 0.34$ g/L

27. (a) $[Ag^+] = [I^-] = 1.2 \times 10^{-8} M$; (b) $[Ag^+] = 2.86 \times 10^{-2} M$, $[SO_4{}^{2-}] = 1.43 \times 10^{-2} M$; (c) $[Mn^{2+}] = 2.2 \times 10^{-5} M$, $[OH^-] = 4.5 \times 10^{-5} M$; (d) $[Sr^{2+}] = 4.3 \times 10^{-2} M$, $[OH^-] = 8.6 \times 10^{-2} M$; (e) $[Mg^{2+}] = 1.6 \times 10^{-4} M$, $[OH^-] = 3.1 \times 10^{-4} M$.

29. (a) 2.0×10^{-4}; (b) 5.1×10^{-17}; (c) 1.35×10^{-4}; (d) 1.18×10^{-5}; (e) 1.08×10^{-10}

31. (a) $CaCO_3$ does precipitate.
(b) The compound does not precipitate.
(c) The compound does not precipitate.
(d) The compound precipitates.

33. $1.42 \times 10^{-9} M$

35. $9.2 \times 10^{-13} M$

37. $[Ag^+] = 1.8 \times 10^{-3} M$

39. 6.2×10^{-4}

41. (a) 2.28 L; (b) 7.3×10^{-7} g

43. 100% of it is dissolved

45. (a) $Hg_2{}^{2+}$ and Cu^{2+}: Add $SO_4{}^{2-}$.
(b) $SO_4{}^{2-}$ and Cl^-: Add Ba^{2+}.
(c) Hg^{2+} and Co^{2+}: Add S^{2-}.
(d) Zn^{2+} and Sr^{2+}: Add OH^- until $[OH^-] = 0.050 M$.
(e) Ba^{2+} and Mg^{2+}: Add $SO_4{}^{2-}$.
(f) $CO_3{}^{2-}$ and OH^-: Add Ba^{2+}.

47. AgI will precipitate first.

49. $4 \times 10^{-9} M$

51. 3.99 kg

53. (a) 3.1×10^{-11}; (b) $[Cu^{2+}] = 2.6 \times 10^{-3}$; $[IO_3{}^-] = 5.3 \times 10^{-3}$

55. 1.8×10^{-5} g $Pb(OH)_2$

57. $Mg(OH)_2(s) \rightleftharpoons Mg^{2+} + 2OH^-$ $\qquad K_{sp} = [Mg^{2+}][OH^-]^2$
1.14×10^{-3} g $Mg(OH)_2$

59. $SrCO_3$ will form first, since it has the smallest K_{sp} value it is the least soluble. $BaCO_3$ will be the last to precipitate, it has the largest K_{sp} value.

61. when the amount of solid is so small that a saturated solution is not produced

63. $8 \times 10^{-5} M$

65. 5×10^{23}

67.

	$[Cd(CN)_4{}^{2-}]$	$[CN^-]$	$[Cd^{2+}]$
Initial concentration (M)	0.250	0	0
Equilibrium (M)	$0.250 - x$	$4x$	x

$[Cd^{2+}] = 9.5 \times 10^{-5}\, M$; $[CN^-] = 3.8 \times 10^{-4}\, M$

69. $[Co^{3+}] = 3.0 \times 10^{-6}\, M$; $[NH_3] = 1.8 \times 10^{-5}\, M$

71. 1.3 g

73. 0.79 g

75. (a)

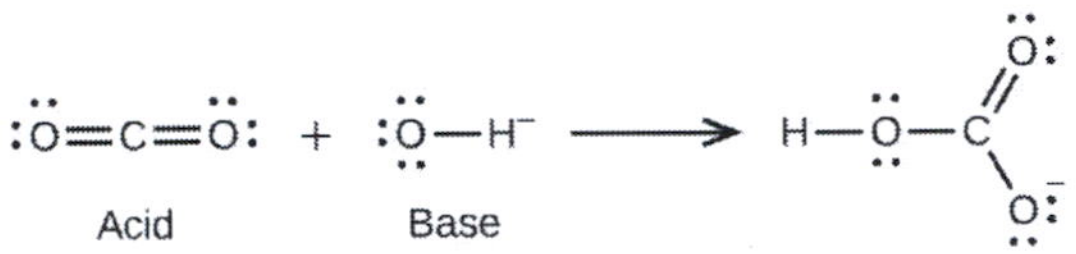

;

(b)

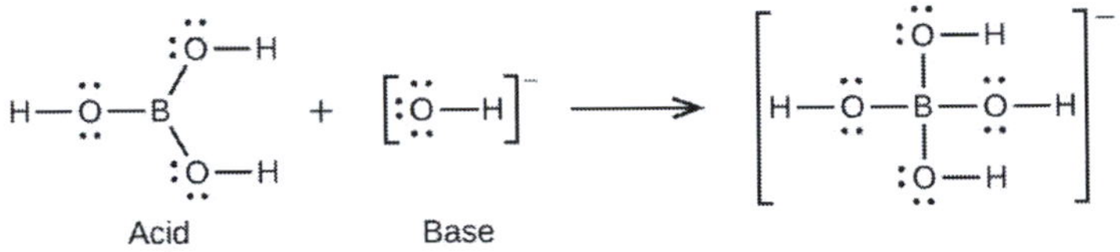

;

(c)

;

(d)

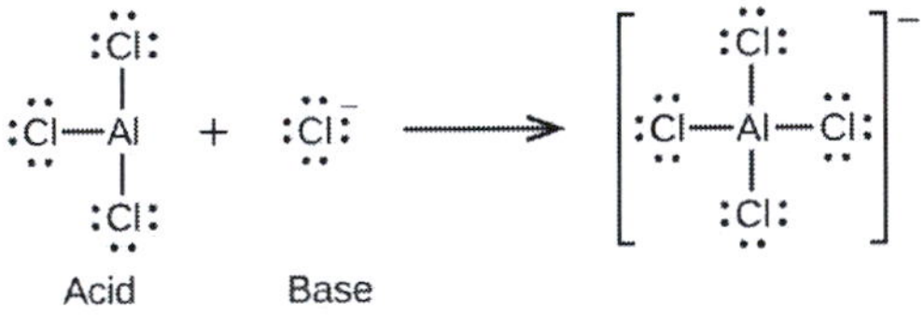

;

(e)

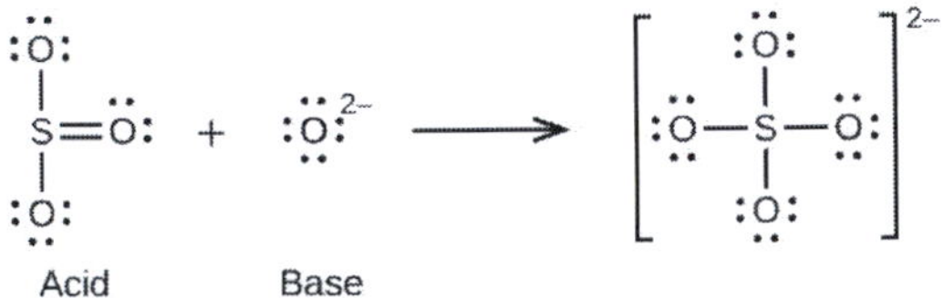

77. (a)

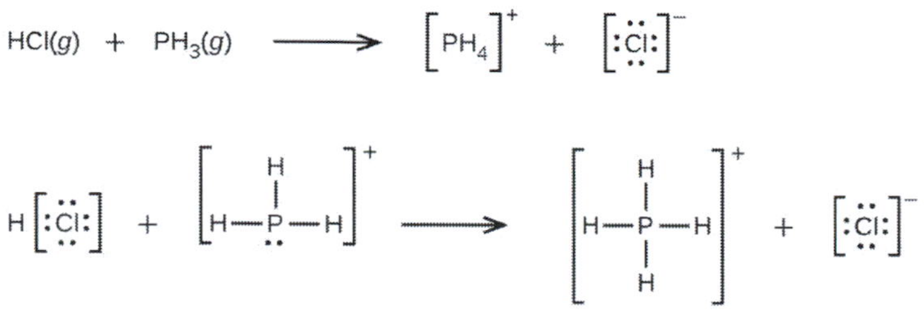

$$HCl(g) + PH_3(g) \longrightarrow \left[PH_4\right]^+ + \left[:\ddot{C}l:\right]^-$$

;

(b) $H_3O^+ + CH_3^- \longrightarrow CH_4 + H_2O$

;

(c) $CaO + SO_3 \longrightarrow CaSO4$

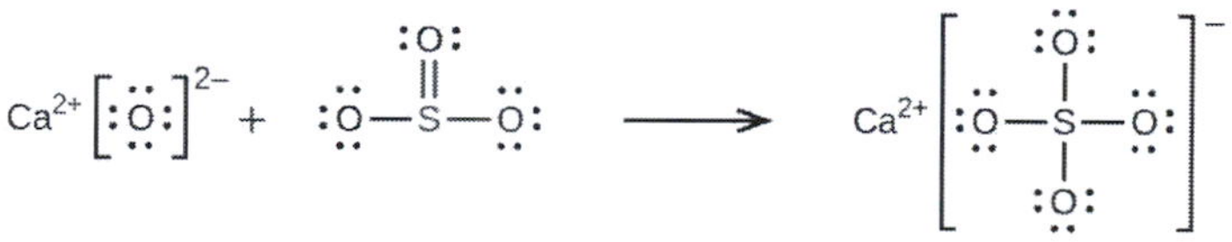

;

(d) $NH_4^+ + C_2H_5O^- \longrightarrow C_2H_5OH + NH_3$

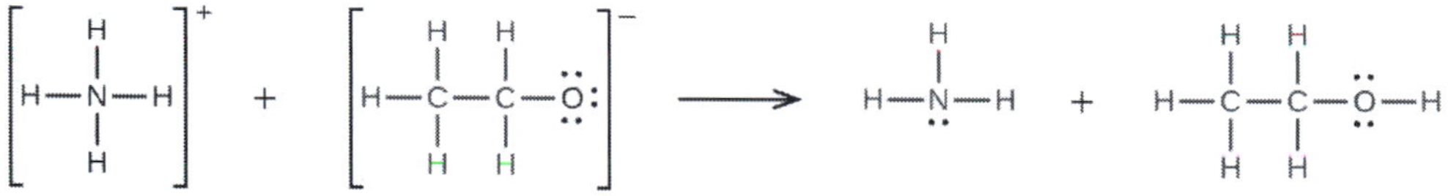

79. 0.0281 g

81. $HNO_3(l) + HF(l) \longrightarrow H_2NO_3^+ + F^-$; $HF(l) + BF_3(g) \longrightarrow H^+ + BF_4$

83. (a) $H_3BO_3 + H_2O \longrightarrow H_4BO_4^- + H^+$; (b) The electronic and molecular shapes are the same—both tetrahedral. (c) The tetrahedral structure is consistent with sp^3 hybridization.

85. 0.014 M

87. 7.2×10^{-15} M

89. 4.4×10^{-22} M

91. 6.2×10^{-6} M = $[Hg^{2+}]$; 1.2×10^{-5} M = $[Cl^-]$; The substance is a weak electrolyte because very little of the initial 0.015 M $HgCl_2$ dissolved.

93. $[OH^-] = 4.5 \times 10^{-5}$; $[Al^{3+}] = 2.2 \times 10^{-20}$ (molar solubility)

95. $[SO_4{}^{2-}] = 0.049$ M
$[Ba^{2+}] = 4.7 \times 10^{-7}$ (molar solubility)

97. $[OH^-] = 7.6 \times 10^{-3}$ M
$[Pb^{2+}] = 2.1 \times 10^{-11}$ (molar solubility)

99. 7.66

101. $[CO_3{}^{2-}] = 0.116\ M$
$[Cd^{2+}] = 4.5 \times 10^{-11}\ M$

103. $3.1 \times 10^{-3}\ M$

105. 0.0102 L (10.2 mL)

107. 5.4×10^{-3} g

109. (a) $K_{sp} = [Mg^{2+}][F^-]^2 = (1.21 \times 10^{-3})(2 \times 1.21 \times 10^{-3})^2 = 7.09 \times 10^{-9}$; (b) $7.09 \times 10^{-7}\ M$
(c) Determine the concentration of Mg^{2+} and F^- that will be present in the final volume. Compare the value of the ion product $[Mg^{2+}][F^-]^2$ with K_{sp}. If this value is larger than K_{sp}, precipitation will occur.
$0.1000\ L \times 3.00 \times 10^{-3}\ M\ Mg(NO_3)_2 = 0.3000\ L \times M\ Mg(NO_3)_2$
$M\ Mg(NO_3)_2 = 1.00 \times 10^{-3}\ M$
$0.2000\ L \times 2.00 \times 10^{-3}\ M\ NaF = 0.3000\ L \times M\ NaF$
$M\ NaF = 1.33 \times 10^{-3}\ M$
ion product $= (1.00 \times 10^{-3})(1.33 \times 10^{-3})^2 = 1.77 \times 10^{-9}$
This value is smaller than K_{sp}, so no precipitation will occur.
(d) MgF_2 is less soluble at 27 °C than at 18 °C. Because added heat acts like an added reagent, when it appears on the product side, the Le Châtelier's principle states that the equilibrium will shift to the reactants' side to counter the stress. Consequently, less reagent will dissolve. This situation is found in our case. Therefore, the reaction is exothermic.

111. BaF_2, $Ca_3(PO_4)_2$, ZnS; each is a salt of a weak acid, and the $[H_3O^+]$ from perchloric acid reduces the equilibrium concentration of the anion, thereby increasing the concentration of the cations

113. Effect on amount of solid $CaHPO_4$, $[Ca^{2+}]$, $[OH^-]$: (a) increase, increase, decrease; (b) decrease, increase, decrease; (c) no effect, no effect, no effect; (d) decrease, increase, decrease; (e) increase, no effect, no effect

115. 9.2 g

Chapter 16

1. A reaction has a natural tendency to occur and takes place without the continual input of energy from an external source.

3. (a) spontaneous; (b) nonspontaneous; (c) spontaneous; (d) nonspontaneous; (e) spontaneous; (f) spontaneous

5. Although the oxidation of plastics is spontaneous, the rate of oxidation is very slow. Plastics are therefore kinetically stable and do not decompose appreciably even over relatively long periods of time.

7. There are four initial microstates and four final microstates.
$$\Delta S = k\ln\frac{W_f}{W_i} = 1.38 \times 10^{-23}\ J/K \times \ln\frac{4}{4} = 0$$

9. The probability for all the particles to be on one side is $\frac{1}{32}$. This probability is noticeably lower than the $\frac{1}{8}$ result for the four-particle system. The conclusion we can make is that the probability for all the particles to stay in only one part of the system will decrease rapidly as the number of particles increases, and, for instance, the probability for all molecules of gas to gather in only one side of a room at room temperature and pressure is negligible since the number of gas molecules in the room is very large.

11. There is only one initial state. For the final state, the energy can be contained in pairs A-C, A-D, B-C, or B-D. Thus, there are four final possible states.
$$\Delta S = k\ln\left(\frac{W_f}{W_i}\right) = 1.38 \times 10^{-23}\ J/K \times \ln\left(\frac{4}{1}\right) = 1.91 \times 10^{-23}\ J/K$$

13. The masses of these molecules would suggest the opposite trend in their entropies. The observed trend is a result of the more significant variation of entropy with a physical state. At room temperature, I_2 is a solid, Br_2 is a liquid, and Cl_2 is a gas.

15. (a) $C_3H_7OH(l)$ as it is a larger molecule (more complex and more massive), and so more microstates describing its motions are available at any given temperature. (b) $C_2H_5OH(g)$ as it is in the gaseous state. (c) $2H(g)$, since entropy is an extensive property, and so two H atoms (or two moles of H atoms) possess twice as much entropy as one atom (or one mole of atoms).

17. (a) Negative. The relatively ordered solid precipitating decreases the number of mobile ions in solution. (b) Negative. There is a net loss of three moles of gas from reactants to products. (c) Positive. There is a net increase of seven moles of gas from reactants to products.

19. $C_6H_6(l) + 7.5O_2(g) \longrightarrow 3H_2O(g) + 6CO_2(g)$
There are 7.5 moles of gas initially, and $3 + 6 = 9$ moles of gas in the end. Therefore, it is likely that the entropy increases as a result of this reaction, and ΔS is positive.

21. (a) 107 J/K; (b) −86.4 J/K; (c) 133.2 J/K; (d) 118.8 J/K; (e) −326.6 J/K; (f) −171.9 J/K; (g) −7.2 J/K

23. 100.6 J/K

25. (a) −198.1 J/K; (b) −348.9 J/K

27. As $\Delta S_{univ} < 0$ at each of these temperatures, melting is not spontaneous at either of them. The given values for entropy and enthalpy are for NaCl at 298 K. It is assumed that these do not change significantly at the higher temperatures used in the problem.

29. (a) 2.86 J/K; (b) 24.8 J/K; (c) −113.2 J/K; (d) −24.7 J/K; (e) 15.5 J/K; (f) 290.0 J/K

31. The reaction is nonspontaneous at room temperature.
Above 400 K, ΔG will become negative, and the reaction will become spontaneous.

33. (a) 465.1 kJ nonspontaneous; (b) −106.86 kJ spontaneous; (c) −53.6 kJ spontaneous; (d) −83.4 kJ spontaneous; (e) −406.7 kJ spontaneous; (f) −30.0 kJ spontaneous

35. (a) −1124.3 kJ/mol for the standard free energy change. (b) The calculation agrees with the value in Appendix G because free energy is a state function (just like the enthalpy and entropy), so its change depends only on the initial and final states, not the path between them.

37. (a) The reaction is nonspontaneous; (b) Above 566 °C the process is spontaneous.

39. (a) 1.5×10^2 kJ; (b) −21.9 kJ; (c) −5.34 kJ; (d) −0.383 kJ; (e) 18 kJ; (f) 71 kJ

41. (a) $K = 41$; (b) $K = 0.053$; (c) $K = 6.9 \times 10^{13}$; (d) $K = 1.9$; (e) $K = 0.04$

43. In each of the following, the value of ΔG is not given at the temperature of the reaction. Therefore, we must calculate ΔG from the values $\Delta H°$ and ΔS and then calculate ΔG from the relation $\Delta G = \Delta H° − T\Delta S°$.
(a) $K = 1.29$;
(b) $K = 2.51 \times 10^{-3}$;
(c) $K = 4.83 \times 10^3$;
(d) $K = 0.219$;
(e) $K = 16.1$

45. The standard free energy change is $\Delta G°_{298} = −RT \ln K = 4.84$ kJ/mol. When reactants and products are in their standard states (1 bar or 1 atm), $Q = 1$. As the reaction proceeds toward equilibrium, the reaction shifts left (the amount of products drops while the amount of reactants increases): $Q < 1$, and ΔG_{298} becomes less positive as it approaches zero. At equilibrium, $Q = K$, and $\Delta G = 0$.

47. The reaction will be spontaneous at temperatures greater than 287 K.

49. $K = 5.35 \times 10^{15}$
The process is exothermic.

51. 1.0×10^{-8} atm. This is the maximum pressure of the gases under the stated conditions.

53. $x = 1.29 \times 10^{-5}$ atm $= P_{O_2}$

55. −0.16 kJ

57. (a) −22.1 kJ;
(b) 61.6 kJ/mol

59. 90 kJ/mol

61. (a) Under standard thermodynamic conditions, the evaporation is nonspontaneous; (b) $K_p = 0.031$; (c) The evaporation of water is spontaneous; (d) P_{H_2O} must always be less than K_p or less than 0.031 atm. 0.031 atm represents air saturated with water vapor at 25 °C, or 100% humidity.

63. (a) Nonspontaneous as $\Delta G^\circ_{298} > 0$; (b) $\Delta G^\circ_{298} = -RT \ln K$,

$\Delta G = 1.7 \times 10^3 + \left(8.314 \times 335 \times \ln \frac{28}{128} \right) = -2.5$ kJ. The forward reaction to produce F6P is spontaneous under these conditions.

65. ΔG is negative as the process is spontaneous. ΔH is positive as with the solution becoming cold, the dissolving must be endothermic. ΔS must be positive as this drives the process, and it is expected for the dissolution of any soluble ionic compound.

67. (a) Increasing P_{O_2} will shift the equilibrium toward the products, which increases the value of K. ΔG°_{298} therefore becomes more negative.
(b) Increasing P_{O_2} will shift the equilibrium toward the products, which increases the value of K. ΔG°_{298} therefore becomes more negative.
(c) Increasing P_{O_2} will shift the equilibrium the reactants, which decreases the value of K. ΔG°_{298} therefore becomes more positive.

Chapter 17

1. 5.3×10^3 C

3. (a) reduction; (b) oxidation; (c) oxidation; (d) reduction

5. (a) $F_2 + Ca \longrightarrow 2F^- + Ca^{2+}$; (b) $Cl_2 + 2Li \longrightarrow 2Li^+ + 2Cl^-$; (c) $3Br_2 + 2Fe \longrightarrow 2Fe^{3+} + 6Br^-$; (d) $MnO_4 + 4H^+ + 3Ag \longrightarrow 3Ag^+ + MnO_2 + 2H_2O$

7. Oxidized: (a) Ag; (b) Sn^{2+}; (c) Hg; (d) Al; reduced: (a) $Hg_2{}^{2+}$; (b) H_2O_2; (c) PbO_2; (d) $Cr_2O_7{}^{2-}$; oxidizing agent: (a) $Hg_2{}^{2+}$; (b) H_2O_2; (c) PbO_2; (d) $Cr_2O_7{}^{2-}$; reducing agent: (a) Ag; (b) Sn^{2+}; (c) Hg; (d) Al

9. Oxidized = reducing agent: (a) $SO_3{}^{2-}$; (b) $Mn(OH)_2$; (c) H_2; (d) Al; reduced = oxidizing agent: (a) $Cu(OH)_2$; (b) O_2; (c) $NO_3{}^-$; (d) $CrO_4{}^{2-}$

11. In basic solution, $[OH^-] > 1 \times 10^{-7} M > [H^+]$. Hydrogen ion cannot appear as a reactant because its concentration is essentially zero. If it were produced, it would instantly react with the excess hydroxide ion to produce water. Thus, hydrogen ion should *not* appear as a reactant or product in basic solution.

13. (a) $Mg(s) \mid Mg^{2+}(aq) \parallel Ni^+(aq) \mid Ni(s)$; (b) $Cu(s) \mid Cu^{2+}(aq) \parallel Ag^+(aq) \mid Ag(s)$; (c) $Mn(s) \mid Mn^{2+}(aq) \parallel Sn^{2+}(aq) \mid Sn(s)$; (d) $Pt(s) \mid Cu^+(aq), Cu^{2+}(aq) \parallel Au^{3+}(aq) \mid Au(s)$

15. (a) $Mg(s) + Cu^{2+}(aq) \longrightarrow Mg^{2+}(aq) + Cu(s)$; (b) $2Ag^+(aq) + Ni(s) \longrightarrow Ni^{2+}(aq) + 2Ag(s)$

17. Species oxidized = reducing agent: (a) Al(s); (b) NO(g); (c) Mg(s); and (d) $MnO_2(s)$; Species reduced = oxidizing agent: (a) $Zr^{4+}(aq)$; (b) $Ag^+(aq)$; (c) $SiO_3{}^{2-}(aq)$; and (d) $ClO_3{}^-(aq)$

19. Without the salt bridge, the circuit would be open (or broken) and no current could flow. With a salt bridge, each half-cell remains electrically neutral and current can flow through the circuit.

21. Active electrodes participate in the oxidation-reduction reaction. Since metals form cations, the electrode would lose mass if metal atoms in the electrode were to oxidize and go into solution. Oxidation occurs at the anode.

23. (a) +2.115 V (spontaneous); (b) +0.4626 V (spontaneous); (c) +1.0589 V (spontaneous); (d) +0.727 V (spontaneous)

25. $3Cu(s) + 2Au^{3+}(aq) \longrightarrow 3Cu^{2+}(aq) + 2Au(s)$; +1.16 V; spontaneous

27. $3Cd(s) + 2Al^{3+}(aq) \longrightarrow 3Cd^{2+}(aq) + 2Al(s)$; −1.259 V; nonspontaneous

29. (a) 0 kJ/mol; (b) −83.7 kJ/mol; (c) +235.3 kJ/mol

31. (a) standard cell potential: 1.50 V, spontaneous; cell potential under stated conditions: 1.43 V, spontaneous; (b) standard cell potential: 1.405 V, spontaneous; cell potential under stated conditions: 1.423 V, spontaneous; (c) standard cell potential: −2.749 V, nonspontaneous; cell potential under stated conditions: −2.757 V, nonspontaneous

33. (a) 1.7×10^{-10}; (b) 2.6×10^{-21}; (c) 8.9×10^{19}; (d) 1.0×10^{-14}

35. Considerations include: cost of the materials used in the battery, toxicity of the various components (what constitutes proper disposal), should it be a primary or secondary battery, energy requirements (the "size" of the battery/how long should it last), will a particular battery leak when the new device is used according to directions, and its mass (the total mass of the new device).

37. (a)
anode: $Cu(s) \longrightarrow Cu^{2+}(aq) + 2e^-$ $\qquad E^{\circ}_{anode} = 0.34$ V
cathode: $2 \times (Ag^+(aq) + e^- \longrightarrow Ag(s))$ $\qquad E^{\circ}_{cathode} = 0.7996$ V
; (b) 3.5×10^{15}; (c) 5.6×10^{-9} M

39. Batteries are self-contained and have a limited supply of reagents to expend before going dead. Alternatively, battery reaction byproducts accumulate and interfere with the reaction. Because a fuel cell is constantly resupplied with reactants and products are expelled, it can continue to function as long as reagents are supplied.

41. E_{cell}, as described in the Nernst equation, has a term that is directly proportional to temperature. At low temperatures, this term is decreased, resulting in a lower cell voltage provided by the battery to the device—the same effect as a battery running dead.

43. Mg and Zn

45. Both examples involve cathodic protection. The (sacrificial) anode is the metal that corrodes (oxidizes or reacts). In the case of iron (−0.447 V) and zinc (−0.7618 V), zinc has a more negative standard reduction potential and so serves as the anode. In the case of iron and copper (0.34 V), iron has the smaller standard reduction potential and so corrodes (serves as the anode).

47. While the reduction potential of lithium would make it capable of protecting the other metals, this high potential is also indicative of how reactive lithium is; it would have a spontaneous reaction with most substances. This means that the lithium would react quickly with other substances, even those that would not oxidize the metal it is attempting to protect. Reactivity like this means the sacrificial anode would be depleted rapidly and need to be replaced frequently. (Optional additional reason: fire hazard in the presence of water.)

49. (a) mass Ca = 69.1 g, mass Cl_2 = 122 g; (b) mass Li = 23.9 g, mass H_2 = 3.48 g; (c) mass Al = 31.0 g, mass Cl_2 = 122 g; (d) mass Cr = 59.8 g, mass Br_2 = 276 g

51. 0.79 L

Chapter 18

1. The alkali metals all have a single s electron in their outermost shell. In contrast, the alkaline earth metals have a completed s subshell in their outermost shell. In general, the alkali metals react faster and are more reactive than the corresponding alkaline earth metals in the same period.

3.
$Na + I_2 \longrightarrow 2NaI$
$2Na + Se \longrightarrow Na_2Se$
$2Na + O_2 \longrightarrow Na_2O_2$
$Sr + I_2 \longrightarrow SrI_2$
$Sr + Se \longrightarrow SeSe$
$2Sr + O_2 \longrightarrow 2SrO$
$2Al + 3I_2 \longrightarrow 2AlI_3$
$2Al + 3Se \longrightarrow Al_2Se_3$
$4Al + 3O_2 \longrightarrow 2Al_2O_3$

5. The possible ways of distinguishing between the two include infrared spectroscopy by comparison of known compounds, a flame test that gives the characteristic yellow color for sodium (strontium has a red flame), or comparison of their solubilities in water. At 20 °C, NaCl dissolves to the extent of $\dfrac{35.7 \text{ g}}{100 \text{ mL}}$ compared with $\dfrac{53.8 \text{ g}}{100 \text{ mL}}$ for SrCl$_2$. Heating to 100 °C provides an easy test, since the solubility of NaCl is $\dfrac{39.12 \text{ g}}{100 \text{ mL}}$, but that of SrCl$_2$ is $\dfrac{100.8 \text{ g}}{100 \text{ mL}}$. Density determination on a solid is sometimes difficult, but there is enough difference (2.165 g/mL NaCl and 3.052 g/mL SrCl$_2$) that this method would be viable and perhaps the easiest and least expensive test to perform.

7. (a) $2Sr(s) + O_2(g) \longrightarrow 2SrO(s)$; (b) $Sr(s) + 2HBr(g) \longrightarrow SrBr_2(s) + H_2(g)$; (c) $Sr(s) + H_2(g) \longrightarrow SrH_2(s)$; (d) $6Sr(s) + P_4(s) \longrightarrow 2Sr_3P_2(s)$; (e) $Sr(s) + 2H_2O(l) \longrightarrow Sr(OH)_2(aq) + H_2(g)$

9. 11 lb

11. Yes, tin reacts with hydrochloric acid to produce hydrogen gas.

13. In PbCl$_2$, the bonding is ionic, as indicated by its melting point of 501 °C. In PbCl$_4$, the bonding is covalent, as evidenced by it being an unstable liquid at room temperature.

15.
$$2CsCl(l) + Ca(g) \xrightarrow[\text{tower}]{\substack{\text{countercurrent}\\\text{fractionating}}} 2Cs(g) + CaCl_2(l)$$

17. Cathode (reduction): $2Li^+ + 2e^- \longrightarrow 2Li(l)$; Anode (oxidation): $2Cl^- \longrightarrow Cl_2(g) + 2e^-$; Overall reaction: $2Li^+ + 2Cl^- \longrightarrow 2Li(l) + Cl_2(g)$

19. 0.5035 g H$_2$

21. Despite its reactivity, magnesium can be used in construction even when the magnesium is going to come in contact with a flame because a protective oxide coating is formed, preventing gross oxidation. Only if the metal is finely subdivided or present in a thin sheet will a high-intensity flame cause its rapid burning.

23. Extract from ore: $AlO(OH)(s) + NaOH(aq) + H_2O(l) \longrightarrow Na[Al(OH)_4](aq)$
Recover: $2Na[Al(OH)_4](s) + H_2SO_4(aq) \longrightarrow 2Al(OH)_3(s) + Na_2SO_4(aq) + 2H_2O(l)$
Sinter: $2Al(OH)_3(s) \longrightarrow Al_2O_3(s) + 3H_2O(g)$
Dissolve in Na$_3$AlF$_6$(l) and electrolyze: $Al^{3+} + 3e^- \longrightarrow Al(s)$

25. 25.83%

27. 39 kg

29. (a) H$_3$BPH$_3$:

;

(b) BF$_4$$^-$:

;

(c) BBr$_3$:

(structure of BBr₃ showing Br atoms bonded to central B with lone pairs)

(d) B(CH₃)₃:

(structure of B(CH₃)₃ showing three CH₃ groups bonded to central B)

(e) B(OH)₃:

(structure of B(OH)₃ showing three OH groups bonded to central B)

31. $1s^2 2s^2 2p^6 3s^2 3p^2 3d^0$.

33. (a) $(CH_3)_3SiH$: sp^3 bonding about Si; the structure is tetrahedral; (b) $SiO_4{}^{4-}$: sp^3 bonding about Si; the structure is tetrahedral; (c) Si_2H_6: sp^3 bonding about each Si; the structure is linear along the Si-Si bond; (d) $Si(OH)_4$: sp^3 bonding about Si; the structure is tetrahedral; (e) $SiF_6{}^{2-}$: sp^3d^2 bonding about Si; the structure is octahedral

35. (a) nonpolar; (b) nonpolar; (c) polar; (d) nonpolar; (e) polar

37. (a) tellurium dioxide or tellurium(IV) oxide; (b) antimony(III) sulfide; (c) germanium(IV) fluoride; (d) silane or silicon(IV) hydride; (e) germanium(IV) hydride

39. Boron has only s and p orbitals available, which can accommodate a maximum of four electron pairs. Unlike silicon, no d orbitals are available in boron.

41. (a) $\Delta H° = 87$ kJ; $\Delta G° = 44$ kJ; (b) $\Delta H° = -109.9$ kJ; $\Delta G° = -154.7$ kJ; (c) $\Delta H° = -510$ kJ; $\Delta G° = -601.5$ kJ

43. A mild solution of hydrofluoric acid would dissolve the silicate and would not harm the diamond.

45. In the N_2 molecule, the nitrogen atoms have an σ bond and two π bonds holding the two atoms together. The presence of three strong bonds makes N_2 a very stable molecule. Phosphorus is a third-period element, and as such, does not form π bonds efficiently; therefore, it must fulfill its bonding requirement by forming three σ bonds.

47. (a) H = 1+, C = 2+, and N = 3–; (b) O = 2+ and F = 1–; (c) As = 3+ and Cl = 1–

49. S < Cl < O < F

51. The electronegativity of the nonmetals is greater than that of hydrogen. Thus, the negative charge is better represented on the nonmetal, which has the greater tendency to attract electrons in the bond to itself.

53. Hydrogen has only one orbital with which to bond to other atoms. Consequently, only one two-electron bond can form.

55. 0.43 g H_2

57. (a) $Ca(OH)_2(aq) + CO_2(g) \longrightarrow CaCO_3(s) + H_2O(l)$; (b) $CaO(s) + SO_2(g) \longrightarrow CaSO_3(s)$;
(c) $2NaHCO_3(s) + NaH_2PO_4(aq) \longrightarrow Na_3PO_4(aq) + 2CO_2(g) + 2H_2O(l)$

59. (a) NH^{2-}:

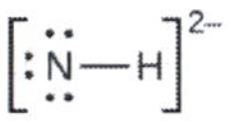

;

(b) N_2F_4:

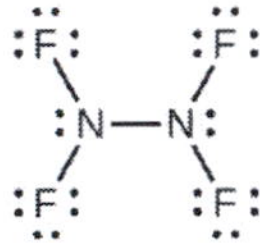

;

(c) NH_2^-:

;

(d) NF_3:

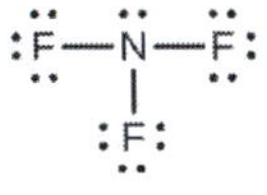

;

(e) N_3^-:

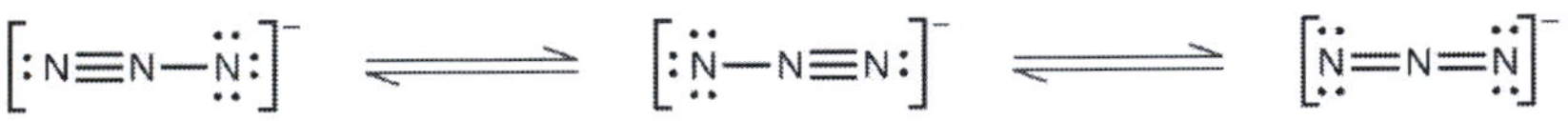

61. Ammonia acts as a Brønsted base because it readily accepts protons and as a Lewis base in that it has an electron pair to donate.
Brønsted base: $NH_3 + H_3O^+ \longrightarrow NH_4^+ + H_2O$
Lewis base: $2NH_3 + Ag^+ \longrightarrow [H_3N - Ag - NH_3]^+$

63. (a) NO_2:

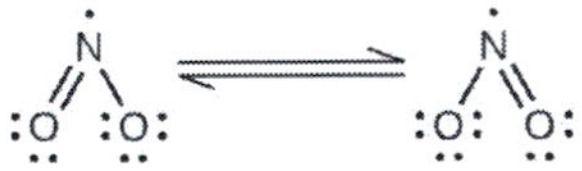

Nitrogen is sp^2 hybridized. The molecule has a bent geometry with an ONO bond angle of approximately 120°.
(b) NO_2^-:

Nitrogen is sp^2 hybridized. The molecule has a bent geometry with an ONO bond angle slightly less than 120°.

(c) NO_2^+ :

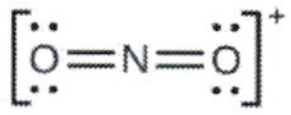

Nitrogen is *sp* hybridized. The molecule has a linear geometry with an ONO bond angle of 180°.

65. Nitrogen cannot form a NF_5 molecule because it does not have *d* orbitals to bond with the additional two fluorine atoms.

67. (a)

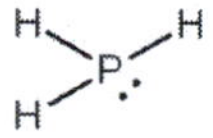

;

(b)

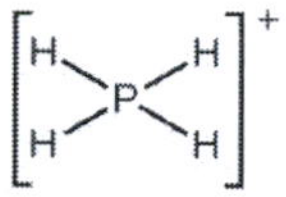

;

(c)

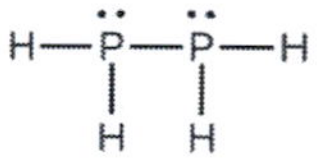

H—P—P—H
    | |
    H H

;

(d)

;

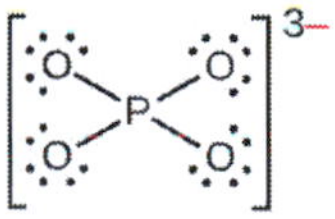

(e)

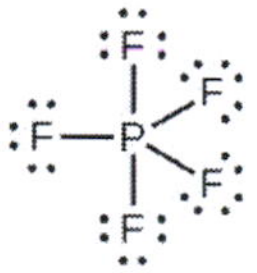

69. (a) $P_4(s) + 4Al(s) \longrightarrow 4AlP(s)$; (b) $P_4(s) + 12Na(s) \longrightarrow 4Na_3P(s)$; (c) $P_4(s) + 10F_2(g) \longrightarrow 4PF_5(l)$; (d) $P_4(s) + 6Cl_2(g) \longrightarrow 4PCl_3(l)$ or $P_4(s) + 10Cl_2(g) \longrightarrow 4PCl_5(l)$; (e) $P_4(s) + 3O_2(g) \longrightarrow P_4O_6(s)$ or $P_4(s) + 5O_2(g) \longrightarrow P_4O_{10}(s)$; (f) $P_4O_6(s) + 2O_2(g) \longrightarrow P_4O_{10}(s)$

71. 291 mL

73. 28 tons

75. (a)

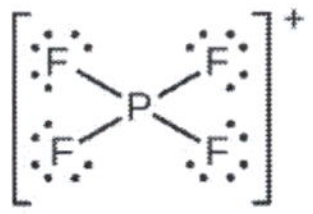

Tetrahedral

;

(b)

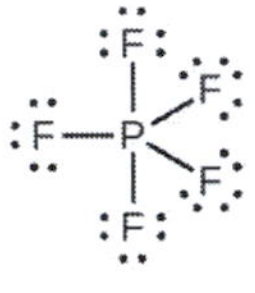

Trigonal bipyramid

;

(c)

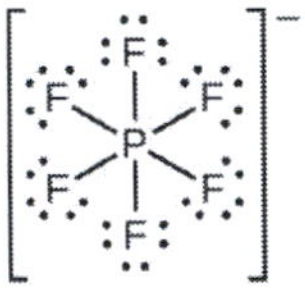

Octahedral

;

(d)

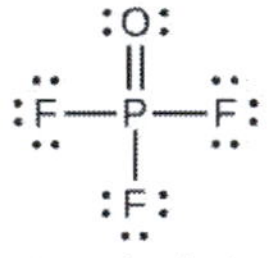

Tetrahedral

77. (a) P = 3+; (b) P = 5+; (c) P = 3+; (d) P = 5+; (e) P = 3–; (f) P = 5+

79. FrO_2

81. (a) $2Zn(s) + O_2(g) \longrightarrow 2ZnO(s)$; (b) $ZnCO_3(s) \longrightarrow ZnO(s) + CO_2(g)$; (c)
$ZnCO_3(s) + 2CH_3COOH(aq) \longrightarrow Zn(CH_3COO)_2(aq) + CO_2(g) + H_2O(l)$; (d)
$Zn(s) + 2HBr(aq) \longrightarrow ZnBr_2(aq) + H_2(g)$

83. $Al(OH)_3(s) + 3H^+(aq) \longrightarrow Al^{3+} + 3H_2O(l)$; $Al(OH)_3(s) + OH^- \longrightarrow [Al(OH)_4]^-(aq)$

85. (a) $Na_2O(s) + H_2O(l) \longrightarrow 2NaOH(aq)$; (b) $Cs_2CO_3(s) + 2HF(aq) \longrightarrow 2CsF(aq) + CO_2(g) + H_2O(l)$; (c)
$Al_2O_3(s) + 6HClO_4(aq) \longrightarrow 2Al(ClO_4)_3(aq) + 3H_2O(l)$; (d)
$Na_2CO_3(aq) + Ba(NO_3)_2(aq) \longrightarrow 2NaNO_3(aq) + BaCO_3(s)$; (e) $TiCl_4(l) + 4Na(s) \longrightarrow Ti(s) + 4NaCl(s)$

87. $HClO_4$ is the stronger acid because, in a series of oxyacids with similar formulas, the higher the electronegativity of the central atom, the stronger is the attraction of the central atom for the electrons of the oxygen(s). The stronger attraction of the oxygen electron results in a stronger attraction of oxygen for the electrons in the O-H bond, making the hydrogen more easily released. The weaker this bond, the stronger the acid.

89. As H_2SO_4 and H_2SeO_4 are both oxyacids and their central atoms both have the same oxidation number, the acid strength depends on the relative electronegativity of the central atom. As sulfur is more electronegative than selenium, H_2SO_4 is the stronger acid.

91. SO_2, sp^2 4+; SO_3, sp^2, 6+; H_2SO_4, sp^3, 6+

93. SF_6: S = 6+; SO_2F_2: S = 6+; KHS: S = 2−

95. Sulfur is able to form double bonds only at high temperatures (substantially endothermic conditions), which is not the case for oxygen.

97. There are many possible answers including:

$$\text{Cu}(s) + 2\text{H}_2\text{SO}_4(l) \longrightarrow \text{CuSO}_4(aq) + \text{SO}_2(g) + 2\text{H}_2\text{O}(l)$$
$$\text{C}(s) + 2\text{H}_2\text{SO}_4(l) \longrightarrow \text{CO}_2(g) + 2\text{SO}_2(g) + 2\text{H}_2\text{O}(l)$$

99. 5.1×10^4 g

101. $SnCl_4$ is not a salt because it is covalently bonded. A salt must have ionic bonds.

103. In oxyacids with similar formulas, the acid strength increases as the electronegativity of the central atom increases. $HClO_3$ is stronger than $HBrO_3$; Cl is more electronegative than Br.

105. (a)

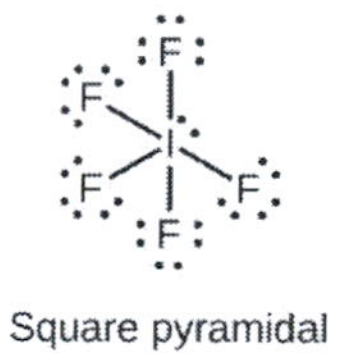
F
F
I
F
F
F
Square pyramidal

;

(b)

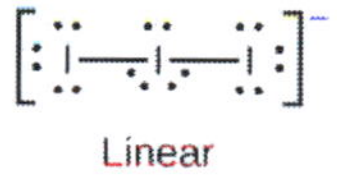
I
I
I
Linear

;

(c)

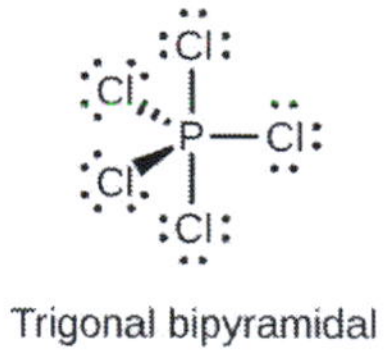
Cl
Cl
P
Cl
Cl
Cl
Trigonal bipyramidal

;

(d)

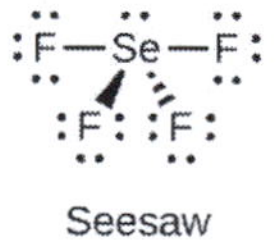
F
Se
F
F
F
Seesaw

;

(e)

T-shaped

107. (a) bromine trifluoride; (b) sodium bromate; (c) phosphorus pentabromide; (d) sodium perchlorate; (e) potassium hypochlorite

109. (a) I: 7+; (b) I: 7+; (c) Cl: 4+; (d) I: 3+; Cl: 1−; (e) F: 0

111. (a) sp^3d hybridized; (b) sp^3d^2 hybridized; (c) sp^3 hybridized; (d) sp^3 hybridized; (e) sp^3d^2 hybridized;

113. (a) nonpolar; (b) nonpolar; (c) polar; (d) nonpolar; (e) polar

115. The empirical formula is XeF_6, and the balanced reactions are:

$$Xe(g) + 3F_2(g) \xrightarrow{\Delta} XeF_6(s)$$

$$XeF_6(s) + 3H_2(g) \longrightarrow 6HF(g) + Xe(g)$$

Chapter 19

1. (a) Sc: $[Ar]4s^23d^1$; (b) Ti: $[Ar]4s^23d^2$; (c) Cr: $[Ar]4s^13d^5$; (d) Fe: $[Ar]4s^23d^6$; (e) Ru: $[Kr]5s^24d^6$

3. (a) La: $[Xe]6s^25d^1$, La^{3+}: $[Xe]$; (b) Sm: $[Xe]6s^24f^6$, Sm^{3+}: $[Xe]4f^5$; (c) Lu: $[Xe]6s^24f^{14}5d^1$, Lu^{3+}: $[Xe]4f^{14}$

5. Al is used because it is the strongest reducing agent and the only option listed that can provide sufficient driving force to convert La(III) into La.

7. Mo

9. The $CaSiO_3$ slag is less dense than the molten iron, so it can easily be separated. Also, the floating slag layer creates a barrier that prevents the molten iron from exposure to O_2, which would oxidize the Fe back to Fe_2O_3.

11. 2.57%

13. 0.167 V

15. $E° = -0.6$ V, $E°$ is negative so this reduction is not spontaneous. $E° = +1.1$ V

17. (a) $Fe(s) + 2H_3O^+(aq) + SO_4{}^{2-}(aq) \longrightarrow Fe^{2+}(aq) + SO_4{}^{2-}(aq) + H_2(g) + 2H_2O(l)$; (b) $FeCl_3(aq) + 3Na^+(aq) + 3OH^-(aq) \longrightarrow Fe(OH)_3(s) + 3Na^+(aq) + 3Cl^+(aq)$; (c) $Mn(OH)_2(s) + 2H_3O^+(aq) + 2Br^-(aq) \longrightarrow Mn^{2+}(aq) + 2Br^-(aq) + 4H_2O(l)$; (d) $4Cr(s) + 3O_2(g) \longrightarrow 2Cr_2O_3(s)$; (e) $Mn_2O_3(s) + 6H_3O^+(aq) + 6Cl^-(aq) \longrightarrow 2MnCl_3(s) + 9H_2O(l)$; (f) $Ti(s) + xsF_2(g) \longrightarrow TiF_4(g)$

19. (a) $Cr_2(SO_4)_3(aq) + 2Zn(s) + 2H_3O^+(aq) \longrightarrow 2Zn^{2+}(aq) + H_2(g) + 2H_2O(l) + 2Cr^{2+}(aq) + 3SO_4{}^{2-}(aq)$; (b) $4TiCl_3(s) + CrO_4{}^{2-}(aq) + 8H^+(aq) \longrightarrow 4Ti^{4+}(aq) + Cr(s) + 4H_2O(l) + 12Cl^-(aq)$; (c) In acid solution between pH 2 and pH 6, $CrO_4{}^{2-}$ forms $HrCO_4{}^-$, which is in equilibrium with dichromate ion. The reaction is $2HCrO_4{}^-(aq) \longrightarrow Cr_2O_7{}^{2-}(aq) + H_2O(l)$. At other acidic pHs, the reaction is $3Cr^{2+}(aq) + CrO_4{}^{2-}(aq) + 8H_3O^+(aq) \longrightarrow 4Cr^{3+}(aq) + 12H_2O(l)$; (d) $8CrO_3(s) + 9Mn(s) \xrightarrow{\Delta} 4Cr_2O_3(s) + 3Mn_3O_4(s)$; (e) $CrO(s) + 2H_3O^+(aq) + 2NO_3{}^-(aq) \longrightarrow Cr^{2+}(aq) + 2NO_3{}^-(aq) + 3H_2O(l)$; (f) $CrCl_3(s) + 3NaOH(aq) \longrightarrow Cr(OH)_3(s) + 3Na^+(aq) + 3Cl^-(aq)$

21. (a) $3Fe(s) + 4H_2O(g) \longrightarrow Fe_3O_4(s) + 4H_2(g)$; (b)

$3NaOH(aq) + Fe(NO_3)_3(aq) \xrightarrow{H_2O} Fe(OH)_3(s) + 3Na^+(aq) + 3NO_3{}^-(aq)$; (c)

$2CrO_4{}^{2-}(aq) + 2H_3O^+(aq) \longrightarrow 2HCrO_4{}^-(aq) \xrightarrow{+2H_2O} Cr_2O_7{}^{2-}(aq) + 3H_2O(l)$; (d)

$6Fe^{2+}(aq) + Cr_2O_7{}^{2-}(aq) + 14H_3O^+(aq) \longrightarrow 6Fe^{3+}(aq) + 2Cr^{3+}(aq) + 2H_2O(l)$

$Fe(s) + 2H_3O^+(aq) + SO_4{}^{2-}(aq) \longrightarrow Fe^{2+}(aq) + SO_4{}^{2-}(aq) + H_2(g) + 2H_2O(l)$; (e)

$4Fe^{2+}(aq) + O_2(g) + 4HNO_3(aq) \longrightarrow 4Fe^{3+}(aq) + 2H_2O(l) + 4NO_3{}^-(aq)$; (f)

$FeCO_3(s) + 2HClO_4(aq) \longrightarrow Fe(ClO_4)_2(aq) + H_2O(l) + CO_2(g)$; (g) $3Fe(s) + 2O_2(g) \xrightarrow{\Delta} Fe_3O_4(s)$

23. As CN⁻ is added,

$Ag^+(aq) + CN^-(aq) \longrightarrow AgCN(s)$

As more CN⁻ is added,

$Ag^+(aq) + 2CN^-(aq) \longrightarrow [Ag(CN)_2]^-(aq)$

$AgCN(s) + CN^-(aq) \longrightarrow [Ag(CN)_2]^-(aq)$

25. (a) Sc^{3+}; (b) Ti^{4+}; (c) V^{5+}; (d) Cr^{6+}; (e) Mn^{4+}; (f) Fe^{2+} and Fe^{3+}; (g) Co^{2+} and Co^{3+}; (h) Ni^{2+}; (i) Cu^+

27. (a) 4, $[Zn(OH)_4]^{2-}$; (b) 6, $[Pd(CN)_6]^{2-}$; (c) 2, $[AuCl_2]^-$; (d) 4, $[Pt(NH_3)_2Cl_2]$; (e) 6, $K[Cr(NH_3)_2Cl_4]$; (f) 6, $[Co(NH_3)_6][Cr(CN)_6]$; (g) 6, $[Co(en)_2Br_2]NO_3$

29. (a) $[Pt(H_2O)_2Br_2]$:

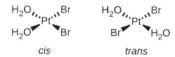

cis   trans

(b) $[Pt(NH_3)(py)(Cl)(Br)]$:

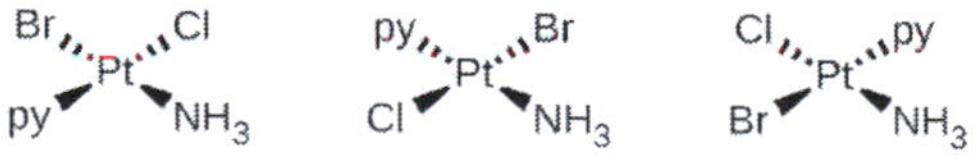

Br,,,Cl   py,,,Br   Cl,,,py
  Pt   Pt   Pt
py   NH₃   Cl   NH₃   Br   NH₃

(c) $[Zn(NH_3)_3Cl]^+$:

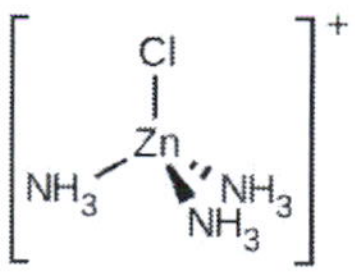

```
        Cl        +
        |
       Zn
   NH₃    NH₃
      NH₃
```

(d) $[Pt(NH_3)_3Cl]^+$:

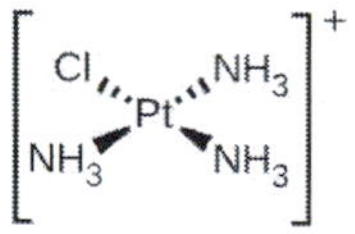

```
   Cl,,,  ,,NH₃    +
      Pt
  NH₃    NH₃
```

(e) $[Ni(H_2O)_4Cl_2]$:

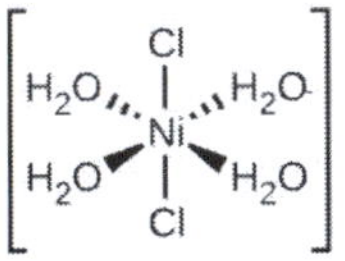

trans 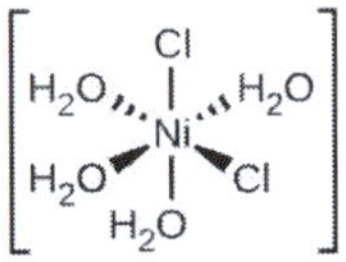   cis

(f) $[Co(C_2O_4)_2Cl_2]^{3-}$:

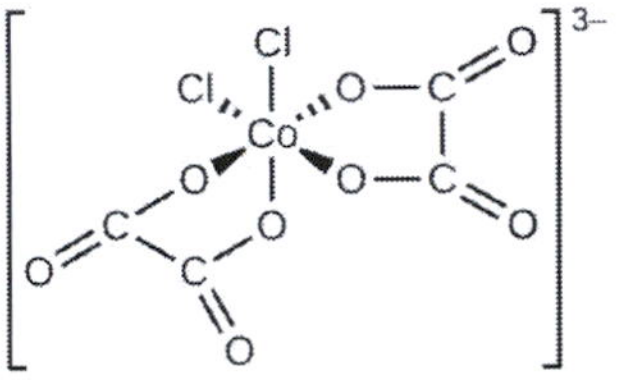

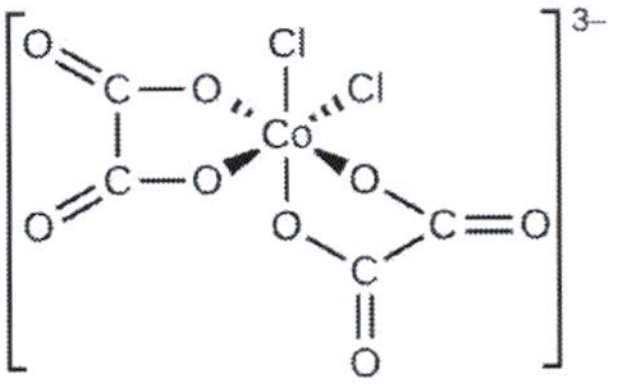

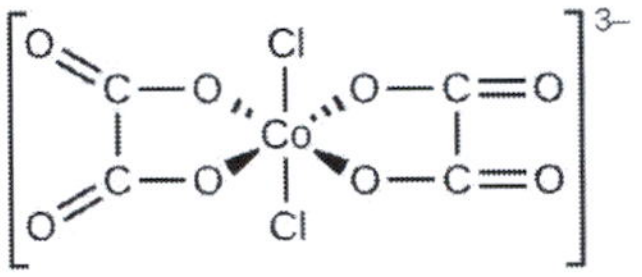

31. (a) tricarbonatocobaltate(III) ion; (b) tetraaminecopper(II) ion; (c) tetraaminedibromocobalt(III) sulfate; (d) tetraamineplatinum(II) tetrachloroplatinate(II); (e) *tris*-(ethylenediamine)chromium(III) nitrate; (f) diaminedibromopalladium(II); (g) potassium pentachlorocuprate(II); (h) diaminedichlorozinc(II)

33. (a) none; (b) none; (c) The two Cl ligands can be *cis* or *trans*. When they are *cis*, there will also be an optical isomer.

35.

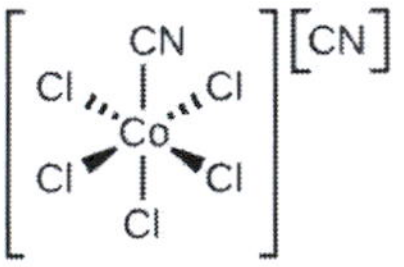

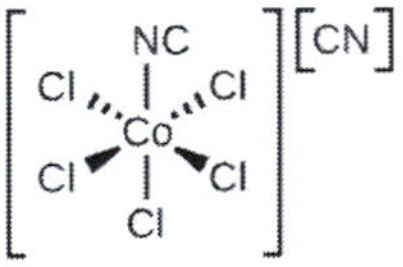

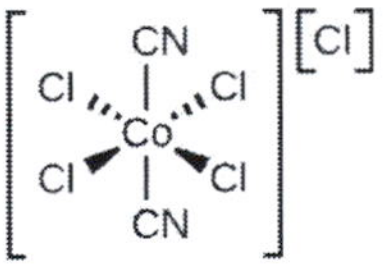

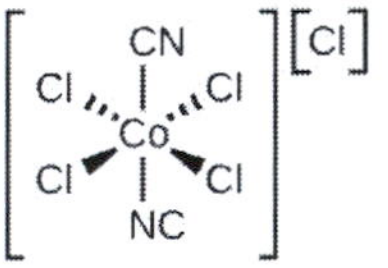

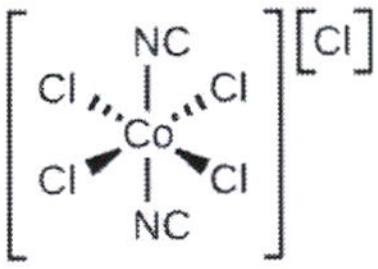

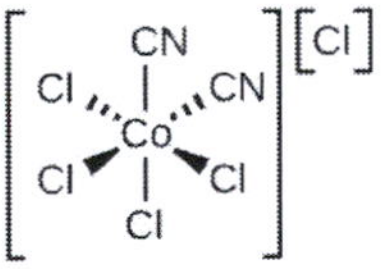

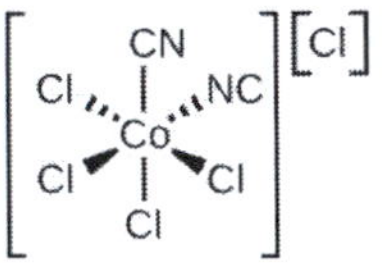

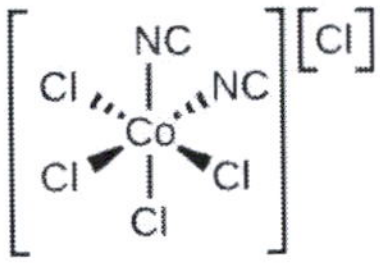

Download for free at http://cnx.org/content/col11760/latest/

37.

$[Fe(NO_2)_6]^{4-}$

Low spin, diamagnetic, $P < \Delta_{oct}$

$[FeF_6]^{3-}$

High spin, paramagnetic, $P > \Delta_{oct}$

39. $[Co(H_2O)_6]Cl_2$ with three unpaired electrons.

41. (a) 4; (b) 2; (c) 1; (d) 5; (e) 0

43. (a) $[Fe(CN)_6]^{4-}$; (b) $[Co(NH_3)_6]^{3+}$; (c) $[Mn(CN)_6]^{4-}$

45. The complex does not have any unpaired electrons. The complex does not have any geometric isomers, but the mirror image is nonsuperimposable, so it has an optical isomer.

47. No. Au^+ has a complete $5d$ sublevel.

Chapter 20

1. There are several sets of answers; one is:
(a) C_5H_{12}

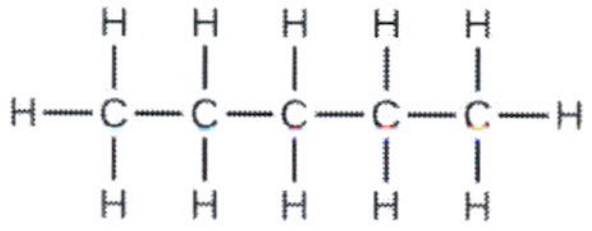

;

(b) C_5H_{10}

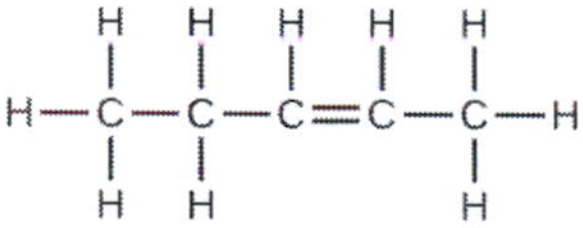

;

(c) C_5H_8

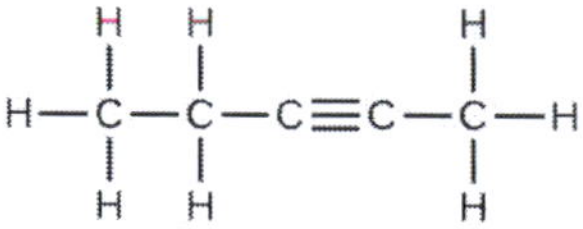

3. Both reactions result in bromine being incorporated into the structure of the product. The difference is the way in which that incorporation takes place. In the saturated hydrocarbon, an existing C–H bond is broken, and a bond between the C and the Br can then be formed. In the unsaturated hydrocarbon, the only bond broken in the hydrocarbon is the π bond whose electrons can be used to form a bond to one of the bromine atoms in Br_2 (the electrons from the Br–Br bond form the other C–Br bond on the other carbon that was part of the π bond in the starting unsaturated hydrocarbon).

5. Unbranched alkanes have free rotation about the C–C bonds, yielding all orientations of the substituents about these bonds equivalent, interchangeable by rotation. In the unbranched alkenes, the inability to rotate about the

$C = C$ bond results in fixed (unchanging) substituent orientations, thus permitting different isomers. Since these concepts pertain to phenomena at the molecular level, this explanation involves the microscopic domain.

7. They are the same compound because each is a saturated hydrocarbon containing an unbranched chain of six carbon atoms.

9. (a) C_6H_{14}

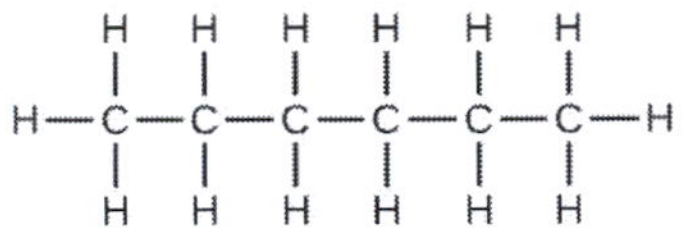

```
     H   H   H   H   H   H
     |   |   |   |   |   |
 H — C — C — C — C — C — C — H
     |   |   |   |   |   |
     H   H   H   H   H   H
```
;

(b) C_6H_{14}

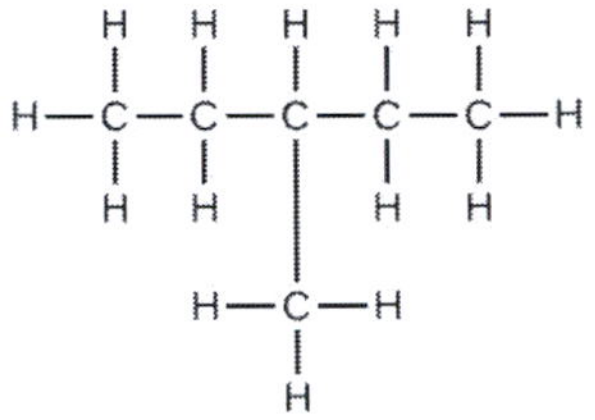

```
     H   H   H   H   H
     |   |   |   |   |
 H — C — C — C — C — C — H
     |   |   |   |   |
     H   H   |   H   H
             |
         H — C — H
             |
             H
```
;

(c) C_6H_{12}

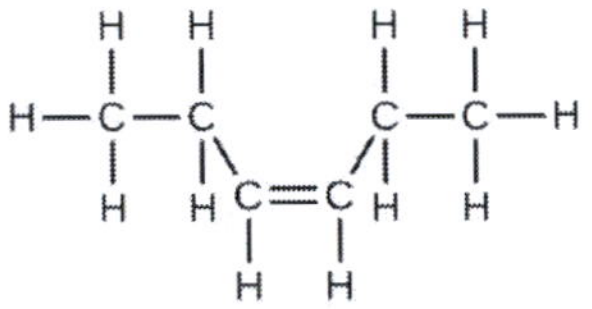

```
     H   H           H   H
     |   |           |   |
 H — C — C           C — C — H
     |    \         /|   |
     H   H C = C   H H   H
           |   |
           H   H
```
;

(d) C_6H_{12}

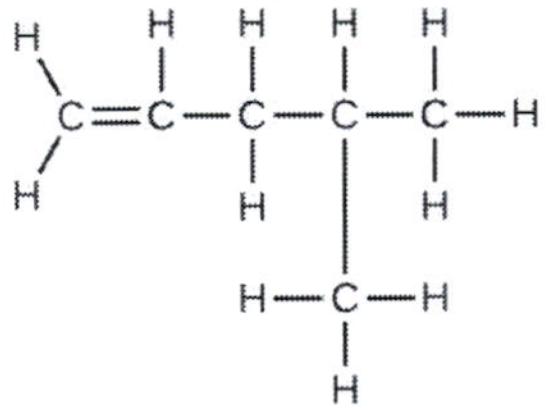

```
  H       H   H   H   H
   \      |   |   |   |
    C = C — C — C — C — H
   /      |   |   |
  H       H   |   H
              |
          H — C — H
              |
              H
```
;

(e) C_6H_{10}

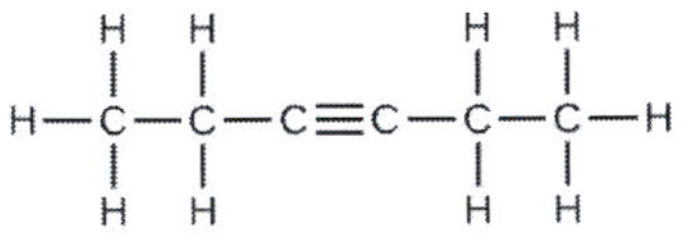

```
     H   H           H   H
     |   |           |   |
 H — C — C — C ≡ C — C — C — H
     |   |           |   |
     H   H           H   H
```
;

(f) C_6H_{10}

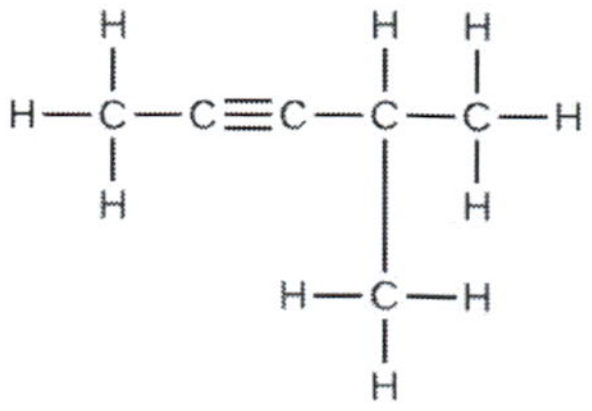

11. (a) 2,2-dibromobutane; (b) 2-chloro-2-methylpropane; (c) 2-methylbutane; (d) 1-butyne; (e) 4-fluoro-4-methyl-1-octyne; (f) *trans*-1-chloropropene; (g) 5-methyl-1-pentene

13.

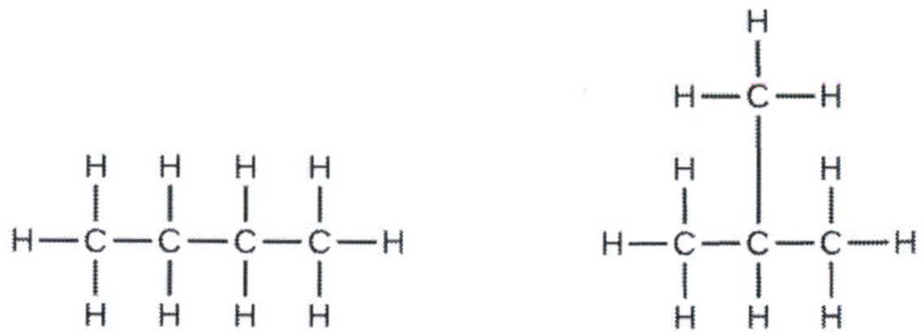

n-butane

2-methylpropane

15.

cis-

trans-

17. (a) 2,2,4-trimethylpentane; (b) 2,2,3-trimethylpentane, 2,3,4-trimethylpentane, and 2,3,3-trimethylpentane:

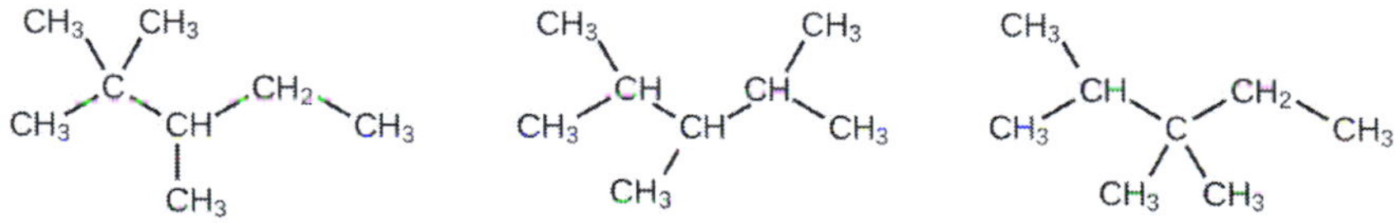

19.

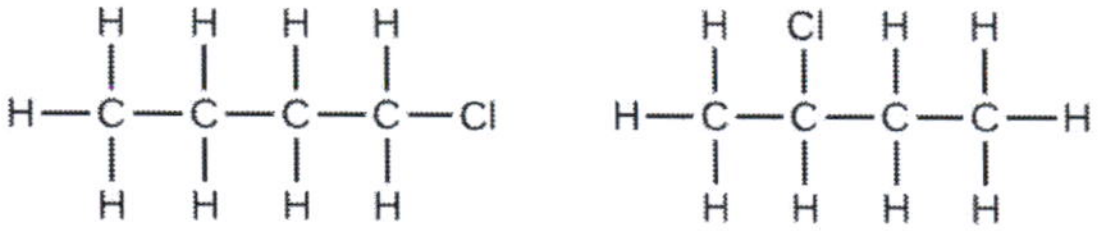

1-chlorobutane

2-chlorobutane

2-chloro-2-methylpropane

1-chloro-2-methylpropane
(1-chloro-2-methylpropane)

21. In the following, the carbon backbone and the appropriate number of hydrogen atoms are shown in condensed form:

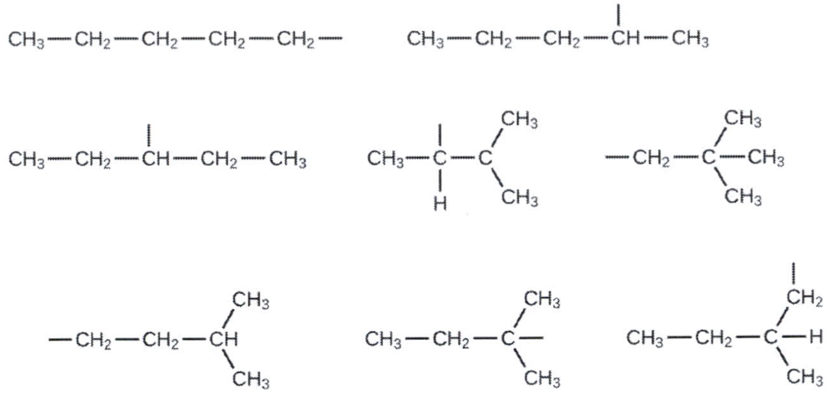

$$CH_3\!-\!CH_2\!-\!CH_2\!-\!CH_2\!-\!CH_2\!- \qquad\qquad CH_3\!-\!CH_2\!-\!CH_2\!-\!\overset{|}{CH}\!-\!CH_3$$

$$CH_3\!-\!CH_2\!-\!\overset{|}{CH}\!-\!CH_2\!-\!CH_3 \qquad CH_3\!-\!\overset{|}{\underset{H}{C}}\!-\!C\!\overset{\diagup CH_3}{\diagdown CH_3} \qquad -\!CH_2\!-\!C\!\overset{\diagup CH_3}{\underset{CH_3}{\diagdown CH_3}}$$

$$-\!CH_2\!-\!CH_2\!-\!CH\!\overset{\diagup CH_3}{\diagdown CH_3} \qquad CH_3\!-\!CH_2\!-\!\overset{CH_3}{\underset{CH_3}{\overset{|}{C}}}\!- \qquad CH_3\!-\!CH_2\!-\!\overset{CH_2}{\underset{CH_3}{\overset{|}{C}}}\!-\!H$$

23.

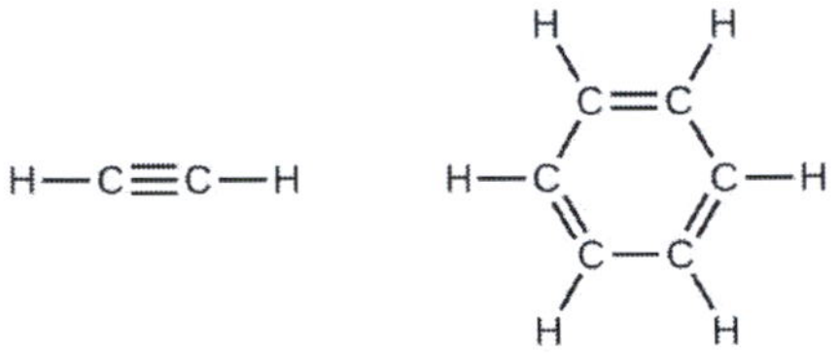

In acetylene, the bonding uses sp hybrids on carbon atoms and s orbitals on hydrogen atoms. In benzene, the carbon atoms are sp^2 hybridized.

25. (a) $CH = CCH_2CH_3 + 2I_2 \longrightarrow CHI_2CI_2CH_2CH_3$

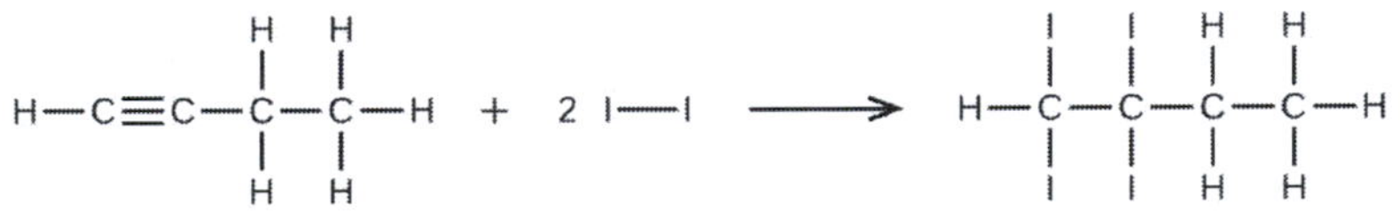

(b) $CH_3CH_2CH_2CH_2CH_3 + 8O_2 \longrightarrow 5CO_2 + 6H_2O$

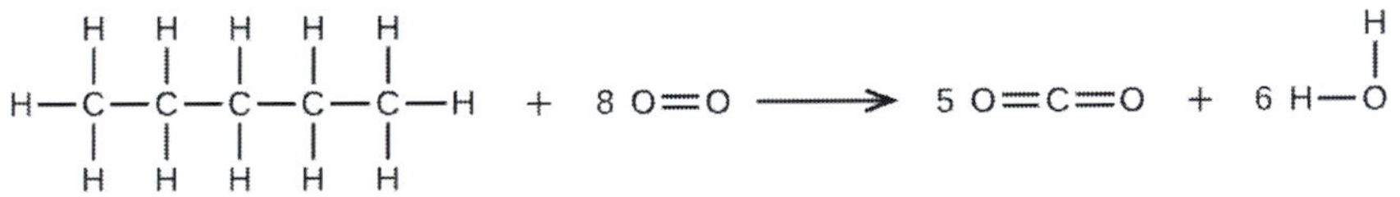

27. 65.2 g

29. 9.328×10^2 kg

31. (a) ethyl alcohol, ethanol: CH_3CH_2OH; (b) methyl alcohol, methanol: CH_3OH; (c) ethylene glycol, ethanediol: $HOCH_2CH_2OH$; (d) isopropyl alcohol, 2-propanol: $CH_3CH(OH)CH_3$; (e) glycerine, 1,2,3-trihydroxypropane: $HOCH_2CH(OH)CH_2OH$

33. (a) 1-ethoxybutane, butyl ethyl ether; (b) 1-ethoxypropane, ethyl propyl ether; (c) 1-methoxypropane, methyl propyl ether

35. $HOCH_2CH_2OH$, two alcohol groups; CH_3OCH_2OH, ether and alcohol groups

37. (a)

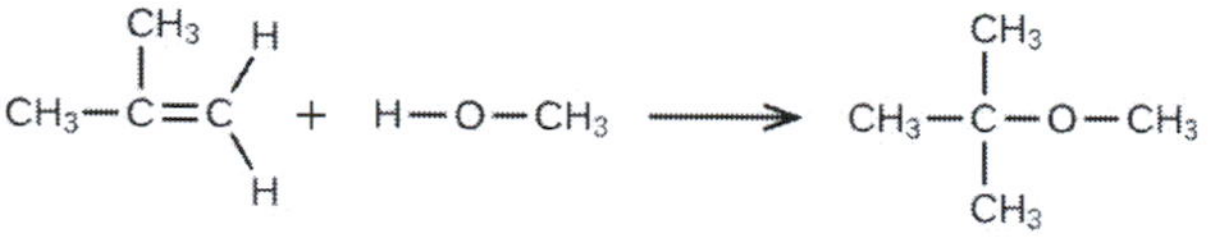

;

(b) 4.593×10^2 L

39. (a) $CH_3 CH = CHCH_3 + H_2 O \longrightarrow CH_3 CH_2 CH(OH)CH_3$

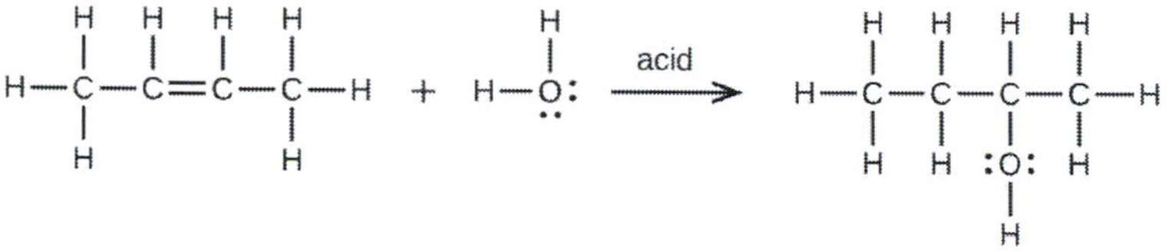

;

(b) $CH_3 CH_2 OH \longrightarrow CH_2 = CH_2 + H_2 O$

41. (a)

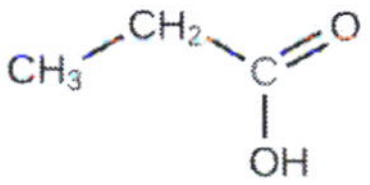

;

(b)

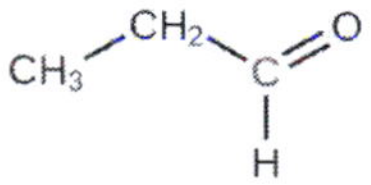

;

(c)

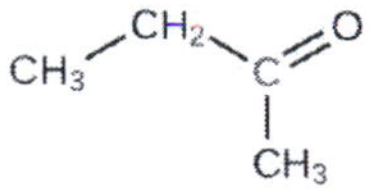

43. A ketone contains a group bonded to two additional carbon atoms; thus, a minimum of three carbon atoms are needed.

45. Since they are both carboxylic acids, they each contain the –COOH functional group and its characteristics. The difference is the hydrocarbon chain in a saturated fatty acid contains no double or triple bonds, whereas the hydrocarbon chain in an unsaturated fatty acid contains one or more multiple bonds.

47. (a) $CH_3CH(OH)CH_3$: all carbons are tetrahedral; (b) $CH_3 C(==O)CH_3$: the end carbons are tetrahedral and the central carbon is trigonal planar; (c) CH_3OCH_3: all are tetrahedral; (d) CH_3COOH: the methyl carbon is tetrahedral and the acid carbon is trigonal planar; (e) $CH_3CH_2CH_2CH(CH_3)CHCH_2$: all are tetrahedral except the right-most two carbons, which are trigonal planar

49.

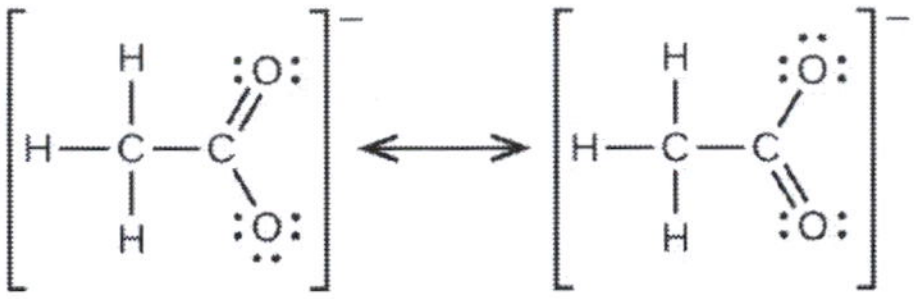

51. (a) $CH_3CH_2CH_2CH_2OH + CH_3C(O)OH \longrightarrow CH_3C(O)OCH_2CH_2CH_2CH_3 + H_2O$:

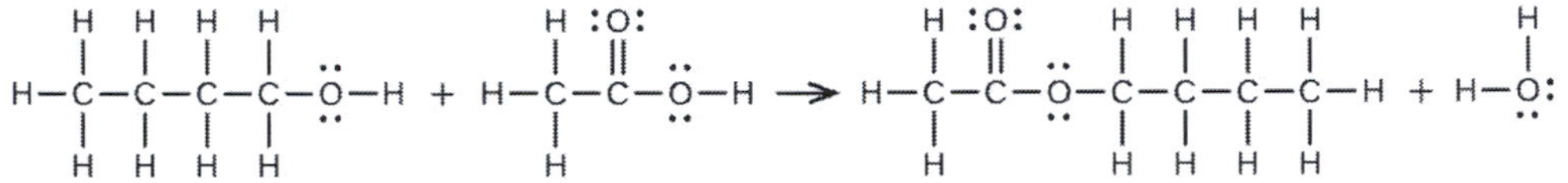

;

(b) $2CH_3CH_2COOH + CaCO_3 \longrightarrow (CH_3CH_2COO)_2Ca + CO_2 + H_2O$:

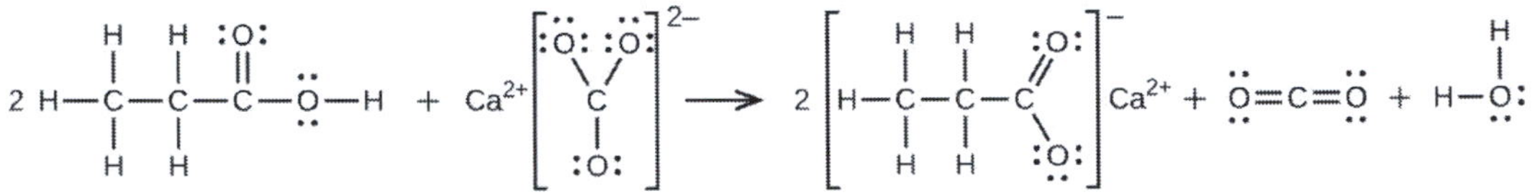

53.

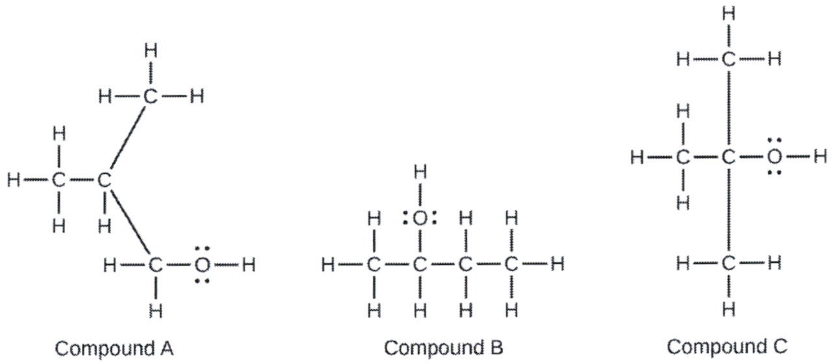

Compound A   Compound B   Compound C

55. Trimethyl amine: trigonal pyramidal, sp^3; trimethyl ammonium ion: tetrahedral, sp^3

57.

pyridine,
trigonal planar, sp^2

pyridinium ion,
trigonal planar, sp^2

59. $CH_3NH_2 + H_3O^+ \longrightarrow CH_3NH_3^+ + H_2O$

61. $CH_3\underline{C}H = \underline{C}HCH_3(sp^2) + Cl \longrightarrow CH_3\underline{C}H(Cl)H(Cl)CH_3(sp^3)$;
$2\underline{C}_6H_6(sp^2) + 15O_2 \longrightarrow 12\underline{C}O_2(sp) + 6H_2O$

63. the carbon in $CO_3{}^{2-}$, initially at sp^2, changes hybridization to sp in CO_2

Chapter 21

1. (a) sodium-24; (b) aluminum-29; (c) krypton-73; (d) iridium-194

3. (a) $^{34}_{14}Si$; (b) $^{36}_{15}P$; (c) $^{57}_{25}Mn$; (d) $^{121}_{56}Ba$

5. (a) $^{45}_{25}Mn^{+1}$; (b) $^{69}_{45}Rh^{+2}$; (c) $^{142}_{53}I^{-1}$; (d) $^{243}_{97}Bk$

7. Nuclear reactions usually change one type of nucleus into another; chemical changes rearrange atoms. Nuclear reactions involve much larger energies than chemical reactions and have measureable mass changes.

9. (a), (b), (c), (d), and (e)

11. (a) A nucleon is any particle contained in the nucleus of the atom, so it can refer to protons and neutrons. (b) An α particle is one product of natural radioactivity and is the nucleus of a helium atom. (c) A β particle is a product of natural radioactivity and is a high-speed electron. (d) A positron is a particle with the same mass as an electron but with a positive charge. (e) Gamma rays compose electromagnetic radiation of high energy and short wavelength. (f) Nuclide is a term used when referring to a single type of nucleus. (g) The mass number is the sum of the number of protons and the number of neutrons in an element. (h) The atomic number is the number of protons in the nucleus of an element.

13. (a) $^{27}_{13}Al + ^4_2He \longrightarrow ^{30}_{15}P + ^1_0n$; (b) $Pu + He^2 \longrightarrow ^{242}_{96}Cm + ^1_0n$; (c) $^{14}_7N + ^4_2He \longrightarrow ^{17}_8O + ^1_1H$; (d) $^{235}_{92}U \longrightarrow ^{96}_{37}Rb + ^{135}_{55}Cs + 4^1_0n$

15. (a) $^{14}_7N + He^2 \longrightarrow ^{17}_8O + ^1_1H$; (b) $^{14}_7N + ^1_0n \longrightarrow ^{14}_6N + ^1_1H$; (c) $^{232}_{90}Th + ^1_0n \longrightarrow ^{233}_{90}Th$; (d) $^{238}_{92}U + ^2_1H \longrightarrow ^{239}_{92}U + ^1_1H$

17. (a) 148.8 MeV per atom; (b) 7.808 MeV/nucleon

19. α (helium nuclei), β (electrons), β⁺ (positrons), and η (neutrons) may be emitted from a radioactive element, all of which are particles; γ rays also may be emitted.

21. (a) conversion of a neutron to a proton: $_0^1\text{n} \longrightarrow {}_1^1\text{p} + {}_{+1}^{\,0}\text{e}$; (b) conversion of a proton to a neutron; the positron has the same mass as an electron and the same magnitude of positive charge as the electron has negative charge; when the n:p ratio of a nucleus is too low, a proton is converted into a neutron with the emission of a positron: $_1^1\text{p} \longrightarrow {}_0^1\text{n} + {}_{+1}^{\,0}\text{e}$; (c) In a proton-rich nucleus, an inner atomic electron can be absorbed. In simplest form, this changes a proton into a neutron: $_1^1\text{p} + {}_{-1}^{\,0}\text{e} \longrightarrow {}_0^1\text{p}$

23. The electron pulled into the nucleus was most likely found in the $1s$ orbital. As an electron falls from a higher energy level to replace it, the difference in the energy of the replacement electron in its two energy levels is given off as an X-ray.

25. Manganese-51 is most likely to decay by positron emission. The n:p ratio for Cr-53 is $\frac{29}{24}$ = 1.21; for Mn-51, it is $\frac{26}{25}$ = 1.04; for Fe-59, it is $\frac{33}{26}$ = 1.27. Positron decay occurs when the n:p ratio is low. Mn-51 has the lowest n:p ratio and therefore is most likely to decay by positron emission. Besides, $_{24}^{53}\text{Cr}$ is a stable isotope, and $_{26}^{59}\text{Fe}$ decays by beta emission.

27. (a) β decay; (b) α decay; (c) positron emission; (d) β decay; (e) α decay

29. $_{92}^{238}\text{U} \longrightarrow {}_{90}^{234}\text{Th} + {}_2^4\text{He}$; ${}_{90}^{234}\text{Th} \longrightarrow {}_{91}^{234}\text{Pa} + {}_{-1}^{\,0}\text{e}$; ${}_{91}^{234}\text{Pa} \longrightarrow {}_{92}^{234}\text{U} + {}_{-1}^{\,0}\text{e}$; ${}_{92}^{234}\text{U} \longrightarrow {}_{90}^{230}\text{Th} + {}_2^4\text{He}$
$_{90}^{230}\text{Th} \longrightarrow {}_{88}^{226}\text{Ra} + {}_2^4\text{He}$ ${}_{88}^{226}\text{Ra} \longrightarrow {}_{86}^{222}\text{Rn} + {}_2^4\text{He}$; ${}_{86}^{222}\text{Rn} \longrightarrow {}_{84}^{218}\text{Po} + {}_2^4\text{He}$

31. Half-life is the time required for half the atoms in a sample to decay. Example (answers may vary): For C-14, the half-life is 5770 years. A 10-g sample of C-14 would contain 5 g of C-14 after 5770 years; a 0.20-g sample of C-14 would contain 0.10 g after 5770 years.

33. $\left(\frac{1}{2}\right)^{0.04}$ = 0.973 or 97.3%

35. 2×10^3 y

37. 0.12 h^{-1}

39. (a) 3.8 billion years;
(b) The rock would be younger than the age calculated in part (a). If Sr was originally in the rock, the amount produced by radioactive decay would equal the present amount minus the initial amount. As this amount would be smaller than the amount used to calculate the age of the rock and the age is proportional to the amount of Sr, the rock would be younger.

41. c = 0; This shows that no Pu-239 could remain since the formation of the earth. Consequently, the plutonium now present could not have been formed with the uranium.

43. 17.5 MeV

45. (a) $_{83}^{212}\text{Bi} \longrightarrow {}_{84}^{212}\text{Po} + {}_{-1}^{\,0}\text{e}$; (b) $_5^8\text{B} \longrightarrow {}_4^8\text{Be} + {}_{-1}^{\,0}\text{e}$; (c) $_{92}^{238}\text{U} + {}_0^1\text{n} \longrightarrow {}_{93}^{239}\text{Np} + {}_{-1}^{\,0}\text{Np}$,
$_{93}^{239}\text{Np} \longrightarrow {}_{94}^{239}\text{Pu} + {}_{-1}^{\,0}\text{e}$; (d) $_{38}^{90}\text{Sr} \longrightarrow {}_{39}^{90}\text{Y} + {}_{-1}^{\,0}\text{e}$

47. (a) $_{95}^{241}\text{Am} + {}_2^4\text{He} \longrightarrow {}_{97}^{244}\text{Bk} + {}_0^1\text{n}$; (b) $_{94}^{239}\text{Pu} + 15{}_0^1\text{n} \longrightarrow {}_{100}^{254}\text{Fm} + 6{}_{-1}^{\,0}\text{e}$; (c)
$_{98}^{250}\text{Cf} + {}_5^{11}\text{B} \longrightarrow {}_{103}^{257}\text{Lr} + 4\text{n}^0$; (d) $_{98}^{249}\text{Cf} + {}_7^{15}\text{N} \longrightarrow {}_{105}^{260}\text{Db} + 4{}_0^1\text{n}$

49. Two nuclei must collide for fusion to occur. High temperatures are required to give the nuclei enough kinetic energy to overcome the very strong repulsion resulting from their positive charges.

51. A nuclear reactor consists of the following:
1. A nuclear fuel. A fissionable isotope must be present in large enough quantities to sustain a controlled chain reaction. The radioactive isotope is contained in tubes called fuel rods.
2. A moderator. A moderator slows neutrons produced by nuclear reactions so that they can be absorbed by the fuel and cause additional nuclear reactions.
3. A coolant. The coolant carries heat from the fission reaction to an external boiler and turbine where it is transformed into electricity.

4. A control system. The control system consists of control rods placed between fuel rods to absorb neutrons and is used to adjust the number of neutrons and keep the rate of the chain reaction at a safe level.

5. A shield and containment system. The function of this component is to protect workers from radiation produced by the nuclear reactions and to withstand the high pressures resulting from high-temperature reactions.

53. The fission of uranium generates heat, which is carried to an external steam generator (boiler). The resulting steam turns a turbine that powers an electrical generator.

55. Introduction of either radioactive Ag^+ or radioactive Cl^- into the solution containing the stated reaction, with subsequent time given for equilibration, will produce a radioactive precipitate that was originally devoid of radiation.

57. (a) $^{133}_{53}I \longrightarrow {}^{133}_{54}Xe + {}^{0}_{-1}e$; (b) 37.6 days

59. Alpha particles can be stopped by very thin shielding but have much stronger ionizing potential than beta particles, X-rays, and γ-rays. When inhaled, there is no protective skin covering the cells of the lungs, making it possible to damage the DNA in those cells and cause cancer.

61. (a) 7.64×10^9 Bq; (b) 2.06×10^{-2} Ci

Index

Symbols

Made in the USA
Middletown, DE
12 August 2017